Dr Vij Sodera FRCS

# ONE small SPECK TO MAN

## the evolution myth 2

# One small Speck to Man
*the evolution myth*

by

Dr Vij Sodera

First Edition. 2004
Second Edition (India). January 2009
Second (Revised) Edition. August 2009

This book is dedicated to
Margaret, Melanie and Lisa

**Vija Sodera Productions**

onesmallspeck.com

Book design by Lisa Sodera

one small Mango .com

onesmallmango.com

All photographs, paintings and artwork by Vij Sodera except where otherwise stated

ISBN 978 0 9531376 5 7



Printed and bound in Malaysia
Managed by Bekaam Printers Pte Ltd

It doesn't matter what you believe, as long as what you believe is true

# Question everything

FRCS
Dr VIJ SODERA
Fellow of the Royal College of Surgeons

# CONT

# ENTS

## SECTION THREE: *all features great and small*



## SECTION FOUR: *the Man, one giant leap for apekind?*



## SECTION FIVE: *from one small speck to Man*

# PREFACE

## Preface to the First Edition

It is taught almost universally that humans have evolved from ape ancestors, and that chimpanzees are our distant cousins. Indeed, many biologists are quite happy to classify themselves as being no more than just another species of chimpanzee. Certainly, to the casual observer, it might seem quite plausible that a few modifications to the brain, face and body of a chimpanzee could result in a human being. However, the question that needs to be addressed is whether or not such modifications can be achieved by the blind processes of genetic mutations and natural selection. Furthermore, few people realise that *to accept that we are modified apes requires us to accept an even more fantastic series of propositions*.

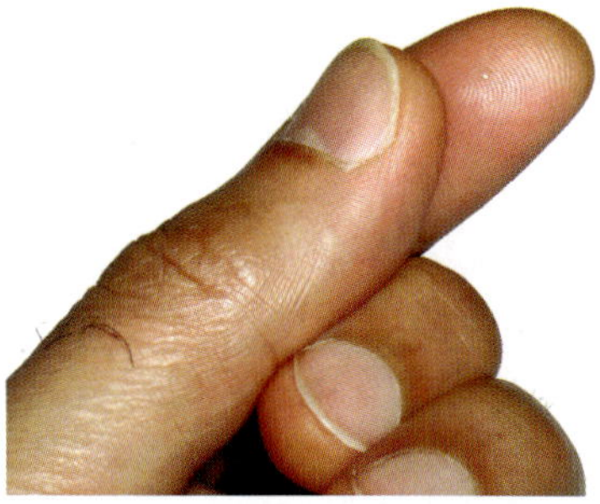

**One small speck**
If it is accepted that humans evolved from simple chemistry and that chimpanzees are our closest cousins, then it must also be accepted that goldfish and even bananas are our very distant cousins.

Look at a speck of dust on the table in front of you. Given the time span of a few billion years, is it possible for such a particle, being composed only of simple molecules, to rearrange its structure, to join with other simple particles and to acquire *consciousness* - ultimately to peer down on another dust particle and to contemplate its origins? Yet inherent in the idea of evolution by natural selection is the assumption, that not only is this possible, but also that it actually *happened*. In other words, spontaneously, without the action of any outside influence whatsoever, simple molecules, dissolved in water, ultimately became living cells. Some of these cells became fishes, some of which became amphibians, some of which became reptiles. Some reptiles became birds and mammals. Some mammals became monkeys, some of which became apes, some of which became you and me.

It is generally assumed that the Earth is so old that there has been plenty of time for Man to have evolved from simple chemicals. Granted, the journey through time from (supposedly) 4.5 billion years ago may seem like a long one, but that journey is like a mere afternoon stroll compared to the magnitude of the (supposed) molecular journey from cosmic dust to cosmic observer.

The fundamental tenet of the idea of evolution is that the natural selection of (mainly) random and simple small progressive changes in genes is all that was necessary to have produced all living things, including all their component parts: for example, the mammalian eye supposedly started out as a simple and primitive light-sensitive pigment in some ancient and simple single-celled organism. But genes and chromosomes are *extremely* complex structures, and no gene can mutate and evolve *unless it first exists*. Furthermore, not all things are *possible*. Since most biological structures are highly complex and are formed under the control of multiple genes, a stepwise change from one structure into another structure (for example, an ape foot into a human foot) - which from a cursory glance and given enough time might appear to be an inevitable progression - can, in fact, be *impossible*. In addition, since many genes are actually inextricably involved in the formation of many unrelated structures, any *significant* change in any one gene can result in wide-ranging and damaging effects elsewhere. Numerous congenital diseases and abnormalities demonstrate this basic principle.

Of course, we know that many genetic changes make *no* overall difference whatsoever to the function of genes. In addition, even serious changes to relatively unimportant genes will have no effect on the chances of an organism's survival. Nevertheless, it remains true that alterations to *important* components of any *vital* gene will have seriously detrimental effects on the organism in which that gene resides. So, significant changes to important genes are likely to be disadvantageous, rather than generating new information so as to result in the production of a *new* structure or a *new* biological system of a type which previously did not exist.

It is important to emphasise that discussing any particular supposed evolutionary relationship without addressing the *overall* picture makes for shaky science. For example, even if it might be *presumed* that the soft tissues (such as ligaments and cartilages) associated with the bony parts of the australopithecine knee joint are only slightly different from the human form, nevertheless in evolutionary terms, ultimately the human knee joint has to be derived from the knee joint of a (supposed) ancestral shrew that lived (supposedly) 65mya and in turn, the shrew knee joint has to be derived from the pelvic fin bones of (supposed) ancestral fishes that lived (supposedly) over 350mya. Since fish fins do not possess the cartilages and locking mechanisms inherent in the human design, and since there is no trajectory from the fish fin to the mammalian knee, then simply discussing the (supposed) evolutionary relationships between australopithecine and human knee joints *in isolation* is meaningless. Do we have evidence of evolution - of one creature, in time, turning into a completely different creature? As we shall see, it is misguided to quote the variety of Galapagos finches, or the (supposed) diversification of mammals following the demise of dinosaurs, as examples of evolutionary progress in response to environmental opportunity. In the first example, the finches represent not the origination of new information that produced new chracteristics, but simply inbreeding of characteristics that were already present in the parent group that first arrived at the island refuge. In the second example, according to a common interpretation* of the fossil record, dinosaurs were dying out before the mammals (supposedly) started to diversify and multiply. Furthermore, mutations in DNA are not at the beck and call of any perceived need or of any environmental opportunity. So it is implausible that *new* mammals came into being just because there were no dinosaurs to contend with. As for intermediaries between apes and humans, we shall see that there is no evidence to support the notion that our ancestors were, at some time in the past, not human.

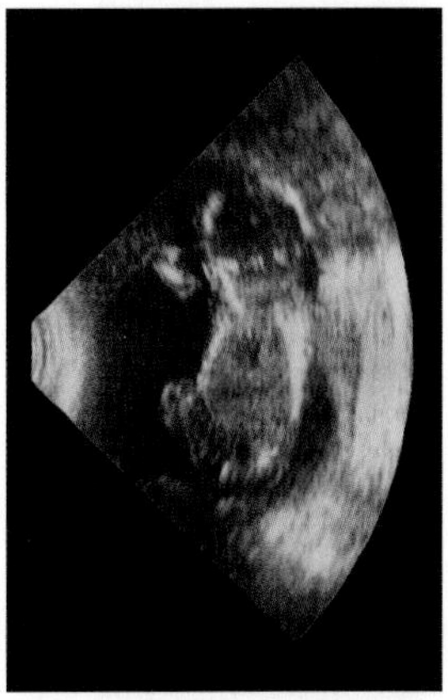

Ultrasound scan of a developing human baby (Anjali, above) at 16 weeks' gestation.

Scan of baby (Arthur-Worsop) at 20 weeks' gestation (below), showing that the face and hands are fully developed.

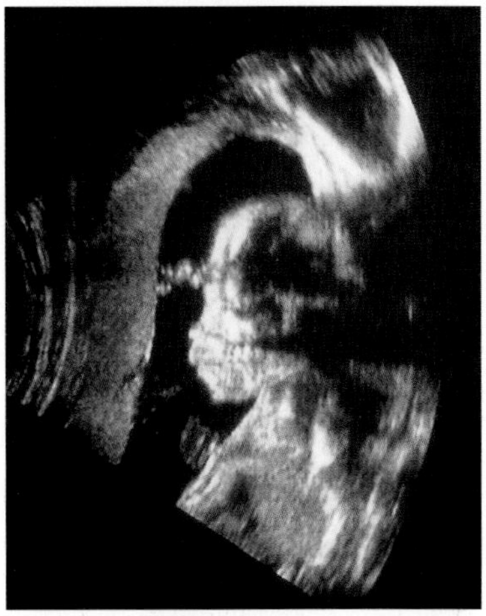

*As we shall see later, there is good evidence to suggest that all creatures could have lived contemporaneously in the past.

# One small Speck to Man ~ *the evolution myth*

The basic principle of scientific enquiry is to test a hypothesis against *all* the evidence. If a hypothesis fails in a number of significant aspects, it is likely that the hypothesis is invalid. Now, starting from only simple chemicals, if it can be shown that chemical laws and biological constraints will not allow the evolution of any number of biological systems, organs, and structures, nor allow the conversion of one creature into a quite different creature (for example, a dinosaur into a bird), then we must face the inescapable conclusion that: no matter how much time you give it; no matter with what energy you supply it; no matter what your interpretation is of fossils or of their supposed dates; no matter how much you dislike the thought of it; *no matter if you have no alternative working model to present in its place*; and no matter what else... the evolution of one type of creature into a different type of creature *did not* occur, and *cannot* and *will not* occur under any circumstances...*ever*.
History shows that regardless of the subject being discussed, any questioning of established and *seemingly* proven dogma will generate intense opposition - opposition which is often simply emotional and without logical foundation. And the subject of evolution and the origin of Man is no exception. However, to those who would say that to question the theory of evolution is to be at once unscientific, foolish and ignorant, I make no apology for this book. To question is the foundation stone of all science, and only by being willing to look objectively at the evidence, and by being willing to challenge established dogma, can we aim to get to the truth of the matter. So, was our great, great…great grandfather an ape? This is *the* question that this book sets out to answer.

October 2003

## Preface
### to the Second Edition

This Second Edition includes 100 extra pages which contain over 500 new images and many new sections, including fossilisation, geology, tetrapod limb origins, vertebrate ear bones, lungless frog, manatees, feathered dinosaurs, bird lung, mammalian diaphragm, eye muscles, retinal Müller cells, Flores fossils, panda anatomy and baculum, knee joint, ape and human genomes, human larynx, circulation changes at birth, and extensive additional information and references. Another new feature includes a number of dissections which give additional insights into the truly awesome structural features found across the whole spectrum of animal biology.
Since the First Edition of this book was released, the human, chimpanzee, and a number of other genomes have been sequenced. What is abundantly clear is that all the new genetic evidence underlines just how markedly different apes are to humans, and how unscientific it is try to hold on to the evolution myth and to the notion that chimpanzees are our distant cousins.
Never before has it been so obvious that THE *Paradigm Lost* can be *Paradigm Found.*

Vij Sodera August 2009

Dr Vij Sodera FRCS
Fellow of the Royal College of Surgeons

Elephant
Ngorongoro crater, Tanzania

# NOTE

Since, as we shall see, the biological evidence contradicts many of the dates given by dating processes, and since evolution is an idea which is not only based on numerous questionable assumptions but is also contradicted by evidence from every aspect of biology, I make no apology for the repeated use of the term 'supposed' when any reference is made to a given paleontological date, or to any assumed relationship (whether ancestral or descendant) between any creatures.

The term 'hominid' has commonly been used to encompass both human beings and (supposed) intermediaries between the (supposed) common ancestors of modern chimpanzees and fully-human-beings. However, the term serves only to mislead since, as we shall see, there is no evidence for any non-fully-human-creature that represents an ancestor to living human beings. Therefore, the term 'hominid' is not used in this book.

To make for more consistent comparisons, all ape and human skulls are shown from the left side. Although in some cases this has necessitated a reversal of the right-sided view, in terms of the anatomy this is of no importance.

No animals were harmed in the course of producing this book. All dissections were performed on animals that had previously died of natural causes or as a result of injury.

**Implausible - Impossible**

Whenever the term 'implausible' is used, it is meant that, from the evidence, it is irrational to think that a particular outcome will occur. So, in context, 'implausible' effectively means 'impossible'.
On many other occasions (where, from a position of sure biological knowledge, certainty must be acknowledged), it can be said, unreservedly, that some things are categorically impossible.

Llama

# To be or not to be (alive): that is the quantum

# PROLO

Gray whale skeleton

*The year is 2009. Quantum Evolutius is a man who believes and teaches that his father is a man, and that his distant cousin is a chimpanzee. Searchio is a handsome youth. He has no direct knowledge of any tree-living distant relation in any central African rain forest. His father taught him when considering any matter, never to be hasty in making conclusions, and never to ignore any obvious discrepancies. We join the two in the whale hall of the Natural History Museum in London.*

**Quantum Evolutius:**

*Leaning over the 'Please do not touch' sign and gently tapping the tail flukes of the model of a bottle-nosed dolphin.*

"...So that's about it. Atoms, being neither particles nor waves, are incomparable to anything imaginable. The material universe, including all life forms within it, is nothing but scattered tiny little bits of something separated by huge amounts of nothing much at all".

**Searchio:**

*He folds his arms for a moment. Then he moves one hand to cup his chin and puts the other in his back trouser pocket like some TV detective as he turns to catch a panoramic view of the great hall with the bones and life-sized models of the huge whales hanging from the ceiling.*

"OK, now let me get this straight...what you're saying is that maybe there was no such thing as the beginning of the universe, since what we see at this moment in time could just be a snapshot of an endless series of cycles in which the universe expands and then collapses upon itself. At the moment we're part of the expanding part of a cycle.

Once upon a time (say about 15-20 billion years ago), very near where we would ordinarily think a beginning ought to have been, all the matter in the universe was condensed into one tiny dot. Then suddenly, as a result of the immense...whatever forces...all the matter was blasted outwards in all directions. That's what you call the 'BIG BANG'. Actually, it was not an explosion as we know it since the tiny dot comprised both all the matter and all the existing space. So there wasn't any huge empty universe of space for the tiny dot to explode into. Time itself started at the Big Bang, and as the *space* itself expanded, matter became increasingly separated. You haven't stated where all the energy that started it all came from, or what existed before time, but only that within the first few microseconds of the 'start' of the universe, the energy of this unimaginably immense 'explosion' formed a number of the simpler atoms*.

*** A big bang can only produce matter and antimatter in equal quantities**

*There must always have been more matter than antimatter in our Universe. If it weren't so, we wouldn't be here. Matter and antimatter would have annihilated each other, leaving only radiation. What caused this asymmetry is still unknown, but it was vital for our existence.*

New Scientist. 15th January 2000. p45.

Much later that same epoch, these bits of matter clumped together to form stars, and clusters of stars became galaxies. Atoms of the heavier elements may have been formed later on within stars. Our sun, just one of the millions of stars in the millions of galaxies, expelled some of its matter which then orbited around it, so forming the planets.

A few aeons later still, one of the planets (Earth) found itself with, amongst other things, ammonia, methane and water vapour in considerable quantities. With the

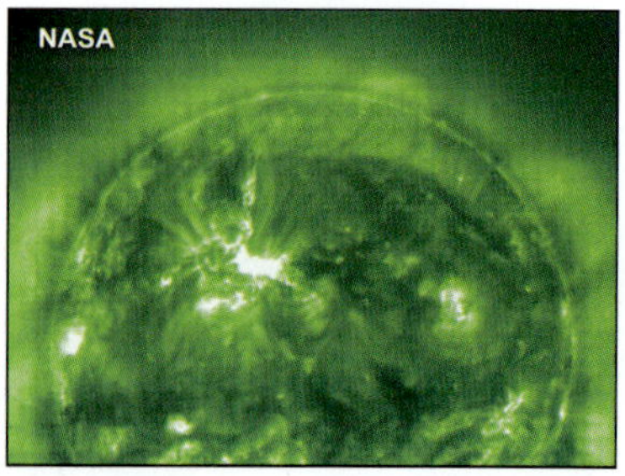


False colour image of solar flare

addition of energy in the form of very frequent lightning and/or intense ultraviolet light, these gases were converted into simple amino acids.
Also about that time, a number of other chemical reactions were occurring, producing fatty acid molecules (fatty acids combine together to form the molecules that we know of as fats, or lipids). These fatty molecules naturally bonded together and formed simple hollow spheres.
Maybe sometime later, other molecules were reacting together and formed the amazing double helix of deoxyribose nucleic acid (DNA). Amazing, because DNA has the property of self-replication. But there's nothing magical in it, that's just the way this molecule behaves chemically. Now, most of these early complex molecules must have broken up again with time, but enough of them must have persisted long enough so that as the molecules became more complex, they bonded together in wonderful and previously unseen ways. New molecules opened the way for newer and more diverse chemical options. Then, the amino acids joined together to produce small proteins which in turn bonded with lipids to form spherical lipo-protein membranes and incorporated the proteins, lipids and nucleic acid fragments inside themselves, along with sodium, potassium and chloride ions, and of course, water molecules.
So, you're saying that such a complex collection of chemicals, the whole encapsulated by the lipo-protein bag, became the first primitive and very simple single-celled living organism?"

**No-one knows how a star is made**
Star formation is shrouded in mystery.

*Generally speaking a star forms when a cloud of gas collapses under gravity. However, if the cloud is too hot, pressure will combat the effect of gravity and prevent the cloud from collapsing. So to form a star, the gas cloud must have a way of cooling down. This isn't as easy as it sounds. In today's Universe this is accomplished by a huge array of molecules which collide and radiate away the heat. However, the atoms necessary for making all but the simplest molecule – molecular hydrogen – have to be made inside stars. It is a chicken and egg situation.*
New Scientist. 7th Feburary 1998. p26-30.

*Nobody knows how a mixture of lifeless chemicals spontaneously organized themselves into the first living cell.*
New Scientist. 12th July 2003. p24-39.

**Quantum Evolutius:**
"I like to use the term *'bio-cule'* instead of the term 'single-celled organism' because the sound of the word 'bio-cule' more accurately reflects the organism's true nature, this being simply an ultra-complex chemical conglomeration that just happens to be able to self-replicate. But ultimately, bio-cules are entirely dependent on and directed by the same chemistry that causes reactions in everyday school test-tube experiments. And just as it is (under certain conditions) in the nature of groups of atoms to form molecules and groups of molecules to form bio-cules, so it is (under certain conditions) in the nature of groups of bio-cules to form polybio-cules (or multi-cellular organisms)."

**Searchio:**
*Stepping backwards and looking directly upwards at the undersurface of the blue whale's huge tail flukes.*
"OK. At some point the sum of the chemical and electrical activity inside these conglomerations could be recognised as what we call *LIFE*. Over the past millions of years, gradual alterations (mutations) in the DNA inside these bio-cules, along with natural selection, have resulted in the evolution of all species from simpler chemical forms. And this whale appeared simply as a result of mutations in the chromosomes of some ancient pig- or dog-like animal. In other words, the increasing complexity of the chemical conglomerations (bio-cules) allowed an increasing complexity and variety in subsequent chemical conglomerations and their possible interactions."

**Quantum Evolutius:**
"Just so. We ourselves are no more than just one type of polybio-cule, a massive conglomeration of molecules. In reality, all polybio-cules only differ from simpler molecules by their size and by the degree of complexity of their interactions."

**Searchio:**
(Aside) *So, ultimately, my great, great...great grandfather was an...amino acid!*
"Fine. So to summarise your thinking, subatomic particles and forces are responsible for the characteristics of atoms and molecules, that is, the reactions of nuclear physics and chemistry. Atomic and chemical forces are responsible for the characteristics of biological cells (the ultra-complex chemical conglomerations that you call bio-cules).
And further, living cells are dependent solely upon atomic, chemical and biological (or bio-cular) forces which dictate each and every one of their characteristics. So, in terms of their physical and chemical attributes and the forces that govern their existence and interactions, there is *no* distinction between particles or molecules described as 'non-living' when they are isolated in a test-tube (for example, water, free amino acids, free proteins and free strands of DNA) and the more complicated chemical conglomerations described as 'living', when these same chemicals come together with other chemicals inside a lipid-protein membrane to form a cell (or bio-cule). Is that it?"

**Quantum Evolutius:**
"No, I mentioned the business of thought, intellect and behaviour."

**Searchio:**
"Ah, yes! What you're saying is that just by *being*, groups of nerve cells naturally create spontaneous electrical discharges between themselves. So thought is as much a natural and *unavoidable* consequence of how nerve cells inter-react as it is with water molecules inter-reacting to form ice if the temperature is right. It's in their nature (whatever that is). And therefore, intellect and behaviour only and merely reflect the degree, complexity and style of the physical/chemical/bio-cular interactions in operation within an ultra-complex chemical conglomeration.
If the brain is big enough you get thought. If the brain is really big as in humans, you get more complicated thoughts, ideas, and dreams...or even a voice from God. You're saying that all this is one and the same process, electro-chemical 'tricks' in the mind (whatever that is)."

Ice crystals

**Quantum Evolutius:**
"And morals...and death?"

**Searchio:**
"Morals?... You have already said that there is no such thing as LIFE as it is commonly understood, but there exist only degrees of molecular complexity. It follows

that if there is no such thing as LIFE, there simply cannot be any such things as morals. How can a (purely and only) physical/chemical unit itself define Right and Wrong in *any* given situation?
Since everything that exists is merely the result of random and arbitrary subatomic, atomic and chemical interactions, you're saying that it is as natural and acceptable to destroy a polybio-cule (for example, a human being) as it is to annihilate a subatomic particle. Both are of no ultimate value. Both comprise some of the matter originally formed after the BIG BANG. Literally, mere stardust. Your view on life implies and requires that anger, love, hate, like and dislike are just *innocent* consequences of fundamental electro-chemical interactions within a complex blob of bio-cules (the brain), and are therefore beyond criticism. Who are you...or more precisely, *what* are you to criticise thoughts inside anyone else's head, or the outworkings of those thoughts reflected in behaviour or action? Thus it is as natural and acceptable for nitric acid (whose innocent behaviour towards other chemical structures is dictated purely by its chemical make-up) to dissolve the skin of your hand, as it is for one man (a particular polybio-cule, whose nerve cell interactions and therefore whose thoughts and behaviour towards other polybio-cules are ultimately dictated purely and simply by its very complicated chemical make-up) to destroy another man.
So, you're saying that when a man deliberately kills a number of innocent people we should not call him evil or deranged. Instead we should accept that this man's brain chemistry is not *abnormal*, but simply *different*. Our ('normal') brains might not *like* what this man has done, and indeed we might in all probability make a judgement and impose a punishment, but ultimately, our views and actions could not be construed as Right, and his as Wrong."

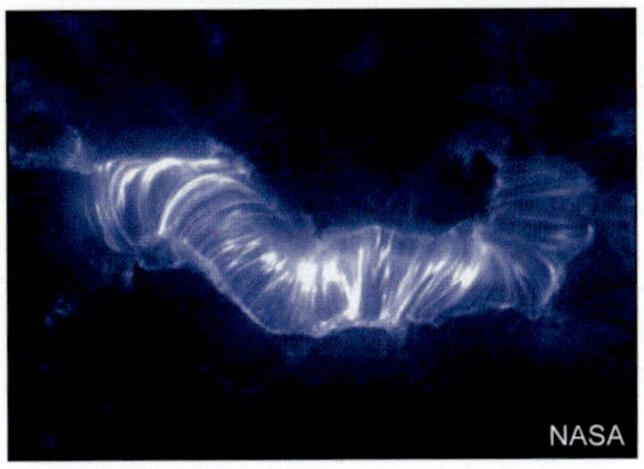

Solar storm: According to evolution, plasma (free nuclei and free electrons) inside the sun turned ultimately into thought.

**Quantum Evolutius:**
"Exactly. There is no difference between a man and chemicals in a test-tube. And if there is no such thing as LIFE, there can be no such thing as DEATH - what we call death is only an alteration of a high energy bio-cular state to simpler chemical structures with a lower energy state. Furthermore, it follows that if there is no such thing as LIFE *during* life, there can certainly be no LIFE *after* life. So you can throw things like religion straight out of the window, along with cathedrals and temples. And if there can be no LIFE in any creature, the distinction between all life forms becomes completely blurred and a man cannot be of any greater importance than any animal. Since apes are self-aware in time and space, can perform complex tasks, have emotions, and can communicate in sign language, they have all the characteristics of a 'person'. And since under this definition of 'person', healthy adult apes exhibit more features pertaining to personhood than healthy human foetuses and healthy new-born human babies, it would be more correct morally to destroy or experiment on human foetuses or new-born babies, than to destroy or experiment on adult apes."

**Searchio:**
"So, if pigs can perform better than chimpanzees in some intelligence tests, maybe pigs should also be considered as persons. Therefore, we should stop experimenting

on pigs or eating them, or using their skins for shoes or coats.
All that heartache about helping the starving humanity in its millions on some distant part of the globe...why bother? They are all merely and only inconsequential stardust. And saving endangered species...save the whale, save the recently discovered *lesser-spotted Outer Mongolian snow kangaroo* etc...mere stardust...
And why worry about a dolphin and not a tuna fish - haven't tuna fish got equal rights in biological terms, because surely any difference in *worth* is arbitrarily imposed by our human classification? So why worry about the extinction of some rare birds because of some man-made ecological change in some estuary? In fact, why be concerned about the destruction of a human foetus or a chimpanzee when both are essentially only a couple of variations amongst billions of molecular conglomerations? Since there's no purpose in anything, why make a fuss at all? When you've seen one clump of atoms, you've seen them all!
So, from ashes to ashes, stardust to stardust, nothing is of any ultimate value..."

(Aside) *Hmm...methinks there is much mischief in such thinking.*

"If we have evolved by a process of survival of the fittest, rather than striving to treat illness and congenital abnormalities, it would be more consistent to let the weakest of us die so as to preserve naturally only the healthiest individuals in subsequent generations. Furthermore, any man-made destruction of ecosystems or man-induced extinctions should not be reviled, since they represent merely the latest in a chain of events that have been going on ever since life began. In this light, man-induced climatic changes, extinctions, suffering and death are entirely *normal* and *natural.*
But...your morals are based on your *assumption* that all life forms *have* appeared as the result of an undirected and unconcerned evolution and that animals are equal to human beings. Yet you have not shown me any convincing physiological, genetic, biochemical, neurological or fossil evidence to demonstrate that human beings *are* simply evolved apes.
Hypothetically, even if, for example, there is found evidence showing conclusively that australopithicines walked fully upright, made sophisticated tools, and communicated by speech - none of this in itself would constitute evidence that ancient apes evolved into humans. It would simply show that humans have not been the only bipedal, tool-making, speaking creatures to have lived. To show that humans evolved from apes requires conclusive evidence that the ape anatomy and brain function *could have changed* (based on data producing a complete logical series of fully functional modifications) and *did change* (based on real fossil, biochemical, genetic and other biological evidence showing a complete series of intermediaries) into the human form.
From the limited evidence you have given, it is quite possible that human beings (despite sharing a common biology with other animals) have no ancestor-descendant or distant cousin relationship to any ape, and may in some unique way be different from other living things. In science you cannot reject a notion just because you *dislike* it...

And another thing: what are subatomic particles made of? And what makes up the things that subatomic particles are made of? And what made the things that make up the things that subatomic particles are made of? How did the *information* by which the very *physics* and *mathematics* on which the universe depends come into being? Is not information *a more fundamental essence* than even matter or energy itself? So, does not the necessity for information to have existed prior to the appearance of the universe demand the pre-existence of an intelligence...?"

We can leave these two to carry on with their discussion at this point. But, is Quantum Evolutius correct when he describes the origin of all current life forms as being the result of sequential changes to ancestral early molecules and primitive cells over unimaginable stretches of Earth time?
In our search for the truth, we must not be hood-winked or blinkered. We must take care to remember that when we talk about a fact in science, this represents simply the accepted or the most fervently promoted wisdom at a particular time, so that if and when more knowledge or insight is gained, yesterday's fact may be discarded in favour of a new wisdom. Therefore, we must not disregard *any* information when formulating our understanding. Everything must be weighed and every inconsistency must be acknowledged.
And we must try resolutely not to be biased.

Gas:
representing the Sun

Water and rocks:
representing the planet Earth

Spider monkey

Evolution requires that, spontaneously, hydrogen gas turned into rock and water, which ultimately turned into monkeys and human beings.

# ATOM

## Blips in the history of matter

Orphan elephants
Nairobi

IC
VIEWS

It's a freezing cold day and looking out of the window, the ground is covered with fresh white snow. But what is snow? Nothing but crystallised water molecules. Below 0°C water is crystalline, above 0°C it is liquid. A slight change in its energy state results in rearrangements of its molecules, but the atoms remain just the same.

Standing in the snow is a bird table. It looks solid and inert and it remains the same day after day. There is no simple way of telling that the wood comprising the bird table was once chemically very active and that it had some control over its own being. However, this wood will never be a tree again. From where we stand, it looks as if this wood always was and always will be just an immobile block composed of billions of molecules, atoms, and subatomic particles.

On the table lie some small round objects. They look solid and inert and could easily be mistaken for small flakes of wood. They, too, look as if they will stay that way for ever, just like the bird table, and you can't easily tell that these seeds are quite different from the wood. Although they are just composed of millions of molecules made up of millions of atoms, made up of millions of subatomic particles, these small round things retain the ability to change into a totally different structure - a self-replicating plant - given the right temperature and a drop or two of melted snow.

But now, suddenly coming into view is a really strange group of atoms. The whole looks solid enough but it is entirely different from anything we have seen so far. This is not inert wood or snow. These atoms seem to have considerable control over their destiny. The atoms are individually moving and oscillating in all directions (as all atoms do, according to their temperature) and yet collectively are all moving in the same direction. And these atoms *themselves* decide which direction to take. Amazing. It's chemistry, but not as we normally understand it. It's a bird.

This sparrow lands on the bird table, flapping its atomic wings, beating its atomic heart, circulating its atomic blood, balancing on its atomic feet and swallowing the atomic seeds. Looking closely, through the feathers and skin and right into the muscles, you can see the molecular ratchet movements of the massive actin and myosin molecules, making and breaking molecular bonds as the muscles contract and relax in response to the intracellular movement of calcium ions, which are themselves triggered to move by electrical changes across nerve fibre walls.

Sparrow

But, amazing and determined as the atoms of the sparrow are, they will not stay this way for ever. One day these atoms will be as unenergetic and purposeless in their movements as the atoms comprising the wood. You could then call the sparrow *dead*. But as Quantum Evolutius suggests, was it ever *alive*?

Turn your eyes to the little girl asleep on the bed, and let's confine our consideration to the head on the pillow. At first glance there seems to be no difference between the pillow and the head and both look solid. At the electron microscopic level the adjacent surfaces of the skin and the pillow-case are as irregular as the branches of the top of a tree that has lost its leaves, and where the skin stops and where the pillow-case begins becomes very indistinct. Deeper inside the head is the bone of the skull. For our purposes here, the bone is mainly an inert crystalline structure of hydroxyapatite (we need not concern ourselves with the cells within the bone architecture), not far different from a sea shell on the sea shore. Solid, but as with everything else, if you look closely enough, you can see the atoms with their electron clouds and the subatomic particles deep within*.

Touching the inside of the skull is the brain. A rather delicate thing with the consistency of blancmange (during brain surgery this material all but falls apart if you try to pick it up with forceps). It is made up of millions of cells comprising countless complex molecules, atoms and subatomic particles. Those of us who have been in the position of operating on the living brain can have nothing but the most profound appreciation of its sheer complexity. Each individual nerve cell has hundreds of fine excitatory and inhibitory electrical connections with other nerve cells and the surface of the brain is alive with electrical impulses. Look carefully - can you see a dream whizz by? Not quite. Let's not get carried away. However, this electron activity in the superficial parts of this child's cerebral hemispheres constitutes the single most awesome wonder within the known universe - the human mind in action. But the mind is not something that we can isolate within the brain, since it seems to be a phenomenon arising from the whole.

*For the purpose of our discussion I have ignored the fact that whatever it is, the atom most certainly cannot be likened to a miniature solar system (even though in high school physics it is often convenient to do so). All subatomic particles exhibit the characteristics of both particles and waves, and the atomic world cannot be accurately compared to *anything* that we are familiar with in our macroscopic world.

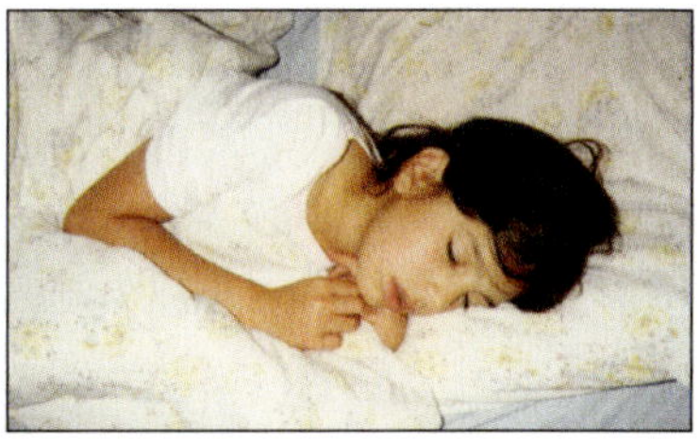

Are both the child and the pillow simply immaterial material?

And stepping back and looking at the scene again, we must ask the question, is this brain only stardust? Certainly, it behaves like stardust. It obeys the same physical laws that are obeyed by every other bit of matter in the universe. Certainly, each atom in this child's brain has a gravitational effect (albeit miniscule) on the most distant star. When this child wakes up and lifts her head off the pillow and consequently moves some of the mass of the planet away from its centre of gravity, the rate of the Earth's spin will be reduced (but you won't be able to measure it) much in the same way as when the body of an ice skater rotates more slowly when the arms are outstretched. But are the thoughts, dreams, aspirations, and the SELF just and only the outward manifestation of the electron cloud movements, or are we missing something? How can we be sure that there is nothing else to consider?

The generally accepted view is that most of the atomic nuclei that exist were formed either shortly following the BIG BANG or later on inside early stars, and therefore all subsequent life forms represent but one aspect of the continuity of matter since the beginning of the universe. You could simply say that the subatomic particles in this child's brain are some of the very same that appeared within a few microseconds of the beginning of the universe and have previously been cycled and recycled through countless gases, liquids, rocks, plants and living creatures, and now here they are inside this head. Over time, these atoms have all, to some degree, been broken down, rebuilt and rearranged, but unlike a second-hand car, these atoms are as perfect as the moment their ancestral component particles were formed. In that sense, they do not age or deteriorate.

So, is this all we are, inconsequential blips in the history of matter?

The *amoeba* is a single-celled animal which may grow to be about 0.1mm in diameter, being just visible to the naked eye. It is thought to represent one of the most primitive of all creatures and indeed, millions of years ago some of its fellows are thought to have been the ancestors of all living animals.

Imagine that you're standing on the muddy floor of a shallow pond. Imagine that you are so small that an amoeba looks as though it is the size of the moon in comparison. It flows silently towards you, its outer cell membrane bulging and changing shape as it unfolds and contracts its pseudopodia, just like a slightly speeded-up film of a

Dr Ralf Wagner

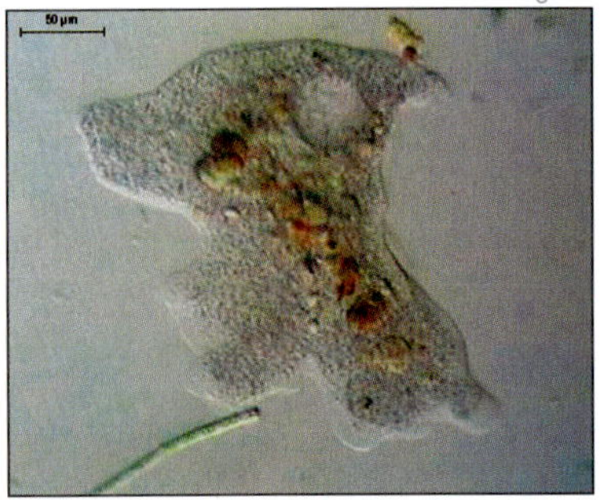


Amoeba

huge fluffy cirrus cloud. As it gets right up close, you can see that its cell membrane is made up of a massive lattice-work of protein and lipid molecules.

Dotted here and there are what look like some holes in the lattice that are just big enough for ions such as sodium and potassium and a number of small molecules like glucose to pass through. But these are no ordinary holes. These holes are actually concavities within the structure of some special proteins embedded within the lattice. If, for example, a potassium ion were to pass nearby, it might be drawn into the concavity and bond onto a part of the protein structure. Immediately, as if by magic, this bonding would cause a dramatic change in the 3-dimensional shape of the protein which would result in the potassium ion being taken through the cell membrane and released on its inner aspect, as if it were on a spring. Simultaneously, a sodium ion would be captured from within the cell and spat out into the outside world. Such might be the nature of the 'sodium pump mechanism' which (without thinking) manages to regulate the concentrations of sodium and potassium ions inside the cell. If you hitch-hike on the next passing potassium ion, you can take advantage of this molecular transport and pass swiftly through the cell membrane and into the protoplasm, a sea of mainly water, potassium, sodium and chloride ions. This sea has a meshwork of special microfilaments which give the cell some rigidity and also has amino acids, enzymes and other molecules for seaweed.

As you swim inwards in this sea of ions and molecules, you pass by what looks like a submarine. This is a mitochondrion, a huge collection of special proteins and other molecules, whose main task is to convert glucose and oxygen into energy which is then used to drive all the processes going on in this sea. Actually, to say that these molecules have a task is quite inappropriate. They simply react with each other as they would in a flask in a chemistry laboratory, and it seems pure coincidence that their combined interactions result in the storage of energy within the phosphate bonds of another molecule called *adenosine triphosphate* (ATP), ready for use in other parts of the cell. Gazing in wonder at the multitude of complicated reactions going on in such a bustling enzyme market-place, there seems to be no obvious overriding control over how or where within the mitochondrion each molecule is guided.

Deeper into the protoplasm and way beyond the mitochondrion you can see an enormous sphere which has a surface lattice much like the outer cell membrane. This is the cell nucleus and as you approach it and pass through one of the pores in the nuclear membrane you find yourself in another sea. Nearby you can see one end of a long double-helical molecule, DNA. This simple chromosome looks huge but compared to human chromosomes it is very small indeed and is but one of some 500-600 in this amoeba. It stretches way out into the distance and you can make out that some segments of this DNA are unzipping, leaving raw molecular surfaces that immediately attract and bond with other molecules called nucleotides which are then strung together to form molecules called RNA.

The DNA is made up of numerous segments called genes, and these genes are codes for particular sequences of amino acids which when linked together produce specific proteins. So here you are watching the initial part of the sequence of molecular events that converts the genetic information in the DNA into a temporary template in the form of RNA, from which chains of amino acids are hitched together to form proteins. These proteins control and direct all the cellular processes that you see everywhere around you. All clever stuff, so it seems, but none of these molecules has a clue about what it is doing, or why.

While you are watching this scene, something new seems to be happening to the DNA strands. Simultaneously all the DNA double helices begin to unzip along their lengths. All of the half-DNA strands attract new groups of molecules and make replicas of their former other halves until they become double helices completely like their original forms. Only now there are double the number, two of each chromosome where previously there was only one. Then, for no apparent reason and without any apparent stimulus, these chromosomes start to coil and shorten. Half the chromosomes move away in one direction and the other half move off in the opposite direction. You sense something dramatic is about to happen and you quickly swim on, until you bump into the far side of the nuclear sphere. Just as you grab hold of the nuclear membrane, it starts to buckle inwards at opposite points until it separates the chromosomes into two identical groups inside two identical nuclei with you inside one of them. You're beginning to wonder whether or not the odd goings-on will cause your own body to split into two! There seems to be nothing controlling what's going on and you have a distinct feeling that it's time to make a run for it. So you clamber through the nuclear membrane, and back into the general cell protoplasm. You can see the two nuclei moving away from each other and you can feel the sound of a low rumble as you witness the very outer cell membrane itself start to buckle inwards between the two diverging nuclei. Not a moment too soon you come up against the inside of the 'back end' of this amoeba. Grabbing hold of a passing sodium ion, you are sucked into the sodium pump and immediately spat out onto the mud floor of the pond. You get up off your knees, wipe the dirt off your clothes and as you look back over your shoulder, you see the two massive clouds of two amoebae flowing silently away from you without the slightest concern for you and without any knowledge that you have just swum around its (or is it *their*) inside.

So what did you see during your little Disney-like adventure? You saw nothing but water and ions and molecules. Perhaps the most important thing is that there was no suggestion of anything that we could call LIFE.

While quantum mechanics tells us that the behaviour of *individual* subatomic particles cannot be precisely known, nevertheless, in general, the way that any atom or molecule behaves (its likelihood of combining with other particles, its stability, its likelihood of breaking down into simpler molecules or atoms, etc.) is determined entirely by its size, shape, and charge, the temperature, and the pressure. So, given any particular set of parameters its chemical behaviour is potentially predictable.

If everything is but atoms and chemistry, then simple chemical options will determine whether or not a group of amino acids will combine with lipids and form simple membranes. Further chemical options will determine whether or not phosphates, sugars and nucleotides will combine to form DNA and whether or not the DNA will sit inside an amino acid-lipid membrane. If we wanted to, we could go on spelling out the chemical options open to each and every aspect of the living cell. Since all particles must simply follow their obligate chemical and physical paths, there is no 'struggle' anywhere, or by any thing. For example, there is no 'struggle' between sulphuric acid and calcium carbonate in a test-tube. And there is no fight between the different nucleotides *cytosine*, *guanine*, *adenosine* and *thymine* when they form interconnecting bonds in DNA. Because of their physical shapes, cytosine *always* pairs with guanine and adenosine *always* pairs with thymine. Cytosine *will not* and indeed *cannot* pair with thymine. There is no struggle, it's just the way it is.

So we could redefine the evolution of life, the origin of species, the survival of the fittest as ultimately...'*the most appropriate chemical conglomeration to remain in its test-tube (the environment)'*.

Cheetah

Look at it this way: a particular part of the DNA in a chromosome will only chemically react in a particular way. There are more ways in which it will *not* react than there are ways in which it will. Similarly, the neighbouring proteins and lipids, amino acids and sugars, intracellular ions, water and all the other cellular components - *at each of their respective chemical interfaces* - will only have certain chemical options open to them. How they then react inside (for example) a brain cell will eventually determine what sort of nerve impulse will be generated, what sort of change in hormone level will be produced etc. These things in their turn will dictate how active, how aggressive, how clever, how determined an animal will be. So those basic chemical options will ultimately dictate whether a particular cheetah is fast enough, determined enough, clever enough, that is - *able* enough - to trip the back legs of a gazelle at 60mph. Furthermore, as he stands panting over his kill and looks over his shoulder, those same types of basic chemical options will dictate whether or not our cheetah will be bold enough, strong enough, determined enough and able enough to keep his prize from the nearby pack of hungry hyenas.

Hyena fossil

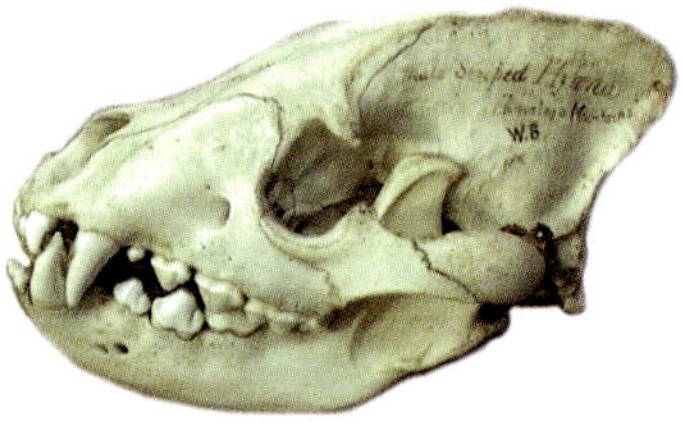

Hyena skull

Spotted hyena

If there occurs an alteration (a mutation) in the chemical structure of part of the DNA, this will necessarily create new options at some of the chemical interfaces pertaining. Such a change might hold the balance between survival and non-survival of the animal in which this mutation occurs. So these basic types of chemical options deep within the mind-numbing complexities of individual cells will ultimately decide the fates of the gazelle, cheetah and hyena. The survival of the fittest, determined by pure chemistry, itself determined by atoms and subatomic particles, *nothing* else. So the idea of evolution would lead us to believe. But is it true that this is all we are: *simply and only inconsequential blips in the history of matter*?

Chance inconsequential blips in the history of matter?
Left: Asian one-horned rhino from Nepal.
Below: Twin Asian elephants recently born in Nepal.

Keith Barnes

CHAP

# The fossil record: *the dead and buried*

# SECTION ONE
# The living and the dead

Allosaurus skull

# THE GEOLOGICAL TABLE

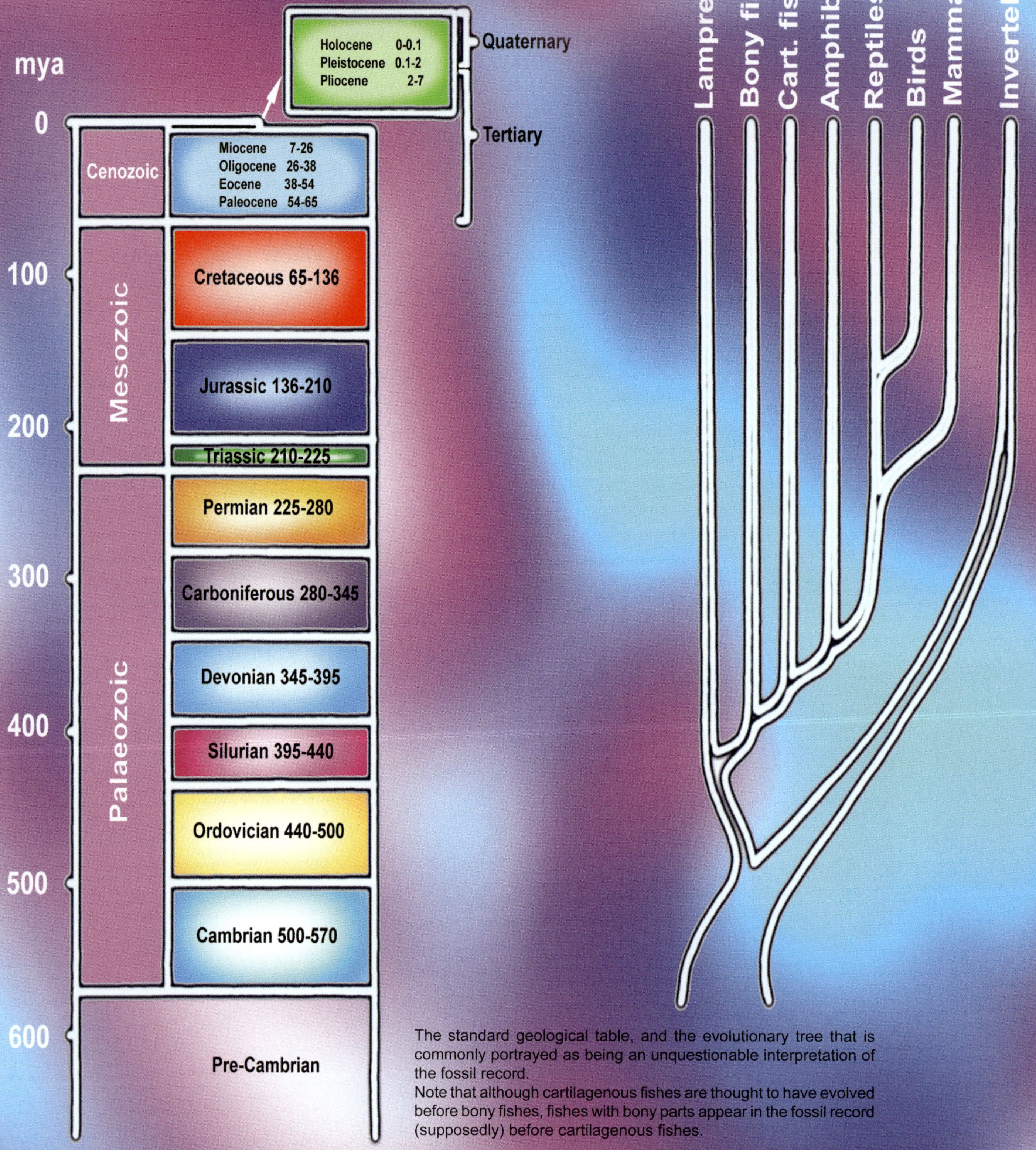


The standard geological table, and the evolutionary tree that is commonly portrayed as being an unquestionable interpretation of the fossil record.
Note that although cartilagenous fishes are thought to have evolved before bony fishes, fishes with bony parts appear in the fossil record (supposedly) before cartilagenous fishes.

Slime growing at the edge of a geyser where the water temperature may approach boiling point. One supposition is that Life started out in hostile places such as this.

It is generally taught that the earth is about 4,500-5,000 million years old (myo), and that the earliest living things were bacteria-like organisms called *prokaryotes* (meaning cells with no nucleus) which appeared about 3,500 million years ago (mya). Some of the early bacteria are thought to have resembled *mitochondria,* which are the intracellular organelles responsible for energy production in all *eukaryotes (*meaning cells with nuclei) which are the cells found in all plants and animals. One theory is that some ancient prokaryotes (which at that time contained no mitochondria) engulfed the mitochondria-like bacteria to form the ancestors of all eukaryotes.

It is thought that once the earth had been shielded from the lethal ultraviolet light by the new layer of oxygen in the atmosphere produced by the early bacteria, early cells gave rise eventually to single-celled animals (*protozoa)* some 2,500-3,500mya, and multi-cellular life started to flourish some 600-1,000mya.

This sea squirt is an invertebrate and has no skeleton. It is supposed that all vertebrates (animals with backbones) once had ancestors that were invertebrates.

*Vertebrates* (animals with backbones) are thought to have originally evolved from *invertebrate* ancestors. The earliest relatives of the vertebrates are thought to be those animals which possess a *notochord*, which is a rod-like structure running along the length of the back of the animal, and which is thought to be the ancient precursor to the spinal column. Although the adult *sea squirts* and the *echinoderms* (starfishes and sea urchins) look unlikely candidates for the job of vertebrate ancestors, the larval forms of these creatures resemble tadpoles in shape and possess a notochord running the length of their bodies. So ancient sea squirts and sea urchins are thought to have diversified to become primitive jaw-less fishes like *lampreys* (500mya); *sharks* (395mya); *bony fishes* (435-395mya); *amphibians* (380mya); *reptiles* (350mya); *birds* (150mya); *mammals* (200mya) ...and *human ancestors* (about 6mya). So the story goes. This *theory* of evolution was in part already emerging by the early part of the nineteenth century but seems to have completely captured people's imagination after the publication of Charles Darwin's *'The origin of species'*:

Starfish. It is supposed that vertebrates (e.g. fishes, frogs, lizards, birds, whales, apes and humans) evolved from creatures like this.

> **Maybe** fish had developed muscular fins and crawled on to land to become amphibians; **maybe** amphibians in their turn had developed water-tight skins and become reptiles; **maybe**, even some ape-like creatures had stood upright and become the ancestors of man. (My bold). Charles Darwin. 'The origin of species', paraphrased by David Attenborough. ' Life on Earth'. Collins/BBC. 1979. p13.

To be precise, Darwin did not in any way *prove* the *origin* of species by the natural selection of inherited characteristics: in fact, he did little more than simply document the commonly observed tendency for living things to show *variation*, and then propose that the principle behind any such variation might be extrapolated so as to become the basis by which ultimately one creature could become another completely different creature.

Certainly, from one point of view the fossil record does seem to *suggest* a series of progressive changes over time, from simple to more complex organisms. And over the last 150 years it has become a common belief that the fossil record shows well- established *proof* of the theory of evolution:

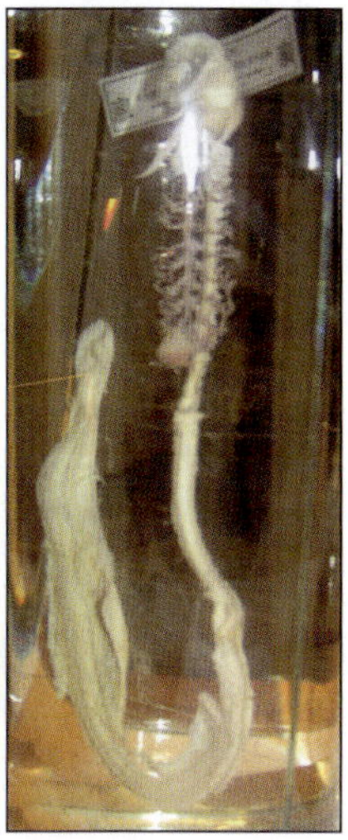

Lamprey skeleton.
Lampreys are jaw-less creatures with eel-like bodies and skeletons made of cartilage. They are blood-sucking parasites with mouths containing tiny teeth, which they use to cling onto other fishes. Lampreys have a flat tail and a dorsal fin but lack both pectoral and pelvic fins.

Although the largest dinosaur eggs were up to 40cm long, most were smaller than an ostrich egg (above) or emu egg (below).

Fossilised dinosaur eggs

Fossilised dropping

Baby mammoth. Whilst in most cases the soft tissues rot to leave only the bones, whole animals may be preserved if buried under ice.

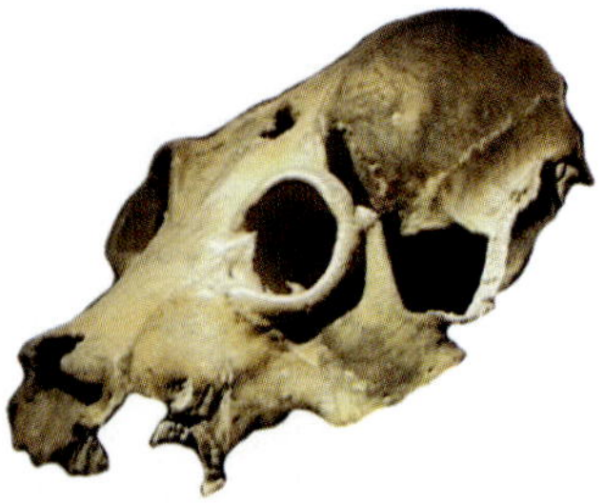

Skull of extinct giant lemur

> Fossils provide the most tangible **proof** of the evolution of living organisms...fossils demonstrate **beyond any shadow of a doubt** that living species are not fixed immutables, but are on the contrary, the product of a very long series of changes whose history, through fossil remains, can now be reconstructed in outline and in several cases, also in detail. (My bold). Paolo Arduni and Giorgio Teruzzi. 'The MacDonald Encyclopaedia of fossils'. Little, Brown & Co. 1993. p15.

But is such confidence justified from the evidence?

### *Fossilisation can occur in a very short time*

Although there is little information about how long is needed to convert a bone into a fossil, it is commonly understood that the process of fossilisation requires millions of years. Most fossil bones have been converted into rock by mineralisation, a process in which the bone has been replaced by a mineral substance that was present in the water permeating the entombing sediment. Calcium carbonate (in the form of calcite) or silicon dioxide (in the form of silica) are the commonest mineral substances found in fossils, although iron pyrites and other minerals are less commonly found.

Any sediment layer in which fossils are found buried is likely to have been deposited in minutes or hours or days at most, rather than over thousands or millions of years as might be commonly supposed. There is no other way to explain the fossils of the large mammoth graveyards; fossils of animals dying whilst fighting, eating or giving birth; and footprints and coproliths (fossilised droppings).

If rock layers truly represent millions of years, then we should not find (as we do) examples of fossil trees embedded upright within thick coal seams whose thickness (supposedly) represents millions of years of time. Clearly, such fossil trees point to millions of tons of plant-matter sludge having been deposited over and through ancient forests in a very short period of time - perhaps in only a matter of days.

The Messel Quarry near Darmstadt in Germany contains many fossils of mammals (supposedly) 35-65myo. Imprints in many of the Messel specimens show great detail of hair and skin, and soft tissues including internal organs and gut contents. And impressions of other soft tissues of animals have been found fossilised; for example:

> One lizard can be identified because its forked tongue is preserved...Fossil chameleon showing head and neck with bones and skin preserved in a Miocene deposit from Rusinga Island, Kenya. LB Halstead. 'Hunting the past'. Hamish Hamilton. 1982. p43.

Elsewhere there are even fossils of fishes showing details of *muscle fibres*, so proving that the entombing sediment containing the fossilising minerals must have hardened considerably before the onset of any decay. So one important question is: just exactly how old are fossils? In fact, contrary to what may be a general understanding, not all fossils are hundreds of thousands or millions of years old. Fossils of extinct lemurs have been found in Madagascar that are very young indeed:

> Some fossils of extinct species found at Amparhinidro in the north-west of the island are only about 3000 years old, and those at Lake Itampola in the south are only about 1000 years old. Steve Jones, Robert Martin, David Pilbeam. 'Cambridge Encyclopaedia of human evolution'. Cambridge University Press. 1996. p180.

Therefore it is quite possible that fossils that are younger than 1000 years exist somewhere but have not yet been found. This would not be all that surprising since we are dealing here with a process which is plain chemistry. If that is the case, then impregnation of minerals into bones *should* take less time than we are commonly led to understand. And, since a bone is not a solid mass of chalky rock but is in fact a honeycomb structure with millions of connections between the outer surface and the whole of the inside, then if a bone were to be immersed in any liquid or semi-liquid fluid, we should expect that liquid or fluid to be able to permeate most of the innermost reaches of the bone material in a matter of minutes, hours, days or months, depending on its viscosity and the pressure applied.

This porosity of bone can be easily demonstrated. The nerves to our teeth pass through canals deep within the bones of the upper and lower jaws. Dental anaesthesia can be achieved by injecting local anaesthetic into the gums *over* the surface of the jaws without having to inject around the main nerves at the back of the mouth. This is because the anaesthetic permeates into the bone and thus anaesthetises the nerves lying deep within. Many of those who have visited a dentist will know that such permeation can take place in less than 15 minutes. Hence it is not out of the question for the bones of a carcass to be replaced by new minerals to form fossils, all in a very short space of time. And indeed there is concrete evidence to show that fossils can be formed in an *extremely* short time:

> A relatively common fossil found on many parts of the coast of northern Australia, *Thalassina* is a ...kind of lobster...Throughout its life, a single *Thallasina* can produce a dozen or more shells...the cast-off **shell can be fossilized in** a very short time, **perhaps less than a year. Some of these fossils are so young that the animals that shed them may still be alive**. (My bold). Arthur Busbey III, Robert Coenraads, David Roots and Paul Willis. 'Rocks and fossils'. Harper Collins. 1996. p96.

Other examples show that the fossilising process must have at least commenced before the onset of any decay:

> Comparisons with the effects of decay in experiments on…fish suggest that phosphatization must have taken place within **a few hours**. (My bold). Stephen Donovan (Ed). 'The process of fossilization'. Belhaven Press. 1991. p129.

At the other extreme, shells that are (supposedly) over *one hundred and forty million years old* can be found without having undergone any chemical change:

> In some cases, the original shell or bone of an organism is preserved **unaltered**...the shells of several ammonites of the genus Quenstedtoceras are preserved [unaltered] in clay. (My bold). Chris Pellant. 'An illustrated guide to fossils'. Dragon's World. 1995. p8.

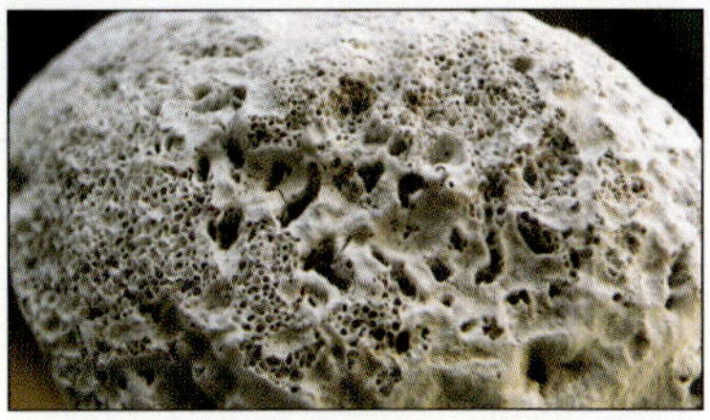

Part of the skull of a gray whale showing honeycomb structure of the bone.

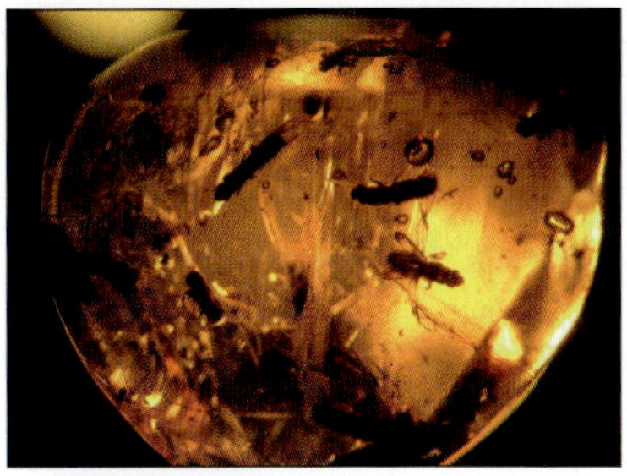

Insects preserved in amber.

Thalassina. A living individual can have shed a shell that is already fossilised.

Above: Dinosaur footprint
Below: Human footprints
It may be thought that it must take millions of years for sediments to become rock. In July 2001, these footprints were found in the Canadian Rockies between the receding lower edge of the Athabasca glacier and the marker indicating the position of the glacial edge in the year 2000.
The interesting thing is that although these prints were made less than one year before this photograph was taken, the ground is not dried mud - it has become solid rock.

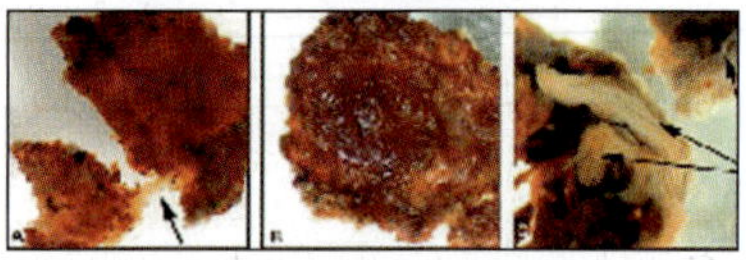

T rex marrow. Tissue fragments from a Tyrannosaurus rex femur are shown at left, when it is flexible and resilient and when stretched (arrow) returns to its original shape. The middle photo shows the bone after it is air dried. The photo at right shows regions of bone showing fibrous character, not normally seen in fossil bone.
*From M Schweitzer et al. 'Soft-tissue vessels and cellular preservation in Tyrannosaurus rex'. Science. 25th March 2005. Vol 307. No. 5717. p1952-1955.*

**Megalania**

Megalania was a 6m iguana-like lizard thought to have become extinct long ago. However, when paleontologist Ralph Molnar found a megalania bone with surprisingly fresh appearances, he suggested that the creature may have lived in recent times:

> *'Here we have part of the hip bone...now, this specimen actually looks as if it came from an animal that died two to three hundred years ago...dry, chalky, this sort of thing...doesn't actually mean that it died two to three hundred years ago...in Montana I've seen dinosaur bones that looked like they came from animals that died two to three hundred years ago...and **I know very well** that they died much longer [ago] than that. So it gives the suggestion, however, that megalania may have been around fairly recently.'* (My bold)
> 'Sightings of huge sea serpents'. Animal Planet Broadcast. 21st May 2002.

Molnar can't seem to make his mind up about when his megalania specimen died, but he cannot have it both ways. How does he know *very well* that dinosaurs died out millions of years ago? If the fresh-looking megalania bone represents an animal that died recently, then (in the absence of any other evidence) the same inference must be made regarding the Montana dinosaur bones he refers to.

The absence of fossilisation in old specimens is not restricted to shells. In 2005 the fossilised bones of a *Tyrannosaurus rex* (specimen MOR 1125) were found in Montana. When a part of the femur containing the marrow cavity was examined in the laboratory after deliberate chemical dissolution of the minerals, the remaining tissue was found to still be elastic, and microscopy showed blood and marrow cellular architecture:

> Removal of the mineral phase reveals transparent, flexible, hollow blood vessels containing small round microstructures that can be expressed from the vessels into solution. Some regions of the demineralized bone matrix are highly fibrous, and the matrix possesses elasticity and resilience. Three populations of microstructures have cell-like morphology. M Schweitzer et al. 'Soft-tissue vessels and cellular preservation in Tyrannosaurus rex'. Science. 25th March 2005. Vol 307. No. 5717. p1952-1955.

This was a surprising finding since the specimen was dated as being (supposedly) 70myo. Some make the comment that we will have to accept that soft tissues can survive for far longer than we previously thought. But, can soft tissue inside a fossilised bone really be 70 *million* years old? A more logical line of thought would be to re-evaluate the dating of the specimen, and this common sense approach is backed up by other evidence. Since the Tyrannosaurus rex find, many other dinosaur fossils have been examined and they show that soft tissue remains inside fossil bones are not rare. Furthermore, a number of other dinosaur bones have previously been found in Montana and the Isle of Wight (for example, an iguanodon), along with the extinct lizard *Megalania* from Australia, that are *unfossilised*, having the look and texture of other bones that have been buried for only a relatively short period of time.
Evidence from wood and other materials also confirms that fossilisation can occur in an extremely short period of time. When it was noticed that branches which had fallen naturally into the silica-rich acidic hot springs (70°C, pH3) at Tatyama, northern Japan, had become fossilised, an experiment was set up in which freshly-cut branches were suspended in the water using stainless steel wire. The branches were subsequently examined over a period of 7 years. After 3 years there was no sign of any change. However, after 7 years, 40% of the wood had become fossilised by the silica salts.

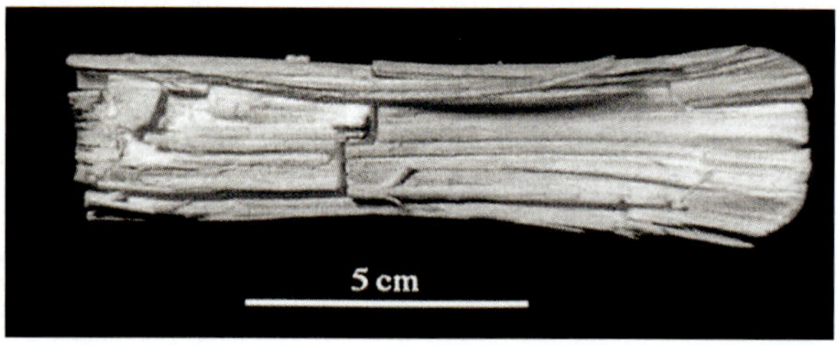


Tatyama experimental wood, 40% fossilised in 7 years.

Furthermore, when cross-sections of the experimental wood were compared to naturally-fossilised wood from Miocene strata (supposedly 7-26 *million* years old) from the Noto Peninsula of north-western Japan, the appearances were identical, showing that both woods had been formed by essentially the same process:

> We conclude that some silicified wood near volcanic or pyroclastic rocks could have formed...during a fairly short period of time, in the order of several tens to hundreds of years, with the same mechanism as described in this study. H Akahan et al. 'Rapid wood silcification in hot spring water'. Sedimentary geology. Vol 169. 2004. p219-228.

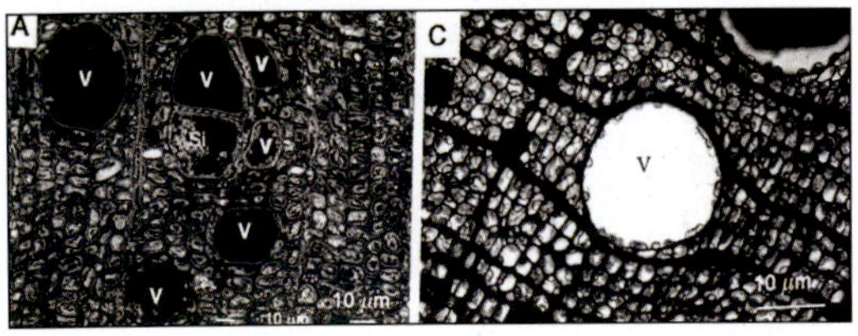


Both the naturally-fossilised (left) and the experimentally-fossilised wood (right) have the same appearances, showing that they were both likely to have been produced by essentially the same process.

Although the authors of the study suggest that complete fossilisation might take 'several tens to hundreds of years', extrapolation of the rate of fossilisation data suggests that the experimental wood may well have become completely silicified within 20 years.

The Natural History Museum in London displays a section of a portion of a tree trunk some 30cm in diameter that has been completely fossilised by copper salts. Since it is known that the wood had been used to prop up the roof of a copper mine 2,400ya, this tree trunk had to have been fossilised in less than that time.

Support poles are regularly placed into the mineral-rich broth that comprises the canals of Venice. When these are removed for any reason, it is found that they tend to be quite brittle as they have become partially fossilised.

To make the point that fossilisation of dinosaur bones would have occurred very rapidly, water (72°C) from the hot springs at Karlovy Vary in the Czech Republic has been sprayed onto toy teddy bears and paper roses, resulting in the items becoming completely per-mineralised within just 2 weeks.

Artificial petrification of teddy bear and paper rose.

So, the evidence shows unequivocally that fossilisation of wood does not take millions of years, but can occur in less than twenty years. It is also a fact that it does not take millions of years to produce oil or coal, but a very short time with high temperatures and pressures:

> We can simulate natural oil generation in the laboratory from organic-rich source rock. New Scientist. 17th August 2002. p17.

This tree trunk has been fossilised in less than 2,400 years.

When support poles in Venice are removed, they are frequently found to be partially mineralised.

**Coal formation is rapid**

It is well described that coal can be formed very quickly in the laboratory:

> *Coal can be formed in a few hours in the laboratory.*
> George Hill. 'Aspects of coal research'. Chemical Tech. 1921. 296.

> *Coal formed from heating wood, clay and water at 150o for 36 weeks.*
> Argonna National Laboratory. Organic Chemistry. 6:483-471. 1984.

Note that since coal seams in Oregon and China are (respectively) 200 and 300 feet thick, this means that at a compression ratio of 10:1 the coal must have formed from a plant layer that was originally 2000-3000 feet thick. This could only have been the result of catastrophic flooding and compositing of huge amounts of plant debris.

Since there is no doubt that fossils can be formed in a very short time, it is quite reasonable to question the generally accepted dates at which extinct creatures lived.

### *All creatures could have lived contemporaneously*

Because the process of fossilisation commonly requires that death is followed by rapid burial in very specific circumstances, it is obvious that there must be a large number of creatures that have never died and never will die in circumstances that have resulted or will result in fossilised evidence of their existence. So, although the world's museums contain millions of fossil specimens, the fossil record represents only a very tiny window into the history of life on Earth. It has been suggested that there are between 5 and 50 million species alive today, and that since life first started, perhaps some 1 to 3 billion species have come and gone. It is supposed that of the 1 to 3 billion species that have existed, only at most a few million (less than 1%) have been fossilised. Of those fossilised, it is suggested that only about 10% have so far been discovered. In particular, fossils of vertebrates are relatively few:

> In general, the fossil remains of higher vertebrates are uncommon for many reasons, the major one being that the land surface is not conducive to preservation. Although they may be abundant in certain places, they are usually difficult to find... Arthur Busby III, Robert Coenraads, David Roots and Paul Willis. 'Rocks and fossils'. Harper Collins. 1996. p236-7.

> In the majority of sedimentary sequences vertebrates are exceedingly rare or absent. Stephen Donovan (Ed). 'The process of fossilization'. Belhaven Press. 1991. p270.

It therefore follows that the absence of vertebrate fossils in sediments of the same (supposed) age as those containing ancient marine invertebrates cannot be taken to imply that vertebrates were not living at that particular time or place:

> Despite the robust nature of dense bone and tooth enamel, many sediments contain little or no vertebrate tissues, but were nevertheless deposited in environments similar to those which today contain many vertebrates. The explanation that vertebrates were absent from the fauna is not usually tenable. Stephen Donovan (Ed). 'The process of fossilization'. Belhaven Press. 1991. p278.

Tuatara. Unlike true lizards, this unique reptile has no external ear, no penis (insemination occurs simply by cloaca-cloaca contact), may live to over 100 years in age, and takes around 20 years to become mature.
Another interesting feature is that its respiration rate may be as low as one breath *per hour*.

Furthermore, since all animal species that are represented in the fossil record must have existed before they died, they must all have been alive some time *before* the time thought to be represented by the sediment in which their bones are found interred. So, just *how long* ago a species lived before its earliest known fossil remains an open question. In addition, any particular animal could have lived at a later date than its youngest fossil.

A number of different creatures demonstrate this last point. The *Tuatara* is a small beakhead lizard found only in New Zealand, and is celebrated amongst biologists because its pineal gland (which lies at the back of the midbrain) acts as a 'third eye' (although, to be accurate, whilst it is true that the pineal gland in the tuatara is sensitive to light, it has no lens or features resembling an eye). Depending on the duration of daylight, the pineal gland of mammals produces varying amounts of the hormone *melatonin,* which has an influence on an animal's behaviour. Despite the fact that the tuatara is clearly alive today, there are no fossils of beakheads found anywhere dated later than (supposedly) 135mya. It is curious indeed that a creature with apparently so little on its side with regard to the struggle for existence could survive so many millions of years *without* significant change and *without* leaving any fossil evidence of its continued presence here on Earth.

Until recently, a tiny 8mm-long member of the wasp family *Xylelidae* was thought to have been extinct for the last 20my, even though:

> The species was known to have been in existence as far back as...200mya.
> Chris Cook. BBC Wildlife magazine. Nov 1995. p25.

However, museum specimens of wasps which were caught *living* in California in 1937 and 1966 have now been identified as the same wasp that has remained unchanged for (supposedly) 200my.

Glass sponge reefs were previously thought to have died out in the Jurassic era (136-210mya), but:

> ... have recently been discovered in British Columbia's Hecate Strait...living reefs of glass, some as high as 21 metres...The (supposedly) 9,000-year-old Hexactinellid sponge reefs...cover about 1,000 square kilometres. (My parenthesis) D Wood. 'Living fossils in peril'. British Columbia magazine. Winter 2007. p17.

Extinct giant elk

Elk from Alberta, Canada

New finds show that the Woolly Mammoth and Giant Elk were alive a few thousand years after the time when they were previously thought to have become extinct:

> Previous assumptions that the megafauna of northern Eurasia had disappeared by the Pleistocene/Holocene transition (0.1mya) were first challenged a decade ago by the discovery that the latest woolly mammoths on Wrangle Island...were **contemporaneous with ancient Egyptian civilization**...The giant deer or 'Irish elk' survived to around...(supposedly) 7,700 yr ago...in western Siberia - **more than three millennia later** than its previously accepted terminal date. (My bold and parentheses).
> 'Pleistocene to Holocene extinction dynamics in giant deer and woolly mammoth'. Nature. Vol 431. 7th October 2004. p684.

Mammoth model

The *Diatomyidae* family of squirrel-sized rodents were thought to have become extinct (supposedly) millions of years ago, but recently a fresh example was found for sale in a Laos meat market*:

> Member of the previously known rodent family Diatomyidae...thought to have died out (supposedly) 11mya. (My parenthesis). Science. 10th March 2006.

Having (supposedly) first appeared about 300mya, the *Coelacanth* was thought to have become extinct about 65mya - until a fresh specimen was caught and identified in 1938. Unknown to the western scientific community, the fishermen of the Comoros Islands in the Indian Ocean not infrequently found the coelacanth amongst

*Some say they are best when lightly fried in garlic butter.

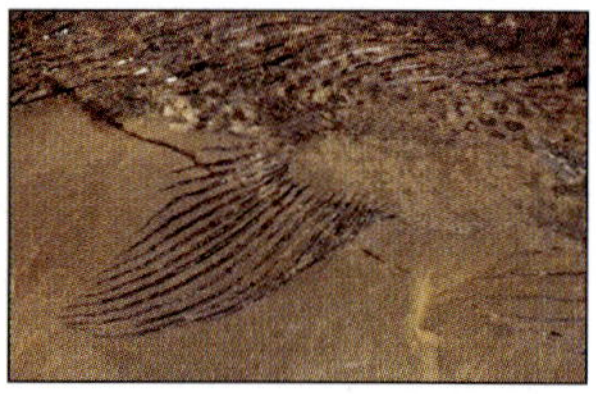

Coelacanth: modern specimen (far right), fossil (above left) and lobed fin close-up (above right)

their catches. More recently, specimens have been found as far away as Indonesia and South Africa.

This creature has (supposedly) lived on Earth for at least 300 million years without any significant change in form, and yet no coelacanth fossils have been found that have been dated younger than 60-70 million years. The coelacanth is about 2m long and has pectoral and pelvic fins which both have stumpy bases. The fins cannot be classed as legs, but the coelacanth's ancestors are believed to have made the transition from the sea to the land and perhaps to have had semi-aquatic behaviour similar to that of modern mudskippers. However, it is notable that the coelacanth lives in very deep water and does not inhabit the shallows *at all*. Furthermore, film of the coelacanth swimming in deep water shows that it does not walk - even on the seabed. (Interestingly, after detecting an artificial electric current, the coelacanth adopts a curious vertical head-down pose while slowly rotating in the vertical axis.) Its limbs are therefore *not* 'on the way' to becoming adaptations to the land. This is an example of how just looking at the fossil bones of an animal can be very misleading as regards not only how and where that animal lived, but also that creature's relationship to other creatures.

On the subject of walking fishes, propelled by its tail and body movements the *mudskipper* waddles across land and uses its pectoral fins as stabilisers. The *climbing perch,* on the other hand, uses the bony spikes on its gill covers to drag itself over land, its pectoral fins being no different from those of any normal fish. Both methods are equally effective. However, since neither the bones of the pectoral fins of the mudskipper and climbing perch, nor the gill covers of the climbing perch resemble amphibian limbs, we should expect that if amphibians have evolved from fishes, then as well as ancient fins evolving into legs, some ancient gill cover spikes should also have evolved into legs. So today, we ought to find some amphibians with legs arising from the side of their necks. In addition, perhaps we should also expect to find amphibians with *three* pairs of legs (arising from the neck, pectoral and pelvic areas).

Returning to gaps in the fossil record, the egg-laying mammals (*monotremes*) such as the *echidna* and *duck-billed platypus* must surely have existed over 200mya if they were ancestral to all other mammals. However, although some say that there is *no* fossil record of monotremes before (only) 2mya:

> There are no paleontological data for monotremes prior to the Pleistocene, even though they display certain very archaic characteristics such as the laying of eggs... Paolo Arduni and Giorgio Teruzzi. 'The Macdonald Encyclopaedia of fossils'. Little, Brown & Co. 1993. p40,

others are confident that:

> ...when early Cretaceous monotreme fossils were found they neatly filled an embarrassing gap in the fossil record! Michael Benton. 'The rise of the mammals'. The Apple Press. 1991. p90.

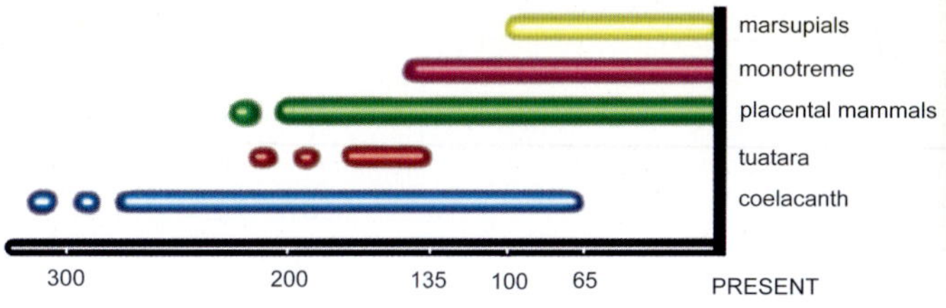


The fossil record does not truly reflect the times during which different creatures lived.
The lines represent the supposed dates of sediments in which the fossils of these animals have been found. For example, no coelacanth fossils have been found which are younger than (supposedly) 65myo.

But 'early Cretaceous' means 100-140mya at the most, so even if we were to accept Benton's view, there would still remain a huge gap of over 60my, since monotremes must have existed over 200mya. As to the *origin* of the egg-laying mammals (presumed to have evolved from egg-laying reptiles), the evidence that exists is frail:

> The body design of the echidna and platypus is undoubtedly of great antiquity, but we have no hard evidence to indicate which fossil reptiles were their ancestors. David Attenborough. 'Life on Earth'. Collins/BBC. 1979. p205.

If monotremes were the earliest mammals, they would have demonstrated some significant advantage over their reptilian ancestors (whatever they supposedly were). And if they had some advantage, they must have been able to breed in large enough numbers to ultimately give rise to more advanced mammals. And if that were the case, there must have been *ample* opportunity over many millions of years for *many* of their kind to have been preserved as fossils. Most certainly, fossils of the duck-billed platypus, which lives in and near water, should be particularly common in ancient sediments. If the fossil record is to be believed, it actually suggests that placental mammals existed many millions of years *before* egg-laying mammals. So, the fossil record of monotremes contradicts the notion that monotremes are *ancestral to* or *more primitive than* placental mammals. Among the primates, fossil bones of *baboons*, *orang-utans* and a small number of *extinct apes* and *ancient humans* have been found, but chimpanzees and gorillas are only represented in the fossil record by a few teeth which, dated around (supposedly) 4myo, do not pre-date *australopithicines*, their supposed descendants.

Many fossils are found in geographically-circumscribed areas. However, this in itself does not mean that the animals did not live elsewhere. For example, the *Antrimpos* is a type of shrimp which is supposed to have lived over a period of 180my, dying out some 70mya. Although fossils of Antrimpos are found *only* in Europe and Madagascar, clearly it must also have lived in areas *between* these two sites. And if it lived between these two locations, it most probably also lived elsewhere, being well able to swim in any direction. Indeed, after 180my, such a free-living marine animal (its favoured climate permitting) *must* have extended its territory practically world-wide. Similarly, *Tenticulites*, which are a type of mollusc with a long straight shell, are supposed to have lived for 155my, dying out some 345mya. Yet tenticulites are only found in Europe and North America.

*Mesolimulus* fossils are found in Bavaria. The mesolimulus looks like a type of king crab and is very similar to the living horseshoe crab, *Limulus*. It is thought that the limulus has been around for 140my but it is only found off the Atlantic coast of North America, and off the Pacific coast of Asia. Since these two populations of limulus are the same, and are very similar to the extinct mesolimulus, we should expect that the observed ranges in which both mesolimulus and limulus are found should be far more extensive than that suggested purely by their fossil record, particularly considering the (supposed) great time-span of their existence.

Thus, the absence of fossil evidence at any site cannot be taken to imply that any particular animal did not live in that particular locality. Furthermore, if a fossil is not found in any particular sedimentary layer, this does not mean that that particular organism was not alive elsewhere at the time supposedly represented by that sediment. It only means that the animal was not involved in the particular interring event represented by that sediment.

But one might argue that the reason, for example, that there are no fossils of tuataras in sediments dated between 135mya and the present is because the environment was different, or there simply was no opportunity for individual tuataras to die in a way conducive to them being fossilised. However, this is not convincing logic. One hundred and thirty-five million years is such an unimaginably enormous period of time that *any* variations in the likelihood of fossilisation for *every* type of creature must surely have been evened out. So we should predict that there *ought* to be fossils of tuataras dating between 135mya and the present, and pro rata, in as many numbers as for time periods prior to 135mya.

Thus, we could state the *Coelacanth Principle*:

In the absence of any direct evidence to the contrary, *any* creature could have lived at *any* time *before* or *after* the time of its earliest or latest known fossil.

This principle forces us to two conclusions: firstly, since there is no way that we can have any direct evidence from pre-history, we are not free to assume the times at which ANY organism lived...whether we are discussing a tree frog, or zebra, or Tyrannosaurus rex, or penguin...or Man. Furthermore, just how long an animal could have existed before the date given to its supposed earliest fossil, cannot be easily defined. Clearly, it could have been a short time or a very long time. If one marine animal such as the coelacanth can (seemingly) exist for 65my without leaving any fossil trace, then it *could* be the case that modern whales were present at least 65my before *their* first fossils suggest. And if 65my, why not 100 or 200my before their first fossils suggest? And since our own existence is subject to the same logic, we are obliged to at least consider the possibility that Man could have existed 20, 30, or 300mya.

Secondly, all living and extinct organisms *could* have existed *contemporaneously*.

**Grand Canyon, USA**

It is questionable whether the fossils found in different sediment layers actually represent the evolution of life forms. The uppermost rocks of the Grand Canyon are dated as being about 200 million years old and contain traces of reptiles and impressions of ferns and insect wings. Further down there are fossils of fishes, and further down still are fossils of shells and worm trails. There are no mammals in the layers of the Grand Canyon, and there are no dinosaurs. In fact, *nowhere* on the planet are there found sequences of fossils in the order invertebrates-fish-amphibians-reptiles-mammals (from deep to superficial) in contiguous layers of sedimentary rocks representing the different ages of the Earth.

### *The fossil record is a composite artefact*

The Grand Canyon is an exceptionally deep, steep-walled canyon in north-western Arizona, excavated by the Colorado River. It is about 349km long, 6 to 29km wide, and more than 1.6km deep, and cuts steeply through an arid plateau region that lies between about 1525 and 2745m above sea level. It is commonly suggested that most of the strata were originally deposited as marine sediment, indicating that for long periods of time the canyon area was the floor of a shallow sea. The uppermost rocks are dated as being about 200 million years old and contain traces of reptiles and impressions of ferns and insect wings. Further down there are fossils of fishes, and further down still are fossils of shells and worm trails. No fossils of mammals or dinosaurs have been found in the layers of the Grand Canyon. As a matter of fact, *nowhere* on the planet are there found sequences of fossils in the order invertebrates-fish-amphibians-reptiles-mammals (from deep to superficial) in contiguous layers of sedimentary rocks representing the different ages of the Earth. Instead, the fossil record is an artificial composite constructed out of comparisons of diverse rocks from all over the world. Fossils of the great dinosaurs are not found lying underneath layers of mammal bones. For example, the mammoths are well known to be found in their thousands in Siberia - but they do not lie on top of layers of dinosaurs. Extinct apes are found in Africa and Europe and Indo-China. But their bones do not lie on top of fossils of primitive mammals, nor the latter on top of dinosaur remains.

Whilst from one point of view it appears that the fossil record shows seemingly primitive or simple creatures being found in sediments supposedly millions of years older than those sediments containing seemingly more advanced or complex creatures, in actual fact the manner in which dead creatures find their final resting places is both complex and diverse. For example, carcasses may be transported over great distances by predators, fluvial systems or ocean currents:

> Post-mortem drifting of floating carcasses may be prolonged…especially in the case of large vertebrates such as baleen whales. It is possible that carcasses may travel tens or even hundreds of kilometres in ocean currents before descending to the sea floor. Stephen Donovan (Ed). 'The process of fossilization'. Belhaven Press. 1991. p275.

Although the Jurrasic cliffs (top) at Lyme Regis, UK, have yielded fossils of diverse marine reptiles, they largely contain huge numbers of ammonites of varying sizes. This boulder (above) contains only ammonites and gives a clear demonstration of hydrodynamic sorting.

Fossilised shoal of fish

And indeed, many fossils are found entombed many miles away from where they lived:

> Post mortem drifting has been elegantly demonstrated in the Lower Oxford Clay (Callovain, Middle Jurassic) of the UK, where several dinosaur skeletons have been reported from fully marine sediments. The nearest landmass to the discovery sites is thought to have been at least 80km distant. Stephen Donovan (Ed). 'The process of fossilization'. Belhaven Press. 1991. p275.

It is also well recognised that during an entombing event dead creatures are subject to some form of segregation (whether by size, shape or density) by the effect of fluid movements, called *hydrodynamic sorting*:

> Sorting of different skeletal elements is, however, an important process… vertebrate accumulations in the Permian Dokum Group of north Texas frequently contain an abundance of palaeoniscoid fishes, acanthodians and less commonly, therapsid reptiles. In a 20m section I counted more than five individual bone beds. One bone bed contained only the spines of acanthodians, while another contained only their dermal denticles. Another bone bed was dominated by palaeoniscoid scales, while a fine sandstone contained scattered scales and some larger palaeoniscoid skull bones. The fifth bed contained teeth and bones of *Dimetroden* and of *Ophiacodon* as well as elements present in all the other bone beds. Stephen Donovan (Ed). 'The process of fossilization'. Belhaven Press. 1991. p281.

Here lies a serious problem. Consider, for example, bony fishes (*Osteichthyes*):

> They first appear as minute scales in microresidues of Late Silurian age [over 408mya]...by the late Palaeozoic [over 230mya] their [complete] remains occur all around the world. John Long (Ed). 'Palaeozoic vertebrate biostratigraphy and biogeography'. Belhaven Press. 1993. p14-16.

It cannot possibly be that at one time sludges engulfed dead fishes in such a way that only their scales would remain entombed, and that many millions of years later sludges were somehow so different that they allowed the preservation of skeletons of complete fishes. This would be bizzare indeed. A more likely explanation is that the scales and the skeletons may represent simply the same individuals with parts of their anatomy having been separated both horizontally and vertically by hydrodynamic sorting.

Clearly, since the finding in a particular sediment layer of a scale or tooth may, in fact, tell us *nothing* about when or where its owner lived, such hydrodynamic sorting has great implications when considering whether a particular creature was either ancestral to or contemporaneous with another.

But, of course, before we can say for *any* fossil when the creature it represents lived, we must have a reliable dating method. The dating of any fossils in a sedimentary layer is in part determined by *comparisons with similar fossils found in other rocks* of *(supposedly) known* age:

> Suppose that a stratum containing a dinosaur bone is sandwiched between two volcanic ash layers. Ideally, an absolute age date could be obtained from each of the ash layers. We would know that the bone was younger than the lower layer, but older than the upper layer. Depending upon how much time separates the two layers, the bone between them can be dated with greater or lesser precision...Though all that looks fine in theory, datable ash layers rarely fall out of the sky because stratigraphers need them! **More commonly, rocks for which absolute ages can be obtained are widely separated, not only in time, but also geographically.** The challenge then becomes correlating the strata of known absolute age with those of unknown age...**For example, if one knows that dinosaurs lived from 228 to 65mya, then any rock containing a dinosaur bone fragment must fall within that age range.** The question is how precise a date can it really give? Biostratigraphic correlation - the linking of geographically separated rocks based upon the fossils they contain - can be very precise...the fact that many species of organisms have existed on earth for 1-2 million year intervals enables them* to be used as powerful dating tools...T. rex itself lived for only about two million years (from 67-65mya). **Thus, if we found a Tyrannosaurus rex fossil, we would know that the sediments** in which it was found were **deposited between 67 and 65mya.** (My bold). David Fastovsky and David Weishample. 'The evolution and extinction of the dinosaurs'. Cambridge University Press. 1996. p26-27.
>
> (* so-called '*guide fossils*')

But there cannot ever be a *guide fossil* unless the time span during which the represented creature lived is known *directly*. Since any animal could be living well before and well after the time supposedly represented by the sediment that interred it, then it follows that if we found an isolated Tyrannosaurus rex fossil in a sediment that could not be directly dated, we *could not* infer an absolute date to this fossil. This principle is exemplified by the huge South American dinosaur *Giganotosaurus carolini* (a larger version of Tyrannosaurus rex), which:

> ...roamed around **90**mya. (My bold). Gabriella Gamini. 'Dinosaurs' lost world comes alive'. The Times. 21st February 1998. p6.

That is (supposedly) *23my* before the earliest Tyrannosaurus rex fossil. Given their great similarity, Tyrannosaurus rex must surely have lived at a time close to, or the same as, Giganotosaurus carolini.

Suppose that we did not know that the coelacanth is still alive, and we found a new coelacanth fossil that was in actual fact only a thousand years old. Suppose that the sediment containing this new fossil could not be directly dated by radiometric means. Since we would assume (incorrectly) that coelacanths died out over 65mya, we would incorrectly date this new fossil as being at least 65 *million* years old.

Tyrannosaurus rex

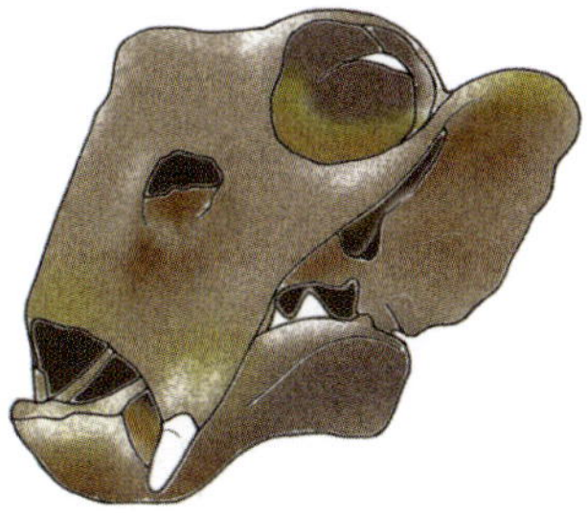

**Lystrosaurus**

This was a mammal-like reptile that (supposedly) lived in the Early Tertiary period (less than 245mya). However, fossils of this creature have recently been found in Upper Permian sediments (supposedly) over 245myo. Lystrosaurus has long been considered a guide fossil so that any sediment found containing it was dated as being Early Tertiary. However:

> *...The occurrence of Lystrosaurus in Late Permian rocks indicates that isolated specimens of this genus should no longer be used for biostratigraphical purposes...*
> M. King & I. Jenkins. 'The dicynodont Lystrosaurus from Zambia'. Paleontology. Vol 40 (1). 1997. p149-156.

Furthermore, the new finds can be interpreted as showing both that the Upper Permian and the Lower Tertiary sediments may have been laid down at the same time in history, and also that Lystrosaurus lived contemporaneously with the less-mammal-like creatures from which it (supposedly) evolved.

So, the use of 'guide fossils' is clearly a flawed method, and although it would be a monumental task, *all* the fossils in the world which have not been and cannot be dated *directly*, and which have been dated *only* by reference to supposed *'guide fossils'*, should be distinguished as such *and left entirely undated.*
But a common understanding is that sediment layers from different parts of the world are united accurately by correlations in radiometric dating. Most fossils are found in sedimentary rocks but since sedimentary rocks cannot be dated directly, their contained fossils are dated by reference to the radiometric dating of any igneous rock found either above or below a particular sedimentary layer. However, this assumes that radiometric dating methods are entirely accurate and trustworthy. Before proceeding further, a little explanatory physics is necessary.
All atomic nuclei (except hydrogen, the nucleus of which contains only one proton) contain protons and neutrons. It is the number of protons within a nucleus which defines any element and which gives an atom its particular chemical properties, so that if the number of protons change, a different element is formed. On the other hand, changing the number of neutrons within an atomic nucleus does not alter the chemical property of that particular atom but it can make the atom *radioactive*. All radioactive atoms are unstable, and different radioactive atoms spontaneously break up (decay) at different rates, but in the presence of the normal physical conditions encountered on Earth, the rate of decay of any specific radioactive element does not vary. In any given sample, the time taken for half of the radioactive atoms to decay is called the *half-life*. For example, the half-lives of radioactive potassium 40 ($K^{40}$) and radioactive carbon 14 ($C^{14}$) are 1.3 billion years and 5570 years respectively. This means, for example, that after 5570 years half of the $C^{14}$ atoms in any sample will have decayed into nitrogen 14 ($N^{14}$). After another 5570 years half of the remaining $C^{14}$ atoms will have decayed into $N^{14}$, and so on. So, if for any rock that contains radioactive atoms, the radioactive decay rate is known and the proportion of parent and daughter nuclei is also known, then this can provide the basis for elucidating how long ago that rock was formed.

According to radiometric dating techniques the stegosaurus family lived about 160mya. But are these techniques reliable?

There are a number of radiometric dating techniques in use, a common one being the *Potassium-Argon* method, which makes use of the fact that one of the products of the radioactive decay of $K^{40}$ is the inert gas argon 40 ($Ar^{40}$). If, during some volcanic activity, rock containing radioactive potassium becomes molten and then hardens, it is assumed that the intense heat would force any argon gas present to escape, and that at the time of its formation the resulting igneous rock would contain no argon at all. Over time, some of the $K^{40}$ atoms in the new volcanic rock would decay into argon atoms which would become locked physically within the rock. By comparing the proportion of radioactive potassium atoms and argon atoms and knowing the half-life of $K^{40}$, the time of the volcanic activity that formed the rock in the first place can be deduced.
Whilst radioactive methods similar to the K-Ar method seem quite valid, nevertheless, there are a number of assumptions that cannot be independently verified. For example, it is assumed that no argon was present at the time of the formation of the rock; that all the argon present in the rock has come from the decay of the potassium atoms within the rock; that no potassium or argon has been either lost or

gained over time; and finally, that the radioactive decay rate has remained constant during the time concerned.

The K-Ar method can only be applied to rock formed in association with volcanic activity or other intense heat, and the length of the half-life of $K^{40}$ further limits its application to dating of material that is (supposedly) many millions of years old.

When material from once-living organisms is found in sediments *unassociated* with any igneous rocks, then the potassium-argon method cannot be applied, and in such cases the *radioactive carbon* (or *radiocarbon*) method is commonly used to analyse the material directly.

High up in the earth's atmosphere, gas atoms are continually bombarded by cosmic rays from space, and as a result, free neutrons are released. These free neutrons collide with other atoms, and when a nitrogen ($N^{14}$) atom absorbs a neutron, its nucleus emits a proton. This reduces its proton number from 7 to 6, transforming it into a radioactive carbon ($C^{14}$) atom. As a result, the carbon dioxide in the atmosphere contains both radioactive and normal carbon ($C^{12}$) atoms, the latter existing in far greater numbers. Since carbon dioxide is absorbed by plants, and plants are eaten by animals, all living organisms necessarily contain both normal and radioactive carbon. During life, all organisms contain normal and radioactive carbon atoms in the same proportion as is found in the carbon dioxide in the atmosphere. However, when an organism dies, its physical remains no longer take in any more carbon and any radioactive carbon present will gradually decay. So the proportion of radioactive carbon in the remains will gradually decline. Thus, by knowing the half-life of $C^{14}$ and comparing the proportion of radioactive and normal carbon in living tissues with that of dead tissues, the time of death of the organism concerned can be deduced.

The main assumptions here are that the rate of formation and decay of $C^{14}$ has reached an equilibrium state and that the rate of formation of atmospheric radioactive carbon has remained constant over Earth's history. However, an equilibrium state is questionable. Furthermore, we know that in the past the Earth's climate has been dramatically different from how it is now, and this could have been associated with quite different rates of radioactive carbon production in times past. If there had been any significantly less production of $C^{14}$ in the past, then for any specimen this would suggest (erroneously) a greater than actual age. Obviously, with all radioactive dating methods the question is one of validity and reliability. Since there are no independent methods of verification, it follows that radioactive decay methods cannot be accepted as foolproof, and therefore it is reasonable and proper to subject the methods to scrutiny. In fact, radiometric dates are susceptible to great error and *do not* necessarily represent absolute age. For example, specimens taken from the newly-formed lava dome which appeared following the eruption of Mt St Helens in 1980, have been dated by the K-Ar method as being 2.4 million years old, showing that at the time of its formation igneous rock is not necessarily devoid of all argon. Since the assumption that, at the time of its formation, there would have been no argon present in ancient igneous rock is false, the K-Ar method cannot be relied upon to give any meaningful results.

Another example illustrates the circular logic of how fossils are used to check on radiometric dates, and radiometric dates are used to check on fossil dates. In 1969, samples of tuff (volcanic rock) from alongside Lake Rudolf (now Lake Turkana) in east Africa were dated by the potassium-argon method as being around *two hundred million* years old. However:

> From these results it was clear that an extraneous age discrepancy was present... F.J.Fitch & J.A. Miller. 'Radioisotope age determinations of Lake Rudolf artefact site'. Nature. Vol 226. 18th April 1970. p226-8.

The results were thought to be wrong because the dates did not fit in with tests done on rocks (assumed to be) of equivalent age from other parts of east Africa. The tests were repeated using less calcified pumice, and it was then concluded that the age of the tuff horizon:

> ... in the Koobi Fora beds from east of Lake Rudolf in east Africa is ... very close to 2.6 my... F.J.Fitch & J.A. Miller. 'Radioisotope age determinations of Lake Rudolf artefact site'. Nature. Vol 226. 18th April 1970. p226-8.

Skull KNM-ER 1470. This fossil most likely represents a large-brained ape.

This represented a discrepancy of some *two hundred million* years between the two analyses. When in 1972, the human-looking fossil skull KNM-ER 1470 was found in Koobi Fora *below* the so-called KBS sediment layer, even the date of 2.6my was deemed too old since the fossil looked too much like a modern human skull, and the standard evolutionary scheme simply did not allow for human beings to be living so long ago. The skull resembled:

> '...something that's so modern it makes your hair stand up.' Donald Johanson & Maitland Edey. 'Lucy, the beginnings of humankind'. Penguin. 1990. p149.

A search was therefore made to see if the date could be revised to an even younger time. In other words, it was hoped to revise the date so as to fit in with the supposed evolutionary scheme. This was done by comparing pig teeth from a site which was also dated as being 2.6myo some 50 miles away, in the Omo valley.
In accordance with the *assumption* that differences in the physical appearances of teeth represent evolutionary modifications over time, the pig teeth from Koobi Fora looked younger than the pig teeth found at Omo, and looked similar to teeth found at other 2myo sites in Ethiopia:

> ...'The Koobi Fora pigs don't look like our pigs from the same time range. Their pigs are too young.' Donald Johanson & Maitland Edey. 'Lucy, the beginnings of humankind'. Penguin. 1990. p150.

As a result the radiometric dating of the KBS tuff was repeated and revised to the younger date of 1.88my:

> It is of interest to note perhaps (for those who put faith only in chemical or physical dating methods) that it was faunal discrepancies between the two sites (the Omo and Koobi Fora) that alerted geochronologists to the problem of the dating of the KBS tuff... Michael Day. 'A guide to fossil man'. 4th edition. Cassell. 1986. p199.

It is important to note that the KBS layer was in most paleontologists' eyes *satisfactorily* dated at 2.6my, before KNM-ER 1470 was found. Therefore, if KNM-ER 1470 had not been found or, more importantly, if KNM-ER 1470 had not looked so human, then the dating of the KBS tuff might never have been questioned, and today it might still be dated at 2.6my. The KBS layer is not alone in all this:

> ...several ash layers of - it now turns out - quite different age had all been mapped (erroneously) as the KBS tuff. Steve Jones, Robert Martin, David Pilbeam. 'Cambridge Encyclopaedia of human evolution'. Cambridge University Press. 1996. p181.

And, if when the original sample of the KBS layer was analysed it was considered contamination-free, and its potassium-argon date was *only* rejected on other (fossil) evidence, then it follows that *no* currently-*assumed*-to-be-contamination-free potassium-argon date for *any* site can be accepted on its own merits alone.
So, it seems that in practice, radiometric dates are accepted when they give results that seem to concur with the *expected* evolutionary pattern, and are rejected when they give results that seem to conflict with the *expected* evolutionary pattern. Therefore, statements such as:

> ...modern chemical analysis...has shown previous field identifications and correlations to be in error. Michael Day. 'A guide to Fossil Man'. 4th edition. Cassell. 1986. p199,

can be meaningless*. Furthermore, as we shall see in the course of this book, there is good evidence to show that dates derived from radioactive methods are quite incongruous with the evidence from biology. In addition, there are many issues that fit neither into the standard evolutionary scheme, nor into its supposed timetable.

***Current dating methods give conflicting results**
Dates of fossils of Australopithecus africanus from South Africa, that were thought to be 3.5myo, have been re-assessed and are now thought to be 2.4myo:

> *"What we have at Sterkfontein is a jumbled-up mess", says Jim Ohman..."what we need is a new absolute dating technique".* Dan Eatherley. 'Is Little Foot only a cousin?'. New Scientist. 5th October 2002. p13.

### *The fossil record does not show a progression in complexity*

Contrary to general opinion, the fossil record *does not* show a consistent progression in complexity. There is a fallacy in accepting firstly, that creatures commonly understood as being 'simple' such as snails or worms are indeed *simple*, and secondly, that they are *significantly* less complex than other creatures generally considered to be more complex. In fact, most organisms, however 'simple', are *incredibly* complex.
Let's look at it this way: should we anticipate that starting *only* with simple chemicals (i.e. with no available DNA or proteins etc.), it would be a simpler matter for us to produce a living snail than it would be to produce a living elephant? Under the microscope both these creatures have virtually identical intracellular features. And both contain nerves, muscles, blood. They also both have seeing, feeling, breathing, digesting, excreting and reproducing mechanisms. Therefore, since all living organisms are *so* complex, we should expect that the production of either a snail or an elephant (starting *only* with simple chemicals) would, for practical purposes, be *equally* difficult.
Furthermore, the complexity of an animal cannot be predicted purely from its bones since the bones themselves are only a very small part of an animal's make-up. For example, sharks can sense electromagnetic fields but goldfish cannot. Goldfish

Trilobites were not simple animals.

Horseshoe crab

Ammonite

**Materpiscis**

Materpiscis is a (supposedly) 380-million-year-old fossilised fish found in the Gogo formation in Western Australia's Kimberley region. What is remarkable is that it was in the process of giving birth, with evidence of an umbilical cord still attached to the offspring:

> *In earlier studies of placoderms, remnants of young have been observed inside adults and it was assumed to be cannibalism.*
> C Dennis. 'The oldest pregnant mum'. Nature. Vol 453. 29th May 2008. p650–652.

The notion of evolution would expect such (supposedly) ancient fishes to have laid eggs, and would not predict such an 'advanced' reproductive process. This proves that (supposedly) ancient creatures were not less complex than (supposedly) less ancient creatures.

Fish such as this Cephalaspid lived (supposedly) 350mya. It had a bony skeleton and bony head plates. Bony fishes such as this are supposed to have pre-dated cartilaginous fishes such as sharks. However, this is incongruous since bone (which is the result of mineralisation of a cartilage scaffold) is a more complex material than cartilage. Thus, contrary to the fossil record cartilaginous fishes should pre-date all bony fishes.

Cephalaspids had no jaws but had small slit-like mouths. The fossil record gives little clue as to the origin of jaws since fishes living (supposedly) 395mya already possessed them.

*Note: the hole on the top of the skull was an opening for the pineal gland (which may have been light-sensitive, as in the tuatara). The eyes were tiny structures very near the front tip of the head.*

can see both colour and infra-red yet dogs (which are generally considered to be less primitive and more intelligent) can manage neither. The common octopus has been shown to be at least as able to find its way around a maze as a rat yet it has not even got a skull to house its brain. Thus you cannot predict the *intelligence* of an animal from the size or shape of its brain, or its bone structure. Therefore, it is not a straight-forward matter to state whether or not a *trilobite* was a less *complex* or a more *primitive* creature than an elephant.

There are many examples of sophisticated animals presumed to have lived millions of years ago. Trilobites are amongst the most well known fossils. They were crab-like animals which had shells with 3 segments to their bodies (hence their name). While most were only about 30mm long, some were over a metre in length. It is thought that they evolved from soft-bellied ancestors which left no fossil evidence of their existence. Like modern horsehoe crabs, trilobites would have had hearts to pump their blood, a well-defined brain and nervous system, and would have been able to periodically shed their shells (*carapaces*) to enable them to grow. The eyes and limbs were all very well developed. Virtually all the enzymes, proteins, internal organs, muscle systems, etc. found in living animals would have been represented in this (supposedly) ancient and 'primitive' animal. And yet it is supposed to have developed all this sophistication and originated some 500 million years ago. The shrimp-like *Antrimpos* (250mya) and *Aeger* (200mya), and the lobster-like *Homarus* (140mya), were no more complex than a trilobite, but are supposed to have appeared only some 140-250 million years ago. Crabs are first seen in sediments 35-55 million years ago with the appearance of animals such as *Harpactocarinus* (which was very similar to modern crabs). Yet since they are of equivalent complexity, one would expect that these animals would have appeared around the time of the trilobites, 250-300 million years earlier. *Ammonites* were squid-like creatures with spiral shells. Some ammonites, *Leioceras*, *Phymatoceras*, *Parkinsonia* and *Haploceras* are supposed to have existed only during the period 100-175 million years ago. However, since their soft bodies were somewhat less complex than the trilobites with their hinged legs, and muscle and tendon insertions, we should expect these ammonites to have appeared *before* the trilobites.

The Devonian Hunsruck Shale of Germany is thought to represent the evolution of spiral ammonoids from straight-shelled *Nautiloids*. The shells are supposed to have evolved a curved, and then a coiled shape in a series of steps over millions of years, and this is taken to represent evolution of one species into another. But a curved shell does not require millions of years to develop from a straight one since selective breeding can easily produce a change within a few generations. Furthermore, since the curvature of the shell says nothing about the animal contained within, the production of a new creature need not necessarily be invoked. Indeed it is more probable that a group of molluscs could divide into a number of breeding populations and produce slightly varying shells and still remain the same creature, much in the same way as various breeds of domestic cattle with straight, curved or spiral horns are still cattle.

Furthermore, are we to suppose that ammonoids developed into other molluscs in the same time that it took fishes to evolve into mammoths? More particularly, are

Background: Tiger shark jaws

we really to believe that some minor alterations in the form of the spiral of the shell could have *determined the survival* of a particular mollusc over its neighbour? For a mollusc, survival or non-survival must surely depend more on a chance entombing deluge of mud than on any features in its shell.

So, the fossil evidence most certainly *does not* describe a transformation from simple living things to more complex types. Furthermore, the common evolutionary scheme contains incongruous steps. We are all generally taught that cartilaginous fishes such as sharks and rays evolved into the more complex bony fishes. However, the fossil record suggests that:

> Forms similar to modern sharks appeared in the Devonian [345-395mya]...fishes that possess an osseous skeleton [originated] in the Silurian [395-435mya]. Paolo Arduni and Giorgio Teruzzi. 'MacDonald Encyclopaedia of fossils'. Little, Brown & Co. 1993. p37-8,

and again,

> The Chondrichthyes [cartilaginous fishes] were abundant in Carboniferous [280-345mya] and Permian [230-280mya] times and include the shark genera Lamna and Charcarodon...Osteichthyes or bony fish, appeared in the Devonian period [345-395mya] and gave rise to many modern fish. Chris Pellant. 'An illustrated guide to fossils'. Dragon's World. 1995. p156-7.

Thus, although these authors differ (*significantly*) in their dating, nevertheless they both agree that the fossil record actually suggests that bony fishes were in existence *before* sharks and rays. So, this suggests that the sharks and rays developed from ancient ancestors which had bony skeletons. But we have seen how powerful

As with all sharks, the dogfish has a skeleton made of cartilage.

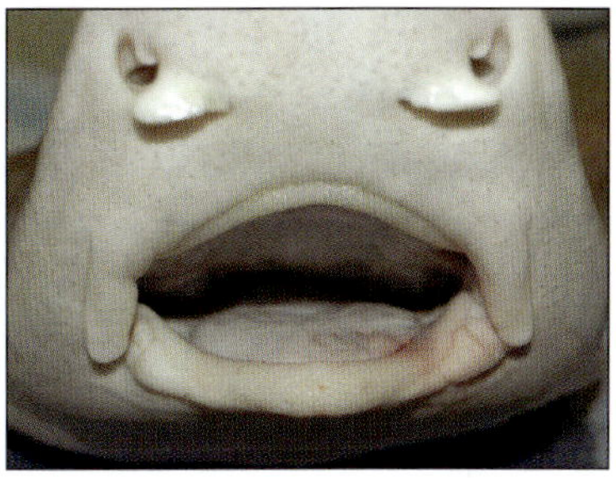

Dogfish jaws

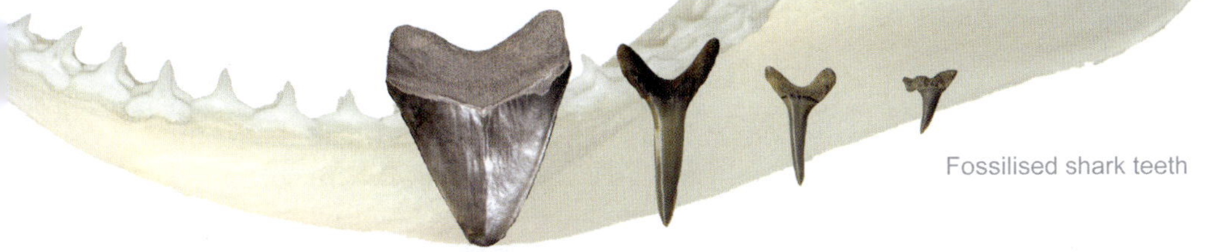

Fossilised shark teeth

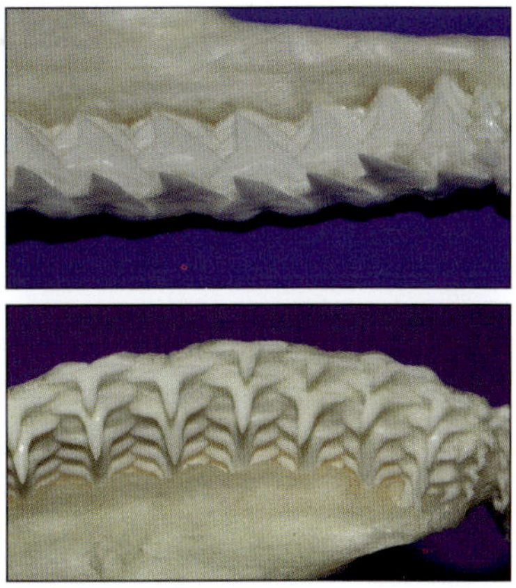

Above: Shark.
Right: Close up of living shark teeth.

Above: Tiger shark teeth and lower jaw (top view from outside; bottom view from inside). Unlike in bony fishes where the teeth are derived from the jaw bones, the teeth of sharks develop continuously from skin tissues of the mouth lining and migrate outwards to replace the outermost teeth once they break off.
Shark teeth are highly complex and are not simple precursors to the teeth of bony fishes.

hydrodynamic sorting is in its ability to sift and displace animal remains. Healthy scepticism must be the order of the day, since:

> In many cases the correlative value of a palaeozoic vertebrate may lie in finding one scale or tooth from a drill-core sample. John Long (Ed). 'Palaeozoic vertebrate biostratigraphy and biogeography'. Belhaven Press. 1993. p xiv.

However, there is no clear reasoning or evidence to show that natural selection could favour bony fishes over cartilaginous fishes or vice versa. The majority of bony fishes are relatively small and insignificant creatures whereas the top marine predators are the (cartilaginous) sharks. Furthermore, whilst sharks and bony fishes are undoubtedly significantly different from each other, we cannot say that one is more complex or more able than the other. Nevertheless, if for any reason having a cartilaginous skeleton did offer some survival advantage to fish that already had a bony skeleton, then we should expect there to be alive today many forms of fishes with diverse mixtures of shark and bony fish characteristics. Most certainly, any bony fish could manage quite well, for example, with multiple gill clefts, or with the unique shark heart which is housed in a rigid chamber.

With a head as much as 1.25m long and a total body length of 3m, the *Mastodon-saurus* was the largest amphibian known to have lived. Its fossils have been dated as between 230 and 200myo. Incongruously, (supposedly) 50my *earlier* lived the 60cm *Seymouria* which had some features more in common with reptiles and which:

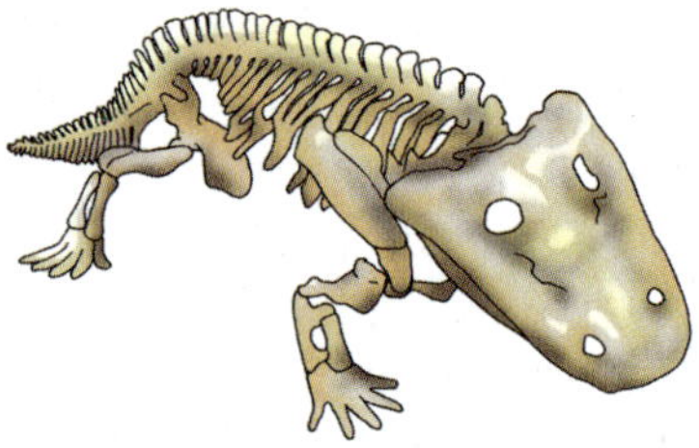

Eryops.
An extinct 2m-long amphibian.

> ...represents one of the most advanced stages in amphibian development. Paolo Arduni and Giorgio Teruzzi. 'MacDonald Encyclopaedia of fossils'. Little, Brown & Co. 1993. p207.

Even more incongruous is the finding that early frogs and toads had short back legs:

> Anurans [frogs and toads] are known from the Upper Carboniferous [290mya] in forms that do not diverge greatly from their modern counterparts, even though their hind limbs are not yet adapted for jumping. Paolo Arduni and Giorgio Teruzzi. 'MacDonald Encyclopaedia of fossils'. Little, Brown & Co. 1993. p209.

Common frog

Which means that it (supposedly) took another *245 million years* simply to lengthen their legs:

> The genus Rana [frogs] first appears in Europe in the Eocene [55 mya], but does not occur in Asian, African or North American sediments until the Miocene [22.5mya]. Paolo Arduni and Giorgio Teruzzi. 'MacDonald Encyclopaedia of fossils'. Little, Brown & Co. 1993. p209.

These dates would suggest that modern frogs evolved *after* the presumed origin of modern reptiles, birds and mammals.

There are three types of living mammals (so-named because they all produce milk from special glands – the *mammae*). The vast majority comprise the *placental mammals* which are born almost fully formed, having developed in the uterus and having derived their nutrition via the umbilical cord and placenta. The *marsupials* (for example, kangaroos, koalas, opossums and the thylacine) are a much smaller group which give birth to very poorly developed babies that continue their development inside a pouch on the mother's abdominal wall. They suckle on milk from teats within the pouch itself. The *monotremes* (whose living examples comprise only the duck-billed platypus and the echidna) are mammals which lay eggs. One theory is that egg-laying mammals, marsupials, and placental mammals all evolved separately from different stocks of primitive reptiles. However, it is generally thought that the egg-laying *monotremes* evolved into the pouched *marsupials* which in turn evolved into the *placental mammals*:

Koala

> ...analyses since 1980 have demonstrated convincingly that the monotremes are indeed closely related to the marsupials and placentals, with a branching-off point possibly in the late Jurassic [around 150mya]. Michael Benton. 'The rise of the mammals'. The Apple Press. 1991. p89.

Early mammal-like reptiles ranged in size from that of a small cat to a rhinoceros. The earliest known true mammal is *Morganucodon*, which is thought to have lived about 200mya. Although the oldest known fossils consist only of teeth, other slightly younger and beautifully-preserved whole fossils show that morganucodon looked very much like a modern shrew. Close examination of the jaws of different sized individuals suggests that morganucodon had two sets of teeth in its lifetime, much like most modern mammals. The fossils, however, give no clue as to whether morganucodon was a monotreme, a marsupial, or a placental mammal.
Until recently, no fossils of monotremes were known that were older than 2my but even so, monotreme fossils are extremely rare:

> Monotremes today are a very small group, and fossils indicate that this has always been the case. Michael Benton. 'The rise of the mammals'. The Apple Press. 1991. p89.

But the problem is that if monotremes evolved into marsupial and placental mammals, then monotremes could *not* have been a small group, either in numbers of individuals or in numbers of types. Evolution could not happen without involving millions of individuals over millions of years. And, by definition, (supposedly) before marsupial and placental mammals ever appeared monotremes would have been the dominant mammalian group. It is implausible that egg-laying mammals turned rapidly into marsupial and placental mammals. Therefore, monotremes must have existed for a very long time before they (supposedly) accumulated the necessary mutations to change them into marsupials and placentals. Therefore, before placentals (supposedly) evolved, monotremes *must* have been plentiful, and although it is true that fossilisation is a rare event, nevertheless, being plentiful they ought to have left plenty of fossils. That no such extensive fossil evidence exists suggests that monotremes were not ancestors to marsupial and placental mammals, nor were they at any time plentiful.

It is commonly stated that before 65mya 'dinosaurs ruled the Earth', and that as a result of this reptilian dominance mammals remained small and insignificant from 200 to 65mya. This does not make sense. Whilst it is true that a small number of dinosaurs were gigantic animals, nevertheless, most were in fact, pig-sized or smaller.

An environment containing very large animals and microbes does not preclude the existence of substantial numbers of creatures of every size in between. For example, the oceans of today contain blue whales which are creatures far larger than any *Diplodocus*. We also have very large cetaceans such as right and humpback whales, and large predators such as the orca and the great white shark. But alongside these large animals live other creatures such as seals and common dolphins, all comprising a variety of sizes and predatory abilities. And these all prey on millions of fishes of even smaller size.

Harbour seal

Blue whales and orcas do not rule the seas any more than elephants rule the plains of Africa. Similarly, nor would the *Apatosaurus* and Tyrannosaurus have ruled the ancient Earth. Creatures co-exist. There is a balance. Hyenas, lions and cheetahs often share the same territory but all three just happen to be placental mammals. So there is no reason why hyenas, lions and cheetahs could not live comfortably alongside large herbivorous dinosaurs and share common prey with lion-sized predatory dinosaurs.

Nor is the pre-existence in a habitat of large creatures a barrier to the appearance of new up-and-coming kids on the block. Whales (supposedly) evolved from dog-sized creatures, and entered into an environment which already contained powerful predators such as the sharks. So, if creatures have evolved, there is no reason why 200mya some of the (supposed) early shrew-like mammals could not have given rise to slightly larger shrew-like mammals, which many generations later (but still only a little later than 200mya) could have given rise to early lion-sized mammalian predators…in the presence of the 'ruling' dinosaurs of the day.

But incongruously, the number of mammalian fossils dated (supposedly) from 208-66mya is *almost non-existent*, since they could be:

> …fitted into a large top hat. This is a hopelessly poor fossil record…
> Michael Benton. 'The rise of the mammals'. The Apple Press. 1991. p36.

*Allosaurus could have hunted alongside large mammals:

> *Not only did large mammals live alongside their giant reptile cousins, but some were big and bold enough to go dinosaur hunting. Repenomamus giganticus and Repenomamus robustus were carnivorous mammals that lived in China about 130mya…R. Robustus died with its belly full of young dinosaur.*
> J Hecht. 'Revenge of mammals'. New Scientist. 15th January 2005. p6.

If a lion-sized dinosaur could have lived in the presence of large dinosaurs, then a lion-sized mammal, be it a carnivore or herbivore, could have too.*

So neither the logic nor the evidence suggests that the presence of dinosaurs prevented mammals from evolving, or that placental mammals evolved from egg-laying ancestors.

### *Constancy of form is evidence against evolution*

The suggestion by the fossil record that over huge periods of time many creatures show a constancy of form is actually evidence against a gradual theory of evolution. Both silverfish and cockroaches demonstrate no particular features that might allow them to survive for so long and in preference to similarly complex and able creatures that have long since become extinct. So, it is important to ask questions such as, why has:

> The silverfish...survived since the early Devonian [345-395mya]...The cockroach remained unchanged since the Carboniferous [280-345 mya]...
> LB Halstead. 'Hunting the past'. Hamish Hamilton. 1982. p196-7.

As we shall see, although DNA replicates with extreme fidelity, and although errors are uncommon, nevertheless errors are *inevitable*. And any such errors (genetic mutations) in the DNA code are far more likely to result in deleterious effects than in any (supposed) beneficial ones. Since mutations are passed on to subsequent generations, it follows that *all* living animals should differ to some considerable degree from any (supposed) very ancient ancestral forms.

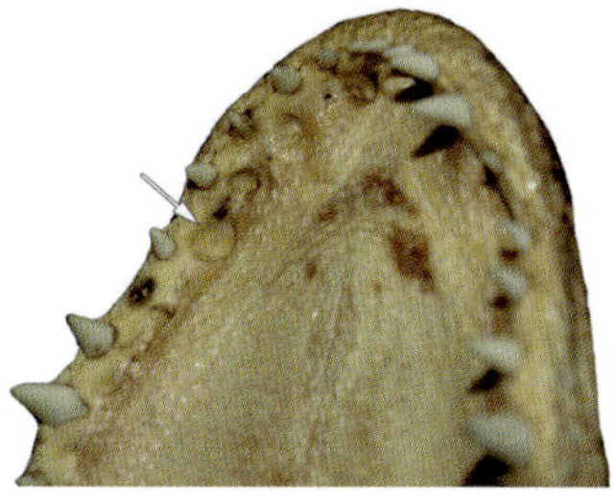

Alligator upper jaw showing socket for 4th lower tooth.

If, in any organism, a system or structure is found which is not essential for its survival, and which if lost would not render the organism less able to survive than a similar organism which retained the structure or system, then losing that structure or system would not in itself cause the organism to become extinct in preference to a similar organism which retained that structure or system. However, over time deleterious mutations of genes coding for that structure or system will *inevitably* arise, and most likely will result ultimately in the loss of that structure or system. It follows then, that if organisms exist today that have any such non-essential structures or systems, and if similar organisms with *the same* structures or systems existed in times past, then the time interval between the present and the time represented by the extinct organisms cannot have been very great.

Above: Crocodile skull.
Below: Alligator skull.

Crocodiles are a good example of a (supposedly) constant form spanning over 200my. All reptiles have only three chambers to their hearts - a right and left atrium and a single ventricle. Crocodilians have the same set-up as other reptiles but their single ventricle is partially divided into two halves by a septum. Therefore, if crocodiles have seemingly evolved a septum in their ventricle, *many* other reptiles (the genes of which must have much in common with those of crocodiles) should by now also have secured the same features by mutation and natural selection.

Crocodile

Alligator

Above: Rear foot of crocodile.
Below: The crocodile's front foot (top) has a complete fifth toe (arrowed), whereas in the rear foot (bottom), the fifth toe is represented only by a short metatarsal bone inside the foot (arrowed).

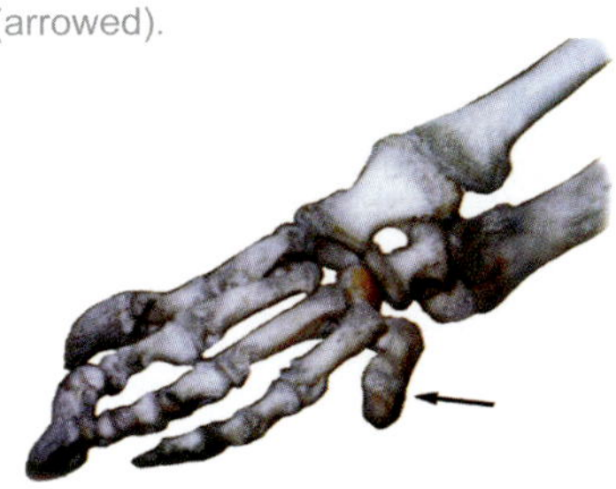

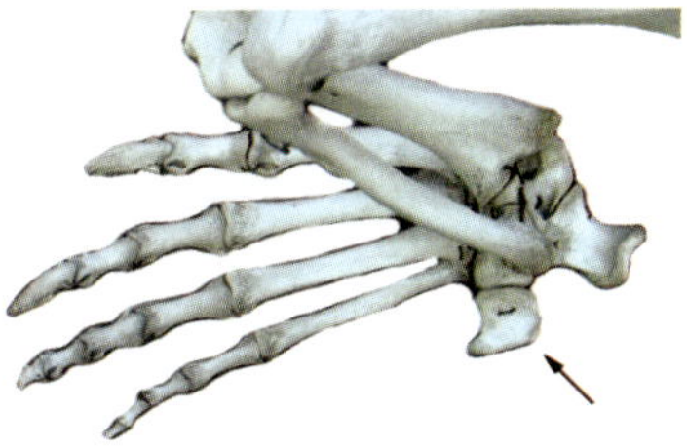

Although amphibians are (supposedly) more ancient animals than crocodiles, this newt has retained five toes on its rear feet.

As in the crocodile, the foot of Tyrannosaurus rex (below) had four toes and a small metatarsal bone. There is no evidence of a T. rex ancestor with five rear toes. (The first toe is hidden on the far side of the picture).

It can only be an advantage for a reptile to have a four-chambered heart, in order to separate completely the flow of oxygen-rich blood (coming from the lungs) and oxygen-depleted blood (returning from the rest of the body), just as in birds and mammals. The fact that no other reptiles have hearts like crocodiles shows, firstly, that non-crocodilian reptiles have *not* accumulated mutations as evolution would predict, and secondly, that their *not having done so* shows that it is implausible that they have been around long enough *to have done so*. In other words, it is implausible that crocodiles and other reptiles have been living for millions of years.

Crocodiles can be readily distinguished from alligators by looking at their jaws: the fourth tooth in the lower jaw of a crocodile is visible when the mouth is closed whereas in an alligator all the teeth fit into sockets and are hidden from view. But since having all its teeth hidden is not detrimental to an alligator, we can safely conclude that it is unlikely to be detrimental to a crocodile. Thus, since the development of the different teeth is governed by different genes, all of which must be susceptible to mutations, then over a long period of time (200my should be more than ample) we should expect to find all manner of tooth arrangements in crocodilians, none of which would be particularly advantageous one over the other, including, for example, alligators with protruding teeth. But this is not the case, again suggesting strongly that crocodilians cannot have been around for millions of years.

Crocodiles have five toes on their front legs, and the rear legs have four toes along with a small (supposedly) '*vestigial*' fifth toe. However, what is notable is that even the earliest known crocodiles (e.g. *Protosuchus* and *Orthosuchus*) of (supposedly) 200mya have *the same arrangement* in their toes. So this leads us to four points. Firstly, since the *earliest* crocodiles had small rear fifth toes, these toes *cannot* be truly vestigial, as there is no evidence to show that any crocodiles at any time in the past had *normal* fifth toes. For example, the 1m-long *Protosuchus*:

> ...is without doubt a very early crocodile...there were 5 toes on each front foot and 4 toes on the hind (the fifth toe of the back foot is represented only by a vestigial metatarsal bone). Rodney Steel. 'Crocodiles'. Christopher Helm Pub. 1989. p164.

As we shall see, horses are supposed to have reduced their toes from four to just the single hoof in a matter of only 20 million years or so. Secondly, therefore, if the Protosuchus' fifth rear toe truly *was* vestigial, then after 200 *million* years many, if not all, crocodiles should surely by now have lost it completely (and possibly other toes too). Thirdly, since evolution insists that mutations *must* occur over time, the fact that crocodile toes have remained unchanged makes it implausible that crocodilians have been around for millions of years.

Finally, if in (supposedly) *two hundred million* years the crocodile has been unable to acquire even *detrimental* mutations so as to alter its anatomy in something as insignificant as a toe, and since evolution is dependent solely on the chance appearance of *beneficial* mutations, in the same time period the incidence of *beneficial* mutations must have been so low as to be considered negligible. Most certainly, then, on the evidence of crocodiles, it is implausible that since the origin of the Earth, chance beneficial mutations could have arisen in any numbers sufficient to provide natural

selection with the raw materials for the formation of the variety of life that exists and which has existed.
But it is often suggested that animals do not change because if a creature is well suited to its habitat, *there is no need to change a 'good design'*. However, crocodilians, sharks and penguins show that there is little force in this argument. We all accept that the crocodilian fifth rear toe is not a design that can be categorised as being perfectly suited to its lifestyle. And many aspects of sharks, for instance, have (supposedly) remained virtually unchanged for 300 million years. But sharks (just like any other animal) have no choice whether their genes mutate or not, and there are *bound* to be a number of changes in their design that could be advantageous, such as a bigger brain, a swim bladder, or sonar. Therefore, evolution should predict that there ought to be a number of *significantly* different features amongst sharks alive today, and that modern sharks should be very different in form to their (supposedly) ancient ancestors.
It is considered that penguins had:

The ancient crocodilian Protosuchus had relatively long legs and a small 5th metatarsal (arrowed) in its rear foot.

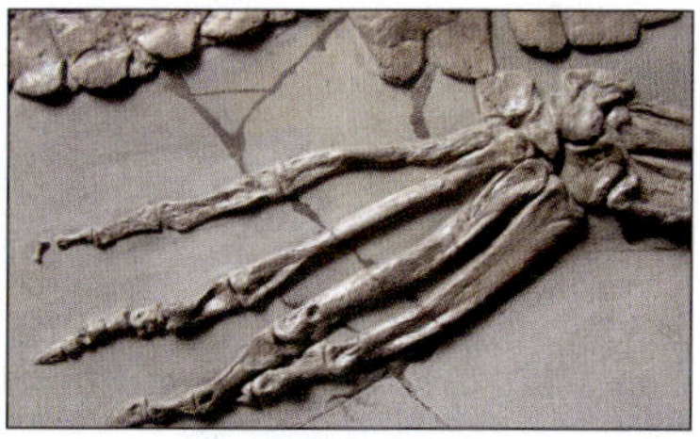

180myo crocodile rear foot.

> ...begun to diverge from petrel-like flying birds over 65mya. By the late Eocene 45-50mya, penguins were clearly differentiated from other birds.
> Jonathan Chester. 'The nature of penguins'. Greystone Books. 1996. p9.

If we were to accept that it took only 15my for a penguin to evolve from a petrel-like flying bird, then we must ask why subsequently the penguin remained unchanged over the next 40my. As with sharks, there are *bound* to be a number of changes in their design that could be advantageous. For example, if we are to accept that toothed whales evolved their teeth in order to catch fish, then we should expect at least some fish-eating birds to also evolve teeth (or not all of them to have lost their teeth). Again, since whales have managed to breed in the water, and even the *Icthyosaurs* gave birth to live young (supposedly) millions of years ago, then by now, we should expect at least some penguins to have evolved so as to become totally aquatic. So, the fact that they have not changed suggests that penguins have not been around for many millions of years.
We have seen that in favourable conditions fossils can be formed in less than a year; that the fossil record is a composite artifact; that all types of creatures (both living and extinct) could have lived contemporaneously in pre-historical times; that there is little evidence for any trend showing that, over time, any organisms have acquired an increase in complexity; and that the constancy of form shown by many creatures is evidence against creatures having lived for millions of years. To this last point the geological strata and geographical distributions of different animals provide yet further supportive evidence.

Humboldt penguin from the western coast of South America.

CHAP

# Event horizon: *a matter of time*

Drumheller hoodoos
Alberta, Canada

Mt. Annapurna. Along the entire range of the Himalayas, the mountain bases are made of granite which is topped by layers of sedimentary limestone containing marine fossils.

### *Going with the flow*

It is commonly thought that the Earth is millions of years old, and that this has been unambiguously confirmed by diverse methods, the most unquestionable one being radiometric dating. Some would say that if we were to question the age of the Earth, why not question the idea of the Earth orbiting the sun? Clearly, they might say, it is an unshakable fact that the Earth is *very* old and to question its age is a sure sign of ignorance. Certainly, at a casual glance the Earth may look old but, as always, we should not be afraid of checking the evidence.

In mountains and cliffs all over the world, we can see horizontal sedimentary strata which clearly suggest that the layers of dirt have been produced following deposition by water. Given that today, everywhere, we see the slow-but-steady accumulations of dust and mud, punctuated by sudden localised flooding events, it might be reasonable to assume that all those sedimentary layers represent multiple depositional events separated by thousands or millions of years.

Olduvai Gorge, Tanzania

Rocky Mountains, Canada

Paul Garner

Grand Canyon, USA

Petra, Jordan

Andrew Dickinson

Gozo, Malta

### *What amount of time does a sedimentary layer represent?*

If, for example, the Grand Canyon walls represent 300my of sequential sediment deposition, with each of the *n* individual layers representing just one deluge event horizon lasting (say) a day, then the layers themselves can only represent *n* days of time. Therefore, it is argued that there must have been *very* long periods of time elapsing *between* the laying down of each individual layer. So, if (for the sake of a simple calculation), there were 300 layers, then, on average, 300my/300 = 1 million years would have passed between the laying down of one layer and the next. This immediately presents a problem since all of the layers are completely horizontal. Over a period of 300my the layers *must* have been subjected to significant erosion, and in addition, thousands of earthquakes would have disrupted the tidy horizontal lines *at varying times* with *varying epicentres,* resulting in randomly-scattered *jagged* discontinuities at *different levels* and *all over* the Grand Canyon site. So, the question is: could the undisrupted horizontal sedimentary layers constitute a series of occurences related to a *single* water-borne event?

Furthermore, the distinct lack of substantial erosion at the layer interfaces shows that very little time had elapsed between the laying down of one layer and the next. For example, by radiometric dating, the gap between the Coconino sandstone and the Hermit Formation limestone underneath it is supposed to represent about 10my of time. However, since there is no significant sign of erosion at the interface of these two layers, the radiometric dating cannot be correct.

This problem with long geological timescales is dramatically demonstrated by fossilised ripples; footprints (e.g. combined dinosaur prints and ripples at Tumbler Ridge, British Columbia; amphibian prints in the Coconino sandstone in the Grand Canyon); and even rain drops (e.g. Mortceau-les-Mines, France). Since delicate features could only have been preserved if there was *virtually no time* between their formation and their covering over by another layer, any significant length of time would have resulted in their partial or complete erosion.

The Glencoul Thrust in Scotland shows the older metamorphic rock (Lewisian Gneiss, L) lying on top of the younger Cambrian rock (C).

Paul Garner

Coconino sandstone. The interface between the Coconino sandstone (light) and the Hermit Formation limestone (dark), Grand Canyon. By radiometric dating, the gap between these layers is supposed to represent about 10my of time. However, since there is no significant sign of erosion between the layers, the radiometric dating cannot be correct.

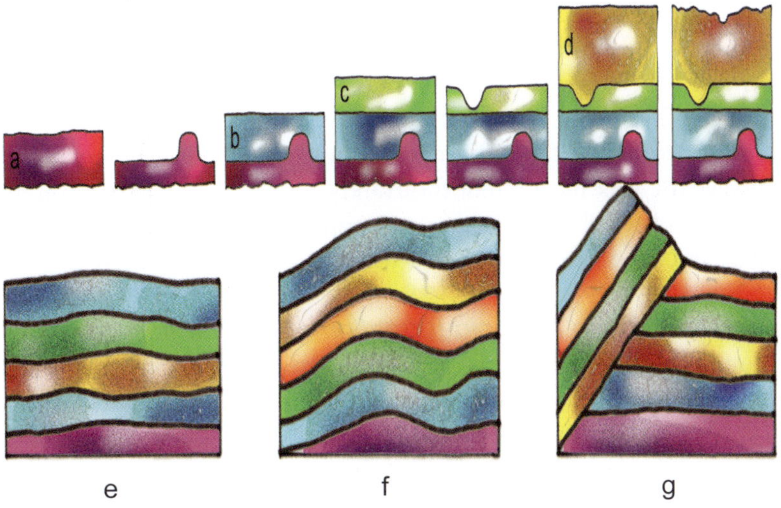


**The sedimentary strata do not generally show signs of erosion**

If sediment is deposited and left for any length of time, we should expect its surface to show signs of erosion or gully formation, which should be visible at the interface with any subsequent deposition.

Above left: This is the sort of appearance that we should expect to be commonly seen if sedimentary layers are the result of multiple depositions spanning hundreds or millions of years. In this example, layer **a** is deposited, hardens and then erodes to leave a promontory. Subsequently, layer **b** is deposited and remains unweathered by the time it is covered by layer **c**, which is itself eroded to incorporate a gully. Finally, layer **d** is deposited and subsequently weathers to leave an irregular top surface.

In contrast, the observed sedimentation patterns (below left) generally show horizontal layers (**e**), buckling of horizontal layers (**f**), or overthrusts (**g**), with little irregularity between the layers. This strongly suggests that multiple planar layers *are not* the result of multiple depositions spanning huge periods of time, as is commonly believed.

Note that overthrusts are usually low angle faults, but are exaggerated in this diagram.

Modern open bivalve and closed fossilsed bivalve.

**Rapid burial of whales in diatomite**

The (mostly) fully-articulated bones of a few hundred whales are buried in sediments packed solid with diatoms in the Pisco Formation (Miocene–Pliocene, supposedly 2-26myo), in Peru. Diatoms are single-celled marine organisms and each cell is encased in a shell made of silicon dioxide. The peculiar thing is that diatoms are not found in concentrations anywhere near those required to bury a whale. Yet there are:

> *346 whales within 1.5 km2…distributed uninterrupted through an 80-m-thick sedimentary section. The diatomaceous sediment lacks repeating primary laminations…There is no evidence for bioturbation…These sedimentary features suggest rapid burial due to high diatom accumulation…Current depositional models do not account for the volume of diatomaceous sediments...*
> L Brand et al. 'Fossil whale preservation implies high diatom accumulation rate in the Miocene–Pliocene Pisco Formation of Peru'. Geology. Vol 32. No 2. February 2004. p165-168.

Clearly, these whales must have been buried during a single catastrophic event of a nature and magnitude that is unparalleled in our modern experience.

In addition, we see fossils of large animals such as sauropods, and even herds of animals which are found buried together *en masse*, showing that whatever the cause, their burial must have been sudden, and involved a huge body of water and dirt. There is a fossil of an *Icthyosaur* in the process of giving birth, and fossils of fishes in the process of eating other fish. Yet another specimen from the Judith Formation in Montana (which may represent a dry burial since there are no signs of cross-bedding) shows a protoceratops and a therapod in very close proximity. Normally, when shellfish die their shells open up, but there are countless fossils of bivalves that have their shells firmly closed, showing that they were buried while still alive. Large numbers of whales have been found entombed in diatomite*, and yet diatoms are not found in concentrations anywhere near those required to bury a whale.
It is also important to note that vast numbers of animals did not live and die where their fossils are found. For example, *Lystrosaurus* fossils predominate:

> ...to a seemingly preposterous extent...in every case it makes up more than nine-tenths of animal remains. This degree of dominance is not typical of natural assemblages of animals... Michael Benton. 'The rise of the mammals'. The Apple Press. 1991. p26.

Also, very little plant material is found in association with the large dinosaur fossils. Since they must have lived amongst dense vegetation, their bones must have been carried away and hydrodynamically separated from the plant debris.
So there is a vast amount of fossil evidence to show that the animals were buried quickly, that many were buried alive by large bodies of entombing slurries, and that the fossil remains do not represent buried ecosystems. This means that the sedimentary layers that we see cannot have occurred very slowly, grain-by-grain, but must have occurred catastrophically.

### *Evidence for rapid formation of sedimentary layers*

We also have field evidence that confirms that each of the sedimentary layers could not possibly have existed for tens of years (let alone thousands of years), before any subsequent layer was deposited on top. In September 1961, *Hurricane Carla* ploughed along the coast of Texas and laid down multiple distinctive layers of sediment in the Gulf of Mexico. However, just twenty years later, the layers were virtually unrecognisable, having been all but eroded away.
Since it is not possible for the surface of a deposit not to become weathered and eroded in just a few years, this means that thousands or millions of years cannot possibly have elapsed between the laying down of consecutive *uneroded* layers of sedimentary strata all over the world, including those of the Grand Canyon.

When coarse and fine dry grains are fully mixed and then poured through a funnel, the different particles automatically separate out and the resulting pyramid is made up of distinct layers.

### *Experimental evidence for rapid formation of sedimentary layers*

Experimental evidence shows conclusively that when particles of different size and density are suspended in water and allowed to settle, separate layers are formed *automatically*.
When experiments are done in long flow tanks with rapidly flowing suspensions, particles are deposited in oblique layers. This can result in the seemingly impos-

This section of rock is 7cm thick. It might be assumed that the distinct layering is the result of a series of sedimentation events over thousands or millions of years.
However, the layers could have been formed in just minutes or hours.

sible situation that a particle B can be laid down *below* a particle A that was laid down *before* particle B:

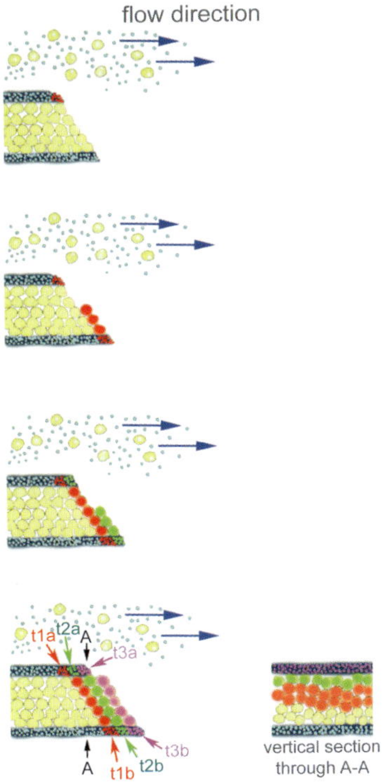


Diagrammatic representation of an experiment examining how sedimentary layers develop as a result of the differential settling of small and large particles suspended in water flowing from left to right.
Because smaller particles have less kinetic energy than larger particles, the small particles tend to settle first with the larger particles rolling over the top of them.
Thus, at any particular time (e.g. t1, in red), the small particles settle first at the top (t1a) and then at the floor (t1b) of the advancing sediment wall. The larger particles then collect above the smaller particles at 1b. The process is repeated and, as a result, the sediment wall advances from left to right.
Since particles at t1a were deposited *before* particles at t2b, this experiment demonstrates that: (i) particles or fossils lying higher up in a sedimentary sequence may have been deposited *before* those found lower down, and (ii) distinct layers (in this case, a large particle layer sandwiched between two small particle layers) can be produced during a single event. More complex patterns of layers are produced when particles of many different sizes are used.

Andrew Dickinson

Cross-bedding rock formations at Petra, Jordan. Cross-bedding is formed when sand is blown or washed up the stoss side (i.e. the back side) of a dune and avalanches down the lee slope (i.e. the front face). As the dune migrates, the lee slopes are preserved as a series of cross-laminations within the sandstone layer. Structures like these can form in a variety of environments, from rivers to deserts to oceans. The crucial factors are a steady supply of sand and high-energy currents (whether of wind or water).

Such oblique layers are *exactly* like those seen in cliffs all over the world, for example, the sediments containing shells in Iceland, as shown below.

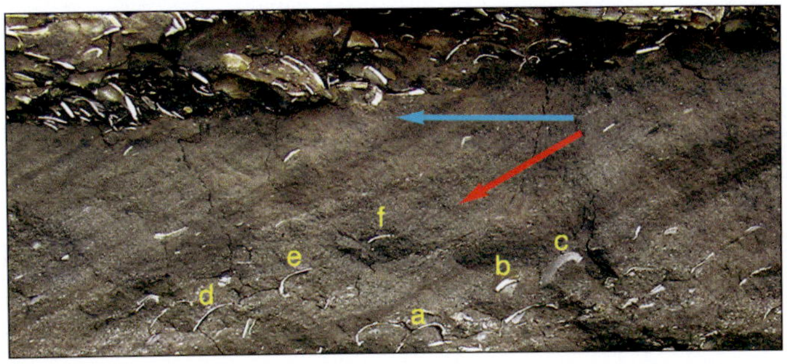


These shells lie in a cliff face north of Husavik, Iceland. The oblique lines suggest that the sediment flowed in the direction of the blue arrow and settled progressively from right to left, parallel to the red arrow. So the shells **a**, **b** and **c** were deposited before shells **d**, **e** and **f**. Therefore, although shell **a** was deposited before shell **f**, shell **c** was deposited *before* both shells **d** and **e,** even though it lies higher up in the sediment.

## *Field evidence for rapid formation of sedimentary layers*

### *a. Mount St. Helens*

Mount St. Helens on 10th April 1980 prior to its eruption.

Following an earthquake at 8:32 am on 18th May 1980, the volcano Mount St. Helens in Washington State erupted, sending a *pyroclastic* blast of molten rock, gases, and pulverised lava northward towards Spirit Lake at speeds of 500mph. At the same time, thousands of trees were sheered away by the wall of destruction and deposited all over the area north of the volcano. The associated landslide sped down the north fork of the Toutle River and filled its valley with up to 600 feet (180m) of debris.

Steve Austin

Effects of the Mount St. Helens eruption.
Above: This photograph, taken in 1989, shows the mini 'Grand Canyon' of the Toutle River which was formed following a mud-flow on 19th March 1982.
Below: This is the Lower Loowit Canyon which lies immediately north of Mount St. Helens.

Steve Austin

The horizontal strata in the block above, and in the middle section of the cliff below, were formed by pyroclastic flows on 12th June 1980, following the eruption of Mount St. Helens.

Steve Austin

Multiple horizontal strata in the Canadian Rockies.

Most of Mt. St. Helens' former north side became a rubble deposit 17 miles (27km) long, averaging 150 feet (46m) thick. As a result of the eruption, the volcano lost about 1300 feet of its elevation.

As this landslide hit Spirit Lake, all the water was temporarily displaced sending a wave about 800 feet high smashing into the north ridge of the lake. As the water fell back into its new basin, it dragged with it thousands of trees that had been felled by the eruption just seconds before.

On 12th June 1980, a massive pyroclastic flow surged down the volcano which resulted in an 8 metre-thick deposit on top of the previous ash deposits.

On 19th March 1982, an enormous mud-flow deposited a dark mudstone layer on top of the previous ash and pyroclastic layers. Simultaneously, in the course of that one day, the mud-flow cut through those previous layers in some places and created a mini 'Grand Canyon' with walls 140 feet high. It was then clear that the pyroclastic flow of 12th June 1980 had spontaneously formed distinct layers. This dramatically proved that layering and canyon-making can occur in a single day, and does not need thousands or millions of years.

Mt. St. Helens is not the only example of rapid sedimentary layering. Following the eruption along a 4km-long fissure under the Vatnajokull glacier in Iceland in 1996, a massive flood burst southwards into the Atlantic Ocean. Sedimentary layers were deposited in less than one day:

> During the outburst flood in Iceland on 5th November 1996, sediment containing around 200 planar rhythmites and 15m thick was laid down in just 17 hours. AJ Russell & O Knudsen. An ice-contact rhythmite (turbidite) succession deposited during the November 1996 catastrophic outburst flood (jokulhlamp), Skeidarajokull, Iceland. Sedimentary geology. Vol 127. 1999. p1-10.

Right: Vatnajokull glacier, Iceland, with close-up (far right).

*b. Mam Tor*

Mam Tor in Derbyshire, UK, is a 100 metre-high cliff comprising around 100 sedimentary layers which are supposed to have been laid down over a period of some 1 million years. If each layer was laid down very slowly, this means that (according to standard radiometric dating) each layer represents 10,000 years. But there is no evidence at all of any erosion between the layers, and in fact, the Mam Tor layers are typical *turbidites* caused, not by gradual silting, but by deposition by very rapidly-moving water. Such turbidites were seen after the Grand Banks earthquake involving 200,000 square miles of the eastern Canadian coast on 18th November 1929. This earthquake (7.2 on the Richter scale) dislodged huge amounts of sediment and carried it into deeper water. Judging by the timing of the severing of a number of transatlantic communication cables, the underwater flow of debris proceeded at some 500mph. So, most likely, all the Mam Tor layers were deposited in a short time in a series of closely-spaced events.

*c. Polystrate trees*

Many examples from coal and other rocks around the world show fossilised trees in the upright position which span many sedimentary layers. Clearly, no such *polystrate* tree could live while being systematically buried over thousands or millions of years. So, these trees must have been completely buried by sediments in a relatively short period of time, and the sediments cannot possibly have taken millions of years to form. In addition, these trees do not usually have root balls, showing that they were buried at a site distant from their growing site.

The polystrate trees at Specimen Ridge in Yellowstone National Park, USA, are also embedded in multiple layers of sedimentary rock.

The usual explanation is that this represents one forest becoming buried and a new forest growing up above the buried forest, with the process having been repeated some 27 times.

However, the Specimen Ridge trees have a number of features* which suggest that they did not grow in situ, and do not represent sequential growths over long periods of time. For example, the fossil wood and leaves belong to trees that are nowadays found in widely differing habitats and environments, showing that they must have been transported by water from diverse locations and lumped together. Many of the supposed soil layers are thin and do not contain enough organic material to qualify as soils. There is an apparent absence of clay in the majority of levels which suggests that normal weathering of soil has not occurred. The petrified trees do not show any signs of decay which would be expected if the trees were partially buried and died over a long period of time. A distinctive feature is that both the vertical and horizontal petrified trees lack bark, which suggest that the trees were severely abraded before they were buried. Although the Specimen Ridge trees have small roots attached, the larger roots are broken off, suggesting amputation prior to burial, and that this is not where they grew. Finally, tree ring analysis shows that the trees from various layers have common ring patterns, showing that they all grew *contemporaneously*, and not thousands or millions of years after each other.

> If trees lived at different times, their tree rings would show entirely different yearly patterns…A recent study of Yellowstone petrified wood [showed] that the trees retained matching signature patterns in the tree rings. Thus, they lived at the same time and were transported and deposited within different strata by successive mudflows. They did not live in successive forests. J Morris & S Austin. 'Footprints in the ash'. Master Books. 2003. p102-3.

A modern example of this is shown by the Mt. St. Helens eruption in which thousands of trees were blown off the ground north of the volcano, with many ending up in Spirit Lake (with their root balls missing). At first, the trees could be seen vertically bobbing up and down on the surface of the lake, with the heavier root ends downwards. In time, the trees started to sink to the lake bed and many have become stuck in the upright position, gradually becoming buried by sediment. The current situation is that vertical trees can be seen to be buried at different levels in the bottom of Spirit Lake.

*For more details, see the excellent article by Harold Coffin: http://www.grisda.org/origins/24002.html

Ian Juby

Polystrate tree from Tennessee, USA.

Ian Juby

Polystrate lycopods (giant reeds) from Joggins, Nova Scotia, Canada.
For more polystrate plant examples, see: http://www.ianjuby.org/jogginsa.html

Steve Austin

Fossilised tree trunks at Specimen Ridge, Yellowstone National Park, USA.
One particular example on the lower slopes of Ramshorn Peak in the Tom Miner Basin is about 15 metres tall and may be the world's tallest standing petrified tree.

Fossil tree-rings

As a result of the Mt. St. Helens eruption, thousands of trees were thrown into Spirit Lake. Since the root ends of trunks are denser than the upper parts, some of the trunks floated upright in the water (right). Other trunks sank to the bottom and were quickly embedded vertically in layers of bark and other debris. In time, the floating trunks will also sink, and eventually the lake bed will show a number of distinct sediment layers with trunks embedded at different levels. The satellite picture (far right) shows Mt. St. Helens (MSH) 5 months after the initial eruption with a smoke flume and the huge pale raft of trees (T) on the surface of Spirit Lake (SL, dark area).

Steve Austin

If seen after a few thousand years, it might be thought that these buried trees were the result of sequential forests, and yet all the trees lived and died at the same time. So, the picture at Specimen Ridge matches exactly the pattern that is emerging in Spirit Lake.

This acute angulation pattern suggests an event which was sudden rather than gradual, and which occurred after the sedimentary rock had hardened. (Cornwall, UK)

### *d. Bent rocks point to soft sediments*

We can also tell that virtually all of the world's sedimentary layers must have been made during the same event, at a time when the layers were still *soft*. When extreme sustained pressure is applied to the ends of a block of solid rock, the material can be made to 'flow' into itself and eventually crumble. When more gentle force is exerted on a rock for a long time, it can bend with a gentle curvature, and this can be seen in very old stone lintels or stone tables. In contrast, when sudden pressure is applied, rock will simply break. So, only when sediments are still soft could extreme bending be tolerated without disruption. It is true that if rock is subjected to extreme pressure and *heat* (for example, deep in the earth's crust) it will become deformed and bent, but in such instances the rock will show obvious evidence of the heating. For example, when it is subjected to extreme heat and pressure, limestone is converted into a rock called *mylonite*.

Therefore, when a marked curvature is seen in solid rock comprising multiple sedimentary layers which have no signs of heating, this is proof that the bending could only have occurred while the sediments were still soft. Therefore, all the layers involved in a bent rock must have been laid down simultaneously, or within a very short time of each other.

Many examples of curved sedimentary rock layers are known from diverse locations all over the world.

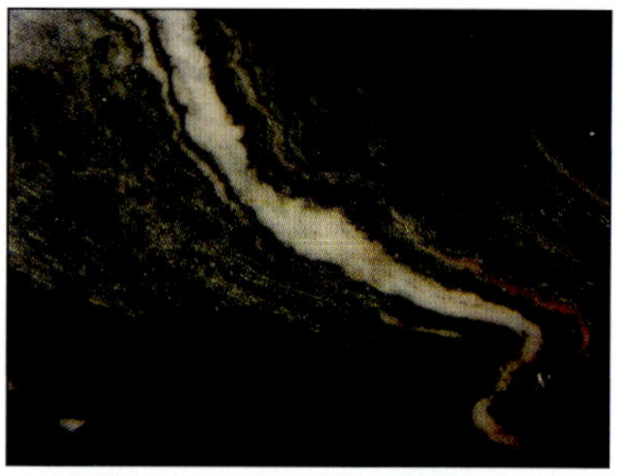

This limestone has been converted into mylonite.

Cave roof at Cabrespine, France. The limestone on the left is compressed while that on the right still shows layering.

Raé Swart

Curved rocks showing that all the layers were bent before the sediments had hardened.
Above: South Africa
Right: Canadian Rockies

So, where does this leave the Grand Canyon, the biggest terrestrial gash in the earth's surface? The common paradigm says that the sedimentary layers took many millions of years to become deposited, and that the canyon system was gouged out by a relatively small amount of water (the Colorado River) eroding its way through for a very long period of time. However, the evidence from all the above strongly suggests quite the reverse: that the Grand Canyon's sedimentary layers were laid down in rapid succession, and that the canyon system was cut through very rapidly by a sudden massive flow of water – *while the sediments were still soft, or partially lithified.* There is evidence that long ago, north-east of the Grand Canyon site, there was a series of large interconnected lakes which were restrained by an uplifted zone called the Kaibab upwarp, through which the Grand Canyon now cuts. One view on how the Grand Canyon was formed is that at one point, the Kaibab upwarp was breached and the lakes drained into the Pacific Ocean, creating the Grand Canyon in its wake. Another view is that the canyon was carved out when the massive inundating waters that laid the sedimentary layers retreated into subterranean cavities caused by immense tectonic activity. Whatever the exact chain of events, the evidence certainly points to the Grand Canyon sediments being laid down collectively over a period of months, and the canyon itself was probably carved in a matter of days rather than millions of years.

Grand Canyon

Paul Garner

Grand Canyon

Some might find it difficult to accept that it is possible for such huge amounts of rock to be eroded away in such a short time. However, the Missoula floods (of supposedly 12,000ya) showed just how powerful moving water can be. Lake Missoula was a glacial lake some 2000 feet deep, with an area equivalent to that of Lakes Erie and Ontario combined. When the ice dam that restrained the waters burst, flood waters equivalent to about 60 Amazon rivers surged with immense force over the land that now comprises the States of Idaho, Oregon and Washington:

> '...bedrock channels can be quickly excavated. The Missoula floods were able to pluck basaltic columns of bedrock, each weighing several tonnes, from stream beds, and carve out canyons through bedrock in the western United States in a matter of days...present day flood events are capable of completely eroding in a matter of days flood plains that may have taken centuries to form.' E.A. Bryant. 'Natural hazards'. Cambridge University Press. 1991. p142.

Lake Coeur D'Alene, Idaho.
This lake is a leftover from the great Missoula flood.

Just as the streams that flow into the canyon north of Mt. St. Helens are only there because the canyon was there first, so the Colorado river is only there because the Grand Canyon was there first. The Grand Canyon is not the only example of a mega-gorge in the Earth caused by mega-volumes of water: the sides of the English Channel at the Dover Straits still have the scar marks of rapid erosion.

The cliffs at Hastings, on the English Channel.

*e. Tilted mountain ranges*

In a number of places around the world, mountain ranges can be seen with vast numbers of parallel *tilted* sedimentary layers. Since the layers must have been laid down by water-borne material, those tilted layers must have all been horizontal and stacked upwards, at some time in the past. Between Lake Castiac and Pyramid

Tilted strata in the Canadian Rockies.

Lake in California, there is a tilted range that extends some *eight miles* (12.6km). Since the vast majority of the oceans are over 3 miles (4.8km) deep, this means that at some time in the past, the water level could have been *far above* the current sea level. Indeed, many tall mountains (including Mt. Everest) have high-altitude sediments containing marine fossils. Such high water levels can only mean that the flooding that caused the highest sedimentary layers could not possibly have been local. Of course, we know that ocean floors have been uplifted in places, but it is impossible for any substantial uplifting *involving hundreds of square miles* to have occurred whilst ensuring that the layers remained *perfectly* horizontal. Furthermore, whatever the event was that resulted in the tilting an *eight-mile-high* block of sedimentary layers, it either must have had *global repercussions*, or was *part of* global tectonic activity, and it must have been of a nature that was unlike any process that is seen going on today.

### *Questions regarding rapid laying down of sedimentary layers*

While the evidence for the vast expanses of horizontal sedimentary layers points essentially to a single huge inundation event, a number of issues might still seem to contradict this view.

#### *a. Stromatolites*

Stromatolites are flat, hemispherical, or toadstool-shaped structures ranging from a few centimetres to 2.5m across. They are made almost entirely of calcium carbonate which has been laid down sequentially, so that a cross-section shows distinct layering. Stromatolites are found today in highly salty environments (such as Shark Bay in Australia, Cuatro Ciénegas in the Mexican desert, and off the Bahamas). Small stromatolites are found growing today in hot springs such as in Yellowstone National Park. Fossil stromatolites are found worldwide in the lower-most geological strata, but are uncommon in the majority of sedimentary layers. Those in the *Awatubi Member of the Kwangunt Formation* in the Grand Canyon (about 635m below the base of the *Sixtymile Formation*) are particularly large.

Although they are commonly thought to have been produced by precipitation triggered by the growth of *cyanobacteria* (also known as *blue-green algae*), this is not universally accepted. For example, stromatolites (supposedly) 1.9byo were considered to be formed purely by physio-chemical effects:

> ...the morphogenesis of all stromatolites can be accounted for by four abiotic processes: fallout of suspended sediment, downslope movement of that sediment (or the mathematical equivalent surface-tension effects in chemical precipitation), surface-normal precipitation, and random effects. M Walter. 'Old fossils could be fractal frauds'. Nature. Vol 383. 3rd October 1996. p385-6.

Since micro-organisms are found everywhere today, then they should have been present on all ancient stromatolites too. If that were the case, it would be difficult to identify whether stromatolites are formed by purely chemical or biological processes, or by a combination of both. Certainly, it can be shown that twigs placed in water containing high concentrations of carbonates can become encrusted with thick layers of precipitations (*travertines*) in a matter of months (e.g. at Havasu Creek, Grand Canyon); so some non-biological process may be involved in stromatolite formation. If stromatolites are formed by bacterial action, then it is assumed that sunlight and photosynthesis are also necessary. Whatever the mechanism of their formation, the inclusion of stromatolites in sedimentary layers may be interpreted as proving that at least a number of years must separate the laying down of a stromatolite-bearing surface from the layer above it.

Apart from the possibility of stromatolites having been transported to their final resting places from elsewhere, there are at least three possibilities to account for the stromatolites that are found between sedimentary layers. Firstly, sedimentary layer 1 is laid down and dries out, a stromatolite grows for a number of years, and then sedimentary layer 2 is laid on top. However, since the stromatolite-bearing sedimentary layers do not have significant weathering, this is unlikely to explain their formation.

Secondly, sedimentary layer 1 is laid down and remains submerged, one or more sedimentary layers are laid down on top, and then the stromatolite starts to grow within the layers. We could predict that, in this case, there would be no weathering of the sedimentary layer 1. However, if all the layers were still soft, a purely chemical action would result in a *sphere* of growth, both down into sedimentary layer 1 as well as up into sedimentary layer 2. Since stromatolites only occur as surface structures, this scenario is unlikely to explain their formation. Although the lack of sunlight might be thought to be a problem, yet in biology, the bizarre and unimagined are always waiting to be discovered. For example, some bacteria have been found in the dark depths of the ocean which can use the dim red glow of a hot hydrothermal vent to photosynthesise:

> The vent, which spews water at 350°C, is nearly 2.5 kilometres deep on the East Pacific Rise, a submerged mountain range that runs south from the Gulf of California...These bizarre organisms convert carbon dioxide into organic compounds by oxidising sulphur compounds. They thrive in oxygen-free environments such as hot springs and can photosynthesise at extremely low light levels, but they have never been found in sunlight-free environments before. 'Life thrives on ocean's faint glow'. New Scientist. 25th June 2005. p22.

Thirdly, sedimentary layer 1 is laid down, the stromatolite starts to grow while still submerged, and then sedimentary layer 2 is laid down. We could predict that, in this case, there would be no weathering of the sedimentary layer 1, and the deposition of sedimentary layer 2 would result in distinguishable flow features in the grains on the 'upstream' and 'downstream' sides of the stromatolite. These flow features could confirm whether or not the stromatolite was fully formed prior to the deposition of sedimentary layer 2. Since large precipitations can occur in a matter of months, and no weathering is seen on stromatolite-bearing surfaces, this last scenario may well represent the true method of their formation, and is consistent with a single inundation event.

These mud cracks on a shore-line appeared just two days after torrential rains, and without any intense sunshine to dry out the ground.

*Note that a single huge flood event could have been subdivided by a number of ebb and flow movements of water.

### *b. Fossilised mud cracks*

A number of rock layers, including some in the 8-mile-long series of tilted sediments between Lake Castiac and Pyramid Lake in California, contain fossilised mud cracks. It might be thought that these were created when a mud surface dried up and cracked, and then the next rainfall filled in the cracks with sandy material to leave, ultimately, sandstone tentacles within a mudstone. Since there must be a drying-out period, it could be thought that these prove that the multiple sedimentary layers could not have been laid down during one devastating flood event*.

However, we do not know what sort of unusual features a truly massive inundation would produce, particularly if associated with underwater volcanic activity that would result in *hot* flood waters. Nevertheless, since we know from the edges of hot water vents that mud can crack *while it is still damp*, it is possible that the fossil mud cracks were made and became infilled with sandstone *while still underwater*. In addition, the surfaces of the layers with mud cracks do not have any erosional features, showing that there was no significant time between the cracking and the infilling.

Damp mud cracks from hot vents in Iceland (above) and New Zealand (below).

### *c. Ripple marks*

Fossilised ripple marks are not uncommon, and it might be thought that ripples must have had to dry out before they could have become covered by sedimentary layers. So it might be thought that ripple marks within sedimentary layers prove that the layers could not have been laid down rapidly. However, the surfaces of the layers with ripples do not have any erosional features, showing that there was only a small amount of time between the ripple formation and their burial.

Furthermore, ripple marks are formed by wave action in either a shore-line or a pond, and commonly persist underwater. With a to-and-fro movement of a large body of water, ripples could have become buried by sediment which might itself develop ripples on its surface, which also might become buried. In this way a number of layers, each with ripples, could be formed and hardened together in a short period of time. As previously noted, dinosaur footprints have been found within ripples, which obviously means that the animals had been walking through shallow waters before a subsequent inundation. However, this does not mean that the ripples had dried out before their subsequent burial.

Fresh ripples on a beach (right), beside a river (far right), and in a still lake (above).

Fossil ripples in Scotland (far left) and Canadian Rockies (left).

## *Radiometric dating revisited*

Many people are conditioned to accept that the radiometric dating methods *prove* that the fossil-bearing sedimentary layers are millions of years old. However, as we have seen, the lack of erosion, combined with the almost universal lack of disruptions amongst the different sedimentary layers, shows that the radiometric dating of those layers *must be completely wrong*. Furthermore, radiometric dating discrepancies can be very serious.

### Potassium-Argon and similar methods

As an example of radiometric dating discrepancy, when the Cardenas Basalt Lavas in the Grand Canyon were analysed by two of the commonly accepted methods, Potasium-Argon and Rubidium-Strontium, these gave *wildly* different results – 516my and 1070my respectively.

> We report a new fourteen-point K-Ar isochron 'age' of 516± 30 Ma which is strongly discordant with the published Rb-Sr isochron 'age' of 1.07 ± 0.07 Ga for Cardenas Basalt. S Austin & A Snelling. 'Discordant Potassium-Argon Model and Isochron "Ages" for Cardenas Basalt (Middle Proterozoic) and associated diabase of Eastern Grand Canyon, Arizona'. Proc. Fourth Intl. Conf. on Creationism. R Walsh (Ed). Creation Science Fellowship. 1998. p35–51.

In addition, the diabase lava at Bass Rapids in the Grand Canyon also yields markedly disparate dates by a variety of methods: Potassium-Argon, 840my; Rubidum-Strontium, 1055my; Uranium-Lead, 1209my; Samarium-Neodymium, 1375my.

> …our new K-Ar, Rb-Sr, Sm-Nd and Pb-Pb radioisotope data from eleven whole-rock samples (eight diabase, three granophyre) and six mineral phases separated from one of the whole-rock diabase samples yield discordant whole-rock and mineral isochron 'ages'. These isochron 'ages' range from 841.5±164 Ma (whole-rock K-Ar) to 1375±170 Ma (mineral Sm-Nd). Although significant discordance exists between the K-Ar, Rb-Sr, Sm-Nd, and Pb-Pb radioisotope methods, each method appears to yield concordant 'ages' internally between whole rocks and minerals. A Snelling, S Austin & W Hoesch. 'Radioisotopes in the diabase sill (Upper Precambrian) at Bass Rapids, Grand Canyon, Arizona: an application and test of the Isochron Dating Method'. Proc. Fifth Intl. Conf. on Creationism. Ed. R Ivey, Jr. Creation Science Fellowship. 2003. p269-84.

Other studies of lava from known eruptions of the volcano Mount Ngauruhoe in New Zealand yielded impossible results:

> The radioactive potassium-argon dating method has been demonstrated to fail on 1949, 1954, and 1975 lava flows at Mt Ngauruhoe...Argon gas, brought up from deep inside the earth within the molten rock, was already present in the lavas when they cooled. We know the true ages of the rocks because they were observed to form less than 50 years ago. Yet they yield 'ages' up to 3.5 million years which are thus false. A Snelling. 'The cause of anomalous potassium-argon "ages" for recent andesite flows at Mt Ngauruhoe, New Zealand, and the implications for potassium-argon "dating" '. Creation. 22(1). December 1999. p18–21.

This means that radiometric dating cannot be relied upon to give true ages.

### *Carbon 14*

It may be surprising to know (but it is not unknown to radioactive dating laboratories) that $C^{14}$ has been recovered from *uncontaminated* rocks (including coals and diamonds) everywhere from the uppermost to the lowest sedimentary layers:

> An astonishing discovery made over the past twenty years is that, almost without exception, when tested by highly sensitive accelerator mass spectrometer (AMS) methods, organic samples from every portion of the Phanerozoic* record show detectable amounts of 14C. J Baumgardner, A Snelling, D Humphreys and S Austin. 'Measurable 14C in fossilized organic materials: confirming the Young Earth Creation-Flood Model'. Proc. 5th Int. Conf. Creationism. Ed. R Ivey, Jr. Creation Science Fellowship. 2003. p127–42.

*Phanerozoic: the last 545my since (supposedly) the first hard-bodies creatures appeared.

> Although tests showed some surface contamination, it was not possible to reach lower 14C levels through cleaning, indicating the contamination** to be intrinsic to the sample...So far, no theory explaining the results has survived all the tests. M Nadeau et al. 'Carbonate $C^{14}$ background: does it have multiple personalities?' Radiocarbon. 43:2A (2001). p169.

**Note that Nadeau et al use the misleading term 'contamination' when this cannot be the case since the $C^{14}$ is intrinsic to the rock.

Since no $C^{14}$ should be found in any material older than 70,000 years, then its finding means that all sedimentary rock must be younger than that. Therefore, any radiometric dating method that gives dates for rocks in the order of millions of years *must* be wrong. And there are a number of other radioactivity-related phenomena which are incompatible with a very old age for the Earth's rocks.

### *Radiohalos*

Granites are commonly thought to have been formed over periods of hundreds of thousands of years, from the cooling of molten rock, deep within the earth's crust. However, the existence of polonium *radiohalos* is strong evidence that granites must have formed in a matter of *days*, and this in turn, suggests that radiometric dating is completely erroneous.

*Zircons* are tiny crystals, some 60 micrometres long, made of zirconium silicate ($ZrSiO_4$) which are found in *biotite*, the dark flakes in a chunk of granite. *Radiohalos* are microscopic spheres of discolouration found in rock crystals caused by the radioactive release of *alpha particles* (helium nuclei). The formation of a single visible radiohalo requires the decay of about 0.5-1 billion parent radioisotope atoms.

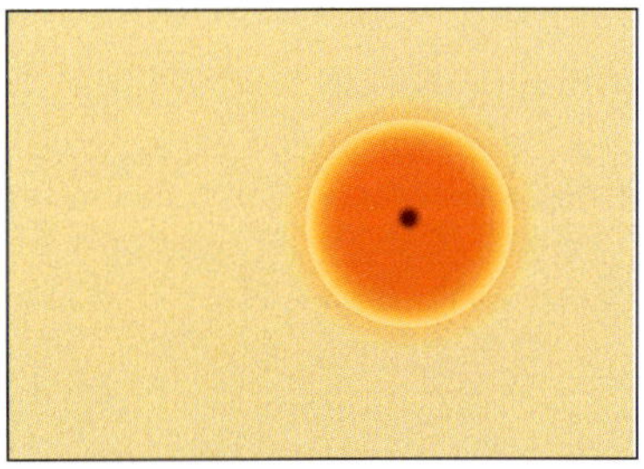

$Po^{210}$ radiohalo with a single ring. (Diagrammatic)

When radioactive elements release alpha particles from inside rock, the energy of the particles creates such distinct radiohalos that the parent element can be identified by appearances alone. Since the alpha particles are released in every direction, concentric spheres of discolouration form in the rock, but when seen under a microscope, these spheres are seen as concentric circles.
When uranium-238 ($U^{238}$) decays to lead ($Pb^{206}$), it does so by a number of steps during which alpha particles are released and intermediary elements are formed, such as polonium (Po), which is the last in the chain before $Pb^{206}$. So, for example, a $U^{238}$ radiohalo will have eight rings (caused by the release of eight alpha particles), that of $Po^{218}$ will have 3 rings, $Po^{214}$ will have 2 rings, and $Po^{210}$ will have 1 ring.
Although the half-life of $U^{238}$ is 4.46 billion years, the half-life of the isotopes $Po^{218}$, $Po^{214}$ and $Po^{210}$ are 3 minutes, 164-thousandth of a second, and 138 days, respectively. This means that after 4 years (10 half-lives), the whole of a sample of $Po^{210}$ will have completely decayed, leaving no polonium at all.

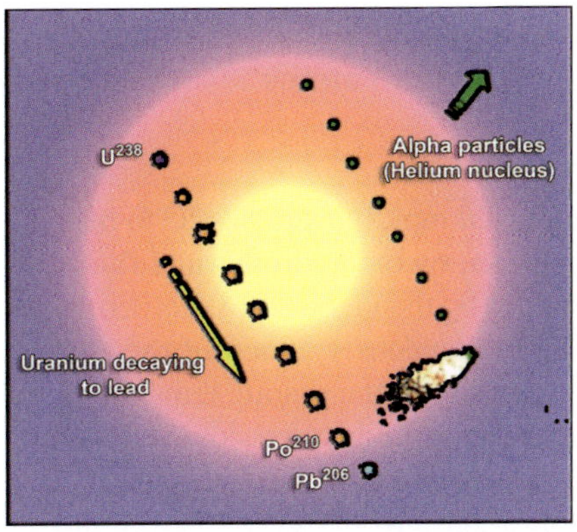


The process of uranium-238 decaying into lead-206, involves the release of 8 alpha particles. The release of the final alpha particle (from polonium-210) is shown.

Now, the interesting thing is that polonium radiohalos are *very commonly* found in biotites *separate from the radiohalos that arise from the decay of the original parent uranium atoms*. In order for *orphan* Po radiohalos to have formed, a number of features would have been necessary:
1. A sufficient number of uranium atoms must have decayed in order to have produced enough polonium atoms (representing about 100 million years of uranium decay) which then must have released enough alpha particles (in the region of 0.5 -1 billion) to have created a radiohalo.
2. In order to have created an orphan radiohalo, the polonium atoms must have travelled away from the parent uranium atoms, and must have been concentrated at a separate site, before the polonium atoms decayed. (The only way that the polonium atoms could have been transported is by the action of hydrothermal fluids, themselves generated by the process of crystallisation of the magma. The polonium atoms could then have been carried along cleavage planes in the biotites, to be subsequently deposited at certain spots).
3. Because of the very short half-lives of polonium isotopes, the time between polonium isotope generation by the parent uranium atoms and the polonium decay at the separate site must be very short, in the region of minutes or days at most.

4. The magma had to be hot enough to produce hydrothermal convection to transport the polonium atoms away from the parent uranium atoms.
5. Before they decayed, the polonium atoms must have been embedded within cooled (below 150°C) and crystallised magma, because at higher temperatures any discolouration caused by alpha particles would have been obliterated by heat, and would not have persisted to become a radiohalo.

Granites are commonly assumed to have taken hundreds of thousands of years to form*, and it might be thought that ongoing uranium decay could have provided the millions of orphan polonium atoms over an extremely long time period, with the polonium halos being formed very gradually. However, if the polonium had been transported away from the parent uranium over hundreds of thousands of years, it is unlikely that the polonium atoms would have become concentrated in exactly the same spot in order to create a radiohalo. In such a scenario, we would expect the polonium atoms to have been transported in many directions with varying concentrations, and to many final resting sites. So, rather than distinct radiohalos, this would have resulted in diverse *streaks* of discolouration, each of varying intensity, leading away from the parent uranium atoms.

** The conventional wisdom has been adamant until recently that granites take millions of years to form...However, [the] long-accepted timescale for these processes is now being challenged, even by conventional geologists[1,2]. The essential role of rock deformation is now recognized. Previously accepted granite formation models required unrealistic deformation and flow behaviors of rocks and magmas, or they did not satisfactorily explain available structural or geophysical data. Thus it is now claimed that mechanical considerations suggest granite formation is a "rapid, dynamic process" operating at timescales of less than 100,000 years, or even only thousands of years.*
A Snelling. 'Rapid Melting of Source Rocks, and Rapid Magma Intrusion and Cooling'. Answers Research Journal 1. 6th February 2008. p11-26.

*1. J Clemens. 'Granites and granitic magmas: Strange phenomena and new perspectives on some old problems'. Proceedings of the Geologists' Association. Vol 116. 2005. p9–16.*

*2. N Petford and J Clemens. 'Granites are not diapiric!' Geology Today. 16(5) 2000. p180-4.*

Furthermore:

> ...unless the granite cooled rapidly from 400°C to below 150°C, most of the...polonium transported by the hydrothermal fluids would have been flushed out of the granites by the vigorous hydrothermal convective flows as they diminished. Simultaneously, much of the energy to drive these fluid flows dissipates rapidly as the granite temperature drops. Thus, below 150°C the hydrothermal fluids have slowed down to such an extent that they cannot sustain protracted flow, and with the short half-lives of the...polonium isotopes, they would decay before those atoms reached the polonium deposition sites...
> Thus sufficient polonium had to be transported quickly to the polonium deposition sites to form the polonium radiohalos, while there was still enough energy at and just below 150°C to drive the hydrothermal fluid flow rapidly enough to get the polonium isotopes to the deposition sites before the polonium isotopes decayed.
> The time "window" is especially brief in the case of the decay of the 218Po and 214Po isotopes (half-lives of 3.1 minutes and 164 microseconds respectively) and the formation of their radiohalos. It would thus be simply impossible for these polonium radiohalos to form slowly over millions of years at today's groundwater temperatures in cold granites. Heat is needed to dissolve the...polonium atoms, and to drive the hydrothermal convection that moves the fluids which transport the...polonium atoms to supply the radiocenters to generate the polonium radiohalos. Furthermore, the required heat cannot be sustained for the 100 million years or more while sufficient 238U decays at today's rates to produce the required polonium atoms to form the polonium radiohalos. Thus the granites need to have crystallized and cooled rapidly (within 6–10 days) to still drive the hydrothermal fluid flow rapidly enough to generate the polonium radiohalos within hours to a few days. A Snelling. 'Rapid Melting of Source Rocks, and Rapid Magma Intrusion and Cooling'. Answers Research Journal 1. 6th February 2008. p11-26.

The natural inference of this is that in order to produce enough Po atoms *the $U^{238}$ must have decayed at a far greater rate than it does today*. This seems to contradict the idea that radioactive decay rates do not vary. However, we must follow the evidence, and not any preconceptions of how we might think atoms ought to behave.
Quite clearly, timescales for the generation of granitic magmas and their intrusion, crystallization, and cooling are quite compatible with the earth being far younger than commonly thought.

### *Helium in zircons*

Helium is formed as a by-product of uranium decay, and because it is so inert, it readily escapes from within crystalline structures. After millions of years of uranium decay and subsequent helium escape, we would expect that zircons should contain only a very small amount of residual helium. However, when the rate of helium diffusion was experimentally analysed in zircon crystals, it was found to be so high that it could not account for the amount of helium that was still trapped inside the crystals. In other words, if all the helium inside the zircons had been produced by millions of years of uranium decay, then after millions of years there would be very little helium left:

> The helium indeed could not have remained in the zircons for even a million years, much less the alleged 1.5 billion years. R Humphreys, S Austin, J Baumgardner, and A Snelling. 'Helium diffusion age of 6,000 years supports accelerated nuclear decay'. CRSQ. Vol 41. No.1. June 2004.

This means that the zircons could not have been discharging their helium for millions of years, and also that the rate of uranium decay was far greater in the past. So, this runs contrary to the accepted praradigm that radioactive decay rates are constant, and strongly suggests that the Earth's oldest rocks are only thousands of years old:

> Combining rates and retentions gives a helium diffusion age of 6,000 ± 2,000 years. This contradicts the uniformitarian age of 1.5 billion years based on nuclear decay products in the same zircons. These data strongly support our hypothesis of episodes of highly accelerated nuclear decay occurring within thousands of years ago. R Humphreys, S Austin, J Baumgardner, and A Snelling. 'Helium diffusion age of 6,000 years supports accelerated nuclear decay'. CRSQ. Vol 41. No.1. June 2004.

Furthermore, if increased decay rates in the past were mirrored in other radioactive elements, then dating methods such as K-Ar etc. would give us false old ages for igneus rocks. This is a big pill to swallow, given the perception that radioactivity decay rates are constant and very well understood – after all, radioactivity is used daily in power stations and in medical treatments. However, whilst it is important to acknowledge that there must be many aspects of nuclear physics that are as yet unknown, nevertheless, on the evidence to date, it would be unscientific to deny the *science* that suggests that the rocks are not millions of years old at all.

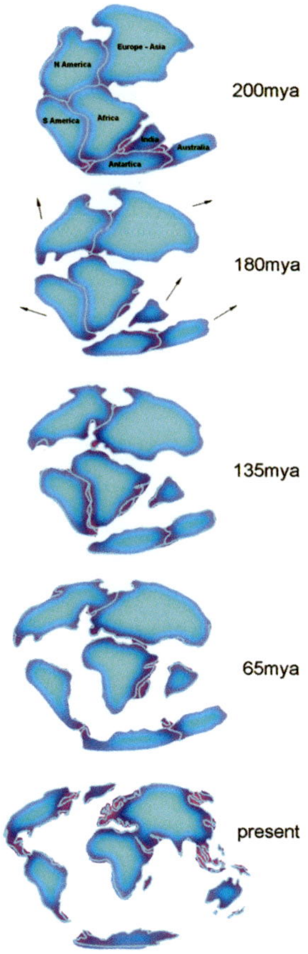


Continental shift. It is commonly thought that (supposedly) 200mya, all the continents were united in a common land mass, called Pangea. Then, (supposedly) between 200 and 165mya, a rift appeared between North America and Africa. Supposedly around 135mya, it is thought that South America and Africa started to split. India is thought to have collided with the Asian mainland plate (supposedly) about 20-50mya.

Sagarmatha, in Nepal, is the tallest mountain in the world. It is also known as Mt. Everest, or by its Tibetan name, Chomolangma.

### *Continental shift*

It is now well accepted that, in times past, the continental land masses occupied different positions on the Earth's surface, and that the continental plates are in a continual state of movement.

This movement of the Earth's crust (commonly called 'continental *drift'*) is responsible for geological phenomena such as earthquakes and mountain formation. The main evidence for continental shifting comes from comparisons of the chemical composition and direction of magnetism in rocks at different locations, along with comparisons of the very shapes of the land masses themselves. For example, the western part of Africa fits into the eastern side of South America; the European coast has identical rock formations to those on the east side of North America; and Madagascar seems to fit neatly alongside the east coast of southern Africa etc. Looking at the map of the world, it certainly does look as though different parts of the land masses could fit together like a giant jigsaw puzzle.

The fossil content of rocks found in the different continents also lends support to the idea of continental drift. For example, fossils of *Mesosaurus,* a small semi-aquatic reptile dated as being 180myo, are found in South America and southern Africa. Fossils of *Lystrosaurus*, a terrestrial mammal-like reptile dated as being over 245myo, are found in Antarctica, India, South America, and southern Africa.

At the moment, the continental plates are gradually moving apart at the rate of a few centimetres per year, so, working backwards, it is supposed that some 200mya, all the continents, including Antarctica, were united in a common land mass which has been called *Pangea*. Gradually, supposedly between 200 and 165mya, a rift appeared between North America and Africa, and produced the embryonic Atlantic Ocean. At around 135mya, it is thought that South America and Africa started to split, and that at about the same time, India broke away from Africa. The north-west part of India is thought to have collided with the Asian mainland plate about 50mya, and different sources suggest that the Himalayan mountain range was formed somewhere between 5-50mya, as a result of the Asian plate being pushed upwards by the force of the Indian plate ploughing into it.

Whilst it is generally thought that continental shifting has *always* occurred at only a few centimetres per year, nevertheless, we are not obliged to accept that the rates of the movements of different parts of the Earth's crust have always been constant, or that they have not, at times, moved very rapidly. For example, previously it was thought that after the north-west corner of the Indian plate collided with the Asian plate, the rest of the northern edge of the Indian plate made impact over a very long period of time. However, recent evidence suggests that the entire northern edge of the Indian plate made its impact with the Asian plate at the same time, and that the granite comprising the base of the Himalayan mountain range may also have been formed:

> ...at the same time, about 20 million years ago, all the way down the line of the collision. Carolyn Fry. 'Everest's rocks reveal their secrets'. New Scientist. 31st May 2003. p14.

Therefore, it is quite possible that the Indian plate/Asian plate collision was far more sudden than is commonly believed. If that is the case, then in times past, the rates of the movements of the continental plates might have been very much more dramatic than we are commonly led to understand.

Lake Thingvallavatn in Iceland was formed as a result of rifting of the mid-Atlantic ridge. In the foreground, the basalt rocks of the Almanngja gorge represent a visible part of the mid-Atlantic ridge between the North American and Eurasian plates.

### ***The distribution of fossils and living creatures do not match continental shift dates***

In fact, the biological evidence highlights a number of inconsistencies in the generally accepted timescale in which the continents are supposed to have moved about with the Earth's surface crust. Throughout this book we will see diverse evidence of this, but here are some examples.

#### *Australian placental mammals*

In 1998 a lower jaw containing a number of teeth and dated as being (supposedly) 115myo was found near Melbourne, Australia. The structure of the teeth suggests very strongly that the shrew-like creature (called *Ausktibuosphenos*) was a placental mammal. The problem is that:

> ...except for bats, placentals are not known to have entered Australia until five million years ago...A good case can be made that it is a placental...If so, it will force us to to rethink basic ideas about the origin and evolution of mammals. Steve Morton. 'A new jaw reopens some old questions'. National Geographic. April 1998.

So, although it is generally accepted that Australia-Antarctica and Africa split around 180mya, the new finding of ausktibuosphenos suggests that Australia was possibly connected to Africa as 'recently' as 115mya. The alternative is that this animal entered Australia about 180mya (before Australia split from Africa) and remained a shrew for an enormous 65 million years (from 180-115mya). On the other hand, if ausktibuosphenos had indeed only entered Australia (supposedly) 5mya along with other placental mammals, then the dating of the rocks by which the ausktibuosphenos fossil was dated, is completely wrong.

#### *Messel Quarry in Germany*

The Messel Quarry has yielded a large number of fossil mammals, dated as being from the Eocene epoch, (supposedly) *35-65mya*:

> Most of the Messel animals fit our expectations for an animal community from Europe or North America during the Eocene epoc, but two of the fossil mammals discovered there are a great mystery. One is the anteater Eurotamandua, a member of a group known otherwise only from South America. The other is the pangolin Eomanis, from a group typical of South East Asia. How these animals came to be living in central Europe in the Middle Eocene is a mystery... Michael Benton. 'The rise of the mammals'. The Apple Press. 1991. p61.

There is a mystery here because Benton *assumes* firstly, that these mammals arose in the last 65my, and secondly, that the dates usually given to the shifting of the

Giant anteater

continental plates are accurate and extremely ancient. However, if we put aside these assumptions, then we might first conclude that *Eurotamandua* lived at roughly the same time in both Europe and South America, and therefore, that the anteaters were living at a time when Europe and South America were united with North Africa and the eastern part of North America. If we assume that the dates assigned to continental shifting are accurate, then this means that the anteaters were living about 180mya. But this conflicts with the general assumption that most modern mammal forms appeared only 65mya onwards.

Other Messel Quarry fossils of animals such as hedgehogs, bats, lemurs, squirrels and tapirs show features that are very similar (if not identical) to modern forms. Since there clearly has not been enough time for living forms to have acquired significant differences to the fossilised Messel creatures, the Messel creatures must have lived quite recently.

While the common hedgehog (above) grows to an average size of 33cm, the African pygmy hedgehog (below) grows to 15-25cm. Interestingly, the spines on the African pygmy hedgehog are relatively blunt.

Pamela Melia

### *Giant Tyrannosaurs*

Until recently, Tyrannosaurus rex was thought to have lived only in the northern hemisphere. However, in the Andean foothills of Bolivia, cement factory workers near the city of Sucra discovered an amazing series of some 15,000 individual prints embedded on a vast vertical limestone wall. Some of the tracks have been identified and show:

> ... tyrannosaurus rex meeting a smaller herbivore... Christian Meyer, quoted by Gabriella Gamini. 'In the footprints of giants'. The Times. 12th September 1998. p30.

Also living in South America was *Giganotosaurus carolini*, a larger version of Tyrannosaurus rex:

> ...probably the largest meat-eater to have roamed around 90mya. Gabriella Gamini. 'Dinosaurs' lost world comes alive'. The Times. 21st February 1998. p6.

Furthermore:

> The Giganotosaurus has a similar brother in Africa, the Carchardontosaurus... Gabriella Gamini. 'Dinosaurs' lost world comes alive'. The Times. 21st February 1998. p6.

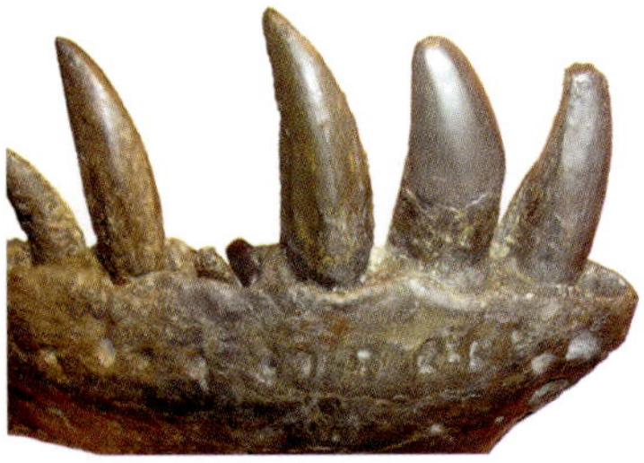

Tyrannosaurus rex teeth

Here again, a problem arises if Africa and South America (supposedly) separated about 135mya. Since it is implausible that these three very similar Tyrannosaurs could have evolved completely separately and independently, their common ancestors must have been present before the continents separated. So, naturally, the discovery of Giganotosaurus:

> ...has led paleontologists to propose a later than supposed link between South America and Africa. Gabriella Gamini. 'Dinosaurs' lost world comes alive'. The Times. 21st February 1998. p6.

## *The distribution of living forms do not match continental shift dates*

### River dolphins and gharials

Evidence for possible serious errors in the dates commonly assigned to the drifting of continents is not limited to fossils. There are five known types of river dolphins. The *Indus* and *Ganges River Dolphins* look identical (although there are some small differences in their skulls) and live exclusively in the very muddy waters of the Indus and Ganges river systems. They are known never to swim into estuarine or coastal waters and they remain entirely within the inland waterways:

Indian river dolphin

> The Ganges river dolphin is widespread along most of the Ganges, Brahmaputra and Karnaphuli rivers of India and Bangladesh. The Indus river dolphin of Pakistan is separated geographically and is classified as a different species, but is **essentially the same animal**. (My bold). Richard Harrison and M.M. Brydon. 'Whales, dolphins and porpoises'. Merehurst Press. 1990. p42.

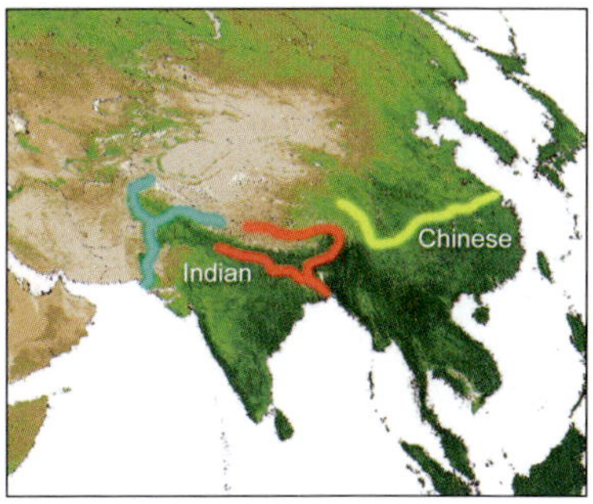


Asian river dolphin distribution

Almost identical to the Indian river dolphins is the Chinese river dolphin, the *Baiji*, which is found exclusively in the waters of the Yangtze river system, and it too never enters coastal waters. On the basis of only minor differences there is no obligation to call the Asian river dolphins different *species*.

In South America there are two river dolphins, the *Boto* and the *Franciscana*. The boto is the largest of the world's river dolphins and the three living populations (which are thought to be genetically identical) are found inhabiting the Orinoco and Amazon Basins, and the Upper Madeira River. These *geographically distinct* varieties are:

> ...separated by mountain ranges but are not sufficiently different to be classified as separate species. Richard Harrison and M.M. Brydon. 'Whales, dolphins and porpoises'. Merehurst Press. 1990. p42.

Boto and franciscana distribution

The boto does not enter estuarine or coastal waters, but unlike all other river dolphins, the franciscana does not live in rivers. Instead, the franciscana lives close to the shore in the shallow (less than 9m deep) eastern coastal waters about 1000km north and south of the La Plata estuary. Size apart, since the boto and the franciscana have only some minor differences in appearance, could they too represent the same creature?

The question arises as to how these isolated populations of river dolphins came into being. It is most unlikely that they were originally simply sea-going animals which happened to find their way into inland waters, since if that were the case, they would surely be found in the oceans as well as in many other river systems, scattered at the very least around the Indian Ocean rim, and all along the eastern South American coastline.

One explanation is that the Asian river dolphins were originally a single population living in a large central Asian river system that perhaps drained into the China Sea as does the present Yangtze River. When the Indian continental plate collided with the Asian plate, this then caused the uplifting of the Tibetan plateau to produce ultimately the Himalayan mountain range. As a result of this, the original large central Asian river system could have been broken up into the Yangtze part which continued to flow eastwards, and a second system which could have flowed southwards. Such a break-up in the river systems would have resulted in three isolated populations of Asian river dolphins, one destined to become the baiji and the other destined to become the Indus and Ganges dolphins.

Similar tectonic events could have resulted in the separation of the South American river dolphins. The idea of *massive* alterations in the direction of flow of major rivers is not a new one, and nor is it incredible. In fact, it is thought that at some time in the past, the Amazon once flowed *in the opposite direction* to which it does now, and drained into the Pacific Ocean.

It could be no coincidence that two identical populations of the *Gharial* (a long-nosed crocodilian) are also found in the same river systems as the Indus and Ganges river dolphins. Since the gharial's weak back legs restrict its movement out of the water, one explanation for this distribution is that gharials, too, were once a single population before the collision of the Indian continental plate with the Asian plate. However, if this process was completed many millions of years ago, this raises the question as to how the Indus and Ganges populations of dolphins and gharials could have remained identical over such a large period of time. Their similarity suggests

Above: Gharial jaw bones.
Right: Gharial.

that their separation, and hence the collision of India with Asia, was far more recent than is commonly supposed.

This idea of river systems being fragmented by tectonic activity and resulting in separated populations of dolphins and gharials has some merit, but the problem remains that the river dolphins of South America and Asia are *also* so similar. However, if, as the geological strata show, the entire globe has been subject to inundation, then this must have a significant bearing on this whole issue.

Gharial fossil

If subsidence of the water occurred rapidly, then canyons and river systems would have been produced very quickly out of unhardened sediments. As a result, any aquatic creatures would have, by chance, found themselves either in the open seas or within river systems of the newly-emerged and still unlithified land masses. Rapid tectonic movements would have separated the continents very quickly, resulting in the current situation, where distinct populations of river dolphins and gharials all retain marked similarities. Simple chance would determine, for example, why gharials are only found in India, and why no river dolphins are found in North

America, Europe or Africa.
Of course, the question arises: if there was a flood affecting the whole globe, what would happen to all the land animals? Here we have to follow the logical path: clearly, without a method of escape, no land animals would survive. The fact is that the geological evidence shows that there definitely was a global catastrophe involving global flooding. Therefore, there *had* to have been a mechanism that ensured the preservation of land creatures. Interestingly, cultures from every corner of the world retain accounts of a global flood, and the preservation of people and land animals in a very large vessel.
We will come back to this later on when we discuss human chromosome 2 in Chapter 12.

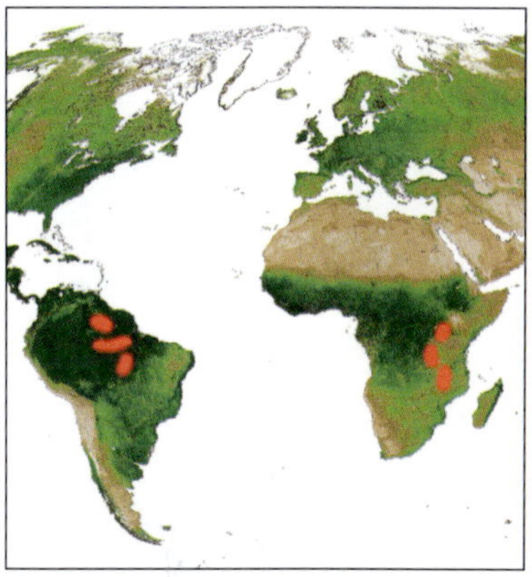
Distribution of some cichlids

### Cichlid and characin fishes

Yet further evidence of a far more recent separation of the continents is provided by the distribution of a number of fishes:

> The Cichlids are a major group of freshwater fishes with a very wide distribution: they are found in the southern USA, Central and South America, Africa, parts of the Middle East, India and Sri Lanka. Brian Ward. 'The Aquarium fish survival manual'. MacDonald. 1985.

A number of cichlids from opposite sides of the Atlantic Ocean (for example, the *Nannacara anomala* of Guyana, the *Aequidens curviceps* of the Amazon, and the *Pseudothropheus tropheops* and *Pseudothropheus zebra* from Lake Malawi) have very similar appearances. If the continental plates of Africa and South America (supposedly) began to separate about 135mya, it is incongruous that on the one hand, after only *thousands* of years, the cichlids of the African lakes have diversified so much, and that on the other hand, (supposedly) after *135 million* years, some African and South American cichlids have remained so similar to each other. Since it is beyond question that these fishes can undergo rapid variation, the logical conclusion is that the Africa-South America continental split may not have occurred *millions* of years ago, but much more recently than is generally thought.

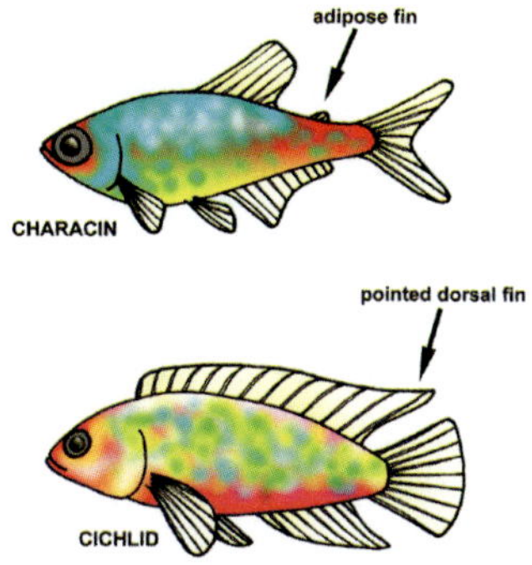

Characin and cichlid fins

The fins of some fishes from both South America and Africa are interesting. The *Characins* are one of the largest group of freshwater fishes and have managed to live in many diverse habitats, except for the sea. They are found mainly in Central and South America, and west Africa, although some are also found in parts of North America, central Africa and east Africa. Since they do not live in the sea, the characins must have comprised a distinct common group *before* the separation of the African and South American continental plates. Characteristically, characins have an *adipose fin* which is a small fleshy structure between the dorsal fin and the tail. A characteristic feature of many of the South American and African cichlids is a pointed or lobed tip at the back end of the dorsal fin.

Piranha showing the typical characin adipose fin.

The problem is that over time, the genes which code for any characteristic are liable to undergo deleterious mutations. If a characteristic is *not essential* to survival, then the lineage of creatures in which that characteristic exists will continue to live on, but eventually, that characteristic is likely to be lost. Clearly, neither the lobed tip of the cichlid dorsal fin nor the adipose fin of the characin serves any function in survival, and we should therefore expect that random mutations over the years would have resulted in their loss, or at least in marked alterations in their forms. However, except in a few cases, many cichlids and most (if not all) characins, from both sides of the Atlantic, have retained the lobed fin and adipose fin respectively. So it is unlikely that characins have existed long enough for their fins to have acquired any significant changes. Certainly, they could not have existed for millions of years. It therefore follows that it is possible that the African and South American continental plates did not separate millions of years ago.

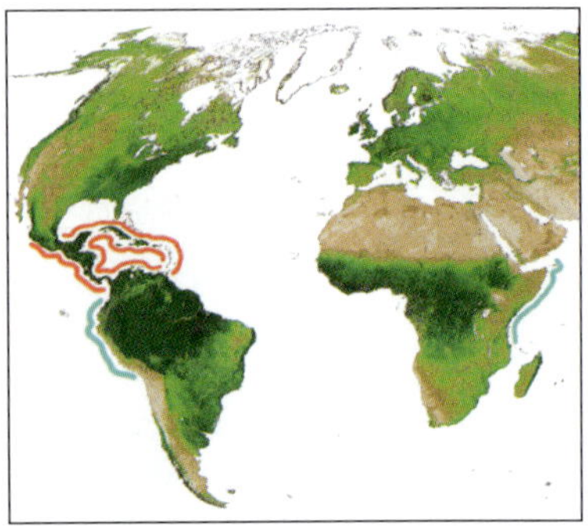

Distribution of cherub fish (red), angel fish and trigger fish (blue).

Above: Octopus camouflaged against coral.
Below: Fossil coral.

### Coral reef fishes

The distribution of a number of marine fishes provides further supportive evidence. Coral reefs only grow in warm shallow tropical and sub-tropical waters. Both *triggerfishes* and *marine angel fishes* live in shallow waters around corals and are not found in open waters. Because of the delicate balance of the ecosystem of a coral reef, it is unlikely that any new colonies would have been founded by individual groups of coral fishes or their fry migrating across thousands of miles of open ocean. More likely, new colonies would have been formed by migration to nearby corals or by the movement of continental shelves, along with their attached whole corals and resident fishes. Given the supposed times for the separation of the continents, those populations of trigger and angel fishes found off the west coast of South America would have become separated from the populations of fishes found off the east coast of Africa some 135mya. Similarly, those populations of trigger and angel fishes found off the west coast of Central America would have become separated from the populations of fishes found off the east coast of Central America at least 2-6mya, when the Panama isthmus is presumed to have closed.

However, the fact is that the angel fishes found off the west and the east coasts of Central America (for example, the *cherubfish* and *bicolour cherub angelfish*) are very similar. And incongruously, the angel fishes found off the west coast of South America and off the east coast of Africa are *identical*. Similarly, the trigger fishes found off the west and the east coasts of Central America (for example, the *clown triggerfish* and the *queen triggerfish*) look very similar. As with angel fishes, the trigger fishes found off the west coast of South America and off the east coast of Africa are *identical*. Since mutations are bound to appear and alter the appearance of subsequent generations of any creature, then, given the similarity of all these fishes, it is quite improbable that the different populations could have been separated by millions of years.

Clown fish

### Birds

Another example is *Archaeopteryx*, supposedly the first bird (see Chapter 8). Despite having no competitors in the air, and despite the plentiful supply of fossil birds found in China (showing that bird fossils are not super-rare), fossils of archaeopteryx are only found in one limited site in Solhofen, Germany. Since archaeopteryx would have been free to expand its horizons all over the world, the limited geographical location of its fossils strongly suggests that it could not have lived for a long enough time to have done so.

Celia Dickinson

Andrew Dickinson

Above left: Lion fish and coral, Red Sea.
Above: Grouper, Red Sea.
Below: Great Barrier Reef, Australia.

Despite penguins (supposedly) having been around for a very long time, surprisingly, the geographical distribution of both living penguins *and their fossils* is completely restricted to the Southern Hemisphere. All penguin fossils:

> ...have been found in the Southern Hemisphere...The Galapagos penguin...is the most northerly of the penguin species, breeding on the Equator. Jonathan Chester. 'The nature of penguins'. Greystone Books. 1996. p9,57.

> Penguins are restricted to the Southern Hemisphere... Ron Freethy. 'Secrets of bird life'. Blanford. 1990. p43.

This distribution is bizarre since the penguin is free to roam anywhere in the seas:

> The king penguin has been seen as much as a hundred miles away from its breeding colonies; the emperor, though it does not venture beyond the cold waters of the Antarctic, **may travel six hundred miles on a single feeding trip**. (My bold). David Attenborough. 'The life of birds'. BBC Books. 1998. p148.

Fishes, which comprise the staple diet of penguins, are found in seas everywhere, and warm seas are not prohibitive to penguin survival: Galapagos penguins live on the Equator, while other penguins live happily in zoos all over the world, and Antarctica would have been closer to the Equator (supposedly) 50mya. Therefore, after 50my we should expect that some penguins would have become separated from their fellows and started breeding colonies *all over* the world. So, it is improbable that the penguin has existed even for a few millions of years, let alone *fifty* million years.

It is commonly accepted that radiometric dating is unquestionably sure evidence that life has existed for millions of years. However, we have seen rock-solid soft-rock evidence that the sedimentary layers around the world were laid down recently under essentially one event horizon. We have also seen that the distribution of both fossils and living creatures, and the lack of change in geographically separated animals, shows that life on Earth could not possibly have been around for millions of years. Furthermore we have seen that the continents must have separated very recently. In the past there was no such thing as continental drift. It was continental *shift*.

King penguin

# CHAP

## Mass extinctions:

*lucky for some*

Cheetah

We have seen that the fossil record does not show consistent and progressive trends in complexity amongst life forms. Instead, in various sedimentary layers different creatures appear in their complete forms and with no unequivocal evidence of any fossils representing intermediaries linking them to their (supposed) less complex ancestors*.

### *Extinction of dinosaurs and radiation of mammals*

In addition to the gaps in the fossil record is the mystery that *all* dinosaurs have become extinct, and yet lizards and crocodiles have managed to survive. This is curious since dinosaurs were of numerous types and sizes, many being not large monsters but animals smaller than pigs. A number may even have been warm-blooded. One theory is that the mass extinction of dinosaurs was caused by a huge meteorite impact 65mya.

*Iridium* is an element which is rarely found in the Earth's crust, but it is often found in meteorites. So, when a relatively high level of iridium was found in sedimentary layers from around the world and dated as being (supposedly) 65 million years old, this was taken to be the result of a huge meteorite impact. It is thought that the offending meteorite landed in what is now the Gulf of Mexico, resulting in a crater which is now largely underwater but involves the Yucatan peninsula at its southern rim.

The Yucatan impact is estimated to have released energy equivalent to many thousands of simultaneous nuclear explosions, resulting in a crater about 160-200km in diameter and perhaps 40km deep. The impact may have caused tidal waves up to 4km high, and as a result of the debris released, most of the land animals over a radius of 5000km may have been killed outright. The huge amount of dust released into the atmosphere would have cut out the sunlight for a very long time, affecting globally the health of both plant and animal life. Proponents of the meteorite theory suggest that it may have taken over one million years for the atmosphere to return to normal.

Basalt columns framing the Svartifoss waterfall in Iceland. Such columns are formed when molten basalt is quickly cooled by contact with cold water.

While the Yucatan meteorite must have been devastating to life worldwide, nevertheless the responsibility for mass extinctions may lie elsewhere. A *flood basalt* is the outpouring from cracks in the earth of molten rock which has originated many kilometres beneath the Earth's surface. The cliffs of the west coast of Scotland, east coast of North America, Greenland and the Giant's Causeway in Northern Ireland are all the result of huge ancient flood basalts. Indeed, the ground comprising a huge portion of the western part of India is the result of basalt floods tens of thousands of times as large as the active flood basalts that are currently seen in Hawaii. Rock containing a thin iridium layer sandwiched between layers of flood basalt, can be seen in various parts of the world, confirming that large flood basalts occurred at times both before and after the impact of the Yucatan meteorite. All flood basalts are associated with large amounts of toxic gases (mainly carbon dioxide and sulphur

* We have seen that hydrodynamic sorting can have a marked effect on the way remains are laid down, and as we shall see, the capability for variation within a creature type is enormous. So, whilst it is true that a number of successive sediment layers may be found that, for example, each contain molluscs with distinct variations in the shell shape, this does not in itself show that the different shapes of shells actually represent different creatures or that the layering represents the evolution of one creature into another.

dioxide), and it is thought by some that the release of these fumes may have been the real cause of mass extinctions:

> It may be that the way life has ebbed and flowed over millions of years has been determined by events that occurred deep in the heart of our planet.
> 'Killer Earth'; Equinox. Channel 4 Broadcast. 22nd September 1998.

Some think that the death of the dinosaurs (which until then were supposed to have *somehow* dominated the land) allowed the little shrew-like mammals that survived to exploit the new environmental niches, and in a relatively short period of time to mutate into all the major forms of mammals seen today. This process has been called *adaptive radiation*. Supposedly, according to this theory, we should not expect to see any transitional forms in the fossil record because the necessary changes happened so quickly that there was insufficient time to lay down fossils.

However, there are a number of problems with the adaptive radiation theory. Firstly, everyone is agreed that dinosaur fossils are found less frequently, and mammal fossils more frequently, in sediments leading up to the 65mya mark, suggesting that dinosaurs were already on the decline, and mammals were already increasing in numbers before the meteorite impact. Secondly, although, following a mass extinction and subsequent breeding amongst any survivors, there would be a greater than usual opportunity for less fit or abnormal individuals to make a living or to avoid predators, nevertheless *the death of the unlucky* does not itself explain how any surviving creature could give rise to a completely different creature. The problem here is that mutations are essentially random and do not occur *because* there is an empty niche.

Although examples such as the Hawaiian finches and African cichlid fishes show that exploitation of new habitats can result in diverse forms, nevertheless, as we shall see, all these animals are still very much animals of the same type (either finches or cichlids). They have *not* become new and different creatures: for example, finches have not become seagulls and cichlids have not become catfishes.

Furthermore, mutations themselves are the result of errors in the replication of the DNA comprising an organism's genes. Apart from the effects of radiation or particular chemicals, these errors occur *independently of the environment*, and normal environmental changes do not and cannot encourage new or more mutations to appear. In fact, serious changes in the environment that might cause mutations are far more likely to result in sickness or death than in any benefit. So there is no reason to presume that new environmental niches which appeared as a result of the death of dinosaurs would have had any significant effect on the stability or accuracy of replication of the DNA inside any shrew-like animals living at the time.

Bats are a good example of mammals that are thought to have appeared 60mya following the death of the dinosaurs. There are about 1,000 known species of bats comprising almost a third of all types of mammals. However:

> The origin of bats is something of a mystery. The fossil record doesn't help much, because the oldest known specimens are strikingly similar to today's bats. 'Bat beginnings'. BBC Wildlife Magazine. Nov 1995. p24.

We know that the death of the dinosaurs was not the *cause* of the *origin* of mammals since the fossil record suggests that mammals have been around for over 200my. So, if shrew-like animals were around long before the dinosaurs, and apart from the *Pterosaurs* the skies were empty for the taking, we should predict that bats would have evolved a long time ago, even 200mya. Bats feed in the air and on bushes and trees, well away from any potential threat that might have been posed by dinosaurs. Since we know that pterosaurs had wings made of thin skin, and if pterosaurs evolved from non-flying reptiles, then we should predict that any small mammals would also have had a similar chance of undergoing mutations to form wings of skin. Thus, if an empty niche is a key factor in evolution, we should conclude that leaving their origins to a time when birds were already occupying the skies, would have made the emergence of bats *less* likely.

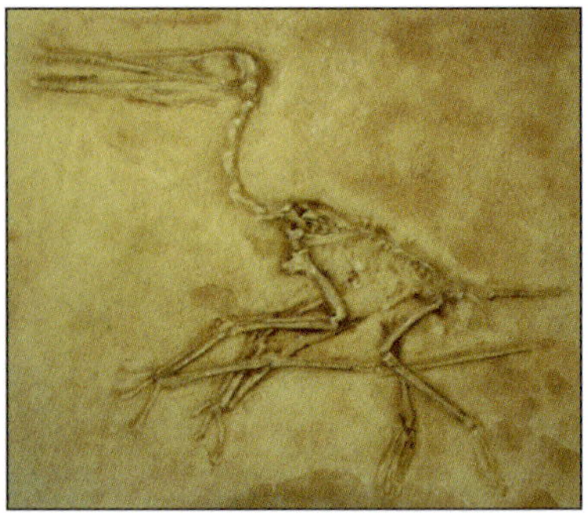
Pterodactyl

Not only could the death of the dinosaurs not be responsible for new creatures radiating from the 'primitive' mammals, but there is also the problem of how previously any shrew-like mammals could have remained static or relatively unchanged for millions of years. From the fossil record, amphibians are thought to have evolved from fish in only 20my. Man is supposed to have evolved from apes in only 8my. And as we shall see later, after only a few years, separate populations of mice living in the same barn can show visible differences in their chromosomes. Therefore, if animals *could* evolve, then for shrew-like animals to have existed *without any significant change* for 135my (from 200mya until 65mya) is *biologically impossible.* However, if we suppose that creatures changed gradually over substantial periods of time, then we should predict that many members within an ancestral line would have had an equal opportunity to exist for long periods of time, and thereby would have had opportunities to be represented in the fossil record in numbers fairly comparable to both their ancestors and their descendants. But, echoing Darwin's thoughts, some people think that the problem of the lack of evidence for intermediaries is not a problem at all:

> ...selection has done its inexorable work in getting rid of intermediates. When we see any structure highly perfected for any particular habit, as the wings of a bird for flight, we should bear in mind that animals displaying early transitional grades of the structure will seldom continue to exist to the present day, for they will have been supplanted by the very process of natural selection. Each new form will tend to take the place of, and finally to exterminate, its own less improved parent. If improvement is great it will not take long to complete the move from old to new. The chances of survival of any ancestor are small indeed. Steve Jones. 'Almost a whale'. Doubleday. 1999. p132.

We should certainly expect that, in general, ancestors would die off and be replaced by their modified descendants which live today, but this does not deal either with the gaps in the fossil record or with the complex practicalities of exactly how one creature could turn into a completely different creature. Take the example of the dinosaur arm being modified into a bird wing where a *considerable number* of changes would have had to be made to the dinosaur skin, bones, muscles and other organs. Since each individual modification would require separate and fortuitous genetic changes to direct its manufacture, and since it is likely that having acquired one change the next appropriate change may not arise for a very long time (if at all), then we can state the *Intermediary Principle*:

If a creature A supposedly evolved into a significantly different creature Z, where Z has a *number* of features which could not have appeared except by *a number* of genetic alterations to A over any number of generations, then a number of intermediary creatures, representing *every one* of the genetic alterations, must have existed in the time between the demise of A and the appearance of Z. And if they existed for equivalent periods of time, collectively the many types of intermediaries would most probably have given rise to far greater numbers of individuals than the numbers of either A or Z alone. Therefore, it follows that in total, more fossils of intermediary forms should be found than of either A or Z alone.

Fossils of transitional forms ought therefore to be at least as common as the known fossils. However, what the fossil record does show is that a particular group of fossils may appear in one layer and then abruptly disappear in successive layers, the latter containing groups of other fossil types. This is contrary to what we should expect from a gradual process. Indeed, instead of the observed distinct groups of different animals that we see, evolution by gradual stepwise changes should predict that *all* extinct and *all* living creatures should show a continuous variation one with another without any stark divisions between them. In other words, contrary to what we actually observe, we should see good evidence of, for example, a giraffe and a *not-quite-a-giraffe*, or a bat and a *not-quite-a-bat,* having lived some time previously, and significant variations of their themes living today.

In an attempt to overcome this obstacle, it has been proposed that changes in the genetic code do occur constantly but that many changes often do not manifest themselves outwardly. However, it is proposed that after many millions of years the very many accumulated mutations could manifest themselves in one fell swoop by a mutation in an appropriate *regulatory* gene (a gene which controls the actions of a group of other genes), resulting in the sudden appearance of a new character or organ. Such a mechanism might result in a *complete* bird or bat from an ancestral dinosaur or shrew-like mammal respectively, without any intermediate transitional form ever having lived. The problem here is that any gene hoping to become involved with some new structure would over time simply acquire random mutations and be thereby rendered functionless long before some hypothetically favourable regulatory gene could make its appearance.

In order to give some idea of the colossal scale of the odds stacked against it happening, the whole process of gaining a series of favourable and mutually complimentary genetic mutations could be likened to a national lottery.

Suppose the lottery gave out prizes instead of money. We know that the chance of any one particular individual winning a prize is extremely remote, but suppose one person actually won a hi-fi amplifier made of biodegradable materials. By itself this amplifier would be of no use in reproducing music. So suppose the amplifier was put away and forgotten as the years passed by. Suppose the owner died and the amplifier was handed down to his children who, over the years, all continued to take

Above: Skeleton of Pliohippus, an extinct horse.

Above: Shire horses.
Below: Shetland pony.

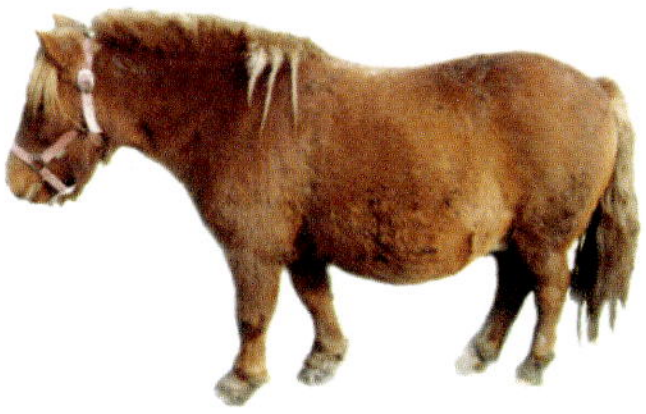

part in the national lottery. Suppose *many* generations later one of the descendants won a biodegradable CD player which was compatible with the amplifier that their great-great-great grandparent had won and that was still in the original packaging and kept in the loft. But of course, without speakers, this descendant could still not use either the amplifier or the CD player to reproduce music. So the CD player, too, was put away in the loft. Then suppose this descendant died and left the amplifier and the CD player to his children. Suppose, many generations later, another descendant won a pair of biodegradable speakers, which were compatible with the amplifier and the CD player that his great---great---great ancestors had won and that were still in the original packaging and kept in the loft. This descendant would now have possession of the amplifier, CD player, and speakers - and might now be able to reproduce music.

The point is, should we consider such a series of lottery wins to be a *realistic* possibility? In fact, we should expect that the prizes comprising the amplifier, CD player and speakers would each be won by *different* and *unrelated* families, separated in both time and space. Therefore each winner, although owning a piece of hi-fi equipment, nevertheless would have no ability to reproduce music. Furthermore, the chances are that by the time someone had won the speakers, the switches and circuitry in the amplifier and CD player might have become inoperable through the passage of time and lack of use. Therefore all the hi-fi components would be useless.

So we can state the *What You Want You Won't Get* (*WYWYWG*) *Principle*:

If a number of genes are together essential for the manifestation of a particular structure or characteristic; and if in the absence of any one of those genes the structure or characteristic could not manifest; and since, if the necessary individual genes were to appear, it is likely that they would appear haphazardly, and within unrelated lineages that would be separated in both time and space: then that structure or characteristic is likely never to manifest - in any lineage.

In other words, it is implausible that a creature will acquire the very mutation that it might hope to acquire so as to continue its (supposed) evolutionary journey: ***W**hat **Y**ou **W**ant **Y**ou **W**on't **G**et.*

Below: Modern Horse.

## ***Transitional forms***

### ***Horses***

It is often suggested that there are plenty of examples of well documented lineages showing transitional forms which confirm evolution as being not just a theory, but *fact*. Perhaps the most often-stated example is that of the horse:

Horse evolution...this story has become one of the classic cases of evolution that is reproduced endlessly in textbooks. Michael Benton. 'The rise of the mammals'. The Apple Press. 1991. p105.

However, things are not so clear-cut at all. Hulbert, in his chapter on horse evolution, admits that:

> Most non-technical surveys of horse evolution present the story as if it were a series of established facts. In reality, it is a web of hypotheses and theories, with many points of contention and controversy. No two paleontologists interpret the fossil record exactly alike, and this is certainly true in the case of horses. Richard Hulbert Jr. 'Horses through time'. Sandra Olsen (Ed). Roberts Rinehart. 1997. p33.

However, Hulbert devotes 21 pages to describing essentially the standard hypothetical story of horse evolution, based entirely on a few selected fossils and tooth patterns. And he does not provide the reader with illustrations or details of whole skeletons. By his own admission there is much controversy, yet his approach is strictly didactic and allows the reader no opportunity to assess his thesis. Furthermore, he presents the reader with no understanding of what objections the known biology of teeth and bones raises against any proposed evolutionary trend. As we shall see when we come to discuss the whale and Man, the genetic control of embryological development is very complex, and many suggested changes necessary for the supposed evolution of the horse have no sound basis. Furthermore, many changes are associated with *diseases*.

Hyracotherium

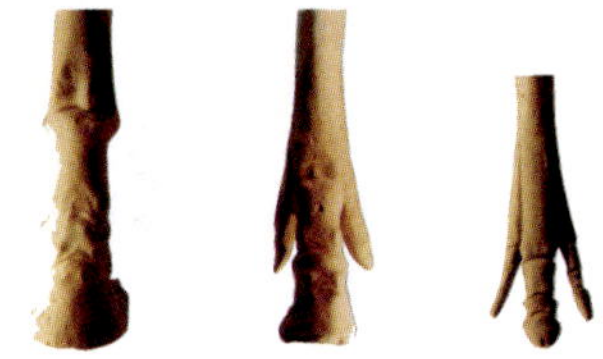

Above: Toes of the living horse (left), the extinct Hipparion (centre) and Merychippus (right). Not to scale.

In essence, the fossil record of the horse (supposedly) shows the evolution of a small ancestral quadruped passing gradually through a number of different forms to produce ultimately the modern horse. Although there are thought to have been a number of 'blind alleys' and variations on the theme, the basic story is that the small *Hyracotherium* (or *Eohippus* = dawn horse) appeared (supposedly) in the Eocene, about 35-55mya. This animal was about the size of a small dog and had a convex spine, with four toes on the front feet and three toes on the back feet. The *Mesohippus* appeared about 25-35mya. This creature was a little larger than the Hyracotherium, and also had a convex spine, but had only three toes on each foot. Later still, over 5mya, the *Merychippus* and *Hipparion* appeared. These animals were like small horses but had three toes on each foot. The large middle toe actually carried all the weight, and there were two very small toes on either side, neither of which touched the ground. About 1.5-5mya, the *Pliohippus* appeared which was like a small horse with variable numbers of toes (see text block overleaf). The modern horse *Equus* (supposedly) appeared within the last 1.5my.

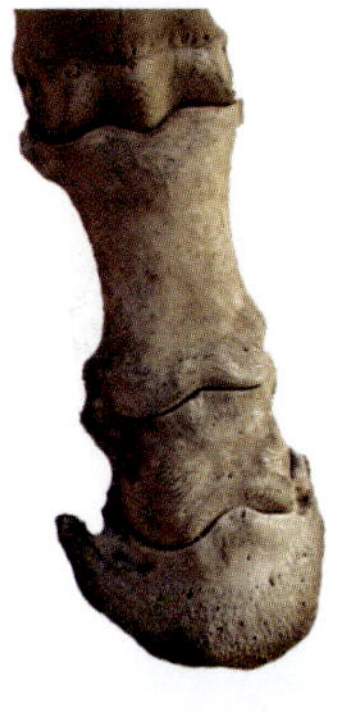

Above: Modern horse foot bones.
Below: X-ray of horse hoof.

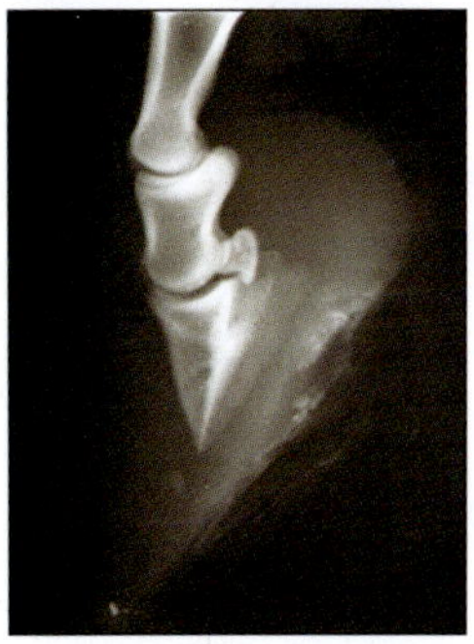

The explanation usually given for this *supposed* lineage is that in ancient times the small forest-dwelling horse ancestors could easily hide, but as the forests receded and savannahs appeared, they were tempted out of the forests by the lush grasses. The horse ancestors then *had* to grow longer legs and become taller in order to see and run away from danger more easily:

> Their limbs became longer and more slender, the number of toes was reduced from three or four on each foot to a single functional middle toe with a large hoof...these changes...large size, long limbs and the single hoof are adaptations for seeing enemies better and escaping from them faster. Michael Benton. 'The rise of the mammals'. The Apple Press. 1991. p105-6.

However, is this simplistic notion supported by the evidence? On the savannah, being

# ANCIENT HORSES

Skulls of a modern horse (right) and of the extinct Chalicotherium (below right). Although the chalicotherium had a horse-like skull, its feet had claws and its front legs resembled more those of a gorilla (below).
Clearly, despite the similarity of its skull, the chalicotherium had no relationship whatsoever to a horse. Similarly, although the skulls and footbones of the hyracotherium, mesohippus, merychippus and the modern horse are often shown in this order so as to suggest an evolutionary trend, such a scheme may be completely invalid.
Furthermore, the oft-repeated story of 'primitive horses' getting taller and losing their toes while moving out of the forests and into the savannah is implausible (see main text).

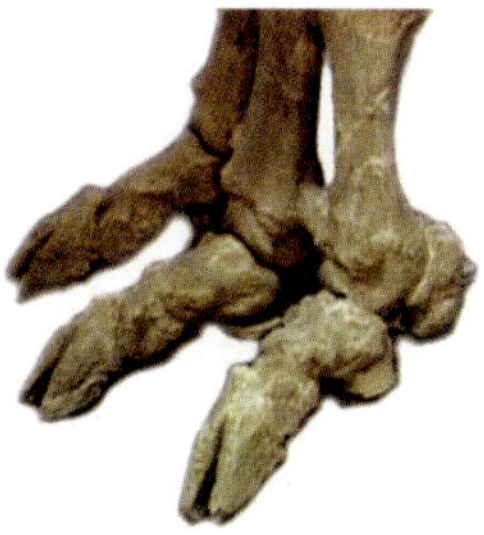

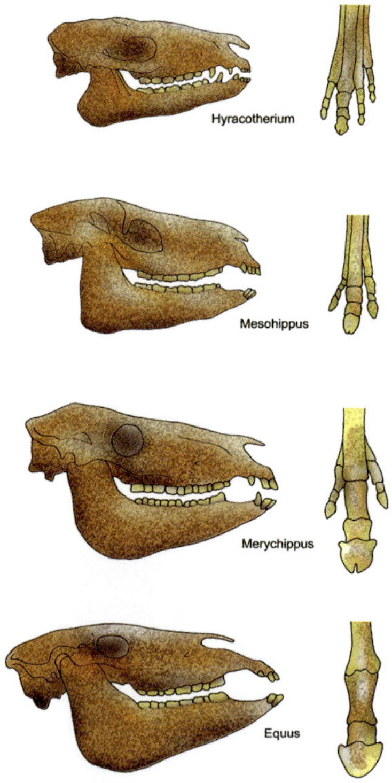


Left: Skulls and feet of some extinct hoofed animals and of the living horse.
Hyracotherium had four toes on its front feet and three toes on its rear feet. If hyracotherium had itself evolved, its ancestors would have had 5 toes on the front feet and four/five toes on the rear feet. No such fossils have been found.
Since many animals exist with two toes, the genetic controls of the development of the first and third toes of mesohippus and merychippus are likely to have been independent of each other.
Therefore, if mesohippus and merrychippus had evolved, some of their descendants would probably have had 2 toes on their feet. No such fossils have been found.

In Ash Fall in Nebraska, USA, some 70 skeletons have been found belonging to five different types of horse. An interesting feature is that a number of pliohippus specimens have three toes while others have only one toe, or very reduced lateral toes. This has been put forward as confirmation that many-toed ancient horses evolved into living one-toed horses.
However, in all other respects (including leg length, tooth form, body shape etc.) the pliohippus specimens were essentially identical. Since all the pliohippus specimens had died in the same site, they most probably lived, ate and mated together within a single population.
As with individuals of a single species of newt living in the same pond, which can vary in the number of their toes yet remain the same creature, so the Ash Fall pliohippus specimens, being the same creature, simply demonstrate variation - those with single toes had not become new and different creatures. In the same way, the rare report of a living horse with three toes or the finding of a dog with five toes on its back feet (see below) does not relate to any other changes in the creatures, showing that the number of toes is genetically unrelated to other features such as type of teeth, or head and body form.
In humans, babies born with extra fingers or toes simply demonstrate defective embryological development and do not point to human ancestors having six digits. In other words, the Ash Fall specimens do not demonstrate the evolution of one creature turning into another.

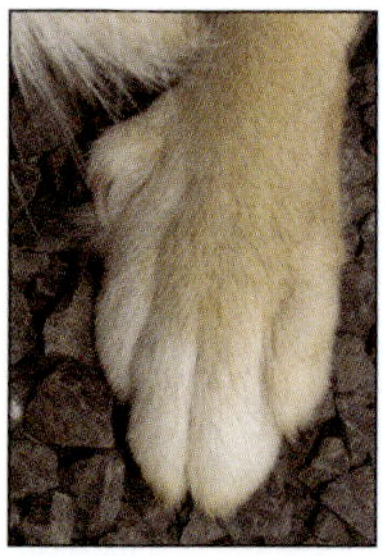

Mongrel dog with five toes on its back foot. Most dogs have five toes on their front feet and only four toes on their back feet.

small and having more than one toe is *not* a serious disadvantage. For example, the Thompson's gazelle (which lives in the same environment as the larger wildebeest and zebra), despite being both small and delicate and having two weight-bearing toes, is quite able to outrun *all* savannah predators, except for the cheetah. And many other small grazing animals which have more than one toe (such as sheep, goats, and deer) continue to happily live and survive on grasslands.

In addition, it does not follow that a taller animal is more likely to survive in a savannah. Indeed, being taller is not always an advantage since it might make the animal *more* obvious to a predator. Being smaller might allow an animal to hide in whatever long grass is nearby. Furthermore, being tall is not so essential when looking out for predators since many grazing animals use their sense of smell, rather than sight, to warn them of danger.

The extinct *Diadiaphorus* from South America is thought to have lived between about 1.5 and 20mya and looked like a slightly-built deer but it had horse-like one-toed feet. Clearly, if the diadiaphorus had itself evolved, its ancestors would also have had many toes. Therefore, if quite different animals such as the diadiaphorus and ancient horses did develop mutations resulting in a single toe, then we should predict that over time other grazing animals *should also* have developed single-toed forms. But since there is no particular advantage in being either single-toed or double-toed (for example, both the single-toed zebra and the double-toed wildebeest live in intermingled herds and manage to survive in the presence of the same predators), then if modern horses did evolve from many-toed ancestors we *should* find at least some types of living horses identical to modern forms but with two-toed feet.

In actual fact, since it is only by a chance detrimental mutation that the development of the non-weight-bearing outer toes could be upset so as to result in shorter toes, a toe would not be lost *in order* for its owner to run faster. Furthermore, since the genes controlling the development of the toes are independent of the genes that control either leg length or the shape of the spine, it does not follow that a reduction in the length of the outer toes would in itself allow a creature to run faster. And the *WYWYWG Principle* demonstrates that it is implausible that a hyracotherium hoping to give rise to a horse would ever manage to acquire the necessary *series* of alterations.

But a hyracotherium would not need to lose most of its toes since most animals run very well with more than one toe. For example, the goat, buffalo, wildebeest, oryx, sable antelope and giraffe all have two weight-bearing toes. Many of the antelopes can run much faster than a zebra or any horse, and wildebeest and large antelopes are well able to run for long distances with equal stamina to that of a horse. In actual fact, a plains animal does not have to run for mile after mile if pursued by a lion or cheetah since both these carnivores can only sprint and give chase for short distances. If the prey is not caught within the first few hundred metres both the lion and the cheetah will give up. The cheetah in particular, has very little stamina. Only African wild dogs and hyenas have the stamina to give chase until a prey animal succumbs to exhaustion. Yet the cheetah (the fastest land mammal), wild dog and hyena all have five toes on each foot. Furthermore, the dog's innermost toe does not touch the ground and so does not bear any weight. Nor has this toe been lost

Wildebeest showing two weight-bearing hooves.

Above: Giraffe showing that height is not restricted by having two toes.
Below: Buffalo showing two weight-bearing hooves.

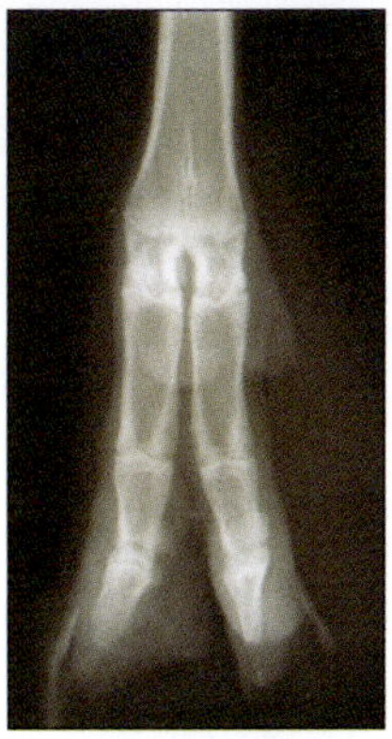

X-ray of goat foot bones.

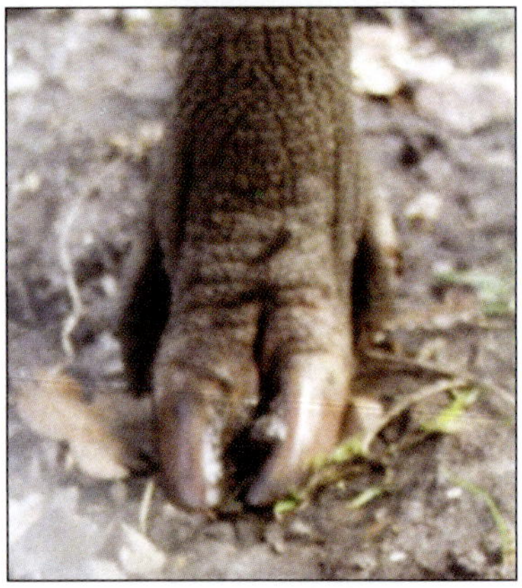

Above top: Warthog showing its four toes.
Above middle: Barburisa, a wild pig from Sulawesi in Indonesia.
Above: The four toes of a barburisa.
Below: Foot bones of domestic pig.

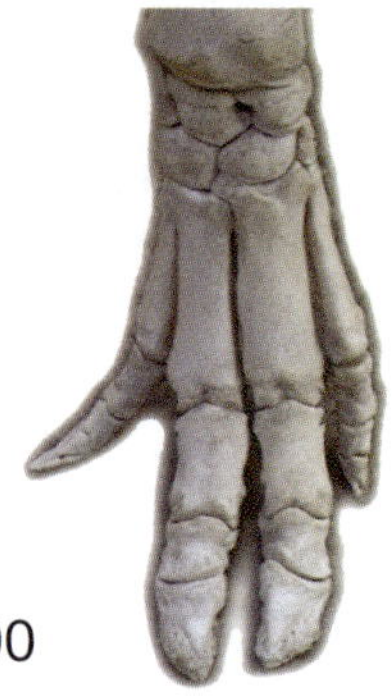

over millions of years. Thus having many toes is not detrimental to either running or survival.

Nor is the loss of toes an essential pre-requisite to an increase in height and bulk. The *eland* and *oryx* are antelopes that are both as large as many living horses, and larger than any supposed horse ancestor, and giraffes are larger than most living horses. Clearly then, since on the grounds of either speed, stamina or size, having extra toes is not disadvantageous to running or to survival in the open grasslands, single-toed-ness cannot have been selected for in preference to two or more toes. And since we have no reason, for example, to presume that in some exclusive way the development of horns must be restricted genetically to those creatures with more than one toe, the theory of evolution should predict that some modern horses should have evolved horns just like deer, antelopes, cows, wildebeest and giraffes (which all have two weight-bearing toes). But curiously (and inexplicably, in terms of natural selection) such animals do not exist - horns are just not found on single-toed ungulates.

If, as is commonly portrayed, we suppose that the horse evolved as it came out of the forests into open savannah, then since there remain plenty of forests today, we should also predict that some of the ancestral forest-dwelling forms should still be alive and living in forests in the same way as modern small deer. Suppose some small four-toed proto-horses spread out, to become over time a number of separate populations living in forests. Suppose one of these populations accumulated mutations that resulted in their having fewer toes. Since (as in domestic pigs) forest pigs have four toes, of which only two touch the ground, it is unlikely that a small forest horse with fewer toes would have any advantage. Furthermore, toe changes will not necessarily result in any reduction in the ability of the differently-toed horses to interbreed. Therefore continued interbreeding would result in individual animals having different numbers of toes, but otherwise remaining the same creature. If, however, there was some resultant difficulty in interbreeding, then distinct populations of small forest horses would exist with different numbers of toes. But these different populations would also still remain the same creature.

As with wild forest-dwelling pigs, there would be no particular reason for *all* those original many-toed forest-living 'horses' to be relatively disadvantaged and die off. Furthermore, the wild pigs that live in open savannah can often outrun lions and have *not* evolved so as to lose their extra toes. If wild forest pigs (which are often hunted by people for food) were to become extinct, their disappearance from the face of the Earth would more likely be as a result of human extermination rather than any evolution into another form. Similarly, since the *Coelacanth Principle* does not exclude the possibility of humans and (supposedly) ancestral horses living contemporaneously, it is not inconceivable that ancient (supposedly) ancestral horses became extinct by virtue of *extermination*, rather than evolution. When we come to examine the supposed evolution of humans, we will see that such a notion is not as absurd as it may at first seem.

One further point, which can also be applied to a number of different animals, is that millions of years is *too long* a time for many aspects of evolution to manifest. We know that following mutations, many anatomical features can change dramatically

Wildebeest and zebra live in mixed herds and are attacked by the same predators. The number of toes gives no advantage to either creature.
Contrary to what the notion of evolution should predict, there are no known three or four-toed wildebeest or gazelle ancestors.

Despite being much smaller and having two toes, the Thompson gazelle (above right) can run faster than a zebra (below).

Sarah McAlpine

in only a single generation. For example, humans can be born with harmless genetic abnormalities affecting a number of different bones, such that collar bones, finger bones, vertebrae, or arm or leg bones can be absent. Thus, the loss of the two small lateral toes in merychippus need not have taken tens of millions of years to happen. In reality, any mutations resulting in changes in toe number could theoretically stabilise in just a few generations, providing the affected animals were grouped into isolated breeding populations. Yet Hulbert writes:

> Throughout the middle Eocene (between 50 and 39mya, **ie 11 million years**) there was **no trend** to increase **body** size, nor any significant advance in **limb** or **foot** structure. The main change was in the dentition. (My bold). Richard Hulbert Jr. 'Horses through time'. Sandra Olsen (Ed). Roberts Rinehart. 1997. p22.

The dik-dik is a very small antelope and manages to run swiftly with two toes. Interestingly, the hooves have a rubbery base which allows extra grip on rocks.

Fossilised horse tooth, split length-wise.

Hulbert describes an *apparent* trend in the change of (relatively pointed) premolars into molar-like teeth over this period of time, and relates this to a change in diet from forest leaves to savannah grasses. However, the glaring inconsistency is that if the environment did change from forest to grasslands, these animals managed to survive for 11 *million* years - *without having to develop longer legs or lose their toes or grow bigger*. Others suggest that despite (supposed) deleterious climatic conditions, leaf-eaters managed to survive for 17 *million* years:

> Grazing descendants of Merychippus diversified greatly in the Miocene (from 25 to 8mya), coexisting for several million years with the last browsing horses. Douglas Futuyma. 'Evolutionary biology'. Sinauer Associates Inc. 1998. p142.

We might accept that natural selection could result in the extinction of leaf-eaters if the process of severe deforestation and the spreading of savannah occurred over a period affecting only a small number of generations. But if browsing 'horses' could survive for millions of years despite the supposed climatic changes which resulted in savannahs and which forced the appearance of merychippus, then this begs the question as to why the browsing horses needed to change at all, and as to why/how they survived for *so* long. When considering periods of time spanning either 20 or 200 million years, there will be a subconscious tendency for the reader not to grasp just how large these time periods are. We have recently come to the start of the third millenium AD: just one million years is the equivalent of *one thousand* millenia. Since even one million years is a very long time, it is implausible that a creature which is battling against environmental changes so deleterious that natural selection should (supposedly) force some of its cousins to change into some completely different creature, could itself survive unchanged for millions of years. So the dating of these 'early horses' is likely to be seriously in error.

Furthermore, there is no evidence to suggest, for example, that when eating grasses instead of leaves, having *four* molars can confer any *significant* survival advantage compared to having only *three*. If having more molars allowed for more efficient processing of food, then we should expect (just as modern children are growing bigger than their parents) that better nutrition should be reflected in larger and healthier animals. But from the fossil record it is supposed that *there was no trend to increase body size, nor any significant advance in limb or foot structure...*over a period of *11 million years*. The great sauropod dinosaurs (the group that includes the diplodocus) also show that changes in tooth structure may simply be aberrations of paleontologists' minds: these animals supposedly roamed the earth for 160my (*one hundred and sixty thousand millenia*) and yet in all that time they *still retained the same peg-like teeth without any modification into molars or premolars*. This is particularly incongruous since the sauropods must have had to eat enormous amounts of foliage to live, and of all animals, surely these giant dinosaurs would have benefited most from evolving molars.

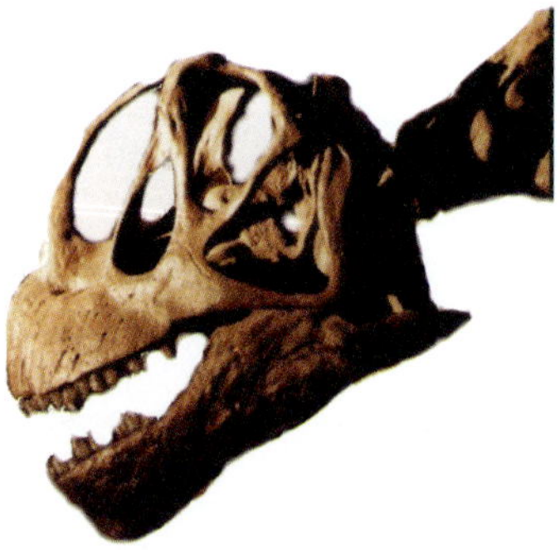

Diplodocus skull showing peg-like teeth. Despite not having molars with which to grind plant material, sauropods managed to grow to gigantic size, proving that if (supposed) early horses came out of forests, changes in their tooth patterns would not have been essential for their survival.

Darwin himself made considerable observations of the horse family and found that many breeds of horse, along with relatives of the horse, are to some degree found with stripes:

> The ass not rarely has very distinct transverse bars on its legs, like those of the zebra...The koulan of Pallas is said to have been seen with a double shoulder-stripe...I have collected cases in England of the spinal stripe in horses of the most distinct breeds...the Kattywar breed of horses is so generally striped...stripes occur far oftenest in duns and mouseduns...the common mule from the ass and the horse is particularly apt to have bars on its legs...I have seen...[that] the pure offspring subsequently produced by a black Arabian sire, were much more plainly barred across the legs than is even the pure quagga. Charles Darwin. 'The origin of species'. Penguin Books. 1985. p198-201. ('Dun' = dull greyish-brown colour)

Donkey showing single stripe over shoulder.

From all this, it might be supposed that all living horses and their living relations once had a common ancestor:

> For myself, I venture confidently to look back thousands on thousands of generations, and I see an animal striped like a zebra, but perhaps otherwise very differently constructed, the common parent of our domestic horse, whether or not it be descended from one or more wild stocks, of the ass, the heminonus, quagga, and zebra. Charles Darwin. 'The origin of species'. Penguin Books. 1985. p201.

Wild asses, showing striped legs.

Such an explanation is certainly plausible. However, even if we were to suppose that modern horses and a number of their modern relatives descended from some common stock, nevertheless this does not permit us to extrapolate further and suggest that living horses arose from creatures that were, at some time previously, *something other than horses*. Indeed, Darwin himself correctly observed that:

> The appearance of the stripes is **not** accompanied by **any** change of form or by **any** other new character. (My bold). Charles Darwin. 'The origin of species'. Penguin Books. 1985. p201.

Thus, we should conclude that although, undoubtedly, there is evidence of variation in the modern horse in terms of colouring, size and shape, there is no obligation to link the modern horse with other extinct creatures, such as merychippus. So, the supposed evolutionary sequence of horses is more likely to be simply an artificial catalogue of animals which have no ancestor-descendant relationship, but which share some similar features. Furthermore, since we can show no clear advantage, one over another, their extinctions cannot be easily explained by natural selection. Thus, we can state the *Horse Principle*:

a. Even if different fossil specimens can be arranged in an order suggesting a *presumed* evolutionary trend, nevertheless no such evolutionary trend need be true.
b. In the absence of any direct evidence, if any number of creatures share any number of characteristics, it cannot be assumed that for *any* of those creatures the shared characteristics reflect any ancestor-descendant relationship.

**The chromosome numbers of different horses**

| | |
|---|---|
| Domestic horse | 64 |
| Przewalski horse | 66 |
| Grevy zebra | 46 |
| Burcheli zebra | 44 |
| Ass | 62 |
| Onager (heminonus) | 54 |

The domestic horse and the Przewalski horse are fully inter-fertile, and are good candidates for having a common ancestor. However, this may not be the case between the domestic horse and zebras, which have a marked difference in chromosome number.

Tiktaalik fossil

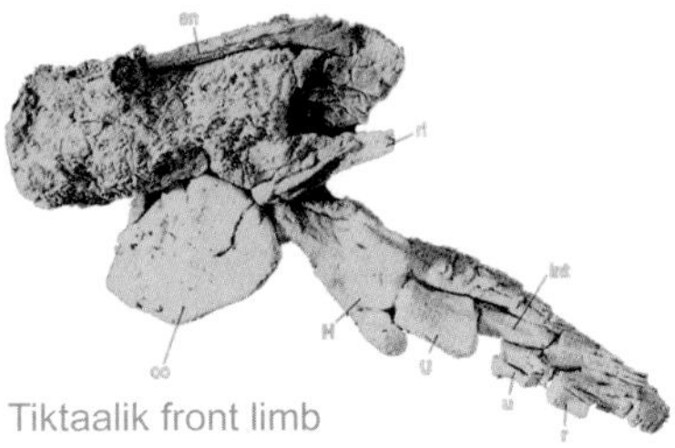

Tiktaalik front limb

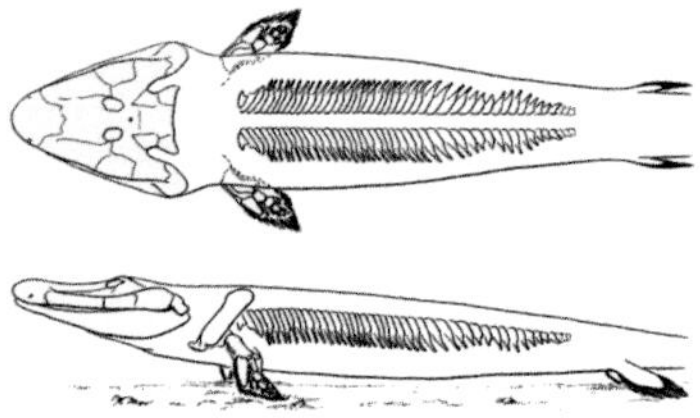

Tiktaalik reconstruction

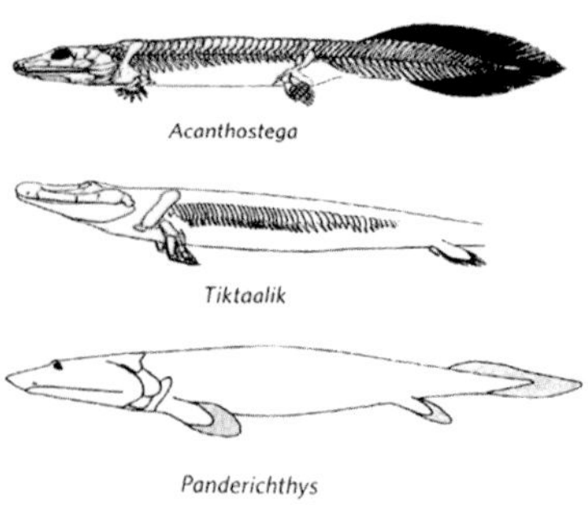


Tiktaalik (middle) and its (supposed) evolutionary relatives.

### *Fish fins and tetrapod limbs*

*Tiktaalik* was a fish with a skull length of around 20cm, whose fossils were found in 2006 in northern Canada, in sediments supposedly 380myo. Tiktaalik was very much a fish with gills, but while in typical bony fishes the skull, gill covers and front fins all have bony interconnections, in tiktaalik the front fins and pectoral girdle were separated from its flattened head. Although the front fins were covered with typically fish-like *dermal rods*, they contained bones that *roughly* mimicked those of land vertebrates. The rear fins were not found. At a superficial glance, tiktaalik seems to share characteristics with other extinct creatures such as the fish *Panderichthys* which lived (supposedly) 385mya, and the 1m-long tetrapod *Ichthyostega*, and so some have made the dogmatic assertion that tiktaalik represents a fish that was on its way to becoming a four-legged terrestrial vertebrate (tetrapod):

> Tiktaalik roseae…This really is what our ancestors looked like when they began to leave the water. Just over 380mya, it seems, our remote ancestors **were** large, flattish, predatory fishes, with crocodile-like heads… (My bold). P E Ahlberg & J Clack. 'A firm step from water to land'. Nature. Vol 440. 6th April 2006. p747-749.

However, we can identify a number of fatal problems with such dogma. Firstly and obviously, before the first land tetrapod (supposedly) appeared, there would have been no tetrapods on the land. So, tiktaalik would have had no advantage in leaving the water since there would have been no land creatures to eat, other than insects and other invertebrates. In the evolution model, even these were rare at the (supposed) time of tiktaalik. Secondly, the *Coelacanth Principle* tells us that pandericthys, tiktaalik and icthyostega could have lived contemporaneously, and so any evolutionary progression is simply supposition, not science. Thirdly, if fishes evolved into tetrapods then there would have to have been the acquisition (by chance) of all the *thousands* (perhaps millions) of interdependent systems and controls that would allow myriad other changes, including the acquisition of lungs and the loss of gills; complete alteration to the heart and circulation (see p273); development of the amphibian ear with loss of the lateral line system; metamorphosis (see p139); and the acquisition of all the countless systems and controls that govern limb embryological development.

Since the main discussion for tiktaalik being an evolutionary intermediate between fishes and tetrapods revolves around limb origination, we need to have an overview of limb embryology.

Small protrusions (the *limb buds*) on the side of the developing human embryo are seen at around four weeks and their very appearance is the result of the activity of highly complicated triggers and cellular migration. The muscles and nerves of the limbs do not arise from inside the limb bud but are actually derived from muscle cells which migrate in from the backbone areas (the *somites*), and from nerve cells which migrate in from the *Neural Crest Cells* of the early spinal cord. To complicate matters, the cells destined to become the tendons develop from within the limb bud *mesenchyme* (non-differentiated embryonic cells) and *separately* from the somites, so abnormal tendons can actually form without any attached muscles.

The early limb bud may look like a simple bump but even at this early stage it has a high degree of regional organisation. In simple terms, the head end has an area of cells called the *apical ectodermal ridge* (AER). This, along with the underlying mesenchyme (the *Progress Zone*, PZ), is responsible for directing limb bud outgrowth, with the tetrapod limb axis being perpendicular to the body. Activity of a small group of cells in the Zone of Polarising Activity (ZPA) in the tail end of the limb bud defines the antero-posterior (head/tail) patterning, with the little finger being posterior and the thumb being anterior. This activity can be mimicked by inserting *retinoic acid*-impregnated beads to the tissues. Dorso-ventral (back/front) patterning may begin even before the limb bud appears and is dependent partly on the expression of the genes *WNT7A* on the dorsal aspect, and *EN1* on the ventral aspect.
The bones of the limbs first appear as condensations of mesenchyme which then form cartilage, which itself becomes calcified to form bone. The limbs grow outwards to produce, sequentially, the humerus/femur, then the radius and ulna/tibia and fibula, then the bones of the hand/foot, and finally the digits. The axis of the limb is defined by digit IV, and the finger/toe bones form in the order IV, V, III, II and I, the last being the thumb/great toe. Controlled cell death (*apoptosis*) then eliminates the tissues between the bones of the hand/foot to produce the individual digits. While specific proteins can be seen to be active in specific places at specific times in growing embryos, these give only a very crude idea of the sheer complexity of what actually goes on during limb development. For example, as the limb bud appears, essential triggers such as FGF4,8 (*Fibroblast Growth Factors*) and *HoxB6* are active. It is found that without the ZPA, the AER degenerates and the limb fails to form. Without the AER the ZPA degenerates, so limb patterning is deranged and the limb is severely shortened and mal-formed. *Sonic* (*Shh*) is a gene that is essential for the proper function of the AER and Shh's protein product can successfully maintain limb growth even if the AER is removed. Conversely, lack of Shh causes truncated limbs. Furthermore, Shh requires another trigger, *Hand2,* without which it is not activated. A number of *Hox genes* (see page 213) are active in determining body patterns and development. For example, *HoxD9* and *10* are active in development of the upper arm/leg. *HoxD11*, *12* and *13*, *Hand2*, and Shh are active in development of the lower arm/leg. *HoxD12* and *13* are active in development of the hand/foot and digits. Many other triggers are also essential including, for example, *BMP2* (*Bone Morphogenic Factor*), which can induce digit lengthening. So, it is not surprising that countless genetic alterations can distort limb development:

> There are over 100 different mutations known in mice that affect development of the limb. C Tickle. 'Limb development: an international model for vertebrate pattern formation'. Int. Journal of developmental biology. Vol 44. 2000. p101-108.

Note that all of the above complexity deals only with the very basic general formation of a limb. None of the above gives *any* understanding as to how the detailed anatomy of a limb (for example, the elaborate architecture of the joints, tendons, tendon sheaths and tendon pulleys) is formed, or why the fingers and toes form on hands and feet.

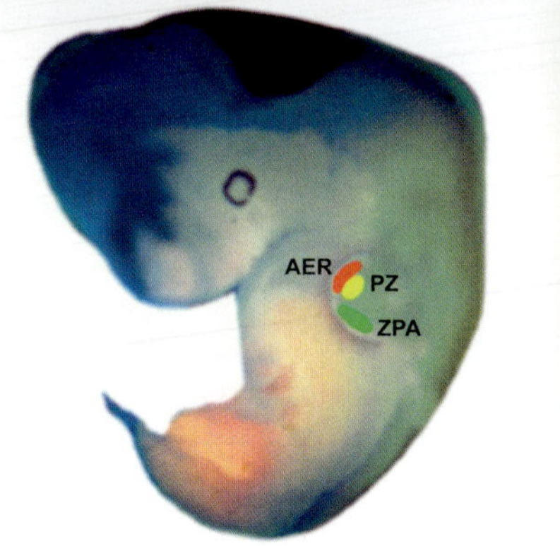


Limb bud showing:
The Apical Ectodermal Ridge (AER, red), Progress Zone (PZ, yellow) and the Zone of Polarising Activity (ZPA, green).
(Diagrammatic)

**Evolutionary changes to the fish head should result in abnormalities seen in the fossil record**

*Mutations of the HOX genes controlling the development of the head produce numerous associated pathologies [abnormalities]. That is how the developmental cascade works. The numerous cranio-facial syndromic anomalies testify to this. You could not have gotten from fish to tetrapod without developmental pathway pathology being registered in the fossil record. If there are mutations occurring to modify the cranio-facial structures in ANY MANNER, pathology is a side product of this action. Therefore, if there is no widespread cranio-facial pathology in the fossil record, there was no cranio-facial evolution.*
Dr Donald Moeller. Maxillofacial Surgeon. Columbus, Georgia. Personal communication. June 2008.

Young coot born with abnormal left leg which flaps behind it. It is unlikely to survive.

The cat front paw: the tendons can be seen to slide through specifically-placed pulleys.

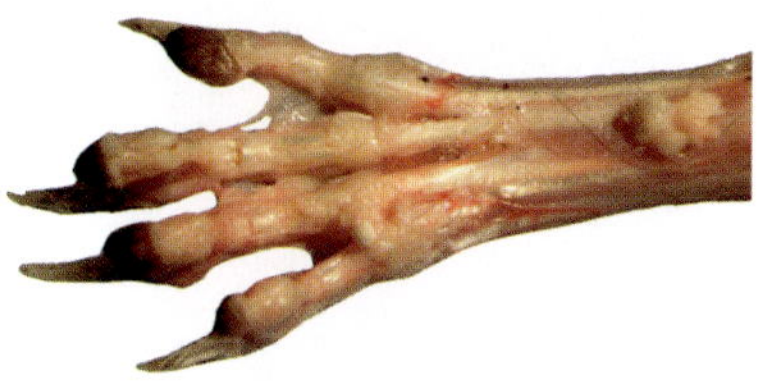

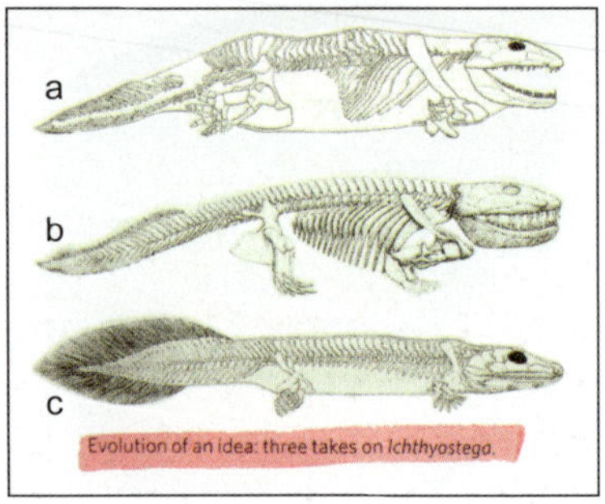


Problems with fossil reconstructions. Of course, anyone can make an honest mistake, but all scientific literature has to be carefully evaluated since serious errors are not rare.
For example, Ahlberg et al's reconstruction of the body of the tetrapod icthyostega (a) in *Nature* (vol 437. 1st September 2005. p137-40) shows that Jarvik's 1996 reconstruction of icthyostega (b) was quite wrong (Ahlberg states that it 'differs radically').

Dogfish pectoral fins

Frogs are not modified fish.

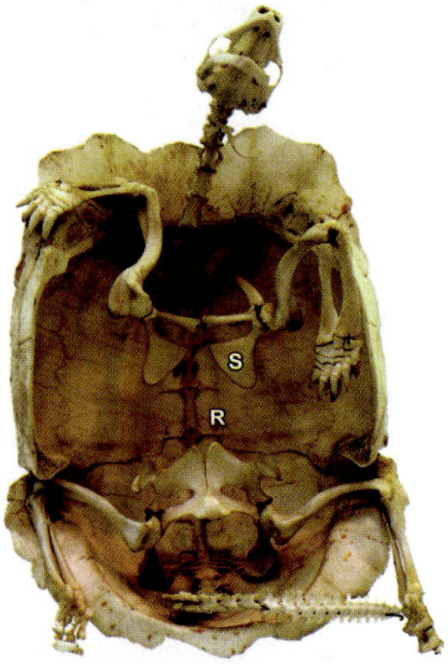


Interestingly, grafting of tissue from within a chick *leg* bud into the tip of the *wing* bud produces toes. So the information for producing toes or fingers is locked *very early* within the cells of the front or hind limb bud, long before the limb is formed..
Now, it has been suggested that the evolution of fishes into ancient tetrapods is *proved* since selected fossils show distinct changes in the number of digits:

> In the course of evolution, animals that begin having eight digits, such as acanthostega here, reduce to seven digits [Icthyostega], to six [Tulerpeton], and to five digits…that may be as good as we'll get in the fossil record in terms of a transition. K Padian. Witness at the Kitzmiller v. Dover trial, USA. 2006.

However, this is clearly nonscience since digit number can be easily changed in one generation (not millions of years) by simply altering the activity of genes within the limb bud:

> An important chicken mutant with respect to antero-posterior patterning is the talpid mutant [in which] the limbs are polydactylous and up to 10 digits develop with uniform morphology. C Tickle. 'Limb development: an international model for vertebrate pattern formation'. Int. Journal of developmental biology. Vol 44. 2000. p101-108.

The interactions of HoxD genes are involved in digit identity. Since the development of digits III and IV are not affected in the absence of HoxDl3, but the other digits are, we should expect that the earliest *fish-hoping-to-become-a-tetrapod* would have had not eight but at most two digits (III and IV). In addition, the (supposedly primitive) dogfish does not seem to have Shh, which is vital in tetrapods. It is also important to note that the dogfish limb axis is *parallel* to the body, compared to it being perpendicular to the body in land vertebrates:

> …we did not detect Shh transcripts in dogfish fin-buds…In S. canicula, the main fin axis seems to lie parallel to the body axis. M Tanaka et al. 'Fin development in a cartilaginous fish and the origin of vertebrate limbs'. Nature. Vol 416. 4th April 2002. p527-531.

If this was the case with (supposedly primitive) cartilaginous fishes, then any bony fish would have to fortuitously acquire Shh before it could develop a limb. But, of course, simply altering the limb axis without there already being in place other triggers and controls to produce a *new* functioning form of fin would only result in deranged shark limb buds. Incidentally, of the two common candidates for being a tetrapod ancestor, lungfish do not have limb-like fins and do not hunt other tetrapods, and coelacanths do not have lungs and do not swim in shallow waters.
So, creatures such as panderichthys, tiktaalik, and icthyostega simply demonstrate different fin, limb and girdle designs. All the evidence shows that tiktaalik is not a fish-tetrapod evolutionary intermediary, and although fish and tetrapods share many of the necessary embryological triggers, the *WYWYWG Principle* makes it clear that a limb *could not have* just appeared in a fish as a result of fortuitous mutations.
So, amphibians are not modified fish.

Tortoise skeleton. The shell is composed of plates of bone which are attached to the ribs (R), and the shoulder blades (S) are therefore *inside* the rib-cage. As with all amphibians, birds and mammals, the shoulder blades of all other reptiles lie *outside* the rib-cage. Since all the evidence shows that randomly interfering with embryological triggers only results in deformities, tortoises cannot be modified lizards.

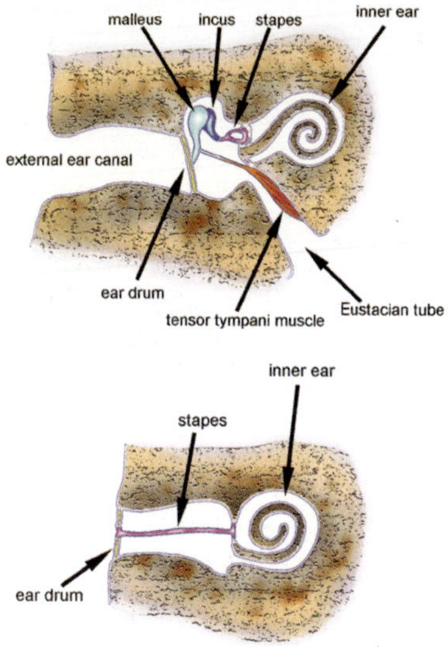


Cross-section through ear of mammal (above top) and of reptile (above). In contrast to the mammal, the reptile has an ear drum which lies on the surface, and it has only one bone (the stapes) in the middle ear.

### *Reptilian and mammalian ear bones*

Another often-cited example of transitional forms that confirm evolution is the (supposed) 'migration' of certain bones of the reptilian jaw to the middle ear in mammals. The mammalian design has a row of three bones which articulate with each other: the *malleus*, *incus* and *stapes*. The malleus is outermost and connected to the ear drum, while the stapes is innermost and is separated from the fluid of the inner ear by a small membrane. In contrast, the reptilian and bird ear design has a single bone, the *columella*, which connects the ear drum with the inner ear. While the mammalian stapes is named because it closely resembles a stirrup, the columella of the reptile is quite different, being just a thin rod of bone. In any discussion about evolutionary advantages of the mammalian ear, it must be remembered that despite their seemingly simpler ears, many birds, such as owls, have far better hearing than do humans.

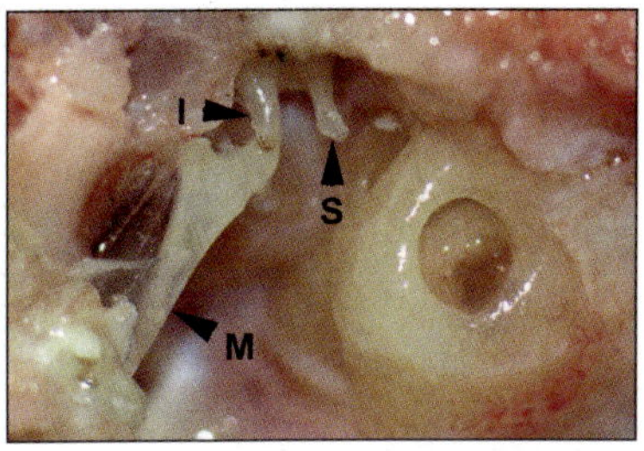


Above: Cat left middle ear cavity showing the malleus (M), incus (I) and stapes (S) bones.

The individual cat ear bones

Crocodile columella

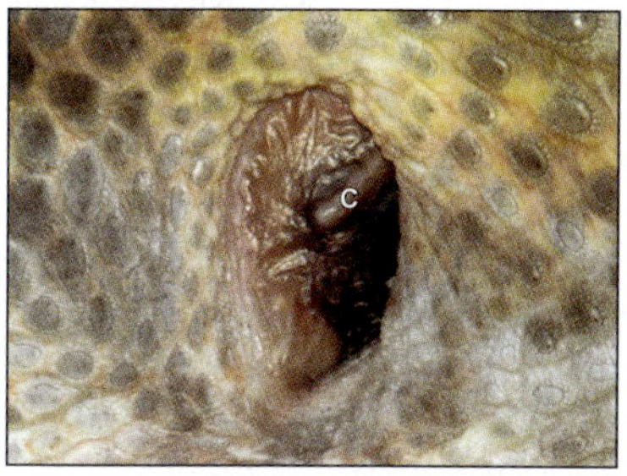


Monitor lizard ear drum from the outside, with the end of the columella (C) showing through.

*Meckel's cartilage* is a block of tissue in the first branchial arch, the most forward of the folds in the developing embryo's neck area. Apart from giving rise to the malleus and incus, it has a role foundational to the proper development of the head, face and neck. Normally, the bulk of Meckel's cartilage is reabsorbed completely and not ossified. It can be observed that the *articular* and *quadrate* jaw bones in reptiles and the malleus and incus ear bones in mammals arise from roughly the same areas within Meckel's cartilage, and that the columella and stapes arise from the second branchial arch.

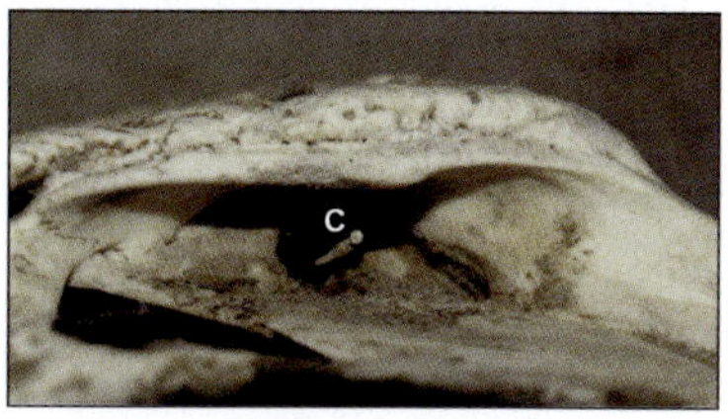


Alligator skull: the left ear opening, with the columella (C) protruding.

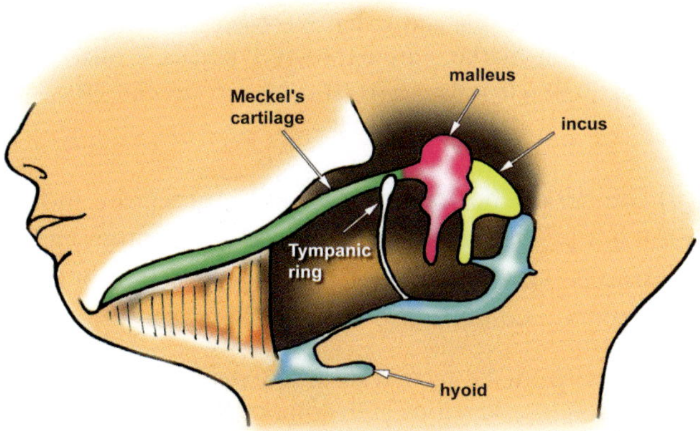


Meckel's cartilage in an 18-week-old human foetus. (After Kolliker)

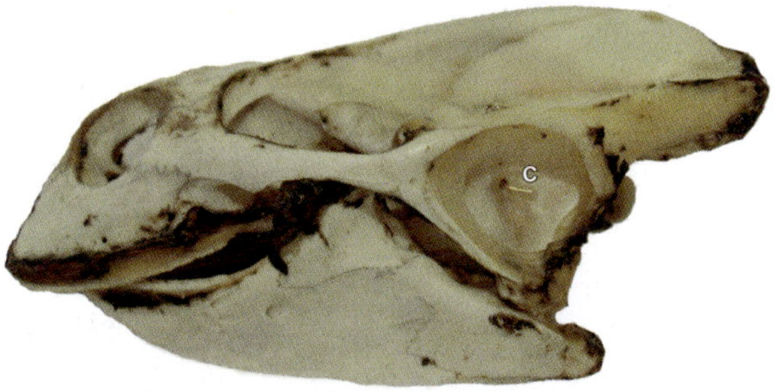


Turtle skull, showing the columella (C)

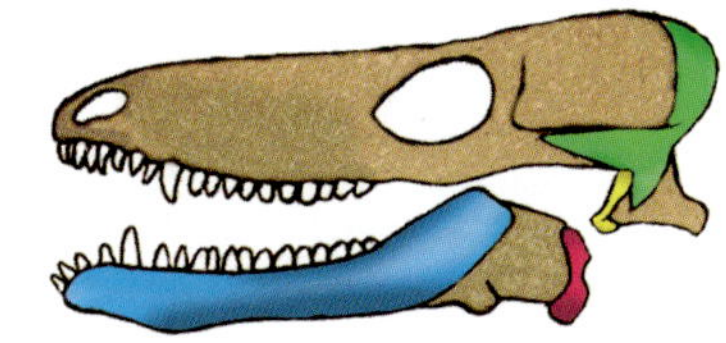

Procynosuchus

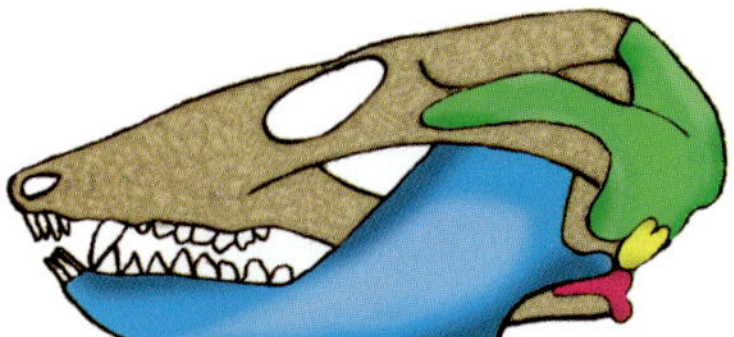

Probainognathus

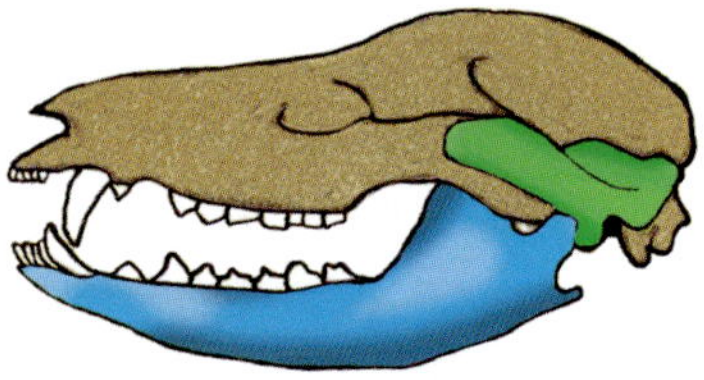

Opposum

In amphibia, reptiles, birds, and mammal-like reptiles (such as Procynosuchus and Probainognathus), the tissues of the upper end of Meckel's cartilage develop into the quadrate and articular jaw bones.
In mammals (such as the opposum), the Meckel's cartilage develops into the malleus and incus bones inside the middle ear.
Blue: dentate (mandible) bone
Green: temporal bone
Yellow: quadrate bone
Magenta: articular bone

A number of different jaw arrangements are found in vertebrates. The format in living reptiles, and also in some extinct mammal-like creatures such as *Procynosuchus* (which, supposedly, lived around 225mya), is that the quadrate is attached to the skull, and the main jaw hinge is between the quadrate and articular. In other extinct mammal-like creatures, such as *Probainognathus* (which, supposedly, lived around 230mya), both the teeth-bearing mandible (or *dentary*) *and* the articular bones are involved in a joint with the skull. In contrast, the jaw joint of living mammals is only between the mandible and the skull.
Because of these findings, the common evolutionary paradigm says that in terms of ear development, animals such as the 10cm-long probainognathus are evolutionary intermediates between animals such as the 60cm-long procynosuchus and placental mammals. In the latter, the quadrate and articular have (supposedly) 'migrated' into the middle ear to become the malleus and incus bones articulating with the stapes:

> ...there has been fossil evidence going back decades that show us animals [such as Probainognathus] with dual pairs of bones in the jaw joints which [are] perfect intermediate form[s]. K Padian. Witness at the *Kitzmiller* v. *Dover* trial, USA. 2006.

Actually, as we shall see, embryological studies from over 60 years ago and more recent evidence show that such an idea is completely mistaken, and that the Dover Court's judgement has compounded the error. It has been shown that unlike the reptilian quadrate and articular, the malleus and incus are *not* formed *purely* from Meckel's cartilage of the first arch, but each have at least two components, as they are *partly derived from the second arch*:

> The head of the malleus, the body and the short limb of the incus arise from the first arch while the handle of the malleus, [and] the long limb of the incus...arise from the second arch. J Orozco et al. 'Ontogenic peculiarities of the human tympanic ossicular chain'. Acta Otorrinolaringol Esp. 2003 Jan;54(1):1-10.

Furthermore, even *before* they become separate entities, the soon-to-be malleus and incus bones have their own distinct molecular controls and triggers:

> Even before a morphological split, we show that the ossicles have distinct genetic identities and joint markers are expressed. S Amin, A Tucker. 'Joint formation in the middle ear: lessons from the mouse and guinea pig'. Dev Dyn. 2006 May;235(5):1326-33.

These markers must be just the tip of the iceberg that determines the individual nature and manufacture of the bones. In addition, the mammalian stapes cannot be the same as the reptilian columella, since the stapes has a quite different structure and is also formed from two separate condensations:

> ...the mass of the stapes arise from the second arch. The vestibular [inner ear] side of the stapedial footplate develops in the otic capsule. J Orozco et al. 'Ontogenic peculiarities of the human tympanic ossicular chain'. Acta Otorrinolaringol Esp. 2003 Jan;54(1):1-10.

Furthermore, the genes that control the development of the joint between the articular and the quadrate are *not* the same as those that control the development of the joint between the malleus and incus:

> In fish, Bapx1 is responsible for the formation of the joint between the quadrate and articular…Contrary to expectations, [mouse] Bapx1 does not affect the articulation of the malleus and incus. A Tucker et al. 'Bapx1 regulates patterning in the middle ear: altered regulatory role in the transition from the proximal jaw during vertebrate evolution'. Development. 2004 Mar;131(6):1235-45. Epub 2004 Feb 18.

All this shows us that the malleus and incus are not *homologous* (being of the same nature) to the reptilian articular and quadrate. Although they arise from roughly the same zone of tissues, nevertheless, in the embryo they are manufactured quite differently.

Also, the common notion that the reptilian jaw 'migrated' to become the middle ear bones in mammals is fundamentally misleading. While the malleus and incus arise from condensations of tissues in the same zones in the first branchial arch as do the articular and quadrate bones, it is quite wrong to think that in the mammalian embryo the Meckel's cartilage starts a developmental trajectory towards forming articular and quadrate bone jaw structures, which *then* become modified to become the malleus and incus bones. Instead, what happens in the early embryo is that the malleus and incus arise *directly* from the condensations of tissues in the Meckel's cartilage.

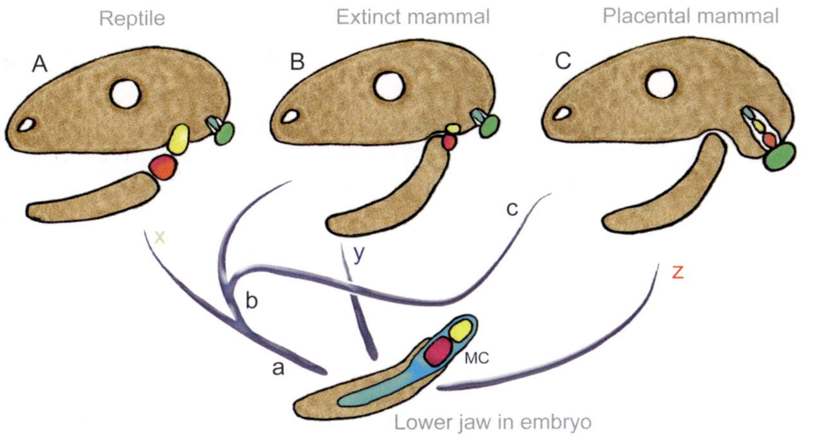


In the ***embryo*** of the placental mammal, the Meckel's cartilage (blue, MC) *does not* take a developmental trajectory (a,b) towards forming articular and quadrate bone jaw structures as in reptiles and extinct mammals, which subsequently take on a different developmental trajectory (c) to become the malleus and incus bones.
Instead, since even in the earliest stage the malleus and incus are seen to arise directly from the condensations of tissues in the Meckel's cartilage, the different forms of jaw in reptiles, extinct mammals and placental mammals must have their own unique develomental trajectories (x,y,z), and therefore, their own unique design plans.
In other words, the mammalian ear bones are not modified reptilian quadrate and articular bones.

Yellow: quadrate bone
Magenta: articular bone
Green: ear drum

In fact, it can be clearly seen in a 12-week human foetus that the cartilagenous shapes of the malleus and incus are *already* fully formed, and at their adult size *before* they start to become bone. The stapes, too, is also virtually fully formed and of adult size at this time:

> …they all commence to ossify at the twelfth week, and at this stage…are already almost *full adult size.* R Last. 'Anatomy, regional and applied'. Churchill Livingstone. 5th edition. 1972. p700.

As with all aspects of embryological development, the structures of the lower jaw and the middle ear cavity are all subject to distinct and highly complex developmental controls. In the embryo, the cells destined to develop into the jaw and middle ear bones actually arise first in the back of the head area (the *cranial neural crest cells, CNCCs*), and then migrate into the tissues destined to become the face and neck. However, in isolation, the CNCCs cannot develop into anything much since their activity is dependent on the overlying epithelium ('skin'):

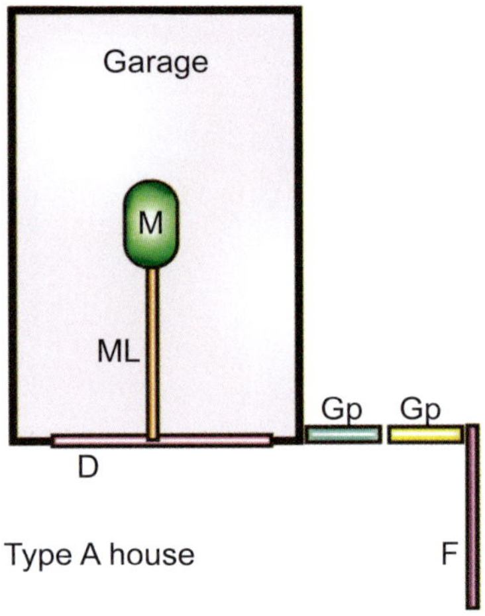


> Although the facial skeleton develops from CNC cells, in isolation these cells cannot enter chondrogenic [cartilage-forming] or osteogenic [bone-forming] differentiation pathways. Specific reciprocal interactions with facial epithelia are required for the developmental potentials of the CNC cells to be turned on according to precise spatial and temporal patterns.
> B Kanzler et al. 'BMP signaling is essential for development of skeletogenic and neurogenic cranial neural crest'. Development. 12. 2000. p1095-1104.

An analogy can be made with two houses (see diagram on left): a type A house with a fence (representing the mandible) with a wooden two-door gate (representing the quadrate and articular), adjacent to a garage (representing the middle ear) which has an 'up-and-over' metal electric door (representing the ear drum); and a type B house with a fence without a gate, along with a complex garage door system in which there are two hinged wooden levers (representing the malleus and incus) *between* the electric motor mechanism (representing the stapes) and the metal door.

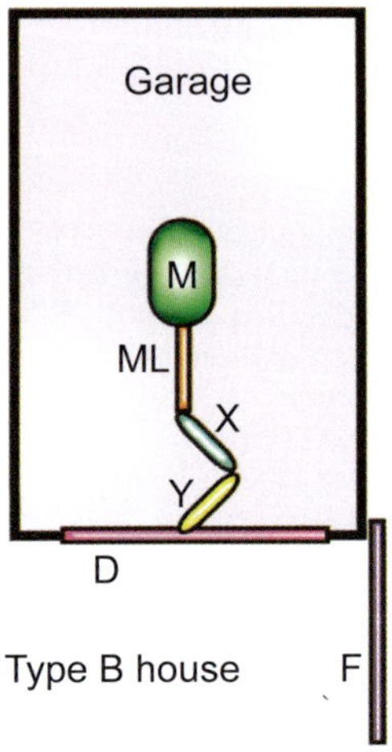


Jaw-ear & garage-gate analogy.
M: motor
ML: link between motor and door
D: door
Gp: gate panels
F: fence
X,Y: new motor-door links
See main text for explanation.

If we wanted to build a type B house, we would not build a type A house first and then modify *that* into a type B house. That would involve dismantling the gate, filling in the resulting space with fence panels, and then modifying the wood of the gate to form the levers, disconnecting the motor mechanism from the metal door and moving the motor mechanism further back into the garage, and then inserting the wooden levers between the motor mechanism and the metal door. Clearly, such modifications could not be done without disrupting the integrity of the fence/gate and also the garage entrance. So, obviously we would build a type B house directly from its component materials, 'from the ground up'. Now, if we did not have the plans, the technology, and the skills to make the type B garage door system, any alterations to the gate of the type A house would simply result in a gap in the fence, and if the gate were to be put onto the motor mechanism this would just jam the door. It would not be a working improvement, and in fact, it would result in derangement of both the fence and the garage.

Similarly, since the malleus and incus are *not* jaw bones, completely different *instructions* and building *capabilities* must exist before either jaw or ear bones can be made. And once a jaw *has been* made, and yet if there are no instructions or capabilities available for making the malleus and incus, the Meckel's cartilage of the first branchial arch *has not even started* on the (supposed) road to being formed into ear bones.

Of course, in addition, a mutation in a reptile that affected the developmental trajectory of the quadrate and articular would result in developmental disaster without a *simultaneous coincidental* alteration in the development of the middle ear cavity to accommodate any Meckel's tissue that was hoping to become a malleus and incus. The reptilian columella would have had to acquire fortuitous mutations so as to attain the mammalian form. Furthermore, as with any situation where a highly complex fully-functioning system is randomly interfered with, we can predict that if genetic mutations were to interfere with the developmental processes directing the Meckel's cartilage to form jaw bones, we should find a whole variety of congenital *abnormalities*, rather than the origination of a new type of ear. And this is *exactly* what we see.

Human congenital abnormalities demonstrate these problems perfectly. The whole external ear, ear drum, middle ear cavity and the jaws *can all be formed perfectly normally*, and yet there can be deafness associated with a variety of abnormalities affecting the bones of the middle ear. For example:

> (i) **discontinuity between the incus and stapes** with mobile stapes, (ii) **congenital fixation of the stapes** with ossicular deformity, and (iii) **congenital fixation of the malleus** and **deformity of the incus** with mobile stapes. (My bold). O Takayuki et al. 'Congenital auditory ossicle malformation without external ear abnormality'. Journal of otolaryology of Japan. Vol 106. No.3. 2003. p199-205.

This confirms that the development of the middle ear cavity is controlled by different signals to those controlling the development of the Meckel's cartilage condensations. Not only can individual middle ear bones be deformed with the others being normal, there can also be a complete *absence* of the one of the middle ear bones, with the other bones developing normally:

> The **malleus and stapes were found to be normal** and mobile on palpation, but there was a complete ossicular-chain discontinuity because the **incus was absent**. The **remainder of the middle ear anatomy was unremarkable**. (My bold). R Ralibar et al. 'Congenital absence of the incus bilaterally without other otologic anomalies: a new case report'. Ear, nose & throat Journal. April 2002 .

This confirms that the development of the individual middle ear bones are controlled independently. Since it is a fact that the malleus and incus are each subject to their own unique developmental controls, if we were to suppose that the reptile form *did* evolve into the mammalian form, we should predict that there should be many living mammals with just two middle ear bones, and also many with the reptilian columella. However, there is not the slightest evidence of this.

Furthermore, since the limb buds develop at the same time as the middle ear, alterations to the development of the quadrate, articular, malleus and incus bones are likely to be associated with abnormal limb development, as highlighted by the effects of the drug *thalidomide*:

> In 4 cases **a fused ossicular mass** was found in contact with a mobile stapes in an oval window. In several cases an ossicular mass, consisting of **a fused malleus and incus**, was found to have **no connection with the stapes**...(My bold). G Livingstone. 'Congenital ear abnormalities due to thalidomide'. Proceedings of the Royal Society of Medicine. 1965. July; 58(7). p493-7.

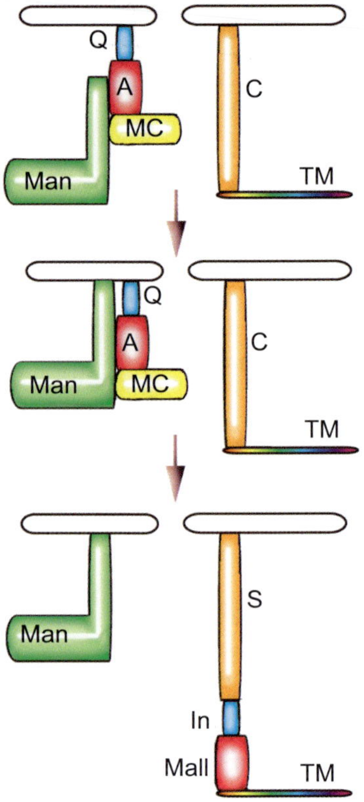


**Don't Mess with It Principle**

Whilst from a superficial view it seems that it is only a simple matter to change the reptilian jaw bone precursors into mammalian ear bones, nevertheless, when the body's hidden developmental inter-relationships come to light (below), it becomes obvious that, as with a tower made of a child's toy blocks, if you pull one block out, the whole system will collapse or any number of connected parts will fall out. Here the blocks represent hypothetical developmental controls and links in different cellular and anatomical systems. So *If you mess with a working complex multipart system controlled by multi-hierarchical activatory and suppressant triggers – you get a mess.* Therefore, the modern mammalian ear bones could not have appeared by stepwise changes to the procynosuchus form.

(Man: mandible; Q: quadrate; A: articular; MC: Meckel's cartilage; C: columella; S: stapes; I: incus; Mall: malleus; TM: tympanic membrane.)

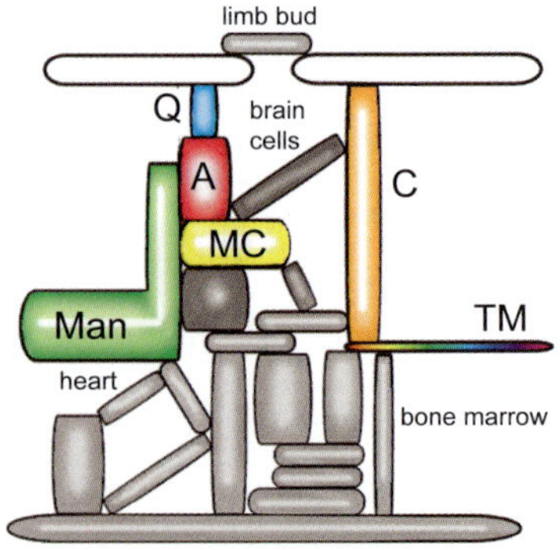


So, interference with the development of the quadrate and articular bones would most likely be associated with diverse bodily abnormalities, rather than the appearance of a new functional design for the middle ear. This is exemplified by the fact that malformations of the middle ear can be associated with the *Split Hand and Foot Malformation*, which itself can be associated with yet other problems:

> ...**clefting of the lip/palate**...profound sensorineural [nerve] **hearing loss**...**congenital vertical talus** [abnormal footbone], and a de novo deletion of the paternal chromosome 7q21.1-q21.3...There may also be additional anomalies typically affecting the **urogenital and lacrimal** [tear duct] systems. Some patients also have **dysmorphic** [abnormal facial development] **facies**...(My bold). Edda Haberlandta et al. Letters to the editor. Journal of medical genetics. Vol 38. June 2001. p405-409.

Mutations affecting the control of pharyngeal arch development also affect the development of the face and head. Much of the research on head and jaw embryological processes is done on the zebrafish, and over 100 mutations have been identified. For example:

> *Hammerhead* [and] *jellyfish* mutants exhibit shortened heads and reduced pectoral fins.
> *Knorrig* embryos [have] tumor-like outgrowths...along the edges of all cartilaginous elements. T Piotrowski et al. 'Jaw and branchial arch mutants in zebrafish II: anterior arches and cartilage differentiation'. Development. 1996 Dec;123:345-56.

> Mice lacking the dHAND branchial arch enhancer died perinatally and exhibited a spectrum of craniofacial defects that included cleft palate, mandibular hypoplasia [underdevelopment] and cartilage malformations. H Yanagisawa et al. 'Targeted deletion of a branchial arch-specific enhancer reveals a role of dHAND in craniofacial development'. Development. 2003 Mar;130(6):1069-78.

No aspect of embryological development is simple. Even just the migration of the CNCCs is a *phenomenally* complicated process, subject to multiple controls and triggers, and interference with just *one* of these can be incompatible with life:

> Chn ['chinless'] mutants lack neural-crest-derived cartilage and mesoderm-derived muscles in all seven pharyngeal arches. T Schilling et al. 'The chinless mutation and neural crest cell interactions in zebrafish jaw development'. Development. 1996 May;122(5):1417-26.

Therefore, what all the evidence *unambiguously* points to is this: any interference with face/head/neck embryological development will *only* result in damaging or lethal effects. So we can state the *Don't Mess With It Principle*:

If you mess with a working complex multipart system controlled by multi-hierarchical acitivatory and suppressant triggers – you get a mess.

It would be negligent if, in discussing the origin of the middle ear bones, we did not look at a number of other important factors, such as the associated joints, ligaments, muscles and nerves, and the teeth.

The bone of the *condyle* of the mandible (which forms the jaw joint with the skull, the *temporo-mandibular joint*, TMJ) is deposited in the embryo by ossification of a *cartilage* matrix (*endochondral* ossification), whereas the rest of the mandible is ossified by a completely different process, directly from *mesenchyme* (*intramembranous* ossification).

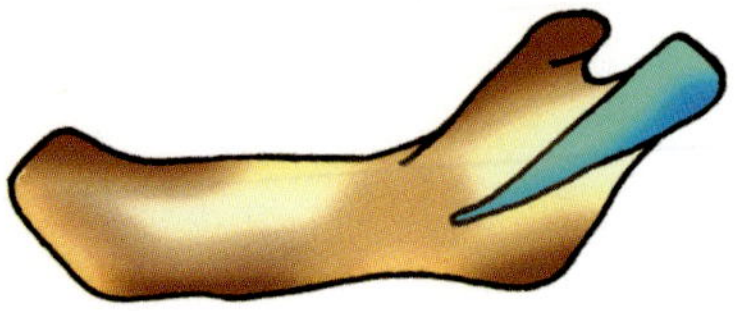

Drawing of an X-ray of a mandible of a mid-term human foetus, showing the distinct carrot-shaped bone of the mandibular condyle, which is ossified directly from cartilage. The rest of the mandible ossifies from mesenchyme. *(After J Scott & A Dixon. 'Anatomy for students of dentistry'. Churchill Livingstone. 1979)*

> The condylar cartilage appears during the twelfth week of development and rapidly forms a cone or carrot-shaped mass [which is] quickly converted to bone by endochondral ossification. A Ten Cate. 'Oral histology'. Mosby. 1998. p45.

This means that the (supposed) transition between a jaw in which the mandible (formed by intramembranous ossification) does not articulate with the skull, and one in which it does, could not have occurred without a *complete* series of *new* ossification and developmental controls being in place to produce the mandibular condyle (formed by endochondral ossification). And these controls could *not* have appeared piece-by-piece. So it is impossible for any modern mammalian jaw to have evolved from a mandible that was completely ossified in membrane.

Furthermore, the human TMJ is not as simple a structure as it might seem. In fact, as well as hinging on the skull, the head of the mandible slides forward as the mouth opens fully. For this to happen, the TMJ contains a highly specialised fibro-cartilage disc that slides during movement of the jaw, and which has a unique crinkled surface which allows it to elongate and recoil in the process. Both the disc and the TMJ develop *simultaneously* in the embryo. The disc is not an afterthought.

In addition, the joint between the reptilian skull and the quadrate is usually a fibrous joint, which has little room for movement. However, just like most joints in our body, those between the malleus, incus and stapes are all *synovial* joints. So a quadrate that was hoping to be an incus would have to acquire a completely different joint structure at its inner end. Also, unlike the quadrate-articular joints, the joints between the malleus, incus and stapes have an exceptional amount of elastic tissue in the associated ligaments. This is because this elastic recoil opposes the action of the two middle ear muscles *tensor tympani* and *stapedius*.

It must also be noted that the malleus, incus and stapes are not arranged so that they act as simple pistons between the ear drum and the inner ear. Instead, their articulations are precisely arranged such that the three bones act collectively like a bent lever. Instead of the base plate of the stapes simply moving in and out, its lower edge rocks on the window leading to the inner ear. All the above arrangements could not occur by simply randomly interfering with the reptilian jaw.

Since the reptile has no malleus, it would have to have evolved a tensor tympani muscle, which sits snugly inside its own made-to-measure canal above the *eustachian tube* (which allows the pressure to be equalised between the middle ear and the back of the nose) and is orientated perfectly so as to be attached to the handle of the malleus. While the tensor tympani is supplied by a branch of the same fifth cranial nerve (the *mandibular nerve, V*) that supplies the *medial pterygoid* muscle (which originates from the roof of the mouth and from the base of the skull and is inserted into the inside of the middle part of the lower jaw), here lies another problem. The nerve fibres to the medial pterygoid direct biting and chewing actions, which are crude and are under conscious control. In contrast, the activity of the tensor tympani muscle deals with *minute* movements and is also completely subconscious. If a little branch of the nerve to the medial pterygoid muscle was 'grafted' across to the malleus during any supposed jaw bone-to-ear bone migration, it would be of no use without major changes in the origination in the brain of its constituent fibres, and without also acquiring previously non-existent functional coupling with the nerve to stapedius, a branch of the seventh cranial nerve (the *facial nerve, VII*), to modulate sonic vibrations.

Sheep mandible showing a fully-formed but unerupted molar.

Because the teeth of mammals are completely formed within the bones of the jaws *before* they erupt, any significant alteration of the angle of the TMJ during any supposed quadrate/articular to incus/malleus evolution would result in skewing of the alignment of the top and bottom teeth. Furthermore, all the experimentation relating to alterations in jaw development result in serious problems:

> The most complex animal used for cranio-facial-mandibular research is the Mouse. MSX-1 and MSX-2 as well as DLX Hox mutations effect the jaws. All such Hox gene modifications produce abnormal 'occlusions' [the way upper and lower teeth fit together] which in no way are improvements over existing occlusions. In some Msx-1 and Msx-2 Hox gene malfunctions, entire groups of teeth are deleted. Dr Donald Moeller. Maxillofacial Surgeon. Columbus, Georgia. Personal communication. 20th July 2008.

The platypus has special electric sensors in its leathery 'bill'.

Recently, the fossil mammal *Yanoconodon*, a tiny 5cm-long creature with an elongated lizard-like body, was found in (supposedly around) 125myo sediments in north-eastern China. It had an ossified Meckel's cartilage attached to the malleus, and the whole jaw/ear bone configuration resembled the embryonic platypus, except that in the latter the Meckel's cartilage does not ossify and is ultimately totally absorbed. In the adult platypus, there is no connection between the malleus and the jaw. The yanoconodon incus had an articular facet for a stapes, although a stapes was not found in the specimen:

> …the middle ear bones are connected to the mandible via an ossified Meckel's cartilage. The connected ear and jaw structure is similar to the embryonic pattern in modern monotremes (egg-laying mammals) and placental mammals… Z Luo et al. 'A new eutriconodont mammal and evolutionary development in early mammals'. Nature. Vol 446. 15th March 2007. p288-293.

So, what advantage was the yanoconodon design? If the mandible was connected directly to the stapes via the malleus and incus, then vibrations from the lower jaw would have been transmitted directly to the inner ear. So, perhaps yanoconodon was able to detect ground vibrations from the movements of insects and worms. It would not be surprising to find that yanoconodon had some unique sonic sensory capability since the platypus itself is endowed with unique electrical sensors in its snout. However, there is another explanation for yanoconodon's curious ear bone configuration when we consider just how complex the whole jaw and ear development really is. Dissolution of Meckel's cartilage can be both promoted and suppressed by the action of various triggers, so we should not be surprised to observe the failure of dissolution appearing in different forms. Persistence of part of the lower section of Meckel's cartilage in an ossified state in humans is not rare, and even duplication has also been observed:

> An exceptional case of a duplicate Meckel's cartilage in a human foetus, with a C-R length of 57 mm… J Rodríguez-Vázquez et al. 'A duplicated Meckel's cartilage in a human fetus'. Anat. Embryol. (Berl). 1997 Jun;195(6). p497-502.

While it is the norm for the malleus and incus to be separate, they are fused together in animals such as the guinea pig, due to failure of the normal joint cavitation process:

> In the adult guinea pig the malleus and incus are normally found as a single complex…In the guinea pig embryo, joint formation is initiated but no cavitation is observed, resulting in a single complex divided by a thin suture [joint line]. S Amin, A Tucker. 'Joint formation in the middle ear: lessons from the mouse and guinea pig'. Dev. Dyn. 2006 May;235(5). p1326-33.

And the joints between the ear bones are formed as a result of *positively* activating or suppressing a range of developmental triggers:

> …The incudomalleal joint region forms by cells in the presumptive joint region switching off cartilage markers and turning on joint markers. S Amin et al. 'Incudomalleal joint formation: the roles of apoptosis, migration and downregulation'. BMC Dev. Biol. 2007 Dec 5;7. p134.

So, yanoconodon may simply represent a lineage with mutational failure of one or a group of suppressant triggers, leading to persistence and ossification of Meckel's cartilage. In other words, the yanoconodon ear configuration may have no special advantage or function at all, but may just represent a degeneration of the platypus' type of ear configuration in which the resulting abnormality is not vital to survival.

Finally, it must be remembered that the *Coelacanth Principle* shows that animals such as procynosuchus, probainognathus, yanoconodon and modern mammals could have all lived *contemporaneously*, and so there is no scientific validity *at all* in stating whether one creature was ancestral to another.

Therefore, the evidence shows that in terms of jaws and ear bones, neither probainognathus nor yanoconodon is an evolutionary intermediate, nor are they any closer to acquiring a placental mammal's malleus, incus and stapes than is procynosuchus. The lesson is that creatures such as procynosuchus, probainognathus and modern mammals simply demonstrate different jaw and ear designs. Others such as yanoconodon may represent inbred abnormalities, a not uncommon finding of which we will see many examples elsewhere.

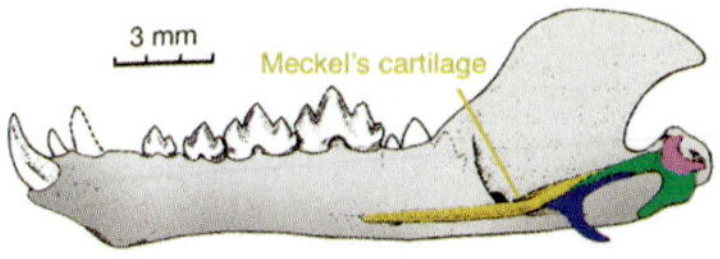

Yanoconodon jaw. (*After Luo*)

Adult platypus middle ear bones. (*After Luo*)

The Yanoconodon jaw has a connection between the malleus and the ossified Meckel's cartilage.
Although the platypus embryo has a malleus which is connected to the mandible with a non-ossified Meckel's cartilage, this does not persist in the adult.

As in living mammals, the skull of Dimetrodon had an opening behind the eye socket. It is because of this common feature that it is commonly believed that creatures like dimetrodon eventually evolved into true mammals - and subsequently into humans.

Thylacosmilus, a sabre-toothed marsupial cat that lived (supposedly) 2mya.

Thylacine, an extinct marsupial from Tasmania.

### *Egg-laying, pouched and placental mammals*

Our fourth example of (supposed) evolutionary transitional forms relates to the origin of mammals. We would be wrong to consider monotremes or marsupials to be either primitive or simple animals. In fact they are both highly complex creatures which are well able to survive in their particular environments. In monotremes the intestinal, urinary and reproductive tubes come together as a single opening (the *cloaca*) and unlike placental mammals, there is no penis, and mating is accomplished by the coming together of the two cloacae. Monotreme eggs hatch after only a 7 to 10-day pregnancy and the babies are born no bigger than rice grains. The duck-billed platypus lays its eggs in a nest and once hatched, the poorly-developed babies cling to the mother's fur and lick up milk that oozes directly from pores in the mother's abdominal skin, the mother having no true nipples. In contrast, the echidna lays eggs but these are carried within a pouch. Adult monotremes have no teeth, although the duck-billed platypus has three tiny ones at birth.

After a pregnancy lasting between 12 and 33 days, marsupials give birth to peanut-sized blobs of tissue which have stubby limbs and hardly any form. These tiny babies make their own way into the mother's pouch and lock onto one of a number of teats with their mouths. Unlike other grazing animals which have teeth that grow continually during life, kangaroos have 4 pairs of molars which erupt in sequence with the new teeth gradually pushing out the old teeth, in much the same way as the six pairs of molars do in elephants.

Mammals are so-named because they have milk-producing glands. But there is no evidence to show *how* a reptilian skin with no sweat or grease (*sebaceous*) glands could develop the milk-secreting glands of the monotremes; *how* the glands and skin of the monotremes could develop into the teats and pouches of marsupials; or *how* a cloaca could be altered to form a penis. Most authors do not even attempt to tackle such issues and remain satisfied simply to *describe* superficially the differences between egg-laying, marsupial and placental mammals and to *reiterate the assumption* of an evolutionary trend. For example, the origin of mammary glands:

> These glands **are** formed from modified sweat glands in the skin. Since *Morganucodon* was an endotherm, it **must have had** sweat glands, in order to dissipate excess heat, and structures **must have been** present in *cynodonts* back to *Thrinaxodon,* at least. Therefore, **there must have been plenty of time for mammary glands to evolve**. (My bold). Michael Benton. 'The rise of the mammals'. The Apple Press. 1991. p33. (Cynodonts such as *Thrinaxodon* were mammal-like animals that are thought to have lived in the Early and Middle Triassic around 250mya.)

Such superficiality is typical of both popular and 'serious' books on evolution. Not only does Benton offer no biological exposition as to how a sweat gland *could* arise within skin that does not have the genes to make one, but he has also missed the point. Any discussion should only deal with what is plausible. Time itself cannot magically make what may be *impossible*, possible.

Certainly, we could programme a computer to show, in an image on screen, a continuous moulding of embryonic tissues, such that the cells within plain skin could be directed to turn seemlessly into a sweat gland, or to fold up to form a pouch, just as we might watch a computer-generated film of Dr Jekyll's face turning into Mr Hyde. But that is an incorrect and an absurdly simplified view of the problem. Evolution by the natural selection of inherited characteristics carries no objective or purpose in mind and does not operate so as to mould a creature in any specific direction. And natural selection can allow to survive only that characteristic which is useful for the present moment.

It is commonly stated (and blindly repeated) that the breast is a *'modified sweat gland'*. Although sweat glands and milk-producing glands share a number of similar features of structure and physiology, this is not evidence to show that one could (or did) ever change into the other. And again, the *WYWYWG Principle* shows how improbable such a series of necessary changes to a reptile would be. The skin cells would have to be reorganised to produce milk glands and teats; the biochemistry of the skin cells would have to be reorganised so as to allow the exudation and active filtration of milk components from the blood; and the baby would have to acquire reflexes so as to be able to clamber towards the milk and to suckle, etc.

But reptiles do not have sweat glands. So before a creature could modify a sweat gland so that it became a mammary gland in the first 'mammal', plain scaly reptilian skin would have to acquire sweat glands. But that would mean that those first hypothetical 'mammals' would have had sweat glands and no milk-producing glands. So they would not have been mammals (since they could not produce milk) nor could their sweat glands have been of any benefit to a baby reptile born immature and needing sustenance. Furthermore, the earliest true mammals were (supposedly) small shrew-like creatures. This raises a problem since small living mammals have very few sweat glands, would manage very well without them, and so would have no advantage in evolving them in the first place. To complicate matters further, a sweat gland is not a simple structure with a simple function. Sweat glands comprise a coiled tubular extension of the epidermal cells deep into the skin. The deeper cells are responsible for allowing the passage of sodium and chloride ions from the surrounding fluid into the sweat gland cavity itself. Cells lining the upper part of the sweat gland are responsible for avoiding excessive salt loss by actively absorbing certain amounts of sodium and chloride ions before the sweat finally escapes onto the skin surface. Both the sodium and the chloride ions are actively absorbed by the action of quite different proteins (called sodium and chloride *channels* respectively), which are embedded in the membranes of the cells in the upper part of the sweat glands.

It is implausible that sweat glands arose piecemeal since, for example, without a chloride channel the incorporation into the developing system of only a functional sodium channel would cause excessive salt loss. This would be of no advantage to a (supposed) *hoping-to-become-a-mammal* reptile. Indeed, it would cause problems: for example, in hot circumstances too much salt loss causes muscle cramps and can lead to low blood pressure and kidney failure. This is why people working or playing in hot conditions should take salt tablets as well as drinking plenty of water. *Cystic fibrosis* is a serious disease caused by a variety of mutations that damage a gene found on chromosome 7, which is responsible for the production of the chloride

Milk production is a highly complex procedure.

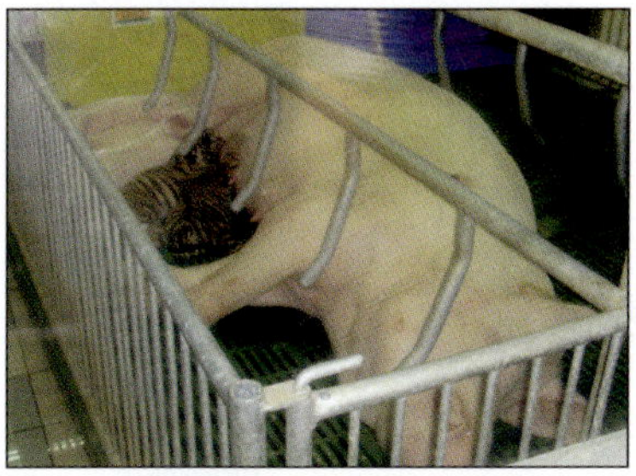

Tiger cubs suckling on a pig.

Tiger cub being bottle-fed.

**A single mutation can result in absent sweat glands**

*Dermatopathia pigmentosa reticularis* is a rare genetic form of ectodermal dysplasia in which a person has missing finger prints, extremely thin hair, brittle nails and mottled skin, in addition to being unable to sweat. The inability to sweat makes the person very vulnerable to overheating, which could be fatal. Since even just a single mutation can prevent *completely* the development of sweat glands, the *WYWYWG Principle* shows that sweat glands could not have arisen stepwise.

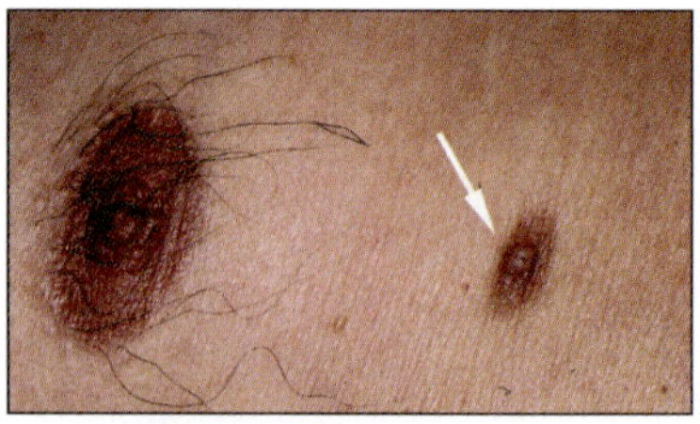

Extra nipple in a man.

The elephant and the hyrax (above) have breasts between the front legs, whereas hoofed animals such as the giraffe (below) have nipples between the back legs. Other animals such the dog (bottom) and pig have a series of nipples along the body.

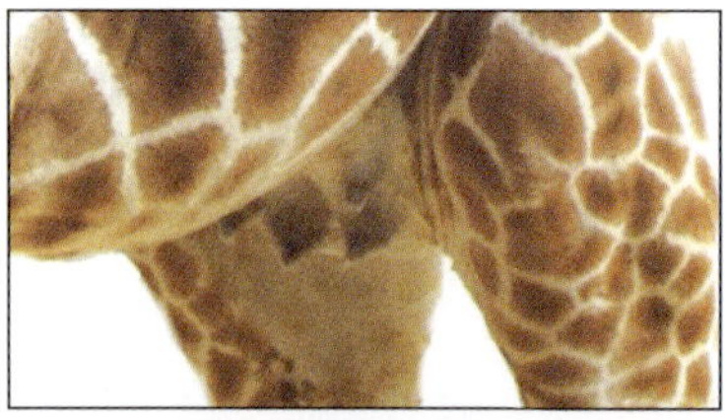

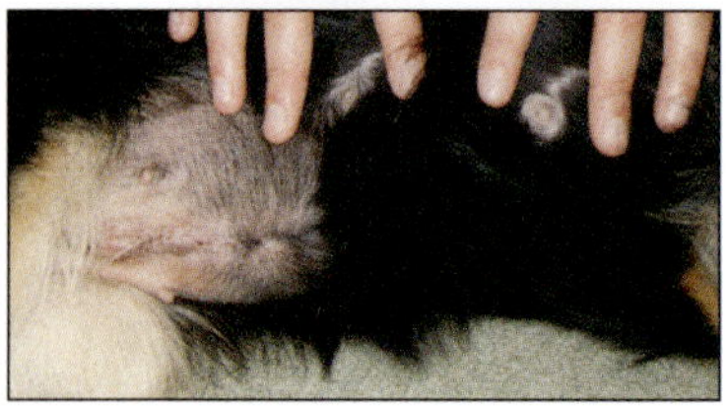

channel. It affects roughly one in 2500 babies. The lungs produce sticky mucus which predisposes to recurrent infections, and the pancreas is unable to produce enzymes resulting in poor digestion and malnutrition. By having faulty chloride channels, people with cystic fibrosis produce excessively salty sweat and this can be used as a test for the condition. Cystic fibrosis shows us that the chloride channel is essential for health and can be damaged by mutations, but this does not show us how the chloride channel could have come into being. In fact, the chloride channel protein is *so* different from other channel proteins (such as sodium and calcium channels), that it could not have appeared by a series of stepwise modifications to some previously-existing channel protein.

Reptiles do not have skin cells sensitive to *prolactin* (the hormone that stimulates milk production). So, having milk glands without hormonal controls to stimulate milk production would be of no use. However, if we suppose that ancestors to monotremes did evolve to produce milk, there is no reason why an ancestral echidna-like animal should mutate to produce nipples over some of its milk pores, rather than produce useless 'dummy' nipples unconnected to milk pores, and sited randomly anywhere on the body (for example, on the back or on the head). All placental mammals have nipples either along the chest and abdomen (along *the primary nipple line*) or between the rear legs. Breasts in the front of the chest are only found in primates and, curiously, in the elephant, manatee and the hyrax (which is a rodent-like creature the size of a rabbit). Since there is no advantage in such arrangements, one over the other, it is impossible for natural selection to have evolved only these forms.

It is not uncommon for humans (both men and women) to have an extra and non-functional nipple on the body, showing that different genes are responsible for controlling the development of the nipple itself and the breast tissue underneath it. In humans, protuberant pigmented moles on the skin can often resemble small nipples and these moles can be found almost anywhere on the body. Thus we should predict that if a nipple appeared by one or a series of mutations, it would not appear *only* where we would think a nipple *ought* to be. And when in humans they appear in the wrong place, extra nipples are not detrimental to survival. Therefore, we should expect to find many creatures that have evolved harmless nipples dotted around all over the body, with only some being associated with milk glands.

Nevertheless, leaving aside these considerable difficulties, consider for the moment that some change did occur in a reptile that caused it to produce sweat. Since sweat glands essentially excrete water and salt, a whole series of changes would have to occur in the cells lining the sweat glands before they could excrete large proteins, fats and other substances i.e. milk. Suppose, however, that the descendants of a reptile produced a milky discharge from some of their glands, but without any changes appearing in the young. As any nursing mother will know, breast-feeding requires a greater food intake on the part of the mother, to make up for the loss to her body of the nutrients in the milk. Therefore, any such reptile would be at a serious disadvantage and would most likely die off. Furthermore, live-born and egg-hatched young reptiles are miniature adults and are perfectly capable of feeding by themselves and *do not need* any milk. In contrast, the baby monotreme is so poorly developed that any reptiles that produced such feeble offspring would certainly die out very quickly. Thus, in order for any reptile to become a mammal, the mother would have to become seriously disadvantaged; and for milk to be of any use, the babies would have to take a huge backward step in development and be born extremely fragile and vulnerable. Natural selection would soon see to the extinction of such creatures.

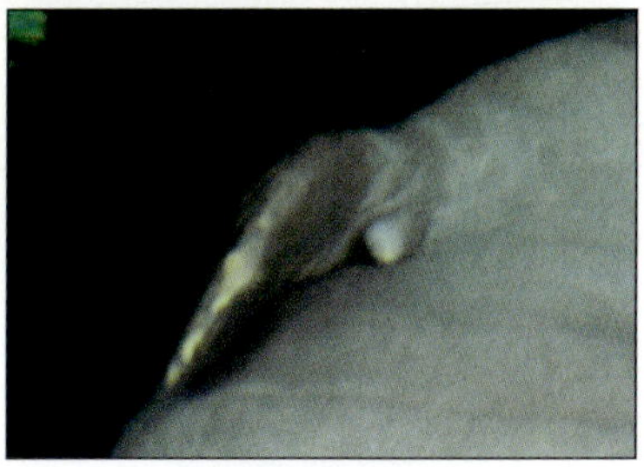

Above: Manatee nipple is located at the base of the flipper.

The same sorts of biotechnical problems would arise in considering the development of a pouch in a marsupial. The genes responsible for the development of nipples and abdominal skin are quite unrelated to those governing the uterus and other organs and structures related to reproduction, and there is no reason to suppose that any pouch would appear *for the purpose* of assisting gestation. If a pouch were to appear in a primitive monotreme, it would most likely arise as a harmless fold of skin appearing elsewhere on the body. *Millions* of years might then elapse before another chance mutation could produce a fold of skin on the abdomen. Even then, such a fold of skin would be of no use unless the developing baby was attracted to it...and unless there was milk inside it...and of course the oozing milk would have to be controlled so that it did not drown the baby etc.

Like all baby reptiles, alligators and turtles (above) come into the world as miniature versions of their parents. They have no need of parental care or nourishment.
In contrast, baby mammals and birds (such as this 3-week-old barn owl, below) are very immature and vulnerable.

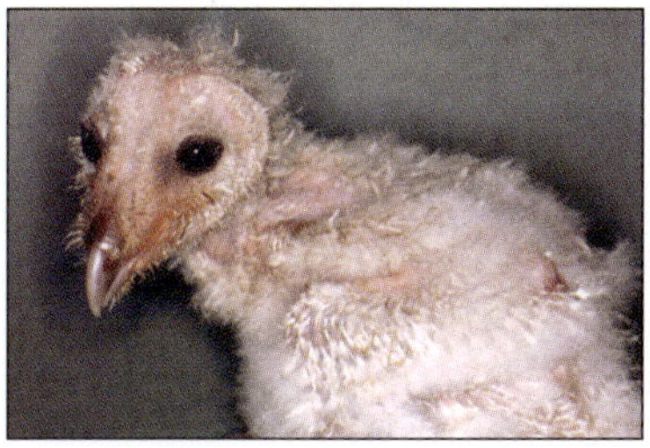

Could we conceive a progression that would allow a penis to appear in a male monotreme instead of a cloaca? In order to produce male external genitalia, a number of new features would have to come into being, with corresponding changes being required *in parallel* in the female. And since mutations resulting in a male with a penis (in whole or in part) *without* the female developing an appropriate vagina (or vice versa) would offer no advantage over the cloaca-cloaca contact, we should not expect any such unilateral changes to be selected for.

The testes of the male embryo produce the male hormone *testosterone*. This acts on receptor sites in cells normally destined to become male genitalia. When these receptor sites are activated, they trigger a programmed cascade of activity to develop the penis and scrotum. In the congenital condition *testicular feminisation syndrome,* the testosterone receptors of the tissues destined to become genitalia are absent, or non-functional. This results in a *genetically-male* individual (complete with a Y chromosome, functional testes and high levels of testosterone in the blood) who fails to develop a penis and scrotum. However, since the testes normally also produce a small amount of female hormones, which are free to exert their female influences without being overidden by testosterone, the individual with testicular feminisation syndrome develops female external genitalia and breasts. Thus the individual is genetically a male, but externally a female. This shows that testosterone acts only as a trigger, or switch. In addition, male-ness is ultimately determined by the small (only a few hundred nucleotides long) *SRY* gene found on the Y chromosome. SRY codes for a protein which triggers a number of other genes, so it acts as a switch that controls a cascade of other actions that determine the sex of the individual. Therefore the real job of directing the production of male organs is carried out by an unseen series of commands and controls, and the *WYWYWG Principle* shows that a stepwise origination of such commands and controls is implausible.

Furthermore, the anatomical arrangements of the testes and kidneys do not support a stepwise evolutionary progression. Fishes, amphibians, reptiles and birds all have testes housed inside the abdomen. In fishes, the testes are elongated structures that

As with this wallaby, most terrestrial mammals have testes inside a scrotal sac. However, in marsupials the scrotum hangs *in front* of the penis.
In addition, there are a number of other major differences between marsupials and placental mammals which do not fit in with an evolutionary progression. For example: unlike placentals, marsupials do not have a *corpus callosum* (a large tract of nerve fibres which link together the two cerebral hemispheres of the brain), and they have *two uteruses* (a right and a left) with a *tripartite vagina* consisting of a mid-line canal and two lateral canals.
Also, many marsupials have the 2nd and 3rd rear toes fused into a single toe with two claws, and a forked penis.

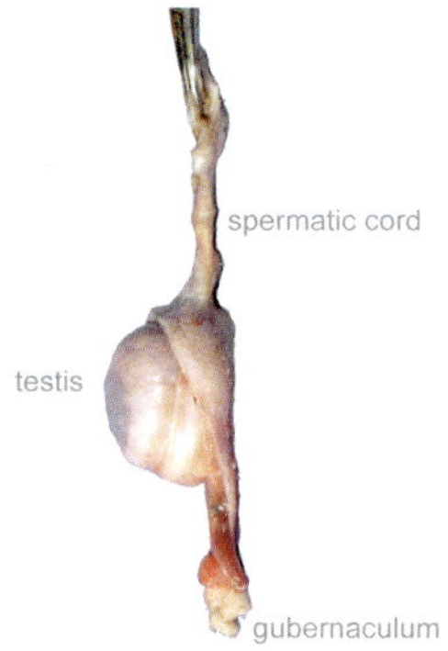


Badger testis

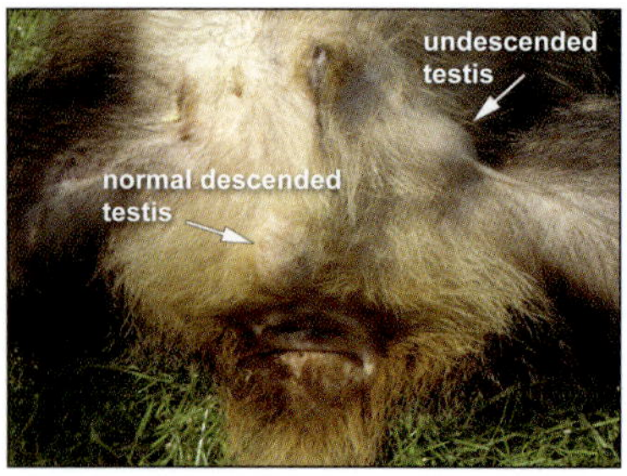


Badger: Undescended testis

can extend the whole length of the abdominal cavity. In amphibians, reptiles and birds, the testes are more compact and are located at the upper end of the kidneys. Monotremes also have intra-abdominal testes (IAT) but, in contrast, their (supposed) descendants, the marsupials, have testes inside a scrotal sac (SS). Most terrestrial placental mammals have testes located in a scrotal sac but the insectivores (such as shrews, hedgehogs and moles), the hyrax, the hippopotamus and elephants all have IAT. In some rodents, the testes descend during each breeding season and ascend again afterwards.

It is generally taught that all modern placental mammals evolved from shrew-like mammals which diversified 65mya. However, shrews have IAT, and since both small and large placental mammals (such as the hyrax, the hippopotamus and elephants) survive very well with IAT, there could have been no necessity or advantage for mammals such as the rat, tiger or rhinoceros to evolve a scrotal sac from their (supposed) shrew-like ancestors. Equally implausible is the notion that shrews (with IAT) evolved from marsupials (with SS), or that marsupials (with SS) evolved from monotremes (with IAT).

Furthermore, the developmental processes that are necessary for the descent of the testis are highly complex. If the location of the testis next to the kidney was the (supposed) primitive condition, then the downward 'migration' of the testis could only have occurred as a result of a series of specific mutations. However, since a testis situated at the pelvic brim offers no selective advantage compared to a testis situated next to the kidney, a stepwise evolution of the testis from high up in the abdomen into a scrotal sac is impossible.

We have seen that testosterone merely acts on a number of commands and controls which must be collectively responsible for directing the developmental processes that manufacture external male genitalia. During the proceedings, the testis itself is 'pulled' out of the abdomen by the contraction of a specific band of tissue called the *gubernaculum*, which is connected both to the lower pole of the testis and also to the tissue that is destined to become the scrotal sac. Clearly, without a gubernaculum any (supposedly) evolving scrotal sac elements would have had no function, and conversely, without a scrotal sac any (supposedly) evolving gubernaculum elements would also have had no function. Furthermore, the lower end of the gubernaculum can (in error) become attached to different sites, resulting in congenital abnormalities, such that a testis can be located in the upper thigh, at the root of the penis or in the groin.

A scrotal sac allows sperm production at a temperature 1-2° below core body temperature, and it is true that, in humans, the condition *undescended testes* (where one or both testes stay inside the abdomen or in the groin and fail to descend into the scrotum) gives rise to subfertility or sterility. However, since a number of mammals and all birds (which have a resting body temperature higher than placental mammals) also have IAT, it is clear that healthy sperms *can* be produced without the testes needing to be housed in a scrotal sac. Therefore, providing that any particular location would not make the testes vulnerable to injury, then, contrary to what we see, an evolutionary descent of the testes should have resulted in animals with a variety of locations of the testes (such as in the thigh or groin).

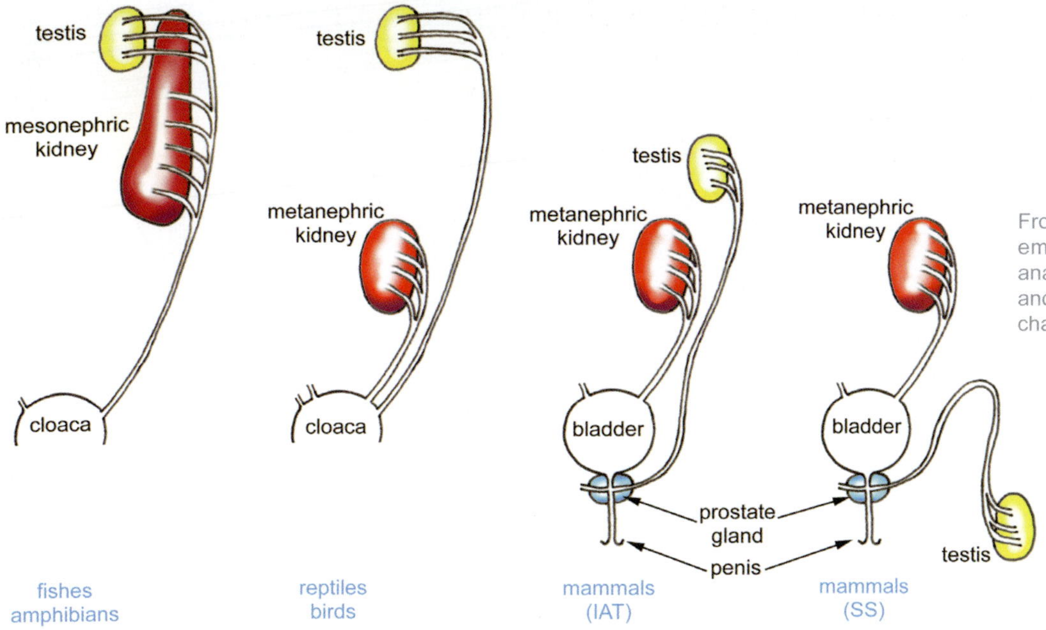


From fishes to mammals, neither the embryological development nor the anatomical arrangement of the kidney and testis reflects a stepwise series of changes.

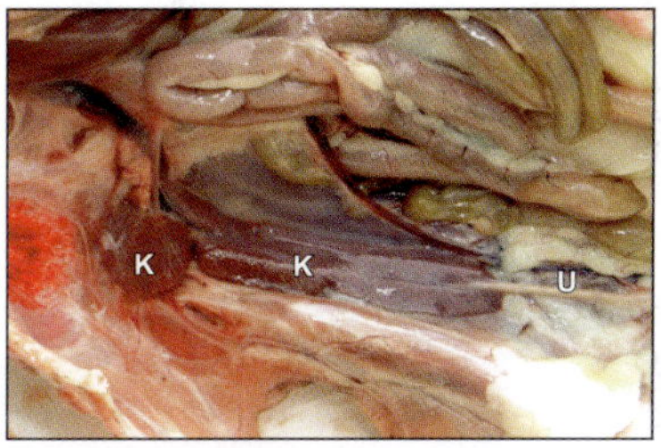


Bird multi-lobed kidney (gannet).
K: kidney
U: ureter

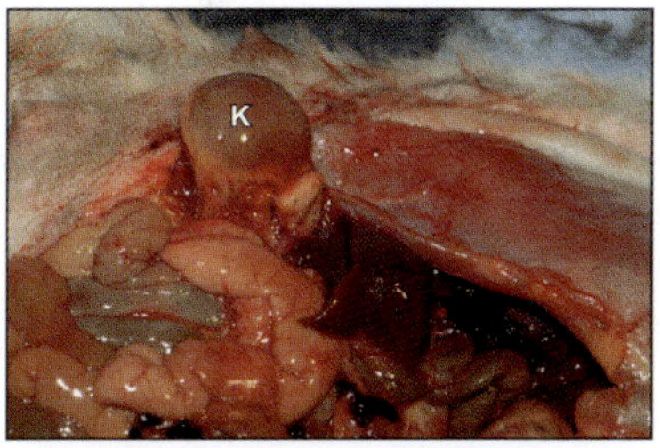


Mammalian single-lobed kidney (squirrel, above) and lobulated kidney (cow, below).
K: kidney

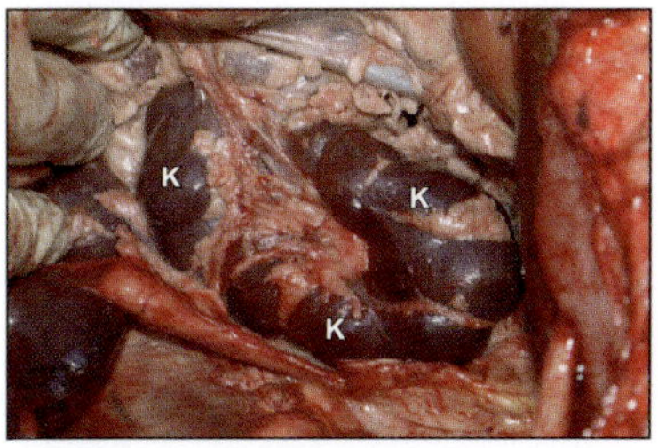


During embryological development, the testis and kidney are intricately linked. In all vertebrates, the testis develops in the cranial (head end) of the abdomen. In fishes and amphibians, the kidney develops from tissue (the *mesonephros*) in the same portion of the abdomen, and the ducts from the testis pass through the kidney tissue and combine with the kidney ducts to form a single tube leading to the cloaca. In reptiles and birds, the kidney develops out of quite distinct tissue (the *metanephros*) which lies in a more caudal (tail-end) position in the abdomen, and the kidney and testis drain into the cloaca via separate ducts. Mammals also have a metanephric kidney but its duct (the *ureter*) drains into a bladder, which itself drains into the *urethra*. The tubules of the mammalian testis join together to form the *vas*, which passes through the prostate gland at the base of the bladder, before emptying into the urethra. These differences in the anatomical arrangements of the testes and kidneys of different vertebrates are substantial and do not show any stepwise progression.

There is no doubt that the mesonephric kidney of fishes and amphibians works quite adequately. So, if evolution *could* occur, and if some improvements to the kidney *were* possible by a blind process, and if any random changes could be *significantly* advantageous, then we should expect that reptiles, birds and mammals should have acquired progressive modifications of the mesonephric system rather than the *de novo* acquistion of a completely new and separate (metanephric) kidney and tubule system.

We have seen in this chapter that while different vertebrates share some obviously similar characteristics with regard to their embryology and anatomy, nevertheless, a critical examination shows that any stepwise evolution, in the order *amphibian-reptile-monotreme-marsupial-placental mammal*, is impossible.

# CHAP

## Variation: *the dice of life*

Flamingos
Lake Nakuru, Kenya

A blizzard may produce the shape of a lizard (Candian Rockies), and erosion may produce the shape of a seahorse (below, Isabella Island in the Galapagos).
Some computer programs have been devised which, when left to run by themselves, produce insect-like shapes. But neither process gives insight into how a real creature is manufactured from its component chemicals.
Yet it is taught that the:

> *'...jet engine ...does not come about by chance but...evolution did come about by chance.'*
> John Sulston. Royal Institute Christmas Lecture. 2001.

We know that organisms vary, and much of this variation is inherited. Those individuals which are most suited to an environment or circumstances will have a greater chance of surviving, resulting in their leaving more offspring. Those offspring will then carry within their genes the coding for any particularly favourable characteristic. During his travels around the world, Charles Darwin had the opportunity to observe the great diversity of form in a number of species. On the basis of his observations on biological variation, he extrapolated and conceived the notion that *all* biological structures and processes were the result of small and gradual changes in previously existing biological structures and processes, and that all organisms shared with each other an *ancestor-descendant,* or *distant cousin relationship*. Extrapolated backwards in time, this view forces the conclusion that our origins lie ultimately in spontaneous pure chemistry, and this is implicit in the stance taken up by many biologists:

> The mere fact that we are here implies that life did get started. Francis Crick. 'Life itself'. MacDonald & Co. 1981. p92.

Yet even though it is admitted that even at the molecular level:

> ...there are complex structures...which have organized complexity - and which cannot have arisen by pure chance. Francis Crick. 'Life itself'. MacDonald & Co. 1981. p52-3,

there remains a supreme confidence in the power of random chemistry and natural selection:

> It might be thought impossible to evolve any large organisms...but one must be careful. The very existence of land animals and plants shows that once a system has progressed some way, natural selection can be very ingenious at surmounting obstacles of this kind. Francis Crick. 'Life itself'. MacDonald & Co. 1981. p61.

But "*the very existence of land animals and plants*" shows only that animals and plants exist. Nothing more. The central question is: do the observed variations offer direct evidence for the origin of life forms, and the evolution of one species to *another*? It is one thing to *suggest* species to species transitions in a few simple words, but it is another thing altogether to define the possible mechanisms and probabilities fairly and squarely...and to also provide *unequivocal* evidence. So, is natural selection some wonderful power that is different from chance acting through random chemical activity?

### *Can natural selection produce new creatures?*

Certainly, it is true that randomly-moving water molecules will produce ordered crystals of ice when the temperature is reduced, and that random atmospheric conditions do produce such phenomena as circumscribed whirlwinds and the Northern Lights. But even the most ardent proponent of evolution will agree that it is quite absurd to think that random activity by itself could produce even the simplest bacterium.
It is true that natural selection can operate as a sifting process on living organisms

by way of the effects of environmental changes or predatory activity. The peppered moth (see later) is a good example of this. In the macroscopic world, there are real birds which have real predatory activity which results in the most camouflaged moths surviving in the greatest numbers.

At the molecular level, chemical laws dictate what sorts of bonds will occur in a given circumstance. Obviously, in any particular situation those chemical bonds which are most stable will persist. Furthermore, the stability of any bond will be determined by other random factors such as temperature, and the passing by of any other atom or molecule that might act as a 'predator' to destroy the bond.

Furthermore, every *possibility* has a different *probability*. Some events are so improbable that their occurence must be regarded, for practical purposes, as *impossible*. For example, the spontaneous generation of a space shuttle from a junk-pile. Conceivably, perhaps a jelly-like substance could be formed by chance - but the question is, could a jelly*fish* be formed by chance?

So, whilst we can observe natural selection operating in living organisms with pre-existing genes, we have no sure evidence that natural selection can say anything about the *origin* of ... *anything*. Indeed, Darwin himself wrote that natural selection is a power which:

> ...acts **solely** through the preservation of variations in some way advantageous...I have **nothing** to do with the **origin** of...life itself. We are concerned **only** with the diversities... (My bold). Charles Darwin. 'The origin of species'. Penguin Books. 1985. p153,234.

But it is still claimed that:

> The difference in fitness among phenotypes is the difference that is not due to chance, but is caused by some characteristic difference between them. Therefore natural selection is the difference in rates of increase among biological entities that is not due to chance. **Natural selection is the *antithesis* of chance**. (My bold). Douglas Futuyma. 'Evolutionary biology'. Sinauer Associates Inc. 1998. p350. (By 'phenotype' is meant the outward characteristics of an organism. In contrast, 'genotype' refers to the genetic constitution of an organism.)

But any difference in fitness can only first arise by *chance* alterations in genes that code for a particular characteristic. And the origin itself of any gene must ultimately lie in the chance interactions of some primordial molecules. Furthermore, even the environment itself is caused by arbitrary and random physical and chemical activity. Thus the *occurrence* of an environment is a chance event in relation to an animal which is itself (supposedly) the result of chance chemical activity. And in relation to that animal any predators that might appear will do so by chance. Therefore, all the involved variables - the internal make-up of an animal and the environmental conditions surrounding it - are associated simply by chance chemical and physical activity. Thus, ultimately *natural selection must simply be the outworkings of pure chance*. Therefore, natural selection can have no inherent creative ability that would make it in some mysterious way, *superior* to pure chance.

Andrew Dickinson

Fried egg jellyfish (Cotyleriza Tuberculata). This is about 30cm in diameter.

Mark Campbell

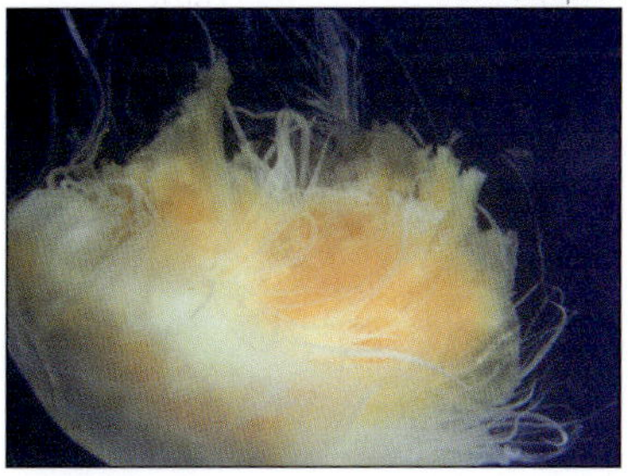

Jellyfish

Jelly-like substance

The polar bear (above top) and brown bear (centre) are varieties of the same creature. The black bear (above) is also found in a white form, the Kermode bear.

**Polar bears are pale brown bears**

In 2006, a 6-year-old white-furred bear was shot on Banks Island in northern Canada. It was obviously the result of the mating of a polar bear and brown bear since it had a shoulder hump and a brown bear-type head. It also had dark rings around the eyes, and instead of a single hind foot pad, this individual's foot pad was divided by a band of fur.

In 1874, in Germany, a polar bear and brown bear mated in captivity and a healthy bear resulted which, incidentally, had no dark rings around the eyes.

The polar bear below, currently in Berlin Zoo, has pale eye rings.

But what then should we make of the very many living examples of evolution quoted in the literature? Before we can proceed, we have to define what exactly a *species* is. In the past, organisms have been classified according to the similarity or dissimilarity of external or internal structural appearances and *in general*, such a method of classification has some value in helping us to distinguish between any two creatures. For example, an elephant looks significantly different from a horse, and in this case the outward appearances correctly reflect significant differences in the anatomy, the genes, and their inability to interbreed. So it makes good sense to say that elephants and horses are different creatures.

However, we know that appearances can be very deceptive, and many creatures previously classified as separate species may represent no more than different degrees of variation of the same animal. For example, the *grizzly bear* and the *European brown bear* are the same animal. And *polar bears* and brown bears are able to interbreed successfully:

> ...in captivity they have mated and produced fertile young... BBC Wildlife magazine. December 1997.

Furthermore, the so-called 'ABC' islands (Admiralty, Baranof and Chicagof) of south-east Alaska are home to a unique population of brown bears. Mitochondrial DNA analysis of these ABC brown bears have shown them to be genetically closer to polar bears than to other brown bears found on the North American mainland:

> ...ABC bears' closest relatives are not other brown bears, but polar bears... Kenny Taylor. 'Discoveries'. BBC Wildlife magazine. March 1999. p59.

Therefore, since they are so similar both in appearance and genetic constitution, to call polar bears and brown bears different species is an artificial and deceptive distinction. In this case it is perhaps more correct to call them off-white and brown varieties of the same bear. In much the same way, the white *Kermode* bear found on Princess Royal Island in British Columbia, Canada, is not an albino but is really a white, *black* bear. There are also, in fact, individual Kermode bears which have coats with a variety of different hues, and black bears, whether brown or white, may have more genes in common than do Great Danes and Pekinese dogs.

It has previously been thought that the *black panther* was a different species to the *leopard.* In fact, the black panther is merely a black variety of leopard and although they are faint, nevertheless the typical broken-ringed black spots of the leopard can be seen all over the black panther's dark coat.

The *slender rasbora* (*Rasbora daniconius*) is a small cyprinid fish from the rivers of South-East Asia and is:

> ...very similar in appearance to *Rasbora einthoveni*... Dick Mills. 'Aquarium fish'. Dorling Kindersley. 1993. p64-5.

In fact these two fish differ only in some slight colour variation. Otherwise they are virtually identical. Similarly, the *great snipe* (*Gallinago media*) of Northern Europe is:

> Slightly larger and stockier than *Snipe* (*Gallinago gallinago*)...with thinner shorter beak. Dr Walter Cerney. 'A field guide in colour to birds'. Octopus Books. 1983. p96-7.

We can go on to compare many other birds but some gulls, woodpeckers and warblers are good examples of (supposedly) different species which may not be different creatures at all:

> The *Mediterranean Gull*...about the size of *Black-headed Gull*...juvenile plumage similar to Black-headed Gull's...Nest and care of brood as Black-headed Gull...
> The *Syrian Woodpecker*...Closely resembles the *Greater Spotted Woodpecker*...
> *Firecrest*...Like *Goldcrest*, but more brightly coloured...Voice resembles goldcrest's... Dr Walter Cerney. 'A field guide in colour to birds'. Octopus Books. 1983. p120,152,192.

The point here is that since we know it is true that great variety can exist within a particular type of creature, it would be logical to say that we are only at liberty to class any two creatures as different species if the chromosomes and genes of both are known to be significantly different. Otherwise the classification simply cannot be upheld. However, even knowing this is not entirely satisfactory in our quest for describing a species since individuals with abnormal numbers of chromosomes do not necessarily constitute different creatures. For example, people with different numbers of chromosomes (such as in *Down's syndrome* with a chromosome number of 47 instead of the normal 46, or in *Turner's syndrome* with a chromosome number of 45) are still clearly human beings. Furthermore, many diverse creatures such as mole rats, house mice, vervet monkeys and wallabies are known to vary in their chromosome number and yet they all remain very much the same type of animal. Despite the great variation that individuals within a species can exhibit, the notion of a *ring species* is often quoted as evidence to support the idea that starting with a single population of a particular species, some individuals can gradually acquire differences which lead ultimately to their inability to interbreed with the rest of the population. The two non-interbreeding populations would then become two species. In this theory, it is supposed that, initially, individuals from within the two new species would still be very similar, but with time the individuals in subsequent generations could ultimately comprise *completely different creatures*:

> The best-known case is *herring gull* versus *lesser black-backed gull*. **In Britain these are clearly distinct species**, quite different in colour. Anybody can tell them apart. But if you follow the population of herring gulls westwards around the North Pole to North America, then via Alaska across Siberia and back to Europe again, you will notice a curious fact. The 'herring gulls' gradually become less and less like herring gulls and more and more like lesser black-backed gulls until it turns out that our European lesser black-backed gulls actually are the other end of a ring that started out as herring gulls. At every stage around the ring, the birds are sufficiently similar to their neighbours to interbreed with them. Until,

Great spotted woodpecker

Polar bear skull (above) and claws (below).

Black varieties of jaguar, leopard and tiger are all found in the wild but are otherwise the same as their normal-coloured relatives.
Above: A black baby jaguar with its spotted mother.
Below: A normal-coloured leopard.

Tim Croad

Herring gull (above top) and lesser black- backed gull (above).

> that is, the ends of the continuum are reached, in Europe. At this point the herring gull and the lesser black-backed gull never interbreed, although they are linked by a continuous series of interbreeding colleagues all the way round the world...The only thing that is special about ring species like these gulls is that the intermediates are still alive. **All pairs of related species are potentially ring species**. The intermediates **must have** lived once. It is just that in most cases they are now dead. (My bold). Richard Dawkins. "Gaps in the mind". In 'The great ape project: Equality beyond humanity'. Paola Cavaleri and Peter Singer (Editors). St Martins Press. 1993.

Such a view can have far-reaching conclusions:

> So if we cannot interbreed with chimpanzees, or with half-human, half-chimpanzee beings who can interbreed with chimpanzees, this **is** merely due to the deaths of the intermediate types. (My bold). Peter Singer. 'Rethinking life and death'. Oxford University Press. 1995. p179.

But the problem is that Dawkins *assumes* that the herring gull and the lesser black-backed gull are indeed two different species, and Singer, following the standard evolutionary theory, *assumes* that humans have evolved from apes. If these two assumptions are false, then Dawkin's views on the formation of new species are false; and Singer's moral extrapolation of evolutionary thinking so as to consider that even in late pregnancy a human foetus is not a *person* (on the basis that as far as we can tell, it is not a conscious being which is able to think, show emotions, have a memory, know self-awareness, or communicate with language), is also false. For example, since:

> ...there is no reason to think that a fish suffers less when dying in a net than a fetus suffers during an abortion, the argument for not eating fish is much stronger than the argument against abortion that can be derived from the possible consciousness of the fetus after ten weeks...Those who oppose late abortions on the grounds of fetal distress will need to be ethical vegetarians if their position is to be consistent and non-speciesist. Peter Singer. 'Rethinking life and death'. Oxford University Press. 1995. p209-210.

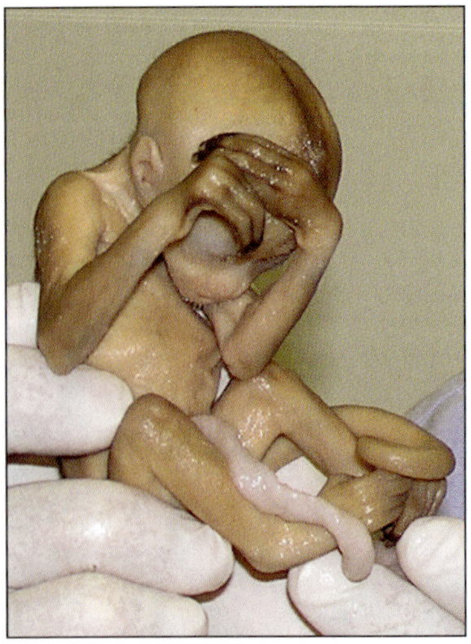

Spider monkey foetus

It will by now be clear to the reader that whether or not human beings are simply one particularly clever variety of ape is a fundamental and essential issue to resolve. Let's spell out the problem: if humans are only evolved apes, then when dealing with morals we should compare and give due consideration and respect, like with like. So, we should perhaps compare an early human foetus to a worm or frog, and a more advanced human foetus with a hibernating beaver. If, on the other hand, humans are *not* just evolved apes then the life of a normal human foetus, which has less brain activity than a baby chimpanzee, might nevertheless and *without exception* be considered more important than that of the chimpanzee. So, before proceeding any further, we need to establish firmly the validity or otherwise of the ring species concept.

Are the herring and lesser black-backed gulls clearly distinct *species*? Before this question can be answered we have to understand bird behaviour. Birds are often

Sarah McAlpine

This starling from East Africa is the same size and shape as its European relatives, but is far more colourful.

highly coloured, or have distinctive patterns, and display many innate and rigid responses to visual cues. For example, the plain-breasted juvenile male robin does not attack a red-breasted adult. However, red-breasted adult male robins defend their territories from other red-breasted males but leave juvenile males alone. Since the adult male robin does not know that it has a red breast but attacks birds (or even fluffy balls) that have the same colouring as itself, its territorial behaviour must somehow be linked genetically to its own appearance. It follows then that other bird behaviours may also be linked inextricably to other features: for example, a bird might only be attracted to and only breed with another individual which shares to some degree its own characteristics, be it colour, pattern or mating call. If the shared characteristics fall below a certain minimum, the mating may be prevented.

It has been shown that, despite some superficial differences, there is no genetic justification for classifying human beings into different *races*. We should also be very careful to apply extreme caution when considering any animal. In fact, there is little evidence to show that the lesser black-backed gull is a different animal to the herring gull. For example:

> Lesser black-backed Gull...Size of Herring Gull...Juvenile...Hard to distinguish from young Herring Gull...Downy plumage like Herring Gull's...Voice: like Herring Gull's... Dr Walter Cerney. 'A field guide in colour to birds'. Octopus Books. 1983. p118.

So, since both herring and lesser black-backed gulls *are* able to mate with other gulls with intermediate features, it is likely that herring and lesser black-backed gulls have such marked differences in appearance that *they are prevented from recognising and being attracted to each other as potential mates*. The recognition of and attraction to potential mates may simply be the result of the imprinting in their brains of their parents' appearances shortly after their eyes first open. Clearly, if there is minimal genetic difference between herring and lesser black-backed gulls, and if they are able to produce fertile young, they should be classified simply as different *varieties* of the same gull. Therefore, if herring and lesser black-backed gulls and the intermediate forms all still represent the same *creature*, then the concept of ring species is fallacious.

And in fact, this fallacy is supported by the plains zebra, the range of which extends from Kenya to South Africa, and which effectively provides us with an example of a 'ring species' which does not form a ring. The stripes of the plains zebra vary:

> ...across its range. In the north, it is most completely and boldly striped... farther south, plains zebra are less boldly striped...with fewer stripes on the belly, hindquarters and legs...the extinct *quagga* had stripes only on its head, neck, and back. This was probably not a separate species, but a subspecies...at one extreme of its geographic range. Jane Wisbey. 'On the Grevy trail'. BBC Wildlife magazine. June 1997. p72.

The *quagga*, which became extinct towards the end of the 19th Century, was a zebra-like creature that had stripes restricted to the head and neck, and to the front half of the body. Despite the fact that in all its visible characteristics, including shape

The quagga (top left) and the plains zebra are the same creature.

Sarah McAlpine

of ears, mane, tail, body and limb form, the quagga was indistinguishable from a plains zebra, it was until recently thought to have been a separate creature to the zebra. However, when plains zebra are selectively bred, the offspring can show all the features of a quagga, thus showing that such marked differences in skin pattern can occur *within* a given type of creature, naturally, in the wild.

If the concept of ring species is fallacious, it follows that in relation to all other distinct creatures (including humans) we have no obligation to suppose that '*intermediates* ***must have*** *lived once*'. And once we can dispense with that obligation we are free to see the relationship of apes to human beings in a different light. In other words, we have no obligation to accept Singer's solid assertion that the difference between humans and chimpanzees '***is*** *merely due to the deaths of the intermediate types*'.

However, we will leave the discussion about the relationship of human beings to other animals until later, once we have examined fully the evidence for the notion that human beings have evolved from ape ancestors. For the present we must get back to the issue of variation and its relevance to speciation.

Clearly, brown bears, herring gulls and zebra can vary markedly in their outward appearance. But does that mean that all creatures that share similar appearances must have arisen from a common ancestor? Lions and tigers can be persuaded to interbreed in artificial circumstances. However, although they both have the same number of chromosomes (38), the resulting offspring (a *liger* when the father is a lion and a *tion* when the father is a tiger) are sterile. Ligers and tions are also more prone to have congenital deformities and abnormalities. But since the lion and tiger *seem* so similar, could it be that perhaps at some time in the past these big cats had a common ancestor?

Normally a population of healthy cats would interbreed and in so doing mix up the genes of the whole population. But consider a population of lions, which become separated into two populations by some geographical divide. Would such a scenario produce, first of all, two populations of lions that could no longer interbreed, and ultimately, one group of lions and a second group which would become tigers? The lions of Nakuru National Park in Kenya are isolated geographically from other populations of lions, and whilst they tend to show quite a marked spottedness, they do not tend to show any degree of striped-ness nor any particular tendency to lose their manes. It is well recognised that as they become geriatric some male lions can lose their manes, and that male lions can be mane-less as a result of a genetic abnormality. But it is unlikely that the genetically mane-less male lion would be so physically superior to a maned lion that it and its male offspring could keep a harem and so perpetuate mane-lessness in a local population. Considering the very short time that any single male lion is actually king of his harem and territory, it is much more likely that if a mane-less lion were king, and his harem taken over by a normal lion, the offspring of the new pride would contain normal maned lions again. Genetically unusual or abnormal forms are usually *recessive* in nature, and mane-lessness is likely to be in this category. Normal forms are almost always dominant. By this is meant that if an individual inherits one recessive gene and one dominant gene from its parents, the individual will show the dominant characteristic. In this case, a mane. Only when an individual inherits recessive genes from both its parents,

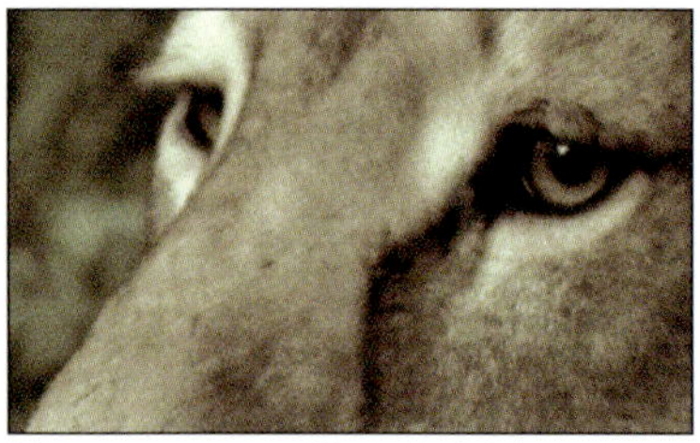

Lions are very active at night.

Tim Croad

Lions showing spots. Above: Lion from Nakuru (central Kenya). Inset: Lion from northern Tanzania.

Tiger

Leopard

Puma (mountain lion)

will the individual show the recessive (mane-lessness) characteristic. In the natural situation, it is likely that the characteristic of mane-lessness amongst male lions would not separate itself into a distinct breeding population for very long. Instead, if mane-lessness is associated with any significant weakness it would be lost in subsequent generations; if mane-lessness is not associated with any weakness, then it will simply appear sporadically in subsequent generations.

It is also notable that despite geographical separation for a long period of time, all African and Asiatic lions are still the same species, and neither one nor the other shows any divergent tendency to become more tiger-like. Furthermore, since there are many forest animals which have plain colouration, and since yellow and black stripes are no more effective as camouflage than a mid-brown coat, we should not expect a lion evolving into a tiger to lose its original colour. And whilst it is commonly thought that lions are diurnal beasts, in fact, like the tiger they are very active at night. Since in moonlight neither a striped nor a plain cat will be easily seen out in the open or in jungle, there is no clear advantage for natural selection to produce striped cats from lions. And, in fact, the finding in the Kerinci Seblat National Park in Sumatra of:

> ...tigers...in the black form. Lindsay Murray. 'Feline find'. BBC Wildlife magazine. June 1997. p58,

demonstrates that since the possession of stripes is not vital for a tiger to survive, neither would it be necessary for a (supposedly) evolving tiger ancestor to acquire them. Finally, since tigers (unlike lions) have a natural tendency to live solitary lives in jungle, we should expect that over the years this very isolation should itself have produced a number of different tiger-like populations scattered amongst the jungles of India and South-East Asia, each with coats varying from fully lion-like to fully tiger-like. But this has clearly not happened. So there is no evidence to show that the lions and tigers that we have today were at any time in the past anything other than lions and tigers. However, even against the evidence, if we suppose that like polar bears and brown bears, lions and tigers did indeed once have a common ancestor (which some think might have been something like the sabre-toothed *Smilodon*), then this would still not comprise evidence that such an ancestor itself had an ancestor that was, in some very distant past, a quite different type of creature - in other words, something that was not a big cat.

Above: Modern tiger skull.
Below: Smilodon, a sabre-toothed cat.

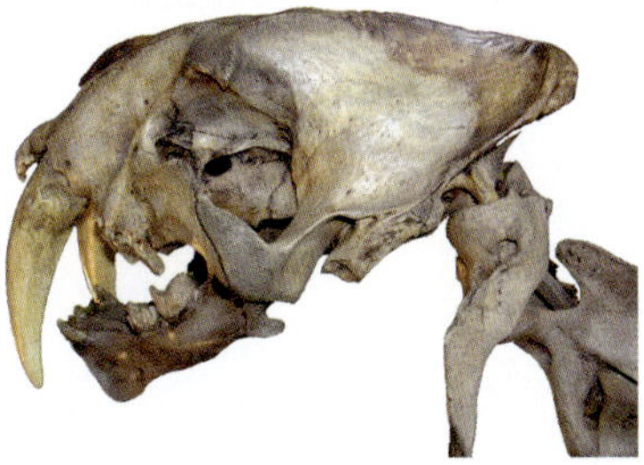

Leopard, Serengeti

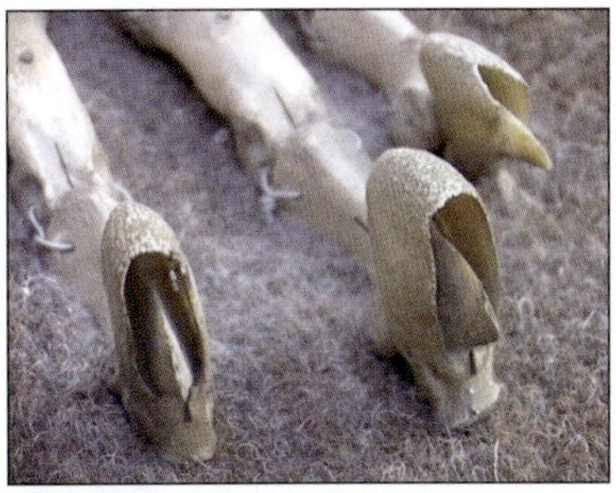

Tiger toes showing retractile claws

Feral kittens (Mi and Ow)

Except in rare circumstances, different breeds or races do not arise by the sudden appearance of *new* genes, but by the shuffling around and concentration of *existing* genes into certain groups of inbreeding individuals. However, thousands of years of selective breeding of domestic animals has not produced any new creatures - only a large number of variations. For example, selective breeding from the wild *rock dove* has resulted in a variety of different shapes and colours of pigeon, with a variety of different shapes of beak, but the common inner-city pigeon is still able to interbreed with the wild rock dove, and still shares the same voice and courtship displays. So, pigeons and rock doves still remain the same creature.
Similarly, thousands of years of selective breeding of ancestral wolves is likely to have resulted in the hundreds of distinctive breeds of domestic dog. As with dingoes in Australia, it is quite plausible that other wild dogs such as coyotes and jackals

Rock doves resting in a cliff

Wolf

Maned wolf

German Shepherd dog

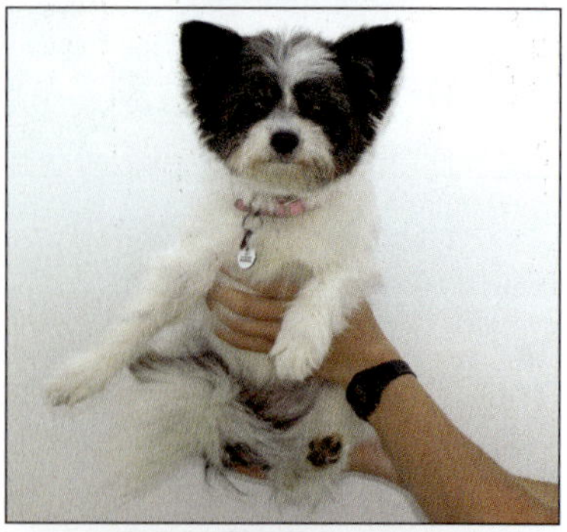

Elaine Dolamore

have also arisen from an ancestral wolf. However, while the German Shepherd dog still looks very similar to its wolf ancestor, the diminutive Pekinese could easily be mistaken for another creature altogether. Indeed, if all there was to go on were some fossilised bones of a Great Dane and a Pekinese, the chances are that they would be mounted in different corners of a room in a natural history museum. Widely different in their appearance they may be, but since there is no difference in their genetic constitution and since they can still interbreed (maybe with some help), the German Shepherd and the Pekinese are still dogs. It remains a fact that thousands of years of breeding has created no *new* genes - all that has happened is that the distribution of certain normal and pre-existing genes (most likely present even in the ancestral wolf) have been concentrated together in different breeds. Thus, if the small Jack Russell terrier and the large Rottweiler were artificially bred, the result would be a mongrel dog. Surprisingly though, in this example no artificial means need be applied, since a female Jack Russell and a male Rottweiler have been known to mate naturally (although the large puppy had to be delivered by caesarian section):

> ...a 12lb Jack Russell, a 133lb Rottweiler and how love led to the patter of not-so-tiny paws. Bill Mouland. Daily Mail. 24th January 1998. p3.

All dogs are descendants of an ancestral wolf. The German Shepherd (above left) closely resembles a wolf, while other breeds vary considerably in their facial proportions (left) and leg length (above). The dog shown above right has normal bodily proportions whereas the dog on the right has the same general shape and colouring but has abnormally short legs. Both the wolf and the Bijon-frise cross (below left) are the same animal.

Over the last one hundred years, the selective breeding of *fruit flies* involving countless thousands of generations has not altered the fruit fly at all. Certainly, there have been some weird morphological forms seen over the years, but they all remain fruit flies, *Drosophila melanogaster*. Nevertheless, it is still suggested that:

> **There are two species** of *Drosophila, D. heteroneura and D. silvestris,* that live on Hawaii. They are very closely related; for example if amino

Inbreeding is always prone to producing congenital diseases by allowing harmful recessive genes to manifest. For example, deafness is common in Dalmations, and Cavalier King Charles spaniels (left) tend to have *syringomyelia*, where swelling within the spinal cord leads to pain, loss of skin sensation, weakness or unsteadiness.

> acid sequence similarity of enzymes is estimated by electrophoresis, they appear to be among the most related pairs of species on the island. In spite of this they look very different, the head of the male heteroneura looking like that of a hammerhead shark. **Athough they rarely interbreed in the wild, they will do so in the lab, giving fertile offspring.** From a quantitative analysis of the progeny it appears that a rather small number of genes are responsible for the difference in head shape, probably about 10. (My bold). Peter Lawrence. 'The making of a fly.' Blackwell Scientific Publications. 1992. p194.

Clearly, these two flies are both fruit flies. The question is: are they different *species*? As with Pekinese and Great Danes, although the heads of these flies look markedly different, the differences are the result of only minor genetic changes. And, close examination confirms that *all the individual head structures are represented*, but *some* of the *proportions* are different. Indeed, such differences in proportion are also seen in some human chromosomal disorders, where the eyes can be set further apart with a broader nasal bridge than normal; the upper eyelids may have an abnormal curve; the ears can be smaller and low-set, etc. We would not classify markedly different humans as different species, and the same principle should be applied to these flies. In fact, as Lawrence submits, they *can* interbreed and yield normal progeny. As with different Hawaiian finches, the fact that they rarely interbreed in the wild most likely reflects simply some restriction in their reproductive behaviour towards each other. Perhaps, for example, the fly pheromones do not gel with each other. These flies then, are not two species, but are simply variations of the same animal.

So we must not confuse the shuffling of existing genes (variation) with the production of new organs or structures, where previously such organs or structures did not exist. Yet despite this, there are a large number of examples of *variations* which are commonly thought to be (and *taught* as) examples of evolution.

Alterations to genes (mutations) or the shuffling of existing genes can give rise to individuals that are just different, resulting in healthy variation. In contrast, when damaging mutations are inbred but are not fatal, this gives rise to abnormal variation. Across the board, however, we see overwhelming evidence which shows that creatures have always been of distinct kinds.

## Healthy variation

### Galapagos tortoises

The giant tortoises from the different islands of the Galapagos archipelago have such distinct features that from the shape of the shell alone, it is possible to identify the island on which an individual tortoise lives. For instance, one variety of tortoise has a very rounded shell and the animal largely feeds on ground vegetation. Another variety has a shell with a distinct lip at the front end that allows the animal to extend its neck vertically to reach foliage that would be impossible for the rounded-shelled variety to reach. Based on these observations Darwin's argument has been summarised as being essentially this:

> All individuals of the same species are not identical. In one clutch of eggs

Dog

Wolf

Barbarissa
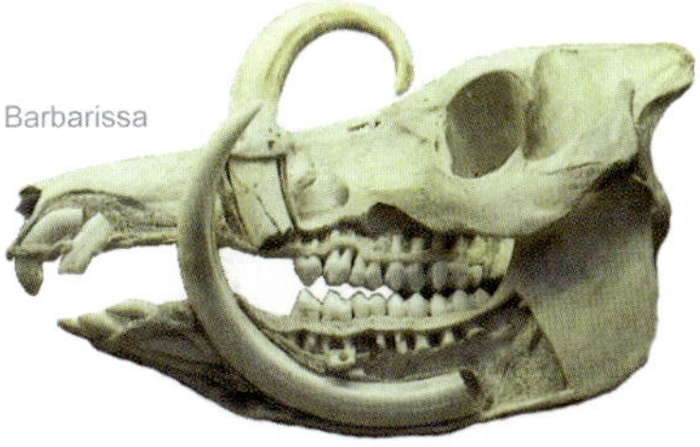

Wild boar, Denmark

Domestic pig, Yorkshire, UK
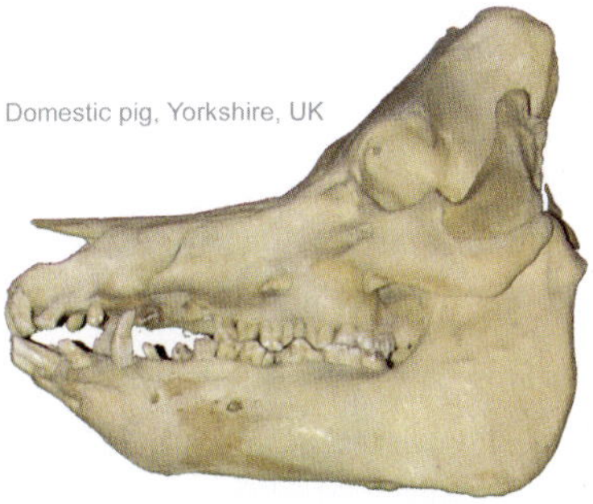

Domestic pig, Denmark
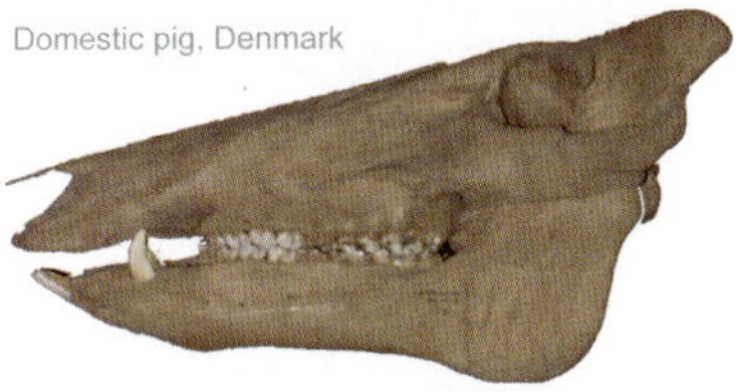

**Healthy variation**
All domestic pigs are derived from wild hogs and, as with dogs and cats, selective breeding results in a variety of skull and body shapes.

Galapagos Islands

from, for example, a giant tortoise, there will be some hatchlings which because of their genetic constitution, will develop longer necks than others. In times of drought they will be able to reach leaves and so survive. Their brothers and sisters, with shorter necks, will starve and die. So those best fitted to their surroundings will be selected and be able to transmit their characteristics to their offspring. After a great number of generations, tortoises on the arid islands will have longer necks than those on the watered islands. **And so one species will have given rise to another**. (My bold). David Attenborough. 'Life on Earth'. Collins/BBC. 1979. p13-14.

It all sounds scientific and quite plausible. But drought is unlikely to affect only one or two of the islands in the Galapagos *to the exclusion* of the others, since the archipelago occupies only a relatively small area (only 7,964 sq km of land spread in an area of ocean less than half the area of the U.K.). Furthermore, since in times of drought water is more important than food, the tortoises tend to seek out roots and underground water and survival is not dependent on long necks and access to foliage. In fact, the Galapagos tortoises can survive for very long periods without *any* food or water:

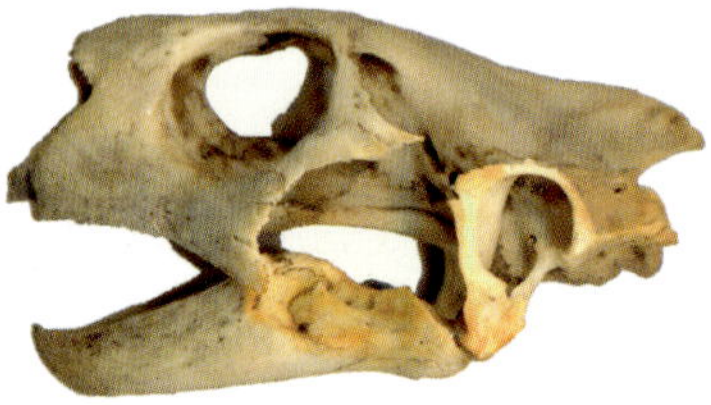
Hermann's tortoise skull

In 1815, Captain David Porter, USN., Commander of the frigate Essex, wrote: ....“Vessels on whaling voyages among these islands generally take on board two to three hundred of these animals and stow them in the hold, where, strange as it may appear, they have been known to live for a year, without food or water...” Kenneth Brower (Ed). 'Galapagos, the flow of wildness'. Vol 2. Sierra Club-Ballantine Books. 1968. p27.

The African leopard tortoise can grow to 70cm in length.

Furthermore, if the differences between the normal- and long-necked tortoises determined their survival simply by the ability or lack of it to reach for higher foliage, then we should find that, given enough time, all the tortoises living on all the more arid islands should be long-necked and all the tortoises living in well-watered islands should have normal necks. However, this is not what we find since there are perhaps fifteen distinct varieties scattered within the archipelago. So, it is likely

A male Hermann's tortoise bites a shoe as if it were trying to stimulate a female.

Giant tortoise with a thick, straight front edge to its shell.

that the observed differences in the tortoises from the different islands are the result simply of inbreeding.

But both Darwin and Attenborough actually argue that the tortoises have become different *species*. Darwin's knowledge was limited and his argument was based purely on outward appearances. With the benefit of modern genetics, Attenborough should be more cautious since to class the tortoises as different species requires at least a demonstrable difference in the chromosomes of the different types. However, as we have seen, even the inability *per se* of two tortoises to interbreed does not constitute evidence that the two animals are two distinct creatures. Actually, some varieties of Galapagos tortoises are in danger of becoming extinct, so there are plans to mate the rarer forms with other varieties that are considered to be closest to them. If succesful this would confirm that Galapagos tortoises are simply variations of the same creature, being no more distinct than greyhounds and terriers.

Giant tortoises from the different islands of the Galapagos vary in neck length and shell shape.

But how did the Galapagos giant tortoises get to be scattered about in the archipelago in the first place? Some suggest that perhaps the giant tortoise was carried to the Galapagos islands on some raft of material from the South American mainland across the 1050km of dangerous Pacific Ocean. If we suppose that, fortuitously, some tortoises *did* manage to get across the open ocean, then after thousands of years many individuals from many of the islands in the archipelago ought to have been successful in getting from one island to another. If that had happened, after so many years we should expect the Galapagos islands to be inhabited by 'mongrel' tortoises, all looking much the same. So the raft theory is rather unconvincing.

What explanation then, can we put in its place? It is possible that the tortoises walked between the islands on land bridges that have long since disappeared. The continental shelf west of Ecuador and Panama actually extends as far as, and beneath, the Galapagos islands. Suppose the continental shelf was once exposed land which was inhabited by giant tortoises. Then when sea levels subsequently rose, the giant tortoises living around the volcanic slopes of the archipelago-to-be would have been left stranded on their particular islands. The same scheme would hold for other creatures such as the marine iguana, and the flightless cormorant.

Short-necked Galapagos tortoises grazing.

The land bridge concept solves the mystery regarding the Galapagos, but the giant tortoise is also found in Madagascar and the Seychelles in the Indian Ocean – halfway round the world from the Galapagos. So how is it that the ancestral giant tortoise managed to get to such distant places on the globe? The answer could be that rather than being ancient mariners transported by driftwood rafts, the animals were living in the prehistoric super-continent of Gondwana before its break-up and were separated into isolated populations *as the land masses on which they lived, separated.*

Suppose that the giant tortoise lived in Africa and moved into the area representing ancient South America and Madagascar. Suppose too that the Seychelles, Mauritius, and other Indian Ocean islands broke off from Madagascar leaving their pockets of giant tortoises. Suppose again that after the land masses had separated, humanity spread into southern Africa and exterminated the African giant tortoises. Suppose that the South American tortoises were exterminated later on as men populated South America. That would leave the little isolated pockets of giant tortoises as we see them today. Of course, such a theory would mean that the world's giant tortoises

were alive before South America and Africa separated (supposedly) 135mya. If the variation that clearly exists amongst the Galapagos tortoises has arisen in a relatively short period of time, then it is incongruous that the Galapagos and the Indian Ocean tortoises are still remarkably similar. So, their very similarity points again to the possibility that Africa and South America did not separate millions of years ago.

### *The peppered moth*

The story of the peppered moth is one of the examples most often quoted as representing evolution in the making. The idea is that the industrial revolution forced the peppered moth to produce two almost distinct populations, with darker moths predominating in the smoke-polluted cities and the lighter moths predominating in the rural areas. This was because the lighter variety was more easily seen and eaten by predatory birds, and the smoke pollution in the cities simply favoured the survival of the better-camouflaged darker individuals.

Peppered moths from 1850 (left) and 1900 (right).

Certainly, the story of the peppered moth demonstrates natural selection inasmuch as the lighter moths declined in numbers, and the dark moths increased in numbers, but the important point is that the dark peppered moth was still a peppered moth. Although in normal circumstances the peppered moth exists in predominantly a lighter variety, darker-coloured moths do also exist, albeit in smaller numbers. So, within its chromosomes lie the genes that make up both light and dark forms, along with any intermediate forms. And in fact, since both light and dark forms of the peppered moth were well known to moth-fanciers *before* the industrial revolution, the dark moths *did not* evolve as a result of the pollution. In fact, all that had occurred was a differential in survival, with a redistribution of genes that *already existed.* So, the dark peppered moth was not a *new* species with some new (previously non-existing) genes, but it was just a *dark* peppered moth. And when the atmospheric pollution receded with the introduction of clean air policies, the darker moths became more conspicuous to predators and the lighter-coloured moths reappeared in increasing numbers. These lighter-coloured moths were not mutations arising from the population of the then predominantly darker moths, but were the descendants of the lighter-coloured individuals that had survived through all those years of pollution. Thus the predominant colour of the moths was determined simply as a result of interbreeding amongst the moths that survived in the greatest numbers.

With clear air, it was the lighter variety; with polluted air, the darker variety. No new animal or characteristic or structure had been formed. Nevertheless, there is still the view that:

> If pollution had continued...the rural moth population would have become entirely light and the industrial entirely dark. Then each population would be subject to somewhat different selective pressures because the two environments vary. In time, the dark and light populations would differ by groups of genes, with each group advantageous locally. The moth populations might eventually become incapable of interbreeding. At this point, natural selection would have caused them to diverge from a common ancestor **into two species**. (My bold). 'Evolution'. Microsoft Encarta. 1995.

We must certainly accept that given enough time and the inevitable mutations that must consequently occur to any existing genes, the separation of two populations of the same animal must eventually result in genotype differences between the two populations. However, we have seen that *even if* the genetic differences render two populations incapable of interbreeding, and the populations have distinctly different outward appearances, this does not oblige us to label the two populations as different creatures. Nor does this constitute evidence that forces us to extrapolate that by the same process and after a very long time, a *completely different* creature might result: for example, that some day one of the two populations of moths might give rise to wasps.

**Peppered moth revisted**

The story of the peppered moth has come under severe criticism, since moths come out at night and do not normally rest on tree trunks during the day. The original research involved *placing* laboratory moths onto tree trunks. This methodological flaw was rectified by experiments in which moths were released over a number of years. Some did indeed come to rest on tree trunks and birds were observed to eat them. Not surprisingly, on dark bark, birds can more easily see light-coloured moths, and vice versa, and so it is claimed:

> *'It is the most simple to understand, visually attractive story of **evolution in action**,' says Michael Majerus, a geneticist at the University of Cambridge.* (My bold)
> J de Roode. 'Reclaiming the peppered moth for science'. New Scientist. Vol 2633. 8th December 2007.

Not so. They were moths at the beginning of the time of observation, and they were moths at the end. The peppered moth story remains simply an example of camouflage and natural predation affecting the survival of subsections of a population of moths, and has no bearing on how one organ or creature could change into a completely different organ or creature.

### *Cichlid fishes*

Cichlids are a large group of fishes which have characteristic spines on their dorsal and anal fins and which share many features with the common perch. The cichlid fishes of the lakes of the African Rift Valley (such as Lake Victoria and Lake Malawi) are often quoted in discussions on the rapidity with which species can evolve. It is thought that the Rift Valley cichlids have evolved in a very short period of time. For example:

> ...many of the endemic species in Victoria apparently evolved in...13,500 years... Curt Stager. 'Africa's Great Rift'. National Geographic. May 1990. p36.

It is commonly presumed that the various different forms of cichlid fishes in the Rift Valley lakes are derived from an ancestral form which is found in nearby rivers.
At this point it is well to understand that apart from size the cichlid fishes are all virtually identical from the head backwards, and it is only in the mouth and face that they really show any great differences. It is suggested that the ancestral cichlids found their way into the early Rift Valley lakes, and being confronted with few competitors or predators, were free to diversify into a variety of species. This is thought to have happened:

> ...by their budding off of distinctive new forms, which would, in turn, have budded off others. The result would be exactly what we see today:

Lake Malawi cichlids.
Top: Labeotropheus trewavasae.
Middle: Lethrinops fucicauda.
Bottom: Pseudotropheus zebra.

> a mixture of generalised ancestral forms and the specialised descendant forms...adaptive radiation typically follows the opening up of environmental space. Goldsmith. 'The new evolutionary timetable'. p119.

So *why* should the formation of a new lake system allow the formation of a number of new species? Whilst it is plausible that different groups arising from the original ancestral stock of Rift Valley cichlids have had the opportunity of finding and staying within diverse environmental nooks and crannies, nevertheless, the separation of different small groups into different habitats within the lakes does not mean that the fishes are different species. And in their natural habitats, the tendency for the different groups not to interbreed does not mean that these fishes have an *inability* to interbreed.

So the question arises: are the different forms of cichlids in the Rift Valley lakes different species or just different inbreeding populations? Suppose that, within a population of apes, some were born with mutations that resulted in a complete failure of their arms to develop, and they could only feed by biting off leaves and fruit from bushes. It would be probable that in order to survive, those unfortunate animals would move into and stay within areas that had plenty of bushes. Meanwhile, any normal apes would be able to use their hands to forage for diverse foods in all manner of places. There would then result a certain segregation, and if the normal apes found the armless apes sexually unattractive, this might result in a (behaviour-induced) reproductive barrier, producing two distinct populations of apes. However, they *could* still interbreed, and they would all still be the same creature.

We have seen that the way animals behave amongst their own kind is very important when it comes to how they reproduce. This also applies to cichlids. Many cichlids are very brightly coloured and their instinctive tendencies would most likely cause fishes of a certain colouring to be attracted to other individuals with similar colours and patterns. Furthermore, many cichlids (from both the Americas and Africa) are extremely aggressive and territorial. For example:

> *Pseudotropheus zebra* (Malawi): ...an exceptionally belligerent fish.
> *Pseudotropheus tropherops* (Malawi): ...this fish is aggressive to its own kind.
> *Melanochromis auratus* (Malawi): Belligerent and territorial...
> *Labeotropheus trewavasae* (Malawi): ...very belligerent and territorial fish.
> *Herichthys cyanoguttatum* (Texas & Mexico): These territorial fish are quarrelsome amongst themselves and others...
> *Paratheraps synspillum* (South America): To its own kind it is very quarrelsome. Gina Sanford. 'An illustrated encyclopaedia of aquarium fish'. Apple Press. 1995. p110-122.

Such aggressive and territorial behaviour must have forced many cichlids to separate into particular parts of any lake, and this would have encouraged inbreeding. Nevertheless, to help explain how the cichlids evolved, Goldsmith goes on to report that 67 bairdiellas (a small marine fish native to the Gulf of California) were transplanted into the Salton Sea, California, in 1950/51. The first successful spawning produced serious abnormalities in 3-23% of the young. Some were blind, others had oddly-shaped heads, or had odd-looking jaws etc. However, by 1954 the abnormalities were only found in 2-3% of young. He then concludes:

> Let us imagine that a vacant body of water invaded by a small population of fishes happens to be somewhat larger and more complex than the Salton Sea. Let us further imagine that, because of divergent feeding preferences or other habits, different forms of fish within the population happen to favour different habitats within the large body of water. Possibly some small schools of closely related and therefore similar fishes become isolated by the flooding or drying up of certain areas. Such events...might lead to the fixation of certain unusual traits within distinct populations. Fixation would occur over the course of several generations, with natural selection guiding much of what happens from generation to generation. **Once distinctive populations became reproductively isolated from others they would represent new species.** (My bold). Goldsmith. 'The new evolutionary timetable'. p121.

We know that large numbers of progeny always result in a large variety of shapes and features. But variety in offspring does not equate with different species. In the first generation of translocated bardiellas, it would be predictable and natural to find a whole spectrum of odd forms, many of which would be feeble. And it is to be expected that after a few generations the feeblest types would have died out - either by being infertile or by being unable to attract mates and reproduce, or by being eaten by predators. It is likely, therefore, that the reshuffling of genes that already existed in the original ancestral group of cichlids has allowed a diverse number of forms to be realised. In exactly the same way the original wolf, from which the modern domestic dog has been bred, had within its genome *at the beginning* of the domestication process, *all* the genes that could make either a Chihuahua or a Great Dane, a German Shepherd or a Pekinese. Of course, unlike the cichlids of the Rift Valley lakes, the isolation of different breeds of dog has been artificially maintained. Nevertheless, any observer can see that the Pekinese and the Great Dane look more dissimilar than most, if not all, of the Rift Valley cichlids. So we can say that there is no evidence that any new or novel feature or structure has been formed in these fishes, and therefore, there is no reason to consider many of the different types of Rift Valley cichlids as different species.

The question remains: are the modern Rift Valley cichlids genetically so different that they cannot be persuaded to interbreed in a tank and remix their genes? Contrary to Goldsmith's enthusiasm for speciation to be *inevitable*, the fact is that the cichlids of the African lakes are far more similar than he would think:

> ...many of the Lake Malawi 'species' are probably just colour varieties, and seem to interbreed happily if none of their own type are available. Brian Ward. 'The aquarium fish survival manual'. MacDonald. 1985.

Indeed, when the DNA of the cichlids was compared:

Variation amongst koi carp. Large numbers of eggs always yield large numbers of progeny with a large variety of colour, shape and size.

Koi carp tooth

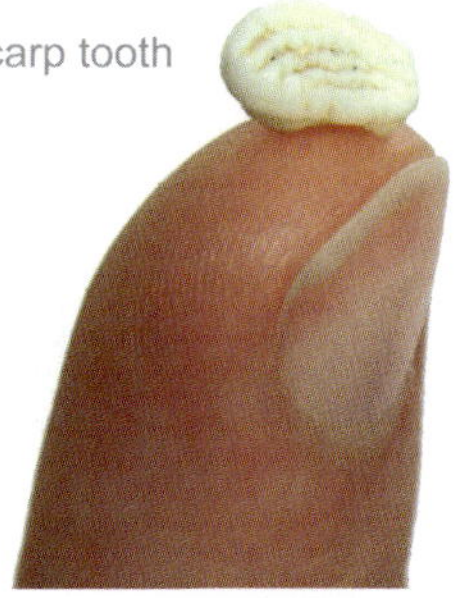

View from above

View from below

Koi carp have curious small molar-like teeth that are found in the back of the mouth.

Trout teeth.
In addition to teeth in the jaws, many fish also have teeth on the tongue.

> The genetic differences between the furu [a type of cichlid] species were extremely small, even smaller than the differences between certain human populations. Tijs Goldschmidt. 'Darwin's dreampond'. MIT Press. 1996. p65.

So, we should predict that when the visual and other cues that serve to prevent the mating between fishes of different appearances are removed, interbreeding should inexorably proceed. And this is exactly what happens. However, Goldschmidt goes on to explain that:

> It is striking that it is precisely the furu that develop new species so readily. The numbers of species among other vertebrates pale in the face of those of the furu. Even other groups of cichlids, such as the *tilapiass* are not species-rich. A living fossil such as the lungfish *(Protopterus aethiopicus)* is believed to have changed little since Lake Victoria came into existence. It has not even split into two species, even though it has spent as much time in fragmented lakes and pools as the cichlids. Tijs Goldschmidt. 'Darwin's dreampond'. MIT Press. 1996. p127.

Here Goldschmidt admits to incontrovertible evidence *that different animal types have different potentials as far as variation is concerned*. But he remains convinced that evolution has occurred. The problem is that early on in his book, when explaining the mechanism of the origin of species, Goldschmidt makes a fundamental error of reasoning, and unwittingly bases all his subsequent thoughts upon this flaw:

> The recipe is simple: allow a saucer to form in the earth's surface... let the saucer...fill up with river water. Ensure that several riverine cichlids, preferably the omnivorous kind, find their way into the virgin lake, and that's it: the rest will take care of itself. After, at most, several hundreds of thousands of years, anywhere between dozens and hundreds of kinds of colourful cichlid species will have developed, closely resembling each other in their basic features but outfitted with anatomical equipment and behaviour to allow them to process totally different kinds of food. Several millions of years later the radiation will have progressed even further: the diversity of shapes, colours, and behavior patterns will have become so great that any amateur could easily learn to distinguish one kind from another. **The differences will have become as distinct as those between a cow and a pig.** (My bold). Tijs Goldschmidt. 'Darwin's dreampond'. MIT Press. 1996. p7-8.

There is no doubt whatsoever that cichlid fishes can diversify. But, the error of reasoning is that we *cannot* simply extrapolate what has happened over thousands of years, so as to declare what is *inevitable* after millions of years. The fact that cichlids can segregate, breed and produce different-looking fishes (*but which have virtually identical genomes*) does not in any way allow us to conclude either that cichlids will eventually diversify enough so as to become as *distinct* as cows and pigs, or (as the theory of evolution insists) that some ancient fishes have *already become* cows and pigs. So, rather than the Rift Valley cichlid forms having evolved *to suit* different environmental niches or feeding options, the genetic variation itself, coupled with associated instincts, causes the different forms to feed in the manner that comes most naturally to them, and hence to separate and inbreed - thereby perpetuating distinct characteristics in different populations of the same creature.

### *Hawaiian honey-creepers and finches*

The Hawaiian Islands are thought to be only about 750,000 years old and are home to a number of different forms of honey-creepers:

> The keratin of the beak, like that of the reptilian scale, seems to be easily moulded by evolutionary pressures. Just how quickly it can is vividly shown by the honey-creepers of Hawaii. The ancestor of these birds was probably a sparrow-sized creature with a short, straight beak that lived in continental America. A few thousand years ago, a flock of them **must have** been carried out by a freak storm. They eventually reached the Hawaiian islands, and there found lush forests empty of other birds for the islands are volcanic and were formed comparatively recently. To exploit the many kinds of food now at their disposal, **they rapidly evolved** into different species, each specialised for a particular diet with the beak shape that was best suited to gather it. Some have short thick bills for seed-eating, others have hooked and powerful ones for tearing carrion. One **species** has a long curving bill for extracting nectar from lobelia blossoms; another has an upper mandible twice the length of the lower which it uses to hammer bark and lever it off in its search for weevils; yet another has crossed mandibles, a form apparently that enables it to extract insects from buds. Darwin had noted similar variations in the bills of the finches of the Galapagos Islands and regarded them as powerful evidence for his theory of natural selection. (My bold). David Attenborough. 'Life on Earth'. Collins/BBC. 1979. p178.

But, the two questions we must ask are: firstly, is it true that the different beaks shown by the Hawaiian honey-creepers are those that are 'best suited' to gather a particular food source, and secondly, are the different *forms* of Hawaiian honey-creeper actually different species? We have already seen that a bird might only breed with another individual which shares some of its own characteristics. And, in fact, the bill and mouth provide very important cues during courting rituals:

> A variety of birds touch bills during courtship, for instance, gulls, ravens, great crested grebes, and some finches. JZ Young. 'The life of vertebrates'. 3rd edition. Oxford University Press. 1994. p382.

Amongst the Galapagos finches (which do not have the bright colourings found in other finches):

> The recognition of individuals of the same species in this case is probably based on the characteristics of the beak; a male will begin to attack an intruder only when the face is seen. JZ Young. 'The life of vertebrates'. 3rd edition. Oxford University Press. 1994. p380.

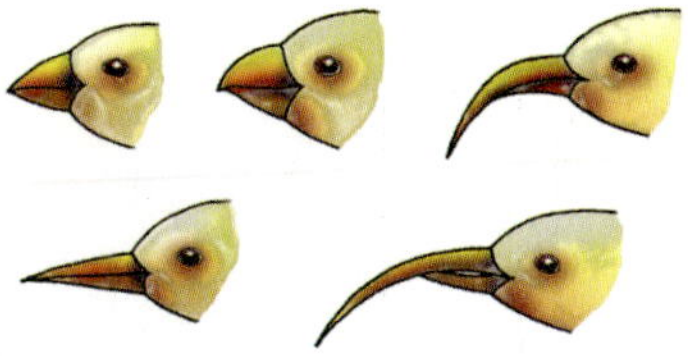

Hawaiian honey-creeper bill shapes.

Above: Galapagos finch.
Below: Hawaiian honey-creeper beak variation.

**Beak variation**

Beak size can be altered dramatically by the activity of calmodulin (CaM), a molecule involved in mediating signalling by calcium ions:

> *...calmodulin...is expressed at higher levels in the long and pointed beaks of cactus finches than in more robust beak types of other species. We validated this observation with in situ hybridizations. When this upregulation of the CaM-dependent pathway is artificially replicated in the chick frontonasal prominence, it causes an elongation of the upper beak, recapitulating the beak morphology of the cactus finches...our results implicate the CaM-dependent pathway in the developmental regulation of craniofacial skeletal structures. A Abzhanov et al. The calmodulin pathway and evolution of elongated beak morphology in Darwin's finches.* Nature. Vol 442. 3 August 2006. p 563-567.

Stork with crossed beak. Crossed bills are not some special design specific to a particular lifestyle. In fact, as with the abnormal beak of the pelican below, they are congenital abnormalities which are not so severe as to render the owner non-viable.

Therefore, it is clear that birds have no choice but to respond automatically to visual cues such as colour and shape. So, it is likely that cross-billed or long-beaked honey-creepers would only be attracted to and preferentially mate with other cross-billed or long-beaked honey-creepers respectively. Thus, rigid bird behaviours can explain how different isolated groups of finches might be found with different shapes and sizes of beak and different colouring. But such inbreeding does not mean, for example, that the cross-billed form is 'best suited' for *any* type of food at all - but simply that it *manages* to get enough food to survive, *despite its inbred congenital abnormality*.

Rather than birds evolving *in order to suit* their environmental niches, any differing types of bill that might result either by the process of normal variation or by means of a mutation of an existing gene, would *force* a bird to feed and behave in different ways. For example, any honey-creeper offspring that had longer than normal beaks would not be able to crack seed pods as well as a bird with a short strong beak (such as the *palila*). Therefore, those individuals with longer and more delicate beaks will be forced to find food elsewhere, for example, the nectar of obelids. Over a period of time the longer-beaked birds, having found and stayed in the vicinity of a suitable food source, would most likely interbreed only with similar longer-beaked birds in the same locality, and produce progeny with long beaks, such as the *akialoa*. So it is not that the akialoa is adapted to probing for insects and to getting nectar from obelids - it's just that that's all it *can* do. It certainly cannot easily crack seed pods so it has to either go after insects and nectar, or die.

Similarly, the beak of the *akiapolaau* has a lower mandible which is about one-third the length of the upper mandible. This odd feature could not have arisen *in order* for it to be able to hunt for insects:

> ...by hammering with its shorter lower mandible and extracting its prey with its longer, curved upper one. William Gray. 'Coral reefs and islands'. David & Charles. 1988. p148.

As with the *akialoa,* it's just that that's all the akiapolaau *can* do. It too can't crack seed pods so it has to go after another food source, or die.

What about the 'woodpecker' form? It is inconceivable that the wood-pecking genes could have arisen *de novo,* since wood-pecking must require a series of special features for protection of the brain against the trauma which would otherwise be caused by the repeated hammering of the bill. Such repeated jarring to the brain would cause damage to millions of brain cells just as it does in human boxers. As is well known, even the jarring caused by a boxer's head repeatedly *avoiding* punches can cause progressive brain damage. And many of us know what happens to a bird when it accidentally flies into a window pane: providing it has not broken its neck, the poor bird often falls to the ground *concussed*, and may take a while to recover before it can fly off.

So these *already perfect* 'wood-pecker' genes, clearly programming for features that cannot have any use except for wood-pecking, must have been present *within the original* honey-creeper genomes. And this is supported by the fact that pecking at hard objects is a common bird behaviour, so much so, that the potential for wood-pecking resides in a wide range of birds. For example, whilst at some times in the year *nuthatches* feed on seeds and nuts, at other times they regularly peck at bark in search for insects. And using their beaks, *jays*:

> ...can cut through even an acorn's armour without any difficulty. David Attenborough. 'The life of birds'. BBC1 Broadcast. 4th November 1998.

Furthermore, the genes coding for, on the one hand, the anatomical protection against the effects of hammering, and on the other hand, hammer bills and hammering behaviour, must occur *together* to be beneficial. Otherwise any bird which mutated so as to acquire (for example) only a hammering behaviour and no protection for its brain, would seriously damage its health. Clearly, such chance sequential appearances and associations would be highly improbable – as shown by the *WYWYWG Principle.*

Over time, it is natural for some variation to arise within the very chromosomes themselves. For example, changes might occur in the chromosome number. If that were the case, then the mating between finches with differing chromosome constitutions might not be able to produce viable progeny. This would further consolidate different populations into distinct breeding groups, and (because of the mutations that must inevitably appear) facilitate their chromosomes in becoming increasingly different.

However, if individuals from distinct breeding groups (either with or without different numbers of chromosomes) were to attempt to mate, then viable and fertile progeny might still occasionally result. However, in such cases, hybridisation simply mixes genes from different groups. Of course, this is rather like mixing the features of a rowing boat and a speedboat, both of which happen to have the same basic hull design. When the two boats are hybridised, the resulting new boat has the benefit of both oars and engine. However, in actual fact, no new type of boat has been produced by the union, and no *new* form of power unit has been manufactured. So, the process of hybridisation does not shed any light on how an oar and an engine could themselves come into being, or indeed, the hull itself. In the same way, neither variation nor hybridisation sheds any light on how a finch came into being. As we shall see when we come to look at chromosomes in more detail, shuffling genes around onto different locations on chromosomes, or changing the size and shape of chromosomes, does not result in the formation of new creatures. The Reeves and Indian Muntjac deer (which, although having identical appearances, have essentially 3 and 22 pairs of chromosomes respectively) provide a good example.

So we can see that what the Hawaiian honey-creepers demonstrate is not the ability of creatures to change *in order to* take advantage of any empty environmental niche, but the fact that mutants or variants must find food somewhere, or perish. Since those individuals with features that are quite inadequate will most certainly die, the Hawaiian honey-creepers, rather than demonstrating the survival of the fittest, actually demonstrate the survival of some of the not-so-fit. Thus, if the environment is not too unfavourable to its particular characteristic, any particular bird type will survive to mate, and so inbreed and perpetuate its characteristic. Such segregation of the initial population of honey-creepers into groups with differing bill sizes and shapes (forced either by innate mating behaviours or by the geographical location of a particular food source, and perhaps maintained to some extent by chromosomal variation preventing certain hybrids being viable) does not mean that the different forms are different *creatures*. The birds are all still honey-creepers and cross-breeding (either naturally or artificially) should once again result in 'mongrel' honey-creepers.

### *The giraffe neck, elephant trunk and stick insects*

The giraffe neck and elephant trunk are commonly put forward as examples of structures that arose from a short neck and a normal snout. There are three possibilities as to their origins: firstly, that the elongation of the neck or trunk was rapid, being the result of one or more mutations that in the embryo increased the rate of growth of the neck bones or soft tissues in the nose/lip respectively; secondly, that the elongation occurred more gradually by the selective inbreeding of the animals which (by virtue of the normal genetic variation found within any sizeable populations) had the longest necks and noses/lips; thirdly, that giraffes and elephants have always been giraffes and elephants.

#### *The giraffe neck*

Darwin himself was convinced he had found the answer to the origin of the giraffe:

> The giraffe, by its lofty stature, much elongated neck, forelegs, head and tongue, has its whole frame beautifully adapted for browsing on the higher branches of trees. It can thus obtain food beyond the reach of the other... hoofed animals inhabiting the same country; and this must be a great advantage to it during dearths... So under nature with the nascent giraffe, **the individuals which were the highest browsers and were able during dearths to reach even an inch or two above the others, will often have been preserved**; for they will have roamed over the whole country in search of food. That the individuals of the same species often differ slightly in the relative lengths of all their parts may be seen in many works of natural history, in which careful measurements are given. These slight proportional differences, due to the laws of growth and variation, are not of the slightest use or importance to most species. But it will have been otherwise with the nascent giraffe, considering its probable habits of life; for those individuals which had some one part or several parts of their bodies rather more elongated than usual, would generally have survived. These will have intercrossed and left offspring, either inheriting the same bodily peculiarities, or with a tendency to vary again in the same manner; whilst the individuals, less favored in the same respects, will have been the most liable to perish...By this process long continued...it seems to me almost certain that an ordinary hoofed quadruped might be converted into a giraffe. (My bold). Charles Darwin. 'On the origin of species'. Modern library edition. Random House. 1993. p227-8.

Could the giraffe have evolved from a shorter, stockier ancestor?

It seems that even today Darwin's logic remains fully convincing to some:

> There is still no better explanation for the giraffe's long neck than this one'. Tijs Goldschmidt. 'Darwin's dreampond'. MIT Press. 1996. p76.

And perhaps most people would offer this same explanation when asked about the possible origin of the giraffe neck. However, Darwin's surmising is based on a number of false assumptions.

Firstly, it is assumed that in times of drought it is *food* and the ability to browse high up which is the main deciding factor in survival. In fact, it is *water* which is the deciding factor. Humans and many animals can survive without food for many weeks, but for most large mammals *in the heat of a drought*, survival without water for longer than one week is most improbable. A large giraffe would need to drink more water than a small one, and since a taller giraffe has a greater surface area relative to its mass than one slightly shorter, the taller animal might be likely to dehydrate more readily. However, it is likely that any size differences are of no practical relevance since if a drought extends over a long time, *all* animals, irrespective of age and sex and height, will die off without partiality. Since the largest individuals in all probability would need the most food and water, in times of drought the *hardiest* animals may not necessarily be the ones with the longest necks. Therefore, when food and water are in short supply, some smaller and shorter individuals might fare better than the larger and taller individuals and hence survive to produce the next generation - of *shorter* animals.

But things are not so simple since hardiness is governed by many factors (not just height and size), so more likely, individual giraffes of *varying* height and size would survive. Hence, interbreeding in the next mating season would once again give rise to a spectrum of shapes and sizes.

Secondly, it is assumed that repeated droughts over long periods of time have provided the opportunity for the natural selection of the tallest 'hoofed quadrupeds'. However, even if we suppose that food *is* the most important factor in surviving a drought, and the tallest animals *are* in fact the hardiest, there is good evidence that droughts in times past are unlikely to have played any part in the origin of the giraffe. In Laetoli, Tanzania, there are some excellently preserved footprints dated as being (supposedly) 3.5myo, belonging to a number of animals including giraffe. From the shape and depth of the footprints there is no suggestion that the giraffes living at that time differed in weight, height or stride length from those living today. Furthermore, since many other animals (such as guinea fowl, hare, baboon, gazelle) and plants (such as whistling thorns) can be identified as being *the same as those living in the same area now,* it is likely that the central African habitat and climate have not changed significantly in the last (supposedly) 3.5 million years. Most certainly, the footprints show conclusively that whatever droughts there have been over this immense time period, they have not had any effect on the size of giraffes or other savannah animals.

Normally a giraffe gets a significant amount of its water needs from foliage. However, in the dry season it has to drink around 60 litres each day.

Thirdly, many shorter animals live side-by-side with the giraffe *in the same terrain*. In addition, female giraffes are generally 30-60cm shorter than the males. So if food had been scarce for any considerable length of time, and if height were the most important factor influencing survival, *all* the shorter tree browsers, along with

Giraffes, Kenya

Giraffe

the young and female proto-giraffes, would have died out before the tallest male proto-giraffes. Therefore, if only the tallest males survived, there would have been no females to mate with. It follows that if females could have survived then so could shorter adult males. Since height is not the only factor that would determine which males would become dominant and hence breed, any subsequent interbreeding would give rise to animals with varieties of heights. Furthermore, since all the other browsers that have lived side-by-side with giraffes would also have been subject to the same climatic changes we should expect all these smaller animals to have evolved a long neck too. And if long necks, why not longer tongues and lips to grasp leaves that would otherwise be that little bit too far away to reach? Of course, if the neck elongated over a long period of time there should have been many transitional forms of giraffes. But no fossils of such animals exist. Since distinct breeding groups of giraffes are found widespread in Africa, even separated by thousands of miles, there is no reason to suppose that *all* previously evolved shorter-necked proto-giraffes should have died out. Thus, if Darwin's speculation is valid, we should have found fossil evidence of extinct intermediaries, and in addition, short-necked and short giraffes should still be found today, scattered in different parts of Africa.

Fourthly, it is assumed that the giraffe is just a tall spotted antelope. Certainly, for any particular animal type, selective breeding of only the tallest individuals results in taller offspring, but there is no evidence to show that lengthening the neck and limbs in an animal would per se result in a *different* animal. (For example, seven-foot tall basketball players are still 100% human beings.) Whilst it might be possible for shuffling and mutations of *existing* genes to produce a longer neck, this does not mean that an ancient giraffe ancestor might have resembled a short-necked antelope. If, for example, there existed only impalas and no giraffes we should expect Darwin's notion of repeated droughts and natural selection to produce simply long-necked impalas which would be more capable of withstanding a water shortage.

We have seen that gradual changes are implausible explanations for the origin of the giraffe neck. But what about more rapid increases in neck length? Certainly, single genetic changes can alter dramatically the development of different parts of the body. For example, in humans, the mutation that causes *achondroplasia* dwarfism results in individuals in whom the trunk and spine proportions are essentially normal, but the limb bones extremely short.

The *okapi* is a forest-dwelling animal that shares many features with a giraffe but in other respects it has the bodily proportions and size of a large antelope such as the *eland.* Although there is *no* fossil record of any okapi, nevertheless, what if the ancestors of giraffes were once like okapis? The okapi survives quite happily in its forest habitat, showing it does not need a longer neck to survive. If it moved out of the forest into the savannah, it would be more likely to survive by eating the same vegetation as large antelopes than by gaining fortuitous neck-elongating mutations or by inbreeding amongst the tallest of its kind.

Okapi. Until the early part of the 20th Century the okapi was unknown to anyone outside the Congo.

It might be supposed that in an animal like the okapi a mutation that simply allowed the neck bones to grow at a faster rate than the other spinal bones would result in a giraffe. However, whilst it is conceivable that a single genetic mutation might be able to cause a normal antelope to develop a very long neck, nevertheless, it would

still be the same antelope.
Furthermore, even if a mutation could increase rapidly the length of its neck, it would still be unlikely to have any effect on the length of its legs (the *gerunuk* from northern Kenya is a good example of a long-necked antelope but in the length of its legs it is no different from other antelopes), or to change its stripes to speckles.
The notion of an impala gaining a very long neck and becoming very tall is not a simple issue since, unless a number of protective anatomical and physiological mechanisms are present, having a *very* tall neck can itself cause serious health problems. Although, in humans, a systolic* blood pressure of 120mm Hg is sufficient to pump blood from the heart up to the brain, this is quite inadequate in a giraffe since its brain is about 2m above its heart. In fact, the blood pressure in the adult giraffe has to be as high as 260mm Hg at the level of its heart:

Despite its long neck, the other bodily proportions of the gerenuk are that of a typical antelope. This shows that neck length is not linked genetically to the giraffe body form. The gerunuk is found in the arid scrublands of northern Kenya.

> Measurements of blood pressures in the giraffe have shown a systolic blood pressure as high as 260mm Hg...This is more than twice as high as the normal systolic blood pressure in humans... Knut Schmidt-Nielson. 'Animal physiology'. 5th edition. Cambridge University Press. 1977. p109.

Without such a high pressure the giraffe would faint from lack of blood flow to the brain. However, whilst the high blood pressure ensures that blood reaches the brain when the animal is standing normally (with the pressure inside the head being around 100mm Hg), nevertheless when the giraffe lowers its head to have a drink, the great and sudden increase in the arterial pressure inside the head (to around 400mm Hg) could easily rupture blood vessels in the brain (causing a stroke) or in the retina (with a risk of blindness). In time, high blood pressure also causes other problems such as heart and kidney failure. So, the giraffe has special valves in the neck blood vessels and special sumps in the head which serve to protect it from massive changes in blood pressure as it lifts or lowers its head. Since, for example, the development of abnormal valves (resulting in congenital varicose veins or heart problems) can arise in otherwise healthy humans, the genes responsible for neck or limb size and vascular valve formation are likely to be independent of each other. So, there is no reason to suppose that any increase in neck length in a supposed proto-giraffe would be likely to produce simultaneously the necessary vascular sumps or valves.
Of course, the problem of the origin of the giraffe neck is shared by many other long-necked animals, the most dramatic example being the great sauropod dinosaur *Diplodocus*, which had a neck many times the length of a giraffe's. It used to be thought that sauropods either walked about submerged in rivers and lakes with just their heads above water, or that they walked on land with their heads held up high. However, by examination of the bones for evidence of how their muscles and ligaments must have been attached, it is now accepted that they carried their heads outstretched, about level with their shoulders. So any explanation for the long necks of sauropods based on the repeated drought theory becomes even less plausible, since they would have eaten leaves found at heights easily accessible to other large-but-not-so-long-necked dinosaurs, and so their long necks would have served as no particular height advantage. And as with the giraffe, so it is with diplodocus or the

Top: Swan with babies.
Above: Cygnets.
Clearly, swan chicks have short necks and tiny wings which enlarge greatly within a few weeks. However, it is also clear that neck length is controlled by different triggers to those which determine leg length.

* Immediately after the heart contracts, the pressure in the arterial system is maximal (the *systolic* pressure). In between heartbeats, the lowest pressure in the arterial system is called the *diastolic* pressure.

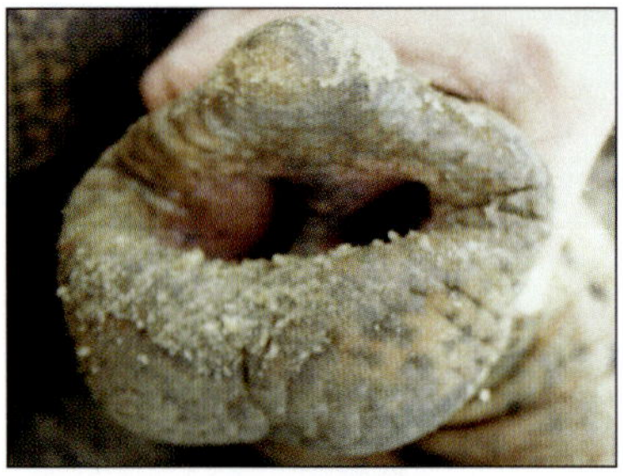

Nostrils of Indian elephant.

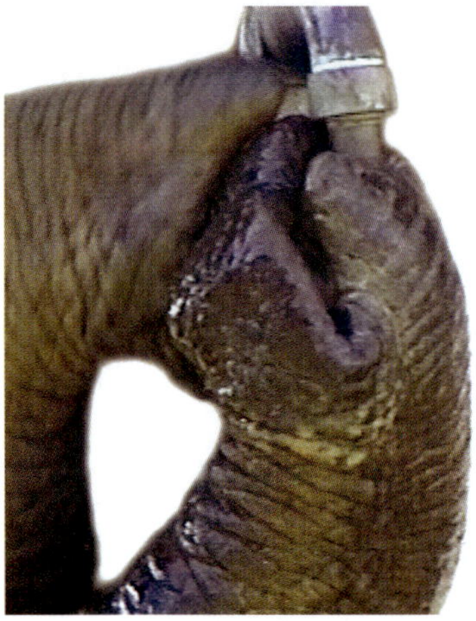

Two Asian elephants drinking from a tap.

The elephant shrew does not use its long snout to manipulate food. Instead it prods the ground with its tongue which is *longer* than its snout.

These African elephants are not using the tips of their trunks as fingers with which to grip food. The trunk on the right is simply trying to reach a branch. The trunk on the left has managed to loop itself over a branch and is now pulling the branch down.

other sauropods - there is no evidence of any ancestral short-necked form.
Thus the notion that the giraffe arose by the evolution of some short-necked ancestral hoofed quadruped is not supported by the evidence.

### *The elephant trunk*

The elephant's trunk is certainly an amazingly versatile structure, being at the same time, an arm and hand, a snorkel, a shower hose, a water bucket, a trumpet and a powerful means of defence or attack. Although the mammoths had trunks much like those of living elephants, some extinct types of elephant-like animals had a variety of different head, teeth and body forms: for example, the *Gomphotherium* had four tusks arising from both the upper and lower jaws, and the *Platybelodon* had spade-like tusks arising from the lower jaws only. However, the question is: is the trunk of living elephants the result of modifications to the snout and upper lip of ancestors which originally had conventional lips and nostrils, or which looked something like a tapir?

It might be thought that an animal with a nose of intermediate length between a 'normal' snout and the living elephant's trunk would be of no use, but it has been argued that, during evolution, perhaps noses of intermediate lengths had functions that were different from those now employed by the full-sized trunk.

There are a number of living animals with long noses. Perhaps with its longer nose the elephant shrew may have a more acute sense of smell than a house mouse, but dogs have an extremely good sense of smell, yet have a 'normal' snout. So, is this increased olfactory sensitivity in the elephant shrew due the extra amount of the soft tissues of its long nose? When we smell something, what happens is that airborne particles are carried into the roof of the nasal cavity where they lodge in the mucus. Extending into the mucus are fine hair-like processes arising from the *olfactory* nerve cells. When a molecule comes into contact with the olfactory nerve endings, electrical impulses are triggered which convey to the brain information about that particular particle. This signal is then perceived as a smell (note that by the time you smell something awful some distance in front of you, some particles from whatever it is that you are looking at are already inside your face). Since the olfactory nerve endings are situated in the roof of the nasal vault (below and between the eyes), increasing the size of the nasal vault to accommodate an increased number of nerve endings would certainly improve the sense of smell. However, simply and only increasing the length of the soft tissues around the nostrils can do little to improve the sensitivity of the sense of smell. So it is implausible that an elephant shrew or a (supposed) ancestral elephant would, by increasing the length of its nose, manage to significantly improve its sense of smell.

If they were not of any benefit in improving the sensation of smell, perhaps during evolution trunks of intermediate lengths had an advantage when it came to making sounds. Certainly, the male elephant seal produces loud bellows with which it fends off rivals and by which it may also attract females. But other large seals with normal snouts such as the walrus manage adequately to mate and defend territory. Furthermore, it is likely that the long nose is not the main mechanism by which a loud deep growl is produced, since lions and tigers manage their powerful roars with perfectly 'normal' snouts. In fact, resonating is accomplished mainly by the use of cavities inside the skull and body. An extreme example is that of the huge curved funnel-like resonating chambers at the top of the heads of some dinosaurs such as *Parasaurolophus*. Smaller examples of resonating chambers are provided by a number of birds. The importance of bony cavities in sound projection can be easily demonstrated by the change in voice caused by a common cold in which the sinuses fill up with catarrh. This is not to say that sound resonance cannot be accomplished by using soft tissues (a good example is the under-chin air bags of some gibbons), but a highly complex structure like a trunk is unnecessarily elaborate for the job. In terms of sound amplification, an increase in size by 10-20cm of the elephant seal's nose would add nothing to its ability to survive or mate. Similarly, an increase in length of the snout of 10-20cm would have been of no use to a proto-elephant.

So, whether for smell, or for vocal resonance, there is no particular advantage of a small or slightly bigger trunk over the 'normal' snout. Thus it is implausible that while they gradually elongated their noses, ancestral elephants might have had uses for their little trunks other than for manipulating food and drinking.

What, then, if the trunk has always had the same function throughout its (supposed) evolution? The question then arises that for use in manipulating food and sucking water, is there a clear stepwise path from a 'normal' snout to the living elephant's trunk, with each step demonstrating a (significant in terms of survival) benefit over its (supposed) shorter predecessor?

Taking platybelodon as an example, the standard explanation for the trunk origin goes something like this: the lower jaw and teeth grew outwards to produce a shovel with which the animal could dig up food. The nose and upper lip elongated to work against the shovel and later this fleshy extension increased in size so as to allow food to be conveyed to the mouth. The nascent trunk then elongated and became more efficient, the lower jaw (being now superfluous for food-gathering) became progressively shortened and the lower incisor teeth were lost altogether:

> ...the trunk, originally evolved to act in unison with the lower jaw, finally reached a threshold at which it was by itself an effective instrument for gathering food, and this was followed by a reversal in the direction of evolution of the lower jaw. John M. Smith. 'The theory of evolution'. Cambridge University Press. 1993. p293.

Since most of the time the tapir's prehensile nose plays no part in feeding, and other large animals such as the Malaysian rhino and the okapi manage without a long nose, a tapir without a protuberant nose would have no difficulty surviving in tropical forests. Furthermore, since tapirs have managed to live quite happily for a very long time (fossils have been found that are, supposedly, 50myo) and there are no tapirs with trunks, it follows that a nose 10cm or 20cm longer would give it no survival advantage. In other words, for survival in a forest, an animal does not need to evolve a trunk.

It is important to understand that any discussion of trunk evolution cannot be valid without considering the teeth (and their genetic controls), and also the neck anatomy which governs the ability (or otherwise) of an animal to lower its mouth to the

Parasaurolophus

Above: Male and female elephant seals. The male has a large expandable snout.
Below: Skulls of extinct (supposed) elephant relatives.

Moeritherium

Phiomia

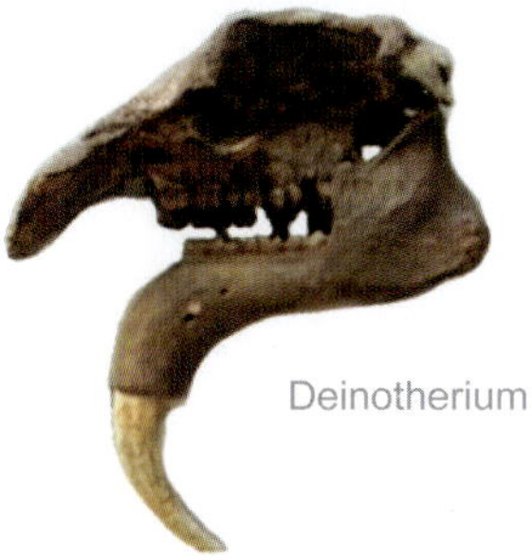

Deinotherium

Mammoth

As with pigs, both the tapir (above) and coati (above right) use their long snouts to sniff for food under leaf litter. They do not use their snouts to grab or manipulate food.

Longitudinal cross-section through a dog skull showing the perforated wall (the *cribriform plate,* arrowed) between the front end of the brain case and the nasal cavity.

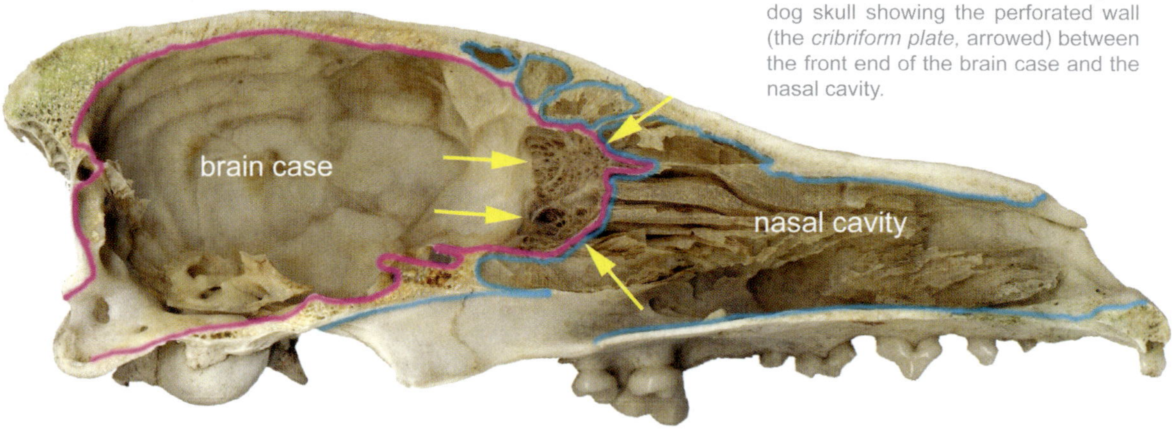


Right: Close-up of the cribriform plate. Embedded within the lining on the nasal side of the cribriform plate are millions of olfactory nerve endings which, when stimulated by inhaled particles, convey impulses through the cribriform plate to the brain where they are recognised as a *smell*. For clarity, only one bundle of olfactory nerve fibres is shown (in yellow).
Note that the vast bulk of the nasal cavity does not have olfactory nerve endings and is *not* involved in the sense of smell.

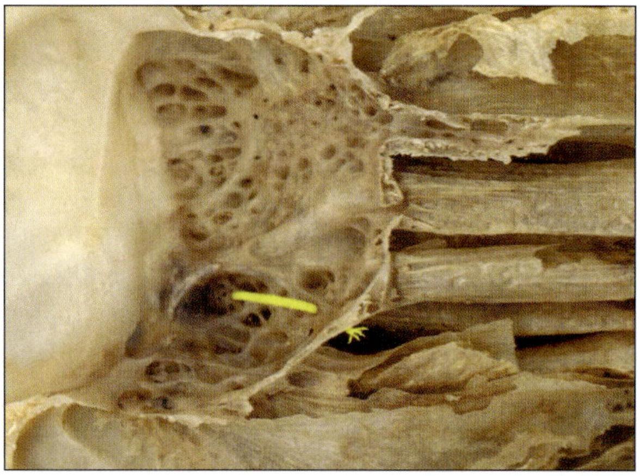

Left: Tapir skull showing that its nasal openings are situated between the eyes as is the case with animals with long trunks, yet tapirs only have short flexible noses.
However, having (supposedly) been around for many millions of years, the tapir is perfectly capable of surviving without a trunk at all. Other supposed elephant ancestors such as moeritherium would have been equally capable of surviving without evolving a trunk.

Tapir skulls.
Left: 50myo fossil.
Right: Modern skull.
There has been no change in (supposedly) 50my.

Big noses are not necessary for making loud noises. The Indian hornbill (left) with its large beak has a horny crest with which sounds can be amplified.
In contrast, the Siamang gibbon from Asia has a throat pouch; and the South American Howler monkey (skull shown below) has a unique *bony* throat chamber (arrowed) with which it amplifies its messages through the jungle.

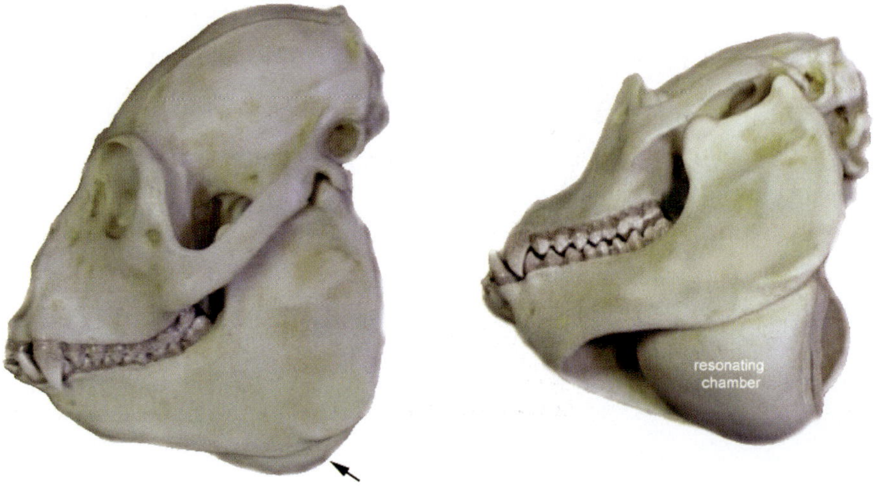

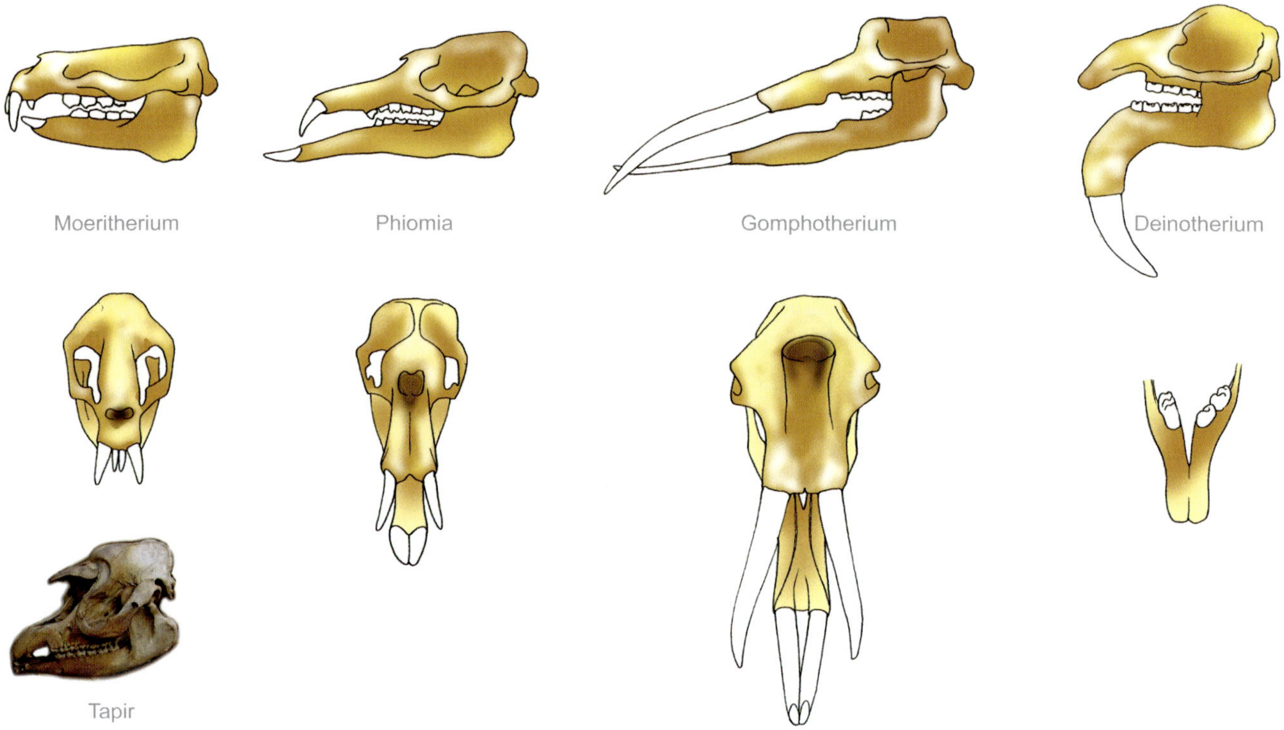


Above: Fossils of extinct large mammals: moeritherium (left), phiomia (middle left), gomphotherium (middle right) and deinotherium (right).
The tapir skull (above, bottom left) shows that its nasal openings are between the eyes as is the case with animals with trunks, unlike in moeritherium (whose nasal openings are situated at the tip of its snout). What sort of short trunk (if any) the moeritherium had is questionable.

Moeritherium
(Natural History Museum, London)

Below: Models of moeritherium and mammoth
(Natural History Museum, Paris)

Above and right: The upper row of illustrations shows common interpretations of the noses of moeritherium (above), gomphotherium (top centre) and deinotherium (top right). However, moeritherium may not have had a tapir-like nose at all. Furthermore, due to the short and relatively rigid necks of gomphotherium and deinotherium, having a short trunk as depicted would have rendered them unable to eat grass or to drink whilst standing by the water's edge. Modern elephants that have had sizeable portions of their trunks amputated are doomed to die from hunger and thirst. Phiomia and platybelodon are not illustrated here but these creatures would also not have had short trunks for the same reasons.
The implausible reconstruction of deinotherium (top right) is redrawn from '*Walking with beasts*' (BBC Books. 2001. p151).
The bottom row of illustrations (right) shows gomphotherium and deinotherium redrawn with more plausible long trunks.

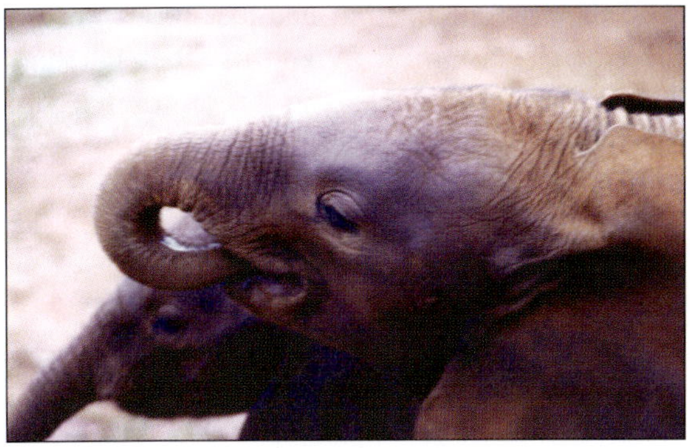

Although an elephant can be very dextrous with the tip of its trunk, it does not regularly pick up large amounts of food in this way. In order for an elephant to eat the huge amounts of food that it needs to keep it alive, it has to grab large portions within a coil of its trunk.

Since an elephant has little movement in its neck, in order for an adult elephant to grasp a clump of grass it requires a trunk that (when fully extended) is some 0.5m longer than the height of its mouth from the ground. Without a long trunk neither a baby nor an adult elephant can both pick up grass, foliage or water *and* place it in its mouth.

Gomphotherium

Deinotherium

ground or to water. Apart from the tusks (which can be thought of as incisors) the living elephants have only four teeth (upper and lower right and left molars), each the size of a house brick. Unlike rhinoceroses, hippopotami, tapirs and all other large terrestrial mammals, elephants have no cutting or chopping teeth with which to bite leaves or graze grasses.

For practical purposes the elephant also has a short rigid neck which means that even if it had the appropriate teeth it could not reach the ground with its mouth to graze. Or to drink. So it is that an elephant that has lost its trunk through some accident or fight is doomed to die of starvation. What this means is that both living elephants and mammoths, with their rigid necks and large molars, could not have had any immediate ancestor that did not have a long trunk.

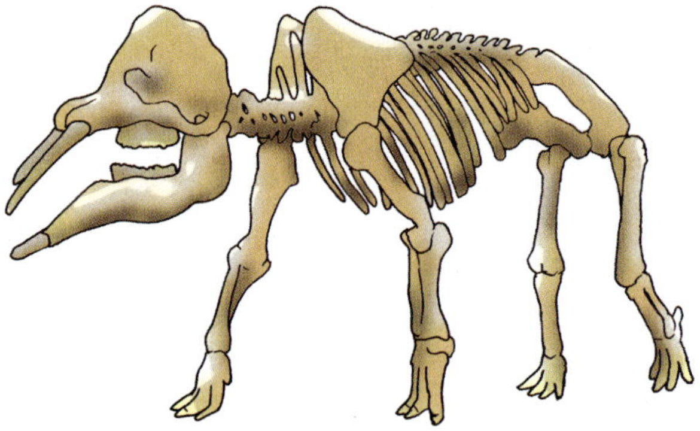

Gomphotherium was the size of a modern elephant. It had a long lower jaw which carried two large incisors at its tip. Like phiomia, platybelodon and deinotherium, it had no cutting or chopping teeth with which to bite leaves or grasses.

Baby elephants

A smaller animal with the same body plan would face the same problem. Found in an island off California are the remains of a diminutive extinct pygmy elephant (standing less than 2m at the shoulder) which show the typical rigid neck of living elephants. Because of the limitations imposed by its teeth and neck anatomy, it too could not have survived without a long trunk.

But what should we make of other extinct elephant-like animals? The *Moeritherium* and *Phiomia* were tapir- to rhino-sized animals with multiple molars and flexible necks, and would have survived happily with or without tapir-like noses. Any animal like the tapir, moeritherium and phiomia which had a flexible neck and conventional teeth would have been able to survive anyway, and so, if by chance an individual developed a longer trunk, this would be of no advantage to it at all.

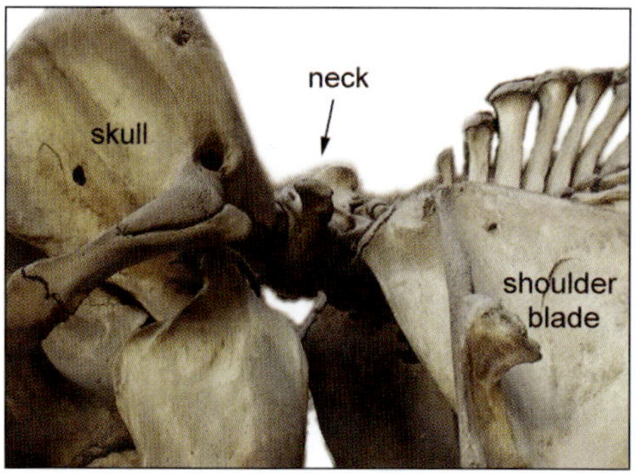


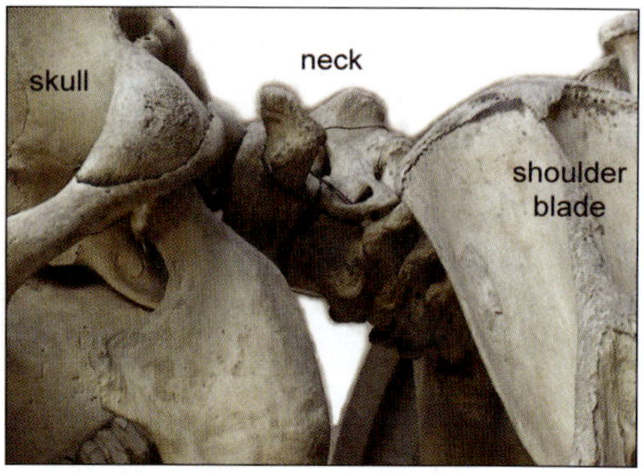


Right: Asian elephant neck.
Far right: African elephant neck.

This African elephant is trying to get at the inside of a fallen tree. The lack of flexibility in its neck is made up for by the length and dexterity of its trunk.

Above: White rhinos.
Below: The extinct hornless rhinoceros Indricotherium was the largest terrestrial mammal known to have lived. At a total height of 5.5m its shoulders were as high as the head of an adult male giraffe, and it may even have weighed up to 30 tons - as much as five adult African elephants.

Model of an Indricotherium leg

Indeed, it would more likely be a hindrance when grazing. Such animals would, therefore, not gradually evolve a longer and longer trunk. And indeed, tapirs (53mya-present), moeritherium (53-36mya) and phiomia (36-23mya) lived (supposedly) over immense time periods, and yet did not show any change in their teeth or any clear change in the length of the nose. Even if a tapir-sized animal grew larger but retained its bodily proportions and flexible neck, a trunk would still be of no survival advantage to it, providing the animal could reach the ground and graze with its teeth and drink water with its mouth. Which is why the largest land mammal that ever lived, the extinct rhinoceros *Indricotherium,* which stood 5.5m (18 feet) at the shoulder, did not have, did not need and did not evolve a trunk during its (supposed) lifetime 36-23mya.

On the other hand, any tapir-like/sized animal that grew larger and developed a rigid neck would not be able to eat and drink without a trunk. However, since nose length is genetically unrelated to neck form, it would be implausible for an enlarging, neck-stiffening animal to coincidentally acquire mutations that would provide it with a trunk. More likely any individual born with a stiffer neck would perish. And as we shall see when we discuss whales, the different types of teeth (incisors, canines, premolars, and molars) are subject to different genetic controls. Even individual teeth (for example, the inner and outer upper incisors) are subject to different genetic controls, so the idea of a lineage of tapir-like animals losing most of their teeth and developing huge molars is quite implausible. The *Gomphotherium* (23-5mya) was the size of a living elephant with four large molars but it had four tusks, one pair in each jaw. The *Platybelodon* (23-5mya) was also the size of a living elephant. It had a number of molars in each jaw, and in addition to a pair of upper tusks, it had a pair of spade-like lower incisors. *Deinotherium* had multiple molars, downward-curving tusks arising from the lower jaw, and lived until 2mya.

Indricotherium compared to an African elephant

Sarah McAlpine

Elephant family, Tanzania

We have already seen that the Laetoli footprints show that deinotherium was living alongside modern east African animals and therefore the climate would have been much the same as it is now. It is quite implausible that animals such as gomphotherium and platybelodon could have lived for 18my without change and then (without any reason for an increased rate of mutations affecting the trunk and molar teeth, and without any increased selection pressure) evolve into modern elephant forms. Indeed, deinotherium, with its multiple molars, was living (supposedly) 3mya so it could not have been ancestral to the modern-like elephants with which it lived contemporaneously.

Except for elephants, all other mammals drink with their mouths. Of course, the elephant does not drink through its trunk but, instead, sucks up water in the trunk and then squirts it into the mouth. Suppose a tapir-like ancestor gained a mutation which slightly altered its body form just enough to make drinking by mouth a little difficult. Since sucking up water into the nose is a very unnatural and dangerous thing to do (since water could be inhaled directly into the lungs), it is implausible that the animal would attempt to do it. In fact, all of its in-built instincts and reflexes would prevent it even trying. Even if, against all reason, it did suck up water into its snout then (unlike an elephant) it would still lack the anatomy and reflex mechanisms that would prevent inhalation of water. Therefore, it is implausible that drinking by use of the trunk has evolved from some previous nose-drinking behaviour.

In conclusion, the fossils simply document the variety of extinct elephant-like animals that have lived and do not constitute evidence that one form turned into another. Thus, there is no evidence that, going backwards in time, any of the ancestors of living elephants had progressively shorter trunks.

### *Stick and leaf insects*

Stick insect

Leaf insect

Could natural selection be such a powerful force that it could *cause* prey animals to ultimately look like their background? The story regarding stick insects is that millions of years ago ordinary-looking beetles crawled around on branches. Some must have been more camouflaged than others and predation by birds favoured the survival of those beetles that happened to have the appropriate genes that made them look just a little more like twigs. Following the natural selection of a series of fortuitous mutations which happened to make the beetle body look more and more like a stick, stick insects acquired the form they have today. It is assumed that the same determinants caused the appearance of leaf insects, so the following discussion applies to both leaf and stick insects.

Certainly, if thinness equates with inconspicuousness, then natural selection might favour a more slender beetle rather than a chunky form, but there is no obvious reason or mechanism by which any change should be towards a beetle looking like the very twigs it lives on. Suppose that the beetles ancestral to stick insects were ordinary chunky beetles and lived on a background of hexagonal nut- and bolt-shapes instead of branches. Are we to suppose that over a period of time gradually- accumulating mutations and natural selection would make the beetle's descendants look more like nuts and bolts? We should consider such a thing to be most unlikely. So for *any* primitive beetle living with *any* background, the chances of it undergoing just the right type of mutation for natural selection to ultimately make it resemble that background must be vanishingly small.

But, it could be argued that there must have lived millions upon millions of insects, from millions of generations, so out of the huge number of possible mutations, those causing their owners to look most like twigs have survived. It might be supposed that insects looking a little like nuts and bolts may have appeared at some time in the past, but since this feature conferred no survival advantage such affected insects therefore died off. It may seem plausible, but this raises a question about all those other countless species of insects which are preyed on by birds and live on plants and yet which look nothing like leaves or twigs. Surely they would all, in time, be naturally selected to look like their background. Surely, insects have enough genome in common for the necessary mutations to arise and be selected for, to make them all resemble sticks or foliage. Yet the reality is that an incredible variety of different insect forms continue to exist on the same plant, sharing the same predators. This suggests that there has always been a number of distinct types of insects (which never had a common ancestor), some of which contain the genetic information to diversify into stick or leaf insects. The same principle applies, for example, to seahorses and pipefish, which are quite distinct from other fishes.

### *Variation and abnormalities*

We have seen examples of variation which result in normal progeny. Now let's look at a few examples of variation or new mutations resulting in *abnormalities*. Having an abnormality does not necessarily mean that a creature will die out since if there is a safe haven, disadvantaged forms may still survive so as to inbreed the condition.

But first we must correct a commonly held notion that the *lack of use* of an organ might, with the passage of time and in subsequent generations, result in degenerate organs or structures:

> As difficult as it is to imagine that eyes, though useless, could be in any way injurious to animals living in darkness, I attribute their loss **wholly** to disuse…Disuse, aided sometimes by natural selection, will often tend to reduce an organ, **when it becomes useless by** change in habits or under changed conditions of life… (My bold). Charles Darwin. 'The origin of species'. Penguin Books. 1985. p178, 451.

Here Darwin is completely mistaken. *Why should* a little-used organ such as an eye or a wing become degenerate in subsequent generations? We have seen that in any given individual, only genes that are detrimental can be selected against. Muscles that are used infrequently may become wasted, but this will not affect the genetic constitution of an individual or its descendants. With improvements in nutrition and health, human beings are, on the whole, getting bigger with each generation. At the same time, however, less and less physical activity is required of most people in earning a living. As a consequence, in general, people are less muscular than previously. However, an office worker who is unfit due to lack of activity does not produce an unfit child unless that child, too, does little physical activity. In time, as more and more machines take the physical labour out of work, it is possible that people will become less and less muscular. However, *unless a deleterious mutation*

Pipefish

Common seahorse

Spiky seahorse

Weedy seadragon

Leafy seadragon

**Pipefish and seahorses**

The common seahorse and the weedy and leafy seadragons are probably a unique group of fishes derived from pipefishes.

Uniquely, the male seahorse fertilises and incubates eggs inside a special pouch at the base of its tail.

The tail of the leafy seadragon is not prehensile as in the common seahorse, and its eggs attach to the wrinkled undersurface of the tail.

*affected any subsequent generation* the genes would continue to retain the DNA responsible for bulking up muscles if they are exercised.

An eye which *squints* is one which fails to properly track the visual point of attention in synchrony with the other eye and is usually caused by an imbalance of the muscles that move the eyeball itself. Initially, this results in the two eyes sending two sets of impulses to the brain that represent two images that do not properly fit together, and this causes double vision. If no surgical correction is undertaken, in order to make sense of the visual world, the brain will in time ignore completely the impulses from the squinting eye in preference to those from the non-squinting eye. We know that although both the eyes and the brain are physically normal, the outcome is *functional* (not physical) blindness in the squinting eye, and affected individuals *will not* give rise to blind children. A similar thing happened to Madeira cows which became blind as a result of being continually locked up in dark sheds (they were confined in this way because of the lack of hill-side space). Their genes code for normal eyes and normal brains and their calves are born normally-sighted.

Madeira. In the past it was common practice to keep cows in dark sheds. As a result they became blind.

In Darwin's time there was no knowledge of DNA and the genetic code, but herein lies the fallacy of his thinking. It is not that the habit of disuse *causes* degeneration of organs or structures, but that degeneration can appear as a result of a random and deleterious mutation. Any such affected individuals will either survive despite the disadvantage (if the resultant abnormality does not have fatal consequences) and live to perpetuate the disadvantaged characteristic in subsequent generations, or succumb and die. In this way, any disuse in the surviving generations is caused by an inability for the characteristic (being degenerate) to be used.

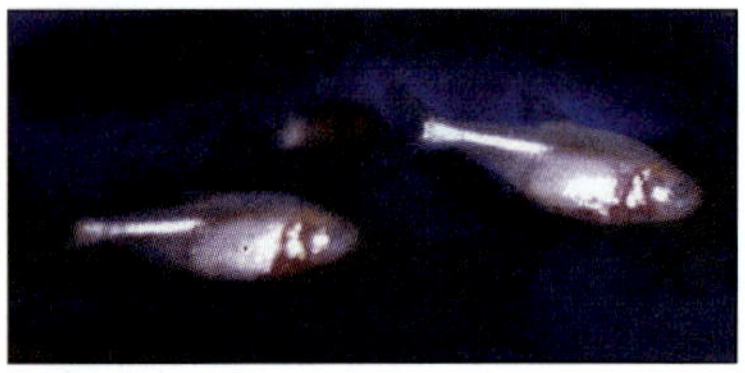

Adult blind cave fishes have degenerate eyes.

### Blind cave fishes

Some cave-dwelling fishes, such as some populations of the Mexican Characin *Astyanax fasciatus mexicanus* (also known as *Anoptichthys jordani*), have almost unrecognisable eye remnants. However, the early development of some of these fishes is quite normal and only the adult fishes have degenerate eyes and a degenerate visual cortex in the brain. Nevertheless, they are said to have evolved their blind nature to *suit* their dark habitats:

> The eyes of this species **have become superfluous** as it navigates in totally dark underground waters using its lateral line system. (My bold).
> Dick Mills. 'Aquarium fish'. Dorling Kindersley. 1993. p87.

Above and below: Blind cave fishes vary in the degree to which their eye components fail to develop.

But in the dark, neither normal nor blind fishes would have any selective advantage since they would both have to rely on their *lateral line system* to sense objects and movement in the water, and on their sense of smell to locate their food (the lateral line system comprises an easily visible line of pores along almost the whole length of a fish, which contain special nerve cells which are sensitive to pressure and movement).

We know that genes are very stable and if the presence of eyes does not become hazardous to survival (and there is no reason why it should), then there would be no opportunity for natural selection to weed out the genes responsible for eye development. Thus simply living in darkness can have no effect on the genetics of the eye. So why are blind cave fishes blind? We have seen that when bairdiellas were transplanted into the Salton Sea, the first spawnings produced a significant percentage of individuals that were blind. After a few generations, blind fishes still appeared but the percentage was greatly reduced. Clearly the blind fishes were at a disadvantage in the open sea and were more easily taken by predators.

Suppose, therefore, that normal-sighted Mexican characins gave rise to a small percentage of mutant blind individuals amongst a large number of normal progeny, then we should expect that all the individuals, both blind and normal, would explore all the nooks and crannies of their surroundings. Since we know that blind characins can locate food (by smell) and avoid obstacles (by use of the lateral line system) as efficiently as sighted individuals, we should expect that both blind and normal individuals might find their way into dark underground streams where there were no predators. The sighted individuals would be attracted by any light and would more likely find their way out again into daylight, whereas those born blind would simply just go to find food where their sense of smell took them. Blind fishes would then become 'lost' in the darkness and unable to find their way out again. Thus, a population of blind fishes, which had originated in open water, might remain in the dark caves. Interbreeding would then perpetuate the mutation and all the progeny would be born blind. As with the bairdiellas, any blind fishes (whether born in daylight or in caves) which find themselves outside in open water would be less likely to avoid predators than sighted individuals. So, the mutation resulting in blindness is not the *result* of the environment and the blind cave fishes have not adapted to the dark. Blind cave fishes have not arisen *in order to* fill an environmental niche, nor have the eyes become degenerate *because they are unnecessary*. It is just that the imperfect blind fishes are not too seriously disadvantaged and manage to survive because the environment in dark caves is simply not so hostile.

Since there is no change in the genome, the animal is still the same animal, but it has a developmental abnormality. (In the same way, a human baby born blind would not and could not be regarded as a different species.) And, indeed, the Mexican blind cave fish can breed perfectly well with the normal-sighted astyanax, showing that it is just an example (albeit a congenitally abnormal one) of variation, and it cannot be considered a separate and distinct creature.

Blind cichlid.
Other blind creatures are also found in caves, such as the blind salamander.

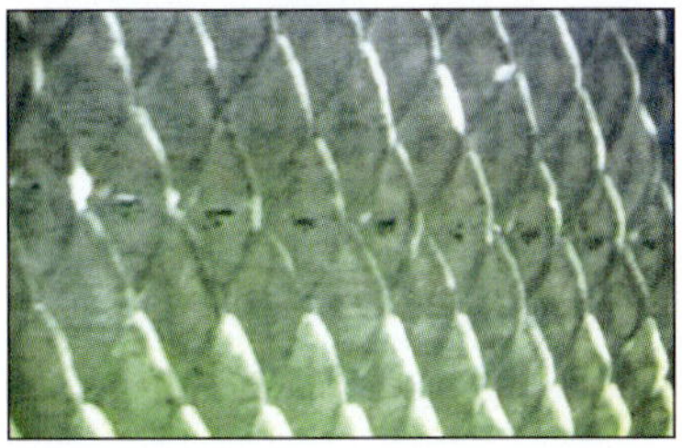

Above: The lateral line system of a fish is composed of a series of small pressure-sensitive pores which run along the whole length of the body. Here a number of pores are seen as dark spots.
Below: The lateral line system enables fish, such as the baracudas, to swim in a shoal that often moves as if it were one large organism.

### Flightless birds

The Galapagos cormorants are flightless and are only found in a limited area on the shores of the island of Fernandina and along the adjacent coast of Isabella. They are exactly the same as cormorants elsewhere in the world, except that they have smaller wing bones, and the flight feathers are abnormally short and do not zip properly. The body feathers are also abnormal but are nevertheless much like those of a normal cormorant. It has been suggested that these birds have evolved flightlessness because they have no need of flight (there are no predators in the Galapagos for them to flee from), and that since they swim excellently, they are still evolving, being perhaps part of the way to becoming something like a penguin. However, unlike penguins which have excellently proportioned wings with which they 'fly' underwater, these cormorants have *mal-developed* wings which cannot be used either for flight or for

Galapagos flightless cormorant showing its abnormally small wings.

Above: Nile cormorant showing normal wings.
Below: Galapagos flightless cormorant showing its abnormally small wings.

swimming. Instead, like other cormorants they swim with their webbed feet and waddle clumsily on land. When they come out of the water, just like their normal flying cousins, the Galapagos cormorants stand in the sun and stretch out their wings to allow them to dry. Clearly they still have the instinct to look after their feathers even though they can no longer fly.

Despite the fact that there are many normal *flying* seabirds in the Galapagos islands, the Galapagos cormorants are commonly believed to have evolved to suit their environment:

> They are clearly adapted to an aquatic life, as penguins and auks are adapted. Having long ago colonized the Galapagos Islands as flying birds, the cormorants found an environment without land enemies, with space, and with abundant food. They lived more in the sea, from which they obtained their living, than on the land, where they occupied only a narrow zone above the tide. Gradually **flight lost its selective advantage**. Genetic changes favourable to a closer relationship with the sea became advantageous, and dominated other less advantageous traits. Feathers became hairlike and **unused wings atrophied**. The cormorants **evolved a more efficient mode of life** for the conditions… (My bold). Kenneth Brower (Ed). 'Galapagos, the flow of wildness'. Vol 2. Sierra Club-Ballantine Books. 1968. p63.

Most birds, whether flyers or not, usually feed on the ground or in the sea, and hardly need to fly anyway. So we must ask the question: why should the genes that produce normal healthy *flying* cormorants become a *disadvantage*? After all, the normal cormorants elsewhere in the world can swim and walk, *and* fly. Is there a simple explanation as to why the ancestors of the Galapagos cormorants lost their ability to fly? We know that a bird which is kept in a small cage and is not allowed to fly will still produce babies with normal wings and feathers. And even after a number of generations, mice which have their tails amputated will still produce offspring with normal tails. Thus, as we have seen with eyes, the lack of use of a structure does not *itself* result in offspring with faulty structures. Therefore, the flying ability could only have been lost by some process that would have prevented the feathers and wings from developing normally.

The most plausible explanation is that a deleterious mutation has been inbred and caused the original Galapagos cormorants to become flightless. Since there are no predators to flee from, the *(congenitally abnormal)* cormorants of the Galapagos

Apart from the cormorants, all other sea-birds in the Galapagos have normal wings.
Below left: Galapagos albatross chick.
Below middle: Galapagos pelican.
Below right: Galapagos blue-footed boobie, which is essentially a variety of gannet.

have been able to multiply and survive. To put it another way, the lack of predators is not the *cause* of their changing from normal birds. Instead, it is the lack of predators that *allows* the abnormal (and vulnerable) Galapagos cormorants to survive. It is important to understand that the Galapagos cormorants represent a degeneration of an existing fully-functional gene that is normally responsible for the proper development of the cormorant's wings and feathers. In other words, the Galapagos cormorants do not represent the production of a new gene, which codes for a new design which previously did not exist.

The Galapagos cormorants, then, are *not* adapted to life in the sea. They have no selective advantage over normal cormorants and they simply manage to survive despite their congenital abnormalities. So, we should not even call the Galapagos cormorants a particular variety or a particular race of cormorants. They are just simply *abnormal*. Thus we can state the *Galapagos Cormorant Principle*:

Disuse *per se* is not the cause of degeneration of a structure or function. Mutations occur as inevitable consequences of random chemical substitutions within DNA. However, if a mutation results in a degeneration of a structure or function which *is not essential for the survival of the affected individual*, then the individual might survive and pass on to its offspring that mutation. The progeny will then have degenerate structures and functions not because of disuse, but as a consequence of the prior mutation.

But what about other flightless birds - how could we explain their inability to fly? Here is one suggestion as to their origins:

> This relapse to a ground-living life is an indication of the great demands that flying puts on a bird's energies and the amount of food it needs in consequence. If life can be led in safety on the ground, then this is a much easier option and the birds take it. David Attenborough. 'Life on Earth'. Collins/BBC. 1979. p198.

Not only is such a notion simplistic, but it is also wrong, since metabolic studies show conclusively that:

> ...for a given body size, flying is a far less expensive way to move to a distant point than running. Knut Schmidt-Nielsen. 'Animal physiology'. Cambridge University Press. 1997. p206.

Examples of recently exterminated flightless birds include the *great auk,* the *dodo* in Mauritius, and the *moa* in New Zealand. Living flightless birds include the *ostrich, rhea, emu, cassawary,* the diminutive *kiwi* and the *kakapo* (a flightless parrot from New Zealand). It seems that whenever birds colonise areas devoid of predators, some flightless forms arise. The *great auk* lived in the North Atlantic and like the Galapagos cormorants, it had small wings and was flightless. If it had not been exterminated by Man in 1844 it would most likely still be found today. Similarly, the dodo and the moa had no natural predators and therefore also survived without the ability to fly until being exterminated.

Moa. Curiously, this huge bird had no wing, wishbone or shoulder-blade bones at all.

Above: Dodo skeleton. Recent DNA studies suggest that the dodo is more closely related to pigeons from South-East Asia than to those from Africa.

We can consider three possibilities as to the origin of flightlessness in any bird. Firstly, could flightless birds have descended from flying bird ancestors? Attenborough's view is fallacious since a flying bird would not change into a flightless bird *because of* energy considerations. If all flightless birds evolved from ancestral flying birds, then as with Galapagos cormorants, they could only have done so as a result of mutations that would have resulted in deficient wings and feathers. If the environment was not *so* hostile such that affected individuals could still survive, then flightless individuals would live to interbreed with other flightless individuals so as to produce a population of flightless birds. And as with the Galapagos cormorants, such flightless birds should be considered as being simply *congenitally abnormal.* Indeed, inbreeding amongst pigeons or chickens yields many unusual forms, and a few are fat and clumsy, just like the dodo and the great auk. These abnormal individuals would still be the same creature as their parents and their flying cousins, and they would still retain the ability to interbreed (either naturally or artificially) so as to produce fertile descendants. So, some flightless birds are likely to be descendants of flying ancestors. The great auk, flightless rail, dodo and kakapo might be included in such a group.

Above: Cassawary.
Below: Emu.

But how should we explain the origin of other flightless birds, such as the *ostrich* and the *penguin*? Could it be that they are all the results of deleterious mutations in ancestral flying birds? The ostrich in Africa, the rhea in South America, the emu in Australia, the cassawary in New Guinea, and the moa from New Zealand are so similar that it is quite plausible that these large flightless birds all arose from a common ancestor that lived at a time before the continents separated (supposedly, over 135mya).

All flying birds have flight feathers with asymmetrical vanes which form an aerofoil in cross-section. Whilst the Galapagos cormorant has asymmetrical vanes, showing that it probably descended from flying ancestors, other non-flying birds like the ostrich and penguin have feathers with symmetrical vanes. As with the Galapagos cormorant, retaining asymmetrical flight feathers would not confer any survival disadvantage on other flightless birds. So, if ostriches, emus, rheas and the like evolved from flying ancestors, whilst in time we would expect further mutations to result in degenerative wings and feathers, we should not expect to find all (if any) populations of these birds to have acquired symmetrical feathers. Instead, we should expect to find these flightless birds still having flight feathers with asymmetrical vanes but with different degrees of feather degeneration. Furthermore, since after countless generations the Galapagos cormorant has not lost its asymmetrical feathers, we have no evidence at all to show that mutations can convert asymmetrical feathers into symmetrical feathers. So, perhaps the ancestors of ostriches and the like were never flyers at all.

The condor is near the upper weight limit for a flapping flyer.

However, if we suppose that birds such as the ostrich originated from ancestors which were at one time flyers, those ancestors could not have been much larger than a swan or a condor since these birds are near the upper weight limit for a flapping flyer. Therefore, linking the ostrich to an original flying ancestor means that there must have existed a number of intermediary ancestral forms that were larger-than-an-albatross-but-smaller-than-an-ostrich non-flyers (these intermediaries representing progeny with congenitally abnormal small wings). But there is no fossil evidence to support the past existence of any such intermediaries. Furthermore, the ostrich, emu and rhea live in open grasslands and any significantly smaller, congenitally abnormal, unable-to-fly bird would be unlikely to survive in the company of so many large predators roaming about, and would quickly become extinct.

Incidentally, if evolution is indeed a process whereby new and more complex structures are added progressively to life forms, and if flightless birds have existed for millions of years, then instead of simply producing degenerative wings, evolution should have, in some of them at least, produced more useful appendages, for example, grasping limbs.

The open savannah is a dangerous place. The ostrich (far left) survives because it is large and powerful. The much smaller secretary bird (left) spends a lot of time on the ground looking for snakes and lizards but is able to fly away from danger.

Not only are the tiny feathers of a penguin hugely different from those of a flying bird or an ostrich, but also their bones are not hollow, and their wings have the form of true flippers (being not simply congenitally deformed wings like those of the Galapagos cormorant). It might be supposed that the tiny feathers and flippers of a penguin can be adequately explained as being simply the result of gradual stepwise modifications of a sea bird such as the Galapagos cormorant.

Penguin

However, the Galapagos cormorant and other diving birds catch fish as effectively as a penguin, propelling themselves through the water with their feet and without using their wings at all:

> Divers...With their wings clasped tightly against their body so that their steamlining is near perfect, they pursue their quarry and often manage to out-manoeuvre them in their own medium. David Attenborough. 'The life of birds'. BBC Books. 1998. p129.

Since there is no advantage to a cormorant or other diver in using the wings for

King penguin preening

Puffin

Stranded juvenile guillemot which is too feeble to escape the incoming tide.

swimming, and there is no evidence to show that mutations affecting a true wing could convert it into a flipper, it is pure speculation to say that flippers could evolve from wings. However, there are other sea birds such as guillemots, puffins and razorbills which, whilst being true flyers in air, also 'fly' underwater using their wings as flippers. So, it may still be supposed that a penguin flipper is only a few progressive changes down the evolutionary line from the wing shape and feathers of a guillimot. If so, could such birds give an indication of what penguin ancestors were like?

The guillemot (with its normal flight feathers) and the Galapagos cormorant (with its deformed feathers) are as able to make a living in the sea as the Galapagos penguin. So there would be no selection pressure for a guillemot or Galapagos cormorant to develop penguin-like feathers. If any mutation *could* (note that we have no evidence to show that it could) cause a Galapagos cormorant's feathers to become like those of a penguin, then, in the safe haven of the Galapagos islands, interbreeding and the lack of selection pressure would simply give rise to a variety of deformed feather types ranging from penguin-like to Galapagos cormorant-like. On the other hand, if any mutation could cause a guillemot's feather to become significantly smaller, then the affected chick would be unable to fly and would be far more likely to plummet fatally from its cliff-side nest, than to survive and breed with other similarly-afflicted guillemots to produce a population of abnormally-winged guillemots. Even if such a chick did survive, since guillemots with normal feathers can out-swim fish, any mutant guillemots with smaller feathers would have no advantage - so there would be no selection pressure for their feathers to become progressively penguin-like.

Furthermore, if we suppose that the penguin originated from ancestors which were at one time flyers, then there must have existed a number of intermediary *no-longer-a-flying-seabird-yet-not-quite-a-penguin* ancestral forms linking the penguin to the original flying ancestor. Again, there is no fossil evidence to support the past existence of any such intermediaries.

The second possibility for the origin of flightless birds is that they could have descended directly from non-flying dinosaur ancestors. In this scenario, flightless birds descended from some dinosaurs which became warm-blooded, developed feathers, gained air sacs and hollow bones, but which did not develop asymmetrical feathers and functional wings. But as we shall see later in the chapter on birds, other fossil and biological evidence suggests that birds *could not* have evolved from dinosaurs.

So, thirdly, it is possible that certain non-flying birds such as the penguin and ostrich are immutable types of birds, having no dinosaur or flying ancestors at all.

### *Lungless frog*

Recently, *Barbourula kalimantanens,* a 2in-long lungless frog, has been found in Borneo. Predictably, since amphibians do not have a diaphragm to separate the lungs and heart from the abdominal contents, the internal organs in these frogs have shifted position to take up the space that would normally be filled by the lungs. As is commonly seen in human congenital abnormalities where abnormal tissue takes on a structure-less form, it was also noticed that there was some abnormal cartilage in the area where the lungs should have been.

Although some biologists consider the lungless frog to be some fantastic find that might shed new light on evolution:

> "this exciting evolutionary event...I think we have a flagship species in these lungless frogs". D Bickford. 'Bizarre frog has no lungs'. LiveScience.com. 7th April 2008,

lungless-ness is also found in a species of *Caecilian* (a limbless creature that looks like an earthworm), and in a number of salamanders*. In fact, lungless amphibians are nothing new – they were well-recognised *over 75 years ago*:

> Any moist thin area of skin or skinlike material is capable of absorbing oxygen (some small land-dwelling salamanders are quite lungless and breathe through their skin...). A Romer. 'Life of vertebrates 2'. (First published **1933**). Pelican. 1963. p326.

The frog 'researchers' conjecture that the loss of lungs might be an adaptation to the cold, fast rivers the frogs live in. If that were the case, then after millions of years, most cold, fast rivers all over the world should be home to lungless frogs. The real explanation is obvious. The frog simply has a congenital disease. If there occurs a mutation resulting in the inactivation of a controlling signal that is active early on in the development of the lungs, the lungs would fail to develop (*pulmonary atresia*). If this were to happen in reptiles, birds and mammals, the embryo would not survive. Although rare, pulmonary atresia is well described in human babies, and is fatal. However, since amphibians can breathe through the skin and via the lungs, we should predict that, in an amphibian, an embryonic defect that results in the failure of the development of the lungs, or that results in impermeability of the skin, would *not* be fatal. Furthermore, if lungless barbourulas are the progeny of normal ancestors and represent simply an inbreeding population of congenitally abnormal frogs, we should expect to find similar but normal healthy frogs not far away. And, in fact, the closest relative of this frog is found in the Philippines – and it has lungs. So, the lungless barbourula is not some exciting new creature, and we should predict that it should be able to breed with its Philippino relatives and produce normal progeny.
Therefore, the kind thing to do would be not to *deliberately* perpetuate the lungless frog's lineage but to let them interbreed with their normal relatives. Similarly, the Galapagos cormorants should be allowed to breed with normal cormorants to produce healthy birds. However, Ecuador has a vested interest (i.e. tourism) in maintaining the abnormal birds, having swallowed the evolution myth that the cormorants represent some wonderful creature evolving its way to ultimately become something like a penguin.

### Axolotl

Just as it is not always fatal for a degenerate cormorant to be unable to fly, so it is not always fatal for an amphibian to be unable to metamorphose. The curious *Axolotl* is a Mexican salamander that normally fails to undergo metamorphosis into the adult

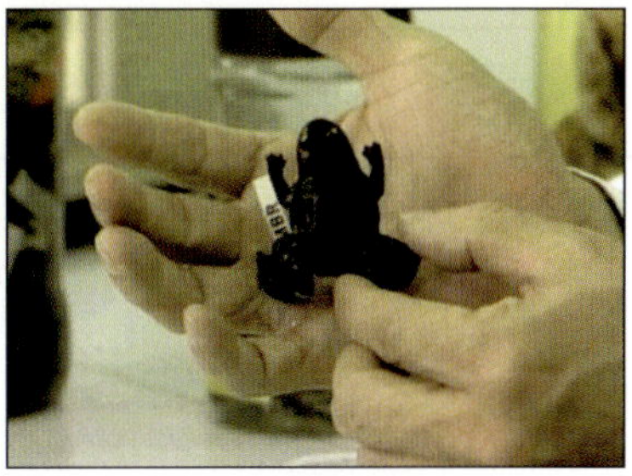

**Lungless frog**
*From National Geographic News. 12th July 2008.*
Although hailed as some fantastic find, this frog simply has a congenital abnormality.
The frog is not unique since it is known that a mutant mouse lacking *Fibroblast growth factor 10 (Fgf10)* does not develop lungs:

> *The [Fgf10] null mouse is characterized by the absence of lung bud development. P Nyeng et al. 'FGF10 maintains distal lung bud epithelium and excessive signaling leads to progenitor state arrest, distalization, and goblet cell metaplasia'.* BMC Dev. Biol. 2008 January. 10; 8:2.

*An example is the slimy salamander from north-eastern USA, which is also interesting because the young do not pass through a tadpole stage.

The axolotl salamander does not metamorphose from its tadpole stage. When sexually mature it still retains its gills. In fact, the axolotl is not unique among salamanders, since a number of *tiger salamanders* (in the United States and Mexico) and the eel-like *olm* (which has only rudimentary limbs and is found in underground rivers and caves in the former Yugoslavia) also remain permanently in the tadpole stage.

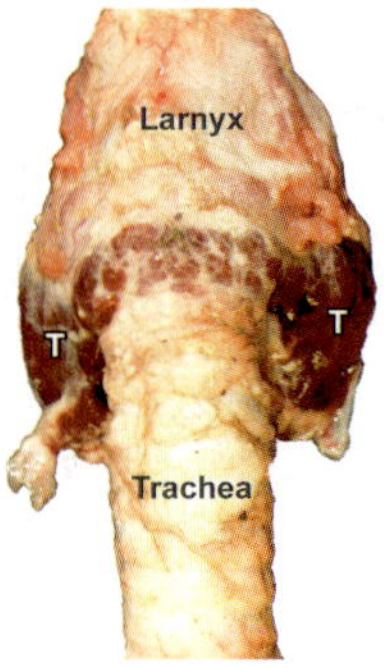

Calf: the thyroid gland (T) lies just below the front of the larynx.

form, and manages to reproduce while it is still in its tadpole stage. This is because it lacks a gene which codes for the production of the hormone which causes *thyroxine* to be produced by the thyroid gland. Thyroxine is itself a hormone which has a pivotal function and without which the metamorphosis process cannot proceed. In humans and other mammals, the thyroid gland is essential for the control of body metabolism, and serious disease can be caused by having either too little thyroxine (causing *hypothyroidism,* in which bodily functions are slowed down) or too much thyroxine (causing *hyperthyroidism,* in which bodily functions are speeded up).

It might be supposed that the axolotl, unlike most other amphibians, has not yet evolved the gene that could allow it to metamorphose. However, this would be bizzare since if the missing step is simple and all other amphibians have accommodated the genetic alterations required, then after millions of years of amphibian existence the mutation should have come about in the axolotl too. In fact, the axolotl is *not* a primitive amphibian which has not yet evolved genes for metamophosis, since if iodine (a vital component of thyroxine) is added to the water or if an injection of thyroxine is given, this will trigger metamorphosis in the creature. Therefore, since the axolotl *can* metamorphose, its genes must have all the coding required to carry out the transformation.

So, how is it that all the complex coding required for metamorphosis could have been selected for and retained in dormant form, while the ancestral axolotl waited for the appearance of the one vital gene that would permit metamorphosis to proceed? The answer to this problem is that the axolotl is not *waiting* to become a proper salamander. It is not waiting to become able to metamorphose. Instead, it has *lost* the ability to metamorphose which it once had. So, the origin of the axolotl can be explained by a *deleterious* mutation, occurring at some time in the distant past, in a small isolated group of salamanders which were previously as fully capable of metamorphosing as any other amphibian. This mutation must have occurred in a gene that regulated the release of thyroxine at the time appropriate for metamorphosis to take place. Such a mutation, by preventing the release of thyroxine and so affecting a key part of the metamorphosis process, would explain why the axolotl has within it the otherwise fully-functional machinery for metamorphosis. If this explanation is true, then we should predict that we should find some wild populations of axolotls which *are* able to undergo metamorphosis in a normal manner, and that interbreeding between those that can and those that cannot undergo metamorphosis ought to be possible. And indeed, this is the case.

So, the axolotl, which lives life completely in the tadpole stage, and which when given thyroxine is able to metamorphose fully into an adult, is simply a creature with a genetic abnormality: a creature which is *not* adapted to an environment, but one which instead manages to survive *despite* its congenital disability.

### *The thymus gland and appendix*

Mention must be made of the *appendix* and the *thymus gland* in humans which have in the past been quoted as examples of vestigial organs, no longer necessary, and which in time might disappear completely.

### *The thymus gland*

The thymus is a gland which is sited in the chest behind the upper part of the breast bone. At birth and during infancy it is quite sizeable but in time it gradually reduces in size, such that in older children and adults it is only represented by a small piece of tissue. For this reason, the thymus gland had been thought for many years to be a degenerate structure left over from some ancestral organ.

In fact, nothing could be further from the truth since the thymus is *essential* for the proper *early* development of the whole of the body's immune system, and without its proper function in early life, serious life-threatening diseases ensue. For example, *Thymic alymphoplasia* is a condition where the thymus and lymph nodes are all but rudimentary at birth. Death usually occurs in the first year of life. In the disease *ataxia-telangiectasia syndrome,* the thymus is poorly developed, there are abnormalities in the circulating white cells associated with abnormal jerky movements (ataxia), abnormally dilated and delicate blood vessels (telangiectasia), and there is also a tendency for affected individuals to develop certain types of cancers.

Surgical removal (*thymectomy*) of the thymus also causes serious disease:

> Thymectomy in rats results in ataxia and low IgA levels. Gavin Forfar & Arneil. 'Textbook of paediatrics'. Churchill Livingstone. 3rd edition. 1984. p728.
> (IgA is one type of antibody.)

As an interesting aside, there is much more to fighting infections than the production of antibodies. Once antibodies lock onto invading bacteria, other chemical signals attract white cells which then engulf and destroy the bacteria. There is a congenital condition in which the white blood cells can ingest invading bacteria, but are unable to destroy the germs since they lack the necessary special enzymes. This is known as the *Chediak-Higashi syndrome*, and it demonstrates the fact that it is *essential* to have a number of different anti-infection systems working alongside antibody production for there to be effective protection against infecting organisms.

In chickens, the *bursa of Fabricius* (which is located near the rectum) acts as a central lymphoid organ controlling the maturation of white blood cells concerned with immunity. To some extent it is similar in function to the thymus of mammals.

### *The appendix*

The appendix is also often presented as a vestigial organ*, an evolutionary relic of no value, which will eventually disappear completely. This impression is not entirely illogical since it is well known that when an infected appendix is removed, its absence causes no illness. However, the appendix does contain abundant lymphoid tissue, and seems to have an important role to play in the proper development of the immune system in the region of the small intestine, during the period shortly after birth. In fact, the immunological function of the appendix was described in textbooks over 30 years ago**, and there is increasing evidence suggesting that the appendix has a significant part to play in the body's immune system:

> '...it achieves its greatest development shortly after birth and then regresses with age, eventually coming to resemble such other regions of GALT

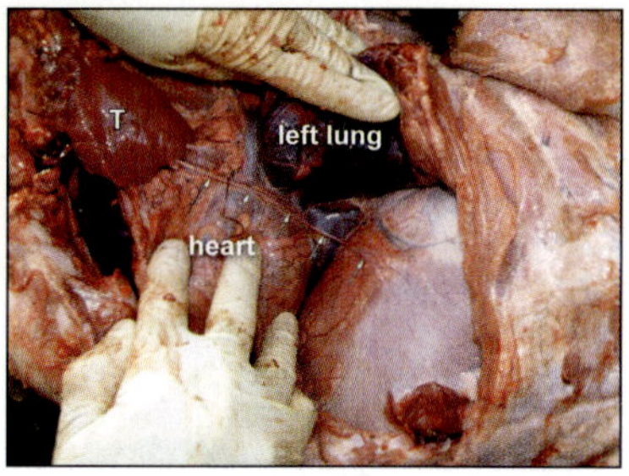


Calf: The thymus (T) lies in the upper part of the chest, just above and in front of the heart.

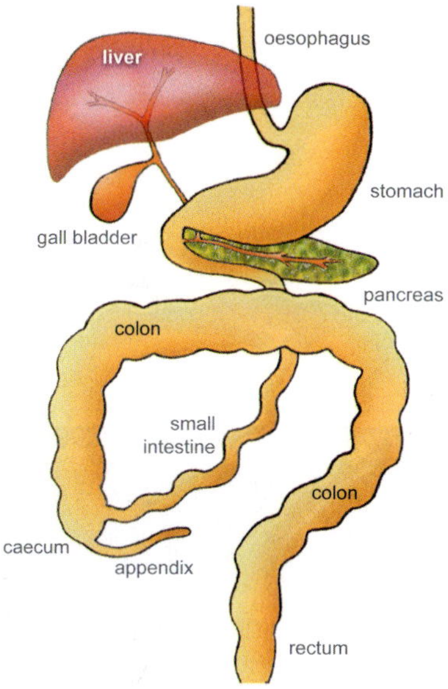


The human appendix is less than 1cm in diameter and about 10cm in length, and is attached to a very short caecum.

***Vestigial organs**

It has been suggested that whether or not the appendix has any function today is irrelevant, because it clearly was originally part of a larger *caecum* in our pre-ape ancestors. And therefore, since it is now very small and no longer functions in fermenting gut contents, it is definitely a vestigial organ.

The problem with this idea is that the wealth of evidence shows conclusively that humans are not modified apes, and have always been human. So, the human appendix cannot be a vestigial part of the caecum.

*** "The Appendix" by G McHardy in 'Gastroenterology 2'. H Bockus (Ed). WB Saunders Co. 1976. p1134-1148.*

***Appendix and ulcerative colitis**

*Appendicectomy itself does not protect against ulcerative colitis... Without appendicitis or mesenteric lymphadenitis, appendicectomy has no impact on subsequent risk [of ulcerative colitis] even when done in childhood or adolescence.*
M Frisch et al. 'Appendicitis, mesenteric adenitis, and subsequent risk of ulcerative colitis: cohort studies in Sweden and Denmark'. British Medical Journal. 4th April 2009. p808-811.

Apart from appendicitis, the appendix is (uncommonly) the site of cancers that have spread from elsewhere (such as breast, lung and stomach). It can also be the site of the rare *carcinoid* tumour.

> [gut-associated lymphoid tissues] as the Peyer's patches in the small intestine.' J Pomerantz. 'Does the appendix serve a purpose in any animal?' Scientific American. Nov 2001. p96.

It may also provide a reservoir of beneficial bacteria that can replenish the intestines after an infection:

> We propose that the human appendix is well suited as a "safe house" for commensal bacteria, providing support for bacterial growth and potentially facilitating re-inoculation of the colon in the event that the contents of the intestinal tract are purged following exposure to a pathogen. R Randal Bollinger et al. 'Biofilms in the large bowel suggest an apparent function of the human vermiform appendix'. J. Theor. Biol. 2007 Dec 21;249(4). p826-831.

Curiously, there is some evidence to suggest that removal of the appendix reduces the risk of developing the serious disease *ulcerative colitis**. This is interesting since ulcerative colitis is caused by some little-understood derangement in immunity. Exactly how the appendix is involved in this is unknown.
While the exact function of the appendix is still unclear, nevertheless, since the wealth of evidence shows conclusively that humans are not modified apes, and have always been human, the human appendix cannot be a vestigial part of the caecum.

### *Conclusion*

We have seen that there is little direct evidence to suggest that all the variety of life on Earth is the result of the natural selection of random mutations over long periods of time. And since we know that a difference in physical form or colouration or behaviour, or even genetic constitution or chromosome number, does not necessarily mean that two individuals or populations are different creatures, there needs to be a revision of what we call a 'species'. We have seen that the man-imposed classification of creatures into different species, whilst providing an excellent (and in general terms, secure) method of distinguishing between living things, nevertheless is at times prone to fail to such a degree as to suggest relationships or distinctions when none, in fact, exist.
Nevertheless, it is generally promoted that the origin of new characteristics, new organs and new creatures can be explained by natural selection. But as Darwin conceded, if it could be demonstrated that any complex organ could not possibly have been '*formed by numerous, successive, slight modifications*', his idea of evolution '*would absolutely break down*'. What we do see is plenty of evidence to show that Darwin's optimism has no foundation at all. Actually, what the evidence points to is that living organisms have within their genes the potential for a wide range of variation; that many organisms have the potential for producing hybrids by breeding with others not of their kind; and that the exact degree to which such things can happen varies amongst different organisms. However, there are limits, since the evidence also suggests that beyond a certain potential for variation and hybridisation, no creature can naturally trespass. I say *naturally* since it is entirely possible, for example, that artificial matings of hippopotami with manatees, or of chimpanzees with human beings, could result in viable offspring, some of which might be fertile.

Doose. This natural hybrid has the head colouration and wing flashes of a male mallard duck and the body of a goose.

Zebroid: a horse-zebra hybrid. This creature is simply the sum total of the mixture of genes from its parents. There is no new gene, organ or structure that the zebroid has that is not present in either horse or zebra.

It is also entirely plausible that someone has already tried the latter.
Hybridisation simply mixes genes from different organisms, and some of the resulting progeny may be viable. So, we might artificially mate a giraffe with an elephant to produce a long-necked plump creature with a trunk. But since, by so doing, no new genes would have been created, neither variation nor hybridisation can produce any new *previously non-existent* structure or function.
However, since the process of hybridisation brings together chromosomes that may be physically different enough to disturb the normal balance during cell division, it is not surprising that developmental abnormalities may be more commonly found in the resultant progeny when compared to the parent stock.
Any understanding of living things must accommodate the fact that individuals can exhibit substantial differences and yet, for all practical purposes, still remain the same animal. For example, house mice can have different numbers of chromosomes yet look exactly the same; the German Shepherd dog and the Pekinese dog have the same number of chromosomes and can interbreed yet have markedly different appearances; herring gulls and lesser black-backed gulls have exactly the same numbers of chromosomes yet have a *reluctance* to interbreed; and the Reeves and Indian muntjac deer look virtually identical, but have very different chromosomes and have an *inability* to interbreed.
Despite the problem with the term, many creatures are labelled as different 'species' based simply on gross appearances. For example, Tanzanian bats (see right) on the Chicago Natural History Museum's website are erroneously given individual species' names based on mainly outward appearance.
Yet these bats clearly have congenital deformities akin to cleft palate and failure of jaw development. For example, *Cardioderma* has a completely missing premaxilla bone and no incisors. In contrast, *Nycteris* has only 4 incisors (yet other bats have 6) and a broader forehead. While they may well have different chromosomal numbers and may not be able to interbreed, nevertheless, they are probably all the same (but defective) animal. They only survive because their abnormalities are not too severe. The Tanzanian bats are not alone with their upper jaw and tooth deficiencies. Another example is the Indian bear (also called the sloth bear) which has missing upper incisors and a defect in the roof of the mouth. Rather than evolving for the purpose, this bear is still able to suck insects out of holes despite the congenital absence of its premaxilla bone and front teeth.
So, in describing and distinguishing life forms, the term 'creature' better suits our purpose than the term 'species'. For example, whether Indian or Reeves muntjac, wolf or Pekinese, herring gull or lesser black-backed gull, polar bear or brown bear, we should classify each of these pairs as the same creature.
All the evidence we have seen so far shows that genes can be mixed together in different combinations. However, this sheds no light whatsoever on exactly how a dinosaur or bird, hippopotamus or manatee, chimpanzee or human being, actually came into being.
We will now step aside from the familiar and visible world and look at the invisible goings-on deep down inside living cells, where atoms and molecules (seemingly blindly) get on with their daily business.

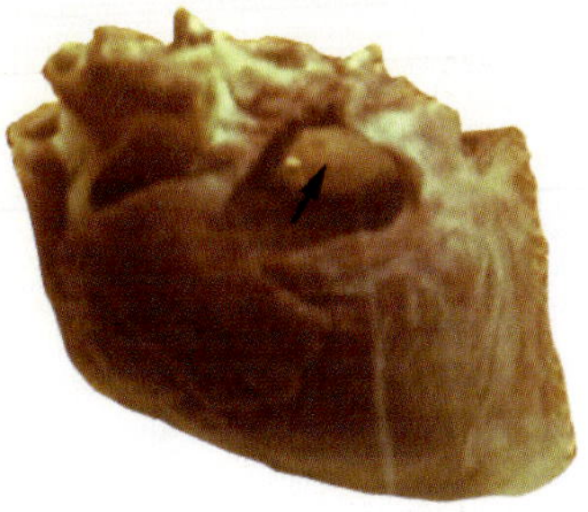

Hybridisation can result in severe congenital abnormalities. This preserved heart of a still-born lion-leopard cross has a hole (patent foramen ovale, arrowed) between the right and left atrial chambers. Such a large defect is incompatible with survival.

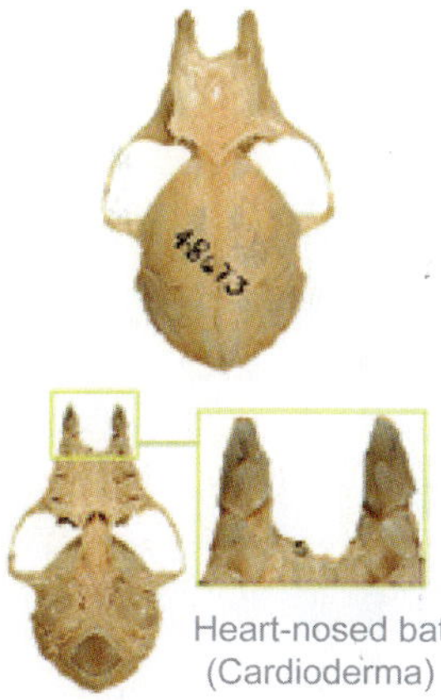


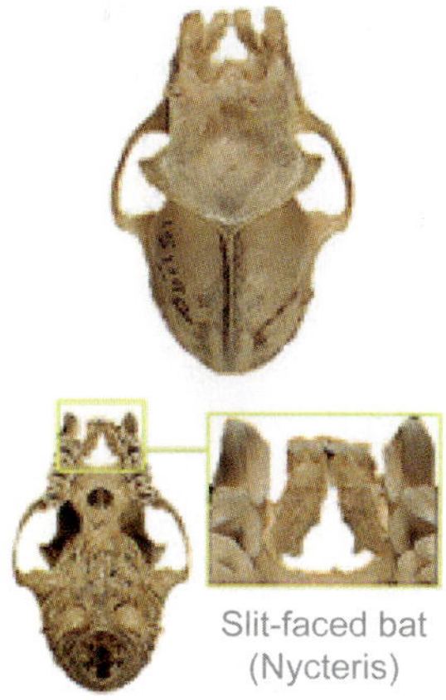


Chicago Natural History Museum.

These bats from Tanzania are not some special designs but simply show congenital abnormalities, resulting in deformed and missing teeth and facial bones.

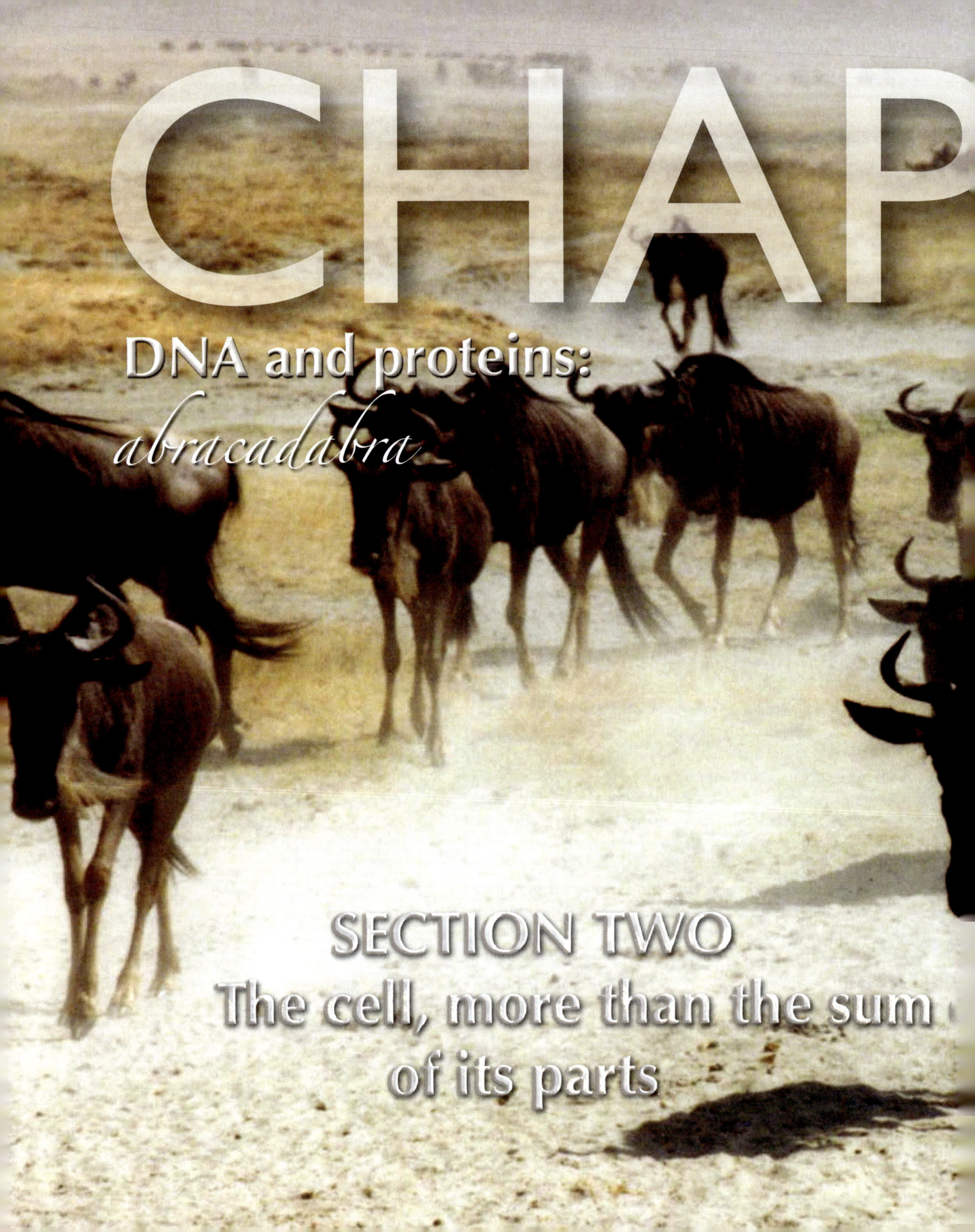

# CHAP

## DNA and proteins: *abracadabra*

## SECTION TWO
## The cell, more than the sum of its parts

Wildebeest
Ngorongoro Crater, Tanzania

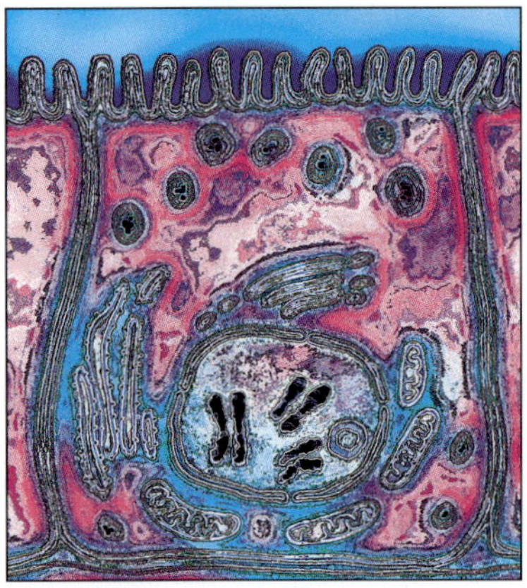

Above: Eukaryote cell (Diagrammatic)

Louisa Howard

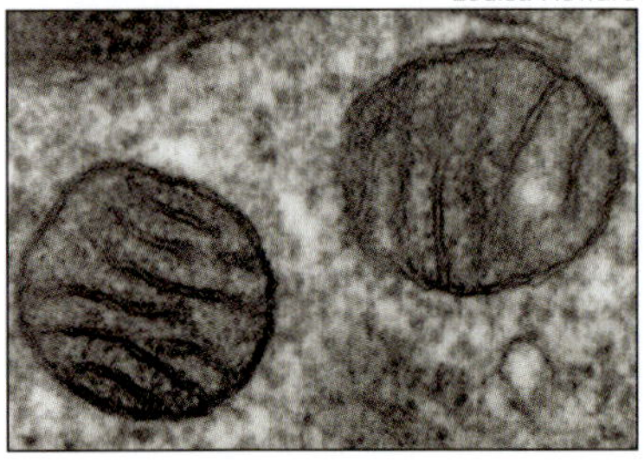

Mitochondria from mammalian lung

In simple terms, a typical cell is a bag of water. However, the cell wall is not some inert plastic but is a highly complex membrane made up of fat and protein molecules. Many molecules can simply drift in and out through the molecular spaces within the membrane, but embedded in the cell wall are numerous specialised proteins which can act as doorways, actively pushing and pulling atoms and molecules in and out of the cell. The watery stuff (*cytoplasm*) inside the cell contains a number of highly complex structures which include the *nucleus*, *mitochondria*, and *ribosomes*. The nucleus itself is a bag whose lining is very similar to the cell membrane. Inside the nucleus lie the *chromosomes* which are made up of coiled lengths of DNA molecules. Mitochondria are found in most eukaryote cells and are the organelles responsible primarily for the chemical reactions necessary for the conversion of the energy, stored in the bonds within glucose molecules, into a form that can be more readily utilised by the molecular goings-on within the cell. The resultant user-friendly energy is carried mainly within the high-energy phosphate bonds in the molecule *adenosine triphosphate* (ATP). Under the electron microscope ribosomes look like rows of beads and are the machines which manufacture proteins.

The cell is often described as a complex chemical factory. Ordered and productive the cell certainly is, but in chemical terms, the organisation and functions of all cells are both mindless and purposeless. So, in discussing any of the molecular systems found in living organisms, we must always remember that it is inherent in evolutionary theory that *all* of the structures and mechanisms that we encounter (whether, for example, bacteria swimming by means of a whip-like flagellum, or chromosomes being pulled to each daughter cell during cell division, or your fingertip being moved by a muscle in your forearm) must have come into being, and continue today, *only* as a result of the random interactions of different atoms and molecules. Indeed, we must accept that all cellular chemistry must have originated likewise. Here, there is no debate:

> If one assumes, according to evolutionary theory, that related species have evolved from a common ancestor, then it follows that each of their proteins must have likewise evolved from the corresponding protein in that ancestor. Donald and Judith Voet. 'Biochemistry'. John Wiley. 1995. p125.

In other words, natural selection ultimately equates simply and *only* with those chemical bonds and reactions that are possible, and which can remain stable (and hence persist) within the milieu pertaining. We have come to understand that the ways in which atoms and molecules interact are very specific and, to a large extent, predictable. For our purposes in this chapter, equally important and predictable are the ways in which atoms and molecules *cannot* interact. For example, every third amino acid in the chain that makes *collagen* (which is the commonest protein found in animals) is *glycine*. This is predictable because considering the way that the collagen chain is coiled, glycine is the *only* amino acid that can fit physically into the spaces available.

Certainly, simple organic molecules can be coerced into being by subjecting even simpler molecules to some form of energy. For example, when an electrical discharge is applied to a mixture of *methane*, *ammonia* and *water vapour*, a whole variety of organic compounds can be produced, including *amino acids*, *formic acid* and *urea*. And if a solution of amino acids is evaporated to dryness and subsequently quenched with water, microspheres composed of a limited number of inter-linked amino acids can be formed. However, a series of reactions in a test-tube requires very careful staging and separation of any component chemicals. In a cell thousands of reactions are carried out in a manner which is supremely organised, and the different stages are separated and ordered by a number of different chemical and structural barriers and constraints, all seemingly governed entirely by the laws of chemistry and physics. So much so, that if we were to focus on *any individual* atom or molecule we would see that its movement would be both very simple and random. Furthermore, isolated from the large number of enzymes, structures and systems within a cell, such individual atomic and molecular movements would have no biological application or purpose. There are literally thousands of aspects that we could look at, but for our purpose of understanding some of the chemical goings-on inside a cell, we shall confine our discussion to the subject of DNA, and the structure and manufacture of proteins. This will be enough to raise questions about the (supposed) evolution of cellular chemistry. But first an overview is necessary.

A *protein* is a complex molecule made up of strings of individual *amino acids* joined end to end in a specific order. Each amino acid-amino acid bond is called a *peptide bond* (hence proteins are also called *polypeptides*). DNA (*deoxy*-ribose-nucleic acid) resides within the cell nucleus and acts as the permanent archive for the code that determines the exact sequences of amino acids which make up any particular protein.

Amino acids cannot spontaneously link up one with another to form a protein chain. Nor can amino acids be strung together by simply aligning themselves to their corresponding codes in the DNA. Proteins are actually manufactured in the cell cytoplasm by an intricate series of chemical reactions involving a highly complex and diverse collection of molecular machines. A temporary template in the form of *messenger ribose*-nucleic acid (mRNA) must first be made from the DNA. This process is called *transcription* and it is from this mRNA template that a protein is manufactured. Once formed, the mRNA passes into the cytoplasm and associates itself with a *ribosome*, which is a complex structure made up of a number of proteins. Individual amino acids are brought to the ribosome by a specific carrier called a *transfer RNA* (tRNA). Each amino acid has to be loaded onto its tRNA by its own specific enzyme (*aminoacyl tRNA synthetase*), and once two tRNAs are locked onto the mRNA, the enzyme *peptidyl transferase* forms a peptide bond between the amino acids that the two tRNAs carry. The process is repeated with fresh tRNAs carrying different amino acids and so the protein chain is constructed.

**Right-handed sugars**
**Left-handed amino acids**

In the lab, left to their own devices, sugars are always produced in equal quantities of right- and left-handed forms. These two forms have the same atoms, in the same relative configurations, with exactly the same chemical properties, but are mirror-image molecules (*optical isomers*). The problem is that in all living cells, sugar molecules are only found in right-handed forms, and the 3-dimensional shapes of cellular enzymes match the shapes of those right-handed sugars.

Similarly, in the lab, left to their own devices, amino acids are always produced in equal quantities of right- and left-handed forms. However, in all living cells, amino acids are only found in left-handed forms. In fact, if a mixture of right- and left-handed amino acids was used to make proteins, this would result in shapeless blobs.

The problem is that there is no natural method by which molecules are only produced in either right- or left-handed forms. And without a mechanism in place to ensure only one form is manufactured, no cell could arise in the first place.

**Lethal oxygen and UV light**

Any supposed early Earth atmosphere could not have had oxygen since that would destroy any supposed pre-life molecules by oxidation.

And if there had been plenty of UV light to supply energy to (supposed) chemical reactions in a primordial soup, the same UV light would have destroyed any biologically-important molecules that (supposedly) might have been made.

## *DNA and genes*

The sequence of amino acids within any protein is coded for by the precise linear order of molecules called *nucleotides* within DNA. Each amino acid is represented by a string of three nucleotides, each triplet being called a *codon*. A complete sequence of codons for a single protein is called a *gene*. An ‘average’ protein of 100 amino

# A gene is a linear arrangement of nucleotide triplets

Within DNA, the sequence of amino acids of any protein is coded for by the precise linear order of molecules called *nucleotides*. Each amino acid is represented by a string of three nucleotides, each triplet being called a *codon*. A complete sequence of codons for a single protein is called a *gene*.

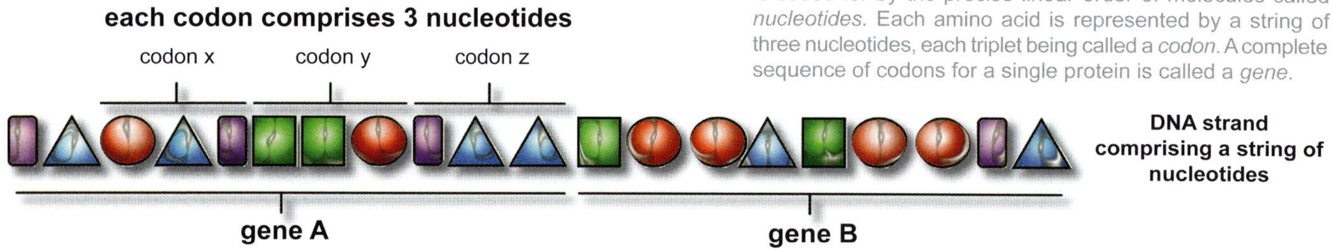


## DNA Transcription

There are three core enzymes involved in the transcription of RNA from the DNA code. *RNA polymerase I* is the enzyme that transcribes genes for the RNAs of ribosomes. *RNA polymerase II* transcribes all genes for mRNAs; and *RNA polymerase III* transcribes tRNA genes.

In eukaryotic cells, the *promotor* is typically a zone comprising a series of nucleotides that are AT-rich (called the *TATA box*), situated some 25 nucleotides upstream of the start of a gene. It is to this TATA box that RNA polymerase II must bind, in order to initiate transcription of the DNA code into an RNA molecule. However, while in bacteria RNA polymerase can itself recognise and lock onto the promoter upstream of a gene, RNA polymerase II cannot do this in eukaryote cells without the help of a number of specific proteins called *transcription factors*, which first bind to the TATA promoter and collectively provide a binding site for RNA polymerase II.

First, a protein called *transcription factor IID* binds to the TATA box, to which other protein transcription factors then bind, with the resulting whole complex providing a binding site for RNA polymerase II. This complete collection of proteins, including RNA polymerase II, is called the *transcription initiation complex.*

RNA polymerase II unzips the DNA helical strands and starts to make an RNA strand complimentary to the exposed nucleotides of the DNA.

At different times during embryogenesis, and subsequently in different cells in the body, the rate of transcription of particular genes is increased by other protein transcription factors called *enhancers*. Yet other proteins selectively switch on, switch off, or suppress transcription.

After the RNA transcript is made, it contains both exons and introns along its length. By a process of *polyadenylation*, a tail of adenine nucleotides is added to the 3' end - the '*polyA tail*' - which may have a stabilising function since RNA is inherently less stable than DNA.

The introns are then excised by a complex made up of a number of proteins called a *spliceosome*, and the remaining exons are bonded together end to end. The introns are recognised by the spliceosome as they all have a GT signal at the 5' end and an AG signal at the 3' end. The RNA molecule is now called a '*processed*' RNA and leaves the nucleus to go to the ribosome in the cytoplasm, in order to be translated into a protein.

Inaccurate splicing of the introns as a result of mutational alteration of a GT signal will result in faulty proteins. One example is β *thalassaemia*, in which the β globin gene is defective, resulting in serious anaemia. Here, the GT of one of the two β globin introns is an AT, which results in a corrupted β globin mRNA which cannot be translated into a β globin protein.

Evolution requires that the whole transcription and translation process is ultimately the result of chance collisions of atoms in an ancient pool, plus some UV light, plus lots of time.

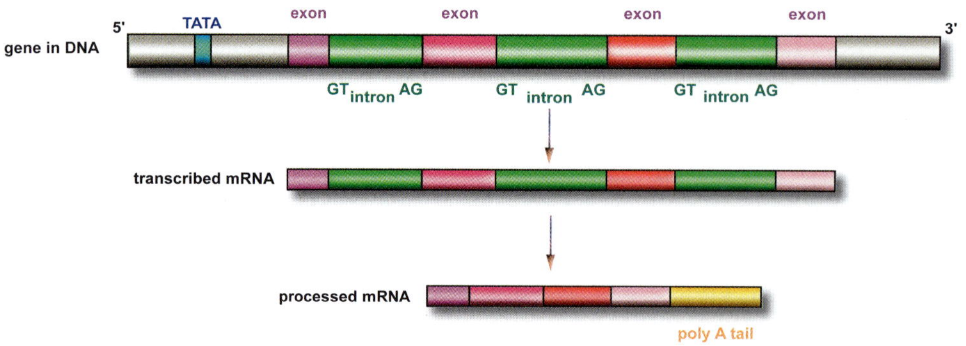

## The genetic code embedded in a gene is translated into a precise linear arrangement of amino acids to form a unique protein

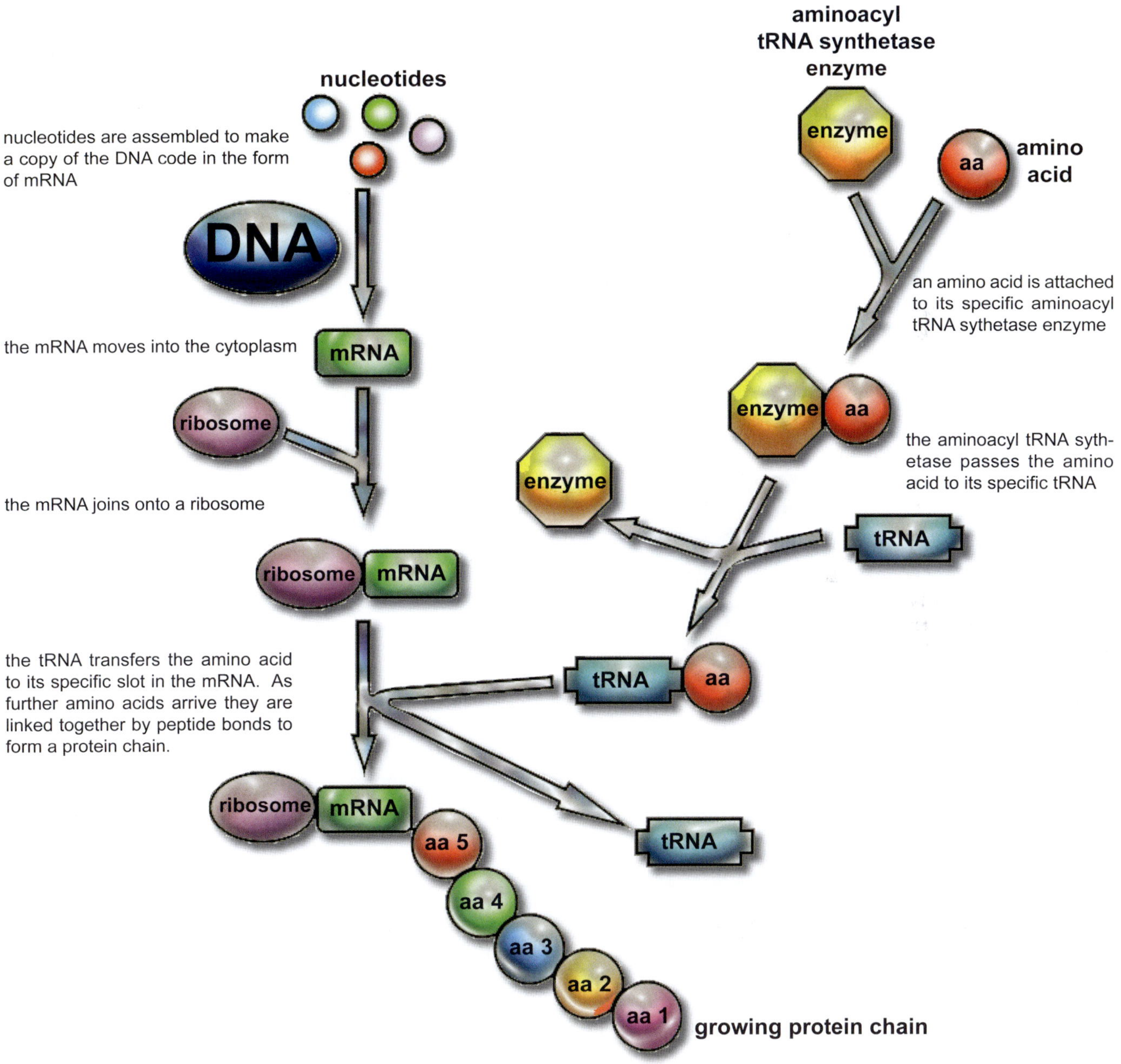

acids requires 100 codons (comprising 300 nucleotides), and some large proteins such as *dystrophin* may have *1000* amino acids. The genes are strung together end-to-end as a continuous line of nucleotides which makes up a long single piece of DNA. DNA exists as twin strands twisted around each other, containing pairs of the deoxy-ribonucleotides, and forming a double helix. In DNA there are four nucleotides, *adenine* (A), *cytosine* (C), *guanine* (G) and *thymine* (T). The nucleotides of one strand of the DNA make chemical bonds with the nucleotides of the other strand. In DNA adenine always pairs with thymine, and guanine always pairs with cytosine. RNA (*ribose*-nucleic acid) exists as single strands containing adenine, cytosine, guanine and *uracil* (U). Unlike in DNA, in RNA adenine pairs up with uracil instead of thymine.

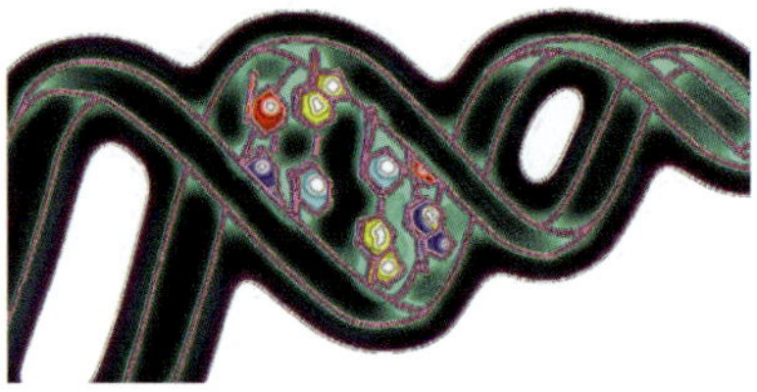

The DNA double helix.

In order to maintain its stability and also to be able to fit physically inside a nucleus, the DNA thread has to be coiled and twisted around special proteins called *histones*.

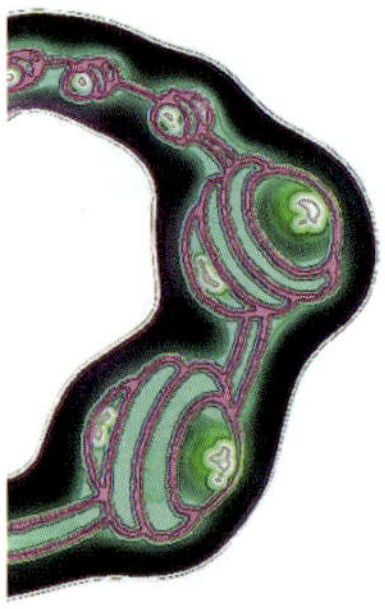

The DNA thread is coiled around histones.

The coiled segments are then coiled further upon themselves many times. The resultant super-condensed double-helical strand is what we see in the light microscope as a *chromosome*.

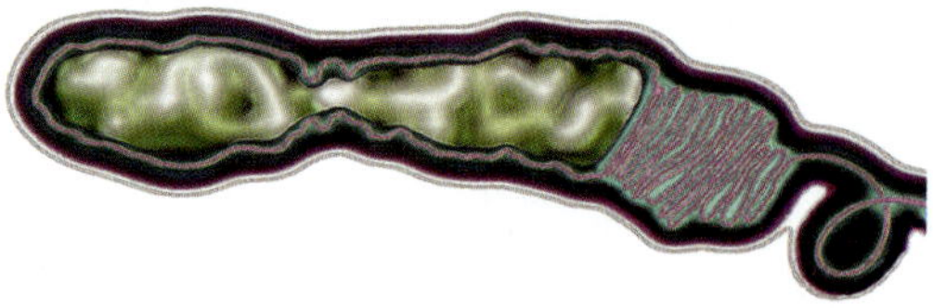

A chromosome is a single tightly-coiled length of DNA.

Every type of creature usually has a set number of chromosomes. For example, humans have 46, arranged in 23 *homologous* pairs, all of which have characteristic shapes and sizes. One chromosome from each homologous pair comes from the father and the other from the mother. The DNA double helix is 20nm (one nannometre = $10^{-9}$ m) in diameter, and one micrometre (one millionth of one metre) of DNA contains about 3000 nucleotide pairs. Thus, although chromosome 21 is one of the smallest chromosomes in humans, it nevertheless comprises an estimated 50 million nucleotide pairs within a single DNA molecule which, if uncoiled, would be some 15mm in length.

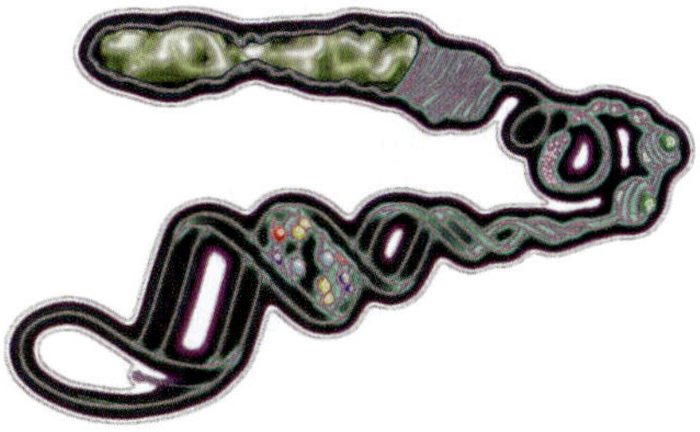

The chromosome unravelled.
The human genome contains around 25,000 genes - only about twice as many as in a fruit fly.

In total, the 23 pairs of human chromosomes contain perhaps 3 billion nucleotide pairs, representing around 25,000 genes. Suppose a protein comprises 100 amino acids. One might expect that the corresponding 100 codons would be all found as a contiguous sequence along a length of DNA. However, this is not the case, since the codons that code for amino acids (called *exons*) are found interspersed between chains of nucleotides (called *introns*) which do not seem to code for any amino acids at all. Indeed, it is estimated that only 3% of the DNA in human chromosomes actually codes for proteins.

Although it was previously suggested that non-coding DNA is *non-functional* (or *junk*), it is axiomatic that *the lack of knowledge about the function or purpose of something does not mean that that something has no function or purpose*. It is becoming increasingly clear that the portions of DNA which do not code for proteins *do not* equal *useless* DNA at all (see *pseudogenes*). In fact, many sequences that were previously thought to be 'junk' have vital roles the in regulation of gene activity. In addition, a certain amount of intron material may be necessary for the stabilising of the DNA molecule during the transcription process, or for facilitating the accuracy of the process itself. Furthermore, the amount of some non-coding sequences does somehow determine both the presence and severity of diseases such as *diabetes, high blood pressure* and *Fragile X mental deficiency*. Other conditions which are similarly affected by the amount of non-coding DNA are bound to come to light with further research.

Although the genetic code is made up of triplets of nucleotides, one cannot simply look at any 3 nucleotides and say that these code for a particular amino acid *x* or *y*. This is because the actual nucleotides are strung together in an unbroken, *continuous* sequence (for example, in RNA it might be: **GCUUGUUUACGAAUU** etc.), there being no molecular markers that specifically define the third nucleotide of one codon and the first nucleotide of the next codon. Sense can only be made of the sequence if the starting point is defined, since moving the starting point just one nucleotide to the right or left will code for a different chain of amino acids. Thus codons have to be read *in phase,* and in general, any phase shift will disrupt the code to a greater or lesser extent. For example, the sequence of nucleotides **GCUUGUUUACGAAUU** can be divided into triplets which code for specific amino acids:

The large-scale study of 1% of the human genome (ENCODE) discovered that vast regions of non-coding DNA (formerly known as 'junk') were transcribed into RNA[1], This included a significant number of pseudogenes [2].

1. E Birney et al. 'Identification and analysis of functional elements in 1% of the human genome by the ENCODE pilot project'. Nature. 2007 447:799-816.

2. D Zheng et al. 'Pseudogenes in the ENCODE regions: consensus annotation, analysis of transcription, and evolution'. Genome Res. 2007 17:839-851.

The *Fugu*, or pufferfish, has the same collection of genes as most other vertebrates, but lacks what was previously called 'junk' DNA. Its DNA comprises 400 million nucleotides compared to the 3 billion in humans.

The idea that all non-protein-coding DNA is 'junk', and is the result of the progressive accumulation of accidental duplications and point mutations, is now known to be quite false. Even if were true, for a fish to have no 'junk' DNA implies that it could not have been around for long enough to have acquired any.

| | | | | |
|---|---|---|---|---|
| **GCU** | **UGU** | **UUA** | **CGA** | **AUU** |
| *Alanine* | *Cystine* | *Leucine* | *Arginine* | *Isoleucine* |

However, if the sequence is read starting with the first C, then the code changes and defines a different sequence of amino acids:

| | | | | | |
|---|---|---|---|---|---|
| **G** | **CUU** | **GUU** | **UAC** | **GAA** | **UU** |
| | *Leucine* | *Valine* | *Tyrosine* | *Glutamic acid* | |

It must be noted that many amino acids within numerous proteins can be freely interchanged without affecting the structure or function of any particular protein. However, any changes resulting in the insertion into a protein of the wrong amino acid has the potential for causing severe problems (for example, *sickle cell anaemia*, see later).

All protein chains in all prokaryote and eukaryote cells begin with the amino acid *methionine,* and in the mRNAs of almost all cells, the *start* signal for protein production is given by the code AUG. The mRNA codons UAA, UGA, and UAG do not code for amino acids but are codes for the *stop* signal. In prokaryotes the start signal can be 1000 nucleotides from the coding section. In eukaryotes this can be *50,000* nucleotides away. So it becomes clear that if the nucleotide pair sequence did not include both a START sequence code and also a STOP sequence code built into the line of nucleotide pairs, codons in the DNA would simply be read as representing a huge long (and nonsensical) protein chain.

Many genes have multiple *identical* copies of themselves repeated a number of times along the chromosome, most often in a head-to-tail fashion. It appears that such *tandemly-repeated genes* are necessary to cope with the high rate of production of their corresponding proteins. Since, for many proteins, any single copy of a gene would be unable to produce enough copies of its mRNA, a number of identical genes doing simultaneously the same job are able to produce more units, in the same way as a number of assembly workers perform identical tasks in a factory. For some proteins it is *essential* that their genes are tandemly-repeated. For example, in early human embryos the rate of cell division is so great that only by having at least *100* copies of the so-called *pre-rRNA gene* can the embryo grow normally. And in *'bobbed'* mutant flies (so-called because their wings are malformed and stubby) there are not enough tandem repeats of the rRNA genes to allow for normal development. And furthermore, for a fly embryo to have less than 50 tandem repeats is *fatal*:

> A bobbed mutation that reduces the rRNA genes to less than ~ 50 is a recessive lethal mutation. H. Lodish et al. 'Molecular cell biology'. Scientific American Books. 1995. p316.

This poses the question as to how any early fly could have existed with less than the minimal requirement of 50 tandemly-repeated rRNA genes. Clearly, since having less than 50 tandem repeats is incompatible with life, then any (supposed) ancestral fly with less than 50 copies would not have lived, so preventing its hypothetical descendants from accumulating stepwise further tandemly-repeated rRNA genes.

Furthermore, if the fly's tandem copies of its rRNA gene have been acquired by evolutionary duplication of a single ancestral gene, then over time, as with all genes, some copies ought to have acquired nucleotide changes; so we should now see, amongst multiple tandem copies, a number of nucleotide variations scattered randomly. Yet the tandem copies seen are all exact replicas, suggesting that the tandem copies have not been acquired as sequential add-ons and that flies have not been in existence for long enough to have acquired mutations - certainly not for millions of years.

Many genes do not function in isolation but are interdependent. Some genes control the activity of a whole group of other subordinate genes (the latter collectively called an operon) and a mutation resulting in the failure of any such controlling genes will result in failure of activation of the operon. This poses the question as to how the genes comprising any operon could have functioned without the corresponding controlling gene. For example, in the bacterium E. coli, synthesis of the essential amino acid *tryptophan* requires five enzymes which are coded for in a continuous stretch of DNA comprising 7000 nucleotide pairs. Tryptophan is an amino acid which is essential for life, and since it is found in most proteins, this raises the problem that if any of the necessary five enzymes were to be absent, tryptophan could not be manufactured and death would ensue. Or more precisely, life could not have begun.

The normal genetic make-up of the DNA of any species is remarkably stable. Indeed, the preservation of any species' line is entirely dependent on DNA's ability to reproduce faithfully, without error, a copy of itself. If it were not stable, there would be little possibility of any subsequent generations being anything like their ancestors. In fact, there would be just random degeneration of all genes over a period of time. Just as in any written paragraph, spelling mistakes are most unlikely to lead to an *improvement* in the way a thought is expressed, or to an expression of a *new* thought, so it is with a mutation in a gene. It is most unlikely to lead to an improvement in its specific protein, or to the production of a *new* and useful protein with a *new* function.

A *mutation* is simply an error in the DNA sequence of a gene, and there are many ways in which the genetic material can be damaged or altered. For example, *point mutations* occur as changes in single nucleotide pairs in the gene, and within each single human cell there occur between 5 and 10 thousand such nucleotide alterations per day. Alternatively, a number of nucleotide pairs can be *deleted* from or *added* to a gene. In more severe cases, the whole gene itself may be deleted or *duplicated.* Gross alterations in the chromosomes can also have a diverse number of effects. For example, *deletions* of anything other than the smallest fragments often result in death of the embryo. *Additions of extra amounts of chromosomal material* can also have severe effects. For example, individuals with Down's syndrome have no abnormal genes, just an extra amount of *normal* genes contained in an extra chromosome 21. *Translocations* can occur when part of a chromosome breaks off and re-attaches itself in a different position or orientation to the same chromosome, or to a different chromosome altogether. In other cases, translocations occur when a complete chromosome attaches itself to another chromosome, making a new

**Mutations are not all random**

It was previously thought that mutations occur completely randomly. However, it is becoming clear that there exist definite 'hot spots' where nucleotide changes occur, and also even where chromosomes break and re-unite.

**Plasmids**

Plasmids are intracellular sections of DNA found in a variety of bacteria, and also in some eukaryotic cells. They are often circular and replicate independently of the DNA of the host cell. Plasmids can be transferred horizontally between two cells without invoking heredity and cell replication, and they include a number of genes that confer antibacterial resistance.

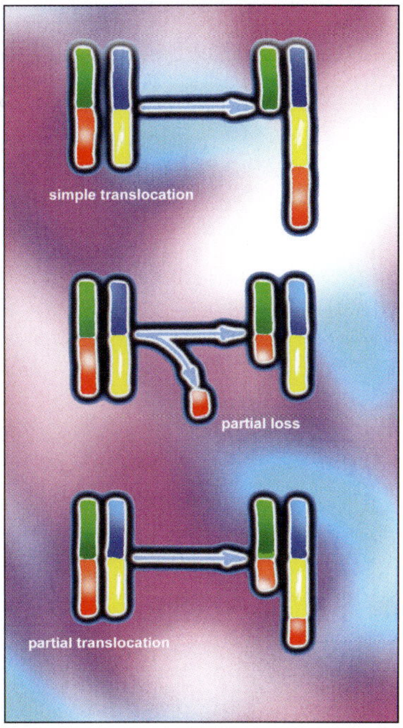


Genes can be translocated in their entirety, lost or broken up. Two chromosomes and four genes are depicted.

larger chromosome. Translocations do not usually produce any problems unless, in the process of breaking up and rejoining, a chromosome loses vital parts of an important gene.

The picture is complicated by the finding that mutations can actually be an active process. Some viruses can insert unstable portions of DNA (collectively called *transposons*) into chromosomes. Such transposons can copy themselves and seed replicas onto different parts of a chromosome, and in the process can induce mutations.

Errors can be induced by a number of factors such as radiation (including ultraviolet light), chemicals and even heat. Errors can also be introduced when DNA is copied by the enzyme *DNA polymerase III* prior to cell division. Whilst errors within the ordinary body cells (somatic cells) can result in faulty transcription resulting in faulty proteins and hence disease, only errors in the DNA replication within cells producing sperms or eggs (also called *germ* cells) can lead to disease or abnormalities in offspring.

Diseases caused by different genetic mutations are numerous and widespread. For example, the commonest serious hereditary condition in humans is *cystic fibrosis* which affects about one in every 1800 live births. This results in a defect in the mucus of the lining of the airways, which becomes more sticky and so causes obstruction in the smaller bronchi and infection. In *haemophilia B,* there is a mutation in the DNA sequence for one of the proteins (called *factor IX*) which is necessary for proper blood clotting. As a result the gene is not transcribed, factor IX is not produced, and this results in a bleeding tendency. *Familial adenomatous polyposis* is a condition in which hundreds of small polyps (growths of tissue) appear in the lining of the colon. Over time, these polyps have a tendency to undergo malignant change. The condition occurs with a frequency of 1:10,000 births and is caused by an abnormal gene which has been found to be located on the long arm of chromosome 5. Some types of human *colon cancer* can be caused by mutations of genes located on chromosomes 17 and 22. The defective genes that are associated with *retinoblastoma* (a type of cancer that arises in the eye), and *Duchenne muscular dystrophy* (causing progressive muscle damage and weakness), have also been identified.

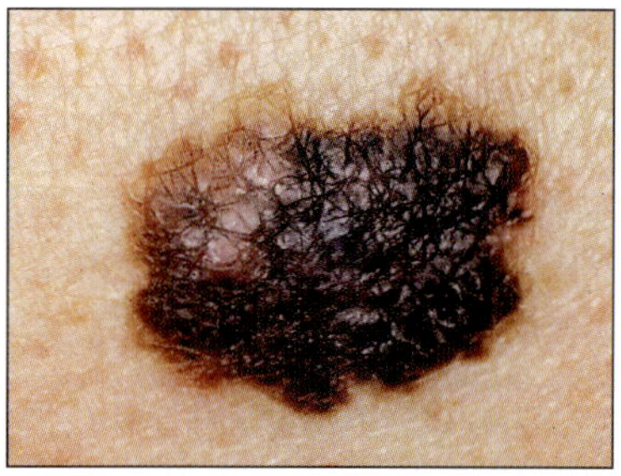

Malignant melanoma. This is a serious form of skin cancer and can be caused by damage to the skin cells by the ultraviolet component of sunlight.
About 70% of all malignant melanomas are the result of a mutation of the gene that codes for the so-called BRAF protein, 80% of which involve just one change (from thymine to adenine) out of the 2200 nucleotides in the gene.

Clearly, if genetic errors:

> ...were left totally uncorrected...cells might accumulate so much genetic damage that they could no longer function. In addition, the DNA in germ cells might incur far too many mutations for viable offspring to be formed. Thus the correction of DNA sequence errors in all types of cells is important for survival... Lodish et al. 'Molecular cell biology'. Scientific American Books. 1995. p385.

DNA errors that occur during the day-to-day activity of cells are to a large part rectified by DNA repair mechanisms. There is, for example:

> ...in E. coli...a remarkably diverse collection of enzymatic repair mechanisms and cellular responses to DNA damage with some **100 genes** participating in these processes. (My bold). Lodish et al. 'Molecular cell biology'. Scientific American Books. 1995. p386.

But if 100 genes are active in the repair kit of a modern E. Coli bacterium, then presumably millions of years ago there would have been a less efficient repair kit. This most likely would have allowed much greater numbers of deleterious mutations to accumulate, so resulting in the rapid degeneration and extinction of any (supposed) primitive cell which had any delusions of grandeur.

An example of disease caused by failure of DNA repair is *Xeroderma pigmentosa*, a rare hereditary disease in which the skin is abnormally sensitive to sunlight and even slight exposure can cause severe sunburn. This is caused by the ultraviolet components of light producing irreversible coupling of some of the nucleotides in the DNA of skin cells. In normal individuals, there is an enzyme repair mechanism which removes the coupled pairs and replaces them with normal nucleotides, thus restoring the DNA to its original form. However, as a result of mutations involving DNA repair genes, this enzyme system is absent in xeroderma pigmentosa and the errors in the DNA cannot be repaired. The abnormal DNA then causes multiple skin cancers.

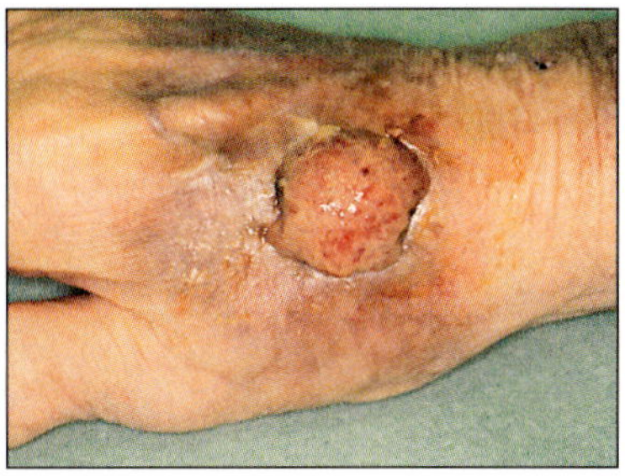
Squamous cell cancer on the back of the hand. The skin also shows severe sun damage.

DNA polymerase III is the main enzyme which catalyses the replication of DNA, and although it can occasionally make mistakes by inserting the wrong nucleotide, the enzyme has inherent proof-reading capabilities that can recognise an error and correct it. Since mutations affecting the gene encoding the delta sub-unit of DNA polymerase III inactivate the enzyme's proof-reading capability, resulting in a:

> ...massive increase in the rate of spontaneous mutations by some $10^5$ fold.
> Lodish et al. 'Molecular cell biology'. Scientific American Books. 1995. p385,

then (as with the absence of DNA repair mechanisms) it is implausible that any (supposed) primitive cell that managed to exist before the existence of the essential delta sub-unit of DNA polymerase III would have survived so as to evolve.

Nevertheless, it is commonly supposed and taught that although mutations are likely to result in deleterious effects, very rarely a mutation might be beneficial to an organism. And following millions of years and millions of very rare but beneficial mutations, then structures, organs and whole creatures could as a result evolve into new structures, new organs and new creatures. So, although it is extremely unlikely, suppose that there might indeed arise a *beneficial* mutation by an error during DNA replication. Since the proof-reading system knows nothing about the usefulness of any nucleotide sequence either for the present or the future, then it is likely that the mutation would be recognised as an error (even though it might potentially be useful) and *corrected back to the original code*, along with the correction of any other mutations (which in all probability would all be deleterious).

And since errors resulting in detrimental mutations would be far more likely to appear than errors resulting in potentially useful mutations, then if DNA repair mechanisms were to fail, they would more likely fail when dealing with a potentially detrimental mutation rather than with a potentially useful mutation. In other words, any (supposed) evolution cannot be considered as being just the natural selection of very rare (but supposedly useful) errors in DNA replication, but as being the natural selection of errors in the DNA that happen to have escaped the DNA repair mechanisms and the almost inevitable corrections back to the original form.

But let us suppose that rare beneficial mutations managed to escape proof-reading and DNA repair mechanisms, and survived into subsequent generations. It is still implausible that evolution could have proceeded. Suppose (although this too, would be most unlikely) that 3 mutations in a single gene over a series of generations 1, 1+*X*, 1+*Y* and 1+*Z* would result in a feather appearing in a dinosaur. Suppose that the first generation starts with the codons ABCDEFG and that the codons A B B D F F C would produce a feather. It might be expected that the changes would occur in a sequence something like this, leaving generation 4 with a feather:

| | | *Codon* | | | | | | |
|---|---|---|---|---|---|---|---|---|
| | | *1* | *2* | *3* | *4* | *5* | *6* | *7* |
| ***Generation*** | 1. | A | B | C | D | E | F | G |
| | 1+*X*. | A | B | ***B*** | D | E | F | G |
| | 1+*Y*. | A | B | ***B*** | D | ***F*** | F | G |
| | 1+*Z*. | A | B | ***B*** | D | ***F*** | F | ***C*** |

But, because of the random nature of mutations, if generation 1+Y is just waiting for a G to C mutation at position 7, it might instead be just as likely to acquire the mutation A to C in the 1st position, such that if a subsequent generation were to acquire a G to C mutation at position 7, the protein would still remain functionless.

| | | *Codon* | | | | | | |
|---|---|---|---|---|---|---|---|---|
| | | *1* | *2* | *3* | *4* | *5* | *6* | *7* |
| ***Generation*** | 1. | A | B | C | D | E | F | G |
| | 1+*X*. | A | B | ***B*** | D | F | F | G |
| | 1+*Y*. | A | B | ***B*** | D | ***F*** | F | G |
| | 1+*Z*. | C | B | ***B*** | D | ***F*** | F | G |
| | 1+*n*. | C | B | ***B*** | D | ***F*** | F | ***C*** |

Of course, in essence this is the *WYWYWG Principle* again. Thus, the dinosaur would most likely miss the feather boat completely.
While on the subject of mutations it is important to clarify a common misconception. Sometimes it is found that two *very* dissimilar organisms both have an identical gene. For example:

> A gene that in humans causes an inherited disorder of the nervous system has an **exact** match in yeast… (My bold). Steve Jones. 'Almost a whale'. Doubleday. 1999. p304.

The misconception is that simply the finding of such similar genes means that in some distant past the organisms involved once had a common ancestor. But, since over time mutations *must* occur, and since all proteins should be able to tolerate some considerable changes in their amino acid sequences, then no gene can exist for any *very* long period of time without acquiring at least some random alterations in its nucleotides. So, if it is found that in two (supposedly) *very distantly* related organisms, a particular gene has exactly the same nucleotide sequence, then this means that firstly, the two organisms cannot have existed for long enough to have acquired any sequence changes (certainly they could not have existed for millions of years), and therefore, secondly, the two (supposedly) *very distantly* related organisms *cannot* have had a common ancestor (since they cannot have existed for a long enough time to have evolved).

### *Protein synthesis*

Although genes code for the sequence of amino acids, proteins cannot be made directly from the DNA template. Furthermore, amino acids cannot be linked together directly from the mRNA template either. To actually manufacture a protein chain, four key classes of molecules are also necessary. One class comprises *transfer RNA* (tRNA). Although it is not quite true, for the purposes of our discussion it is fair to say that there are 20 different tRNA molecules, each of which has an *anti-codon* end which, like a lock and key, fits perfectly with and specifically to its complementary code in the mRNA. The other end of each tRNA has a non-specific site which can bind with any of the 20 amino acids. The purpose of a tRNA is to bring a specific amino acid (the one complementary to its anti-codon site) to an mRNA so that amino acids can be linked together in the specific order that is dictated by the code on the mRNA. If there was no more to the system than this, then the specificity of the DNA code would be of no use since any amino acid could bind with any tRNA (whether complementary to its anti-codon site or not). Furthermore, a tRNA cannot force an amino acid to attach itself to its non-specific site. The job of ensuring that only the complementary amino acid binds with the non-specific site on any tRNA is undertaken by its *complementary aminoacyl tRNA synthetase enzyme*, the second key molecule.

For practical purposes, there are 20 different aminoacyl tRNA synthetases, each capable of binding specifically with one of the 20 amino acids. Each aminoacyl tRNA synthetase collects its specific amino acid and then binds with its comple-

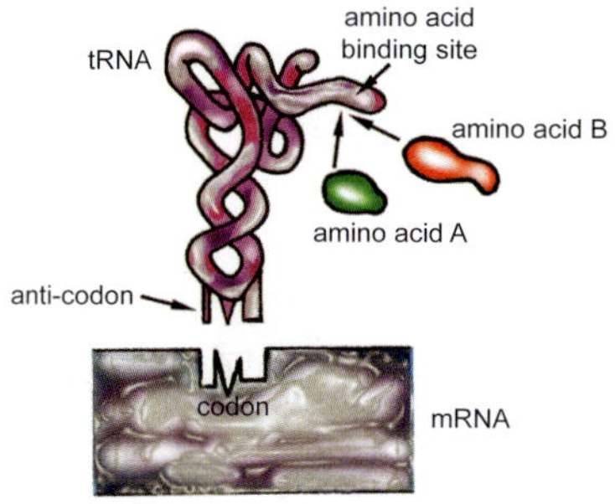


Although the anti-codon site of a specific tRNA fits only with its complementary codon on the mRNA, the amino acid binding site on any tRNA can bind any amino acid (in this case, both A and B). To ensure that only the correct amino acid (the one complementary to the codon of the mRNA) binds to the tRNA, a specific aminoacyl tRNA synthetase enzyme is capable of recognising and locking onto both a specific tRNA and its complementary amino acid.

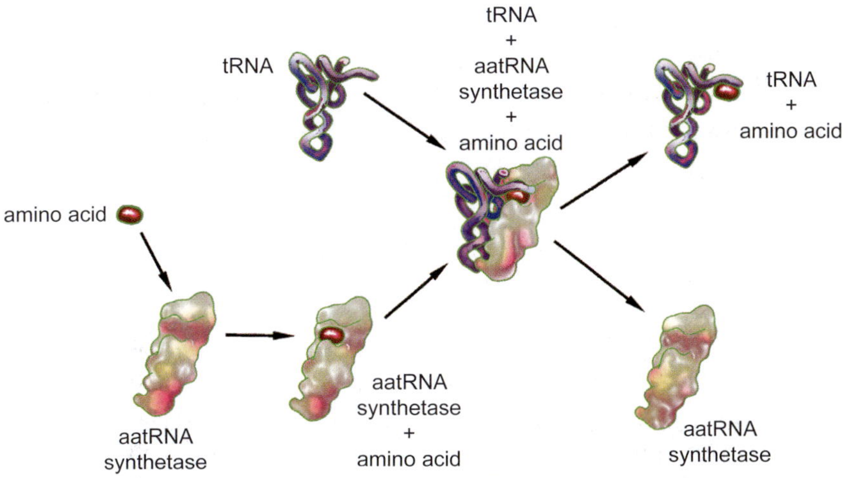


When an aminoacyl tRNA synthetase molecule carrying its specific amino acid comes into contact with its complementary tRNA, the amino acid is caused to bond onto the free arm of the tRNA. The aminoacyl tRNA synthetase then separates to leave the tRNA free to carry its amino acid to a ribosome, for subsequent incorporation into a growing protein chain.

mentary tRNA. In the process the amino acid is transferred to the tRNA, and the aminoacyl tRNA synthetase breaks away, and is then free to pick up another amino acid 'load'. Once primed with its complementary amino acid, the tRNA locks onto its complementary codon site on the mRNA.

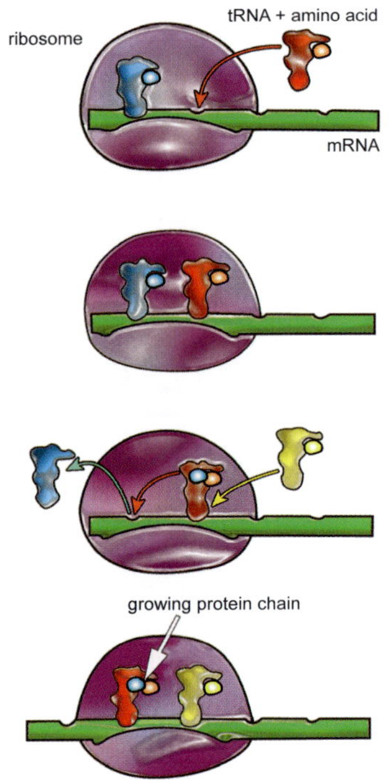


As the ribosome (purple) moves from left to right along the mRNA (green), different amino acids (shown here as blue, orange and yellow circles) are joined together by the enzyme peptidyl transferase.

The ribosome is the third key molecule (actually it is an intricately-linked collection of proteins) and functions as a sort of workbench on which the protein is constructed. There are two sites in the ribosome-mRNA complex that can accept a tRNA. The first tRNA occupies the *P-site* and a second incoming tRNA occupies the *A-site*.

When two tRNAs are in position in the ribosome, the fourth key molecule, the enzyme *peptidyl transferase*, makes a peptide bond between their two amino acids. As the amino acids are joined together the first tRNA is set free from the P-site and floats away, leaving the first amino acid bonded to the second amino acid, the latter being still attached to the second tRNA at the A-site. The ribosome then moves one codon length along the mRNA and in so doing, the second tRNA (carrying its two amino acids) moves into the P-site. The A-site is then free to accept another tRNA, carrying with it a third amino acid. As the process is repeated and other specific tRNAs and aminoacyl tRNA synthetases come and go, the amino acid chain is built up to produce ultimately the whole protein.

In fact, the whole process is far more complicated than this, and involves a large number of other associated molecules and enzymes without which the rate of formation of peptide bonds would be very slow. However, the ribosome protein-making factory can speed up reactions a million, or even a million million times. In a typical mammalian cell, more than one million peptide bonds are formed each second:

> Small proteins of 100-200 amino acids are…made in a minute or less. However the largest known protein, *titin,* found in muscle, has 30,000 amino acid residues and requires 2-3hr to make. Thus the machine that accomplishes this task must be precise and persistent. James Mauseth. 'Botany, an introduction to plant biology'. Saunders College Publishing. 1991.

So how could this machine evolve? An aminoacyl-tRNA synthetase molecule could by chance be complementary to the shape of a particular tRNA, but for it to be *also* complementary to the exact amino acid for which that particular tRNA's anticodon refers, is an enormous coincidence. Clearly, without the specificity with which any single aminoacyl-tRNA synthetase molecule can recognise and bond to a specific amino acid *and* its complementary tRNA, chaos would ensue. Thus the aminoacyl-tRNA synthetase is a protein molecule which has to be in existence *before* it itself, or most other proteins, can be made.

There is another problem raised by aminoacyl-tRNA synthetases. Since they have the same function, and if they did evolve from a common ancestral molecule, it would be expected that all aminoacyl-tRNA synthetases should have the same basic structure, individually modified to suit a specific amino acid. However, aminoacyl-tRNA synthetases:

> …form a diverse group of enzymes…over 100 such enzymes have been characterised…there is little sequence similarity among synthetases

specific for different amino acids. Donald and Judith Voet. 'Biochemistry'. John Wiley. 1995. p973.

In other words, aminoacyl-tRNA synthetases could not have had a common ancestral molecule. Since it is stretching the imagination and the statistics (see later) to expect even one aminoacyl-tRNA synthetase to appear by a random process, involving modifications to an ancestral protein fragment that did not itself have some aminoacyl-tRNA synthetase capabilities, it is implausible that all the aminoacyl-tRNA synthetases could have arisen independently from different ancestral molecules.

There are a number of other essential proteins involved in the process of protein manufacture, and abolition of the function of any one of these proteins causes cell death by blocking protein synthesis. This shows that a stepwise evolution of the protein synthesis machinery is implausible. For example, in the ribosome, the formation of peptide bonds between two amino acids is catalysed by the enzyme peptidyl transferase. This enzyme is blocked by the antibiotic *chloramphenicol*. The movement of a ribosome along an mRNA strand is facilitated by the protein *eukaryotic elongation factor 2* which is inhibited by the *Diphtheria toxin* (diphtheria used to be a common and fatal disease before the advent of antibiotics). The binding of a tRNA to the A-site in a ribosome is prevented by the antibiotic *tetracycline*.

So we have here a minimal complexity within the protein synthesis system without which a cell would die. Since, at some early stage in history, all (supposed) ancestral chemical conglomerates would have lacked this minimal complexity, it is implausible that living cells could have arisen by the natural selection of any favourable chemical interactions, the latter appearing by chance.

The production of a protein is not completed only by the linking together of its component amino acids. The specificity of an enzyme is to a large extent determined by its size and shape since the enzyme molecule has to be able to fit accurately (and preferably *only*) around or alongside its intended substrate (a substrate being the atom or molecule which the enzyme specifically acts upon). But, left to its own devices, immediately a protein chain has been manufactured, it would simply be an arbitrarily-folded ribbon of amino acids. So, before it can function effectively, a newly-formed chain of amino acids has to become folded into its own unique 3-dimensional shape (its *'native conformation'*).

Obviously, if an enzyme folds such that its active site is either distorted or is not formed at all, then as a consequence the enzyme will be less efficient or it will fail completely. An example of faulty folding is the protein *$PrP^{Sc}$*. Normal brain cells contain a protein called *PrP* which, when left to its own devices, folds in a particular way and is called $PrP^{c}$. However, when it comes into contact with an abnormally folded form called $PrP^{Sc}$, the PrP protein folds into the $PrP^{Sc}$ form and produces large clumps. These clumps of $PrP^{Sc}$ cause severe brain damage and death. The condition is called by different names in different animals: *scrapie* in sheep, *mad cow disease* in cattle and *Jacob Creutzfelt* disease in humans. Ingestion of meat containing the abnormal protein carries a risk of the disease, although reported cases are rare.

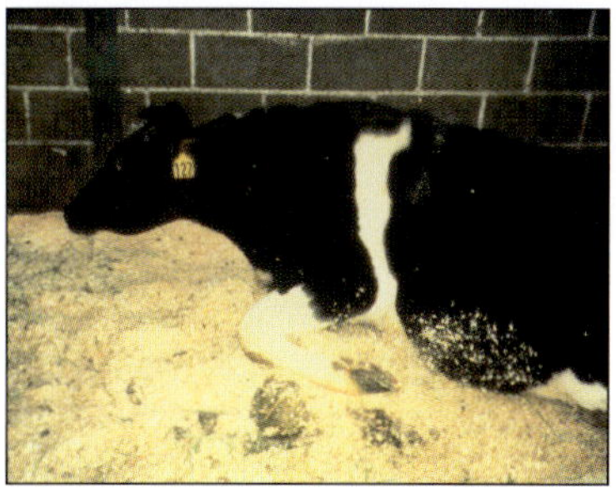

Cow suffering from bovine spongiform encephalitis (mad cow disease). The cow is digging frantically for no reason.

Since amino acids have both charged (*hydrophilic* – attracted to water) facets and uncharged (*hydrophobic* – repelled by water) facets, just as charged particles tend to be soluble in water and uncharged particles tend to be insoluble in water, so the uncharged aspects of any amino acid molecule tend to tuck inwards and away from the surrounding water molecules, while the charged aspects are attracted outwards to make weak bonds with the surrounding water molecules. So it might be thought that a protein achieves its native conformation simply by the outworking of the diverse attracting and repelling forces.

But this is not how it happens. A protein does not simply fold into its native conformation by a *random* process that explores all the myriad permutations of bends and twists, and tucks and folds available to it after the amino acid chain is manufactured. For a rather small protein comprising 100 amino acids, the time it would take would be in the region of:

> ...$10^{87}$ seconds, which is immensely more than the apparent age of the universe (20 billion years, 6 x $10^{17}$s). It would obviously take even the smallest protein an absurdly long time to explore all its possible conformations. Donald and Judith Voet. 'Biochemistry'. John Wiley. 1995. p194.

In the test-tube, the process is inefficient since most proteins fold slowly in a matter of minutes to days, with:

> ...low efficiency into quasi-stable non-native conformations or non-specific aggregates...Yet [inside a cell] many proteins fold to their native conformation in less than a few seconds. Donald and Judith Voet. 'Biochemistry'. John Wiley. 1995. p198,194.

Therefore, *inside a cell*, the folding processes must be directed by factors other than just the amino acid sequence. In fact, proteins fold into native conformations while they are in the process of being synthesised in the ribosome, where a number of *accessory proteins* assist in the folding process. Some accessory proteins are actually synthesised *integral* with the main protein itself. For example, the cow protein *bovine pancreatic trypsin inhibitor* (BPTI) is synthesised with a special 13-amino acid section tagged onto one end of the molecule. This short sequence of amino acids acts to aid folding of the main protein, and once folding is complete, the tagged section is excised by yet another protein, and discarded. As we shall see time and time again, parts of proteins are generally of little use, and indeed altering just one amino acid in the tagged section of BPTI (for example, by replacing *cysteine* with *alanine*) *abolishes* its folding activity. Without the tagged section of BPTI, and in particular the essential cysteine component, BPTI could not be formed. So, it is implausible that BPTI could have evolved by a stepwise process from a simpler ancestral molecule.

In addition, cells also contain essential *chaperone* proteins which prevent clumping and faulty folding of a newly-formed protein. Normal unfolded proteins have numerous hydrophobic points along the chain which cause the protein to fold into useless aggregates. Such clumping is prevented by chaperone proteins which bind temporarily with the hydrophobic points.

So, from the genetic code embedded in DNA to the manufacturing process in a ribosome, the production of a functional protein requires the existence of an intricate series of complicated and interdependent processes. This demonstrates the *Irreducible Complexity Principle*:

If, for it to function, a system requires a minimum number of interdependent components, and if no component could have had any function outside that system, and if without all the components being present that system cannot function *at all*, then those components could not have evolved from simpler entities and that system could not have evolved from any fewer number of components.

Proteins carry out thousands of essential structural, triggering and catalytic functions within living cells:

> Proteins are **designed** to bind every conceivable molecule - from simple ions to large complex molecules like fats, sugars, nucleic acids, and other proteins. They catalyze an extraordinary range of chemical reactions, provide structural rigidity to the cell, control flow of material through membranes, regulate the concentrations of metabolites, act as sensors and switches, cause motion, and control gene function. The three-dimensional structure of proteins **has evolved** to carry out these functions efficiently and under precise control. (My bold). Lodish et al. 'Molecular cell biology'. Scientific American Books. 1995. p52.

A catalyst is a substance which helps to make a chemical reaction proceed more efficiently than it would in the absence of the catalyst. An enzyme is a specific form of catalyst, made up of one or more protein chains. The specificity of an enzyme for a particular chemical reaction is determined by its unique 3-dimensional shape, which is itself in part determined by the linear order in which the amino acids are arranged. The 'average' protein may comprise a chain of around 100 amino acids, with the whole chain bent and twisted into a complex three-dimensional shape, itself made up of sub-sections such as a helix, a thread or a random coil, or any combination of these, along with other forms such as a globule. Each of the different specific sub-sections within a protein is called a *domain*, and the shape of each domain is determined by the linear order of its specific amino acid sequence. Many domains comprise utilitarian shapes and are therefore found in a number of different proteins.

So, the individual shape and identity of a protein is provided by the stringing together, in different permutations, of a number of different domains, in association with other sections of amino acid sequences which are unique to a particular protein. It is only when a protein has the correct 3-dimensional structure that it is able to function effectively, so the form and function of a protein are inseparable. If a mutation affects part of a gene relating to an *essential* aspect of a domain, this will, most likely, result in a serious defect in the domain, and if the domain is crucial for the function of a protein, there might also result a failure of function of the whole protein molecule.

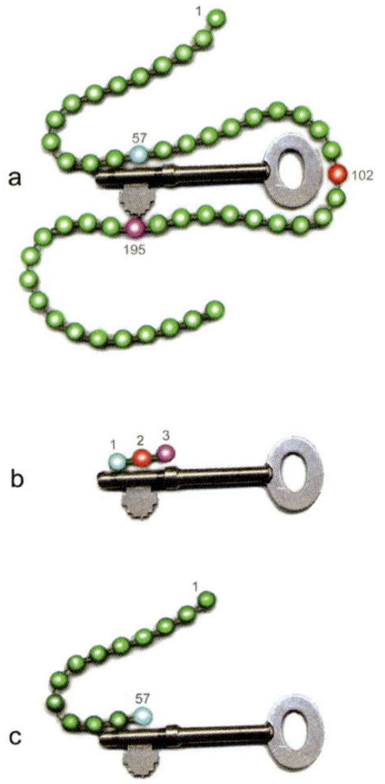


**Chymotrypsin**

Chymotrypsin gives an example of the relationship between the essential amino acids comprising the active site of an enzyme and its substrate* shape. Chymotrypsin has three essential amino acids at positions 57, 102 and 195 (**a**). If any one of these amino acids is not present the enzyme cannot function:
1. Without a chain of other amino acids forming a backbone within which these three essential amino acids can be embedded, the enzyme cannot function.
2. A chain comprising the three essential amino acids alone cannot make up an active site (**b**).
3. There must exist a certain minimum number of amino acids in the chain so as to allow, after folding, for the formation of the correct spatial positioning of the three essential amino acids at the active site.
4. Since a minimal size of chain and a specific spatial separation of the three essential amino acids are essential for the enzyme to function *at all*, it is impossible for the enzyme to have evolved by sequential additions of the correct amino acids in the order 1 to 241. (For example, the first 57 amino acids would have no enzyme activity (**c**), and the first 100 would only contain one of the three essential amino acids and so would still have no enzyme activity.)
These same problems apply to the supposed evolution of all other enzymes.

* A substrate is the specific molecule on which an enzyme acts.

### Proteins lacking essential parts are of no use

Not only does the manufacture of proteins require a number of processes, each showing irreducible complexity, but proteins themselves have a minimal complexity without which they cannot carry out their specialised functions. Within the protein molecule, certain sequences of amino acids relate to essential parts of the molecule and these *cannot* be altered without seriously altering or destroying completely the function of the molecule. Other sequences of amino acids relate to the non-essential parts of the molecule and these *can* be altered without significantly affecting the overall function. Now, it is *inevitable* that over time, the genetic code for any given protein will undergo mutations on a random basis. Those mutations which relate to non-essential aspects of a protein molecule will be able to persist in subsequent generations since function is maintained and no harm is caused to the individual. However, any mutations which relate to the essential aspects of a protein molecule that are vital to life will, most likely, cause an impairment to the individual of varying degrees, or death. As a result, such mutations will tend not to be passed on to subsequent generations.

### *Chymotrypsin*

The enzyme *chymotrypsin* is found in the intestines of animals, and is responsible for the breakdown of the proteins in food into smaller sections that can be subsequently absorbed into the blood stream. Contained within its amino acid sequence:

> The three residues (histidine, aspartate, and serine) are **crucial** for catalytic activity... (My bold). Lodish et al. 'Molecular cell biology'. Scientific American Books. 1995. p77.

The amino acids *histidine*, *aspartate* and *serine* form the catalytic region of the active site of chymotrypsin. Although these amino acids are widely separated along the protein chain (being at positions 57, 102 and 195 respectively), nevertheless:

> ...they are brought together into close proximity in the native folded conformation... Lodish et al. 'Molecular cell biology'. Scientific American Books. 1995. p77.

Because of both the physical size and shape of the cleft at the active site in the chymotrypsin molecule, and also the way in which charges are exposed in the area, chymotrypsin can cleave protein chains by acting specifically on any *phenylalanine tyrosine* and *tryptophan* amino acids present:

> ...the residues lining the cleft are in the right position to participate in hydrophobic interactions with the large hydrophobic side chains of [the amino acids] phenylalanine tyrosine and tryptophan... Lodish et al. 'Molecular cell biology'. Scientific American Books. 1995. p77.

Since, out of the 241 amino acids comprising chymotrypsin, three widely separated amino acids are essential for its function, it is implausible that any ancestral molecule comprising *significantly* smaller portions of the complete protein chain, could have allowed the three essential amino acids to come together to maintain their active site.

So, incomplete ancestral molecules could not have had *any* function at all. Without any function, any non-functioning (even if correct) part sequences could not have been selected for. So, chymotrypsin could not have evolved.

### Histones

*Histones* are the most abundant proteins found associated with the DNA of eukaryotes, and are essential for maintaining the structural integrity of the DNA. There are 5 major types of histones (H1, H2A, H2B, H3, and H4) and it is remarkable that out of the 103 amino acids in the H4 histones of cows and peas, the molecules differ only at *two* sites in the amino acid sequence; the histones of cows and yeasts also differ at only *two* sites; and the H3 histone of the sea urchin differs from that of cattle by only *one* amino acid. In fact, there is little difference between the histones of *all* organisms:

> Such rigid evolutionary stability implies that the...histones have critical functions to which their structures are so well tuned that they are all but intolerant to change. Donald and Judith Voet. 'Biochemistry'. John Wiley. 1995. p1125.

But the reason very different creatures have virtually identical histones is that most changes would result in a useless molecule. Thus the crucial nature of their amino acid sequences means that histones could not have been produced piecemeal by small stepwise changes to any other (supposedly) ancestral polypeptide sequences.

### Cytochrome c

*Cytochrome c* is an essential enzyme which is found in virtually all living cells, and which takes part in the highly complex series of chemical reactions (collectively called the *electron transport chain*) that results ultimately in the production of high energy phosphate bonds in adenosine triphosphate (ATP):

> It is believed that the electron-transport chain took its present form between 1.5 and 2 billion years ago as organisms evolved the ability to respire. Donald and Judith Voet. 'Biochemistry'. John Wiley. 1995. p126.

In vertebrates, cytochrome c contains 103-104 amino acids, and being an essential enzyme, it can only tolerate very small changes in its amino acid sequence:

> ...cytochrome c is an evolutionary conserved molecule. A total of 38 of its 105 residues...are invariant. Donald and Judith Voet. 'Biochemistry'. John Wiley. 1995. p126-7. (The term 'Residue' = amino acid)

Amongst all living things these 38 amino acids are invariant at the positions shown below in red:

1 2 3 4 5 6 7 8 9 10 11 12 13 14 15 16 17 18 19 20 21 22 23 24 25 26 27 28 29 30 31 32 33 34 35 36 37 38
39 40 41 42 43 44 45 46 47 48 49 51 52 53 54 55 56 57 58 59 60 61 62 63 64 65 66 67 68 69 70 71 72 73
74 75 76 77 78 79 80 81 82 83 84 85 86 87 88 89 90 91 92 93 94 95 96 97 98 99 100 101 102 103 104

Donald and Judith Voet. 'Biochemistry'. John Wiley. 1995. Table 6.4. p129-130.

This means that over the last (supposed) 1.5-2 billion years, mutations in the cytochrome c gene that altered any of the invariant amino acids have resulted in non-functioning molecules and have been selected against. Of course, this is because 38 of the amino acids are *essential* for cytochrome c's function, and so no (supposed) primitive fragment missing *any* of those 38 amino acids could possibly function *at all*. For example, with regard to the amino acids *histidine* at position 18 and *methionine* at position 80, it can be shown that:

> ...the substitution of **any** other residues in these positions inactivates the protein. (My bold). Donald and Judith Voet. 'Biochemistry'. John Wiley. 1995. p127.

From this we have the *Essential Sequence Principle*:

No enzyme (nor any of its hypothetical, shorter ancestral molecules), missing any of the amino acids that are *essential* for the integrity and minimal function of the enzyme's active site, could possibly function at all.

And of course, the lack of function of a molecule, system or structure means that natural selection cannot act so as to perpetuate it. And without natural selection there can be no evolution.

Of course, if cytochrome c is so crucial to living things then it *ought* to be extremely similar in all living things. And indeed, whether we look at yeast, a pumpkin or a sunflower, a fruit fly or a donkey, a whale or a human, 38 of the amino acids are always found in exactly the same position in the sequence. Although cytochrome c of horses and monkeys differs from human cytochrome c in 12 and 1 amino acid sites respectively, at first glance perhaps suggesting a closer evolutionary relationship between monkeys and humans than between horses and humans, the sequence of the *great grey kangaroo* is closer to that of humans than it is to that of the horse. But since placental mammals are supposed to have evolved *from* marsupials, the kangaroo sequence should by now show far more differences than it does. Thus, the sequence of cytochrome c does not necessarily show an evolutionary trend - it simply shows more similarities between some groups of organisms and less in others.

### *Haemoglobin*

*Haemoglobin* is the molecule which transports oxygen around the blood and which gives red blood cells their colour. Each haemoglobin molecule is made up of two $\alpha$ chains (each comprising 141 amino acids) and two $\beta$ chains (each comprising 146 amino acids). The protein chains are associated with 4 non-protein components called *haem* and at the centre of each haem is an iron atom. Both the haem group and the protein chains are together necessary for the job of binding oxygen. *Myoglobin* is the molecule that temporarily stores oxygen within muscles and occurs as a single protein chain with a single haem group. The sequences of sperm whale myoglobin (which has 153 amino acids) and the alpha chain of human haemoglobin (which has 141 amino acids) have 38 amino acids in common. It is thought by some that the differences:

> ...**presumably** represent insertions or deletions of nucleotides in the DNA sequences during the evolution of myoglobin and the haemoglobin α chain from a common ancestor... (My bold). Lodish et al. 'Molecular cell biology'. Scientific American Books. 1995. p69.

If that were the case it would mean that even after millions of years, myoglobin and haemoglobin have not been able to tolerate any change in 38 of the amino acids in their sequences. This suggests that any mutations affecting these key positions would result in the polypeptide chain undergoing marked configurational changes, such that the protein would not be able to bind oxygen to *any* valuable degree. So it could not have evolved stepwise from a simple ancestral molecule.

Haemoglobin raises another problem since it is thought to have arisen from an ancient globin gene that first appeared (supposedly) over a billion years ago. This ancient gene is thought to have become duplicated and changed over time to give rise to a number of haemoglobins (e.g. α, β, γ, δ), each of which have slightly different degrees of oxygen-binding capacity. Over the course of time the patterns of haemoglobins found in modern animals, from sharks to mammals, are thought to have slowly changed:

> The different β-globin genes **probably** arose by duplication of an ancestral gene...Over evolutionary time the two copies of the gene that resulted accumulated random mutations; beneficial mutations that conferred some refinement in the basic oxygen-carrying function of hemoglobin were retained by natural selection. Repetitions of this process are **thought to have resulted** in the evolution of the contemporary globin-like genes observed in humans and other complex species today. (My bold). Lodish et al. 'Molecular cell biology'. Scientific American Books. 1995. p314.

But, in terms of both time and genetic space, both the present day shark and the present day mammal are equally distant to their (supposed) common ancestor, an ancient shark. Over a long period of time we should expect that the *essential* sequences of haemoglobin would remain completely unchanged in all generations and in all subsequent lineages, be they fish or reptiles or mammals. And, over a *very* long period of time we should expect that the non-essential sequences might be *completely* (or very significantly) altered in a random manner. Therefore, over a very long period of time *all* breeding groups of animals, be they sharks or reptiles or mammals, should have retained all of the essential sequences in common, with the non-essential sequences showing a whole spectrum of variation. After 250 million years the only common patterns should be those of the essential sequences. Yet this is not the case. And there are other irregularities. For example, the lamprey is classified as a primitive animal which lacks a true backbone but it is thought to resemble an ancestor of all vertebrates. However, at 81%, 78% and 76% respectively, the frog, chicken and kangaroo share a greater percentage of the lamprey's haemoglobin amino acid sequence than does its (supposed) closer evolutionary relative, the carp, which shares only 75%.

NIDDK

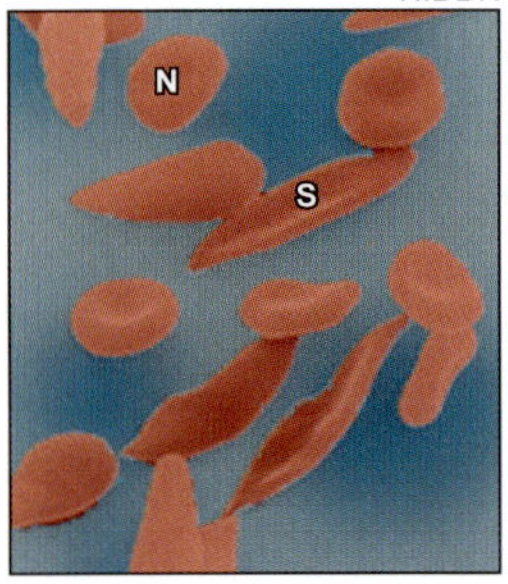


Normal red blood cells (R) are seen as biconcave discs. The sickle cells (S) are abnormal and elongated. (False colour)

### *Sickle cell anaemia*

We have seen that proteins do not tolerate any alterations in their essential amino acid sequences. But, nevertheless, a seriously altered protein can sometimes survive. For example, a simple substitution of the amino acid *valine* for the normal amino acid *glutamic acid*, in the sixth position of the β-chain in the human haemoglobin molecule, causes the serious disease *sickle cell anaemia*. In this condition the red cells readily deform into stiff crescent shapes, and being inflexible they tend to get stuck in and block up the smallest blood vessels, the capillaries. Such blockages to the blood supply result in sickle 'crises' which cause severe and extensive damage to the affected tissues, and are associated with excruciating pain.

Compared to the cells of most other tissues in the body, red blood cells have limited metabolic capabilities, and so naturally degenerate with a life expectancy of around 120 days. Old red blood cells are continuously removed from the circulation by the spleen and their loss is balanced by new red blood cells made by the bone marrow. In sickle cell anaemia, however, the abnormal red cells are far more fragile and the life expectancy of an affected red cell is about half that of normal. Since the bone marrow is unable to keep pace so as to make good the excessive loss of red cells, affected individuals suffer with chronic *anaemia*, with reduced numbers of red blood cells. Affected individuals are ill and lacking in energy, and have approximately a 600-fold increased incidence of serious infection (such as meningitis, septicaemia, bone and intestinal infections) when compared to normal people. Before the second half of the 20th Century the majority of individuals with sickle cell anaemia did not survive into adulthood.

Since it is a recessive condition, sickle cell disease only affects *homozygous* individuals (those who carry abnormal genes from both parents). Although their red cells have a slightly shorter life span than normal, *heterozygote* individuals (those with an abnormal gene from only one parent) can lead essentially a normal life. Malaria is still probably the world's most problematic infectious disease:

> Of the 2.5 billion people living within malaria-endemic areas, 100 million are clinically ill with the disease at any given time and around one million, mostly very young children, die from it each year. Donald and Judith Voet. 'Biochemistry'. John Wiley. 1995. p124.

It is because heterozygotes seem to have some greater protection against malaria than individuals with normal haemoglobin, that HbS is often cited as an example of a beneficial mutation. And so, it is believed that although most mutations are deleterious, nevertheless, given sufficient time, enough beneficial mutations can arise to result ultimately in new structures and systems:

> … sickle cell anaemia provides a classical Darwinian example of a single mutation's adaptive consequences in the ongoing biological competition among organisms for the same resources. Donald and Judith Voet. 'Biochemistry'. John Wiley. 1995. p125.

But does sickle cell anaemia provide a classical Darwinian example? Is the (abnormal) HbS simply extending the evolutionary journey of the haemoglobin molecule so that will it go on, at some distant time ahead, to become a new and better breed of haemoglobin, or even some other hitherto unknown molecule with a hitherto unknown function? Or is HbS, in its heterozygous form, (in the same way that being deaf, while in itself being disadvantageous, might nevertheless protect an affected individual from being killed in a war, by excluding him from being drafted into the army) just a lucky coincidence?

In fact, HbS is found in many areas where malaria is not endemic so its persistence in such areas cannot be ascribed to some protection against malaria. Haemoglobin S occurs and persists in these areas as a result of population migrations and due to the fact that, in heterozygous form, the presence of the HbS gene is not lethal. And since normal individuals *also* survive very well in endemic malarial areas (60% of individuals in endemic malarial areas have normal haemoglobin), this shows that for survival in endemic malarial areas, the notion that HbS is superior to normal Hb is not clear-cut.

Furthermore, many tribal populations that live out their lives within endemic malarial areas, with few marriages outside each tribe, do not have HbS *at all*. From this we can say unequivocally that HbS is *not* essential for survival in endemic malarial areas. So HbS does not provide an example of evolution at work at all. And, of course, the question arises: if altering just *one* out of its 141 amino acids causes such damage to the haemoglobin chain, how did the haemoglobin β chain function before it ever evolved the correct amino acid in the (vital) 6th position?* Thus, HbS should be seen not as a new protein perhaps on its evolutionary journey towards making a 'better' haemoglobin, but rather as an *abnormal* form, which *manages to persist simply because, in heterozygous form, it is not seriously detrimental*. Thus the persistence of HbS is, to some extent, the biochemical equivalent of the Galapagos cormorant which survives despite its abnormal wings.

***The original haemoglobin**

Since individuals with sickle cell anaemia have *abnormal* haemoglobin β-chains, at some time in the past, their ancestors would have had *normal* haemoglobin b-chains. Since evolution requires stepwise changes over time, then further back in time *none* of the ancestors of those with normal haemoglobin would have had the correct and crucial glutamic acid in the 6th position. This means that at some time in the past, there would not have been ANY normal haemoglobin b-chain in ANY organism. We do not see, and we do not expect individuals with sickle cell anaemia to spontaneously acquire a reversal (ie. valine back to glutamic acid) of the 6th position mutation. Which means that ***all*** our distant ancestors would have suffered from sickle cell anaemia (and worse), and died off. So, normal haemoglobin could not have evolved from abnormal haemoglobin.

The only logical conclusion is that *from the beginning*, all haemoglobin b-chains were normal and complete, and what we see in sickle cell anaemia is a degeneration of that previously perfect state.

### *Dystrophin*

*Dystrophin* is a huge protein that provides a link between certain proteins inside a muscle cell and other proteins that lie across its cell membrane. One of its functions is to maintain the structural integrity of the muscle cell, and without it the muscle cells degenerate. The dystrophin gene is located on the X chromosome and, containing over 2 million nucleotides, is the largest gene known in humans. As a result of a defect in the dystrophin gene part of the dystrophin protein cannot be manufactured, and this causes muscle damage and weakness. Mutations in different parts of the gene result in different degrees of defect, and some mutations have little effect at all. For example, *Duchenne muscular dystrophy* is a progressive degenerative condition affecting the muscles of about 1 in 3500 boys. In a milder form of the disease known as *Becker muscular dystrophy*, partially functioning dystrophin molecules are produced. Nevertheless, certain parts of the dystrophin molecule are *essential* for function since any mutations in the *carboxy-terminal* region almost always result in severe muscular degeneration.

It follows, therefore, that any (supposed) 'ancestral' dystrophin molecule fragment could not have had any function without first having the complete carboxy-terminal section, along with the necessary minimal length of protein chain.

### *Drug resistance does not produce new organisms*

The emergence of drug resistance amongst micro-organisms is often cited as evidence in support of evolution. However, there are two important points to note: firstly, the resistance to drugs occurs only as a result of changes to an existing protein, and this tells us nothing about how that protein came into being in the first place. Secondly, mutations that result in alterations of biochemical processes in micro-organisms *do not* arise *in response* to the antibiotics that are used to kill them:

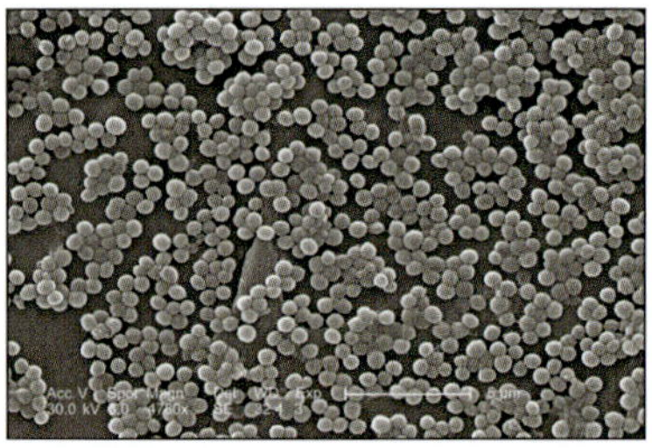
Scanning electron micrograph depicted numerous clumps of methicillin-resistant Staphylococcus aureus bacteria (MRSA).

> The…point to make is that resistant genes and mechanisms existed long before antibiotics were used. For example, antibiotic resistant bacteria have been isolated deep within glaciers in Canada's high Arctic regions, estimated at 2000 years old. C.A Hart. 'Antibiotic resistance: an increasing problem?' British Medical Journal. Vol 316. 25th April 1998. p1255-6.

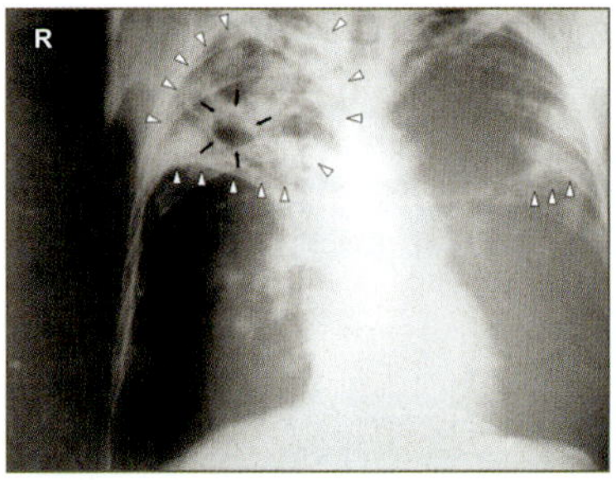

Chest X-ray showing lung damage (white arrows) with cavitation (black arrows), due to tuberculosis.

And although some examples of drug resistance are well-known, for example, the methicillin-resistant *Staphylococcus aureus* (*MRSA*, which causes boils, carbuncles and also serious infection of the blood), the penicillin-resistant *Pneumococcus* (which causes pneumonia), and resistant *Mycobacterium tuberculosis* (which causes tuberculosis), nevertheless, just as we have seen with larger organisms such as fishes and birds, these are examples of *variation* and not *speciation*. And as with larger animals, some micro-organisms are more likely to undergo chromosome changes and show marked variations than others. So, some germs are *unable* to produce resistant forms. Whilst, for example, *Staphylococcus aureus* is becoming increasingly resistant to penicillin, many other bacteria demonstrate little such ability:

> ...Streptococcus pyogenes remains sensitive to penicillin, as do Chlamydia trachomatis to tetracyclines...and most anaerobes to metronidazole. Dr Ronald Salmom. 'Antibiotic resistance: an international problem'. Update magazine. 6th May 1998. p857.

By way of examples of mutations in micro-organisms, let us consider drug resistance in some viruses, and human immunity to a bacterium.

#### *Human immunodeficiency virus*

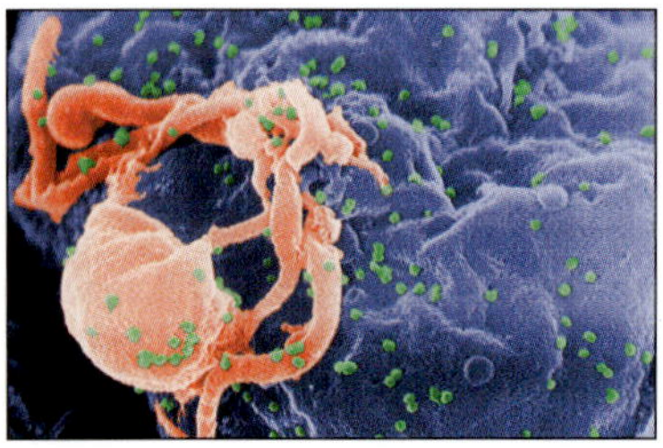
Scanning electron micrograph of HIV-1 particles (green) budding from a white blood cell (purple). (False colour)

AIDS (Acquired Immune Deficiency Syndrome) is caused by the *Human Immunodeficiency Virus* (HIV), which damages cells of the immune system. The condition is fatal if left untreated. Upon entering a cell, viruses hijack the host machinery to produce copies of their own kind. The virus uses its own *reverse transcriptase* enzyme to cause the host DNA to make replicas of its own (viral) RNA. Amongst the drugs used for treatment of HIV are *reverse transcriptase inhibitors*, which are designed (by reacting chemically with the reverse transcriptase enzyme) to interfere with the viral replication. The HIV reverse transcriptase has a sequence of over 250 amino acids. As individual codons representing these amino acids by chance mutate, so a particular drug might become no longer able to react chemically with some particular mutant reverse transcriptase. As a result the drug will not be able to render the enzyme ineffective. Any virus particle containing such an altered enzyme would thus have some degree of resistance to attack by a particular drug. All that 'resistance' means in this case is that being unaffected by the drug, the virus with the altered enzyme will still be able to replicate. However, it is important to understand that the mutant virus just has a slightly altered but fully functional *reverse transcriptase*, and that the virus is still very much an HIV. No *new previously-unknown enzyme* or *gene* or *virus has been produced*.

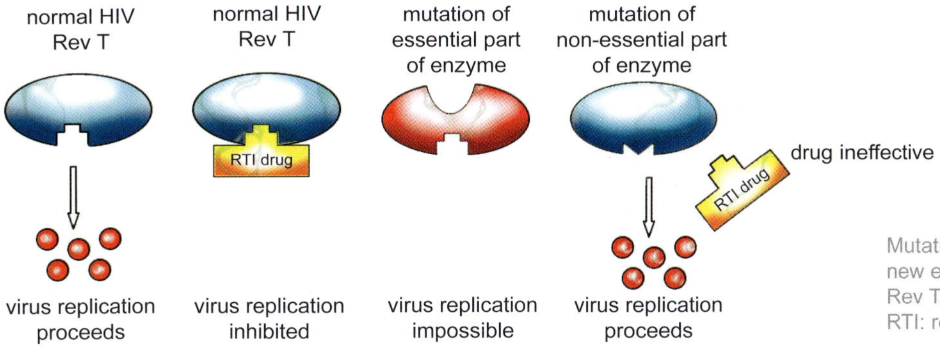


Mutations of the HIV do not produce new enzymes.
Rev T: reverse transcriptase
RTI: reverse transcriptase inhibitor

## *Herpes simplex virus*

The Herpes Simplex virus (HSV) causes diverse problems from cold sores to genital infections. Unlike the HIV, the HSV shows a high degree of fidelity in replication, and therefore, mutations are relatively infrequent. However, after prolonged treatment with drugs such as *acyclovir*, resistant mutant strains do appear.

Before it can act to block virus replication, acyclovir has to be first activated (involving a process called *phosphorylation*) by the virus's own enzyme *thymidine kinase*. Although other mutations are seen, most resistant strains of the HSV survive by having mutant thymidine kinase that is sufficiently faulty so as to be non-functional. Thus the resistant virus survives by being *unable to activate* the very drug that is designed to destroy it. Again, no new enzyme, gene or virus has been formed.

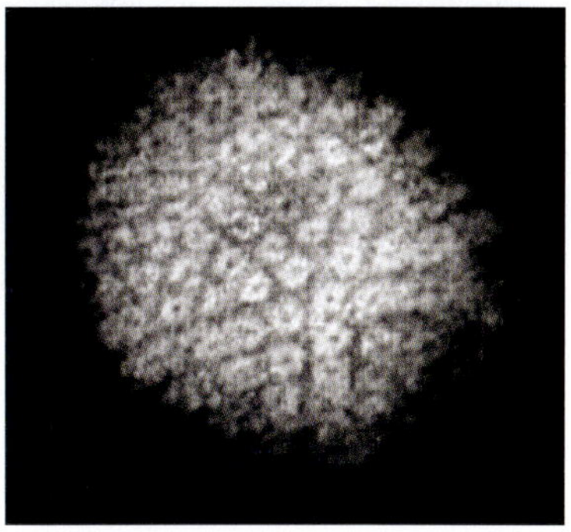

Herpes simplex virus

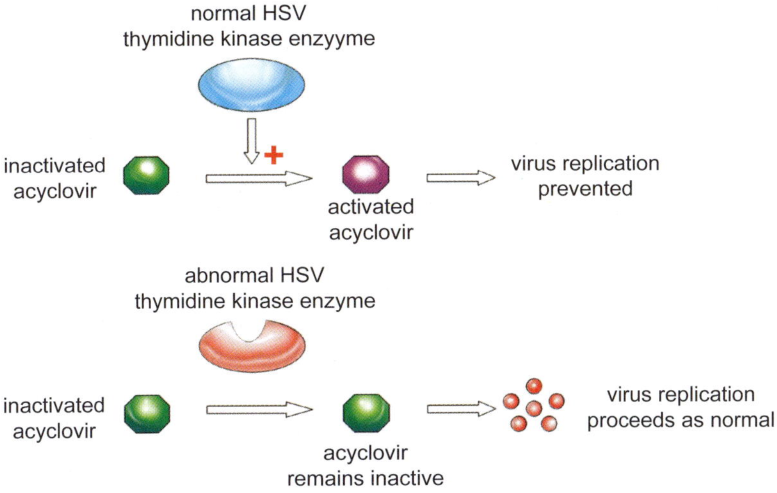


Mutations of the HSV do not produce new enzymes.

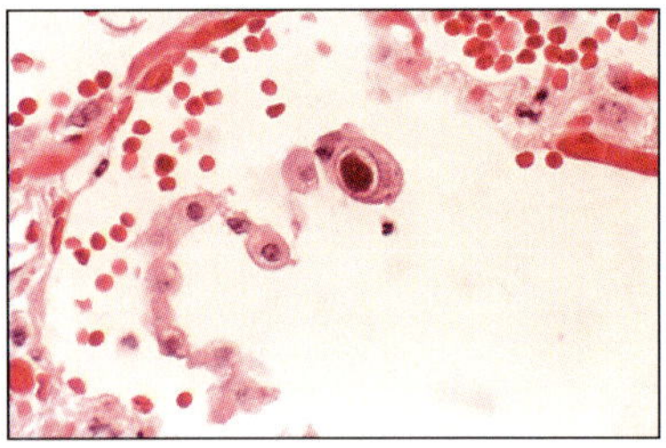
Cytomegalovirus infection of a lung cell. The central cell displays the dramatically enlarged nucleus characteristic of cytomegalovirus infection.

### Cytomegalovirus

The *Cytomegalovirus* (CMV) can cause a glandular fever type of illness. When the CMV invades the host body it uses its own *DNA polymerase* enzyme (a protein comprising around 1000 amino acids) to replicate itself. The drug *Ganciclovir* acts by interfering with the proper function of the CMV's DNA polymerase. Of the few resistant mutations of the virus, one type results in it being resistant to ganciclovir by the substitution of just one amino acid (*glycine* by *alanine*) in position 987 of the DNA polymerase amino acid sequence. However, the altered DNA polymerase is still the same enzyme with the same function, and the CMV is still the same virus, so no new enzyme, gene or virus has been formed.

### Salmonella typhimurium

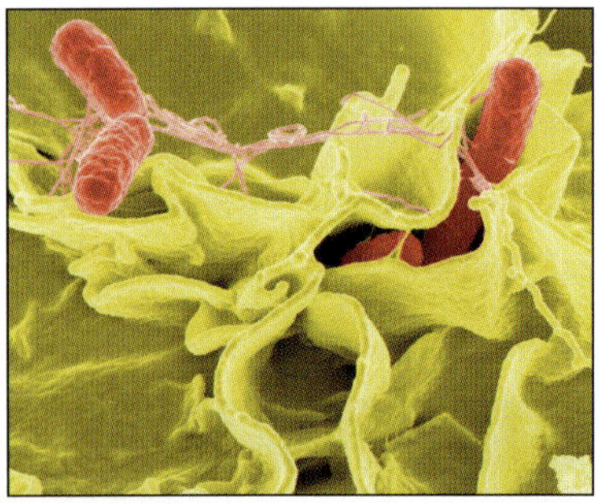
Color-enhanced scanning electron micrograph showing Salmonella typhimurium (red) invading cultured human cells. (False colour)

*Salmonella typhimurium* is a bacterium that causes food poisoning, resulting in diarrhoea and vomiting. The bacterium normally has two surface proteins (called H1 and H2) which are coded for by different genes and which are located at distant sites on its single chromosome. Human resistance to this bacterium usually occurs through producing antibodies against these surface proteins. The H2 gene is coupled with an adjacent gene (called rH1) which inhibits the activation of the H1 gene. Thus, although *both genes are present in all individual salmonella cells*, any individual bacterium normally only manufactures one of these proteins. So, a person immune to a bacterium which manufactures one protein may be susceptible to infection by another bacterium which manufactures the other protein. The two genes (H2 and rH1) are activated simultaneously by a *promotor* gene in such a way that when the H2 gene is activated, the H1 gene is automatically suppressed. Occasionally, the promotor is *inverted* and as a result the DNA code has to be read in the opposite direction, causing activation of the H1 gene and suppression of the H2 and rH1 genes. So, for example, people who have antibodies that protect them from infection by a bacterium with the H2 protein, will be susceptible to infection by a bacterium expressing the H1 protein, and vice versa. These bacteria have not *evolved* resistant strains since nothing *new* has appeared. All the genes already exist and the bacterium is still the same salmonella.

So, from the above, we see that the examples of enzyme mutations resulting in the emergence of drug-resistant or immune strains of micro-organisms, simply demonstrate variation within existing organisms, without any new class of enzyme or new type of organism being formed. In other words, drug resistance does not give us a window into the process of the (supposed) evolution of one creature into another.

## Probabilities and impossibilities

Since evolution knows no plan or direction, we cannot leave the subject of protein synthesis without examining the statistical probabilities of functional proteins arising randomly out of the countless billions of permutations of amino acids that are possible. Some are adamant that there is no room for any debate:

> Let us hear the conclusion of the whole matter. The essence of life is a

> statistical improbability on a colossal scale. Whatever is the explanation for life, therefore, it cannot be chance...**Cumulative selection, by slow and gradual degrees, is the explanation, the only workable explanation that has ever been proposed, for the existence of life's complex design**...No matter how improbable it is that an X could have arisen from a Y in a single step, it is always possible to conceive of a series of infinitesimally graded intermediates between them. **However improbable a large-scale change may be, smaller changes are less improbable**...**we shall be able to derive anything from anything else**...**if there has been sufficient time**...And also only **if there is a mechanism for guiding each step** in some particular direction, otherwise the sequence of steps will career off in an endless random walk. (My bold). Richard Dawkins. 'The blind watchmaker'. Penguin. 1991. p317.

But rather than providing an explanation, Dawkins' bold 'conclusion on the matter' does nothing but open up the question. He also makes a number of false assumptions: firstly, that *a change involving a series of small steps is more probable than a change involving a series of larger steps;* secondly, that *if small enough, any step is ultimately possible*; and thirdly, that *natural selection is a guiding force different from pure chance*.
We have already seen in the previous chapter that ultimately, natural selection is no different from pure chance. Yet the argument of 'cumulative selection' is still put forward as *the* answer and is often presented something like this:
Consider a sentence:

The rain in Spain falls mainly on Manchester.

Starting with a board with 44 slots (one slot for each letter and space position), a blind monkey is allowed to insert letters into the slots in a random manner. When a correct letter is placed into any slot the slot is closed, so that the letter cannot be removed and neither can it be substituted with another letter. The process is then repeated over and over again. It will be seen that by such a process the target sentence will not take millions of monkey insertions to appear.
But as an analogy to natural selection this monkey model is entirely fallacious, for three reasons. Firstly, while natural selection does not work towards some future goal, in contrast, the target phrase is predetermined by a human mind. Secondly, unlike in DNA where every correct nucleotide is susceptible to being eliminated by another subsequent mutation, here, there is a mechanism in place that retains permanently any letters when they are inserted in the correct place. Thirdly, the order of the letters is not analogous to the function of a protein since, for there to be any natural selection, whatever 'the sentence' reads, every step in 'the letter insertion process' would have to have some practical benefit or function in a partly-formed protein. In this particular case, even though it only differs from the target phrase in three letters, the order of letters *'The rain in Spain falls mainly on Chichester'* conveys *nothing* whatsoever of the meaning of the target phrase, since the whole point of the sentence is a joke*.

* *'The rain in Spain falls mainly on the plain'* is a well-known English phrase. For our purposes *'on the plain'* has been substituted by *'Manchester'* since it always seems to be raining there, hence the joke. In contrast, Chichester generally enjoys much better weather, and so the phrase *'The rain in Spain falls mainly on Chichester'* is not a joke at all.

Consider the *by chance* process of a simple thing A giving rise to a very complicated thing Z, occurring in one transitional move. We might say that the probability is so remote that the transition would be considered impossible. But what if the simple thing A could give rise to the very complicated thing Z by an enormously large series of very small intermediary steps A-A1-A2-A3....B1....X5...Z, as Dawkins and others would plead? Surely such a stepwise process would make the transition more likely. In actual fact, when all the parameters are independently variable, the probability of either process ending up with Z would be exactly the same. The mathematics can be easily tested:

Suppose we throw up 100 dice, hoping for them all to land with a 6 facing upwards. The probability for one die landing with a '6' facing upwards is 1 in 6. The probability for all 100 dice is 1 in $6^{100}$ (or 1 in $10^{78}$). Not a very good chance. What if we throw the 100 dice *one by one*, will we be more likely to succeed? No. Since *the probability of a number of independent events occurring is given by the sum of their individual probabilities*, the chance of us getting 6 showing on every single die will be exactly the same.

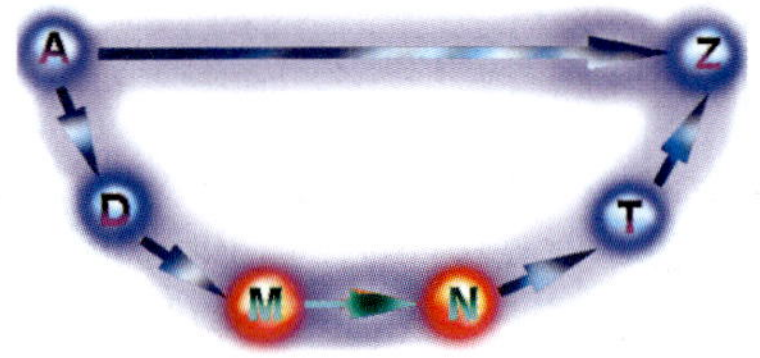


A change from A to B to C....to Z has the same probability as a change directly from A to Z.
If an hypothetical change from A to Z includes a very small step M-N that is not only *essential* but is also *impossible*, then both the change from A to Z directly, and the changes from A...to M to N...to Z, would be *impossible*.

Dawkins also assumes that *any* transition (relevant to the trend towards the end result in question) is *possible*, and that, from any position, any very small change is more probable than a large change. But, if an hypothetical change from A to Z includes a very small step M-N that is not only *essential* but is also *impossible*, then both the change from A to Z directly, and the changes from A...to M to N...to Z, would be *impossible*.

For example, the group A streptococcus bacterium is involved in some 30% of sore throats (the remainder mainly being caused by viruses). Yet despite 80 years of antibiotic use:

> A study of samples of group A streptococcal strains collected over 80 years showed **no** development of resistance to penicillin. (My bold). Merec Bulletin. National Prescribing Centre. Volume 10. No.11. 1999. p43.

What this means is that despite this bacterium having had, in order to ensure its survival, the 'necessity' to 'work out' some different chemical pathway to avoid the action of penicillin, it has been *unable* to do so. Perhaps, it could be argued, that eventually, some day, the group A streptococcus may succeed in generating penicillin-resistant strains. However, the fact is that the group A streptococcus has not been able to resist the action of penicillin, and 80 years of intensive antibiotic use against the bacterium must be equivalent to many millions of years of natural confrontation. So it is more likely that the group A streptococcus *cannot* make a few small steps to make itself resistant to penicillin.

As another example, in the chapter on eyes we shall see that during (supposedly) many millions of years, the eye of the nautilus seems to have been incapable of making even a few 'simple' changes in order to enhance its efficiency.

So, what this means is that we must be willing to accept that there are some impossibilities in this universe. Thus, we have the *Small Step Principle,* which states that:

Any small step, even if infinitesimal, need not be *simple*, and even if a small transitional step may be *imagined*, this does not mean that that particular step *ever happened*, or is *possible*.

We have also to distinguish between *conceptual* and *physical* modifications. For example, whilst it is true that the evolution of the motor car has, over the years, been the result of human ingenuity modifying the *principle* of the horse and cart, nevertheless, most of the important modifications to the horse and cart have been conceptual modifications. Although both the horse and the internal combustion engine can provide motive power, and although it must have occurred to early inventors that perhaps the horse could be substituted by an engine, nevertheless, *not one* component of the internal combustion engine has arisen as the result of direct modifications to the horse. Nor have the mechanisms by which an engine is linked to the motor car's road wheels arisen by stepwise modifications to the harness of the horse. And similarly, the wood-grain-effect plastic of the modern dashboard has not evolved from modifications to real wood. Other non-directly-physical modifications are numerous. So a motor car is not the result of a physical stepwise evolution of a horse and cart.

For it to have occurred, the evolution of living things necessitates *physical* precursors to each and every biological structure or system that exists, or which in times past, has existed. Yet we have seen that DNA cannot directly cause amino acids to bond together, and amino acids do not spontaneously form peptide bonds without the assistance of RNA and aminoacyl tRNA synthetase. Furthermore, in the process of making a protein, it might be possible to unite a number of amino acids, but this is likely not to confer any function while the crucial parts of the molecule remain incomplete. Thus, no organism could exist and function while any *essential* steps are being accumulated. For example, we have already seen that histones are special proteins that hold most DNA in an inert form, protecting the molecule from break-up. They are so essential to life that, most certainly, these proteins could not have arisen in the first place piece-by-piece-by-piece. So, the origin of histones demands a certain number of steps which are likely to be insurmountable by a process of evolution.

We have seen strong evidence that there exists a minimal complexity without which cells and their subcellular systems cannot function. But does this mean that without a minimal complexity cells could not have come into being in the first place? Clearly, if cells *did* evolve from simple molecules, then the obligate interactions and the interdependency of cellular components that now exist, did not always exist. So, some would argue that the problem of minimal complexity is not a problem at all:

> Nobody suggests that technological society cannot have arisen by gradual evolution just because we all need each other today. Take a simple case. A blacksmith shapes iron using iron tools: he needs iron instruments to make iron instruments. So where did the first iron instruments come from? Must they have been handed down from on high, ready made? Of course not. Early blacksmiths might have used stone clubs, for example, or other metal tools to shape the first instruments. It is possible to evolve sophisticated technological cycles in many different ways from fumbling

beginnings, but once a cycle is established, it rapidly becomes refined. Today's organisms are full of high-tech chemical cycles that **must** somehow have emerged from long-discarded molecular groping. (My bold). Paul Davies. 'The fifth miracle.' Penguin Press. 1998. p93.

But the fallacy of Davies' logic is obvious. His example of blacksmiths evolving increasingly sophisticated tools is not, in actual fact, random since it involves the activity of human intelligence with definite objectives in mind. Thus, for example, the origin of the screwdriver was not the result of an accident, but of forethought and design. An atom or molecule has neither intelligence nor an objective. Nor is there within the physical laws and mathematical constraints which an atom is subject to, or within the nature of protein molecules, any identifiable organising principle that directs simplicity to become complexity. Some declare that:

> **Life started out** with simple microbes. If it was to go anywhere it would inevitably be in the direction of greater complexity…**Obviously**, if you start from some simple state, **even a random excursion will likely take you in the direction of greater complexity**, at least at first… Stephen Jay Gould has explained this point well by using the analogy of a drunk leaning against a brick wall, who then begins stumbling about blindly, and eventually ends up in the gutter. The drunk reaches the gutter not because he is seeking it out and methodically moving towards it. He is, in fact, staggering randomly: **at any given time,** the drunk is as likely to be moving towards the wall as away from it. The point is that, because the wall bounds his motion in one direction, he is obviously on average likely to be found somewhat away from the wall, and in due course he is going to encounter the gutter simply by chance...**If life on Earth began 'at the wall'**, i.e. with the simplest cells, and then evolved at random, the average complexity would **inevitably** increase…**Most** biologists are of the opinion that any growth of complexity can be satisfactorily accounted for by the drunken walk effect. (My bold). Paul Davies. 'The fifth miracle.' Penguin Press. 1998. p221-4.

But this biologist most certainly does not agree. And in science (as with anything else), a majority view does not automatically equate with truth. Anyone who has seen someone who is completely intoxicated will know that such an individual is completely oblivious to his actions and does not behave like a ball that, having hit a wall, would simply bounce away from it. Being at the wall, a *completely drunk* drunk is likely to slump against the wall *and stay there*. Even if he slumps slightly away from the wall, his next slump is just as likely to bring him back in contact with the wall. *Providing that each slump results in only a tiny movement towards or away from the gutter, and the width of the pavement is such that a large number of slumps would be necessary to get to the gutter*, it will be found that the drunk will most likely *not* reach the gutter - *ever*. By way of example, suppose we toss a coin. If the coin lands with one side uppermost, this would represent a slight movement of the drunk away from the wall, and if the coin lands with the other side uppermost, this would represent a slight movement of the drunk towards the wall. By tossing a coin, it can easily demonstrated that there will be no net movement either towards or away from the wall.

There is another fallacy in Gould's example of the situation faced by any simple molecule because, *for the drunk*, to be in the gutter is *not* a more complex state than being against the wall. In the living cell, huge numbers of biochemical reactions only proceed into highly organised systems with the input of energy and information. So, a more realistic analogy would be that of a drunk inside an industrial chimney 50 metres high, on the inside wall of which is attached a ladder. The ladder extends to the top of the chimney but the completely drunk drunk cannot reach the bottom rung unless he stands on three stools, one on top of the other. Conveniently, on the floor a number of stools are strewn about. Unfortunately, they are all *one*-legged stools (*very* unstable). Acting randomly, truly oblivious to anything, the drunk would have to balance the three one-legged stools one on top of the other, in order eventually to climb up the ladder. For the drunk, standing on top of just one stool would indeed be a more complex state than standing on the ground. Relating this analogy back to chemistry, having climbed up such a 'ladder', a simple molecule would indeed have attained a state of greater complexity. But we should be realistic and say that even if the drunk pushed one stool directly underneath the ladder, *since he would act without co-ordination, purpose or understanding* he would be unlikely to make it balance, and any subsequent contact with the stool would be likely to push it away from its position under the ladder. Furthermore, it is unlikely for a stool to be so favourably positioned again.

So, increasing chemical complexity *cannot* be explained by the drunken walk effect.

The *WYWYWG Principle* applies here, too. Since the (supposed) occurrence of events which would produce the primitive chemicals necessary for life from simple chemical precursors is associated with an astronomically huge improbability, it follows that if an hypothetical 'towards-life-event' did occur, then another additional and *useful* 'towards-life-event' is most unlikely to occur nearby and in the same lineage of molecules. In other words, if any number of useful 'towards-life-events' were to occur, then we should expect them to occur randomly and to be spatially separate. And we would expect them *not* to come together to produce a cumulative effect.

But what sort of events should we consider likely and what should we consider unlikely? Some examples are necessary to put things into perspective. In any class of 30 school girls, more than one will develop breast cancer. Yet such a risk does not leave women in perpetual fear. One in 700 babies born will have an extra chromosome 21 resulting in Down's syndrome, but we would consider the risk to any individual couple of having a child with Down's syndrome to be extremely small. Thyroid cancer is a well recognised condition but is so rare (it affects only around one in every million individuals) that many general practitioners are unlikely to see a single case amongst their own patients in their working lifetime. The most likely cause of death in healthy adults in the UK is a road traffic accident. Yet none of us consider it to be a probable event when we drive off in the morning. Although thousands of people are killed on the roads in the UK every year, *any single* individual that we might focus on is most unlikely to have a fatal accident. Although it is *correct* to say that it is *possible* for two small meteorites to burn through the

atmosphere and land *simultaneously* on the noses of identical twins, one in London and one in New York....on their birthday...we should sensibly consider such a possibility to be astronomically (literally and mathematically) unlikely. So, we most certainly have to make some sense out of biological and biochemical events by distinguishing between *realistic* probabilities and *unrealistic* probabilities. With that in mind, let's just look at the probabilities associated with the (supposed) evolution of specific and functional proteins.
We know that in a ribosome, with all the necessary cellular machinery at hand and in working order, the elongation of a polypeptide occurs:

> ...at a rate of up to 40 residues/s... Donald and Judith Voet. 'Biochemistry'. John Wiley. 1995. p994.

So, let us say that it takes 0.025 seconds for the completion of one peptide bond between any two amino acids. The average molecular weight of amino acids in most proteins is 115. This means that 115g (1 *mole*) of amino acids of average molecular weight would contain $6 \times 10^{23}$ molecules (which is *Avagadro's Number,* i.e. the number of particles in 1 gram molecular weight [1 mole] of substance).
The mass of the Earth is about $10^{27}$g. If the *whole* of the Earth's mass were a soup comprising *only amino acids* of average molecular weight, there would be:
$\frac{10^{27}}{115}$ moles of amino acids, comprising $\frac{10^{27}}{115} \times 6 \times 10^{23}$ molecules.

If all these amino acids formed protein chains, each comprising 100 amino acids, there could be formed a maximum of:

$$\frac{10^{27}}{115} \times \frac{6 \times 10^{23}}{100} = 5.2 \times 10^{46} \text{ protein chains.}$$

If we suppose (purely for the sake of simplicity, since there are far too many variables to make any other calculations valid) that all the peptide bonds between all the amino acids for all the 100 amino acid proteins could link together *simultaneously*, then all $5.2 \times 10^{46}$ protein chains could be formed in 0.025 seconds. Over time, these proteins could break down (by a process called *hydrolysis*), and the amino acids could reunite in the form of different sequences, thus producing different proteins. However, once formed, proteins do not readily break down spontaneously:

> ...the activation energy for uncatalyzed hydrolysis of peptide bonds in a neutral aqueous solution at room temperature is so high that **little or no hydrolysis occurs even after several months**... (My bold). Lodish et al. 'Molecular cell biology'. Scientific American Books. 1995. p77.

Again, to make it more simple, let us assume that it takes *negligible* time for peptide bonds to break down, so that a completely new $5.2 \times 10^{46}$ set of proteins could be formed *every* 0.025 sec. Now, since the Earth is thought to have originated 4.5 billion years ($1.4 \times 10^{17}$ sec) ago, then a maximum of:

$$(5.2 \times 10^{46}) \times 40 \times 1.4 \times 10^{17} = 3 \times 10^{65}$$

different permutations of 100-amino acid protein chains could have formed in the time available since the origin of the Earth. But these figures are exceedingly over-generous, not least because only a very tiny proportion of the material in the primitive Earth could ever have been available to (supposedly) form amino acids with the help of the energy from the sun's UV light and spontaneous electrical discharges.
Now, many medium-sized proteins are composed of about 100 amino acids. Since there are 20 amino acids to choose from, the number of permutations of the 20 amino acids that can be arranged to form 100-amino acid proteins is $20^{100}$.
Since $X^Y = 10^{(\mathrm{Log}\,X).Y}$, then the chance of any one specific protein of 100 amino acids arising by random interactions of individual amino acids is one in $10^{130}$:

> Clearly, nature can have made only a fraction of the possible different protein molecules. Donald and Judith Voet. 'Biochemistry'. John Wiley. 1995. p57.

Thus, even these simple calculations show that there has simply not been enough time since the presumed origin of the Earth to ensure that even *one specific* 100-amino acid protein could have been formed by chance. Furthermore, most proteins contain far more than just 100 amino acids - for example, the actin molecule is composed of 375 amino acids, for which the number of possible permutations is $10^{487}$! And there are some 25,000 genes in the human genome, each coding for a specific protein. Since the probability of any two independent chance events occurring together is the product of the probabilities for each event, then the probability of 25,000 specific proteins - each containing 100 amino acids - arising by the chance interactions of individual amino acids is one in $10^{3,250,000}$.
It is simply not possible for us to even begin to appreciate just how large these figures are. Some estimate the total number of atoms in the universe to be 'only' in the region of:

> ...$9 \times 10^{78}$. Donald and Judith Voet. 'Biochemistry'. John Wiley. 1995. p57.

Even if any two atoms could interact and form bonds every one billionth ($10^{-9}$) of a second, then, since the Big Bang (supposedly less than $10^{18}$ seconds ago), a maximum of only

$$10^{78} \times 10^{9} \times 10^{18} = 3 \times 10^{105}$$

chemical events could have occurred in *the whole universe* since the (supposed) beginning of time. Thus the chance of all 25,000 proteins - each comprising 100 amino acids - arising *de novo* must be accepted as *impossible*. So, the spontaneous origination of Life on Earth is implausible.
However, it has been suggested that, since there are many billions of stars in space and therefore since there must be billions of planets out there with an Earth-like atmosphere and climate, the chances of alien life being present elsewhere in

It might be argued that discussing the improbability of protein synthesis in this manner is simply restating the notion that a space shuttle is unlikely to result from the random effects of a tornado passing through a junkyard; and that, instead, evolution works by step-by-step modifications to individual facets, with each modification serving some advantage over its ancestral form.
However, such a criticism is flawed, since evolution requires that every amino acid in every position in a protein has to be specified and coded for, and that every single proto-protein (ranging from an initial solitary amino acid through to the complete modern protein) must have had both a function and an advantage over its (supposed) ancestral molecule. To arrive at a modern protein, ancient amino acids would have had to bond with each other by random forces (with or without the help of other primitive molecules acting as catalysts), following which natural selection would have had to preserve 'useful' conglomerates.
It is impossible for such specific and progressive step-by-step improvements to have occurred within the (supposed) $10^{17}$ seconds available since the origin of the planet.

The Northern Lights seen from Husavik, Iceland. The total number of atoms in the universe is estimated to be in the region of $9 \times 10^{78}$.

## Card dealing sleight of hand

Some people criticise the statistical argument against the origination of functional proteins by drawing an analogy with the dealing of a pack of cards. The reasoning behind the Card analogy contains a fundamental assumption - that since very rare events do happen all the time, rare amino acid arrangements can also happen. Therefore, given enough time, functioning proteins will inevitably form:

*One of the mathematical tricks employed by Intelligent Design involves taking the present day situation and calculating probabilities that the present would have appeared randomly from events in the past. And the best example I can give is to sit down with four friends, shuffle a deck of 52 cards, and deal them out and keep an exact record of the order in which the cards were dealt. We could then look back and say 'My goodness, how improbable this is! We could play cards for the rest of our lives and we would never, ever deal the cards out in the exact same fashion'. You know what, that's absolutely correct. Nonetheless, you dealt them out, and nonetheless you got the hand that you did.*

*K Miller. Horizon. BBC2 Broadcast. 2006.*

On the surface, that seems a reasonable suggestion. However, right at the beginning we have a problem, since before the game can start, the cards themselves need to be manufactured, packed and purchased, and brought to the game table. By chance.

What would happen to the game if half of the cards were dealt face upwards? Obviously all the players would know the whereabouts of half of the cards, rendering the game unplayable. In the same way, we know that any protein comprising 50% of right-handed amino acids (equivalent to cards facing upwards) will render a protein useless.

In addition, when placed on the table the cards must stay put next to each other, and not drift apart and mingle with the other hands. In the cell, however, amino acids do not bond with each other without outside influence.

In the card game only the 52 cards in the pack are necessary for producing the hands. However, pre-existing proteins (equivalent to a hand of cards that is in addition to and separate from the pack being dealt) are essential for amino acids to bond together, and also for the resultant protein to fold in the correct manner. Since a newly-formed protein will not serve any useful purpose until it is correctly folded, then, in order to make the card analogy more accurate, a blank set of cards should be dealt, which then acquire their numbers and patterns after making contact with a separate group of 'helper' cards. In some cases (as exemplified by the 13-amino acid tag on the end of the bovine pancreatic trypsin inhibitor protein), yet more cards have to be dealt from a separate pack, and subsequently discarded in order for a valid hand to be left.

What about the purpose of dealing out the hands – the game itself? If evolution is true, then every protein must have had a useful function at every stage from a single amino acid to the complete chain. So, if the card analogy is valid, a valid game should be playable with any number of cards, from only one card up to 52, without an intelligence inventing unique games for every number (between 1 and 52) of cards dealt out, and before the next card is dealt. However, examples such as Cytochrome C show that for many proteins, there are vital amino acids at specific points along the chain, without which there can be no function at all.

The reader would have noticed that the card-dealing analogy is no different from, for example, the unique spatial and temporal orientations of millions of raindrops falling out of a thunder cloud, or of the dust particles ejected by a volcano: these unique trajectories and patterns do occur. So we see that the card analogy is simply a sleight-of-hand and that the occurences of rare events have nothing whatsoever to do with real biology. Time and chance do not make impossible things happen.

However, having said that, Miller is right on one thing: that having once dealt a particular hand, it is improbable that it will happen again (not even in $10^{17}$ seconds). And that is where the card analogy fails completely. Our being here is not dependent on one particular protein arising by chance, but on the tens of thousands of specific proteins that are in our cells, along with the myriad other cellular structures and functions.

the universe must be so high as to be *inevitable*. However, if we were to suppose that there might be around a billion billion ($10^{18}$) Earth-like planets out there, then, on the basis of the previous calculations, there would still only have been a maximum of:

$$(3 \times 10^{65}) \times 10^{18} = 3 \times 10^{83}$$

different permutations of 100-amino acid protein chains that could have been formed throughout the whole universe in the time available since the origin of the Earth. This could still not ensure even one specific 100-amino acid protein chain being produced. If (as is likely) any distant planet is subject to the same physics and chemistry as here on Earth, the spontaneous origination of Life elsewhere is equally implausible. If Life were to be found in some distant galaxy, it is likely not to have evolved spontaneously from simple chemicals.

In a recent trial relating to the deaths in suspicious circumstances of two baby brothers, the jury was told by the prosecution that:

> ...the chances of such a double tragedy were 75 million to one. Nigel Bunyan. Daily Telegraph. 10th November 1999. Front page.
> *(The babies died within 14 months of each other, aged 8 and 11 weeks.)*

The mother was found guilty and sentenced to life imprisonment, but she was later acquitted. But clearly, we have to be consistent in our response to statistics. So, if we accept the jury's view that an event which has a one in 75 million chance* of happening is *so* unlikely that it can safely sentence a person to life imprisonment, then we must accept that the chance of even one specific protein arising since the dawn of time is so unlikely that we can safely convict evolution and sentence it to eternal imprisonment in the dungeon of myths. If on the other hand we accept that no matter how small, no possibility can be discounted, and so against all the odds, Life on Earth could have originated and evolved spontaneously (and supposedly did), then we must reject the jury's initial judgement, and declare on statistical grounds that, compared to the negligible possibility of the spontaneous origin and evolution of Life, the possibility of the defendant being innocent is so great that the conviction is unsafe. Indeed, by extrapolation, in the absence of any direct evidence or witnesses, *all* convictions are unsafe.

*Some time later it was pointed out that the chances of two cot deaths arising within the same family is much higher than one in 75 million, but that the jury made its decision on evidence other than the statistics.
Although the prosecution gave an incorrect probability, this does not undermine the notion that a chance of one in 75 million is by all accounts vanishingly small odds.

The only way out of this impasse is to use common sense. We have seen that amino acids left to themselves do not readily form peptide bonds, and that in the cell, highly organised chemical machinery exists to cause a protein chain to be produced. Furthermore, we have seen that not only is the correct sequence of amino acids required for a protein to achieve a specific 3-dimensional structure, but in addition, a number of accessory proteins and chaperones are essential. We have also seen that large sections of proteins are essential for proper function, and that mutations affecting such essential parts are likely to result in severe loss of function. In the light of all this, and also taking into account the rates of peptide bond formation, and the enormous number of permutations that are possible both in the sequence of amino acids and also in the folding options, we have to conclude that it is implausible that even *one* cellular protein comprising 100 amino acids could possibly have appeared by any random chemical processes, in the time available since the presumed origin of the Earth.

We have seen that the evidence from DNA, from the process of protein synthesis, and from the specificity of proteins themselves, reveals an awesome complexity, comprising a large number of molecular systems and demonstrating a minimal functional complexity. By all sense of reason then, we must conclude that it is implausible that even the loudest 'ABRACADABRA!' could have caused the biochemical machinery inside cells to have evolved stepwise and spontaneously from simple chemicals.

CHAP

# Molecular movement:

*it's chemistry, but not as we know it*

Kreb's Bi-cycle

Even as recently as 150 years ago, with the limits imposed by low-power microscopes, it was thought that the cell was just some very simple blob of jelly with nothing much inside it. Therefore, it was not an unreasonable thought that perhaps something like a cell could arise spontaneously, and from it in due course of time complex life could easily evolve. However, we have seen that far from being a simple blob, a cell and the tiny structures and systems contained within it demonstrate the most *awesome* complexity.

In order to get some further insight into this complexity, in this chapter we will turn our attention from pure chemistry (which is the result of the random movements of atoms and molecules and their interactions) to an *epiphenomenon* - the *purpose*ful movement within cells of structurally organised combinations of molecules. By that I mean that molecules can be seen to be involved in movements that relate far beyond the level of the making or breaking of their own chemical bonds: molecular movements that relate ultimately to purposeful movements of whole systems of giant molecules, whole cells, and even beyond that to the purposeful movements of whole animals. Or to put it another way, unknown to the process itself, in living things chemical activity has dynamic applications that go far beyond the pure chemistry itself.

Cilia on the surface of the cells of the respiratory epithelium.

Recall the sparrow landing on the bird table. If we accept life forms to be just *'biocules'* - extrapolations of pure physics and chemistry - then this forces us to the conclusion that *ultimately*, as a result of its atoms following an astronomically huge series of physical/chemical options, the sparrow is by *obligation* attracted towards the seed, although not quite in the same way that a charged ion is by obligation attracted towards an electrode. Both the sparrow and the ion can do nothing else but move, but the question we have to ask is: does the evidence show that the sparrow is nothing more than just complicated pure chemistry?

### *Flagellae and cilia*

*Flagellae* (long) and *cilia* (short) are rod-like structures that are attached to the surface of many eukaryote cells. They beat rhythmically and either cause an organism to swim (for example, the flagellae in *Chlamydomonas*, a single-celled plant-like organism) or cause movement of fluid adjacent to the cell (for example, the cilia lining the lung passages which force secretions upwards so they can be coughed up as sputum).

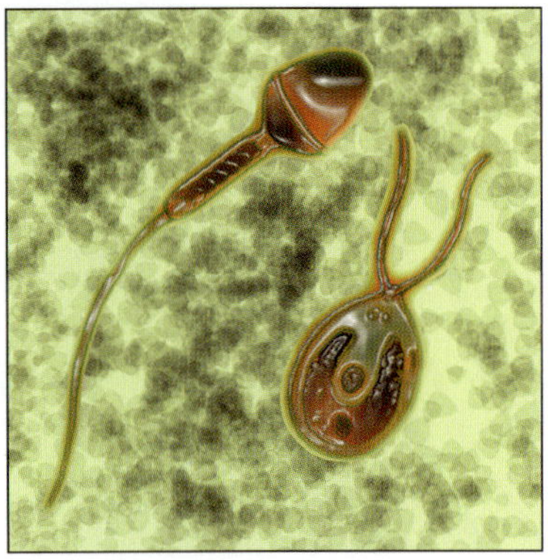
Eukaryote flagellae:
Sperm (above left), and Chlamydomonas (below right).
(Diagrammatic)

In some cells such as sperms, the flagellae beat symmetrically, producing movement along a curved path. In other cells such as those lining the air passages, the cilia have a whip-like effective stroke followed by a limp recovery stroke. In chlamydomonas, the twin flagellae allow swimming by virtue of a symmetrical butterfly stroke followed by a recovery stroke.

Whilst the following discussion refers only to flagellae, the details and principles apply equally to cilia. (The flagellae of eukaryote cells should not be confused with the flagellae of bacterial cells, the latter being totally different structures). In virtually all organisms, a flagellum is made up of a central pair of filaments, which runs along the whole of its length and which is surrounded by an array of nine similar doublets. The whole 9+2 arrangement is called an *axoneme* and is contained within a sheath of membrane continuous with the main cell membrane. At the base of the axoneme and within the cytoplasm are attached the *basal bodies*. The basal bodies are short cylindrical structures which also have the 9+2 arrangement, and resemble the *centrioles* which are involved in cell division, which we will turn to later.

Each of the doublets is composed of A and B filaments, each of which is made up of polymers of the protein *tubulin* coiled in an helical form. Between the doublets are a number of different proteins which provide stabilising scaffolding and which are arranged either as radial spokes or as connections between the doublets. Special motor proteins called *dyneins* are attached to both the A filaments and the B filaments. The dyneins bridging the gaps between the A filament of one doublet and the B filament of the adjacent doublet are arranged into two groups, called inner arm dyneins and outer arm dyneins. The dyneins attached to the outer doublets are found *only* in flagellae and are not found anywhere else in the cell. In fact, there are a number of different dyneins in flagellae. They are all very complex protein molecules comprising a number of component chains linked together. For example, the dynein heavy chain component can be 4500 amino acids long. All in all, in addition to the A and B tubulins, there are about 200 other distinct proteins present within flagellae. As with all proteins, the exact sequence of amino acids is responsible for the three-dimensional shapes of all these proteins, and the shapes and charges in turn dictate how chemical bonds can be translated into mechanical movement of the flagellae, and hence of the whole cell itself.

One end of a dynein molecule forms stable bonds with the A filaments and the other end forms unstable bonds with the B filaments. With the ubiquitous adenosine triphosphate (ATP) molecule providing the necessary energy, the dyneins make and break their bonds with the B filaments and, as a result, the A and B filaments slide past each other, and this causes the bending movements of the whole flagellum. An influx of calcium ions alters the timing and direction of flagella bending, but the exact mechanism of such control is unknown.

Bearing in mind that each and every evolutionary molecular step must allow for some function, and must be of sufficient additional benefit to be selected for, in preference to the original situation, can we show a valid stepwise sequence from simple molecules to the fully functional flagellum?

Here we come to some problems. Consider an isolated mixture of dyneins and A filaments: the dynein forms a fixed stable bond with the A filaments, but since there are no B filaments there can be no bending or sliding movements. Consider an isolated mixture of dyneins and B filaments: loose unstable bonds form between these proteins but again, no discrete structure would be formed and there can be no sliding or bending. Consider a mixture of isolated dynein and isolated A and B filaments, from which the scaffolding proteins have been removed: there is no activity until ATP is added as an energy source. When ATP is added, not only is there no beating, but also, having no stabilising proteins, the doublets simply break up by sliding apart. Once separated, the A and B filaments do not and cannot reform. So, for doublets to retain their integrity, the other flagellar elements are necessary. Furthermore, the site of action of ATP is a very specific part of the dynein molecule, and without this specific site in the protein ATP, it cannot have any effect.

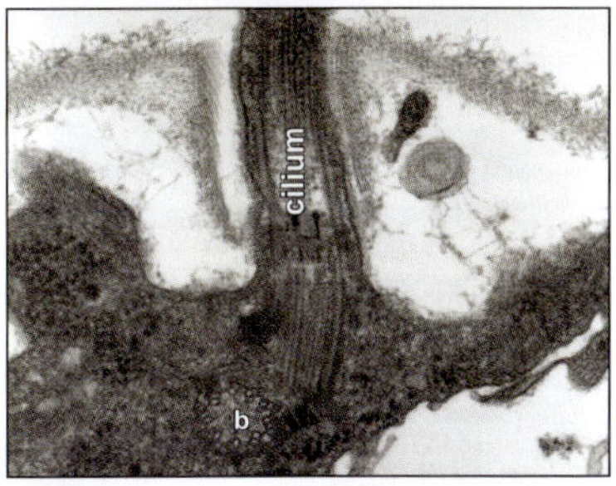


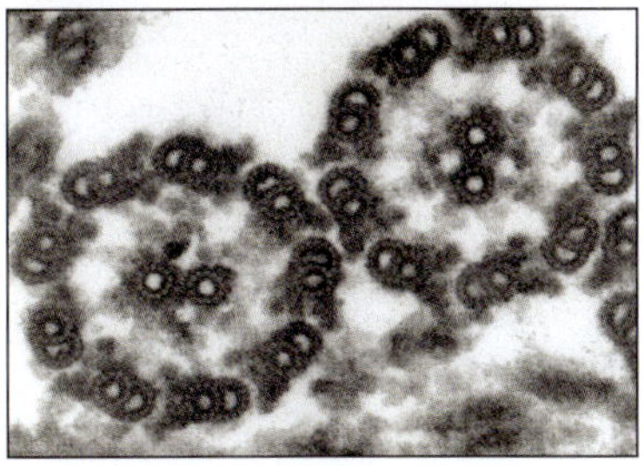

Cilia.
Top: Cilia on wall of a bronchus. (Diagrammatic)
Middle: Close-up of the lower part of a cilium, with a basal body (b, in cross-section) at its base. Basal bodies have a similar structure to centrioles.
Bottom: Cilia seen in transverse section.

If A and B filament doublets were coded for but could not be held together, they would be functionless without the binding proteins. The binding proteins could appear first but would be of no use without the A and B filament doublets. Partial binding proteins and partial A and B filament doublets would also be of no use. Thus, for any useful function, the flagellar system must first exist with a crucial minimum number of features. And since any cell which has an incomplete, non-functional flagellum (and is therefore non-motile) cannot have any selective advantage over an otherwise identical cell which has no flagellar components whatsoever, the flagellar system could not have evolved stepwise from within a situation where no flagellum existed. Of course, even if some (supposed) incomplete primitive flagellum could come into being, the *WYWYWG Principle* shows that it is implausible that a whole functioning flagellum could ever arise.

It could be argued that the flagellum is not really a structure exhibiting irreducible complexity since many of the flagellar proteins may be found having functions elsewhere in the cell, and components could be recruited into an evolving flagellum in a stepwise fashion. However, this is unconvincing. As we have seen, the dyneins attached to the outer doublets are unique and are *not* found elsewhere in a cell; the concept of recruitment of a protein from another site in a cell does not explain the origin of that protein in the first place; and the absence of essential components results in no function at all:

> ...the absence of all of the inner arms renders the flagellum **immotile**... (My bold). Lodish et al. 'Molecular cell biology'. Scientific American Books. 1995. p1086.

Such facts are often ignored and it is still argued that surely a primitive and simple flagellum that allows only a tiny bit of movement would be an improvement on no movement at all, and so an early cell might gain some survival advantage with an incomplete flagellum. But here we come again to the point that any advantage must be *significant*. A tiny movement of a tiny cell will get it nowhere, and without a mechanism for its movement to be favourably directed, any movement will be random. And, of course, the moment the first primitive cell (supposedly) gained its first flagellar component, there would have been no actively motile predatory cells to escape from.

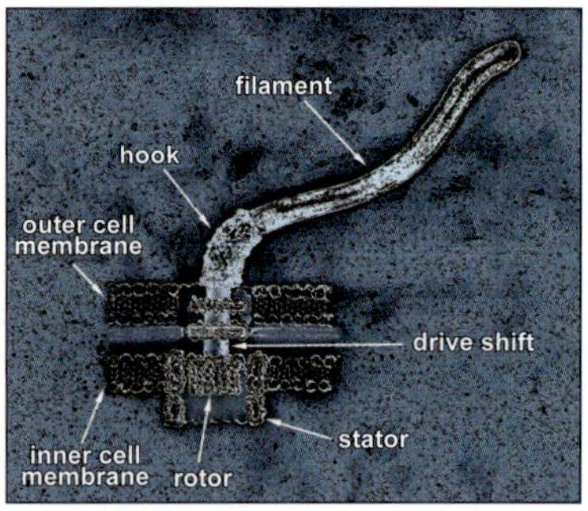

Bacterial flagellum, a molecular motor. (Diagrammatic)

### *Bacterial flagellum*

The bacterial flagellum is quite a different structure altogether from the flagellum of eukaryote cells. Under the electron microscope it is seen to be comprised of a single filament (which acts as the propeller) attached to a universal joint (the *hook*), which is attached to a rod which lies within a number of globular proteins (which act as rotors and stators) embedded in the cell membrane.

The propeller filament is actually an helical structure composed entirely of sub-units of the protein *flagellin*. It has been found that when a flagellum is 'glued' to a microscope slide by the use of anti-flagellin antibodies, the attached bacterial cell slowly rotates about the flagellum, thus showing that the flagellum exhibits true rotary motion and can in many ways be compared to a flexible rod attached to the spindle within an electric motor.

## Mousetrap – flagellum

It has been claimed that a standard mousetrap is made up of a minimal number of components, without any of which the mousetrap would not work at all. Hence it is claimed to be an irreducibly complex structure. With that principle in mind, it is argued that the bacterial flagellum is also an irreducibly complex system comprising some 50 essential proteins. In other words, if any of those 50 essential components are missing, there is complete failure of the whole system. Therefore, it is claimed, the bacterial flagellum cannot have arisen stepwise from simpler molecular components.

The counter argument to this is that prior to the appearance of the bacterial flagellum, a primitive structure (comprising less that those 50 proteins) could have had a totally different function to motility, and that this structure was subsequently added to, ultimately to produce a new function and structure – the modern flagellum.

Supporters of the counter argument show that it is possible to produce ingenious designs of simpler mousetraps that utilise only some of the parts of the standard mousetrap. Furthermore, the Type Three Secretory System (TTSS) is a molecular device situated on the surface of a bacterium that can inject toxic chemicals into another living cell. It is made up of just 20 of the proteins that are found in the bacterial flagellum, and therefore, it is argued, the bacterial flagellum is not an irreducibly complex system, and could have evolved from a TTSS.

However, the flaws in the simpler mousetrap argument are firstly, that any conceived simpler mousetrap is the result, not of chance chemical activity, but of determined intelligence to devise it; and secondly, that playing about with mousetrap designs has nothing, whatsoever, to do with real molecular biology.

The flaws in the bacterial syringe argument can be shown by predictions: if the bacterial flagellum is the result of evolution, then all the hypothetical specific intermediary structures between 20 and 50 protein components must have had beneficial functions (otherwise they could not have evolved). Therefore, many of those intermediaries should not have been lost over time, but retained (because every intermediary must have had an essential function) so as to be functional and still identifiable within cell walls (from bacteria to mammals) today. However, such a variety of structures does not exist. Furthermore, other ancient intermediaries ought to have suffered disabling-but-not-eradicating mutations, so that degenerate, non-functional TTSS-to-bacterial flagellum structures should still be identifiable within many cell walls (from bacteria to mammals) today. But such redundant structures do not exist either.

Furthermore, unlike the bacterial flagellar genes, the guanine/cytosine (G/C) content of genes for the TTSS proteins are found in different proportions to the G/C of the rest of a bacterial genome. This suggests that the genes for TTSS and the bacterial flagellum are different, and those for the TTSS might be transferred between bacteria by plasmids. Other evidence suggests that the TTSS may actually have arisen from the bacterial flagellum, as a result of degeneration:

*Analysis of the type III system and flagellar export genes in the obligate parasites of the family Chlamydiales suggests that the type III system arose from the flagellar export system in chlamydiae or related bacteria. JF Kim. 'Revisiting the chlamydial type III protein secretion system: clues to the origin of type III protein secretion'.*
*Trends Genet. 2001 Feb;17(2):65-9.*

So, in fact, the evidence does not support the notion that the bacterial flagellum evolved stepwise from simpler components.

What then is the mechanism that drives such a molecular motor? One unique feature is the ability of the bacterial flagellum to rotate and beat without the need of ATP or any other external energy source, since it is dependent only on the movement of extracellular protons to power it. So, one model suggests that as individual protons pass from the surrounding fluid into the cell cytoplasm via channels in the basal proteins, slight configurational changes result such that a small rotational movement occurs between the rotor and the stator proteins. It is estimated that some 1000 protons need to pass through the multiple protein channels in order to cause one complete revolution of the flagellum. Typically, the flagellum rotates about its long axis at a rate of about 10 revolutions per second.

It has been argued that a stepwise evolution of the bacterial flagellum is demonstrated by the fact that bacteria vary, for example, in the number of rings in the lower part of the flagellum. Therefore, so it is said, it is easy to see how a primitive bacterium with only one ring might, in time, give rise to descendants with two, three or more rings. All that would be required would perhaps be a duplication of the gene responsible, or a mutation of the gene so that, as a result, its protein product formed clusters of two or more. But since neither of these scenarios actually deals with the

**The bacterial flagellum has a clutch**

It has been shown that the protein EpsE inhibits the rotation of the whip of the bacterial flagellum, and yet the base assembly can still rotate relative to the bacterial body:

*EpsE arrested flagellar rotation in a manner similar to that of a clutch, by disengaging motor force-generating elements in cells embedded in the biofilm matrix. The clutch is a simple, rapid, and potentially reversible form of motility control.*
K Blair. 'Molecular clutch disables flagella in the Bacillus subtilis biofilm'. Science. 2008 June 20;320(5883):1636-8.

appearance of anything *new*, nor explains how the gene arose in the first place, this is not evolution, but molecular variation. Having four molecular rings instead of two is no different from having an extra finger, and an extra finger does not tell us anything about how the other fingers themselves came into being.

So, it is quite clear from its structure that for it to function, the bacterial flagellum system must first exist with a certain minimum complexity. For example, even if any bacterium acquired a flagellar filament attached to its cell membrane, this would be of no benefit without the special rotor and stator proteins that create the rotary movement (and vice versa), and of course, nothing would work without the drive shaft to link the filament and the rotor. Any bacterium which has a partially complete but non-functional flagellum (and is therefore non-motile) cannot have any selective advantage over an otherwise identical bacterium which has no flagellar components whatsoever. The *WYWYWG Principle* shows that the bacterial flagellar system could not have evolved in a stepwise manner.

### *Muscle contraction*

Muscle cells are very different from other cells in the body which are, in general, well-circumscribed discrete entities. In contrast, a muscle is made up of a number of bundles of *muscle fibres*, most of which extend the whole length of the muscle. As a result of the fusion of a number of cells during embryological development, each muscle fibre is essentially a single giant cell containing multiple nuclei. At each end the muscle fibres are attached to tendons which are themselves attached to bones either side of joints. Within each muscle fibre, the cytoplasm contains many parallel *myofibrils*. The myofibrils themselves are made up of complex structural units called *sarcomeres,* millions of which are arranged lengthwise, end to end.

Under the microscope, muscle tissue has distinctive light and dark bands which are caused by the partial overlapping of thick and thin filaments inside the sarcomere.

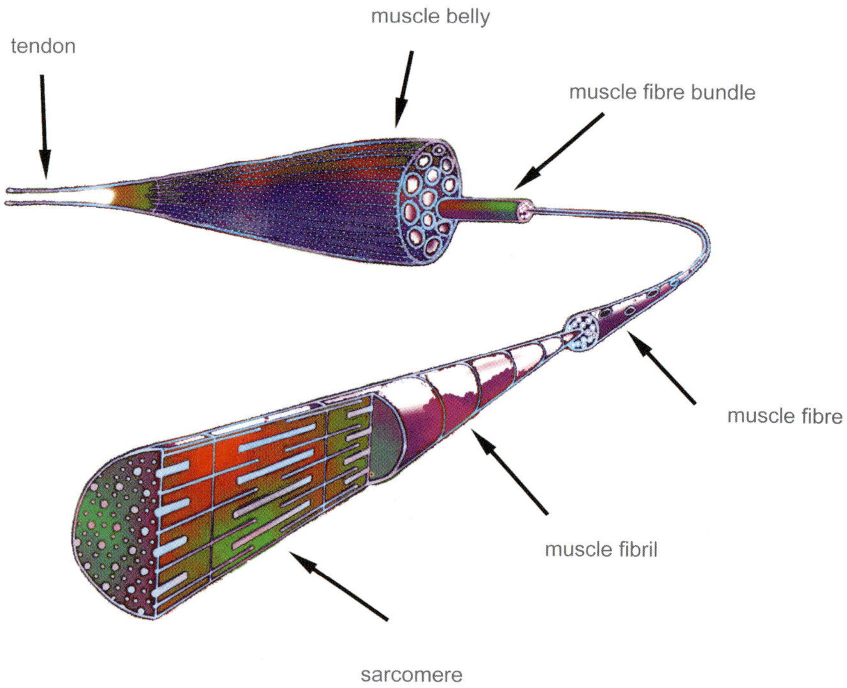


Skeletal muscle: diagrammatic view of muscle fibres.

These thick and thin filaments are composed of the proteins *myosin* and *actin* respectively. Separating each sarcomere is a *Z disc*, either side of which the actin filaments are anchored and aligned in opposite directions. The thick filaments have a central thin portion (called the '*bare zone*') either side of which the myosin molecules are aligned in opposite directions. It is the formation of bonds between the actin and myosin molecules which result in configurational changes that cause the myosin and actin filaments to slide past each other. Since the actin filaments are attached to the Z discs, this action results in the Z discs be-pulled closer together, and ultimately in shortening the muscle during contraction. Other proteins make up cross-linkages to give (amongst other things) stability to the whole formation.

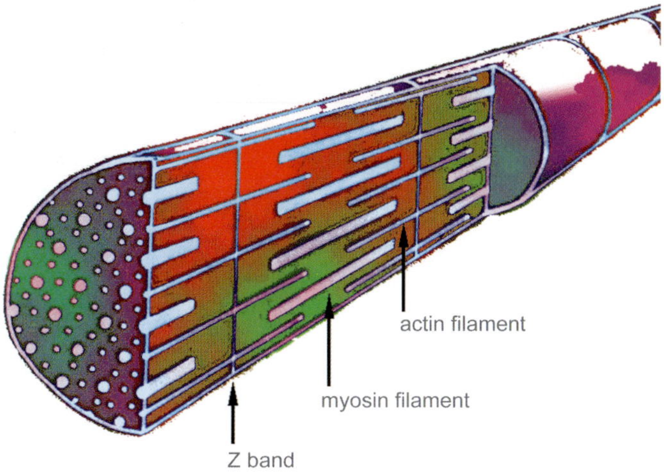


Sarcomere detail

Actin is a large protein that forms filaments only in solutions that mimic the concentration and nature of the fluid within living cells. In solutions of different ionic concentrations, actin assumes a globular form (called G-actin) which is of no use in muscle contraction. Myosin is a huge molecule made up of 6 distinct protein chains, resulting in a complex structure with a shape that is crucial to its function as a molecular hinge during muscle contraction. One part of the myosin molecule (*light meromyosin*) is able to form filaments but it cannot react with ATP and cannot bond to actin. Another part of the myosin molecule is *heavy meromyosin* which reacts with ATP and can bind to actin, but which cannot form filaments. As with actin, myosin forms chains only in the physiological conditions that exist in a living cell. When actin and myosin are mixed together in solution they form a viscous 'soup' of *actomyosin* threads. When ATP is added, the actomyosin threads simply break up into actin and myosin fragments.

The myosin thick filament has a middle '*bare zone*' only when seen in muscle. In the test-tube, myosin molecules form simple filaments without bare zones. Thus the bare zone component is dependent on the living cell for its origination. With all other things present, but without the crucial bare zone uniting two oppositely-orientated myosin filaments, the action of ATP would simply result in the sliding of myosin filaments towards the Z plates, without any muscle shortening.

As with flagellae, a number of secondary proteins are also essential for any muscle function, and contraction itself is dependent on specific triggers. It is only when the muscle is stimulated that the head end of the myosin molecule undergoes a hinge-like transformation, which results approximately in a 10-nannometre relative movement between the actin and myosin filaments. When a rapid series of bonds between actin and myosin are repeatedly made and then broken again at a rate of around 50-100 cyles per second, the molecular hinge movements are converted to the movements which we observe as muscle contraction. But, of course, such molecular movements have no *purpose* except when seen in association with tendons and joints and a nerve supply in the intact organism.

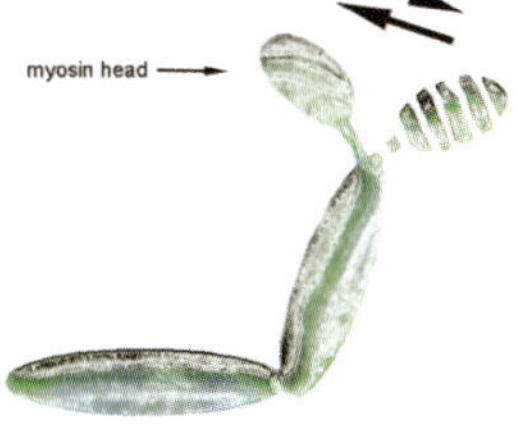


**Myosin hinge**
The heads of the myosin molecules bend to and fro and make bonds with the actin filaments.
A number of myosin molecules joined together make up a myosin filament.

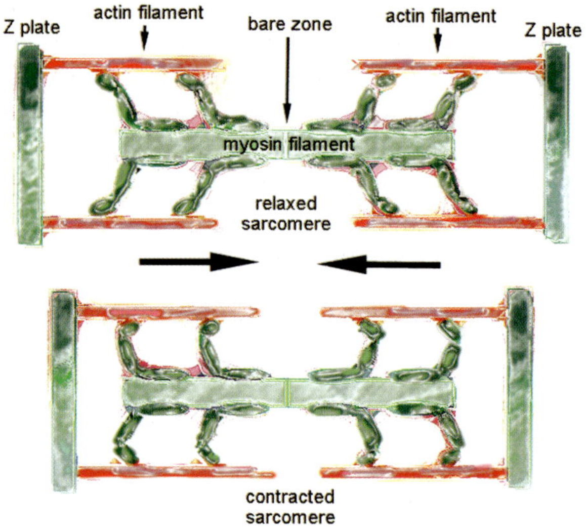


Sarcomere contraction.
As the heads of the myosin molecules move, the actin filaments (with their attached Z-discs) are pulled closer together. As a result the muscle shortens.

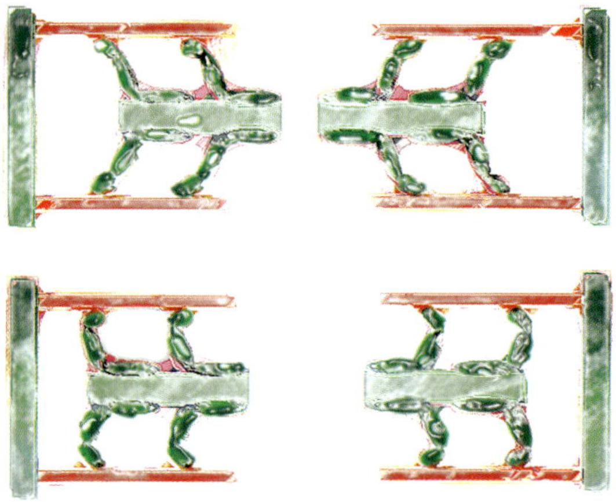

Without the bare zone uniting two oppositely-orientated myosin filaments, the myosin filaments would simply slide towards the Z plates and there would be no muscle shortening.

As with flagellae, it is quite clear that for it to function, the muscle sarcomere must first exist with a minimum complexity. Clearly a sarcomere can be assembled within a living cell - after all, it happens with every new birth. However, these structures cannot self-assemble from disintegrated whole cell components. Thus there must be a mechanism inherent in the living cell that ensures that the components are not only manufactured, but also assembled in the correct manner. And since any organism which has a partially complete, non-functional sarcomere cannot have any selective advantage over an otherwise identical organism which has no sarcomere components whatsoever, a sarcomere could not have evolved.

### *Cell division*

Most living cells grow and divide. But before a cell divides into two new daughter cells, the chromosomes have to be duplicated and then separated, so that both new cells acquire complete working copies of the whole genetic code for that particular organism. For example, a normal ape has 24 pairs of chromosomes. Each pair consists of two *homologous chromosomes*, one of which has come from the sperm and the other from the egg of the parents. Normally each sperm and egg only have half the normal number of chromosomes (24 single chromosomes) so that when fertilisation occurs, the new individual will have the normal full complement of 48 chromosomes (24 pairs). For normal health, a complete pair of homologous chromosomes has to be present in the embryo. When, for example, one of a pair of homologous chromosomes is missing (called *monosomy*), the embryo is likely to be non-viable. If, as in *Down's syndrome*, an extra homologous chromosome is present, making 3 such chromosomes in all (called *trisomy*), the embryo may well develop, but the individual is likely to be seriously abnormal.

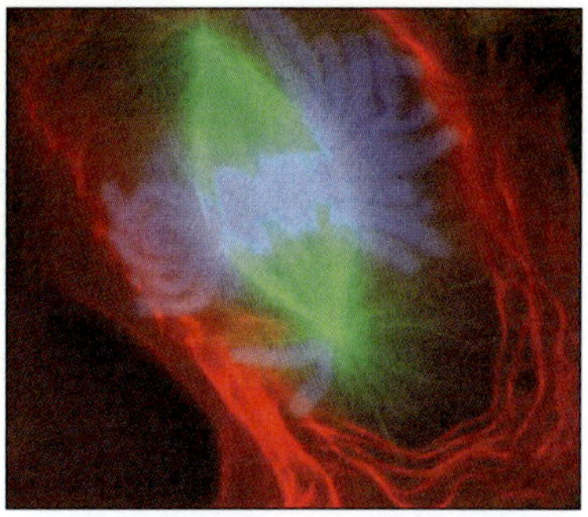

A newt lung cell stained with flourescent dyes undergoing mitosis, showing mitotic spindles (green), cell membrane and some cytoplasmic components (red), and the chromosomes (purple).

In the absence of specific triggers, there is no *chemical* reason why chromosomes (DNA molecules) should start to replicate at any particular time. And if they were to migrate into the halves of a dividing cell, there is no *chemical* reason why all chromosomes should not just move randomly and end up in any daughter cell. But, in fact, we know that chromosomes *do* replicate themselves at specific times, they *do* arrange themselves in an ordered fashion prior to a cell dividing, and they *do* (except in error) divide themselves equally between the daughter cells. Chromosomal separation and cell division are very carefully controlled and executed processes involving a huge number of different molecular components. However, for the purpose of this discussion only, the main characters in the play are the centrioles with their *spindles*, and the chromosomes and their *kinetochores*.

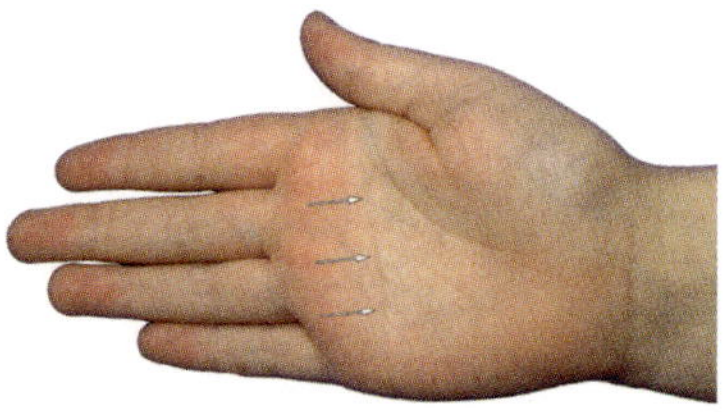

A transverse crease in the palm of the hand is a feature of Down's Syndrome, but is also occasionally found in normal people.

The centrioles are a pair of permanent structures which resemble short sections of a eukaryote flagellum. At the time of cell division the centrioles themselves duplicate, separate and migrate towards opposite poles of the cell. Microtubules then sprout outwards from the centrioles, and these are seen under the microscope as the spindle. The microtubules are not fixed cellular structures but are constructed during, and dismantled after, cell division.

Each chromosome has a small narrowed portion within its length called the *centromere*. The centromere itself has a special part called a kinetochore, to which microtubules attach and by which the chromosomes are pulled to one pole of the dividing cell. At least 4 proteins have been identified within the kinetochore plate. A particular protein called *CBF3* is *essential* for the binding of microtubules to the kinetochore plate, and mutations that affect this gene cause problems with the proper sorting of chromosomes during cell division. For example, in the serious disease *Scleroderma*, for some unknown reason the body produces antibodies to its own kinetochore proteins and this results in cell damage and scarring in tissues.

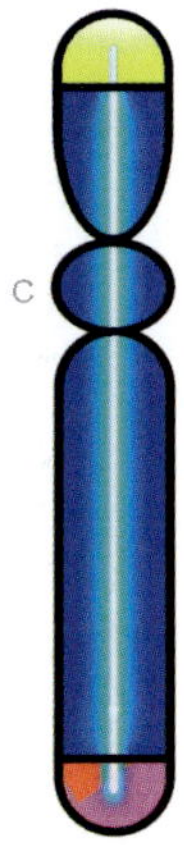


Chromosome and centromere(C). (Diagrammatic)

Once the chromosomes have duplicated, the centrioles at each pole of the cell start to produce microtubule filaments which elongate (as a result of polymerisation of the microtubule proteins) and project towards the centre of the cell. The exact direction of the projection of the microtubules is entirely random but when a microtubule touches a chromosomal kinetochore, special binding proteins cause the microtubule and the kinetochore to become locked together, and the chromosome is 'pulled' towards one centriole. The pulling is effected by the shortening of

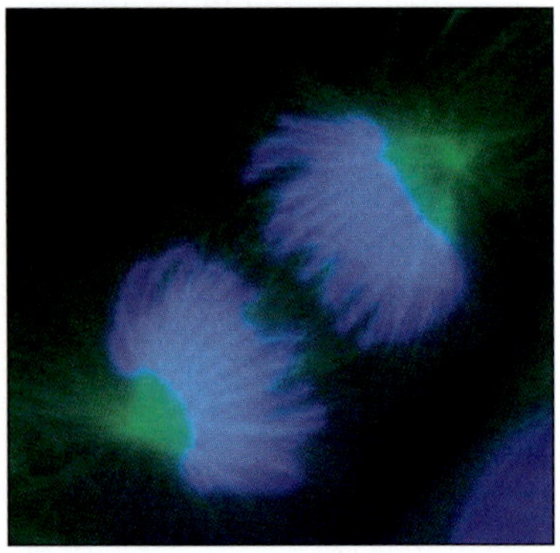
A newt lung cell stained with fluorescent dyes undergoing mitosis showing the mitotic spindles (green), and the chromosomes (purple). The cell membrane is not visible.

the microtubules, which is itself caused by the process of depolymerisation of the microtubule proteins.

Any chromosome fragment that lacks a centromere section cannot be pulled by a microtubule and its genetic material will therefore be lost. Loss of even small fragments of chromosomes can cause serious illness, for example, the *Cri du chat* syndrome (where there is loss of a small part of chromosome 4 resulting in severe mental retardation).

Of course, there is much more to movement in cell division than what has been described so far. For example, the cell itself has to be cleaved into two daughter cells. This is made possible by a contractile ring that appears in the equatorial plane of the spindle apparatus. The position of the contractile ring is in fact determined by the position of the centrioles. Experiments can be done in which the centrioles are moved with ultra-fine needles, and any movement of the centrioles results in movement of the site of the contractile ring. The actual cleavage furrow (which will eventually separate the two daughter cells) is produced by the activation of a special assembly of actin and myosin molecules (this assembly has nothing to do with muscle cells) which are attached in a circular manner to the inside of the cell membrane.

Thus many essential controlling and structural aspects of cell division must be activated at very specific times and sites. Working backwards, there can be no cleavage without actin and myosin. There can be no activation of the actin and myosin contractile ring without some signal from the centrioles. There can be no spindle apparatus formation without the activity of the centrioles. There can be no spindle without the microtubule proteins. There can be no attachment of microtubules to the chromosomes without there being a complex kinetochore within the centromere. And the kinetochore is composed of essential proteins such as CBF3 without which the process of chromosomal capture is severely damaged. Since the microtubule sproutings are random then we should expect that individual chromosomes would be pulled to each pole of the dividing cell in a random fashion. But this is not the case since the beauty of cell division demonstrates the most impressive degree of order - order without which all cells would fail to replicate, resulting in death. All this and much more has to be coded for by the genes within the chromosomes *before* cell division can be undertaken with the fidelity that it necessarily requires.

So, cell division contains yet another collection of structures and processes which requires a minimal structural and organisational complexity without which it cannot operate *at all*.

Thus, by looking at the above *purpose*ful movements of molecules inside cells, not only do we come face to face with the awesome complexity of a living cell, but we are obliged to concede the following: firstly, to have *any* function, molecular machines and systems require a minimum degree of complexity. In other words, any molecular machines and systems that lack *essential* parts can provide not proportionately *less* function, but *no* function at all. Secondly, the appearance of any non-functional part-system can confer no advantage, and so natural selection cannot operate; and without natural selection there can be *no* evolution. And thirdly, if, fortuitously, a non-functioning *part* of a system could appear, then the *WYWYWG Principle* shows that the system is likely *never* to become functional.

## *Intracellular molecular movement*

The development of an embryo is an astronomically complicated process and as is often the case, uncovering one layer of knowledge opens up a Pandora's box with more and more layers of complexities and associated questions within. The manufacture of a butterfly out of the liquidised caterpillar inside the chrysallis is just one such example:

> ...the caterpillar's body now begins to function as a living butterfly blueprint...controlling the liquid flow of cells with military precision. These cells are regimented and re-ordered into new shapes - some into wings, others into legs or huge new eyes. It's a complex process...**It all seems so unlikely, like dropping a pile of bricks onto a sheet of plans and expecting St Paul's Cathedral to materialise. Yet it happens every time.** (My bold). David Attenborough. 'Butterfly metamorphosis'. Wildlife on One. BBC Broadcast. 1997.

Painted lady butterfly

Painted lady caterpillar

Embryological development is best understood in the fruit fly *Drosophila melanogaster*, on which research has been going on continuously for at least the last 100 years, involving the observation of countless thousands of generations of flies.

It is important to note from the outset that (as with all biological processes) there are no 'stages' as such, since the whole process of fly development unravels as a continuum of inter-related phenomena. The fruit fly starts life as a *syncytium* - a large mass of cytoplasm, within the centre of which are a number of nuclei. These nuclei divide every 9 minutes or so, and are free potentially to move about in any direction throughout the cytoplasm. After about an hour there are about 128 nuclei. About 18 of these move towards the rear end of the insect-to-be and continue to divide at a different rate to the remainder of the nuclei; these nuclei will later become the germ cells (sperm or egg) of the adult. After a time, most of the remaining nuclei migrate outwards to the periphery of the cytoplasm to form a monolayer of nuclei just underneath the cell wall. These nuclei are destined to make up the external aspects of the insect. A small number of nuclei remain in the centre of the cytoplasm and these will later develop into deeper tissues, including muscle. As a result of some unknown trigger, the cell membrane then extends inwards to encircle each nucleus so as to produce distinct and separate cells.

The cells then divide many times and after much complex folding and segmentation, the familiar fly body starts to appear with all its organs and appendages. The body is divided essentially into 14 specific segments, within which the organs and structures develop. For example, the *antennae* form in the portion called T1; the *wings* in T2; and the *balancers* (also called *halteres*) in T3.

But our interest here is to do with the movement of molecules. Let's look again at the developing fly syncytium and magnify the picture so that we can see atomic detail. We can see just molecules in a 'soup' made up of water and ions and mi-

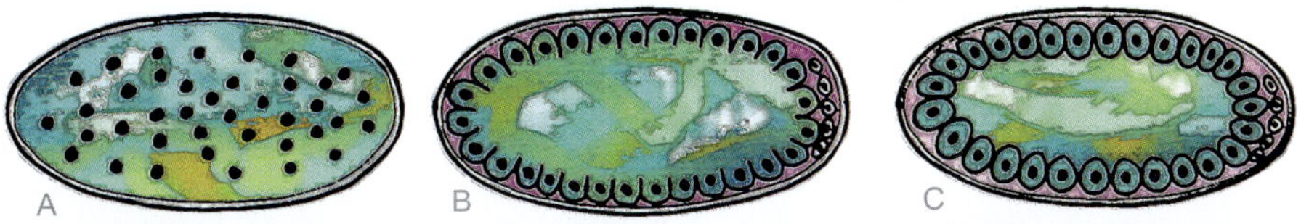


Early development of the fly embryo. See main text.

Andrew Dickinson

**Painted lady metamorphosis**

1-4: The caterpillar skin splits and rolls upwards to reveal the chrysalis within. This process takes about 10 minutes. Inside the chrysalis, the caterpillar cells become liquidised.

5-8: After 7-10 days, the liquidised caterpillar has been completely remodelled and the butterfly emerges over a period of a minute or so.

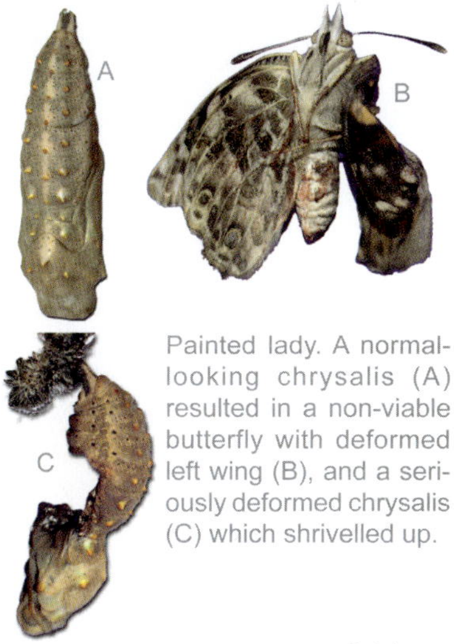


Painted lady. A normal-looking chrysalis (A) resulted in a non-viable butterfly with deformed left wing (B), and a seriously deformed chrysalis (C) which shrivelled up.

**Balancers**

It is commonly suggested that balancers are rudimentary wings. However, balancers have a necessary gyroscopic function and allow the fly to accurately alter its flight path at speed.
Furthermore, it has been shown that in the fly there is no direct eye-to-wing nerve stimulation. Instead, impulses are first sent from the eye to the balancers, and subsequently relayed to the wings. This arrangement of wing control of flight suggests strongly that balancers are not wings that have become degenerate secondary to some gene mutation, but are important structures

Balancers.
Above: Housefly.
Below: Cranefly.

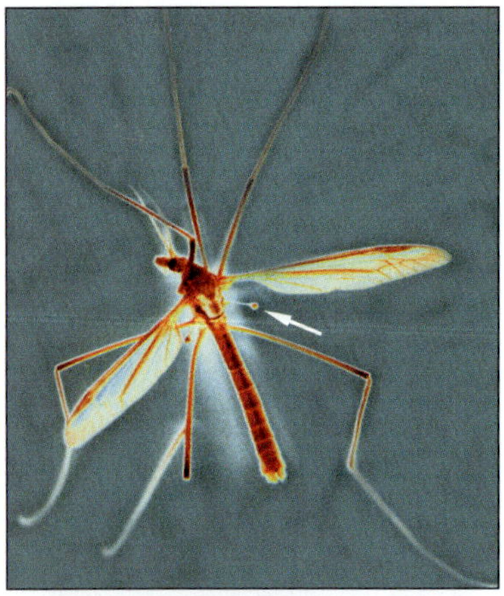

However it does it, a bee can fly very well.

crofilaments, along with ribosomes, lipoprotein membranes, strands of DNA and RNA, and a thousand other essential cellular elements.

If we were to run a film of the developmental process slowly, and if we were able to carefully watch every molecule and particle, and observe every chemical concentration gradient, and see the cascade of chemical reactions and molecular movements that result after every gene is activated, then, strictly speaking, we would see only a whole series of interwoven triggers and events: some resulting in molecules being attracted, others resulting in molecular bonds breaking, yet others resulting in molecules being repelled etc. Most certainly, we would see only pure chemistry.

So what organises the whole process? If we were to suppose, for example, that the information template for the movement of the nuclei and every other aspect of the development of the fly is held solely in the genes of the proto-fly, such a scheme could not evolve since if the first attempt by trial and error failed (as most likely it would), there would be no more genes to mutate, and no more proto-flies to 'trial and error' with.

So, how do the chemicals in the original syncytium 'know' what to do, and when and where to do it? We know that cells can 'communicate' with each other, but any communication is simply the result of passive outworkings of their chemistry. For example, many cells are known to release different molecules into their surroundings. Some of these chemicals act so as to inhibit or stimulate the growth of other nearby cells. For example, in addition to *neurones* (nerve cells), the brain is made up of a number of other essential cell types. So as to ensure that not all the cells in the developing brain become neurones, once formed, embryonic neurones produce chemicals which inhibit other nearby cells from also developing into neurones, thereby facilitating neighbouring cells to develop into other cell types instead. Another example is the developing eye bud, which grows out from the front of the developing brain and releases chemicals which stimulate the cells of the overlying skin to develop into the lens. Other chemical signals direct the subsequent migration of the lens into the developing eye ball. And chemical gradients are involved in causing specific cell death to allow the shaping of tissues. For example, the hand is in part carved out of a solid mass of cells, some of which are destined specifically to be killed off leaving behind the living tissues to form fingers. In this example, if the specific cell death triggering system fails to operate, then some of the fingers may develop with skin bridges, or may even be completely fused together.

So cells *do* interact, but chemical gradients can only produce geometric patterns of cell arrangements; and the simple recognition that chemical gradients and triggers exist, does not explain how growing tissues inside an embryo came to be organised in the sophisticated way that they are. In fact, we know nothing about exactly what stores the information for the actual manufacture of any structure, be it a wing or your finger tip. But the reader may protest that surely it has been well established that DNA holds *all* the information necessary for making living things, and that the recent recognition of a number of specific genes called *homeobox* (often abbreviated to *hox*) genes confirms both the supreme organising ability of DNA and also its evolution.

### *Hox genes*

So what exactly is a hox gene? Some flies which have a mutation of the *ultrabithorax* gene (*Ubx*) develop an extra pair of fully formed wings in the place of the balancers on the T3 segment. Other mutant flies with the mutant gene *antennapedia* (*Antp*) are found to have fully formed legs in place of the antennae. But neither the Antp nor the Ubx gene 'knows' how to manufacture either a wing or a leg since both genes act simply as 'master control' genes, whose protein products influence the expression of a variety of commands which all play a role in the normal development of a range of embryonic structures.

Another example of a hox gene in the fly is the so-called *ey* gene* which can be shown to cause the development of an eye in different sites on a fly's body, and flies with a mutation of this gene develop without eyes. When transplanted into the developing fly, the mammalian *equivalent* of the fly's ey gene can induce the production of eyes at different sites on a fly's body, for example, on a leg:

> ...the mammalian gene can substitute for the fly gene in causing the formation of ectopic eyes in fruit flies. Gerald Karp. 'Cell and molecular biology'. John Wiley. 1996. p559. (The term 'ectopic' refers to a different site from normal.)

*Also called the *eyeless gene.* It has been suggested that this gene is very old and that in 'primitive' organisms, perhaps:

> *...a cluster of light-sensitive cells [were] organised through the regulatory effects of a primitive eyeless gene.*
> Peter Snusted et al. 'Principles of genetics'. John Wiley. 1997. p630.

Of course, this gives no insight as to how the eyeless gene itself arose in the first place, without which a hypothetical *'cluster of light-sensitive cells'* could not have been organised *at all.*

But the mammalian equivalent of the ey gene does not produce a *mammalian* eye in the fly showing that the ey gene is acting simply as a trigger, which switches on a cascade of subordinate genes which are themselves directed by some other manufacturing blueprint. It is found that a specific portion of the amino acid sequences of the Antp, Ubx and ey genes is identical, and it is this common sequence which is called the homeobox. In fact, the DNA from beetles through to humans has a variety of genes containing sequences very similar to the homeobox, and there is no doubt that hox genes are responsible for the development of the body segments in creatures as diverse as mammals and insects:

> ...just as in the case of the fly, the homeobox genes are involved in controlling the basic segmental character of the vertebrate body. Gerald Karp. 'Cell and molecular biology'. John Wiley. 1996. p558.

It has therefore been suggested that this confirms a common ancestry for all living creatures:

> Studies on the mouse indicated that there were four distinct clusters of homeobox-containing genes located on four different chromosomes... This observation suggests that the ancestor of both arthropods and vertebrates, which are thought to have diverged approximately 700 million years ago, must have possessed a similar cluster of genes that determined its body plan during embryonic development. Gerald Karp. 'Cell and molecular biology'. John Wiley. 1996. p557.

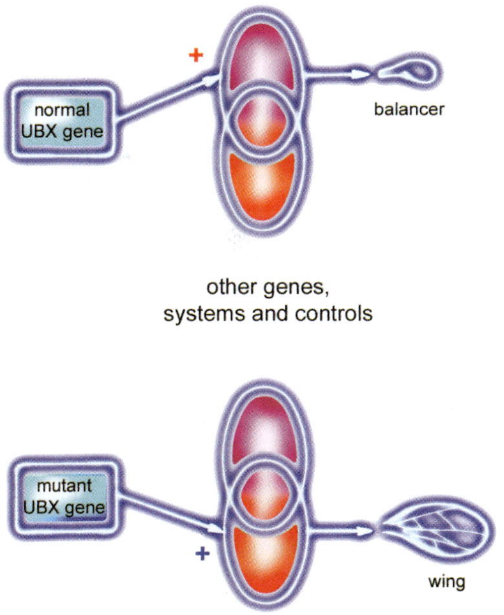


Since different hox genes utilise virtually the same proteins in the manufacture of different structures, hox genes must depend on additional information in order that developmental processes can be directed in different directions.

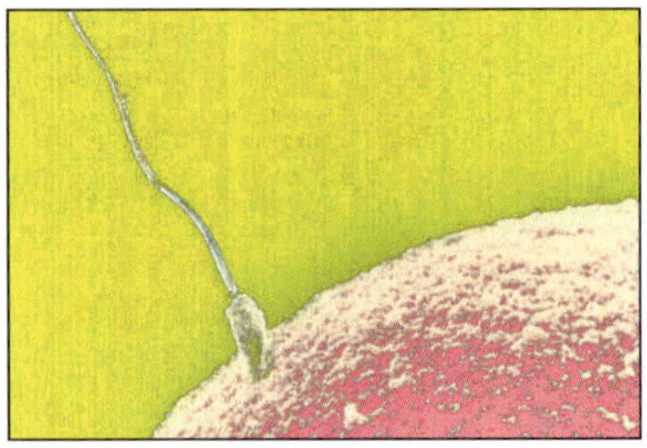

A sperm that is about to fertilise an egg.

In actual fact it confirms nothing of the sort. All it shows is that all organisms have a lot of things in common. For example, most vertebrates have two eyes, two kidneys, intestines, a liver, a heart, four limbs etc., so large numbers of the genes that help to manufacture and maintain their bodies are likely to be similar. At the cellular level, most cell types, whether belonging to a whale or a beetle or a banana, have similar intracellular structures and biochemical systems, and so large numbers of the genes that help to manufacture individual cells in all living things are also likely to be similar. So we should not be surprised to find that hox genes are also similar in diverse creatures.

In contrast to Karp's suggestion that arthropods and vertebrates had a common ancestor, comparisons of the early embryological development of invertebrates and vertebrates do not support such a relationship. Once a fertilised egg divides in both invertebrates and vertebrates, it multiplies and forms a hollow ball (the *blastula)*. With further replication, the surface cells migrate inwards to form a hollow structure composed of two layers of cells (the *gastrula*) with an opening at the point of inward migration. It is at this very early stage that the invertebrate and vertebrate development differs dramatically.

For example, in invertebrates the dividing cells of the blastula arrange themselves in a *regular* spiral manner; the two layers of the gastrula are formed by *immigration* of cells from the surface; the initial opening develops ultimately to become the *mouth;* and a later perforation at the opposite end to the initial opening becomes the *anus*.

In vertebrates, on the other hand, the dividing cells of the blastula arrange themselves in an *irregular* manner; the two layers of the gastrula are formed by *invagination* of the cells from the surface; the initial opening develops ultimately to become the *anus*; and a later perforation at the opposite end to the initial opening becomes the *mouth*.

There are other major important differences which show that the development of invertebrates and vertebrates is directed by quite different instructions. For example,

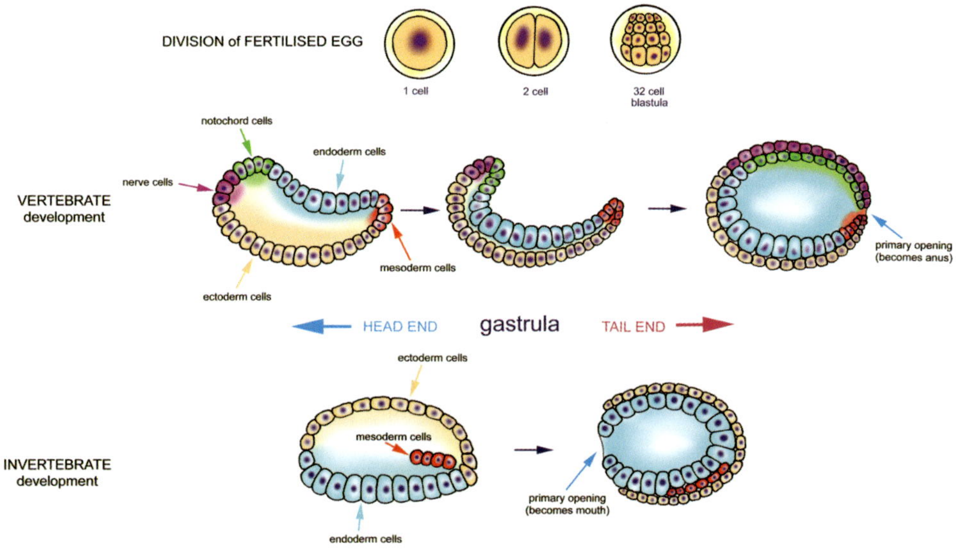


**Egg development in vertebrates and invertebrates**

The fertilised egg divides to become a multicellular blastula. Even at this early stage cells at different sites clearly have different fates, and invertebrates and vertebrates develop in quite different directions. Very soon, distinct front and back ends become discernible.

In invertebrates the dividing cells of the blastula arrange themselves in a regular spiral manner; the two layers of the gastrula are formed by immigration of cells from the surface; the initial opening develops ultimately to become the mouth; and a later perforation at the opposite end to the initial opening becomes the anus; the cells destined to form muscles and other deeper tissues (*mesoderm*) arise within the gastrula (red).

In vertebrates, on the other hand, the dividing cells of the blastula arrange themselves in an irregular manner; the two layers of the gastrula are formed by invagination of the cells from the surface; the initial opening develops ultimately to become the anus; and a later perforation at the opposite end to the initial opening becomes the mouth; the mesoderm cells arise from the surface of the gastrula.

Since the early embryological development of invertebrates and vertebrates proceeds in quite different directions, then despite any similarity in their hox genes, it is impossible for insects and humans to have had a common ancestor.

*Note that this illustration is highly diagrammatic - the developmental process is extremely complex, and the differences between the fate of the initial and secondary openings in the invertebrate and vertebrate are* **not** *due to simple reversal.*

the cells destined to form muscles and other deeper tissues (the *mesoderm*) arise amongst the surface cells in the vertebrate, whereas they arise within the gastrula in the invertebrate; and the spinal cord lies along the back in the vertebrate, whereas the main nerve trunks lie on the underside of an invertebrate.

Since embryological development is the result of a highly complex weave of triggers and responses, and since there are no known intermediary stages between invertebrate and vertebrate developmental processes, the differences (for example, in the fate of the initial opening, whether mouth or anus) are likely not to be the result of simple reversals. It is, therefore, implausible that the invertebrate developmental process reversed itself so as to branch off and produce vertebrates. In other words, the evidence from early development suggests that invertebrates and vertebrates arose separately and are totally unrelated. This shows that claims of evolutionary patterns between diverse organisms based only on genetic similarity can be highly suspect or even invalid.

Nevertheless, it is argued that if changes in hox genes can produce a leg where there was previously an antenna, then why couldn't a hopeful dinosaur acquire a hox gene mutation that in its progeny would convert a front leg into a bird wing in just a single generation:

> ...homeobox genes regulate the sequence and mixture of those materials, thereby determining body shape. A slight variation in one of these genes **can** make the difference between an embryonic limb bud developing into a fin or a foot, an arm or a wing. (My bold). Peter Martin. 'History. The beginning.' The Sunday Times magazine. 12th September 1999. p21.

It is assumed that such eventualities would not require any intermediary forms, and so the glaring absences of intermediary forms of creatures in the fossil record no longer become a problem. So, it is suggested that we should not be surprised to learn of:

> ...the four-*legged* hoofed creature that, a couple of sudden morphings down the line, had turned into a whale. Peter Martin. 'History. The beginning.' The Sunday Times magazine. 12th September 1999. p21.

And for a knuckle-walking ape to turn into a fully upright biped would be:

> Just another random consequence of a sudden all-or-nothing shape-altering mutation. Peter Martin. 'History. The Beginning.' The Sunday Times Magazine. 12th September 1999. p21.

This of course supposes, for example, that any dinosaur that might hope to become a bird already has within its make-up the blueprint and the manufacturing mechanisms that could produce a wing, upon which some newly-mutated hox gene could act. But it is unlikely that hox genes carry within themselves the blueprint for manufacturing either a fin or a toe or a wing, and they most likely act only as master controlling genes whose job it is (somehow) simply to switch on some other genes at key points during the development of living things.

**Finger elongation: mice**

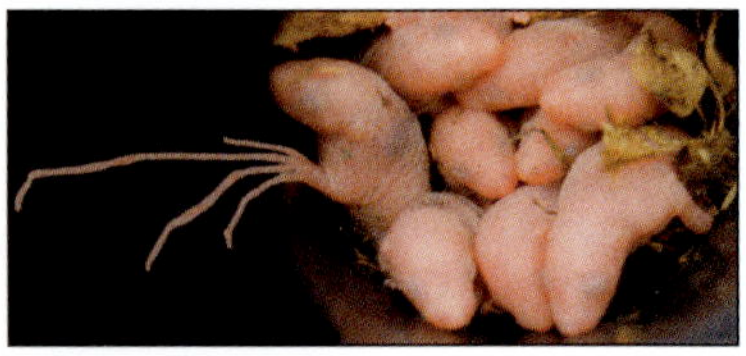

Computer image showing what a rat with long digits would look like.

Recently, it has been shown that digit length can be easily increased by the action of just one protein:

> *We find that the early embryonic bat digits (immediately after the formation of joints) are similar in size to those of mouse embryos of a comparable age...We show that bone morphogenetic protein 2 (Bmp2) can stimulate cartilage proliferation and differentiation and increase digit length in the bat embryonic forelimb.* K Sears et al. 'Development of bat flight: morphologic and molecular evolution of bat wing digits'. PNAS Vol 103. No.17. 25th April 2006. p6581–6586.

From this came the dramatic claim:

> *A change to a single gene **allowed** bats to grow wings and take to the air... (My bold).* 'Rogue finger gene got bats airborne'. New Scientist. 3rd November 2004. p16.

Clearly, if all the digits of a mouse suddenly became very long (like that of the aye-aye), the animal would still be completely a mouse, and be seriously impeded so that it could not run or burrow. It would, therefore, die out rather than evolve into a bat. The claim that Bmp2 activity 'allowed bats to grow wings' is simply unscientific.
Furthermore, the *WYWYWG Principle* shows that if some ancient mice did acquire Bmp2 activation, it would most likely involve bones other than the forelimb digits (for example, the tail or shoulder blades), where it would certainly be of no use as an intermediary step towards flight.

We know that a mutation in a hox gene can result in the manufacture of a balancer instead of a wing, and that virtually the same pool of proteins is used to make both structures. But unlike in the abnormal haemoglobin HbS, where a single point mutation results in a deformed red blood cell, a balancer is not a deformed wing but a structural entity in its own right. So, for two completely different structures such as a balancer and a wing to be manufactured, there must exist two distinct cascades of commands and controls that direct the common pool of proteins to become either a wing or a balancer. So it is not surprising that:

> Molecular biology has not yet helped us to understand how a single protein present in a cell can change it from making part of one pattern (eg a wing) to making part of another (a haltere). Peter Martin. 'The making of a fly.' Blackwell Scientifc Publications. 1992. p117.

It is therefore unlikely that hox genes can explain the gaps in the fossil record. Although we have seen that it is implausible, nevertheless, *if it were* possible for a simple hiccup in a hox gene to occur such that, as a result, it could code for the manufacture of a wing instead of a foreleg, we should also ask: why should a hox gene inside the DNA of a *hoping-to-become-a-bird* dinosaur mutate so as to code for a wing, rather than a hox gene sited in the DNA *inside any other creature* living at the same time, or at any time before or after that hopeful dinosaur?
Look at it this way: a hox gene is simply a series of nucleotides strung together in a portion of DNA. At any moment there will exist billions of hox genes in billions of portions of DNA inside billions of cells inside billions of creatures. For our purpose here we can view all of the world's hox genes as being simply molecules spread out all over the planet. The fact that the molecules are inside individual creatures is of no importance. So it is improbable that *the very* molecular alteration that would code for a wing instead of a foreleg, would occur in a hox gene inside the DNA of the *hoping-to-become-a-bird* dinosaur. Instead, *if it were* possible, such a mutation would, in all probability, occur in some other fragment of DNA somewhere else on the planet and might be inside, for example, some worm or jellyfish. In such a location, the mutated hox gene would find no application and would subsequently become degenerate and lost. Furthermore, having in the history of the planet appeared once, it is likely that that particular hox gene mutation would *not* arise again. In fact, if mutations in hox genes could create wings instead of forelimbs and without the necessity for intermediaries, then we should find the fossil record full of hopeful monsters which have acquired characteristics which for them are useless, but which are still compatible with survival: for example, a compound eye in a mole, or antlers in a sperm whale.

### *The template for making bodies is not in the genes*

Here we are left with perhaps the most important question: can the nature and structure of an organism be dictated *purely* by its genetic code and the manifestation of that code via the operatives of specific proteins? The reader may protest that surely the dogma of the centrality of DNA and genes in the defining and determining of practically all structural and organisational aspects of any organism, is the most revered tenet in modern biology. Surely, since there is *complete agreement* that genes are responsible both for expressing variety, and for the proper development and formation of all living things, then DNA *must* comprise the blueprint for the manufacture of living organisms. Surely the centrality of DNA's position is unshakable and unquestionable.

However, consider a motorcycle. Obviously, working plans for manufacturing all the *components* of a motorcycle (for example, screws, bolts, alloys, plastic, rubber etc.) must exist. But under no circumstances will simply producing all the thousands of components enable those parts to be integrated in the correct order and manner to manufacture the motorcycle. Although a particular screw may well fit exactly into a particular hole, actually placing it into that hole at the right time in relation to the placing of the rest of the motorcycle's components requires information about timing; a mechanism for ensuring that the screw is transported from its site of manufacture to its final destination (via any intermediary resting places); and a mechanism for its correct insertion into its proper hole. For all that to happen requires *additional information* which is *not inherent in the shape or substance of the screw itself.* Nor is that information within the plans which detail the manufacture of the screw itself. Of course, the information detailing how a motorcycle is manufactured is organised in the design stage, and housed in a set of plans which has nothing to do with either the manufacture of the components or the raw materials of which any components are made.

Just as the recipes for making any of its components do not contain the information for making a motorcycle, the recipes for making any protein do not contain the information for making a body.

Similarly, DNA may be the working drawing for the manufacture of proteins, but any protein can only act in a relatively small number of ways: for example, as a structural entity (e.g. in a cell wall), a catalyst (an enzyme), a chemical trigger (e.g. a hormone such as insulin), a switch (e.g. affecting the development of cells in an embryo), or a chemico-mechanical machine (e.g. the myosin hinge in muscle). Proteins are only chemicals, and function only as chemicals, and can do nothing except that which their size, shape, charge distribution and make-up *forces* them to do. So, no protein or group of proteins, nor any hox gene carries with it the information to make, for example, a finger. And by implication *nor does DNA itself.* More than that, the information template for making any organism is .... a complete unknown. Furthermore, it is likely not to be contained within any cellular system currently proposed.

We must therefore come to the conclusion that: it is improbable that the chemistry and physics which we observe now could have been, at some time in the past, responsible *singularly* for the origin and perpetuation of any living thing. So there is more to Life than the sum of an organism's constituent atomic parts. And DNA's position is likely to be subordinate to some higher order.

**DNA is subordinate to other controls**

It is a fundamental fact that the genes contained in the egg and the sperm are not solely responsible for the development of the embryo since the egg already has fully functioning intra-cellular structures and machinery (including mitochondria) which it has acquired from the mother. It is the egg itself that selectively activates the genes within the nuclear chromosomes as and when their protein products are needed, so that the nuclear DNA is not in overall control at all.

# CHAP

## The whale: *something fishy*

# SECTION THREE
## All features great and small

# TER 7

**Orca, Vancouver Island**
Melanie Sodera

According to the Natural History Museum in London, the Mesonyx is the ancestor of whales.

The commonly held view is that whales, being mammals, must therefore have evolved from some ancestral terrestrial quadruped mammal. The evidence is extremely patchy but, based on comparisons of teeth, one theory is that the ancestor was a *Mesonyx* which (supposedly) lived about 60 million years ago. This creature was a carnivore with a skull rather like a modern wolf and, curiously for a predator, it had on its feet five small hooves instead of claws.
However, there remains an extremely high degree of speculation amongst authors:

> The origins of cetaceans are still **uncertain**. Many aspects of cetacean anatomy...**indicate little about ancestry**, but biochemical and genetic studies suggest cetaceans are related to hoofed mammals...the *Mesonychidae*. Cetacea **probably** arose from a small form...**we can speculate** that one of these **possibly** fish-eating animals became adapted to feed on abundant food in shallow waters...and quickly took to an amphibious life...the tail **probably** became adapted for vertical beating very quickly, but **we do not know how** this affected the use of the hind limbs...**or how** the tail became flattened. While the early whales **probably** bred on land for many millions of years, there were **probably** rapid physiological changes to life in water. (My bold). R Harrison and M.M. Brydon (editors). 'Whales, dolphins and porpoises'. Merehurst Press. 1990. p14.

Others consider that since dolphins have more than one stomach, the whale ancestor may have been more like a cow or a goat or a sheep. However, multi-compartmental stomachs are also found in completely unrelated animals such as the *sloth*, the *langur monkey*, and some marsupials such as the rabbit-sized *quokka.*
Although the teeth of hippopotami show no similarity to those of whales (whether extinct or living) and although they are herbivores which show no tendency to chase fish, nevertheless, certain aspects of the DNA of hippos share some similarity to that of whales. From this some consider the hippopotamus to be the closest living relative to whales. For example:

> The DNA responsible for the proteins in milk and in blood-clots **shows** whales to be closely related to hippopotami. (My bold). S Jones. 'Almost like a whale'. Doubleday. 1999. p20.

To the unwary reader such a dogmatic statement, coming from a geneticist, might sound entirely scientific and convincing. However, Jones refers to an article by Montgelard and is clearly discriminative with the information he shares with his readers. Although:

> ...three genes (the mitochondrial cytochrome b and two nuclear milk caseins) indicate an association between Cetacea and Hippopotamidae. C Montgelard et al. 'Phylogenetic relationships of artiodactyls and cetaceans as deduced from the comparison of cytochrome b and 12S rNA mitochondrial sequences'. Molecular biology and evolution. Vol 14. 1997. p550-559,

this is only half the story since the second half of the same paragraph continues:

> ...three others (the mitochondrial 12S rRNA and two nuclear hemoglobins) **do not** support such a relationship. (My bold). C Montgelard et al. 'Phylogenetic relationships of artiodactyls and cetaceans as deduced from the comparison of cytochrome b and 12S rNA mitochondrial sequences'. Molecular biology and evolution. Vol 14. 1997. p550-559.

The interesting thing about cytochrome b is that the 243rd amino acid is *threonine* both in hippopotami and in the 35 cetaceans that have been studied. Jones states that this is conclusive evidence of hippopotamus and whale cousinship, but he fails to acknowledge that a number of other mammals also have threonine in the same position:

> ...threonine...replacement at position 243 in the cytochrome b sequence... is shared by only 11 distantly related mammals (six primates, four rodents, and one sirenian) out of 199 comparisons. (My bold). C Montgelard et al. 'Phylogenetic relationships of artiodactyls and cetaceans as deduced from the comparison of cytochrome b and 12S rNA mitochondrial sequences'. Molecular biology and evolution. Vol 14. 1997. p550-559.

So, contrary to Jones' selective statement, a number of mammals as diverse as sea cows and monkeys are also entitled to have some claim to being whale cousins. The last quote also shows how important it is to read very carefully any scientific statement. Montelard states that the six primates, four rodents, and one sirenian are only '*distantly related*'. There is a big assumption here: that all creatures *are* the result of evolution and that they *do have* varying degrees of relatedness according to how long ago their ancestors diverged. Clearly, if no evolution has taken place then any assessment of relatedness, whether close or distant, may have no validity whatsoever.

In stark contrast, still others dogmatically consider that the whales evolved from a very small shrew-like animal (like the modern insect-eating *tupaia* of Malaysia), at a time when there were no living medium-sized or large terrestrial mammals:

Tupaia

> ...could these immense animals really be descended from a tiny creature like a tupaia? It is difficult to believe, and yet **the logic** of the deduction **is undeniable**. Their ancestors **must have** entered the sea at **a time when the only mammals in existence were the little insectivores.** But their anatomy...gives **no clue as to how the move into the sea was made**... (My bold). David Attenborough. 'Life on Earth'. Collins/BBC. 1979. p242.

Clearly, such a marriage of confident assurance and absolute speculation makes for very poor science, and such assertions must be properly investigated.

### *Whales and the fossil record*

Undeniably, whales are a unique group of completely aquatic creatures and have many features that distinguish them from other mammals. The toothed whales generally have a large number of conical teeth, although some such as the beaked whales are almost toothless and any teeth they do have vary from being peg- to spade-like. The baleen whales lack teeth (except in embryonic form) but have

Gray whale

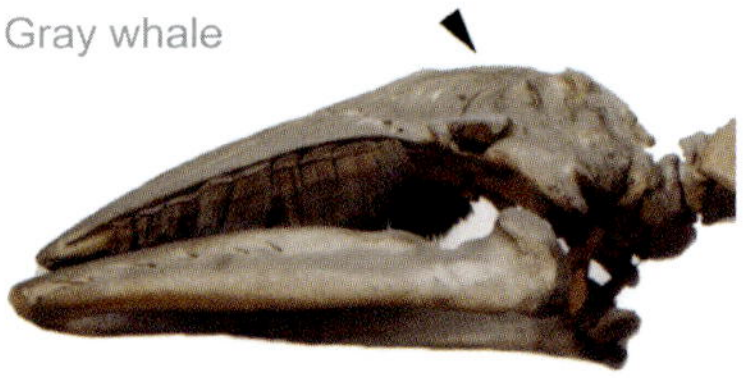

Sperm whale

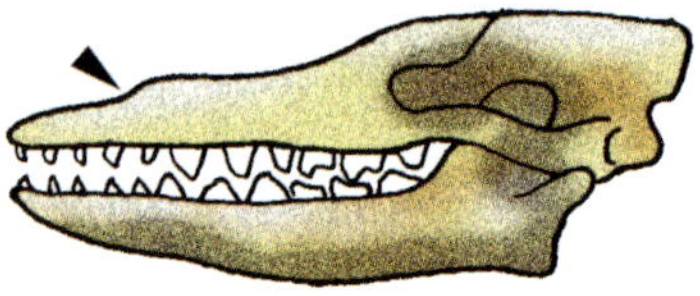

Pakicetus skull

Basilosaurus

Dorudontine

Mammalodon

The arrows point to the position of the blow-holes.

instead fibrous plates which hang like the bristles of a comb from the upper jaws. The toothed whales have an asymmetrical skull with a single blow-hole (usually offset to the left), whereas the baleen whales have symmetrical skulls which have a mid-line double blow-hole. All the toothed whales and baleen whales have their blow-holes situated roughly above their eyes, the exception being the sperm whale which has its blow-hole at the upper front left end of its head.

Toothed whales can (supposedly) be traced back to a small incomplete skull found in Pakistan belonging to the *Pakicetus,* an animal about the size of a large otter, which is presumed to have lived more than 50mya. Pakicetus was clearly a terrestrial carnivore with its long legs and triangular, pointed teeth.

The *Protocetus* is also known only from one small skull found in Egypt, which is thought to be about 50myo. It had nostrils behind the tip of the long upper jaw and its teeth were triangular and pointed. Although there is no evidence with which to formulate an idea of the shape of the rest of the body, and rather than saying simply that we can say *nothing* about the body shape, it is commonly speculated that it had rudimentary hind legs and a dolphin-like body with tail flukes.

The *Basilosaurus* is thought to have lived between 38 and 45mya. A number of specimens have been found in Alabama in the USA, the largest being 21 metres long. Basilosaurus had flippered forelimbs with hinged elbows, a well-developed femur, and nostrils half-way along its muzzle. The teeth were of mixed type and included incisors, canines and cheek teeth. It is possible that basilosaurus could have hauled itself out of the water with its front legs, in the manner of seals. The tiny femur was joined to the pelvis with a well-developed ball-and-socket joint. But it is not clear whether the femur projected out far enough to form simple stumps or small paddles: Benton (in '*Rise of the mammals*') illustrates basilosaurus with complete but tiny hind limbs whereas in contrast, Harrison and Brydon depict basilosaurus with only a vestigial pelvis and femur and no other hind leg bones at all. Such highly speculative illustrations serve only to confuse the issue.

The dolphin-like *Dorudontines* are thought to have lived about 30mya. They were about 5 metres long, had nostrils half-way along the muzzle and their teeth were triangular and pointed. A number of other extinct toothed whales are known from a variety of skulls and teeth. The *Squalodontids* were dolphin-like creatures with shark-like teeth. The *Kentriodontids* were similar to small modern dolphins, although their skulls are described as being more primitive. Both these toothed whales are presumed to have lived between 5 and 25mya. Some presumed ancestral forms of beaked whales had many teeth in both the upper and lower jaws, whereas more recent fossil skulls such as that of *Mesoplodon*, thought to be 5-10 million years old, are almost toothless.

Baleen whales can be traced back to *Cetotheres* which were a group of small to medium-sized extinct whales presumed to have lived between 3 and 30mya. The *Mammalodon* is thought to have lived about 24mya. It was a small animal and may have had baleen-like fringes between its large cusped teeth. Its blow-hole was half-way along its muzzle.

Of living whales, modern sperm whales are known from fossils (supposedly) 22 million years old; modern dolphins are known from fossils (supposedly) 15 million years old; and fossils of modern-looking gray whales have been dated as being 20myo.

So, just how did an ancestral terrestrial mammal become a whale? What genetic and biological mechanisms provided the platforms for the necessary changes? The many popular and specialist books and literature all weave a similar story studded with a variety of suppositions and punctuated with very little hard evidence. The casual reader may be deceived into the notion that there is *no question* about the origin and subsequent evolution of the whale. For example:

> It has been **suggested** that zooplankton-rich deposits in the Oligocene strata indicate conditions that **may have** favoured the evolution of baleen and the filter-feeding mode of life...**While the toothed whale skull was becoming modified** to contain the acoustic apparatus and its jaw and dentition developing for the capture of fast-moving fish, **the baleen whale skull became modified** to feed upon concentrations of plankton. (My bold). P Evans. 'Natural history of whales & dolphins'. Christopher Helm. 1987. p28-29.

David Darling uses more poetry than science as he explains the origin of the whale from a primitive land mammal:

> Unknowingly, the shy creature...Threatened no longer by fearsome predators...ventured into the daylight. Ten million years on, it had blossomed into a dazzling multiplicity of forms. Slipping into the sea, it grew to be a giant - the whale. D Darling. 'Deep time'. Bantam Press. 1989. p98.

Recently, it has been discovered that the ankle bones of the extinct aquatic animals *Artiocetus* and *Rodhocetus** are similar to those of even-toed ungulates (artiodactyls).

* Both supposedly 47 myo.

Left: The skull of the Artiocetus.
Below : Rodhocetus skeleton.

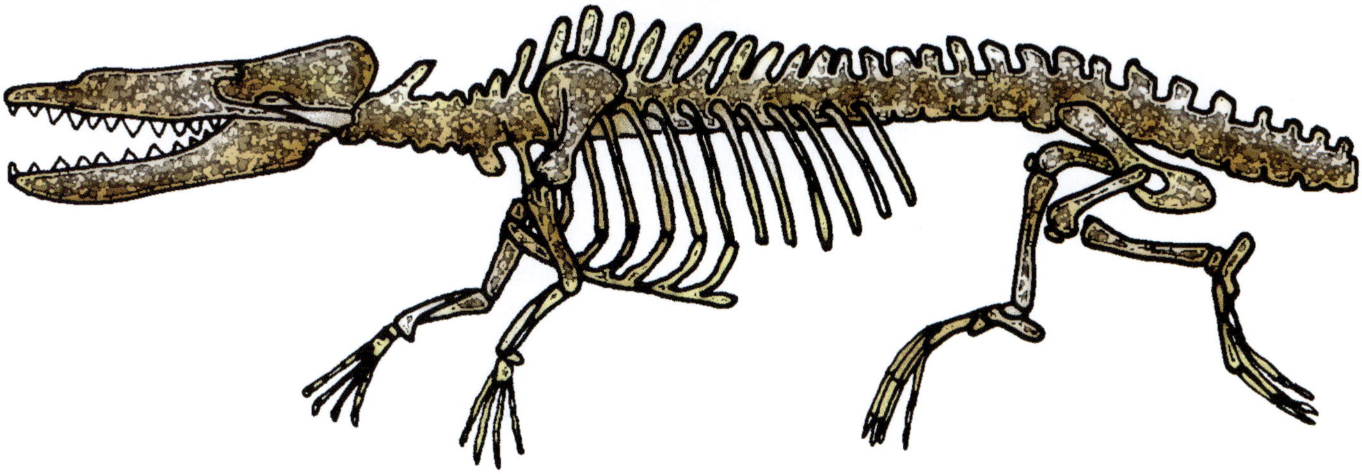

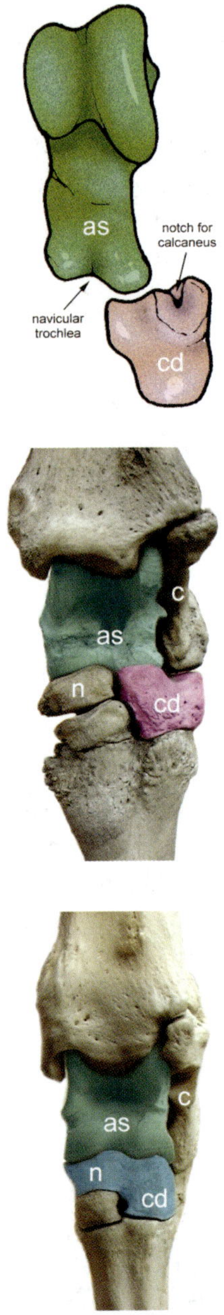


The astralagus (as, green) and cuboid (cd, pink) bones in the Artiocetus (top), camel (middle), and red deer (bottom).
The astralagus of artiocetus has a well-developed concavity at its lower end (*navicular trochlea*), and the cuboid bone has a marked notch for articulating with the calcaneus (c) bone.
Note that the cuboid and navicular (n) bones are fused together in the red deer, showing that artiodactyls vary in the configuration in their ankle bones.

For example, as in ungulates, the *astralagus* (equivalent to the *talus* bone in humans) of both artiocetus and rodhocetus has a well-developed concavity (the *navicular trochlea*) at its lower end, and the *cuboid* bone has a marked notch for articulating with the *calcaneus* bone (the heel bone in humans). These features have led to the current enthusiasm for the claim that artiodactyls were the ancestors of whales:

> The most important evidence comes from the shape and orientation of joint surfaces of several ankle bones in the new fossils. These specialized features, typically associated with adaptation to running, have only been observed in artiodactyls and are widely considered diagnostic of the order... K Rose. 'The ancestry of whales'. Science. Vol 293. 21st September 2001. p2216-7.

Since there remains the important question as to how ungulates themselves arose, what Rose should have said was that '*These features have only **previously** been observed in artiodactyls*'. As Rose admits, we have no understanding of:

> ...the origin of artiodactyls themselves. K Rose. 'The ancestry of whales'. Science. Vol 293. 21st September 2001. p2216-7.

Therefore, since artiocetus and rodhocetus clearly were aquatic animals, we could just as easily (and with as little validity) suggest that the ankle bones of artiocetus and rodhocetus show specialised features, typically associated with adaptation to *paddling*, and that ungulates arose from aquatic mammals.

Artiocetus and rodhocetus had a number of features that separated them greatly from both ungulates and whales. The teeth were not those of plant eaters and were unlike those of living whales. The neck was mobile, unlike in basilosaurus or living whales (excluding river dolphins). Unlike in whales, the nostrils were located near the end of the snout. The spines over the vertebrae in the shoulder area were long, suggesting that artiocetus and rodhocetus had large muscles attached to their front limbs and the neck. This would have allowed them to hold their heads up, and to haul themselves out of the water. Since whales do not leave the water, they have no need for such relatively large neck and shoulder muscles.

Whales swim by using up and down movements of the tail flukes which generate a massive amount of thrust, and front flippers are necessary to control the recoil from such undulations. In contrast, artiocetus and rodhocetus had broad paddle-like front and rear feet, and the limbs could not be extended sideways as flippers. Therefore, it is unlikely that they had tail flukes, and they probably swam by paddling with the feet, and moving the tail sideways like an otter.

Therefore, in order for creatures such as artiocetus and rodhocetus to have given rise to living whales, a number of features would have had to change: for example, the teeth, the nostrils, and the flippers. Furthermore, the back limbs and pelvis would have had to be lost in parallel with gaining tail flukes. We do not know how artiocetus and rodhocetus reproduced but the animals would have had to acquire the ability to give birth wholly in the water. In addition, some of the descendants would have had to acquire echolocation. Since ungulates are generally relatively simple-minded herbivores, this presents a serious difficulty when accounting for how and why some should have become very intelligent predatory swimmers. With all the differences that exist between artiodactyls and whales, the *WYWYWG Principle* shows that it is implausible that a terrestrial ungulate could have evolved into a cetacean. It is likely, therefore, that both artiocetus and rodhocetus simply represent extinct creatures unique in their own right.

On the basis of *only* a skull of protocetus, Fordyce (in 'Whales, dolphins and porpoises') depicts a hairy animal with webbed digits arising from stocky forelimbs, along with *horizontal tail flukes*. Michael Benton on the other hand, at a time when pakicetus was known only from its *skull*, depicts an animal with four stubby limbs and a stocky tail *without flukes*:

> ...The skeleton of Pakicetus is **unknown**, but a **very tentative** reconstruction shows a semi-aquatic coast dwelling carnivore which could still move about on land. (My bold). M Benton. 'Vertebrate paleontology'. Unwin Hyman. 1990. p289.

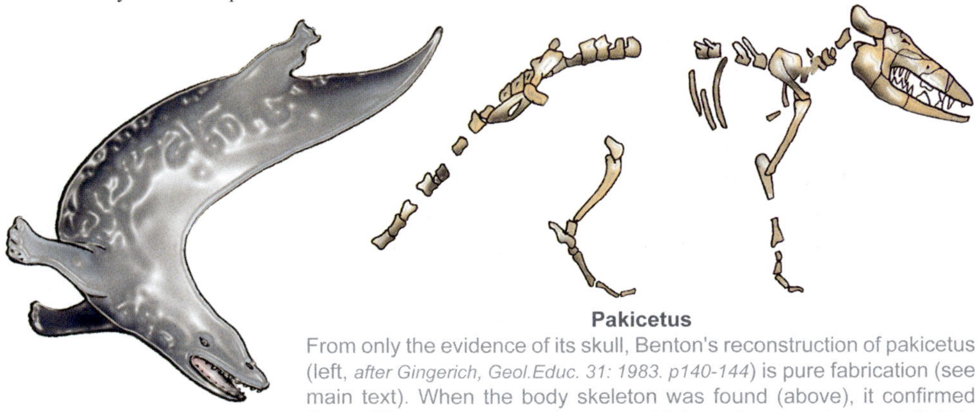

**Pakicetus**

From only the evidence of its skull, Benton's reconstruction of pakicetus (left, *after Gingerich, Geol.Educ. 31: 1983. p140-144*) is pure fabrication (see main text). When the body skeleton was found (above), it confirmed that pakicetus was a running, terrestrial carnivore, and most definitely not a seal-like animal.

Interestingly, the pakicetus skull was found in deposits associated with *terrestrial* creatures:

> ...Pakicetus, known from a skull found in a riverine deposit along with terrestrial animals. It is very similar to a mesonychid, but the ear bones are intermediate between those of *mesonychids* and those of cetaceans. D Futuyma. 'Evolutionary biology'. Sinauer Associates Inc. 1998. p196.

But clearly, the finding of only a skull whose teeth resemble that of a known terrestrial quaduped, in association with other terrestrial creatures, suggests that pakicetus could therefore just have been a creature that was different from other living and extinct terrestrial quadrupeds. And of course, a supposed intermediate form of the ear does not prove that pakicetus was (in evolutionary terms) in any way an intermediary creature. Since new fossil findings* have now confirmed that pakicetus was a *fully terrestrial mammal* with tapir-like *thin legs* which enabled the creature to *run*, then no matter how 'cetacean-like' its ear bones were, it certainly used them while running. Whether it spent any significant time in the water, we do not know.

**WARNING!**
**IMPORTANT INFORMATION LEFT OUT**

Natural History Museum's book deliberately perpetuates (after Gingerich, 1983) the *well-known* false depiction of pakicetus.

*"With four paddle-like limbs and a relatively short tail, this early ancestor of whales probably both swam in shallow water and walked on land."*

So states page 92 of the Natural History Museum of London's 2008 book '*99% Ape: How evolution adds up*'.
Yet, despite the book's claims that:

> *"leading experts...guide the general reader through the latest evidence",*

and despite the authors being fully aware of what the body of pakicetus actually looked like:

> *"the finding in 2001* of a more complete skeleton",*

there is a deliberate failure to inform readers that the skeletal findings, which have now been known for *more than 8 years,* confirm that pakicetus ran on its toes (and did not have paddle-like feet), had long legs, and a long tail.

**J Thewissen et al. 'Skeletons of terrestrial cetaceans and the relationship of whales to artiodactyls'.* Nature. Vol 413. 20th September 2001. p277-281. See also p259-260.

**Unscientific naming**

Note that with the assumption that whales *must have* evolved from a terrestrial ancestor, and despite there being no scientific evidence to support any whale kinship, a number of creatures have been named with the suffix *'-cetus'* (meaning 'whale'). For example, protocetus ('early whale'), artiocetus ('even-toed whale') and pakicetus ('whale from Pakistan'). Since none of these creatures have any relationship to whales, such unscientific naming serves only to misinform. All these creatures should be renamed appropriately.

It would also be well to remember that the natural world includes such a variety of forms (some very bizarre), that we should be wary of constructing a whole skeleton, or of predicting the behaviour of an animal, from only a skull or fragments of bones. For example, when the body of a duck-billed platypus was first brought to London it was thought to have been a hoax - a composite made up from a number of different animals. The coelacanth was thought to be a fish that used its stumpy limbs for walking in the shallows or on the sea bed, until recent film of live specimens showed that they swam like normal fishes. A large bony spike found with the bones of the large dinosaur *Iguanodon* was originally thought to have been a horn on the top of its head, and was illustrated as such in early drawings.

Iguanodon hand showing its thumb spike.

Above: Modern horse skull.
Below, in sequence: Chalicotherium skull, artistic reconstruction, and front toes.

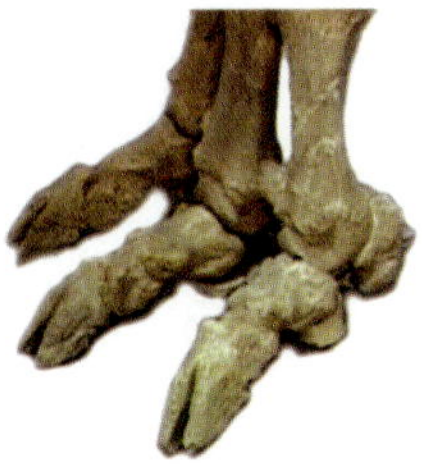

However, when more complete specimens were found, it was clear that the spike actually belonged to the thumb! The *Chalicotherium* was a most unusual creature that supposedly lived between 23 and 5mya and well demonstrates the principle that life forms on earth can be *more* bizzare than we can imagine. The animal had a head that was very horse-like, a sloth-like rear half, stocky legs and claws on the toes:

> This looks like an imaginary mammal composed of the rear parts of a sloth, the arms of a gorilla and a horse's head. But it was real enough, as shown by numerous complete fossil skeletons from the Miocene period of Europe and Africa, and also by relatives from North America and Asia. Michael Benton. 'The rise of the mammals'. Apple Press. 1991. p108.

Many specimens of chalicotherium have been found so there is no doubt about its bony anatomy. However, if for example, only the skull or teeth were found, or if only the front limbs were found, any tentative reconstruction of the rest of the body would be likely to be seriously inaccurate, to say the least.
In the Natural History Museum in London lie the fossil bones of the *Paleoparadoxia*, a creature thought to have lived in Japan between 25 and 6mya. The animal was originally known *only* from a few fossil teeth and a few bone fragments, which were found in 1940. From the appearance of the teeth, the animal was thought to have been a form of sea cow. However, everyone was surprised in 1941 when a complete specimen of paleoparadoxia was found, showing it clearly to have been a cow-sized *terrestrial quadruped*. In fact, the animal has *no* relationship *either ancestral or descendant* to *any* known creature. In the Museum's own words, the paleoparadoxia:

> ...are fossil creatures which occupy a group with no obvious ancestors or successors, they remain a scientific puzzle. Paleoparadoxia exhibit, Mammal Room. Natural History Museum, London. 1997.

In fact, the paleoparadoxia is by no means unique in having no known ancestors or descendants. For example, the rhinoceros-sized mammal *Arsinoitherium*, whose remains have been found only in sediments (supposedly) 35myo and limited to North Africa, also has no known relations. Thus we can state the *Chalicotherium Principle*:

Purely on the basis of its teeth or skull, for *any* fossil: (i) The nature of the body skeleton cannot be predicted or assumed, and (ii) No animal can be assigned to any ancestor-descendant or distant cousin relationship.

Nevertheless, incredible though it may seem, standard works on fossils still fail to learn this lesson, or to even to clarify the point. For example, having laid the reasonable foundation:

> A reconstruction of an extinct animal is only a model of what it might have been like. We can never be really sure... Arthur Busbey III, Robert Coenraads, David Roots and Paul Willis. 'Rocks and fossils'. Harper Collins. 1996. p100,

Busbey et al. go on to add:

> From a fragment such as this shark tooth, a whole animal **can** be reconstructed. Many fossil sharks are known **only** from their teeth. (My bold). Arthur Busbey III, Robert Coenraads, David Roots and Paul Willis. 'Rocks and fossils' Harper Collins. 1996. p100.

But as we have seen, from the examination of just a newly-found tooth we cannot know for sure whether the animal was even aquatic or terrestrial, let alone formulate any suppositions about anatomical details or any evolutionary relationships. Clearly, to ascribe any ancestor-descendant relationship, the body skeletons of both the animal in question and its proposed ancestor or descendant must be known. And even then, the *Horse Principle* shows us that even if different fossil specimens can be arranged in an order suggesting a *presumed* evolutionary trend, no such evolutionary trend need be true.

So, applying the lessons from these examples it is clear that any similarity between the teeth of the mesonyx and protocetus, or comparisons between the position of blow-holes in different ancestral whale-like creatures, *cannot* be taken as evidence of *any* ancestral-descendant evolution. Therefore, it would be logical to conclude that just like the paleoparadoxia, the mesonyx was just an unusual animal, which simply and only had some features in common with some other mammals, and that the mesonyx did not evolve into any cetacean.

Since the fossil evidence gives us *no* idea as to how a terrestrial quadruped could have evolved into a whale, we need to look at other biological evidence. Certainly, Darwin did not provide any great thought on the matter. In fact, Darwin provided no answers at all. His reference to mustela* (mink) simply and only describes a creature that was able to take the opportunity to feed on the ground and in the water, *but which remains the same creature.*

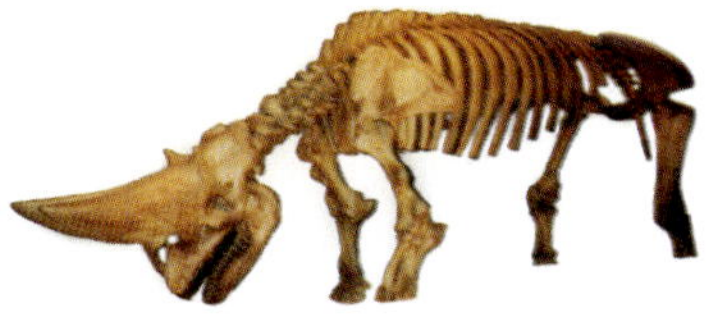

Arsinoitherium. Fossils of this animal have only been found in Egypt, and dated as being around 35myo. Unlike the horns of a rhinoceros (which are made of matted keratin), the twin horns of arsenotherium were made of bone. Both its ancestors and descendants are unknown.

**It has been asked by opponents of such views as I hold, how, for instance, a land carnivorous animal could have been converted into one with aquatic habits; for how could the animal in its transitional state have subsisted? It would be easy to show that within the same group carnivorous animals exist having every intermediate grade between truly aquatic and strictly terrestrial habits; and as each exists by a struggle for life, it is clear that each is well adapted in its habits to its place in nature. Look at the Mustela vison* [mink] *of North America, which has webbed feet and which resembles an otter in its fur, short legs, and form of tail; during summer this animal dives for and preys on fish, but during the long winter it leaves frozen waters, preys like other polecats on mice and land animals. If a different case had been taken, and it had been asked how an insectivorous quadruped could possibly have been converted into a flying bat, the question would have been far more difficult, and I could give no answer...* Charles Darwin. 'The origin of species'. Penguin Books. 1985. p211-2.

Manatee.
Despite having a large layer of fat, manatees cannot tolerate cold water for long. That is why they congregate around fresh water springs with temperatures around 22°C, and only venture out to the open sea in summer.
Since evolution would predict that manatees would be well and truly cold-adapted after 50my, the fact that they are not shows that they cannot have been around for millions of years.

### *Were sea cows once land cows?*

It is commonly stated that the fossils of *sea cows* (dugongs and manatees) demonstrate some of the best examples of the (supposed) evolutionary transition from a terrestrial to a fully-aquatic mammal.

It is true that a number of fossils of sea cows have been found with rear limbs of different sizes containing various numbers of bones. The earliest (supposed) 'proto-sea cow' is *Prorastomus*, which comprised a skull found in Jamaica, and described in 1855:

> ...portions of a skull...An occipital portion...a tolerably capacious brain-case...a facial portion, with the fore part of the lower jaw... Owen. 'On the fossil skull of a mammal (Prorastomus Sirenoïdes, Owen), from the island of Jamaica'. Quarterly Journal of the Geological Society. 1855. Vol 11. Issue 1-2. p541-543.

However, since we know that the *Paleoparadoxia* had a skull that resembles that of the extinct sea cow Halitherium, but was a completely unrelated cow-sized animal, then simply from its skull, any reconstruction of the body of prorastomus is pure conjecture. It may have been fully terrestrial, or it may have been fully aquatic.

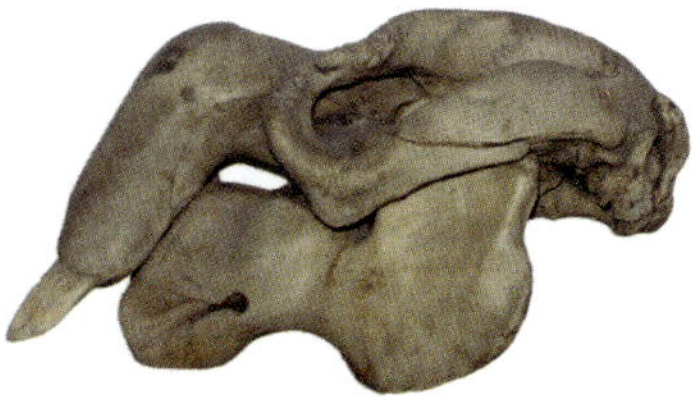

Halitherium: a fully aquatic creature.

Paleoparadoxia: a fully terrestrial creature.

Above: Dugong skull.
Below: Manatee skull.

*Halitherium* and other almost complete fossils (for example, *Metaxytherium*, which was much like a living manatee) show that extinct sea cows did have rear limbs of varying sizes.

It is quite clear that both halitherium and metaxytherium could not have lived on land and were fully aquatic creatures. More recently, a (supposedly) 50myo skeleton of *Pezosiren* was found in Jamaica that was, in terms of its body, a 7-foot-long (2.1m) modern manatee, but it did have a reduced pelvis and four well-developed limbs.

Metaxytherium.
(After Domning, Journal of Vertebrate Paleontology. 1988)

Pezosiren. The white bones are missing.
(After Domning. Nature. 2001)

Domning's reconstruction shows an animal with a relatively long body and short limbs, but he suggests that it was well able to walk about on land:

> Pezosiren... had...four still fully functional legs. D P Domning. 'The earliest known fully quadrupedal sirenian'. Nature. Vol 413. 11th October 2001. p625-627.

While manatees can come out of the water to some degree, they are unable to haul themselves out completely.

It is notable that one of the neck bones, the fibula, most of the foot-bones, and most of the tail bones are missing. So, as Domning accepts:

> The length of the tail, and the form and posture of the feet are partly conjectural. D P Domning. 'The earliest known fully quadrupedal sirenian'. Nature. Vol 413. 11th October 2001. p625-627.

But, more than conjectural, Domning's reconstruction seems ungainly, a somewhat forced interpretation of a terrestrial animal - perhaps because a terrestrial precursor for sea cows is demanded by the notion of evolution. The body seems far too long to allow any significant mobility on land. From its skull and teeth, the animal most likely had the diet of a living manatee, so it most likely was essentially aquatic. The tail has an unnatural directional change where the fossil bones meet the hypothetical tail, and the tail vertebrae seem far too large for a 'dangling' type of tail. Instead, the tail most likely continued horizontally and had flukes. The angulation of the pelvis may have allowed the rear limbs to be turned further backwards, and the feet themselves may not have been feet at all, but flippers. A more natural reconstruction (with the feet left out altogether) is suggested in the illustration below.

Pezosiren: modified reconstruction. Note: The foot bones are missing in the actual specimen.

Now, if a creature such as pezosiren acquired a damaging mutation that affected the development of its rear limbs, it would look like metaxytherium, which, if it lost its rear limbs altogether, would look very much like a modern manatee, as in the following illustration.

Pezosiren: modified reconstruction with pelvis and rear limbs removed.

So, the fossils do not show a series of animals from fully terrestrial ancestor to fully aquatic sea cow.

In addition, we know that since the bony anatomy and muscular and nervous con-

Manatee fluke

trols are all dramatically different, a horizontally-moving tail cannot change into a vertically-moving fluke by a stepwise process. Therefore, if it was ancestral to living sea cows (which have flukes), pezosiren must also have had flukes, and so did every one of its ancestors. We have also seen that mutational interference with limb development only causes derangement or loss, and so forelegs could not turn into flippers. For example, mutations in the *Msx genes* cause serious problems, rather than producing new functional limb designs:

> Mice that are null for both Msx1 and Msx2 develop limbs with lack of antero-posterior polarity, absence of radius [forearm bone], marked inhibition of apoptosis [resulting in webbing] in the interdigital regions, and polydactyly [extra digits]. S Alappat et al. 'Msx homeobox gene family and craniofacial development'. Cell research. 2000. Vol 13. p429–442.

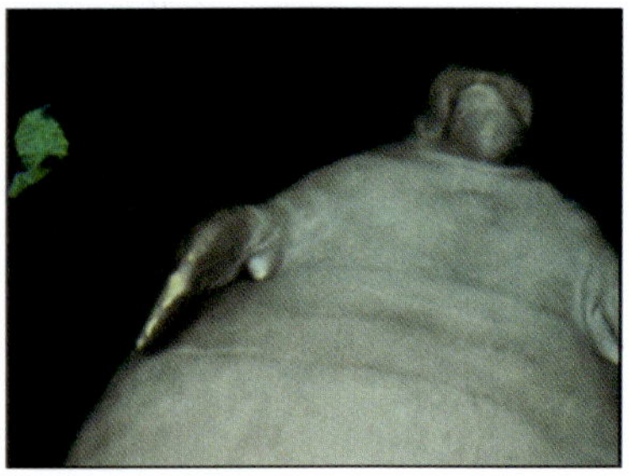
Manatee nipples are uniquely situated at the base of the fins.
The joints of the flipper, with the toe-nails facing tail-wards and the 'palm' sides facing the body, are arranged in a completely different orientation to those of a (supposed) ancestral terrestrial foreleg, which could not have been folded back parallel to the body.

Therefore, since living sea cows have front fins, and if pezosiren was ancestral to living sea cows, then pezosiren must have had front fins. As with all fossils, the bones can only provide a very limited amount of information about a creature. Sea cows are unique in a couple of interesting ways. Firstly, manatees have lungs that are long and thin and extend above the abdominal cavity, almost as far backwards as the anus. Secondly, they have a simpler system of alveoli (air sacs) than that found in all other mammals. Since the manatee lung system is radically different and less efficient compared to the terrestrial mammalian system, it could not have evolved stepwise.

As we have seen, terrestrial mammals have nipples along the primary nipple lines, which extend on both sides from the mid-point of the collar bone to the genitals. The manatee is unique in that its nipples are found at the rear end of the base of the flipper. This makes it easy for the mother and baby to suckle without much manoeuvring. Of course, there is no reason why a terrestrial animal that lost its rear limbs and had to swim for a living would alter its breasts to so favourable a position. Therefore, since manatees have unique nipples, then (most likely) so did every one of its ancestors.

So, how can we make sense of the fossil and living sea cows? The *Chalicotherium Principle* tells us that completely unrelated creatures can have very similar anatomical features. So, the logical explanation for the observed sea cow fossils is that some fully terrestrial creatures unrelated to sea cows (such as paleoparadoxia) had heads that resembled those of the fully aquatic sea cow halitherium. Other (possibly) semi-aquatic sea cows (such as pezosiren) lost their rear limbs through damaging mutations, but their descendants still managed to survive quite adequately since they could get about using their tail flukes and flippered forelimbs. In other words, the ancestors of living sea cows were either fully- or semi-aquatic sea cows, which had flukes and flippers along with rear limbs. Sea cows were never 'land cows'.

### *What would cause a terrestrial mammal to become a whale?*

I'm walking on the beach with my dog and I throw a stick into the sea. The dog rushes forward and dives into the waves, does her doggie-paddle, grabs hold of the stick in her mouth and swims back to me. She shakes herself free of most of the water and waits for me to throw the stick again. As I'm playing this game I wonder just what sort of persuasion would cause a dog or hippopotamus or tupaia to become a whale. How long would it take? How many generations would have to appear before changes in its ability to drink sea-water, its shape, and its behaviour (such as mating, rearing young etc.) would allow for a totally marine existence? If, for example, whales evolved from creatures such as the mesonyx, it is critical to define exactly what sort of environmental change could have occurred that forced such a wolf-like animal to leave its normal habitat and prey, in favour of the open seas and fishes. It is not enough simply to say that natural selection happened. We have to discuss real habitats and real animals with real physiological requirements. And, of course, nothing can change without there first being real mutations.

There are many animals alive today which make their living on the shores of rivers, lakes or seas. Bears in North America commonly perch on a convenient rock and catch salmon as they swim and jump upstream on their way to breeding grounds. Polar bears live almost exclusively on fish and seals, but have evolved no significant marine features. Otters spend much of their time in the water and live on fish and shellfish, yet retain their legs and fur. Whilst the mesonyx is supposed to have been tempted to hunt in the sea, it should be noted that in general, wolves, wild dogs, hyenas, and both big and small cats *do not* have any tendency to hunt for fish. Lions are known to extend their territories to some coastal beaches in Africa, but are not known to venture out into the sea in order to catch fish, even in a drought.

If whale ancestry began on the land, then by definition, any presumed whale precursor would have been a successful terrestrial animal. And therefore, the initial survival and success of a mesonyx or wolf-like animal would have necessarily been determined by characteristics *already present* and operating in its genome. The animal which was the most agile and the strongest would have been the one most likely to survive any environmental change or circumstance that would force a water-dwelling lifestyle. However, these same characteristics in an individual wolf-like animal would most likely have made it well able to compete with other predators on land, so negating any pressure to enter the sea.

More importantly, there would not and could not be any natural selection of flippers, flukes, blow-holes, mating in water, giving birth in water etc., until the appropriate mutations had first by chance occurred. Mutations do not appear *in order* for a creature to survive better in its environment; they occur randomly, *irrespective of an animal's requirements or circumstances*. In other words, if competition on land was very great but there were fishing opportunities in the sea, a dog would not simply go and live in the sea, and then, in order to swim better, grow tail flukes. And even if it spent more time in the water, this behaviour would not cause its teeth or nostril position to change. Neither seals nor penguins nor otters nor polar bears have 'needed' to change into fully aquatic animals despite the bulk of their food being in the water. Or to put it another way, none of these animals have experienced and retained mutations for a fully marine existence. Since no such mutations have arisen in these animals, why should they have arisen before, to produce whales?

But, let's suppose a surface-swimming whale-like creature did appear. Initially, it

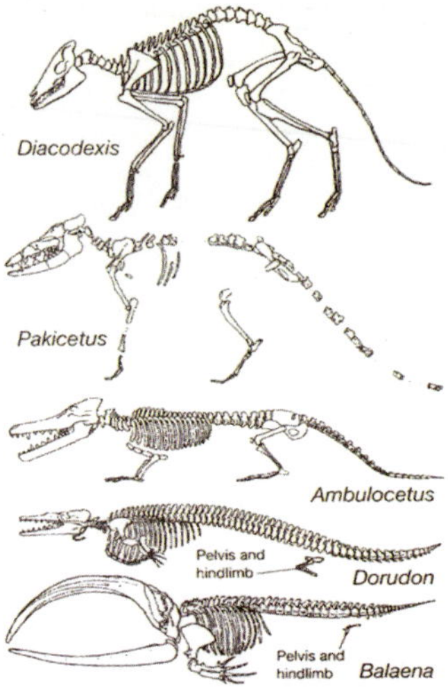

The evolutionary route to a whale.

*All of the mammals that existed… some 65-50 million years ago - lived on land. So it has always been clear that aquatic cetaceans must have evolved from terrestrial mammals.*
C de Muizon. 'Walking with whales'. Nature. Vol 413. 20th September 2001. p259-60.

This supposition is not supported by the evidence: see main text.

Whale precursor?

Beluga whale

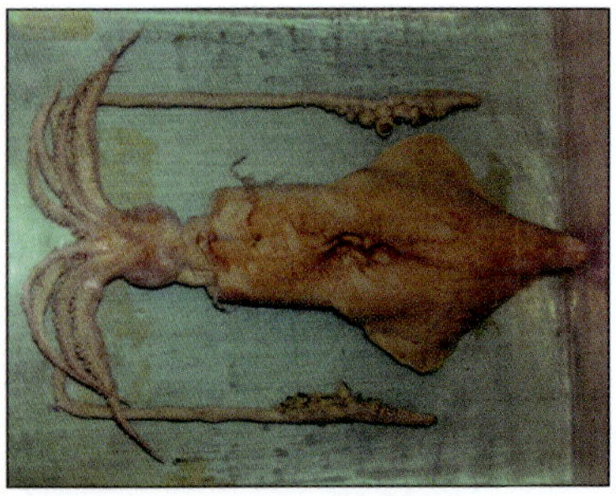
Squid

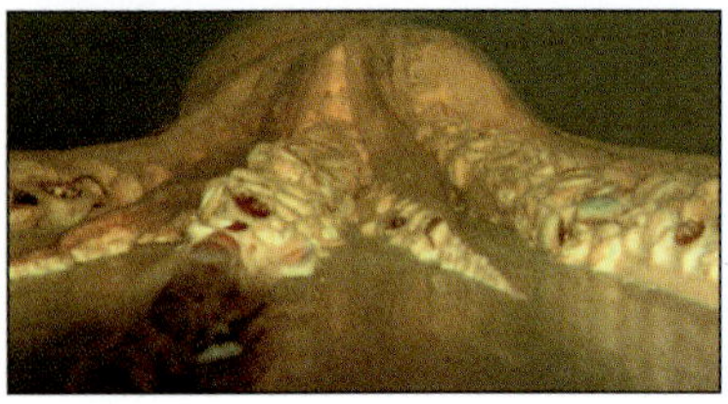
Giant squid viewed from the ends of its tentacles.

ought to have been able to catch fish near the surface, otherwise it would not have survived. There is no reason to expect this creature to be able to dive to any significant depths. The sperm whale can dive to depths of over 2000 metres and stay underwater for up to 2 hours. One of its main foods is the giant squid which is only found in the extreme depths of the seas. How should we suppose that a previously shallow-swimming whale with no ability to withstand the extreme water pressures of the depths, and with no metabolic process to enable it to survive without oxygen for so long, could evolve into a deep-sea-diving sperm whale? There is a huge difference in ability between diving down to 20 metres and diving down to 1000 metres, which is where the giant squid are found. Any primitive whale that managed to dive even to 300 metres would find no squid, and would gain nothing either from trying again, or from managing to dive another 50 metres deeper. So there would be no selection pressure in favour of those (supposed) proto-whales that could dive just a little bit deeper. Thus the problem arises that if the giant squid is only found in extremely deep water, how then would any (supposedly) primitive whale develop a taste for squid, rather than stay near the surface and catch easier prey, as do its many surface-swimming fellows?

A sudden environmental change that forced a wolf-like animal into the sea would not have allowed for any adaptation and the creature would have simply become extinct. If it did venture into the sea it would most likely swim back to the shore with or without a fish. If it stayed in the water, whether or not it managed to catch any more fish, it would have to drink some sea-water at some stage. Salt overload results in an increase in the volume of the blood and a loss of fluid from within the cells of the body. Unless reversed, this results in progressive fatigue and death. Both human and canine kidneys are quite incapable of adequately excreting the salt load caused by drinking sea-water. Our wolf-like animal would have desperately tried to find some land, but finding no land it would have drowned from exhaustion or from salt intoxication. So, it is more likely that a significant predator such as the wolf-like animal would have been as able as other similar predators to continue to make a living on the land, without the necessity for the massive alterations in anatomy and physiology that are required for aquatic life.

Instead of modifying existing kidneys to excrete a greater salt load (as the notion of evolution should predict), a large number of terrestrial animals have unique glands that fulfil this function. The living saltwater crocodile (similar to the marine crocodiles which are supposed to have lived *200mya*) is the largest of all crocodilians, and it can remain in the sea for very long periods. However, like the Nile crocodile (which is known to venture temporarily a short way out to sea) and indeed like all reptiles, its kidneys are *unable* to excrete the excessive salt that results from drinking sea-water. So how does the saltwater crocodile survive at sea? These animals have *certainly not,* over millions of years, evolved a progressive change in the kidneys to accommodate an increasing marine habit. Instead, the saltwater crocodile has some 30-40 unique glands in the back of the tongue which excrete concentrated salt to maintain the normal blood levels. Similar salt-excreting glands are found under the tongue in sea snakes. Marine birds have salt-excreting glands in a shallow depression above the eye sockets. Marine iguanas have salt-excreting glands

Galapagos marine iguana

in the anterior parts of their nasal cavities. This explains why the Galapagos marine iguana periodically 'sneezes'. In doing so, it is simply expelling its salt excretions. In contrast, other reptiles such as marine turtles have salt-excreting glands within the eye sockets and therefore 'cry' salty tears.

It is a fact that while marine fishes show an effective capability to enter into and survive in fresh water, in general, freshwater fishes seem unable to cope with sea water. However, marine and freshwater fishes are generally confined to either seas or rivers respectively. Evolution should predict that by gradual changes in kidney function over millions of years, a whole host of marine animals (such as sharks and rays) should have adapted to living in freshwater rivers and lakes (and vice versa), but except for a few examples, such as the Amazonian stingrays (that live upstream in some South American rivers), eels, lampreys and salmon, this is not the case. Since there is no clear evidence for any terrestrial creature having modified its existing kidneys to excrete the salt load caused by drinking sea-water, and since even after (supposedly) millions of years freshwater fishes are generally unable to handle the salt load of sea-water, it is likely therefore that the capabilities of the kidneys of whales and other sea mammals are *not* the result of gradual changes in a terrestrial ancestor.

Humpback whale bubbles

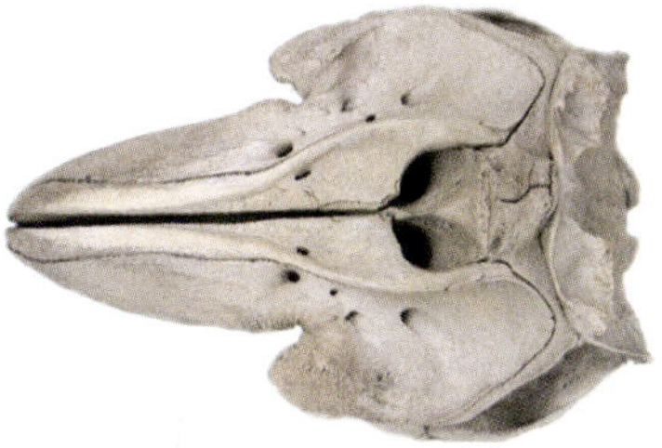

Skull of a gray whale (above top) and orca (above) showing blow-hole (both viewed from above - the snout is to the left).

### ***Nostrils and blow-holes***

The notion that an ancestral dog-like animal with primitive tooth forms of incisors, canines, and triangular teeth developed into a bottle-nosed dolphin requires, for example, that the animal first gained a long snout and nostrils at its tip, and that the nostrils then migrated half-way along the muzzle, resulting finally in a long-snouted creature with a blow-hole and conical teeth.

But we can say categorically that after many thousands of years of selective breeding in dogs, involving millions of individual animals, *all* dogs still produce only dogs. The snub-nosed Pekinese may be physically unable to mate with the wolf-like German Shepherd, but their chromosomes are the same and they can be artificially interbred. And the result of such a mating would be a mongrel *dog*. Furthermore, although there are indeed differences in size and shape, nevertheless, in essence, the various parts of the body remain unaltered. Indeed, the Pekinese dog (with its face

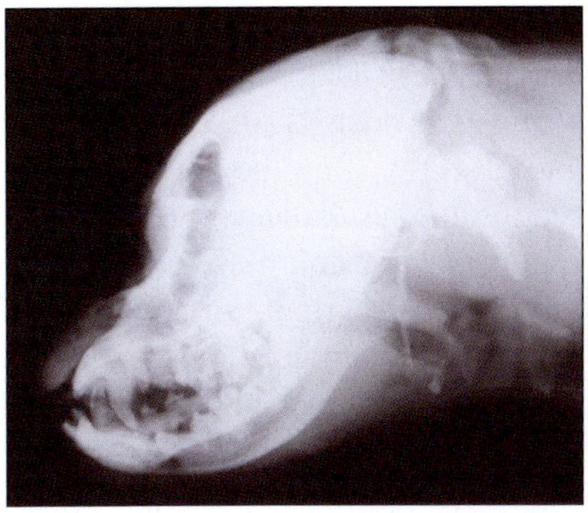

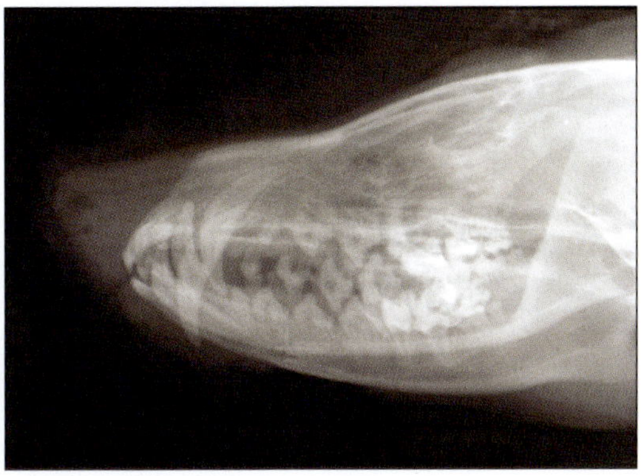

Changing the length of the muzzle of a dog does not alter the relative position or nature of the nostrils.

**Snout length in dogs**

It is found that activity of the gene Runx-2 is involved in determining snout length in dogs:

> *Long-nosed breeds such as greyhounds had repeats within the gene that code for an unusually high number of copies of the amino acid glutamine, followed by unusually few copies of the amino acid alanine. Pug-nosed breeds such as boxers had the reverse.*
> *Canine muzzle points to speedy evolution.*
> New Scientist. 18th December 2004. p10.

more resembling a large vampire bat than a dog) still has nostrils at the end of its (very) short snout. Thus shortening of the muzzle does not give rise to an alteration in the *location* of the nostrils.

Since we know that changing the head shape or the muzzle length of a dog or any other animal by selective breeding does not alter the relative position or form of the nostrils (which remain at the tip of the muzzle), both the nostril position and its form must be locked genetically to the tip of the muzzle, whatever its length. Therefore the migration of the nostril half-way along the muzzle requires a genetic unlocking of this relationship, and in addition there must be *a demonstrable selective advantage* from such a short and insignificant change. It is not permissible to simply accept that the known fossil forms imply that in extinct mammals, the nostrils migrated and developed progressively into a blow-hole. Most certainly, seals and other swimming mammals have normally-positioned nostrils and are well able to close them when diving. If the nostrils did subsequently move backwards and upwards to develop a blow-hole, then totally different and new genes must have come into play to drastically alter the nature and position of the nostrils, the skull, and the airways down to the larynx.

It is supposed that the modern dolphin is the result of a series of changes to the mesonyx via intermediaries such as the dorudontine.
Note: Except for the durodontine, these pictures are diagrammatic and do not represent real fossils.

It is generally assumed that a (supposedly) primitive creature such as a mesonyx, with its nostrils at the end of its muzzle, evolved into creatures such as a protocetus or a dorudontine which both had nostrils some way along from the end of their snouts, and that these in turn evolved into whales and dolphins with blow-holes above the level of their eyes. But what might appear to be a simple progression does not really make any sense: if the mesonyx snout elongated prior to any change in the position of the nostrils, then we should expect to find evidence of a dorudontine-type form which had nostrils on the tip of its snout (M1 in figure on left). If on the other hand the mesonyx nostrils migrated backwards prior to lengthening of the snout, then we should expect to find evidence of a mesonyx-type form but with nostrils set back from the tip of its snout (M2).

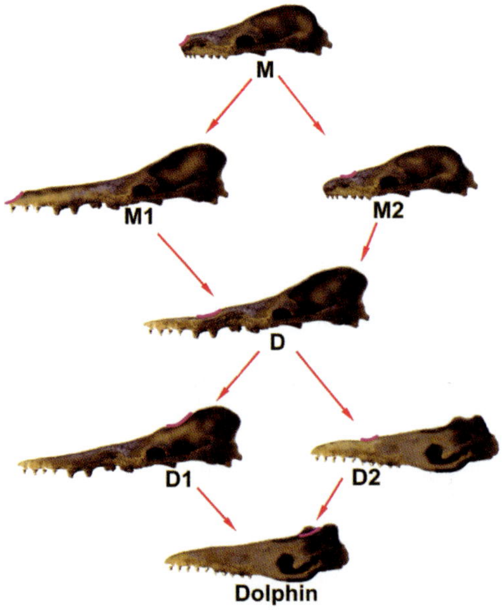


The implausible intermediaries from mesonyx to a modern dolphin. (The nostril positions are shown in pink).
Note: Except for the dorudontine (D), these pictures are diagrammatic and do not represent real fossils.

Similarly, if the dorudontine nostrils migrated backwards towards the skull without any change in the snout, then we should expect to find evidence of a dorudontine-type form but with nostrils on the top of the skull (D1). If on the other hand the dorudontine skull and teeth changed into a more dolphin-like form but without any change in the nostril position, then we should expect to find evidence of a dolphin-like form which had nostrils set back from the tip of its snout (D2). Thus, both the evidence from selective breeding, and the absence of evidence for M1, M2, D1 and D2 forms, strongly suggest that the (supposed) evolution from the mesonyx to the dorudontine form through to the modern dolphin form did not occur. It is of equal importance that there is no evidence to show that any (supposed) intermediary would have had any survival advantage, one over the other.

The question must be asked: is it even necessary for the nostrils of any whale to be on top of the head? All land animals have nostrils on the tips of their snouts. All semi-aquatic air-breathing animals such as hippopotami, seals, otters, beavers, polar bears, frogs, newts, penguins, turtles, anacondas, and crocodiles have nostrils on the tips of their snouts. The largest recorded elephant seal was 6.9m (22ft 6in) long and weighed 4-5 tonnes. The elephant seal is far larger than most dolphins and yet it does not have a blow-hole. And curiously, while the male elephant seal has a proboscis which in effect lowers rather than raises the nostrils, and which also increases the distance between the nostrils and the lungs, this does not cause any problems with breathing. And totally aquatic air-breathing animals, such as dugongs and manatees, have nostrils on the tips of their snouts. The giant *Liopleurodon* was a huge predatory marine reptile rather like a giant plesiosaur. It had a head around 1.5m long* yet it, too, did not have a blow-hole.

Extinct aquatic air-breathing animals such as the *Plesiosaurs* had nostrils on the tips of their snouts, and the *Icthyosaurs* and *Mosasaurs* had nostrils further back along their snouts. So the evidence shows that apart from whales and dolphins, all extinct and living animals had, or have, nostrils *on* their snouts. The only two other exceptions to the rule are the *Machaeroprospus* (an aquatic predator from the Triassic, resembling modern crocodiles, but not closely related to them) and the giant sauropod *Diplodocus*, both of which had nostrils on the top of the head. The other sauropods such as *Shunosaurus* and Apatosaurus all had nostrils on the ends of their snouts. It is significant that a large number of sauropod fossils have been found from diverse parts of the Earth, and that the sauropods lived (supposedly) for

Above: River otter.
Below: Sea otter.

Mosasaur, an extinct marine reptile.

Left: Diplodocus.
Below: Plesiosaur.

*In 1999, the BBC's 'Walking with dinosaurs' book and TV programme claimed that Liopleurodon had a 5m skull, was around 25m long, and weighed 150 tonnes. However, this is likely to be an exaggeration. A juvenile specimen found in Mexico in 2002 was thought to be 15-18m long, but this was only based on the size of some vertebrae and limb bones. A mandible at Oxford University Museum of Natural History, originally described as Liopleurodon, suggests a head length of almost 3m.

Harbour seal

*160 million years* (far longer than whales are supposed to have been around for) yet the diplodocus nostrils are curiously unique, there being no evidence of any gradual upward migration in a series of similar but more ancient dinosaurs.

If underwater breath-holding times are compared, bottle-nosed dolphins (10 mins) fare very unfavourably with the sea-lion (30 mins), the Wedel seal (60 mins) and even the humble beaver (15 mins). Dugongs and manatees can stay underwater for up to 20 minutes. The sperm whale can stay underwater for almost 2 hours. The underwater breath-holding world record for a man is over 7 minutes. Yet river dolphins rarely dive for more than 30-60 seconds. Thus the presence of a blow-hole is not related to diving or breath-holding times.

Nor are blow-holes necessary for 'porpoising'. A dolphin would have the same time in which to exhale and inhale, irrespective of whether its nostrils were on top of its head or on the tip of its snout. Penguins also 'porpoise' but do not have blow-holes. It might be argued that penguins are so short that it does not matter where their nostrils are, but in fact the largest penguins, the king and emperor penguins, are not much shorter than an adult Hector's dolphin, which can be as short as 1.2 metres. Whilst it is common for dolphins to surface with *only* their blow-holes above the water, and in general there is no necessity for dolphins to breathe *so* quickly as to avoid both the beak and head breaking above the surface of the water, nevertheless, they also frequently surface with their heads raised out of the water in a very similar way to sea-lions and seals. Orcas are often seen cruising along whilst exposing a

Above and below: Sea-lions. Although sea-lions are larger than the smallest dolphins, they have conventional nostrils.

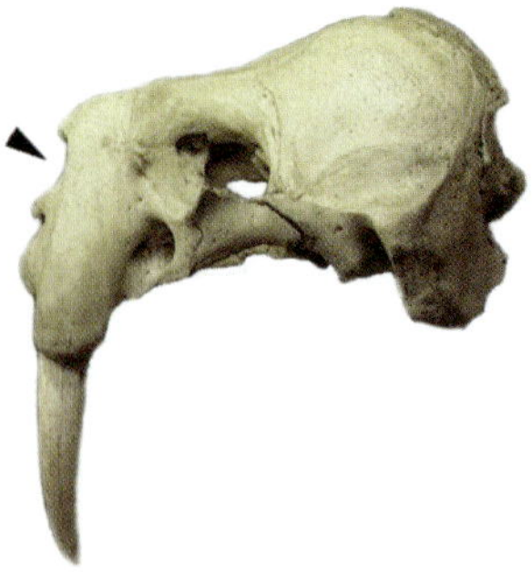

Walrus skull showing nostril position.

Humpback whale

large portion of their body, from the head to behind their dorsal fin, above the water. Although orcas often surface with just the blow-hole and dorsal fin exposed, just as frequently they are seen surfacing in an arc (first with their 'nose' followed by their blow-hole and finally, their dorsal fin), or 'spy hopping' (in which they surface vertically with the entire head clear of the water, as if to look around). When rising to breathe, river dolphins with their long beaks frequently surface at an angle of about 45 degrees. Yet crocodiles with equally long snouts manage to breathe perfectly adequately without such antics. It may be argued that being a shorter distance from the chest, perhaps the position of the blow-hole allows for a more efficient exchange of gases to and from the lungs. But the difference in distance from the top of the head to the lungs, and from the front of the head to the lungs, is insignificant in both dolphins and the larger whales. Furthermore, both the giraffe and the elephant manage to breathe perfectly well with very long tubes connecting their lungs to the outside air. Similarly, the great sauropod dinosaurs must have managed perfectly adequately with their extremely long necks.

It would certainly appear that amongst different air-breathing aquatic creatures, the difference in time taken between the different methods of surfacing and breathing is of no significance. Indeed, although the large whales tend to surface with just their blow-holes above the water, they do so in a casual manner - even when being chased by a whaling ship. Whilst it may be suggested that keeping the eyes below the surface and breathing through a blow-hole allows a creature to keep a look out for underwater predators, this is a weak and unconvincing argument for the evolution of a blow-hole, for two reasons. Firstly, since initially, *no* (supposed) wolf-like creature would have had a blow-hole and since the normal variation in nostril position or muzzle length would have been very limited, then if underwater predators were a significant problem, no particular individual wolf-like creature would have had any significant survival advantage. Secondly, the (supposed) wolf-like creature would have given rise initially to a sea-lion-sized aquatic animal, and yet sea-lions, seals, manatees and dugongs all manage to breathe without blow-holes, and at the same time, avoid predators.

We have seen that nostrils on the top of the head are not essential for a whale lifestyle; and that nostrils on the tip of the snout would cause no risk to a whale, nor cause any impediment in moving about in the water. And, furthermore, there does not seem to be any selective advantage in any of the living whale forms over any of the extinct forms since all seem perfectly equipped for survival. In addition, the variety of animal forms that today live out an aquatic existence, and yet retain ordinary nostrils, testifies to the fact that blow-holes are not necessary for an aquatic lifestyle.

So it can be seen that there is no real evidence for any *trend* in changes in the position of nasal openings - different ancient whales just had different positions of blow-holes, along with other differing features.

Since mutations occur by chance and a mutant individual must somehow survive despite the mutation that it carries, we should not be looking for ways in which a terrestrial quadruped could, for example, lose its legs *in order* to become a whale. Rather, as with the Galapagos cormorants and blind cave fishes, we should be able

Orcas often cruise with the head and dorsal fin exposed above the water (above) or 'spy hop' (below).

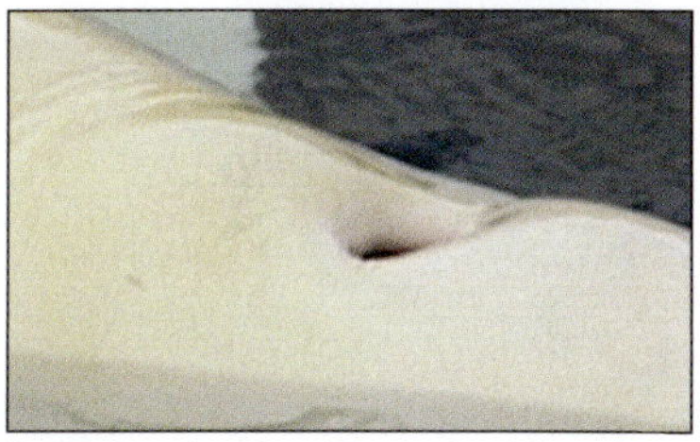

Beluga whale with open blow-hole (above) and closed blow-hole (below).

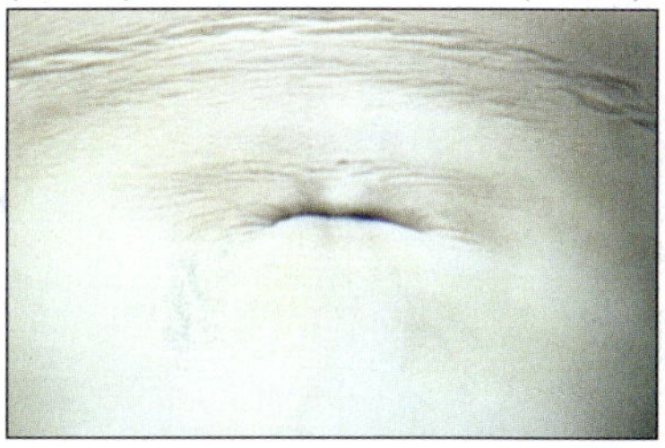

White-sided dolphin with open blow-hole (above) ane closed blow-hole (below).

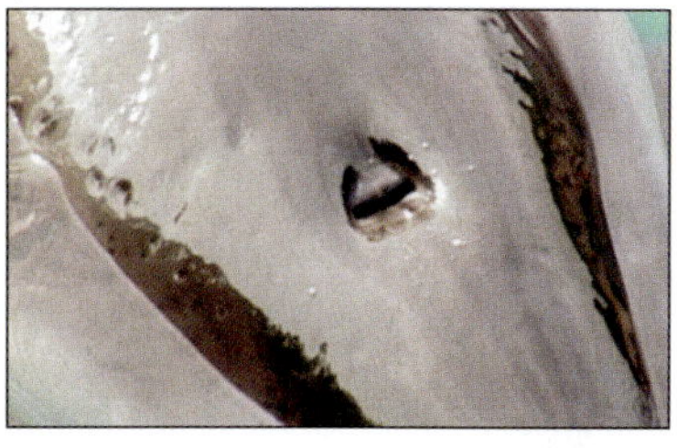

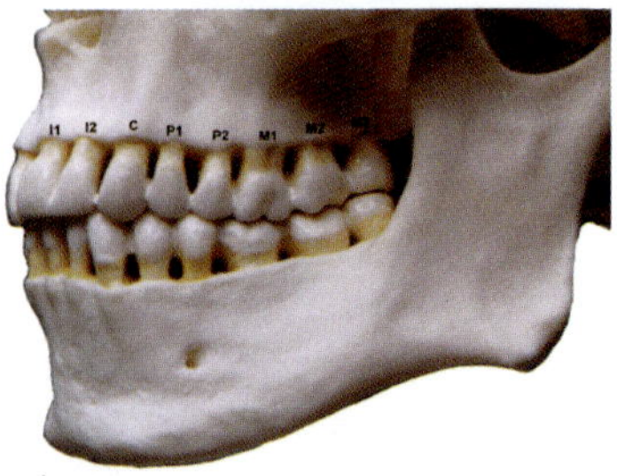

Human teeth

to show that a (supposed) mutation forced a new way of life and at the same time led to the demise of the (non-mutant) ancestral individuals. In other words, a (supposedly) evolving terrestrial carnivore would not have transformed its 'standard' teeth into conical teeth or baleen plates *in order* to go fishing, but *despite* the loss of its normal teeth, the creature would have managed to survive by catching fish, instead of its normal terrestrial prey.

### *Teeth*

The notion of evolution and the study of animal relationships are quite often based on comparisons of tooth anatomy. But if whales evolved ultimately from shrew-like animals 65mya, is there any good evidence to show that the necessary changes in the teeth of a terrestrial quadruped were possible, or that the necessary changes did actually occur? Before we can accept any notion that differences in tooth form can define ancestor-descendant relationships, we must first have some insight into the genetics of tooth development.

On either side of both upper and lower jaws, the normal adult human complement of teeth includes two incisors (I), one canine (C), two premolars (P), and three molars (M). Most mammals have the same tooth format but with a variation in the number of premolars and molars. However, elephants have only one molar on each side of both the upper and lower jaws, in addition to the tusks. Six sets of molars appear sequentially during an elephant's lifetime, and after the 6$^{th}$ tooth falls out no further teeth appear. Kangaroos have only 4 pairs of molars, which erupt in sequence throughout the animal's life. Adult monotremes (the duck-billed platypus and echidna) are toothless. In the toothed whales, all the teeth are identical, being generally conical, and these teeth lack the curvature of normal canines. The beaked whales have only one tooth and this is found in the lower jaw. The tooth varies in shape between different types of beaked whale, and at 10cm the *Ginko beaked whale* has the widest whale tooth. The *beluga* and *narwhal* have an almost identical appearance in silhouette, but the narwhal has a characteristic tusk arising out of the left side of the upper jaw.

The orca has up to 26 conical teeth in each jaw, the bottle-nosed dolphin up to 50, and other dolphins up to 120. The sperm whale has about 50 teeth in the lower jaw,

Above: Sea-lion teeth.
Below: Orca skull showing teeth.

Right: Baleen plates (Gray whale).
Far right: Bottle-nosed dolphins.

the upper jaw being toothless. The baleen whales have sieve-like baleen plates on each side of the upper jaw which can number up to 400 in the blue whale, and through which the whale can filter krill and small fishes from mouthfuls comprising thousands of litres of water. The lower jaws of all baleen whales have no teeth or baleen plates. Baleen plates are not modified teeth and although they are made of keratin (as are rhino horns and finger nails), they are unlike any structure in any other animal.

Whether whales evolved from descendants of the mesonyx or some hippo ancestor, ultimately whale teeth and baleen plates must have evolved from the ICPM format of the early shrew-like mammals of 65mya – since these were (supposedly) the ancestors of all living mammals.

There is more information available about tooth disorders in humans than in animals, and so it is appropriate to look at congenital dental abnormalities in humans as a basis for trying to understand the difficulties that natural selection would have to face in order to change the dentition of animals.

We know that the development of different teeth (ICPM) are controlled by diverse triggers*. For example, the outer milk incisor of the upper jaw is the tooth most commonly absent, and this is not associated with any abnormality of the remaining teeth. The second premolar of either upper or lower jaw can also be absent without any changes in the other teeth. And since some congenital abnormalities only affect the permanent teeth, this shows that the milk and permanent teeth are controlled by different genes.

In addition, the teeth and jaws are controlled by different genes. For example, 90%

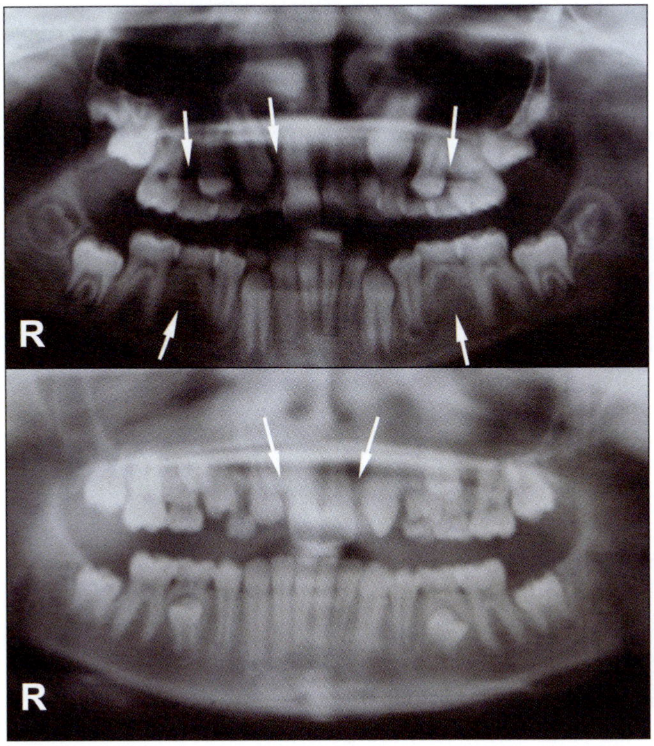


Absent human teeth.
Left: X-ray showing absence of all four permanent 2nd premolars and of the upper right permanent lateral incisor. It is probable that the left upper permanent lateral incisor is also absent.
Below left: X-ray showing absence of both right and left upper permanent lateral incisors, and also of the right upper canine.
The arrows point to where the missing teeth should be.
Below: 9-yr-old boy with absent upper lateral incisors.

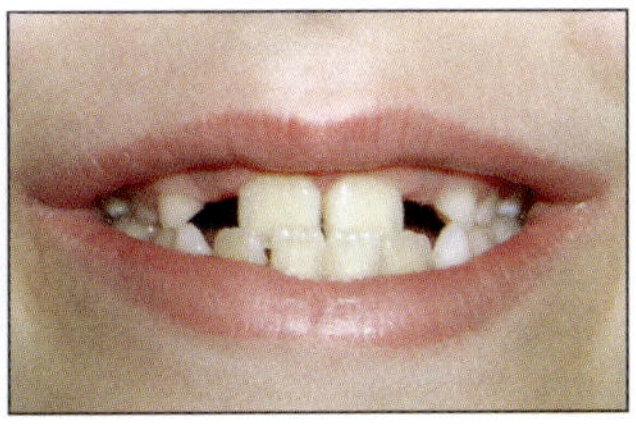

*** Changes to Hox genes only make crude changes to tooth development**

*Hox genes are not very precise in their ability to make gradual changes in dental morphology. Hox gene knock-out mice (Msx-1 and Msx-2) have shown loss of entire groups of teeth, not minor changes in specific teeth. The gradual small changes to various classes of teeth (molars-premolars-canines-incisors), which are required by the notion of evolution, are not possible with the Hox gene models. There are also over 200 genes controlling dental-craniofacial anatomy. In any genetic model (either large group - balancing selection, or small group - founder populations), any advantageous changes in tooth structure would be so insignificant as to 'fall below the natural selection radar'. Therefore the vertebrate cranio-facial-dental complex is very stable and resistant to mutational change.*
Dr Donald Moeller. Maxillofacial Surgeon. Columbus, Georgia.
Personal communication. June 2008.

**The upper and lower incisors have different developmental controls**

*The dHAND gene was expressed in the mesenchyme of the presumptive incisor region of the lower jaw at an early stage and in the mesenchyme of the lower incisor tooth germ at a later stage. However, the dHAND gene was not expressed in the upper incisor region or the upper and lower molar regions during jaw development.*
M Abe et al. 'Tooth-type specific expression of dHAND/Hand2: possible involvement in murine lower incisor morphogenesis'. Cell Tissue Res. 2002. Nov;310(2): 201-12.

of supernumerary (extra) teeth occur in the upper jaw and they often occur in normal individuals. However, in *Gardner's syndrome*, supernumerary teeth are associated with polyps in the intestine, tumours in the bones, and multiple cysts in the skin. Since there is no associated change in jaw size to accommodate them, supernumerary teeth can result in impaction, pain and abscesses.

A rare toothless crocodile from Florida. Note that the mutation that has resulted in the loss of teeth has made no difference whatsoever to the form of the jaws and skull, or the body shape. This is a captive individual - in the wild, a toothless offspring of any toothed animal would most likely have a serious difficulty in feeding and die.

The complete absence of all the teeth is extremely rare, but reduced numbers of teeth, absent teeth, or abnormally-shaped teeth are found in a number of genetic conditions and are associated with a wide spectrum of serious skin and soft tissue problems. For example, reduced numbers or absent teeth occur in the *Ectodermal dysplasia syndrome*. However, any teeth that remain may be conical. This condition, which is linked to the X chromosome, affects mainly males. There is a smooth dry skin, with partial or complete absence of sweat glands. In addition, the sebaceous glands and hair follicles are often absent although the hair follicles on the moustache and beard areas are usually unaffected. The bridge of the nose may be depressed, the lips may be protuberant, and about 25% of affected individuals have a type of *anaemia* which is very difficult to treat. In addition, an abnormally high palate and cleft palate may be present. However, despite all these features, the growth of the lower jaw is unimpaired, showing again that the genetic control of tooth development is independent of the genetic control of the development of the jaws. Other conditions where there are reduced numbers of teeth include the *Cleidocranial dystosis syndrome* (where there are associated abnormalities of the skull bones, facial bones, shoulder and pelvic girdle) and the *Oral-facial-digital syndrome* (where there is cleft palate, small lower jaw, sparse hair, malformed digits, and trembling movements). In the *Chondroectodermal dysplasia syndrome*, abnormally-shaped teeth occur in association with short limbs, short stature, extra digits, joining together of the bones in the hands and feet, and congenital heart disease, and 50% of affected individuals die in infancy. The condition is a recessive trait and has a high incidence in small communities such as the Amish people living in the USA. Notably, the skull, spine and pelvis are all normally developed.

Crocodile 4th lower tooth.
In crocodiles, a new tooth (t) grows *within* the root of the old tooth (T) that it will eventually replace. The tooth replacement process is quite different in mammals.

Since we know that small or large lower jaws can occur with normal upper jaws (with a resulting mal-alignment of the upper and lower teeth) and since some genetic abnormalities affect the upper and not the lower jaw (and vice versa), it is clear that the development of the upper and lower jaws is controlled independently by different genes.

Both the greyhound with its long thin muzzle, and the Pekinese with its very short muzzle, represent the same animal and there is no practical difference in their teeth and nostrils. Therefore, we should predict that if it were to elongate its muzzle, a primitive whale ancestor would retain its ICPM tooth format. Since the different teeth are controlled independently by different genes, we should also predict that a primitive whale ancestor would be unlikely to undergo changes in *all* of its teeth.

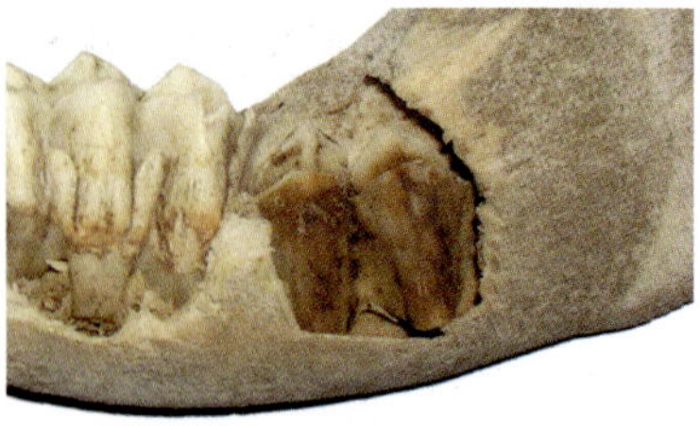

Lamb lower jaw bone.
As with all the teeth, this third molar was fully formed before it erupted and before it could make any contact with its fellow on the upper jaw. Since different teeth are subject to different developmental controls, any mutation affecting the development of the cusps of one tooth will result in mismatching of the upper and lower teeth.

So, since the mesonyx had triangular serrated teeth suggesting a carnivorous eating habit, it is pertinent to ask how its dentitian could change to an overtly peg-like or baleen-type form in the course of time. It has to be borne in mind that since bears, seals, dolphins and beaked whales can all catch and eat fish of similar size, we should expect there to be no selection pressure for any terrestrial carnivore that (supposedly) evolved into a whale, to change its dentition. Furthermore, teeth are *not essential* for an animal to catch such a slippery character as a fish. Fishing birds such as kingfishers, pelicans, penguins and cormorants manage perfectly well without teeth. The beaked whales feed on fish and small squid, yet the single tooth of the male cannot be of any practical use since the females are often toothless. In fact, living toothless whales are not in some unique way 'better' at making a living than toothed whales, but instead, are likely to be simply the result of inbreeding amongst congenitally abnormal ancestors. In other other words, they survive because they are not fatally disadvanged. Even the fossils of the earliest known beaked whales are virtually toothless:

> There are possible *ziphids* [beaked whales] dating back 22 million years and *ziphids* are **common** in marine sediments 5-10 million years old. These later beaked whales...like their living relatives, are **nearly toothless**. (My bold). Richard Harrison and M.M. Brydon. 'Whales, dolphins and porpoises'. Merehurst Press. 1990. p21,

which presents the problem that if beaked whales evolved from very ancient fully-toothed ancestors, resulting in nearly toothless beaked whales living (supposedly) 5-10mya, then by now, beaked whales *should have* lost *all* their teeth. The fact that they have not provides strong evidence against their having lived for millions of years.

We have seen that it is probable that any changes to the teeth are likely to be associated with other changes to the body, many of which are likely to be seriously detrimental. Furthermore, since the individual teeth, the jaws, the collar bones and shoulder blades, the fingers, the hands, the front legs, the toes, the feet, the back legs, the pelvis, and the tail (to number but a few anatomical features) are all coded for by separate and mutually *independent* genes, if whales evolved from a terrestrial quadruped, then, contrary to what is found, a whole variety of intermediate fossils showing all manner of permutations of tooth and body variations should be in evidence.

We have seen that the different teeth are under different genetic controls, and that a mutation resulting in the failure of growth of one type of tooth is unlikely to affect the other teeth. But, could teeth change in their actual form? We have seen that although baleen plates are made of the same basic material as teeth, nevertheless they are quite unlike any type of tooth. So, could baleen plates have evolved from the standard ICPM mammalian set?

It has been suggested that the extinct mammalodon had small 'proto-baleen' fringes between its large cusped teeth, and that it may represent a transitional form leading to modern baleen whales. However, mammalodon had large teeth in both upper and lower jaws, and whether or not it filter-fed, its tooth arrangement must have

Mesonyx teeth

**Mammalian teeth**

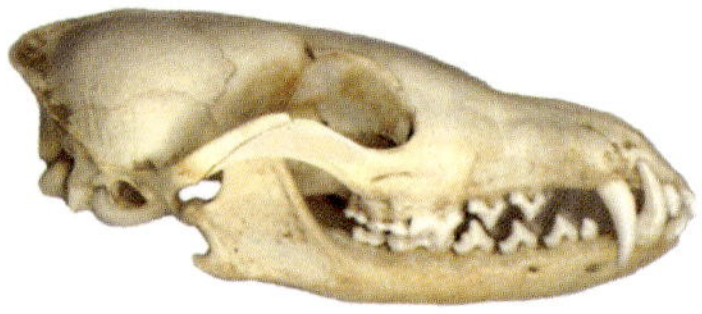

Fox skull showing the typical forms of mammalian teeth.
This fox was the victim of a road accident, hence some of the upper teeth are missing.

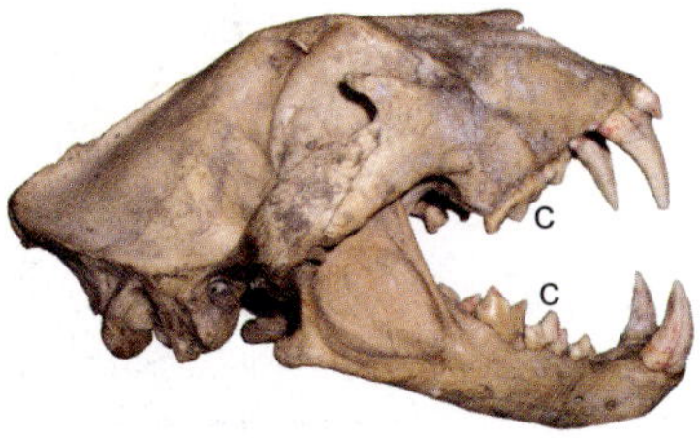


Instead of grinding molars, the tiger has three *carnassial* teeth (c), which slice the food.

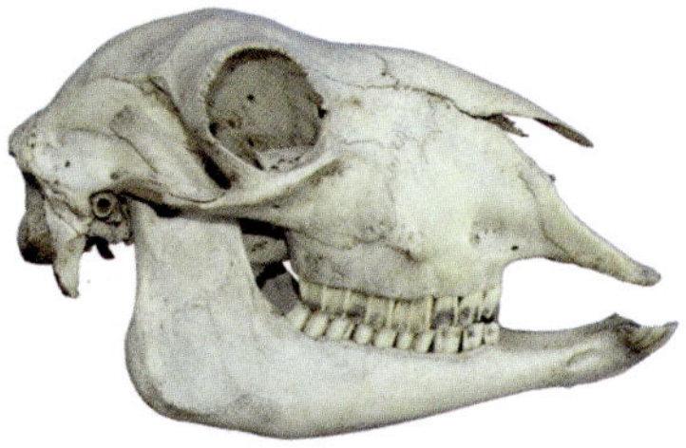

Like the sheep, many herbivores have a large gap between the incisors and the molars, and do not have upper incisors.

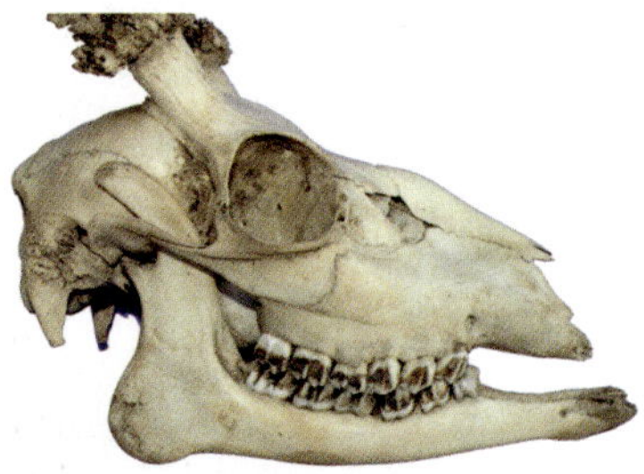

Deer skull showing its molars. The tip of the upper jaw and the lower incisors are missing from this specimen.

Lisa with Anna the Asian elephant.
Whipsnade zoo.

a
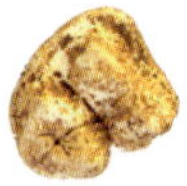
b

c

d
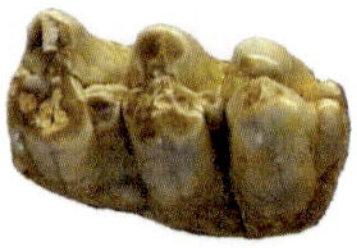

been perfectly adequate to sustain it, so there would have been no reason for it to (supposedly) evolve baleen plates.

The same sort of problem arises with the (supposed) evolution of the unique teeth of elephants. (Supposed) elephant ancestors like moeritherium had six molars on each side of the upper and lower jaws. However, the modern elephant has only one molar tooth (each the size of a house brick) on each side of the upper and lower jaws. This single molar, when worn, is replaced sequentially until a total of six molars have been worn away. The sixth tooth appears at around the age of 25 and may last for some 50 years. When the sixth and final tooth is worn away, animals living in the wild are doomed to die of starvation. It is a massive leap indeed for (supposed) elephant ancestors with the normal mammalian dentition to collect a series of viable mutations that would result not only in a number of the molars not appearing *at all*, but also in six molars being transformed into huge grinding blocks that appear sequentially.

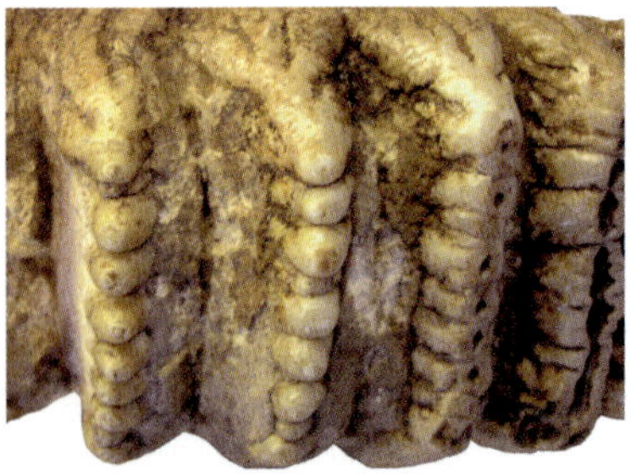
e

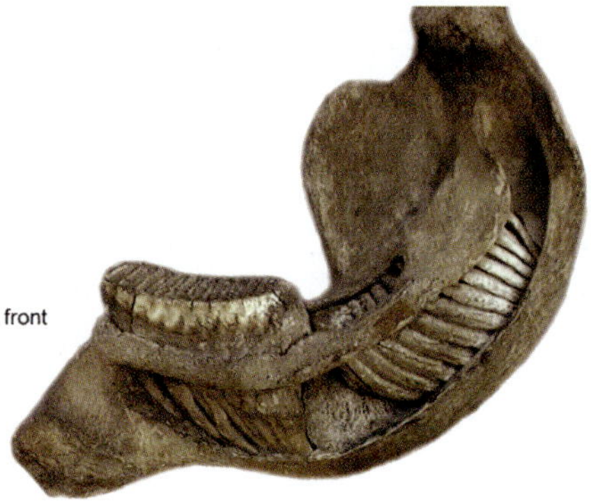


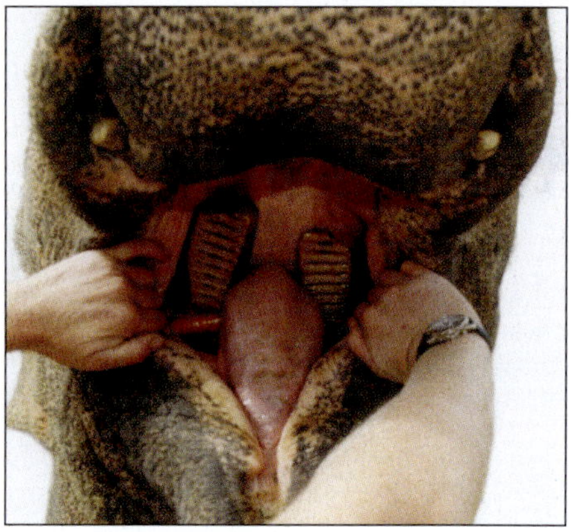

f

Left: The teeth of extinct (supposed) elephant ancestors compared to the teeth of living elephants. From top to bottom *(only roughly to scale)*:
a. Moeritherium
b. Phiomia
c. Deinotherium
d. Gomphotherium
e. Stegodon
f. Mammoth
g. African elephant
h. Asian elephant

Above left: Lower jaw of Asian elephant showing unerupted molar behind the molar in current use.
Above: Mouth of Asian elephant showing the upper molars. A carrot is seen under the right upper tooth.

g

h

We know that the gomphotherium, platybelodon and deinotherium were practically as large as modern elephants, and managed to survive for (supposedly) *millions* of years with a number of smaller molars and without having the large molars of living elephants or mammoths. Clearly then, the smaller molars of gomphotherium, platybelodon and deinotherium were very efficient and quite sufficient to process enough plant material to maintain their large size. This sufficiency of smaller molars is supported by other evidence: firstly, if the teeth of modern elephants were more efficient at mastication than smaller molars, then, contrary to the evidence, we should expect living elephants to be significantly larger than their (supposed) ancestors with their smaller molars. Secondly, the extinct indricotherium was the largest known land mammal (it was as big as five adult African elephants), yet it had a conventional set of molars. So, a number of smaller molars are not less efficient than single large ones. Furthermore, if a mutation caused one of the molars to enlarge to any significant degree without simultaneously removing other teeth to make room for it, this would result in the risk of impaction, tooth decay, infection and death, and so the original form would be selected for.

On the other hand, suppose a mutation (or series of mutations) caused the molars to erupt in series, one at a time. Then the animal would be left with only one small molar in each jaw to do the grinding work, where previously there were three or more. Since elephants have to daily eat vast quantities of vegetable matter, individuals with only one small molar would be doomed to starvation and would be selected against.

It is therefore implausible that in creatures such as gomphotherium, platybelodon and deinotherium, alterations in tooth size and in the scheme of tooth eruption would occur in harmony or offer an advantage so as to evolve the modern elephant dentition. Despite the implausibility of such a thing, suppose the elephant's teeth *did* evolve from a previously 'standard' mammalian ICPM set, then, contrary to the evidence, we should firstly expect a whole variety of transitional tooth forms in elephants. Secondly, since rhinoceroses and hippopotami have managed to grow large and healthy in the same environment as elephants, we should expect some elephants to have retained completely the ICPM format.

So, we have seen that while teeth can be congenitally absent, or abnormal, there is no compelling evidence to show that the dentition of any creature can dramatically change from one type to another completely different type. Therefore, we must conclude that it is improbable that either toothed whales or baleen whales or elephants arose from ancestral creatures that had the ICPM format.

Above: Rhinoceros skull.
Below: Hippopotamus skull.

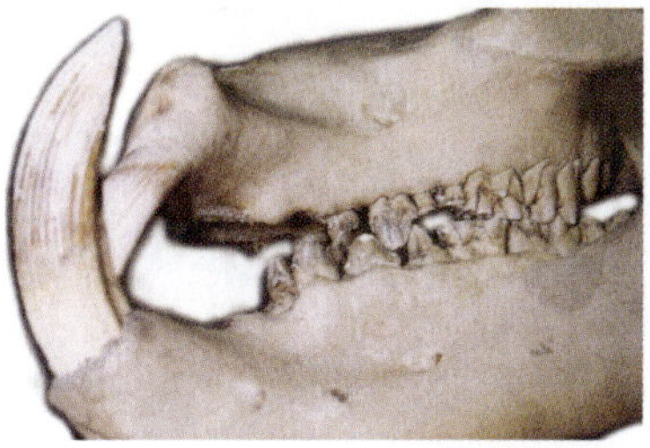

Above: The skull of an extinct form of hyrax showing numerous teeth quite unlike those of the modern elephant. Hyraxes are said to be the closest relatives of the modern elephant. Since it is impossible for the teeth of modern elephants to have evolved (see main text), hyraxes are not related to elephants.
Below: Living hyrax.

## *Flippers and limblessness*

### *Whales*

The sum total of the explanation of the evolution of a whale's flippers from the legs of some (supposed) terrestrial ancestor amounts, in fact, to very little explanation at all:

> The invasion of the water is seen in the bones. There are no hind legs - legs would get in the way and produce friction. In large whales the hind legs have not been completely lost - a single bone of the back leg is still present... Things similar to a dog investigated the sea a bit more, and eventually made the transition into whales. Prof. Simon Morris. Royal Institute Christmas lectures. BBC Broadcast. 1996.

Clearly, this is not a scientific explanation yet it was presented as an unquestionable fact to an impressionable young audience of schoolchildren. Others take the same stance:

> The forelimbs **have become** paddles. The rear limbs **have been lost** altogether, though there are a few small bones buried deep in the whale's body **to prove** that the whale's ancestors really did, at one time, have back legs. (My bold). David Attenborough. 'Life on Earth'. Collins/BBC. 1979. p242.

Gray whale showing tiny bones in the pelvic area.

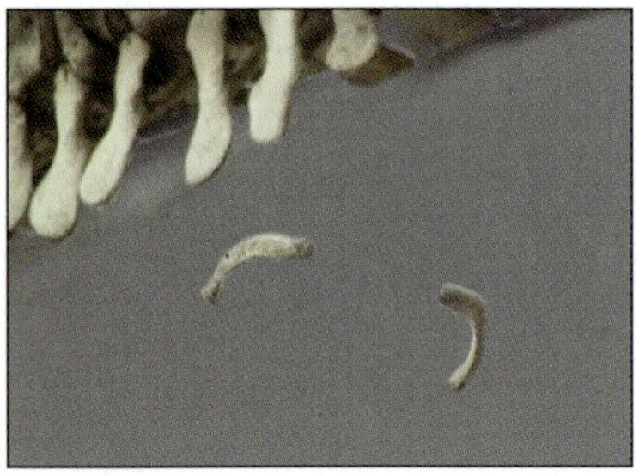

Pelvic bones of bottle-nosed dolphin (above) and right whale (below).

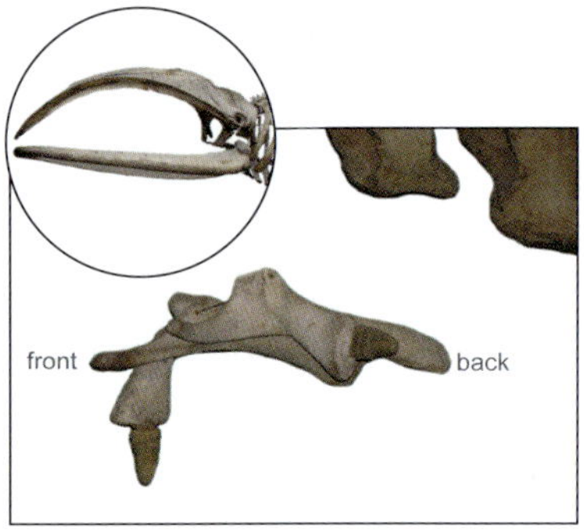


And despite the lack of logic and evidence, the origin of other marine mammals is dealt with in the same manner:

> The sea cows evolved forelimbs which could be used as paddles and they took to the sea. William Gray. 'Coral reefs and islands'. David & Charles. 1993. p98.

As is often the case, the unwary reader is given the impression that creatures evolve *in order* to be more capable of coping in a particular environment. As we have seen when discussing variation amongst Hawaiian finches, nothing could be further from the truth. An animal which has undergone a mutation will just simply either manage to survive, or die.

The forsaking of terrestrial quadrupedalism is not a prerequisite for making a living from the sea: many animals, such as the otter, polar bear, beaver, hippopotamus, marine iguana, and crocodilians manage to live in or around water and yet retain completely their four limbs. And as is shown by icthyosaurs, plesiosaurs and even the huge 25m liopleurodon, all of which had pectoral and pelvic paddles, contrary to Morris' concerns about friction or impediment there is no problem *at all* in having back limbs or fins.

The point is worth emphasising that evolutionary loss of limbs would not have occurred *in order* for creatures to survive at sea, and a terrestrial creature would not have become a marine creature *in order* to find new food sources. So, could a mutant terrestrial creature which was ill-fitted to survive on land be more able to survive in the sea? Since the genetic control of the development of the hands and forelimbs is independent of the feet and hindlimbs, the involution of front and back limbs would also be independent of each other. Furthermore, any (supposed) terrestrial proto-whale would not lose its legs unless a series of mutations first appeared that could cause it to happen. Since mutations are random, the *WYWYWG Principle* shows that a proto-whale which by chance lost its legs would in all probability never acquire the mutations that could change its nostrils or teeth. In addition, it is implausible that a legless proto-whale would have acquired by accident the very mutations that could result in fully functional tail flukes. Indeed, the origin of tail flukes is a complete

Beaver. Some claim that the vertical movement of the whale's tail is confirmation of their land ancestry, since running mammals bend their spines in an up-and-down manner. However, no running mammal uses its tail for locomotion, and the tail essentially swishes from side to side without any significant vertical movements. Even the beaver, otter and platypus use their webbed toes for swimming, not their tails. Although the beaver does slap the water surface with its flat tail, it does not move the tail up and down while swimming.

mystery since all land vertebrates that swim do so either by using their legs or by *horizontal* movements of their tails. The vertical beating of whale flukes has no parallel amongst any land mammals, whether extinct or living. And since a slight bump would have no selective advantage, a dorsal fin is unlikely to have arisen by first being a slight elevation of the skin of the back. Most certainly, the idea of evolution offers no explanation as to the origin of whale tail flukes:

> ...dorsal fins and tail flukes are...folds of skin and hard connective tissue but [with] no skeletal support, radical innovations indeed. It is not easy to imagine by what alterations of habit an early eutherian could come to develop fins out of such folds of skin. J Z Young. 'The life of vertebrates'. Clarendon Press 1991. p498.

A whale can swim perfectly well with a tiny dorsal fin (above, humpback whale) or a large bent dorsal fin (below, orca).

Furthermore, it must be remembered that initially, any proto-whale must have bred on land and would have required the ability to return to the land in order to breed. So, any loss of the ability to breed on land *must* have coincided with fortunate and *additional* new mutations that would have enabled offspring to be born directly into the water and to be able immediately to swim, breathe, and suckle etc., without the need to return to the land.

Basilosaurus had a small well-formed femur with a functional ball-and-socket hip joint. If it (supposedly) existed for a few million years, clearly the presence of the small hind paddles could not have been lethal, nor even a hindrance, nor did they *need* to be lost altogether. Furthermore, slightly larger, slightly smaller, or even absent hind paddles could not conceivably constitute either an evolutionary advantage or an impediment to a basilosaurus. Thus there could have been no selection pressure for basilosaurus to lose completely its minute hind paddles. And if an individual basilosaurus did manage to lose its hind paddles completely, that would not necessarily prevent it mating with other (normal) individuals, and thereby mixing genes to result once again in individuals which retained their hind paddles. Of course, if basilosaurus is the whale equivalent of the Galapagos cormorant, then there should be evidence of a basilosaurus ancestor with 'normal' hindlimbs.

Tail flukes of a humpback whale. The vertical movement of the whale tail has no parallel amongst any land mammals.

Furthermore, since pelvic paddles could not have been essential to icthyosaurs or plesiosaurs, we should expect these creatures to have lost completely their pelvic limbs over the (supposed) millions of years of their existence. That they did not lose their pelvic paddles prior to their extinction means that they did not exist for long enough for random detrimental mutations to cause their pelvic fins to be lost.

There remains the problem that even when mutations in single genes result in changes of limb shape and structure, such mutations result either in changes in length but with retention of the normal overall form, as in *Achondroplasia*, in which there are short limbs associated with a normal torso; or in changes which are associated with a number of other abnormalities, as in *Arachnodactyly,* in which the digits, and to a lesser extent the limbs and trunk, are excessively long and slender, along with abnormal laxity of joint ligaments, dislocation of the lens of the eye, rupture of the aorta, and heart wall defects.

Short stubby limbs associated with normal head and body form are often deliberately inbred in dogs. Achondroplasia is a common cause of dwarfism in humans.

When there is a failure of the normal development of the digits, the hand and foot may be represented by a variety of different structures, ranging, for example, from

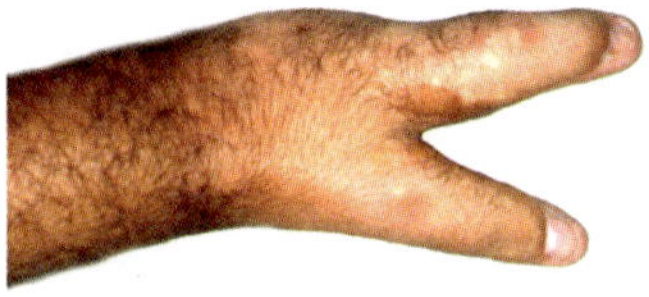

Congenital cleft hand

a flat bony plate to partial fusion (*syndactyly*) of two or more metacarpals and fingers. A particular variety of syndactyly is the so-called *lobster hand*. This is usually inherited as a dominant trait, and the syndactyly is associated with a deep central cleft which divides the hand into two parts.

It is well known that when taken during pregnancy, many substances can give rise to serious congenital structural abnormalities (which are not the result of genetic abnormalities). The drug *Thalidomide* interferes with embryological development and causes limb deformities resulting in severe reduction in or absence of the bulk of the limb, such that the hands and fingers, or feet and toes, arise directly from the shoulders or pelvis respectively. Since thalidomide causes defects in embryological growth but does not affect the genes, children of thalidomide victims have normal limbs. However, thalidomide does demonstrate the possibility that when there is interference with genes acting at any particular time during embryological development, severe abnormalities can result. And indeed, such a situation is exemplified by *Robert's syndrome* which is a genetic disorder resulting in thalidomide-like limb abnormalities (it is also associated with severe abnormalities of facial development and growth deficiency).

### *Flipper heat exchanger*

A look at the internal anatomy of flippers shows that they are not just modified legs. Since whales, sea cows and seals do not have insulating blubber over their flippers, it might be thought that they would have severe problems with heat loss in cold water. We would also think that water birds with their thin legs would have the same problem. However, both flippers and bird legs have a special arrangement of their blood vessels that elegantly deals with the issue: the arteries taking the blood towards the tip of the flipper and the bird leg are completely surrounded by a network of veins, so that the whole arrangement acts as a *counter-current heat exchanger*.

By this arrangement, as the arteries carry blood out towards the tip of the flipper and the bird leg, they lose heat to the surrounding veins which therefore carry pre-warmed blood back to the heart, minimising heat loss to the water.

Below: Sea-lion rear flipper.
Below right: Heron standing on ice.

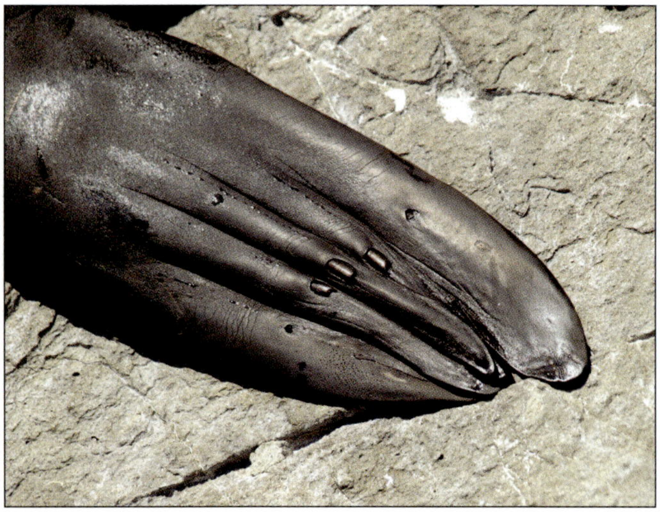

Humpback whale

Walrus

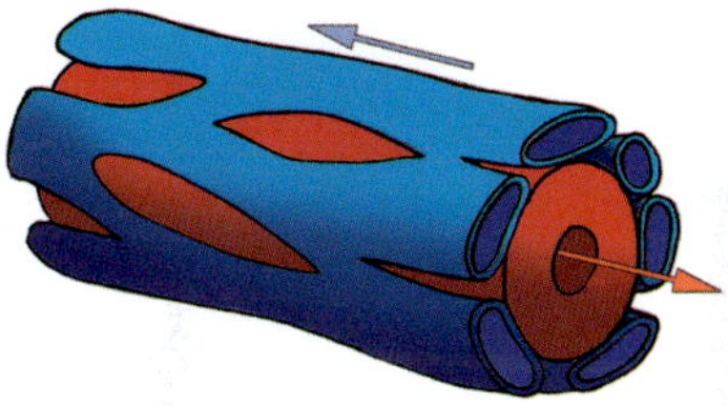
Counter-current heat exchanger: since the blood flow in the artery (red) is in the opposite direction to that in the veins (blue), heat in the arterial blood is progressively passed across to the veins so that little heat is lost in the extremity.

The system is so efficient that, for example, when a gull was placed with its feet in icy water for 2 hours, it only lost an insignificant:

> 1.5% of its metabolic heat production from the feet. From P Scholander. 'Evolution of climatic adaptation in homeotherms'. Evolution. Vol 9. 1955. p15-26.

It is true that arteries are usually adjacent to veins but here we see special arrangements at the very sites where they are necessary. Since it is implausible that a counter-current heat exchanger would appear just where it is needed, the whale flipper and wading bird leg could not have evolved stepwise from structures that could not *effectively* control heat loss.

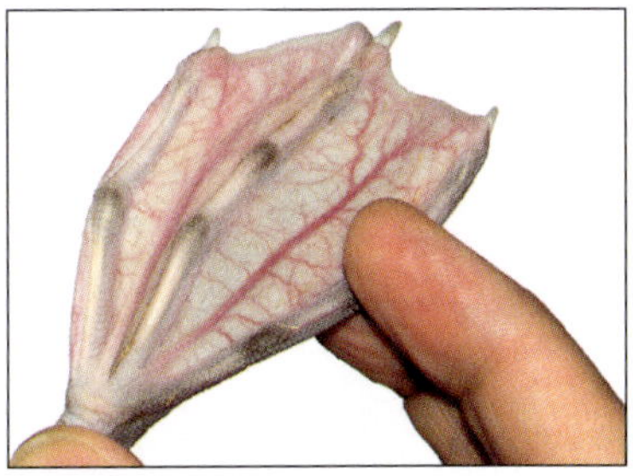
Fulmar foot showing blood vessels.

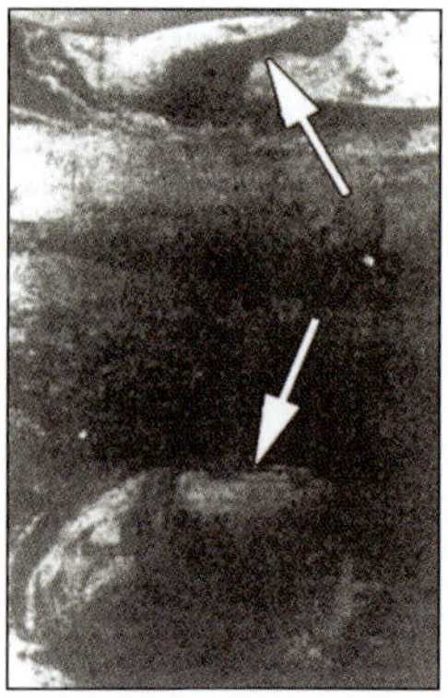

Sperm whale with rear flippers.

### *Whale rear flippers*

It is true that modern whales are occasionally seen with structures projecting outward where we would expect rear flippers to be found. For example, at the beginning of the 20th Century, a sperm whale was caught which had what looked to be distinctly deformed rear flippers. Another sperm whale had what looked to be femur, tibia and fibula-like bones, reportedly some 6 feet long. Recently, a Japanese fishing crew caught an adult bottlenosed dolphin which had definite but malformed rear fins. A number of other examples undeniably show that whales must have the cellular instructions for building rear fins. However, as we discussed with manatees, since it is not possible stepwise to transform a horizontally-moving tail into vertically-moving tail flukes, then whales must always have had flukes. And since they always had flukes, they must have always been virtually obligate swimmers, and could *never* have been terrestrial runners.

Before we put all this into perspective, we need to look at snakes and lizards.

Skull of a mosasaur (viewed from below) showing extra rows of teeth in the palate.

## Snakes

The matter of limblessness also arises in a number of animals other than whales, and is most dramatic in snakes. One common theory is that snakes arose from ancient lizards. Some suggest that monitor lizards (with their snake-like forked tongues) and snakes had a common ancestor. Others suggest that snakes descended from mosasaurs (with their python-like extra rows of teeth in the hard palate), which were large crocodile-like marine reptiles that (supposedly) lived until some 65mya:

> A hundred million years ago…limb reduction took place among a group of ancient lizards. Its consequence was the appearance of the snakes. …**No one doubts**…that snakes once had legs. Indeed a whole group of them, the pythons and boas, still retain internal relics of their hip bones, and show external signs of them - two spurs on either side of the vent. (My bold). David Attenborough. 'Life on Earth'. Collins/BBC. 1979. p164-7.

Spurs (arrowed) are modified scales and are found in constrictor snakes (such as the royal python, below middle) and in legless lizards (such as the Sheltopusik, below right), as well as in many legged lizards such as the plated lizard (below left). Some tortoises such as the spur-thighed tortoise (right) have spurs at the base of their rear legs.

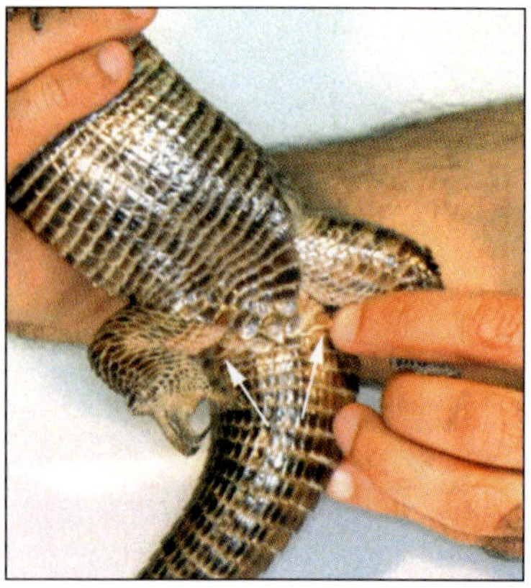

Immediately, a number of questions arise: firstly, do spurs represent vestigial claws; secondly, do these small 'rear limb' bones represent vestigial legs; and thirdly (even if some snakes once had legs) what is the evidence that snakes were once lizards?
Near the ends of the 'rear limb' bones of pythons and boas is a small keratin spur which projects from the skin surface, one on either side of the anal vent. In some snakes the spurs (which are involved in copulation) are found only in males, in some they are present in both sexes, and in others they may also be absent altogether.
Since legless lizards such as the *Sheltopusik* also have spurs, it might indeed be thought that spurs represent the vestiges of hind limb claws. However, spurs are also found in a number of normal lizards in which they are seen as distinctly separate structures quite unrelated to the back legs. So the spurs in snakes may not represent claws at all.
Some living snakes, such as the pythons and boas, have small bones embedded within the body in the pelvic region. The size, number and shape of these bones can vary considerably both between and within different types of snake. In living snakes, the bones seem to resemble leg bones but the bones are never seen to project out of the body. However, there is at least one case of a garter snake which had a number of external rear claws.
The oldest fossil snakes are *Haasiophis* and *Pachyrachis,* which were found in sediments (supposedly) 95myo. These were both aquatic snakes with flattened tails, but they had many features resembling living pythons. Before proceeding, as always when interpreting the fossil record, we must remember that the *Coelacanth Principle* shows that these fossil snakes could well have lived contemporaneously with 'advanced' snakes such as cobras and vipers, and with modern lizards and mosasaurs, and so may not necessarily represent any ancestral form of snake *at all*. Furthermore, it is commonly suggested that 'rear limb' bones and egg-laying are primitive characteristics and so, because of their 'rear limb' bones, pythons and boas are said to be more primitive than other snakes such as cobras*. However, most boas (supposedly primitive snakes) have 'rear limb' bones yet produce live young, whereas cobras (supposedly advanced snakes) have no 'rear limb' bones yet

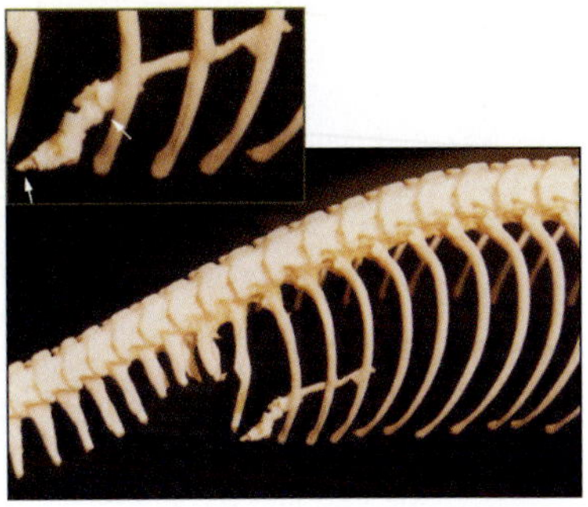

The pelvic bones of an Indian python. It can be seen that the bones are not attached to the spine, and that the claw-like spur (lower arrow) is attached to a bone which has a ball-and-socket joint (upper arrow).

* Chameleons, along with anacondas, boas, green cobras and rattlesnakes, all have 36 chromosomes while iguanas have 38. If the 'primitive' python-like Haasiophis existed 95mya, then by now there should be major differences in chromosome number between pythons and lizards. The fact that there are no large differences suggests that snakes and lizards have not existed long enough to have acquired significant changes. In other words, it is impossible for them to have existed for millions of years.

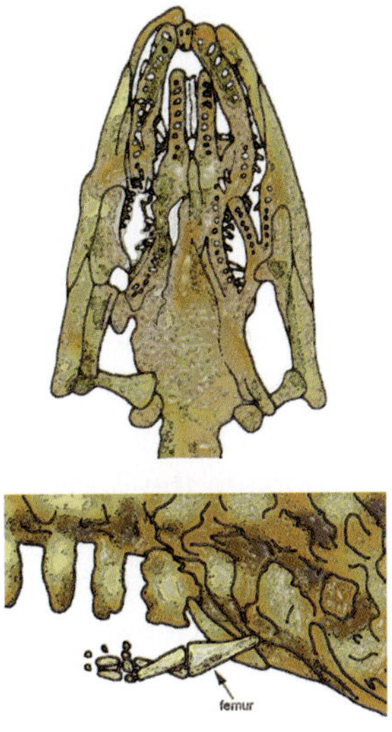


Left: The skull (viewed from below) and the 'rear limb' bones of the fossil snake Haasiophis. As in living pythons, the skull shows extra rows of teeth in the palate. It is not clear whether or not the 'rear limb' bones protruded from the body surface. If living pythons are the descendants of haasiophis, we should expect that after 95my pythons should have lost their 'rear limb' bones completely. Since whales are supposed to have lost their 'rear limb' bones in less than 65my, and since apes supposedly evolved into humans in only 6-8my, the fact that pythons still have 'rear limb' bones suggests that they have not existed for millions of years.

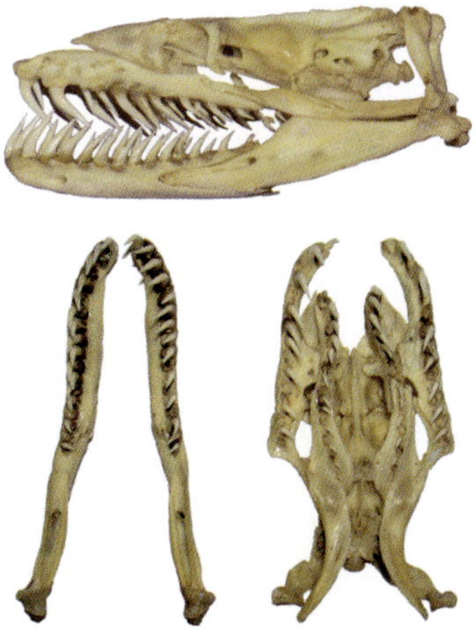

Python skull showing extra rows of teeth in the palate (note that the premaxilla is missing).

Below: Still-born baby boa with umbilical cord attached.

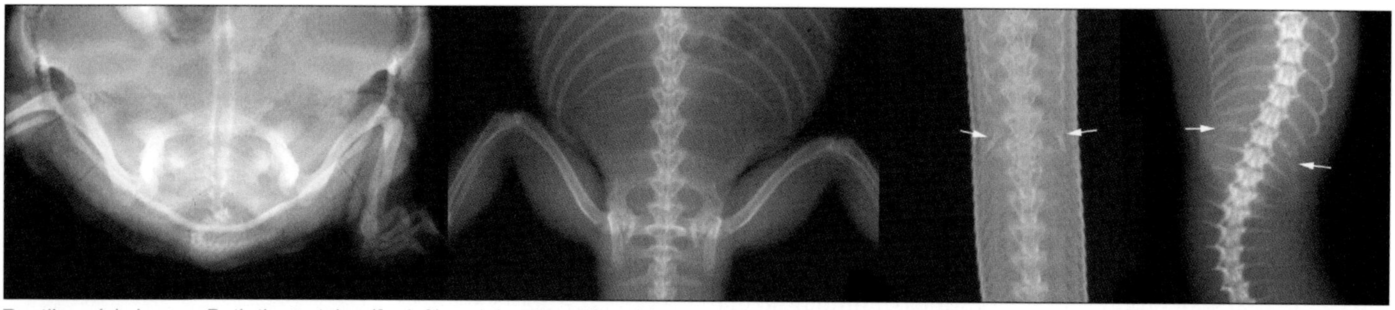

Reptile pelvic bones. Both the tortoise (far left) and the plated lizard (middle left) have a well-developed pelvis and normal limbs. The plated lizard also has spurs next to the outer edge of its anal vent. The sheltopusik legless lizard (middle right) and anaconda (far right) have no signs of any pelvis, but have some tiny bones (arrowed) lying deep in the tissues near to the spurs, which are next to the outer edge of the anal vent.

X-rays of reptile skulls. The tortoise (far left), gecko (middle left) and sheltopusik legless lizard (middle right) all have typical lizard jaw hinge joints with the right and left halves of the lower jaw being fixed together.
In contrast, living snakes (far right) have specialised jaw hinge joints with each half of the lower jaw being separate.

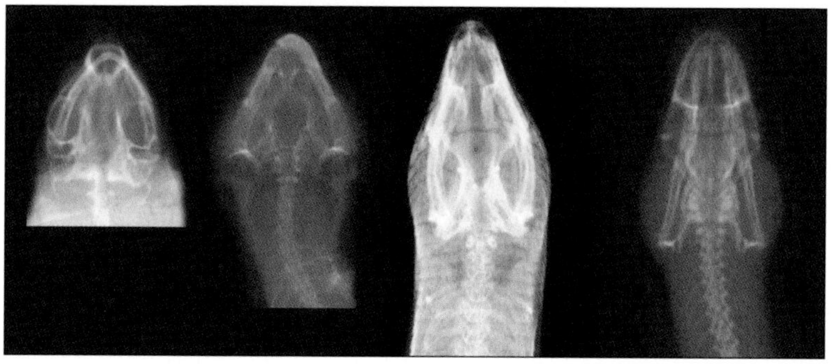

lay eggs. Furthermore, although *pipe snakes* are supposed to be some of the most primitive of all snakes and spend most of their time underground, nevertheless they too produce live young.

As with living pythons, both the fossil snakes haasiophis and pachyrachis had extra rows of teeth in the palate and fully mobile skulls and jaws. Their bodies were elongated like living snakes, although pachyrachis had 18 neck-like vertebrae. Both these ancient snakes also had bones in the pelvic area which closely resemble rear leg and foot bones. It is not clear from the fossils of haasiophis and pachyrachis whether or not their 'rear limb' bones projected from the body surface. As already noted, in living snakes, all the 'rear limb' bones (including those that might resemble foot bones) are found *inside* the body. This is unique since, in contrast, limb reduction in *skinks* (lizards) and *salamanders* (amphibians) results in stumpy *external* rear limbs. A similar appearance is seen in human congenital deformities, in which any short and abnormal lower leg and foot bones are found on the ends of external upper leg bones and are not found embedded *within* the soft tissues of the pelvic area. It is therefore reasonable to suspect that the 'rear limb' bones of some snakes may not represent the vestiges of legs at all.

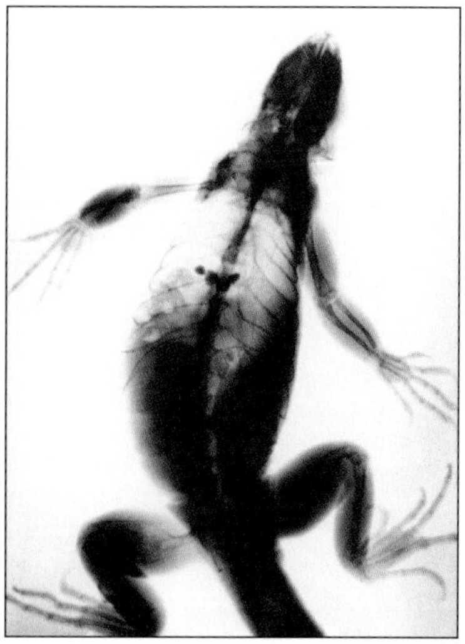

X-ray of iguana

However, in 2000, *Eupodophis,* an 85cm fossil snake, was unearthed in (supposedly) 92myo sediments in Lebanon. The hind limb (which clearly shows a fibula, tibia and femur) is only 2cm (0.8in) long. Although it was clearly outside the body, it could not have been of any use for locomotion. More recently, another fossil snake *Najash* was found which also had rear limbs extending outside its body. It also had a pelvic girdle attached to the spine. Although the fossils haasiophis, pachyrachis,

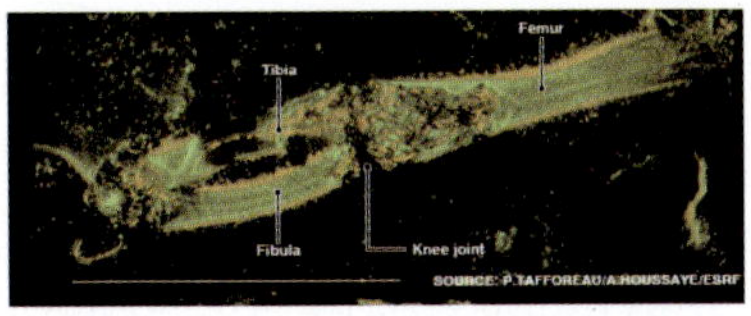


Rear limb of Eupodophis.
*From J Rage, F Escuillié. Un nouveau serpent bipède du Cénomanien Crétacé. Implications phylétiques, Comptes rendus des'séances de l'Académie des Sciences, Paris, Series IIa, 330 (7) 15 April 2000. p513–520.*

Hip region of Najash showing the rear limb bones.
*From S Apesteguía. H Zaher. 'A Cretaceous terrestrial snake with robust hindlimbs and a sacrum'. Nature. Vol 440. 20th April 2006. p1037-1040.*

and eupodophis were all found around the Mediterranean basin, suggesting that maybe there was some local environmental issue affecting limbs in snakes, najash was found in (supposedly) 90myo sediments in Patagonia, showing that a local trigger is unlikely. What is clear is that the legs in fossil snakes are degenerate and useless for locomotion, and found in creatures that are otherwise essentially like modern snakes.

Recently, some genetic evidence has been interpreted as showing that snakes were once four-legged reptiles. In the developing chick embryo, the genes HoxC6 and HoxC8 are active along the chest region and are involved in the development of the thorax. It is found that in snake embryos, these same genes are active along the entire length of the spine from the back of the skull to the anal region, excluding the tail.

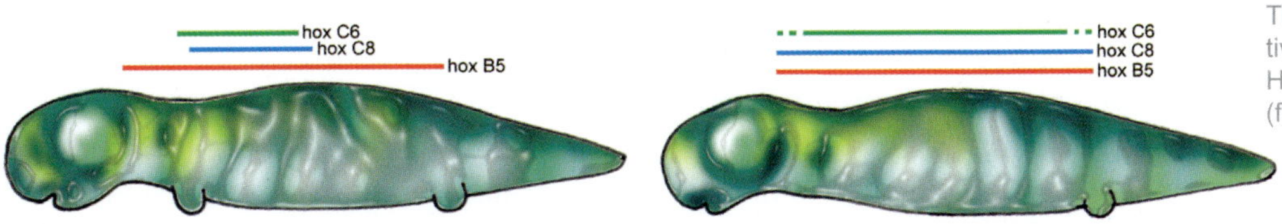


The lengthwise distribution of the activity of the genes HoxB5, HoxC6, and HoxC8 in the chick (right) and python (far right) embryo.

The idea has been put forward that the extension of the activity of HoxC6 and HoxC8 beyond the thorax region somehow inhibits the development of the legs. It has therefore been suggested that mutations might have occurred in some pre-snake lizard resulting in the extension of the regional activity of HoxC6 and HoxC8 to both the neck and lumbar regions, causing the neck and abdomen to develop ribs, and in addition, causing failure of the legs to develop.

However, the sheltopusik (which has ribs in its lumbar region but is otherwise fully a lizard) demonstrates that any mutation resulting in trunk elongation is itself unlikely to also result in changes in the neck, skull, teeth, ear, jaw, ventral scales, or in the ability of the tail to separate etc. Furthermore, any such modifications of a mosasaur or terrestrial lizard would not necessarily confer any advantage and would, more likely, lead to extinction. For example, if, in a mosasaur, alterations in the activities of Hox genes could have caused ribs to develop in the neck and lumbar regions, and also failure of limb development, this would only have given rise to a very vulnerable mosasaur. Detail about mosasaur reproduction is lacking, but if they needed to get onto the land in order to lay eggs, such mutations would render them incapable of doing so.

If in a mosasaur (below left) alterations in the activities of Hox genes could have caused failure of limb development, the abnormal animal (below right) would have had difficulty in changing direction, and would have been unable to get out onto the land to lay eggs. It would, therefore, most likely have died off.

In contrast, the sheltopusik demonstrates conclusively that terrestrial lizards can lose their limbs and still survive, without needing to evolve by altering *any* other lizard features.

We have no evidence of an ancient snake with either mosasaur-like or terrestrial lizard-like form, and the fossil snakes are very similar to living snakes. Since those fossil snakes were no closer to lizards than are living snakes, it can therefore be concluded that:

> Haasiophis and Pachyrachis have no particular bearing on snake-mosasauroid relationships or snake origins. E. Tchernov, O. Rieppel and H. Zaher. 'A fossil snake with limbs'. Science. Vol 287. 17th March 2000. p2010-12.

Legless lizards. The sheltopusik (left) and the slow worm (right) have eyelids and external ears, both of which are typical lizard features.
The sheltopusik here shows its nictitating membrane half way across its eye.

Furthermore, whilst limb loss can be explained in terms of mutations resulting in damage to important genes that control embryological development, this tells us nothing about how the phenomenal number of unique structures and controls that make up an arm or a leg could ultimately have arisen within a lineage, starting with some (hypothetical) ancient pre-fish creature which would have had neither fins nor limbs. The *WYWYWG Principle* shows that the acquisition of such a series of structures and controls is implausible. And, although some fossil snakes definitely had external leg bones, it is questionable that the bones inside the pelvic regions of living snakes truly represent bones comparable to the limb bones of other reptiles. As with the 'pelvic' bones of some whales, it is plausible that the 'rear limb' bones

Below: Skinks are a group of lizards which often have tiny legs.

The scales on the underside of both legged and legless lizards are fixed, do not overlap and are the same circumferentially around the body. In contrast, snakes such as this python have wide single scales which extend across the whole width of the underside and are mobile. In addition, whereas the body of the legless lizard is essentially circular in cross-section, the snake has a flat underside.

The reticulated python can grow to 33 feet (10m) in length.

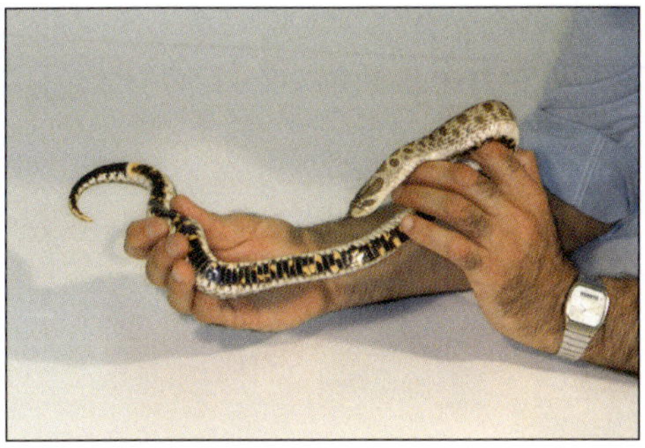

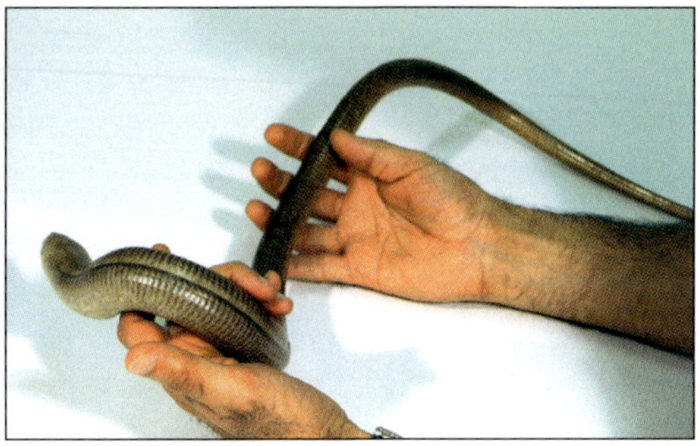

Far left: Hognose snake. Snakes have very flexible bodies and glide by pushing with their flexible underbelly scales. Snakes have unique single underbelly scales which extend right across the underside. Most snakes have relatively short tails (my right thumb is near the anal vent).
Left: Sheltopusik. Legless lizards have relatively stiff bodies and move by pushing their *sides* against any available resistance. The scales on their bellies are small and fixed and the same as those on the back and sides. This individual is moving by rubbing its right side against my fingers. Most lizards (including legless ones) have relatively long tails (my right little finger is near the anal vent).

*inside* the body of some snakes simply constitute a unique design.
A number of lizards are known to have no legs and their study ought to give us some insight into snake origins. Whilst legless lizards (*anguids*) may at first glance look as though they are snakes, nevertheless they no more cease to be lizards as a result of their leglessness, than a dog born without a limb ceases to be a dog. In fact:

> Anguids are elongate lizards that superficially resemble snakes – until they begin to move. The awkward, clumsy, broadly S-shaped curves employed in the movement tells any observer that **this is no snake**. (My bold).
> R D Bartlett and Patricia Bartlett. 'Lizard care'. Barron's Educational Series Inc. 1997. p123.

Above: Gaboon viper.
Below: The hollow fang of a gaboon viper. At around 25mm it is the longest fang of any snake.

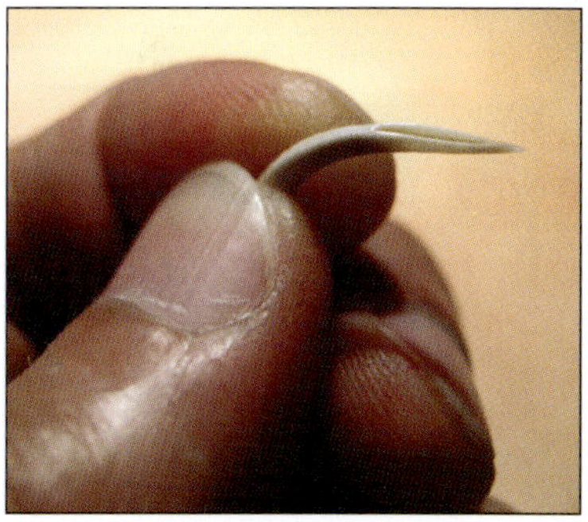

The marked difference in locomotion between snakes and legless lizards is because snakes have unique, wide, transverse, overlapping scales which are found right along the length of their underbelly. Muscle action in the body wall allows these scales to gain purchase on any surface irregularity. In contrast, the scales of legless lizards (like legged lizards) do not overlap, are immobile, and are the same all around

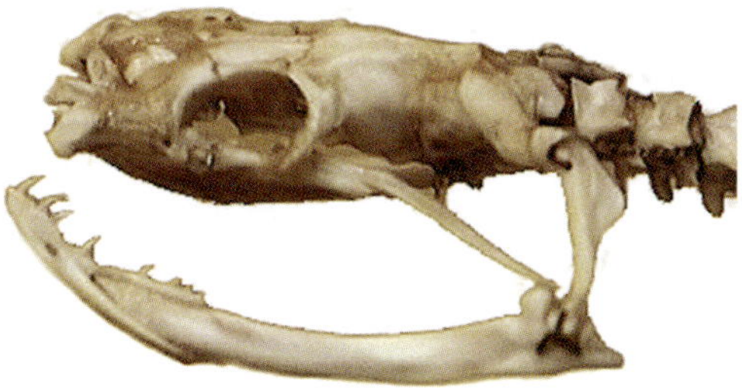

Snake jaw flexibility.
Left: Cobra skull.
Right: Boa.
Unlike all lizards, both the upper and lower jaws of most snakes are only loosely connected, which allows a great amount of flexibility when swallowing large prey.

Slow worm eating slug.

Above: A lizard can shed its tail without any contact, simply in response to a threat. The tail does not tear *between* adjacent vertebrae, but the plane of separation lies *through* one specific vertebra. When the tail separates, any transected blood vessels constrict so there is minimal blood loss and the raw end heals quickly.
Below: The scales at the end of this slow worm tail shows how sharp the separation junction is.

## Some important differences between snakes and lizards

| | SNAKES | Legless LIZARDS | Legged LIZARDS |
|---|---|---|---|
| **HEAD** | | | |
| Eyes | no eyelids | eyelids present | eyelids present |
| | no nictitating membrane | nictitating memb. present | nictitating mem. present |
| | focusing with spectacle | focusing with ciliary muscle | focusing with ciliary muscle |
| Ears | no external ear | external ear present | external ear present |
| Heat pits | present in some | no heat pits | no heat pits |
| Jaws | highly flexible jaw hinges | normal lower jaw hinge | normal lower jaw hinge |
| | R+L halves separable | rigid mandible | rigid mandible |
| | many with extra row of teeth | single row of teeth | single row of teeth |
| Poison | many poisonous types | no poisonous types | only two poisonous types |
| | hollow fangs | solid teeth | solid teeth |
| | hinged fangs in vipers | normally-set teeth | normally-set teeth |
| **BODY** | | | |
| Lateral groove | no lateral groove | lateral groove often | lateral fold often |
| Tail | cannot break off | can break off | can break off |
| | supple | relatively stiff | relatively stiff |
| | relatively short | long | long |
| Ventral scales | unique mobile scales | identical circumferentially | identical circumferentially |
| Movement | supple flexible body | relatively stiff body | stiff body |
| | gliding using ventral scales | twisting using body sides | motion using legs |
| Pelvic bones | present in some snakes: boas and pythons | present | present with legs |
| Pelvic spurs | present in some | present in some | present in some |
| Lungs | single lung in most types | two lungs | two lungs |
| | two lungs in: boas and pythons | | |
| **FOOD** | mainly vertebrates | mainly insects or carrion | mainly insects or carrion |
| **PREY CAPTURE** | capture using mouth in most snakes | capture using mouth only | capture using mouth only |
| | capture by constriction in: boas and pythons | no constrictors | no constrictors |

Regrowth of tail in a gecko and an iguana. After a lizard sheds its tail, a new one will grow but it never matches the original in size, shape or colour. The junction between the old and new tail is at a-b.

Unlike lizards, many snakes have heat-sensors with which they can locate warm-blooded prey (even in total darkness). Those in this python (above far left) are seen as pink pits situated around edge of the upper jaw. Snakes such as pit vipers have a single heat-sensitive pit below each eye.
Most snakes eat vertebrates. In contrast, like most lizards, this chameleon (above left & above) eats insects, although larger lizards eat carrion or vertebrates.

their circumference. The difference in their movement is particularly noticeable when they are being handled: snakes glide gracefully through the hands by pushing themselves along with their underbelly scales. In contrast, a legless lizard twists and gyrates in a clumsy manner as it tries to push the *sides* of its body against any available irregularity.

Close examination shows that apart from having no limbs, legless lizards have no significant features that separate them from legged lizards. In fact, legless lizards are as different from snakes as are legged lizards (see table overleaf). For example, unlike snakes, all anguids have relatively immobile skulls and jaws and have external ear openings. Most anguids also have a longitudinal groove or expandable fold on each side of the body. They also have functional eyelids and quite different eyes to those of snakes (we shall discuss the differences between snake and lizard eyes in a later chapter).

Lizards and snakes often live in similar surroundings. So, given the same sorts of environments and predators, if any mutations were beneficial to a snake ancestor, then, given a long period of time, other lizards ought also to have benefited from and retained similar types of changes. Some lizards should therefore have acquired mutations that would have given them, for example, uniquely mobile jaw joints and stretchable jaws (useful for swallowing any large prey) and hollow poisonous fangs (whether folding fangs as in vipers, or fixed fangs as in cobras). In fact, only two lizards are known to have poisonous bites, the *Gila monster* and the *Mexican beaded lizard.* Both have rear poison fangs, but neither has any snake-like characteristics.

Unlike lizards (both legged and legless types), snakes have a relatively short tail which they cannot shed in response to danger.

The Gila monster, one of only two living poisonous lizards, has solid fangs at the back of the lower jaw.

The point must be stressed that mutations are essentially random and do not conveniently appear just when a particular creature could benefit from a particular change. Furthermore, there are so many *substantial* differences between snakes and lizards that the *WYWYWG Principle* shows that it is implausible that any lizard could acquire the very *series* of mutations that would be necessary for it to change into a snake. So, given the large number of dissimilarities between snakes and lizards, the fact that pythons have some very small bones in the pelvic area *does not* allow the conclusion that snakes were once lizards.

Regarding the ancestral burrowing pre-snake idea: if, in the distant past, burrowing lizards lost their legs and became snakes, and if, today, a large number of snakes are not burrowers, then the notion of evolution should predict that, after 95my, some snakes at least should have acquired mutations that would re-establish their legs or develop new types of limbs. However, no such legged snakes exist. Furthermore, just as not *all* light-coloured peppered moths died off during the industrial revolution, in the same way, many (supposed) pre-snake lizards ought to have survived to show a variety of combinations of snake-like characteristics today: for example, a lizard with no external ears, and in which the right and left halves of the jaws can separate, or a lizard with an extra row of teeth in the hard palate and with snake-like underbelly scales. Such lizards do not exist.

The *olm* is a salamander which lives in caves. It has tiny legs and retains its gills throughout its life.

Lizards are not the only animals that can have legless forms. The adult *Oedepema salamander* is about half a metre long and has puny little legs, which are so small that it looks and acts like a huge worm. Since the oedepema salamander exists today, we know that it is possible for a mutant salamander to survive without proper legs, *provided that the environment is favourable, and there are few predators*. Thus in oedepema, as with the Galapagos cormorant, we are looking at the survival of a creature with genetic abnormalities, which ordinarily would have resulted in its extinction. It is likely that inbreeding of these mutant salamanders has resulted in populations with reduced limbs. Just as a person with thalidomide-associated limb abnormalities is still without question a human being, so we should not call a legless salamander anything other than a salamander.

### *Rib locomotion and limb loss*

Expansion of the ribs during breathing is largely a subconscious thing, although we humans are consciously able to increase or decrease the rate and even to stop breathing for a while. Whether lizards and snakes can consciously control their breathing is unclear. However, in humans and in both snakes and lizards (whether legged or legless, breathing occurs by the expansion of all the necessary ribs collectively as a whole - the ribs do not move independently of each other. Whilst locomotion in snakes and legless lizards is clearly under conscious (to whatever level the reptile brain can make decisions) control, the outworking is by the subconscious ordered movements of the ribs in a head-to-tail sequence, which in some ways mimics the progressively head-to-tail contraction (peristalsis) of the intestines, which causes ingested material to be moved along. In contrast, in legged lizards there is no peristalsis-like movement of the ribs.

In the legless lizards and snakes, the ribs must be able to move outwards to allow the lungs to expand, and also forwards and backwards to allow locomotion. To allow these different rib movements, the shapes of joints between the ribs and the spine have to be different from those of legged lizards. In addition, the muscles have to be arranged appropriately, and there have to be special nerve structures and controls in place to direct rib movements.

Snakes are able to climb vertically without any limbs.

Since rib movements for breathing are entirely collective and automatic, no amount of conscious effort will cause ribs to move independently of each other in a head-to-tail wave. In the same way, since intestinal movements are entirely automatic, no amount of conscious straining will cause the intestines to contract. This is exemplified by *Hirshsprung's disease*, which is caused by a congenital deficiency in certain types of nerve cells (ganglion cells) in the wall of the rectum. This deficiency can severely reduce proper peristalsis and result in chronic constipation or even complete bowel obstruction. Just as without specifically organised nerve networks, there cannot be any peristalsis of the intestines, so also, without specifically organised nerve networks, there cannot be any rib locomotion.

Now, here's the issue. If a legged snake or lizard lost its legs, it would not be able to survive (since it would need to move adequately to feed and avoid predators) unless it already had the nerve/muscle/joint arrangements in place to allow rib locomotion. Since it is impossible for 'peristaltic' rib movements to appear simultaneously with leg loss, both snakes and legless lizards must have always had the ability for rib locomotion.

A snake with rib locomotion does not need legs.

Tiny legs (cgi*) in snakes would be of no use for locomotion but would not be fatal.

A lizard with rib locomotion does not need legs.

Tiny legs (cgi*) in lizards with rib locomotion would be of no use but would not be fatal.

A lizard with no rib locomotion needs its legs.

Tiny legs (cgi*) in lizards without rib locomotion would be fatal.

*cgi: computer-generated image

So we come to the logical conclusion that there are possibly a number of fundamentally different types of lizards and snakes:
1: legless snakes with rib locomotion with ancestors which never had legs.
2: legless lizards with rib locomotion with ancestors which never had legs.
3. legged snakes with rib locomotion with ancestors which were never lizards.
4. legged lizards with rib locomotion (with no ancestors that did not have rib locomotion).
5. legged lizards without rib locomotion (including those with congenitally abnormal limbs).

### *Why do some creatures lose their limbs?*

As we have seen, it is a curious fact that many animals have either diminished or absent limbs. Some lizards such as skinks have stumpy limbs, the uselessness of which creates no problem in survival. Other lizards, such as the sheltopusik and slow worm, do not have either front or back legs and still manage to survive very well. The ancestors of some snakes had rear limbs of varying functionality, and living snakes manage to survive very well without limbs at all. Some bipedal dinosaurs, such as T rex, had tiny useless front limbs but managed to survive very well since feeding and running were in no way dependent on the front legs. Similarly, some flightless birds, such as the Moa, had no front limbs at all and also managed to survive very well. Some whales and sea cows once had rear limbs but this loss has not affected survival since their mobility is determined by the tail flukes.

A whale with flukes does not need rear flippers.

Rear flippers (*cgi) in whales are of no use for locomotion, but would not be fatal.

Clearly, survival with a non-lethal abnormality is determined not by the abnormality itself, but by whether or not the affected animal can eat and escape predators – *despite the abnormality*. It is likely that in the wild, otherwise healthy legged

*cgi: computer-generated image

lizards and crocodiles, and running mammals born with stunted or absent front or back legs, will quickly succumb to predation. While bipedal dinosaurs born without back legs would not have survived, those born without front limbs could still have managed to survive providing they could feed and escape from predators. The same applies to birds.

What is clear is that having diminished or absent limbs is not part of some evolutionary process in which a creature somehow 'improves itself' or 'adapts' to an environment, but is purely and simply a degenerate process in which healthy animals with fully-functioning controls for limb embryological development give rise to progeny with congenital abnormalities. In other words, they are *diseased* offspring, as a result of genetic mutations.

Mutations happen all the time. We have seen that, except for those affecting non-vital sections of a protein, all mutations interfere with the proper working of molecular control systems to some degree, and many result in developmental problems. What mutations *do not* produce is a *new* structure or function, but always some form of *derangement*, which is often (but not always) associated with diverse abnormalities way beyond the limb itself. We saw this when discussing fish and tetrapods (page 94) and ear bones (page 97), and mutations of Msx genes are another good example:

> Msx1/Msx2 double mutant mice exhibit synergistic defects in calvarial [skull], tooth, ear, limb, hair follicle, and mammary gland development.
> S Alappat et al. 'Msx homeobox gene family and craniofacial development'. Cell research. 2000. Vol 13. p429–442.

If a mutation is not so problematic as to result in death, then an individual may survive and perpetuate that mutation. So, if a mutation interferes with development of the limb bud, but does not affect other vital developmental processes, then the individual will be born with no limb. If a mutation interferes with limb development after the limb has started to form, then the individual will be born with varying degrees of limb mal-development.

Leaving aside animals such as seals whose unique rear limbs function very well, the likelihood is that all the extinct and living examples of creatures with stunted limbs had ancestors with healthy functional limbs. Therefore, we should predict, for example, that T rex ancestors had longer functioning front limbs, but were otherwise fully T rex. So, as with the Galapagos cormorant, T rex and its cousins like allosaurus were suffering from congenital diseases.

Evolutionists claim that snakes with tiny legs and whales with rear flippers demonstrate intermediaries between modern whales and creatures that were in times past neither snakes nor whales. However, the *Degeneration Principle* shows us that this is a misinterpretation of the evidence. Since a limb could not appear in an animal whose ancestors did not have the cellular instructions for making limbs, it follows that whenever we see stunted or deformed limbs which are otherwise essentially complete, then these must be *degenerate limbs* in creatures whose ancestors must have had full limb-making instructions. In other words, they are not signs of evolution, but of congenital abnormalities.

If it is suggested that whales descended from terrestrial mammals, then it follows that reptiles such as the extinct plesiosaurs and icthyosaurs must also have descended from terrestrial ancestors. However, although icthyosaur fossils are found in sediments between (supposedly) 65 and 200myo, even the oldest examples were already fully icthyosaurs and they:

> ...give few clues to the earlier origin of these reptiles. David Norman & Peter Wellnhofer. 'The illustrated encyclopedia of dinosaurs'. Salamander Books. 2000. p368.

Icthyosaurs gave birth to live young and have no known ancestors.

This man's middle finger was lost in an accident long ago. He manages very well without it, showing that a pentadactyl limb can have no selective advantage over a limb with four or six digits. (The recent injury to the thumb is incidental.)

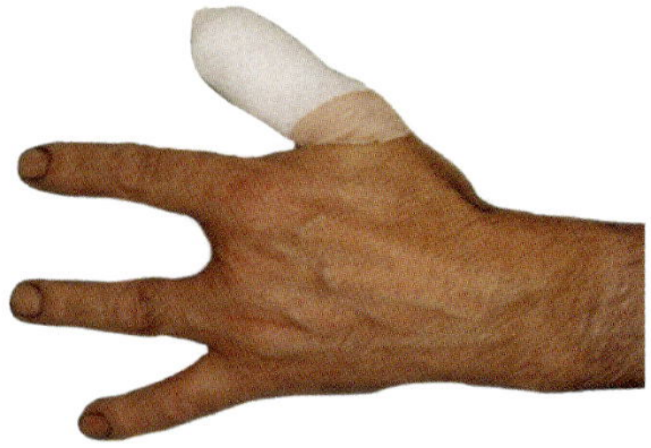

In fact, there is no real evidence for any totally terrestrial animal changing into a totally different and aquatic creature.

If ancient fishes (with no digits at all) evolved into amphibians, then we may here raise the question as to why the limbs of terrestrial vertebrates are based on the *pentadactyl* (five-fingered) format. Since we know that, from frogs to humans, there is no advantage in having five digits compared to having either four or six, then, contrary to the evidence, the notion of evolution should predict that fishes should have given rise to land vertebrates with a whole variety of numbers of digits.

In the typical bony fish pattern (unlike the shark or amphibian form), the front fins arise from the side of skull structures (the front fins are commonly called 'pectoral' fins but this is a misrepresentation since bony fishes do not have a pectoral [or shoulder] girdle). However, since a number of fishes manage to 'walk' on the sea bed or on land by using their gill covers (such as the *climbing perch*), or by using appendages that arise separately from the front fins (such as the *shanny*), we should expect that at least some amphibians should have evolved with front legs derived from tissues on the side of the neck or head, rather than from the pectoral girdle.

**Fish head and tetrapod neck**

This tuna shows the typical bony fish pattern in which the pectoral fin arises from the side of skull structures. All living and extinct amphibians have front limbs arising from the pectoral girdle, which is situated at the junction of the neck and chest. Contrary to evolutionary theory, there are no extinct or living amphibians that show a front limb attached to the back of the skull:

> *The changes in the neck anatomy which would be necessary in the 'fish to tetrapod' transition have never been documented in HOX gene lab animals. The precision which is needed to guide the gradual loss of existing fish 'neck' supportive structures while gradually establishing a tetrapod neck (involving new developmental cascades producing entirely new structures) is scientific fantasy.*
> Dr Donald Moeller. Maxillofacial Surgeon. Columbus, Georgia. Personal communication. June 2008.

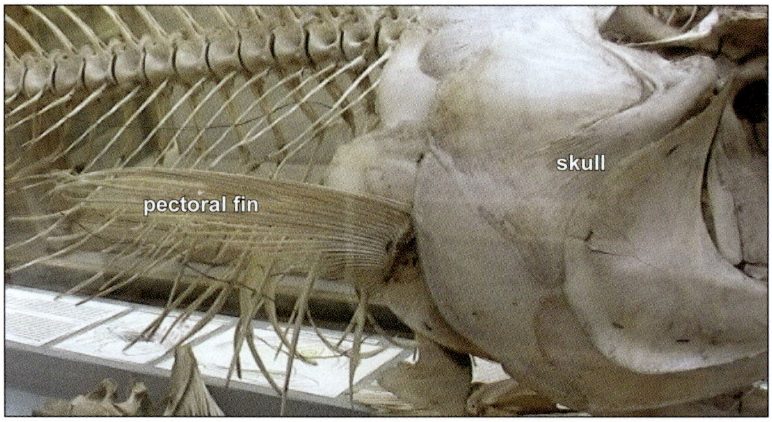

In addition, we might also expect to find that some amphibians had evolved with three pairs of legs arising from the neck, pectoral and pelvic areas.

Whilst the gunard (far left) walks on the sea bed by using the three spines arising from each pectoral fin, the shanny (left) walks using spines (arrowed) which are separate from the pectoral fin.

To round off the subject of limbs, what should we make of the common notion that *homology* (the finding of similar structures in diverse creatures, for example, the *humerus*, *radius* and *ulna* bones of a limb) confirms evolution from a common ancestor?

Flying gunard with greatly enlarged pectoral fins.

> The front leg of a crocodile, the wing of a bat, the flipper of a whale, the wing of a bird superficially look quite different from each other. On closer examination, these differences begin to disappear. For example, the arrangement of bones in the front limbs of these four animals is identical, even if the shape of those bones varies greatly. This common arrangement of bones, which extends also to humans, indicates that these limbs **have all evolved from the same basic plan**. (My bold). Arthur Busby III, Robert Coenraads, David Roots and Paul Willis. 'Rocks and fossils'. Harper Collins. 1996. p95.

If homology truly reflects a common ancestry, then firstly, we should find good evidence for intermediaries between one type of creature and another (for example, between quadrupeds and whales, lizards and snakes, terrestrial reptiles and ichthyosaurs, dinosaurs and birds [see next chapter], and shrews and bats). Secondly, we should be able to demonstrate an increasing and biologically significant advantage within a hypothetical series of intermediaries. Thirdly, we should be able to identify genetic models that demonstrate that such intermediaries *could* have arisen. However, on none of these counts do we see any convincing evidence to support the notion that any creature type has ever changed into another completely different type.
On the contrary, we have seen good evidence that suggests that whales have always been whales, that elephants have always been elephants, and that snakes were never, in the distant past, lizards.

# CHAP

## The bird:

## *a flight of fancy*

# PTER 8

Osprey

Emu feet

Birds are not lizards. Everybody knows that, but some people believe that birds evolved from a small bipedal dinosaur, and made their first appearance on the Earth about 150mya:

> Let's cut to the chase: Birds are dinosaurs. David Fastovsky and David Weishample. 'The evolution and extinction of dinosaurs'. Cambridge University Press. 1996. p293.

> ...I am awakened by a flock of dinosaurs shouting at one another in a dead tree...The dinosaurs are brightly colored burrowing parrots, which like all birds, are now thought by most experts to be living members of the dinosaur clan. James Shreeve. 'Uncovering Patagonia's Lost World'. National Geographic. December 1997. p135.

Bird feet and legs certainly resemble those of theropods (below), but did birds evolve from non-flying bipedal dinosaurs?

> It seems difficult to believe that birds, whose movements are freer than all other animals, could have descended from the traditionally ponderous and cold-blooded dinosaurs, and yet there is now widespread support for such an ancestry...Because it had feathers, *Archaeopteryx* was undoubtedly a warm-blooded animal and is regarded as the first bird...as far as the evolution of birds is concerned, the crucial development of warm-bloodedness may have already taken place in dinosaurs millions of years before they became birds...**Once small dinosaurs evolved feathers** for insulation, it is possible to imagine how flight may have developed...by flapping and leaping and the gradual process of natural selection **whereby their skeletons and general form were modified**, they acquired the ability to fly... (My bold). Stephen Dalton. 'The miracle of flight'. Sampson Low. 1977. p68-71.

The reasoning behind the dinosaur theory comes mainly from comparisons of various parts of the skeleton, coupled with the presupposition that birds *have* evolved, and if so, that they must have evolved from a previous non-bird creature. For example, birds have an upper and lower opening on the side of the skull, and are therefore *diapsids* (meaning two arches). This feature is also shared by lizards and snakes. However, birds also possess an opening in front of the orbit and this makes them *archosaurs* (archo = ruling), a group that includes crocodiles, dinosaurs, and pterosaurs. Birds also have hind limbs that resemble those of a number of bipedal dinosaurs.
However, when looking at structural similarities between any creatures, we should be careful not to read too much into such observations since the *Horse Principle* reminds us that:

In the absence of any direct evidence, if any number of creatures share any number of characteristics, it cannot be assumed that for *any* of those creatures the shared characteristics reflect any ancestor-descendant relationship.

Yet despite the fact that there is more that goes to make up a bird than just its skull and bones, it is commonly promoted that terrestrial reptiles became birds, hence the common belief amongst paleontologists that we eat roast dinosaur at Christmas. But as we shall see, such ideas hinge more on hope and belief than on science. Fastovsky and Weishample ask three all-important questions: *Where do birds come from? Where do feathers come from? How did avian flight evolve?* However, after spending 26 pages essentially and only *describing* the bony features that birds have, the bony features that dinosaurs had, and the bony features that birds and dinosaurs have in common, these questions were answered as follows:

> **Where do birds come from?** Our answer is, of course, dinosaurs.
> **Where do feathers come from?** ...is not easily answered in a mechanistic way...they are **presumed** to be an outgrowth of archosaurian scales, but to date the fossil record is silent on how or when.
> **How did flight evolve?** Again, the answer is not fully laid out in the fossil record...**Birds simply co-opted primitive limb proportions for their own use**... (My bold). David Fastovsky and David Weishample. 'The evolution and extinction of dinosaurs'. Cambridge University Press. 1996. p320-1.

Baby blackbirds or baby black dinosaurs?

To state that *'Birds simply co-opted primitive limb proportions for their own use'* is not an answer at all. What is noticeable is that the text offers no examination of the molecular, genetic, and embryological *mechanisms* by which birds could have evolved from dinosaurs, and instead the quasi-religious approach is taken. In other words, despite their own admission that there is no real evidence at all, the authors *believe* in evolution and plead that the few observed similarities between dinosaurs and birds *prove* that birds evolved from dinosaurs. And to the unwary reader, they present it as a fact:

> So, **in fact**, birds are not some separate biological entity, distinct and apart from "reptiles". **Birds are dinosaurs.** (My bold). David Fastovsky and David Weishample. 'The evolution and extinction of dinosaurs'. Cambridge University Press. 1996. p320-1.

Of course, since there is no doubt whatsover that birds do share a number of features with some bipedal dinosaurs, the real issue here is *not* whether or not to classify birds under the umbrella 'dinosaur', but whether or not the evidence supports the notion that birds have evolved from some dinosaurs which were, at some time in the past, not birds.

## *Archaeopteryx*

It is generally accepted that *Archaeopteryx* was the first feathered flyer, the link between true reptiles and true birds, and that its ancestors were the small group of bipedal dinosaurs known as *theropods*. The first partial fossil of archaeopteryx was found in Late Jurassic limestone (dated as being 150myo) in Bavaria in 1861. Then in 1877 an almost complete specimen was found, confirming that archaeopteryx was a bird about the size of a pigeon. To date, there are only seven known fossilised skeletons of archaeopteryx, all having been found in Bavaria. Although there is no way of confirming the association, the single feather (60mm long by 11mm wide) which was found in the same area in 1860 has always been attributed to archaeopteryx. Logically, there is the possibility that the feather could, of course, have belonged to another flying creature, even to a true bird.
Some of the features of archaeopteryx are undeniably reptile-like and this is reflected in the way that specimens of archaeopteryx have sometimes been mis-identified.

**The archaeopteryx brain**

CT scans of archaeopteryx show that it was well suited to flying:

> *...the size and shape of the delicate semi-circular canals in the inner ear, which are crucial for balance. They were highly arched like those of modern birds, a trait associated with an acrobatic or aerobatic lifestyle. As for the brain itself, archaeopteryx had massive bird-like visual centres jutting out from either side of the brain, and the apparatus for a superb sense of hearing. But perhaps the most notable feature was a hefty cerebellum - the brain's 'autopilot', where sensory information is coordinated and integrated. Its relative size was far larger than even the birdiest dinosaurs, which include velociraptor.*
> New Scientist. 21st May 2005. p34.

Note that since similarity does not necessarily mean relatedness, the features of the archaeopteryx brain do not prove that it was a bird.

Above: Archaeopteryx.
Below: Compsognathus.

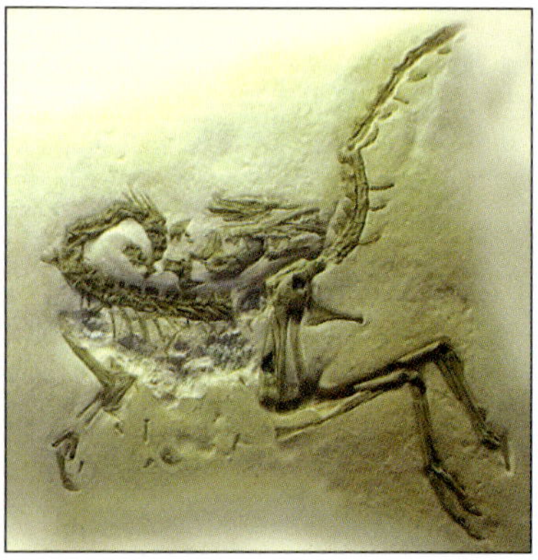

For example, the Teyler specimen (found in 1855) was initially thought to be a *Pterodactyl*, and the Eichstatt specimen (found in 1951) was initially thought to be the small bipedal dinosaur *Compsognathus*. Both the upper and lower jaws of archaeopteryx had many small sharp teeth, and it had a long, distinctly lizard-like tail which had around 22 vertebrae. As in all reptiles (excepting pterosaurs and some bipedal dinosaurs) but unlike modern birds, the bones did not contain large air spaces. In the shoulder girdle there were both bird and reptile characteristics. Although the collar bones (*clavicles*) were fused to form the 'wishbone' (or *furcula*) as in modern birds, unlike living birds the breast bone (*sternum*) was made of cartilage and not bone. There were sharp long curved claws at the end of each of the three long fingers on each wing. Curiously, *whatever the position of the wings*, the anatomical arrangements of the joints would seem to prevent the claws from being projected forward, so it does not seem as though the claws could have been used to capture prey. It has been suggested however, that with the wings held, as it were, at the bottom of their downstroke, the claws could have been in a position to help archaeopteryx to climb up small trees, much in the manner of coconut tree climbers. But since the distance between its shoulder joints was less than 3cm, the only 'trunks' it could have climbed in this way would have belonged to bushes. And since climbing up the centre of a bush does not leave much of an opportunity to fly off, how archaeopteryx used its wing claws is still a mystery.

As in many modern birds, one of the four toes pointed backwards. The sharp claws on the feet suggest that archaeopteryx was well able to perch, but compared to the toes of modern birds the backward pointing toe was relatively short, which suggests that much like a chicken or pheasant, the creature probably spent a considerable time on the ground.

Details of archaeopteryx: the long tail, the three-clawed front limbs and the pelvis are all typical dinosaur features.

Unlike in mammals, the ends of the long bones of amphibians, reptiles and birds are composed of cartilage and not bone. Since the cartilage is not represented in fossils, the detailed form and function of fossilised joints in non-mammals is difficult to deduce. So it is with archaeopteryx. However, it seems that the bones making up the wrist joint were arranged in such a manner that perhaps archaeopteryx could not flex the wrist sideways so as to fold the wing feathers against the body. Therefore it is not clear if the wings themselves could have been folded up as in a modern bird.

Archaeopteryx lacked a keel on its sternum which in modern birds serves to anchor the flight muscles (the *pectoral* muscles for the downstroke and the *supracoracoideus* muscle for the upstroke). It also lacked the *acrocoracoid process* which provides a pulley around which the tendon of the supracoracoideus muscle runs (see later). Some see these two features as prerequisites for flapping flight and so there is some argument over whether or not archaeopteryx was actually capable of flying like a modern bird. However, since the asymmetrical vanes on its feathers are just like those of modern flying birds (modern *non-flying* birds have symmetrical vanes) and since, although they lack a keel on the sternum, bats are well able to enjoy powered flapping flight, it is quite plausible that archaeopteryx was indeed a true flyer, albeit perhaps not so adept as a modern bird.

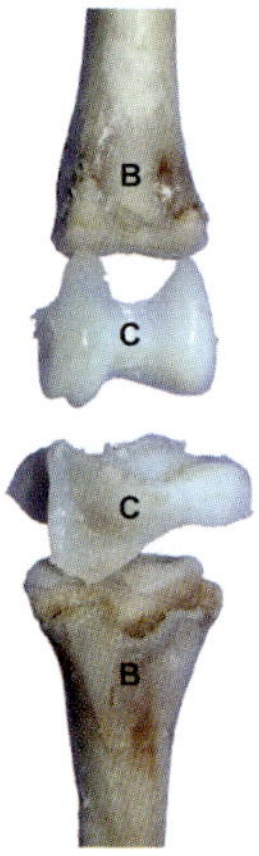


Chicken knee joint. The shapes of the joint surfaces of birds, amphibians and reptiles are formed by the cartilage (C), and are different from the shapes of the ends of the bones (B).

From a distance the living adult *Hoatzin*, a bird the size of a chicken and found in the swamps of Guyana and Venezuela, might be passed off as an unremarkable bird.

Adult hoatzin

However, the young hoatzin has two small claws on the front edges of its wings, quite different in size, shape and anatomical position from those of the archaeopteryx, but claws nonetheless. Also unlike the archaeopteryx, by the time the hoatzin becomes an adult it has lost its claws completely. The hoatzin is not unique in this respect since the young of the *Turaco* of Africa also have claws on their wings. It is because of their claws that the hoatzin and touraco are often mentioned alongside the archaeopteryx, as examples of birds which have retained some of their (supposedly) ancestral reptilian characteristics. Yet birds with claws on their wings are not rare - adult emus and ostriches have one and two claws respectively on each wing, and even the wings of young *moorhens* and *coots* have claws.
But was the archaeopteryx an evolving dinosaur, with its:

> ...tail and teeth showing reptilian ancestry, being half-bird, half-reptile...
> David Attenborough. 'Evolution weekend'. BBC2 Broadcast. 29th March 1998.

Adult emu wing with single claw.

We have seen that there is no evidence to show that the egg-laying echidna and duck-billed platypus are ancestors to any other mammal, or that monotremes are half-mammal and half-reptile. From the evidence we can only say that monotremes represent simply a unique form of mammal. And as we have seen, the chalicotherium and the paleoparadoxia show that it is possible for one creature to have a head and teeth that completely resemble another creature, yet for it not to have *any* relationship ancestral or descendant to the second creature. So if these principles are applied to the claws of the hoatzin and turaco, then this demonstrates *simply and only* that living birds that have claws on their front limbs *exist* - nothing more. And the claws, teeth, tail and breastbone of the archaeopteryx demonstrate *simply and only* that creatures with such features once lived and are now extinct.
It is also quite plausible, in the same way that bats are flying mammals, that some dinosaurs were feathered and that archaeopteryx represents one of their kind that flew. However, in the same way that we have no obligation to consider bats as representing evolved shrews (there is no fossil record of any supposed bat ancestor that was not completely a bat), so we have no obligation to consider that if archaeopteryx was truly a flying feathered dinosaur, the ancestors of archaeopteryx were at one time non-flying or even non-feathered. Logically, the existence of archaeopteryx does not oblige us to consider that feathers evolved in scaly dinosaurs, or that birds evolved from archaeopteryx. In other words, it is plausible that archaeopteryx was not half-way to becoming a modern bird. It was simply and fully an archaeopteryx.
But, it might be said, a number of other fossils of flying feathered creatures have recently been found - surely these confirm that birds are evolved dinosaurs? In actual fact, recent feathered fossils give no real insight into bird origins for two main reasons. Firstly, whatever their supposed age, all the birds that are found in the fossil record appear as complete birds. And all the dinosaurs that are found with hints of feathers are complete dinosaurs. As with the archaeopteryx, none of them are half-way to becoming something else. Secondly, there is far more that makes up a bird than just its bones. If birds evolved from dinosaurs, then all the differences between birds and reptiles, whether structural, biochemical, or behavioural, must ultimately be accounted for in terms of genetic possibility and probability. As

Adult turaco

Turaco hatchling showing claw on its wing.

Adult moorhen

**Archaeopteryx is probably not a bird**

1. Archaeopteryx had a typical dinosaurian head and no beak. All birds have beaks.
2. Archaeopteryx did not have a supracoracoideus pulley, without which its pectoral muscles could not elevate the wing. All birds have a supracoracoideus pulley.
3. Archaeopteryx had a typical dinosaurian tail. All birds have short, stubby tail bones.
4. Archaeopteryx had a dinosaurian pelvis. All birds have a backwards-directed pubic bone.
5. Archaeopteryx did not have a breast bone. All birds have a breast bone.
6. Archaeopteryx did not have special intra-costal rib joints. All flying birds have jointed ribs.
7. Archaeopteryx had typical dinosaurian forelimb digits 1, 2 and 3. Bird forelimbs utilise digits 2,3 and 4.

So, archaeopteryx was a flying, non-bird dinosaur.

The question is not whether birds are or are not dinosaurs, but whether non-flying bipedal dinosaurs, or flying non-bird dinosaurs (such as archaeopteryx) evolved into birds. The evidence shows that they did not.

we have seen when discussing statistics, many of the things that are possible may seem quite improbable. Furthermore, many things are just *not* possible. But let's deal first with the extinct birds.

### *Fossil birds*

It is a fact that fossils of ancient birds are extremely uncommon, and the dating of those fossils which are known show no clear evolutionary trend. Therefore, it is quite improper on this paucity of evidence to infer *any* firm views about any supposed ancestry.

We should not be surprised to find that the head, jaws, limbs, and tails of some extinct birds were quite different from those that exist today, not least because even amongst living classes of animals there exists a great diversity of form and structure. We could compare, for example, the vastly different teeth of marine mammals such as dugongs, dolphins, baleen whales, and seals; or of terrestrial mammals such as the elephant, rhinoceros, dog and horse. Alternatively, we could compare the prehensile tail of a New World monkey with the non-prehensile tail an Old World monkey and the absence of a tail in an ape. Yet again, we could compare and contrast the differences in the feet and toes of a cheetah (with its non-retractile claws) and a lion, a horse and an antelope, or an ape and a human being.

We should therefore ask: what exactly is a bird? It is important to remember that the classification of a bird (or any other class of animal for that matter) is to some *significant* extent an *artificial* label, which can act in the mind so as to impose some limit on what can be accepted or understood as a particular creature type. In other words, we have to ask: why can't a dinosaur have wings and feathers and still be a 'true' dinosaur; and why should it be felt necessary to extrapolate that such a dinosaur, sharing some similarities with birds, actually became one of them?

Recent fossil finds show that in addition to archaeopteryx, a number of other unusual feathered creatures have lived in the past. Among them are three dated as being 120myo from China: the *Protarchaeopteryx* and *Caudipteryx* (both with symmetrical feathers and teeth), and the *Confuciousornis* (with asymmetrical feathers, and a normal beak). Two specimens from Spain show a nestling from 135mya (with modern wings along with teeth) and the *Eoaluavis* from 115mya (which is the earliest bird known to have an *alula*, a feather on the thumb which is specialised for controlling airflow over the wings so as to reduce the risk of stalling during slow flight).

With regard to modern-looking birds, fossils of grebes have been found in sediments (supposedly) up to 26myo, vultures and storks in sediments (supposedly) up to 38myo, and flamingos and gulls in sediments (supposedly) more than 65myo. Giant flightless birds such as the 2m tall *Diatryma* were roaming the ground (supposedly) between 45 and 60mya:

> ...the oldest grebe fossil is Miocene in age...shorebirds, gulls, auks and flamingos are said to have arisen in the Late Cretaceous...Storks date back to the Oligocene. The oldest New World vultures are only Miocene in age, although ancestral forms from the...Oligocene of France...have been reported... Michael Benton. 'Vetebrate paleontology'. Unwin. 1990. p221-3.

Vulture

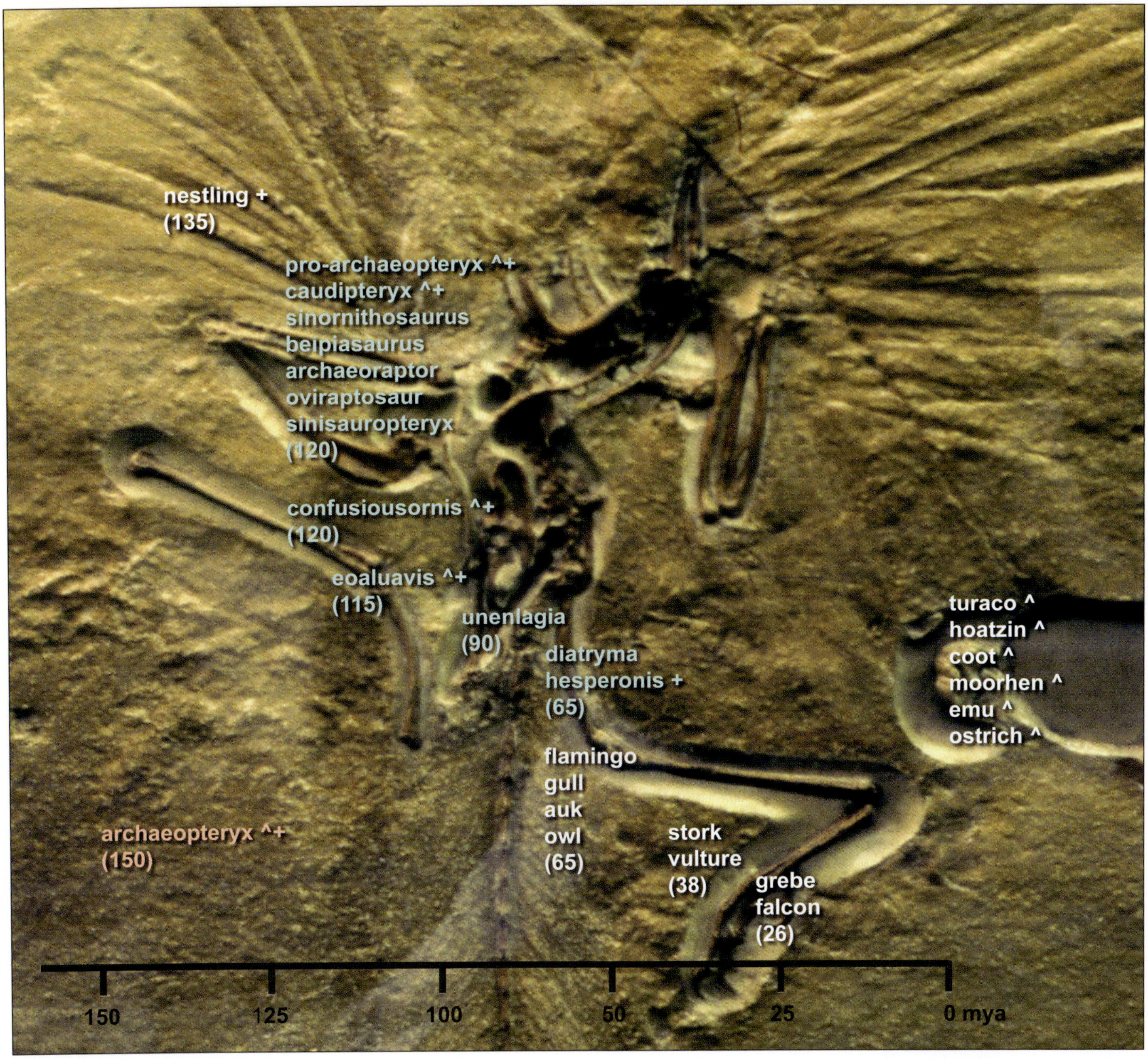


The fossil record shows flying dinosaurs (such as archaeopteryx), non-flying dinosaurs with filamentous skin covering (such as Beipiasaurus), flying birds and non-flying birds. There is no convincing evidence that dinosaurs evolved into birds.
Key:
Fossils resembling living birds are shown in white.
^ claws on wings
+ teeth

What does this actually mean? Does it mean that confuciousornis took 30my to lose the teeth of its archaeopteryx ancestors? And does it mean that vultures are very late arrivals on the evolutionary scene, and that there were none to be found at the time of the Late Cretaceous flamingos and gulls? If bills can change so dramatically in only a few thousand years, as shown by the Hawaiian honey-creepers, we should surely not expect that storks somehow took another 30 *million* years to evolve after flamingos and gulls appeared in the Late Cretaceous. Equally incongruous is the fact that:

> The oldest falcons are Miocene [up to 26mya], while the oldest owl is Paleocene (65mya) in age... Michael Benton. 'Vertebrate paleontology'. Unwin. 1990. p223.

Does this mean that falcons took *40my* to evolve after the Paleocene owls? Does the absence of very old humming bird fossils mean that they have only just recently evolved? Then again, (supposedly) *55 million years* after the appearance of the beaked confuciousornis, the large flightless diver *Hesperonis* had still not lost its teeth.
It is suggested that:

Humming birds

> ...the earliest fossil birds were flamingos, loons, grebes, cormorants and rails. It may be significant that many of these are aquatic and marine animals and hence more likely to be preserved... J Z Young. 'The life of vertebrates'. 3rd edition. Clarendon Press 1991. p390.

However, it is more likely that no individual storks or hesperornis happened to have become fossilised in sediments similar to those of gulls or flamingos or archaeopteryx. The *Coelacanth Principle* shows that a creature can live for (supposedly) 64my without leaving a fossil. If we apply this principle to birds, then flamingos and gulls could have lived 65 + 64 = 129mya, which would have been before most of the (presumed) primitive birds such as caudipteryx, and before any of the well-known large dinosaurs. But then of course the *Coelacanth Principle* must also be applied to the caudipteryx, so *it too* could have lived at least 64my before the date of its first recorded fossil 120mya - i.e. 184mya! And archaeopteryx could have lived 150 + 64 = 214mya! In fact, as we have shown before, the *Coelacanth Principle must* be applied to *every* fossil and to *every* living or extinct creature, such that we are not at liberty to exclude the contemporaneous existence of any living and extinct creatures at all times prior to the historically-documented past. And in fact, the geographical distribution of archaeopteryx is a curious anomaly since it is:

> ...known only from specimens found in late Jurassic limestones of Solnhofen, Bavaria. Michael Benton. 'Vertebrate paleontology'. Unwin. 1990. p209.

What is incongruous is that the barn owl, whose:

> ...earliest fossils so far unearthed date back (supposedly) only 12 million years... (My parenthesis). Ron Freethy. 'Secrets of bird life'. Blanford. 1990. p57,

has managed to extend its range so as to have:

> ...a wider distribution than any other species of land bird... Mike Reed & Jake Allsop. 'The barn owl'. Blanford. 1994. p14,

Barn owl chick, 24 days old. The barn owl is the most widespread of all land birds.

despite competition from many other contemporaneous predatory birds. So, if archaeopteryx was an ancestor to modern birds, it *must* have had very few competitors in the sky, and after only a few hundreds (let alone millions) of years it surely *must* have had the opportunity of extending its range *way beyond* Bavaria. So the fact that

archaeopteryx has only been found in an extremely limited geographic area suggests that archaeopteryx *could not possibly* have been around on earth for millions of years. Furthermore, whatever its biological category, it is plausible that archaeopteryx may have been restricted in its range by other feathered flyers (dinosaurs and/or birds).
Archaeopteryx leads us to a major misconception about the subject of missing links. As previously discussed, the genetic coding for any new structure or system must first appear by a *random* process before it can be selected for. So it is important to make clear that neither feathers nor any other bird characteristic could have evolved for any *reason* or *purpose*. Therefore, for a dinosaur to have acquired all the features that make a bird a bird, there must have first evolved a large number of intermediary forms. Furthermore, each subsequent intermediary form must have had at least some significant survival advantage over its immediate ancestor, along with an opportunity to reproduce in numbers sufficient so as to be realised subsequently as fossils.
We have previously stated the *Intermediary Principle:*

If a creature A supposedly evolved into a significantly different creature Z, where Z has a *number* of features which could not have appeared except by *a number* of genetic alterations to A over any number of generations, then a number of intermediary creatures, representing *every one* of the genetic alterations, must have existed in the time between the demise of A and the appearance of Z. And if they existed for equivalent periods of time, collectively the many types of intermediaries would most probably have given rise to far greater numbers of individuals than the numbers of either A or Z alone. Therefore, it follows that in total, more fossils of intermediary forms should be found than of either A or Z alone.

Thus, if dinosaurs evolved into birds, we should find a whole series of intermediary forms - *either side* (both in time and in form) of archaeopteryx. And it becomes apparent that archaeopteryx cannot be an ex- ‘missing link’, now found. Well, not in the sense that people would normally think. Since evolution by natural selection is dependent upon gradual genetic changes occurring and becoming consolidated in subsequent generations, all creatures that have ever lived were fully *those* creatures. If we were to come across any of them, we would not consider that any of them were ‘partly formed’, nor would we (without hindsight) be able to recognise that some were on their way to becoming something else.
So, suppose just for the sake of argument, we assume that birds have evolved from dinosaurs, and that a small *Velociraptor* was *the* ancestor. Then, in time, velociraptor would have given rise to a slightly-altered-velocirapter. Much later, the latter would have given rise to a slightly-altered, slightly-altered-velociraptor. And so on, until millions of years later the now much-altered velociraptor was in fact an archaeopteryx. The story does not end there since the archaeopteryx would have gone on to give rise to a slightly-altered-archaeopteryx. Later on, the latter would have given rise to a slightly-altered, slightly-altered-archaeopteryx. And so on, until millions of years later the much-altered archaeopteryx would have become a modern bird. All the individuals would in fact have been slightly-altered versions of their immediate ancestors. So, all creatures would be links of some form. So velociraptor would not have been any more a distinct entity than any of its slightly-altered descendants.

And in this scheme of things, velociraptor would itself have been nothing but a slightly-altered velociraptor-ancestor.

So, in keeping with the *Intermediary Principle*, fossils of neither velociraptor nor archaeopteryx should be any more rare or any more common than any other creatures in their particular lineage. Therefore, the fossil record ought to just as likely show evidence for the creatures velociraptor (Vo) and its descendants V+1,V+3, and V+8, and not for archaeopteryx, unlike the observed record. Or alternatively, the fossil record ought to just as likely show evidence for velociraptor ancestors V-1,V-4,V-7, and V–8, and no velociraptor or archaeopteryx at all. Thus, we should find in the fossil record that there are at least as many good examples of 'links' between the creatures that we might consider to be the main groups, as there are of creatures representing those 'main' groups.

If any particular creature (such as the archaeopteryx) has evolved from a markedly different ancient ancestor (such as a velociraptor), and has in its turn given rise to a markedly different descendant (such as a modern bird), then there *should* be evidence for a continuum of changing forms linking the present form to the very ancient.

But it might still be argued that fossilisation is a rare event and gaps in the fossil record are *bound* to exist, and are unlikely ever to be filled. If so, then we should be surprised that we have fossils of an archaeopteryx rather than one of its many (supposed) ancestors which would have had much the same form but no feathers *at all*.

Across the whole spectrum of known fossils, we find 'complete' creatures and no unambiguous series of intermediaries. For example, we have evidence of large sauropod dinosaurs from (supposedly) 220mya (e.g. *Plateosaurus*), from (supposedly) 150mya (e.g. *Brachiosaurus*), and from (supposedly) 100mya (e.g. *Titanosaurus*). It is considered that none of these three sauropods were directly related, so the notion of evolution must require that they were preceded by a series of ancestral forms each being similar to their immediate descendants, but slightly different.

Since the fossil record of sauropods does (supposedly) span over 120my, then we should most certainly expect that over such a long time period, fossils of neither plateosaurus nor brachiosaurus nor titanosaurus should be any more rare or any more common than any other creatures of that type. Therefore, the fossil record ought to *just as likely* show evidence of ancestral forms of those three sauropods with stout bodies and not-so-long necks, rather than *only* the long-necked ones that we have evidence of. Or alternatively, the fossil record ought to just as likely show, scattered throughout the 120my time interval, evidence for *only* the remnants of the stout-bodied-not-so-long-necked ancestral forms and *no* plateosaurus, brachiosaurus or titanosaurus at all. This complete lack of unambiguous evidence for 'intermediaries' in the fossil record suggests very strongly that no such intermediaries ever existed. And so, it follows that it is improbable that any such evolution ever took place.

### *Models for the origin of bird flight*

Despite the great differences between birds and dinosaurs, it is surprising that any discussion about the evolution of birds from dinosaurs usually revolves simply around bones and the origin of flight. Nevertheless, if we suppose that birds *did* evolve from dinosaurs, how then could these terrestrial reptiles have started to fly?

> 'The wing **is** a perfect example of how nature incorporates attributes that came about for unrelated reasons into a new structure', explains Luis Chiappe of the American Museum of Natural History. Feathers that evolved to keep a ground dweller warm, for example, might give it the ability to glide and better catch prey, leading to more feather development. (My bold). Jennifer Ackerman. 'Dinosaurs take wing'. National Geographic. July 1998. p91.

> Over a long period of time, some reptiles began to develop long back legs and short front legs. They could stand upright on their back legs. When the bipedal reptiles ran, they used their front legs to keep their balance. Some of them **may have** flapped their front legs, like tiny wings. **They developed feathers** instead of scales. With the appearance of feathers, came the first birds. They **probably** took to the trees for shelter. They were not able to fly very well, and could **perhaps** only glide by jumping from trees. (My bold). 'The super book of birds'. MacDonald Educational Ltd. 1977.

Two main theories have commonly been suggested to explain how bird flight originated. The first theory involves the *running-flapping model* in which it is thought that perhaps a small terrestrial bipedal dinosaur started to leap into the air in pursuit of flying insect prey. This idea is based largely on the fact that a number of therapod dinosaurs had bird-like legs. The pre-existence of bipedality meant that the forelimbs were free of any locomotor function and so were conveniently available for modification as wings. Thus, during the (supposed) process of change, alterations in the forelimb would not have adversely affected the ability of the dinosaur to move about.

It must be understood that it could not have been *when* bipedal dinosaurs started to leap about that they developed bigger scales and ultimately feathers. Rather, if any dinosaur acquired a mutation that left it with more frayed or fluffy scales, such affected individuals might have found that in the normal process of lungeing after any flying insects, they were able to become that little bit more airborne than their normal cousins. If, in the mutants, this ability allowed them to feed more effectively than their normal contemporaries, then the mutants would have survived to breed in greater numbers and so to perpetuate the mutation. The standard evolutionary stance is that after a number of further fortuitous mutations, the small dinosaur's descendants developed feathers and wings which assisted their leaps until eventually true flight became possible.

There are four main problems with this idea. Firstly, these dinosaurs must previously have managed to survive very well on ground insects. And if they failed to catch any flying insects, then, just like their contemporary dinosaur cousins, they would *still* be able to continue to survive on the plentiful supply of insects on the ground. In fact, the vast majority of birds feed in the water, on the ground, or in folliage, eating:

> ...fish, nuts, nectar, insect larvae, sugar-laden fruit... David Attenborough. 'Life on Earth'. Collins/BBC. 1979. p178.

Even birds that are commonly associated with trees actually feed on the ground:

> You may think of them as woodland birds but green woodpeckers are open ground specialists. They hop about, using their long, sticky, barbed tongues to fish ants out of their nests. 'Birds of healthy places'. RSPB Bird magazine. Summer 1998. p79.

The fact is that the food sources of most birds can be harvested by swimming, running, or climbing, *without* the necessity of flight. Furthermore, only a small minority of birds actually catch *flying* insects, and only occasionally do they do so just above ground level. Therefore, overall, there is no real evidence to suggest that a low-level flying insect food source could tempt dinosaurs into flapping and leaping.

The second problem is that it is unlikely that an animal that could not overtake its prey on foot would be more successful leaping and flapping after it. Supporters of the scales-to-feathers theory assume (and portray in drawings) that the arrangement of (supposed) early frayed scales in a dinosaur would mimic the flight feathers of a bird. But this is unlikely since the whole of a reptile's limb is covered in similar scales. If a mutation caused frayed or longer scales to develop, then it is likely that *all* the scales, all around the limb, would become frayed or longer. Some idea of the circumferentially-arranged long scales that might result is shown by lizards such as the *sungazer* (*Cordylus giganteus*) and the *blue spiny lizard* (*Sceloporus serrifer cyanogenys*). In fact, *whatever* the arrangement of any (supposedly) 'early large and frayed' scales, they would cause a serious problem - *drag*. Furthermore, any *lift* gained would act so as to reduce the grip of the feet on the ground. So leaping and gliding would result in two unavoidable adverse effects: loss of ground speed and loss of manoeuvrability. Therefore, such an animal would be better off feeding on ground insects. It is relevant to note that the *goshawk*, as well as being a superb flier, is well known to *run* after small mammals in thick woodland. This it does with its wings *completely folded*. Clearly, extending its wings would cause problems in thick brush, but the point is that such a fast runner as a goshawk does not *need* any feathers or wings in order to catch a meal. And since the fast bipedal dinosaurs must have been equally capable on the ground, there is no clear advantage that any 'frayed scales' might impart, on which natural selection might act.

The third problem is that, in addition to the normal forward and backward shoulder movements found in reptiles, any proto-bird dinosaur would have had no ability to lift its forelimbs above the back. Furthermore, as with all reptiles, it would have had puny breast muscles. Whilst a bird can take off by running and flapping, nevertheless, without the ability to lift the forelimbs high above the back (the upstroke), and without powerful breast muscles to produce an effective downstroke, take-off would have been impossible. Apart from the 2.5m-long *Unenlagia* (which was clearly fully a bipedal dinosaur, too large to become airborne, and too large to bother with prey as small as insects), no other dinosaur is known to have had the necessary anatomical arrangement at the shoulder to lift the forelimbs upwards. To what use unenlagia put its mobile shoulder joints is unknown, but since unenlagia lived 30my *after* archaeopteryx, in all respects that issue is likely to be of no relevance to the origin of birds.

There are two other anatomical features to mention with regards to birds, the *furcula* (or wishbone) and the *supracoracoideus* muscle. The furcula is formed by the fusion together of the collar bones (clavicles) and provides anchorage for the origin of the supracoracoideus muscle:

> [The furcula] is a rather distinctive feature in birds, particularly in flying ones (it can be reduced or absent in non-flying birds)...virtually no other organism has a furcula; it occurs in all birds and is known in two oviraptorids...*Oviraptor* and *Ingenia*. David Fastovsky & David Weishampel. 'The evolution and extinction of dinosaurs'. Cambridge University Press. 1996. p297.

However, recently a number of terrestrial dinosaurs with skin covered with fibres or feather-like filaments have been found with a furcula, such as the eagle-sized *Sinornithosaurus* from China. Since all other bipedal dinosaurs managed without a furcula, we can safely conclude that as they ran about on the ground, it could not have been essential either to oviraptor, ingenia or sinornithosaurus. It follows that since a furcula could have been of no particular advantage to a running dinosaur, there would have been no mechanism for it to have arisen by natural selection.

The downward movement of both the bird and bat wing is effected by action of the large *pectoralis major* muscle. However, the function of their *pectoralis minor* muscle (in birds it is called the *supracoracoideus*) differs. As in other mammals, the bat's pectoralis minor muscle is inserted into the coracoid bone and so it plays no part in elevating the wing: this is effected by the action of the muscles overlying the shoulder blade and the back. In contrast, in birds, the tendon of the supracoracoideus muscle winds round a unique pulley made by the coracoid and clavicle and

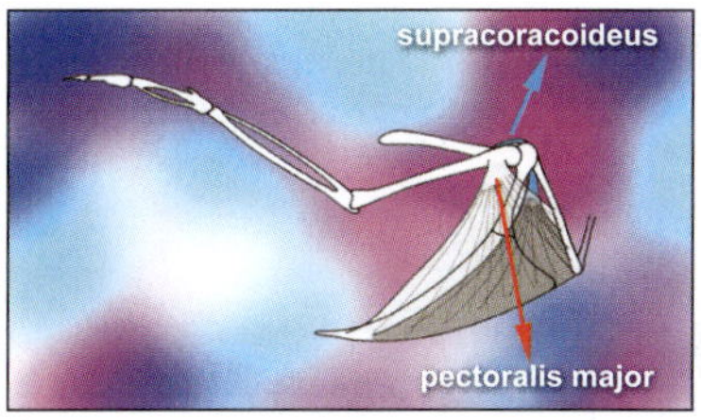


Bird wing showing action of pectoralis major muscle (which pulls the wing downwards) and supracoracoideus muscle (which elevates the wing).

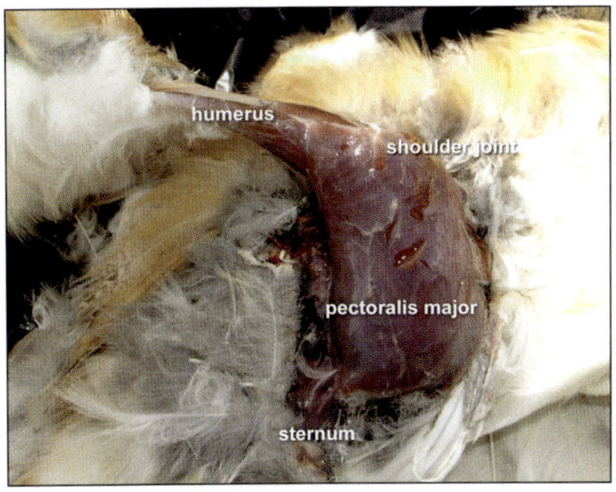


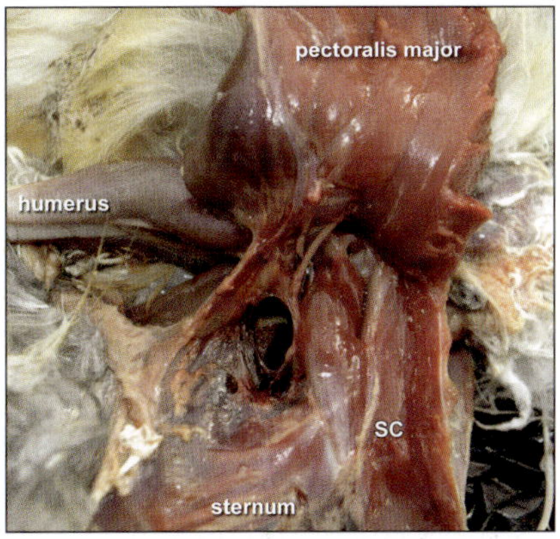


Barn owl dissection.
Left: Showing the right pectoralis major muscle in place. Right: Pectoralis major muscle retracted upwards to reveal the supracoracoideus muscle (SC) underneath.

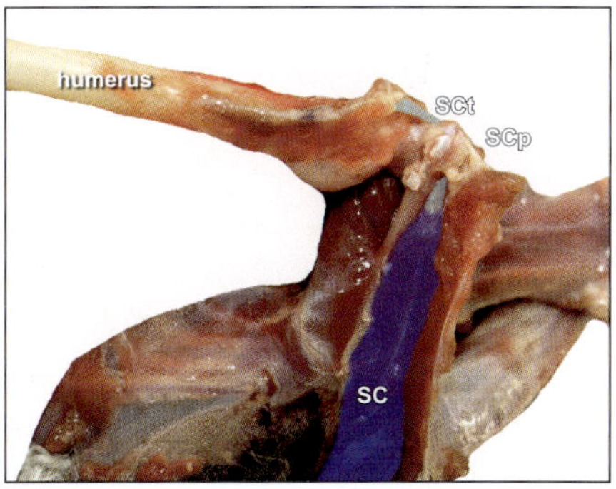


Barn owl dissection showing the right supracoracoideus muscle (SC, purple) and its tendon (SCt, pale blue) passing through the supracoracoideus pulley (SCp).

**Different muscles have different types of nerves**

There are a number of different types of muscles in the body. For example, the small muscles inside the hand have what are called 'white fibres' which are not powerful, but they do allow the fingers to move rapidly and repeatedly. That is why a person can play very rapid sequences of notes on the guitar over a long period of time, without fatigue. In contrast, white muscle fibres are of no use, for example, in the back, where powerful, slow or persistent contractions are necessary to maintain posture. Here, 'red fibres' are found. Actually, the fibres inside the muscles related to posture take turns to contract, so that there are always some fibres that are contracted while others are relaxed. The important point is that whether a muscle has white or red fibres is not dependent on the location or function of the muscle, but is dependent on the nerve that supplies the signal for it to contract. This can be proved by simply dissecting a nerve from a muscle containing white fibres, and re-routing and transplanting its end into a muscle with red fibres (or vice versa). What happens is that the red fibres gradually acquire the appearances and physiological characteristics of white fibres. The opposite happens if the experiment is reversed.
Now, this has crucial importance when we come to discuss bird origins. Even if a small, hollow-boned dinosaur managed to acquire a series of mutations such that its forelimbs developed into wing-like structures, and even if, in addition, it managed to acquire a series of mutations such that its skin developed true feathers, it would still not be able to fly.

*(Continued in opposite margin)*

is inserted into the upper part of the *humerus* (the upper arm bone).
Whilst the supracoracoideus muscle is not essential for flapping flight, nevertheless the ability to take off is severely affected without it:

> Modern birds can fly if that tendon is cut; however, as noted by ornithologist R. Raikow, take-off is greatly impaired. David Fastovsky & David Weishampel. 'The evolution and extinction of dinosaurs'. Cambridge University Press. 1996. p298.

Clearly, if birds evolved from some dinosaur which did not have a supracoracoideus pulley system, and in which the pectoralis minor was inserted onto the coracoid bone, we must ask: how could such creatures fortuitously gain mutations so as to produce the pulley system and cause the pectoralis minor tendon to migrate so as to become inserted into the humerus? Actually, we have no evidence *at all* to show that either one or even a hundred mutations would be necessary for such a change to occur. In fact, it is pure speculation to suggest that such a change is even *possible*.
Whether archaeopteryx (by virtue of its limited shoulder anatomy) only fluttered around on the ground, or whether it managed to take off from trees, the very presence of its fossils shows that it was able to survive without needing to be a proper flyer, and it certainly managed without a supracoracoideus pulley system.
We have seen that the *Galapagos Cormorant Principle* states that:

Disuse *per se* does not result in degeneration of a structure or function. Mutations occur as an inevitable consequence of random chemical substitutions within DNA. However, if a mutation results in a degeneration of a structure or function, *and if that structure or function is no longer essential for the survival of the affected individual*, then the individual might survive and pass on to its offspring that mutation. The progeny will then have degenerate structures and functions not because of disuse, but as a consequence of the prior mutation.

Thus, rather than fortuitously acquiring mutations that might code for the manufacture of the supracoracoideus pulley system, in time archaeopteryx would more likely experience deleterious mutations so that its pectoralis minor muscle, having no pulley system to elevate the wing, and being unnecessary for the downstroke (provided adequately by the pectoralis major muscle), would become increasingly wasted in subsequent generations. In other words, it is unlikely that descendants of archaeopteryx could have evolved a supracoracoideus pulley system. It is more plausible that natural selection would have used, and modified slightly, muscles and systems that already existed. In the case of elevating the wing, 'all' that would have been needed in a supposed ancestral dinosaur was for some mutations to make bigger a few of the muscles overlying the shoulder blade and which attach to the humerus. However, there is no evidence that this ever happened either.
Bats manage without a supracoracoideus muscle because they take off from the hanging position, and a bat on the ground is a very vulnerable creature. So, even if we suppose an oviraptorid was duck-sized and acquired feathers, without a supracoracoideus muscle it would most likely not be able to take off from the ground.

And with its back legs and feet designed for running rather than climbing, it could not have been able to climb a tree in order to propel itself into the air.

There would also have been no purpose in a dinosaur *flapping* its front legs. Flapping requires movement of the forelimb in an inwards and outwards manner - which would have been virtually at right angles to the forelimb movements in either a quadruped or a biped dinosaur, where for the purposes of either weight-bearing or grasping prey, the main movements at the shoulder joint would have been forwards and backwards. And, from the bone structures, we can tell that the pectoral muscles of terrestrial dinosaurs were relatively puny and that they had no significant role to play in locomotion. Indeed, the pectoral muscles are relatively puny in all non-flying animals. In birds:

> The chief muscles of the shoulder are...a massive set serving to raise and lower the wing. There is little development of the muscles present in other vertebrates for the purpose of balance and drawing the limb backwards and forwards for standing and locomotion... J Z Young. 'The life of vertebrates'. 3rd edition. Clarendon Press. 1991. p335.

A number of other alterations to the forelimb bones, muscles and ligaments of a bipedal dinosaur would have been needed to convert it into the wing of an archaeopteryx. Whilst the forelimb of a bipedal dinosaur hung down flexed at the elbow and at the wrist, and extended forwards in use, in contrast, in the resting position, the bones of the bird wing are folded and held neatly against the body. When in use, the wings of a bird are extended *sideways,* and importantly, once extended, are held stable in the flying position by special ligaments and without any further muscular effort.

The completely terrestrial running dinosaurs oviraptor and ingenia both had a furcula but neither had a supracoracoideus pulley system or feathered wings; archaeopteryx had a furcula and feathered wings, but did not have a supracoracoideus pulley system; and the Galapogos cormorant has useless stubby wings with asymmetrical feathers, but its mutation has not interfered with the formation of either the furcula or the supracoracoideus muscle, both of which are retained as normal structures. Therefore, the furcula, the supracoracoideus pulley system and feathered wings are all *genetically independent* of each other. In addition, other bones and ligaments are controlled by many sets of genes, many of which act independently of each other. This being so, the *WYWYWG Principle* shows that the bird wing is unlikely to have evolved in a stepwise process.

The fourth main problem with the running-flapping theory is that from a statistical point of view, if proto-feathers were ever to have arisen, the chances are that they would most likely have appeared in a creature that could have derived no benefit or advantage from them. Then, using parsimony, having arisen once, and since feathers are very complicated structures, they would most likely *never* appear again, in any other creature (whether or not that creature would potentially benefit from feathers). Even in the hypothetical scenario where the only food left comprised flying insects, that circumstance *per se* would not, and could not, induce the necessary genes to

*(Continued from opposite margin)*
The reason is that its slow-twitching pectoral muscles (with red fibres) would never be able to twitch rapidly enough to flap, because their nerve supply would remain the same, being geared up for relatively slow postural movements.

For the dinosaur pectoral muscles to become rapid twitchers, their nerves would have to change. However, if an (implausible) series of chance mutations altered the nerves to the pectoral muscles before the dinosaur acquired its wing-and-feather-developing genes, there would be no advantage, and in time, the nerves would acquire degenerative mutations, so there could be no evolution.

Similarly, if a dinosaur, through an (implausible) series of chance mutations, acquired wing-and-feather-developing genes before it acquired mutations to cause the pectoral nerves to stimulate the pectoral muscles to become rapid twitchers, there would be no advantage, and in time, wing-and-feather-developing genes would acquire degenerative mutations. So, again, there could be no evolution. Therefore, on the basis of nerve-muscle function alone, a dinosaur could not evolve into a bird.

Grey squirrel

appear, so as to allow flight. It would be just as likely that (for example) crocodiles would occasionally produce mutant offspring with feathery skin instead of scales. And most likely, such skin would do them no harm, would provide no advantage, nor would it be selected against, since after all, neither feathers nor hair cause a problem to swimming birds and mammals. So, whilst in all manner of books, the running and flapping scenario is repeated and promoted as an explanation for the origin of bird flight, it remains a fact that:

> ...scientists...have been unable to satisfactorily model the running-to-flight transition in early birds. David Fastovsky and David Weishample. 'The evolution and extinction of dinosaurs'. Cambridge University Press. 1996. p313.

The second popular theory regarding the origin of bird flight holds that perhaps a number of mutations may have resulted in primitive feathers appearing in some arboreal dinosaur. Perhaps after initially jumping from branch to branch, or from a branch to the ground, its progeny became first gliders and later on, true flapping fliers. The attraction of the gliding theory is two-fold: firstly, that an animal at every one of the hypothetical stages from walking through to flying (i.e. plummeting, parachuting, gliding and flapping) could function fully and without impediment; and secondly, that by leaping off some elevated position, the problem of take-off is taken care of.

At the core of the gliding theory is the notion that a *gliding* animal represents an intermediate between a *fully terrestrial runner* and a *fully flapping flyer*. But such a theory has a number of flaws.

Firstly, all known gliders are *quadrupeds* and have skin (a *patigium*) stretched between their otherwise conventional front and back limbs. In fact, as with their non-gliding cousins, modern gliding animals have no significant modifications to their bones, joints or muscles at all. For example, the *colugo* is a rabbit-sized glider from the Philippines, and has a membrane which stretches between all four limbs and the tail, between the toes and also between the front limbs and the neck. Some think that the colugo gives an insight into how gliding animals evolved by gradually lengthening their fingers and developing membranes between the digits and so eventually becoming bats. However, the bones of the colugo's toes are normal and not elongated, and this animal has not even started on the road to becoming anything like a bat. The difference between a terrestrial-quadruped-that-sometimes-glides and a bird that sometimes glides is enormous. So much so that the modifications that would be required to make a terrestrial-quadruped-that-sometimes-glides into a bird are as great as the modifications that would be required to make a plain-terrestrial-quadruped-that-does-not-glide-at-all into a bird. So in effect, a gliding quadruped has not even started on the road to becoming a bird.

Secondly, in order for a glider to become a bird, it would have to attain powered flapping flight. However, in order for gliding to be effective the patigium has to be held taut, and flapping simply destroys the effectiveness of the membrane resulting in a reduction of lift, and stalling. So, while birds and bats can easily glide as a break from the rigorous demands of flapping, gliding animals *cannot* flap. Thus the glider

Gliding vertebrates.
Above top: Colugo.
Above middle: Lizard (*Draco*). Some extinct gliding reptiles, such as the *Coelurosauravus*, also had elongated ribs as in the draco.
Above: Flying squirrel.
Unlike the Egyptian fruit bat (below), none of these gliding animals have any modifications of their front limbs.

that hopes to evolve into a powered flapper is confronted with an insurmountable barrier.

Thirdly, since there are far more tree-dwelling creatures which do not glide compared to those that do, when it comes to getting food or escaping from predators, in practical terms, a gliding animal has no particular advantage over its non-gliding relatives, nor is gliding essential to their survival. We know that a number of animals manage to glide with the help of flaps of skin (e.g. the flying squirrel and the flying phalanger, a type of bush baby) or flattened bodies (e.g. flying snakes) or large webbed feet (e.g. parachuting frogs), but none of these animals catch insects 'on the glide'. Modern gliding animals only take to the air to find a new position or occasionally, perhaps to avoid predators. One common view is that a slight increase in either the bushiness of the tail or the size of a skin flap between the front and back legs would allow an individual to jump a slightly greater gap in the trees. In the normal course of getting around in the trees or escaping from predators, it is thought that any small increase in gliding distance might make the difference between life and death. As a result, some argue that those individuals which could glide the farthest (i.e. those with the bushiest tail or largest skin flap) would more likely survive to pass on their genes to subsequent generations, resulting inexorably in progeny with ever bushier tails and larger skin flaps. In this way it is believed that, for example, ordinary squirrels have evolved into flying squirrels and lemurs have evolved into fruit bats.

The webs between the toes of a gecko would help to soften its landing should it fall from a height. We should expect that those geckos which by normal variation have the most expansive webs would more likely survive a fall than those with lesser webs. However, such parachuting has no bearing whatsoever on the origin of bird wings, feathers and flapping flight.

Whilst at first glance such an idea might seem plausible, the idea assumes that every little increase in gliding distance *will* be translated into an increased ability to survive. Or to put it another way, the idea assumes that the *main* reason why tree-living animals die is because they fail to leap across a particular gap, and so either end up dead on the ground or dead in the grasp of a predator. These can easily be shown to be false assumptions: most other arboreal creatures living in the same habitats as flying squirrels, marsupial sugar gliders and the like, and which are at risk from the same predators, are not gliders. Since successful non-gliding forms of all these animals exist, gliding cannot have any particular overall advantage either in the trees or on the ground. Clearly, gliding is not essential for survival, and since a variety of members of a population, whether gliders or not, would survive to breed, there would be a mixing of genes and there would be no inexorable trend towards gliders in subsequent generations.

Whilst, within a population, inbreeding of a variable characteristic (such as bone length) can result in progeny with different bodily proportions, nevertheless, the progeny remain fully *the same creature* and have no *new* structures. For example, inbreeding amongst tall people can result in individuals with long fingers - but those fingers will retain the same general proportions in relation to other bones. For a colugo to have progeny with massively long fingers with markedly disparate lengths (as in a bat), and a membrane with aerofoil (as opposed to parachute) characteristics, would require a whole series of fortuitous mutations. The *WYWYWG Principle* shows that this is implausible. So, on all counts, the idea of a gradual increase in gliding ability resulting ultimately in true flapping flight is flawed.

I have here discussed the gliding-to-flapping theory because without proper critique

it appears repeatedly in popular writing and on television. However, since we have seen that a bird wing evolving from a dinosaur with a patigium (whether plain or feather-covered) is implausible, it is likely that neither extinct nor living gliding animals have any relevance to the problem of bird origins. Feathers and wings are a completely different matter which will be addressed later.
Finally, there is no way that a typical terrestrial bipedal dinosaur could get up a tree in the first place, and furthermore, there is no fossil evidence to show that any bipedal dinosaur ever had any tree-climbing characteristics:

> There is...no evidence for an arboreal proto-bird, no evidence for climbing adaptations, and no evidence in the skeleton of any non-avian therapod for arboreal habits. David Fastovsky and David Weishample. 'The evolution and extinction of dinosaurs'. Cambridge University Press. 1996. p313.

A look at the pelvic bones of dinosaurs gives one example of the problem. Dinosaurs are classified into two main groups: *Saurischia* (reptile-hipped) and *Ornithischia* (bird-hipped), depending on the shape of the pelvis.

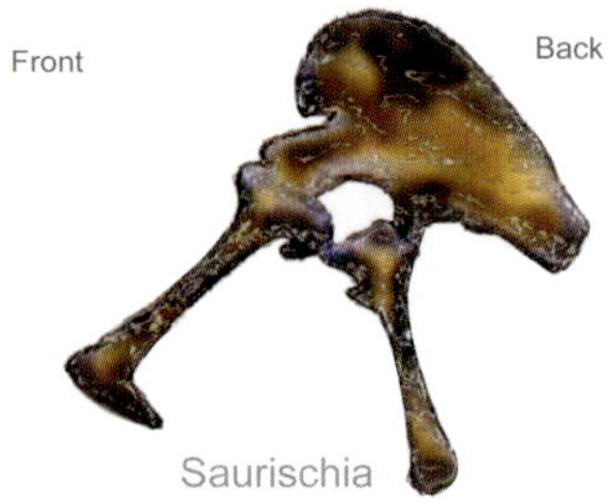


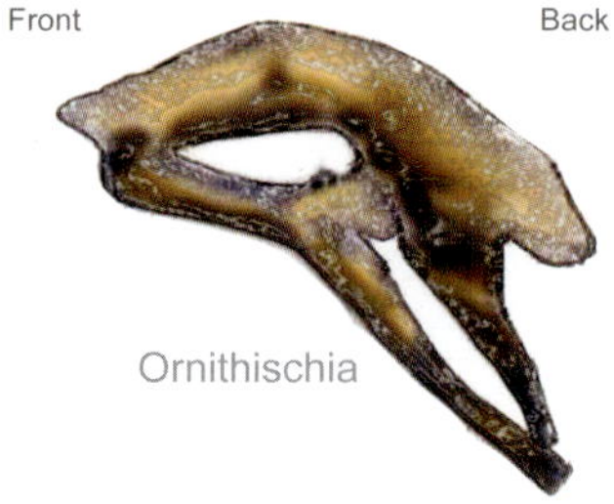


The Saurischia (reptile-hipped) and Ornithischia (bird-hipped) pelvic forms.

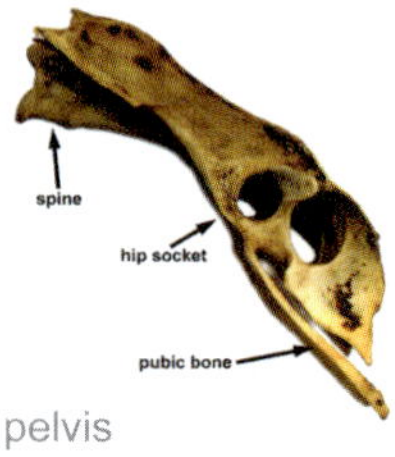


Barn owl pelvis

Emu pelvis

In its head, pelvis and tail, the archaeopteryx (above left) closely resembles a bipedal dinosaur such as Compsognathus (left).
The pelvis of living birds such as the pigeon (top right), barn owl (top left) and emu (bottom left) are quite different.
(Pelvic bones are shown in red, pigeon pygostyle in green)

It is thought that since birds are bipedal, their ancestry must lie with the therapod bipedal dinosaurs. However, therapods belong to the reptile-hipped group, and the dinosaurs with more bird-like hips comprise bulky quadruped animals such as *Triceratops*, *Camptosaurus* and *Stegosaurus* - clearly a group which has nothing whatsoever to do with birds.

So both the gliding and the jumping theories have little to commend them at all. One other theory could be mentioned here, simply to show that once it is accepted (without any question) that birds evolved from reptiles, people are prone to come up with *any* idea as to how the transition was made. If early birds are assumed to have first evolved feathers only on their front limbs:

> A recent theory is that the enlarged feathers on the back of the hands may have served as nets to assist in catching insects. J Z Young. 'The life of vertebrates'. 3rd edition. Clarendon Press. 1991. p390 (see also fig 17.4, p391).

Clearly, feathers allow flight because air *cannot* pass through them. So, without gaps in their structure to allow air to pass through, any (supposed) early proto-feathers would have acted as *fans* rather than *nets,* and as a result, any flying insect would simply be blown away rather than caught. Of course, the theory requires that early in their evolution, the primitive feathers *would have* allowed air to pass through and so could have served as 'nets'. However, the problem is that once the feathers had evolved just a step further, yet at a time when useful flight was still unattainable, then the 'nets' *would* have become 'fans' and the creature would have become once again no better at catching flying insects than its ancestors. And because of the resultant drag the creature might be less able to escape predators. So, having no survival advantage, such a creature could not have evolved further.

Since neither flapping of the front limbs and running, nor climbing and gliding - practised for however long a time - will make any necessary anatomical or physiological changes appear so as to allow flight, then discussing the origin of birds by simply discussing fossils or the mechanistic origin of flight is totally missing the point. Birds have a number of unique features, and feathers represent simply one facet of what makes a bird a bird. In order for a dinosaur to *become* a bird, a *vast* amount of remodelling would need to be done to its anatomy and physiology.

We could enumerate a number of distinct changes that would have had to be made to convert a terrestrial dinosaur into archaeopteryx (for example, feathers, a furcula, large breast muscles and flapping ability, and tree-climbing ability), and further changes that would have had to be made to convert archaeopteryx into a modern bird (for example, the supracoracoideus muscle and pulley system, fusion of wrist bones, loss of wing claws, and loss of teeth). In addition, the modern bird has a 'through flow' lung system and a 4-chambered heart. No discussion about bird evolution can have any validity without accounting for these all-important features, which we will examine after looking at the evidence for feathered dinosaurs.

Left: Allosaurus
Right: Moa

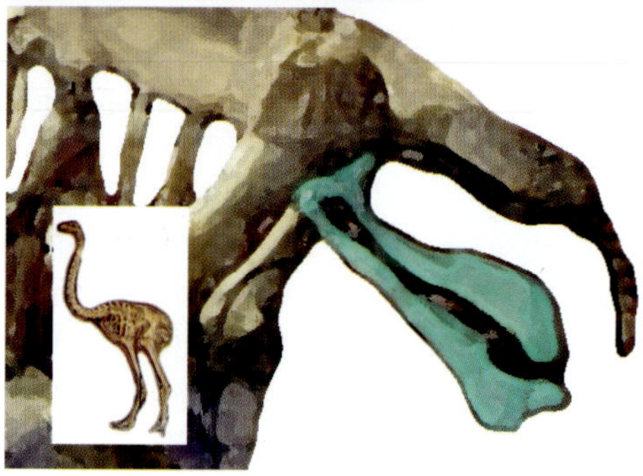

Left: Iguanodon
Right: Dodo

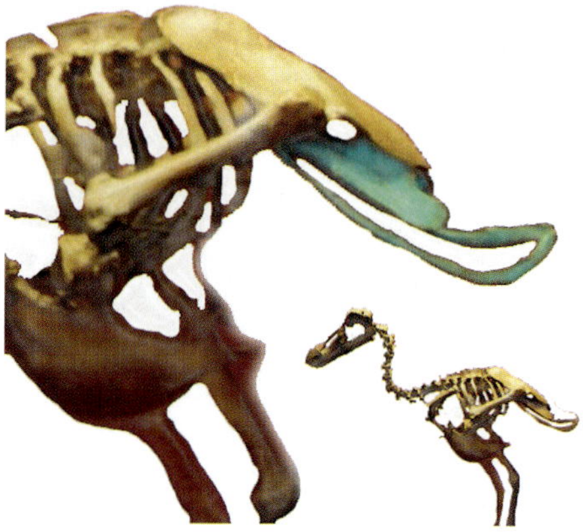

## Dinosaur pelvis types

Gallimimus

The only dinosaurs with pelvic bones that in any way resembled bird hips were the bird-hipped dinosaurs such as Stegosaurus (below), none of which are candidates for having been bird ancestors. Bipedal dinosaurs and the huge diplodocus all had pelvic shapes which were quite un-bird-like (series on left).

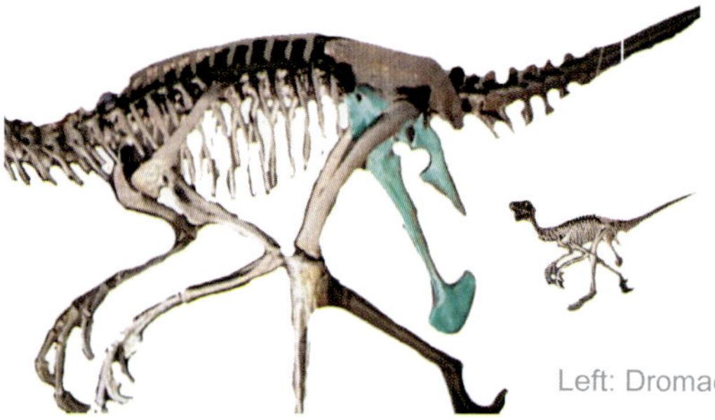

Left: Dromaeosaurus

Stegosaurus

Left: Diplodocus

### *A brief history of feathered dinosaurs*

A number of fossils have now been found that are reported as showing that dinosaurs had feather precursors, or even true feathers. It must be noted at the outset that whenever something is described as a 'proto-feather', this is not based on evidence that demonstrates that such a thing evolved into true feathers. Instead, it is an *assumption* based on the notion that 'since we (evolutionists) know that evolution happened, structures simpler than feathers must have given rise to feathers'. Also recall that the *Coelacanth Principle* shows that no evolutionary trend can be made from the fossil record. So the finding of 'proto-feathers' in a dinosaur found in sediments that are (supposedly) older than those in which a creature with true feathers is found, does not mean that the creature with 'proto-feathers' was ancestral to the creature with true feathers.

In addition, since the anatomy of flightless birds is obviously closer to the terrestrial dinosaur than that of flying birds, in some cases 'dinosaurians' with unequivocal, actual (as opposed to presumed or interpreted) feathers may just be extinct flightless birds. Therefore, we can't always be sure exactly what they were without more fossil material and better preservation in some cases. It is entirely possible that some of these creatures were even a separate and distinct group: in other words, neither like typical birds, nor like dinosaurs.

Furthermore, selling fossils is Big Business and it is very tempting for fraudsters to take advantage of people who are keen to make a name for themselves or sell magazines. With this in mind, any fossil from China and, in particular, any of the following fossils that purport to be of a flying or feathered dinosaur, demand *very* careful scrutiny. Here is a brief summary of a selected number of (supposed) 'feathered dinosaurs' that have been found:

**1996 Sinosauripteryx**, a turkey-sized dinosaur found in (supposedly) 140myo deposits in Laoning, China, was hailed as the first running dinosaur to be found with 'proto-feathers'.

**1999 'Archaeoraptor' fraud.** National Geographic (November 1999) proudly published a major article on the finding of the fossil of a dinosaur from Laoning which was called 'archaeoraptor'. It supposedly had the arms of a primitive bird and the tail of a dinosaur. However, after the publication blitz, the fossil underwent further examination and was found to be a clever fraud involving the glueing together of a number of different and unrelated dinosaur and bird fossils. (See New Scientist. 29th January 2001).

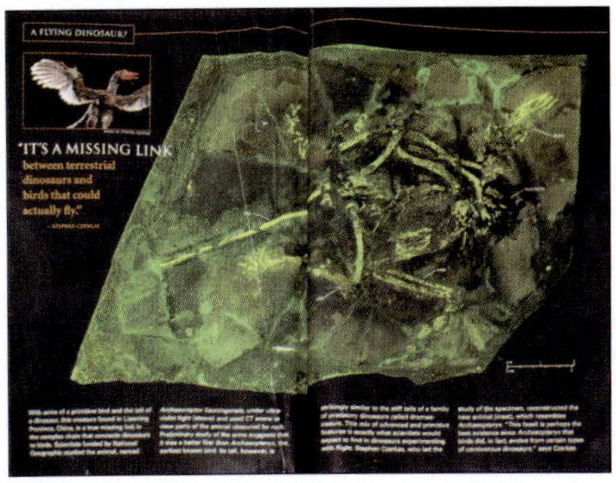


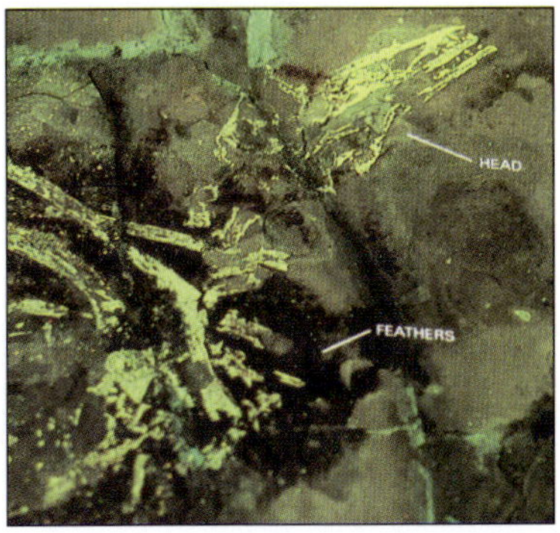


'Archaeoraptor' fraud, 1999 (far left), with close-up of the feathers (left).

**2000 Microraptor**, a small adult dinosaur at 39cm long, of which 24cm was a long straight tail, was found in the (supposedly) 124myo Yixian formation in China. It had long, thin filaments preserved around its body, described as 'dinofuzz'. Both halves of a slab were found with the bones sandwiched between them, so it seems genuine. The foot structure suggests microraptor could grasp branches. (My parenthesis). From: Jeff Hecht. 'Micro-raptor'. Newscientist.com. December 2000.

Juvenile dromeosaur

**2001 Unnamed** (supposedly) 130myo fossil of a juvenile dromeosaur a half-meter long. Downy fibres covered its head and tail, and tufts of filaments sprouted from other parts of the body. Branched structures resembling modern feathers grew on the backs of the animal's arms. Again, both halves of a slab were found with the bones sandwiched between them, so it seems genuine. (My parenthesis). From: Jeff Hecht. 'Warm and fluffy'. Newscientist.com. 25th July 2001. Referring to description in *Nature*. Vol 410. p1084.

**2002 BPM 1 3-13**, a pheasant-sized dromeosaur fossil from the Jiufotang rock formations near the town of Chaoyang, China. The fossil is reported to have 'unequivocally modern feathers, with a single rachis down the middle'. From: Jeff Hecht. Dinosaur reveals fully formed feathers. Newscientist.com. 6th March 2002. Referring to description in *Nature*. Vol 416. p36.

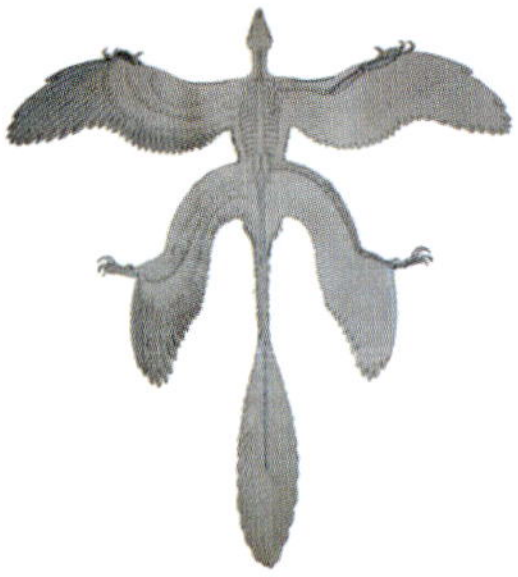
**Microraptor gui**
The arrangement of the leg feathers seems to be incapable of providing lift, and the feathers would also get in the way when walking or clambering on branches. This naturally calls into question the authenticity of the fossils. Also, in order for it to spread its legs out, its (supposed) terrestrial bipedal ancestor would have had to acquire substantial (but implausible) alterations to the hip joints.

**2003 Microraptor gui**. A set of six fossils (supposedly) around 124 million years old, from Liaoning, China. The fossils clearly show a small dinosaur (77cm) that had flight feathers covering its legs, as well as its tail and arms, forming an extra pair of wings, with the most extensive coat of feathers ever seen on a dinosaur. The feathers at the ends of its arms and legs are twice as long as those close to its body. Usually animals have gliding surfaces that narrow, rather than widen, the further they are from the body. The feathers on the legs are particularly baffling, as they cannot spread wide enough to generate much lift. (My parenthesis). From: Jeff Hecht. 'Four-winged dinosaur makes feathers fly'. Newscientist.com. 22nd January 2003. Referring to description in *Nature*. Vol 421. p335.

The arrangement of the leg feathers seems to be incapable of providing lift, and since the feathers would also get in the way when walking or clambering on branches, this naturally calls into question the authenticity of the fossils.

**2004 Dilong paradoxus**, a tiny, earlier cousin of Tyrannosaurus rex sported at least a partial coat of hairlike feathers, found in 130myo deposits from Liaoning. The dinosaur chased prey and roamed the lakeside forests of Liaoning Province in northern China (supposedly) some 130 million years ago. (My parenthesis). From: John Roach. 'New Dinosaur Discovered: T. Rex cousin had feathers'. National Geographic News. 6th October 2004.

**2005 Hagryphus giganteus**, a 7-foot tall dinosaur found in (supposedly) 75myo deposits in Grand Staircase-Escalante National Monument, USA. It had a strong toothless beak and large feathers grew on its back end. (My parenthesis). From: 'Feathered 7' dino'. Associated Press. 4th April 2006. Referring to description in: L Zanno &. S Sampson. 'A new oviraptorosaur (Theropoda; Maniraptora) from the Late Cretaceous (Campanian) of Utah'. Journal of vertebrate paleontology. 25 (4). December 2005. p897-904.

**2007 A velociraptor** specimen IGM 100/981 found in the Gilvent Wash locality near Ukhaa Tolgod, Mongolia, in 1998, was recently re-examined and the ulna bone seems to show clear signs of the insertion pits for long feathers, there being:

'...six low papillae on the middle third of the caudal margin of the ulna. These are regularly spaced about 4 mm apart...these papillae correspond to the quill knobs in living birds'. A Turner et al. 'Feather quill knobs in the dinosaur Velociraptor'. Science. 21st September 2007. Vol 317. No.5845. p1721.

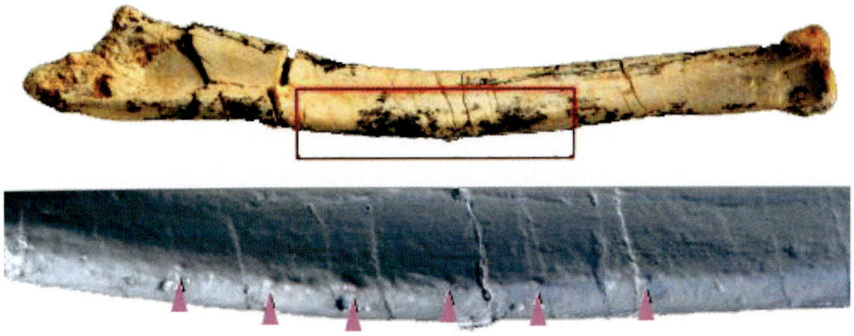

Velociraptor ulna from specimen IGM 100/981, showing signs suggestive of feather quill pits.

With all those reports of the findings of more and more 'feathered dinosaurs', the reader might be led to think that it is a *fact* that a number of dinosaurs had feathers. The American Museum of Natural History admits that 'feathers turn out to be surprisingly complicated' but then continues with a fairy-tale regarding how feathers originated:

Gannet humerus floating on pond water. Bird bones are hollow and some, like this humerus, contain air sacs, making them very light.

> This complicated structure evolved in multiple stages over many millions of years.
> **STAGE 1**: thin, hollow filaments appeared over 150 million years ago.
> **STAGE 2**: tufts of filaments that somewhat resemble down feathers.
> **STAGE 3**: numerous filaments sticking out from a central shaft.
> **STAGE 4**: shaft located off-center; these feathers provide the aerodynamic lift needed for flight.
> (American Museum of Natural History online. 16th April 2008)

The point is that we *know* that is *not* how feathers develop (see later) and so the above 'stages' have *no foundation*, *whatsoever,* in biology. Furthermore, in a new study published in the Proceedings of the Royal Society B, the fossil of sinosauropteryx was re-examined. The findings suggested that the fibres seen in the fossil and previously described as '*proto-feathers*' were probably just probably collagen fibres that supported a frill. The authors of the study, T Lingham-Soliar and A Feduccia, said:

> 'What we have shown is that there's **absolutely no evidence whatsoever** that protofeathers existed in dinosaurs, period.' (My bold). From: Stefan Lovgren. ' "Feathered" dinosaur was bald, not bird ancestor, controversial study says'. National Geographic News. 1st June 2007.

So, the findings definitely show that some dinosaurs had fully-formed feathers, but none show evidence for structures that are '*hoping-to-become-feathers-one-day*'. Despite these findings, museums continue to promote the *filaments-equals-proto-feathers* myth. For example:

> *Sinosauropteryx* had a coating of thin, hollow filaments on its body. These **primitive feathers** may have served to keep the animal warm, much as hair does on mammals. (My bold). American Museum of Natural History online. 16th April 2008.

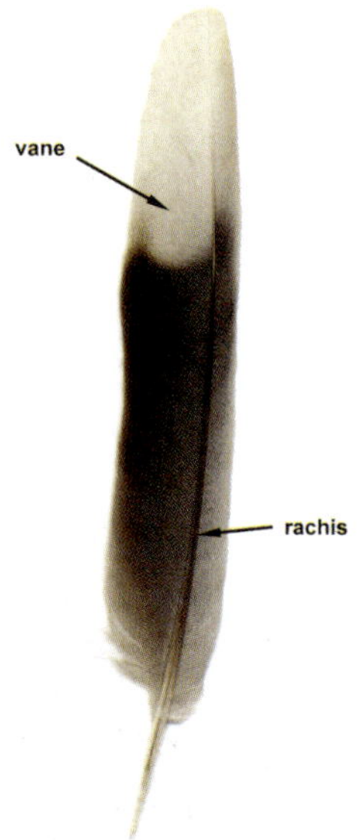


A flight feather showing asymmetrical vanes.

So, what should we make of the many other dinosaurs that appear to have been feathered? It is possible that some of them may just have been non-flying birds:

> Feduccia says these creatures are actually descendants of birds that lost their ability to fly: 'When they become flightless, they superficially resemble small dinosaurs'. S Olson, the curator of birds at the Smithsonian Institute, added: 'The whole notion of feathered dinosaurs is a myth that has been created by ideologues bent on perpetuating the birds-are-dinosaurs theory in the face of all contrary evidence.' From: Stefan Lovgren. ' "Feathered" dinosaur was bald, not bird ancestor, controversial study says'. National Geographic News. 1st June 2007.

In summary, it is clear that there is much money to be made from fossils, and once they found out that they would be paid for the *number* of fossil pieces found, some of the Chinese fossil diggers actually began to break whole fossils into parts to attract more payments. In addition, there is the temptation for fraud (as with the pseudo-creature archaeoraptor) and also for dramatic press releases before a fossil has been properly evaluated, when there is the possibility of associated fame. Nevertheless, while a number of 'feathered' dinosaurs may be frauds, a number *did* have some sort of fibre-like covering (either the remains of decayed skin collagen or hair-like structures). Still others seem to have had true feathers. For example, it seems clear that archaeopteryx is not a fraud and may have been a true feathered flying dinosaur. The feathered velociraptor was clearly not a bird and it clearly did not fly. Finally, it is possible that some fossils described as 'feathered dinosaurs' may, in fact, represent non-flying birds.

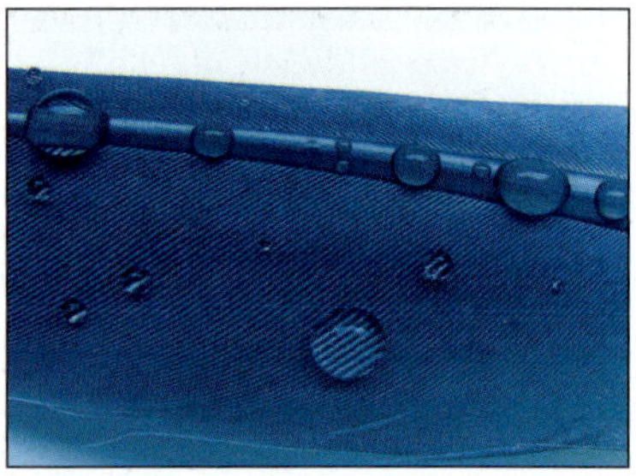

A flight feather.

So, did the feathers of archaeopteryx evolve from the filaments of a non-flying dinosaur, and did those filaments, in turn, evolve from reptilian scales? To answer that we need to look at feathers in more detail.

Juvenile gannet feathers under the wing.

## *Feathers*

Some think that the bird's scaly foot is evidence enough of its reptilian origin. Sure enough, as you follow a bird's foot upwards it is evident that the scales disappear and in their place are found feathers. But in isolation, this simple observation does not mean that feathers arose from scales. As to just exactly how feathers did evolve, the reader will be hard pushed to find even some basic science on the subject. Yet despite this some writers are quite satisfied that:

> ...**the question is**...**easily answered**. Reptilian scales appear to have

> gradually elongated and flattened to support the weight of a flying creature, eventually becoming finely divided (feathery, in fact) to enable them to absorb high pressures without permanent damage. (My bold). Ron Freethy. 'Secrets of bird life'. Blanford. 1990. p14.

Others simply contradict themselves, first stating dogmatically that feathers evolved from scales, then in the same sentence accepting that how this could have happened is a complete unknown:

> Feathers **undoubtedly** evolved from reptilian scales, but lacking fossil evidence, the mode of transition from scale to feather remains a subject for **conjecture**. (My bold). 'Bird'. Microsoft Encarta. Microsoft Corporation. 1994.

So we have to be very careful what we use as our starting point. If we were to *assume* that dinosaurs evolved into birds, then most certainly, as a result of random mutations, scaly skin would somehow *have had to* evolve into skin with feathers. However, if we make no prior assumption, and accept the *fact* of the complete lack of any evidence to show how scaly skin *could* acquire feathers, then we should question seriously whether birds evolved from dinosaurs *at all*. It does not matter for this discussion whether feathers evolved initially for the 'purpose' of flight, or as a means of conserving bodily heat, since the real issue concerns the *mechanisms* (genetic and embryological) of the physical origin of feathers in creatures that previously did not have them, and (more importantly) did not *need* them.

It would be a mistake to consider that the reptilian scale is a simple structure. Reptilian scales are formed and continually replaced by the constant outward growth of the cells of the uppermost part (*epidermis*) of the skin. The whole process of the transformation of the delicate basal skin cells into properly formed scales requires the concerted action of many different genes.

As the cells in the epidermis divide and move outwards, they become dessicated and flattened. However, unless growing and dividing skin cells retain proper cohesion one to another, then the skin surface will simply flake and crack up. And this is exactly what happens in human skin diseases such as *psoriasis* and *eczema* which are in part the result of poor skin cell cohesion.

But the shapes of reptilian scales are not determined purely by the action of local movements of the body and so are not formed simply like cracks in sun-baked earth. Instead, many reptilian scales have very specific shapes which vary from site to site on the body, and which are unrelated to body movements. For example, at some sites scales overlap each other, and at other sites they simply abut against each other. And since the skull and upper jaw of a lizard essentially form a rigid structure, the different shapes and sizes of scales around the mouth margin, eye, upper jaw and skull case cannot be formed as a consequence of any movement of the skin. Indeed, the scales on the head are often bone-like, being partly calcified. The scales along the mid-line of the back and along the tail are often raised, and different from those on the rest of the body. Most lizards have a distinct *lateral fold*, where the scales along the side of the body are arranged in a longitudinal ridge.

It follows that the skin on different sites of the reptile body must have its own spe-

Some think that the bird's scaly foot is evidence enough of its reptilian origin. Sure enough, as you follow a bird's foot upwards it is evident that the scales disappear and in their place are found feathers. However, as in this barn owl, it can be seen that the very tiny feathers on the toes arise from *between* the scales and not from the scales directly.

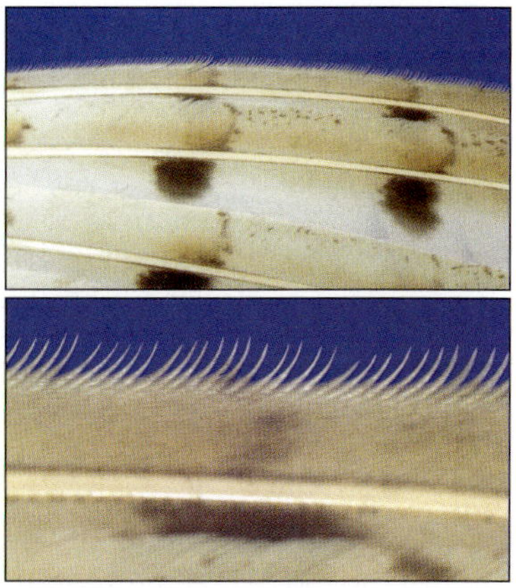

Barn owl feather: the leading edges of the flight feathers of owls are serrated. This allows for quieter flight.

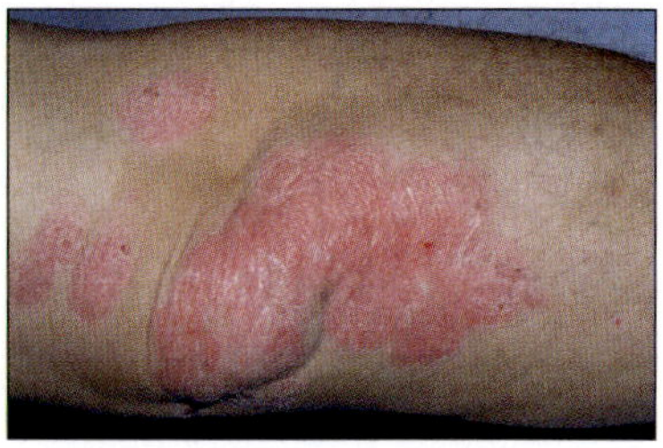

Psoriasis on the elbow. Without treatment, the skin continually flakes off and cracks, leading to sores and infection.

Left: Flamingo feather.
Right: Guinea fowl feather.

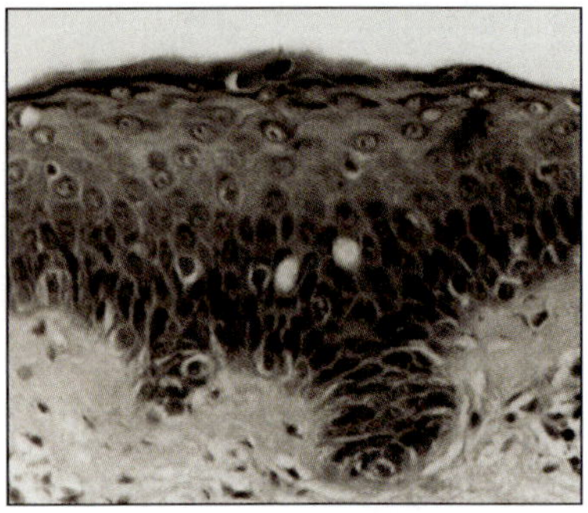

Section through normal human skin showing the gradual flattening of the epidermal cells as they migrate towards the skin surface.

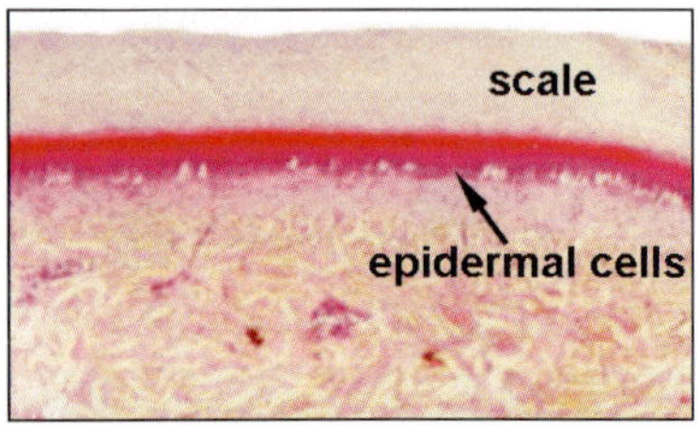


Section through a scale from the leg of a pheasant.

cific genetic programming. The same thing is also true in humans where different areas of skin are affected by different diseases; for example, psoriasis and eczema do not affect the face, and in *hand, foot and mouth disease* (a virus infection) only the lining of the mouth and the skin of the hands and feet are affected. So different genetic controls for the skin of different parts of the body are not confined to reptiles. Indeed, it is likely that the skin of different parts of the body of all vertebrates (including birds) is subject to the same type of diverse genetic control.

But if reptile scales are under diverse genetic control, then we should be able to show, for any particular size and shape of head scale, some significant advantage of one type of scale over another type. For example, we should be able to show that small scales over the brain case are less advantageous than the larger ones that are normally found. Clearly, if the observed patterns of scale sizes and shapes found on different sites of the reptilian body do not have distinct survival advantages, then those distinct patterns *could not* have arisen as a result of the natural selection of scale-less amphibians. Of course, if they had evolved, we should find a whole variety of different scale shapes and sizes randomly distributed on the body of all reptiles.

And if, for example, in a reptile, the skin of the head is under a different genetic control from the skin of the limbs and body, then we should not necessarily expect that a mutation which (supposedly) produced feathery scales on the front limbs of some dinosaur would also produce feathers on its head. Furthermore, if feathers did first appear only on the front limbs of a dinosaur, then since having on the head feathers instead of scales is of no particular advantage to a bird (vultures have bare heads and live and fly perfectly well), there is no obvious selective pressure that would force a (supposedly) evolving bird (such as archaeopteryx) to acquire head feathers. So if birds are evolved dinosaurs, we should find that most (if not all) birds have featherless heads.

Turning back to the origin of feathers, it is still claimed that:

> … the feather…has the same fundamental structure and development as a reptilian scale… Douglas Fatuyma.'Evolutionary biology'. 3rd edition. Sinauer Associates Inc. 1998. p153.

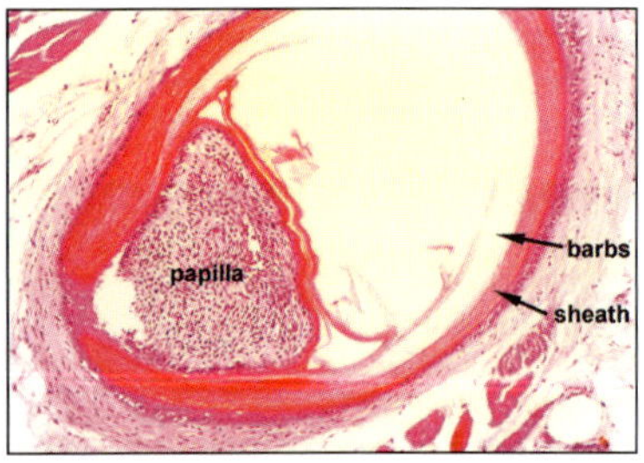


Above: Transverse section through feather sheath.
Below and below right: Longitudinal section through feather sheath.

The truth is that whilst a scale is made of the same material (*keratin*) as a feather, so is a rhinoceros horn, and a finger nail, and a tortoise shell, and a beak. And whilst a scale and any of these other structures might share some common developmental biology, the point is that if it did evolve, a feather would not have evolved *from* a

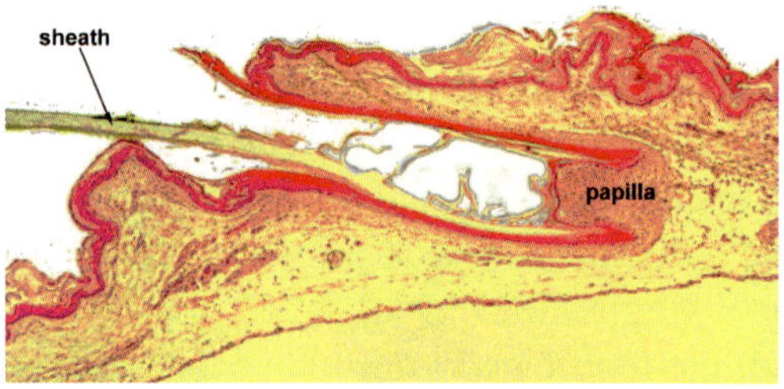


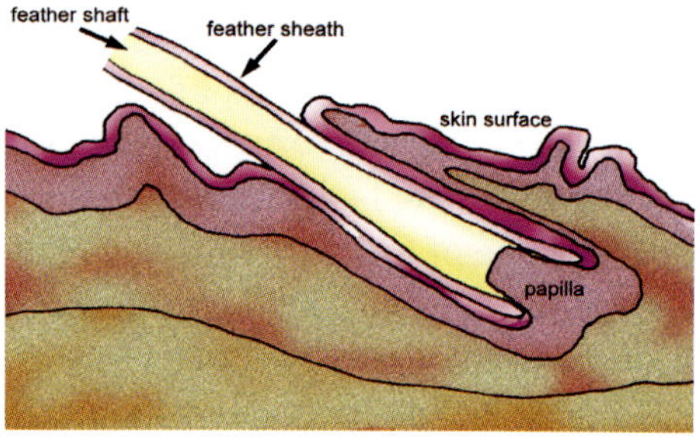


Right and far right: Longitudinal section through feather sheath.

scale. A feather would have arisen as a result of both changes in the materials that make up the scale and changes in the way those materials are utilised. In the same way, although both a moulded plain rectangular bathroom tile and a moulded tile with ornate edges may be composed of exactly the same materials, the ornate tile is not manufactured by first making and then modifying the plain rectangular tile. Whilst it is true that the rectangular and ornate tiles share some common preparatory procedures, it is possible to make an ornate tile *only* by making a unique mould to the specifications required by its design.

So, *if* it did evolve, a feather would have arisen as a result of, on one hand, alterations to genes that code for the utilisation of some of the basic materials that go to make up scales, and on the other hand, the appearance of *new* genes that would code specifically for the completely new structure which is the feather. This seemingly simple distinction is of *fundamental* importance.

The keratin protein actually has a number of forms, coded for by a number of genes, which in humans are located on chromosomes 12 and 17. Keratins are rod-like filaments which are anchored to structures called *desmosomes* which are located along the cell walls of adjacent cells. It is the cohesion between desmosomes of adjacent cells that gives skin (and other tissues) its integrity.

Since there exist within any individual a number of distinct keratins which are manufactured only in certain types of cells, mutations in the corresponding genes can result in diverse patterns of diseases. Furthermore, as we have seen many times before, a mutation of any functioning gene is, if anything, likely to result in some degree of failure of function, or degeneration. So if, for example, a gene might facilitate the utilisation of certain materials causing skin cells to become hardened scales, a mutation of that gene would most likely produce fragile scales, or fragile skin *and* fragile scales. Such skin might be more likely to crack and become infected. Again, the skin diseases psoriasis and eczema give a good idea of how serious such a mutation might be to a reptile.

Even supposing that some individual reptiles acquired some slight 'feathering' to their scales, affected individuals would then still breed with normal individuals, and so mix up the genes. Therefore, inbreeding would most likely perpetuate both the 'feathery' scales and the normal scales within the population.

If the similarity between scale and feather is as great as some make out, then it would be expected that authors provide a number of possible step-by-step models of the process of transformation. However, the reader will not find anywhere any discussion on how the transition was (supposedly) made. As we shall see, a scale is *not* a simple genetic hop away from becoming a feather. Compared to a scale, a feather is an *incredibly* complex structure.

The feather *vane* is curved to form an aerofoil, and in all flying birds (including archaeopteryx) the vanes are asymmetrical, the leading vane being narrower than the trailing vane. In all flightless birds the wing feathers are symmetrical. The vanes themselves are made up of hundreds of *barbs* each carrying thousands of *barbules* which in turn each carry a number of *barbicules*.

In a typical flying bird, the feather-tip side of the barbs carries barbules with 4 or 5 hooked barbicules, while the feather-quill side of the barbs carries barbules which

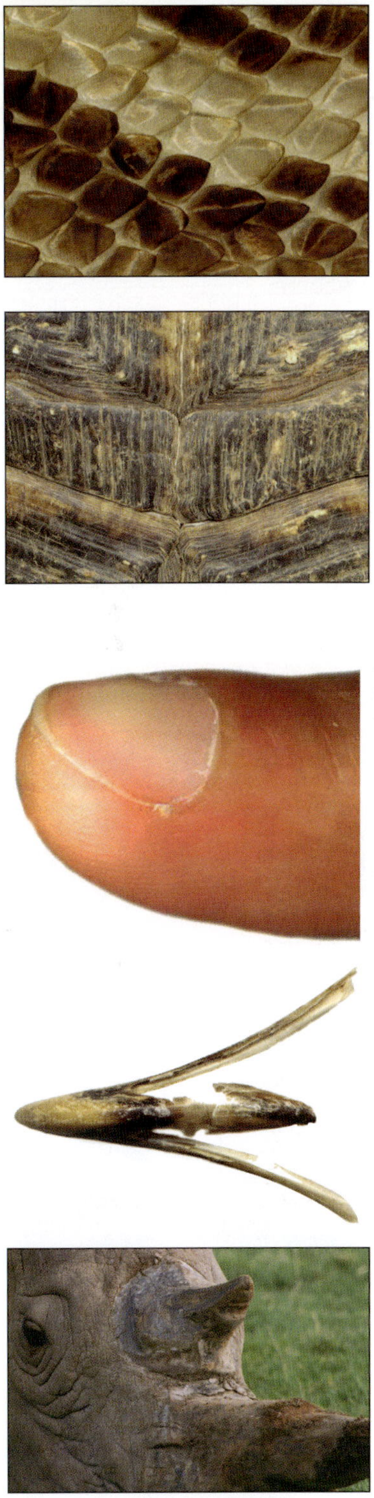

Structures made out of keratin. From top to bottom: python skin, tortoise shell, finger nail, bird beak, rhino horn.

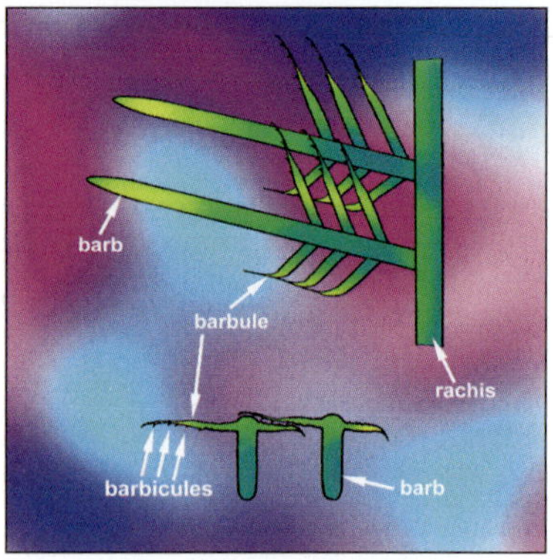


Diagram showing barbs, barbules and barbicules.

have small ridged barbicules. Like *'Velcro'*, the hooked and ridged barbicules from adjacent barbs interdigitate with each other to form a continuous surface. Flightless birds such as the *rhea* and *ostrich* have very few barbicules, so the barbs cannot unite to form a vane.

Unlike the skin over a lizard's forelimb, the roots of the flight feathers are actually *attached to the bones* of the wing. In contrast, the follicles of all the other wing feathers and all body feathers lie only skin deep. So, in some (supposedly) pre-archaeopteryx proto-bird, the follicles of the primary flight feathers would have been unattached to bones and the feathers would therefore have been *flail*, and so would not have allowed for either flapping or gliding flight.

Rhea feather: the barbules cannot zip together.

Gannet ulna showing markings where the flight feathers are attached.

Unlike a single membrane (such as in the bat or pterodactyl wing) which could tear and thus be rendered useless for flight, the vane of the feather has the advantage that if torn, it can be easily restored by bringing the barbules closer together so that they interdigitate once again. This is what a bird does when it is preening. To keep them in good condition, feathers also need to be kept oiled. Most birds have a special gland (the *uropygeal gland*) sited just above the tail from which they can gather oil with their beaks to apply to the feathers. Incidentally, reptilian skin has no glands at all, so the origin of the uropygeal gland is itself inexplicable by natural selection.

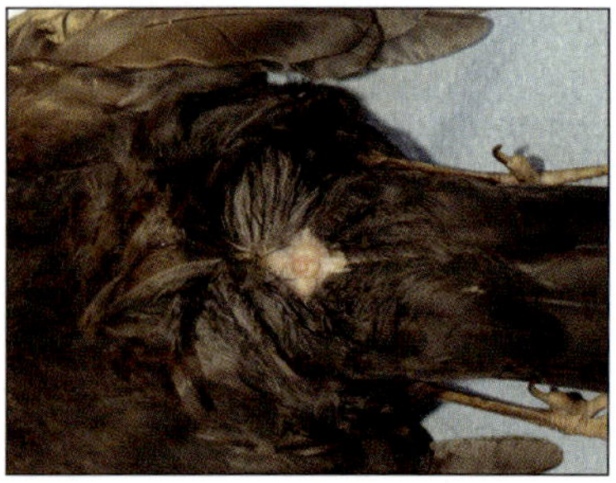

Uropygeal gland in a blackbird.

So how does a feather develop? As we have seen, the reptilian scale is formed from layers of skin cells which grow upwards and become flatter and more dehydrated in the process, until the cells die leaving scales composed mainly of the protein keratin. Although keratin is the main component of both reptilian scales and bird feathers, there the similarity between the structures starts and ends. At first glance, a feather *could be* likened to a modified hair since they both arise out of pits in the skin that are lined with epidermal cells. However, from the cells within the follicle, the feather grows folded as a tube within an outer sheath of keratin (which looks rather like clear plastic). As it grows, the keratin sheath gradually breaks up and the feather unfurls along its length. Although the feather initially develops as a simple tube, even at an early stage it is asymmetrical with its top and underside, leading and trailing edges, all well defined.

Flight feather showing vane with some barbs unzipped.

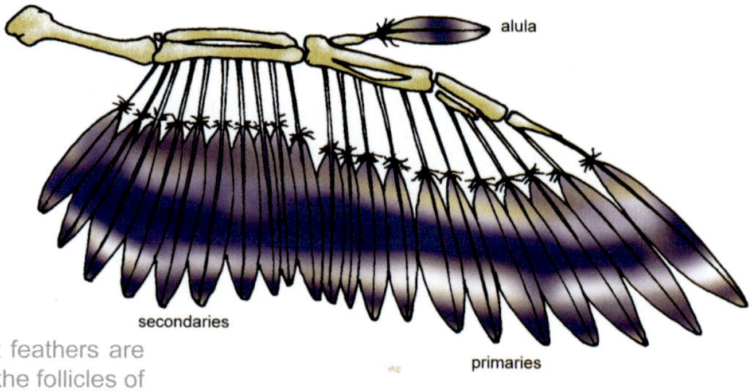


A typical bird wing. Unlike hairs, the roots of the flight feathers are actually *attached to the bones* of the wing. In contrast, the follicles of all the other wing feathers and all body feathers lie only skin deep.

Baby crow with the bases of its flight feathers yet unfurled inside their sheaths.

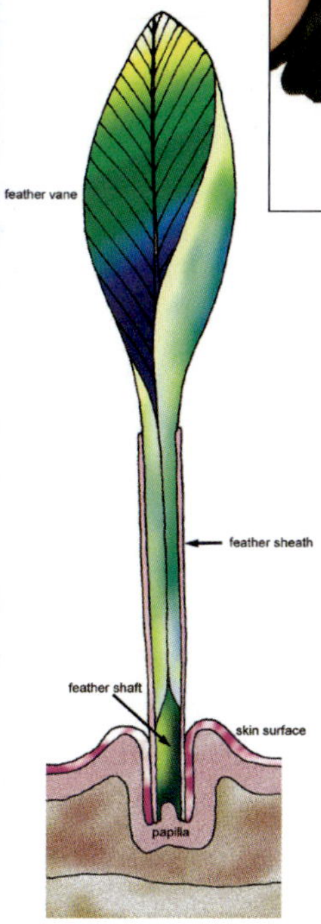


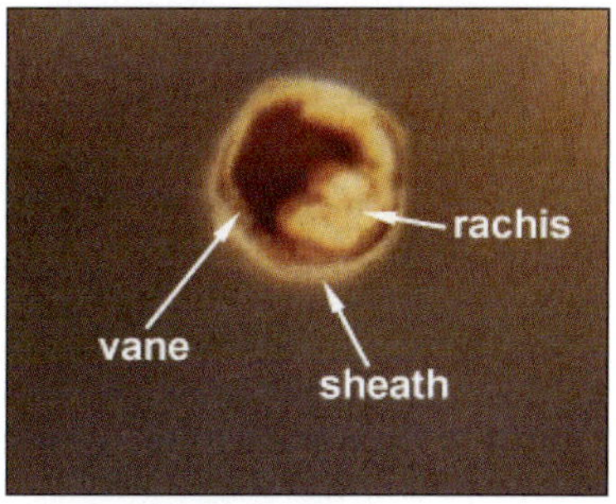


Left, far left and above: The developing feathers emerge by unfurling from within their keratin tubes, which look like plastic sheaths.

Barbs and barbules *do not* develop as a series of branches growing sequentially from a central stem. In fact, the highly specific shapes of the barbs and barbules are the result of differential *cell death* and differential *cell cohesion* within the epidermal cells. The three-dimensional shapes of barbs, barbules and barbicules are, as it were, cut out of the tissue that is the feather tube, much like the cutting of a jig-saw puzzle. In other words, just as highly controlled and specific cell death and cell cohesion (or lack of cell cohesion) result in the formation of fingers out of what was previously a paddle-like sheet of tissue in the embryo, so it is with the formation of the barbs, barbules and barbicules of a feather. So the barbules do not grow out of the barbs like leaves on a branch but both the barbs and the barbules are formed *simultaneously*. Thus the problem of feather development is, in part, one of how to program the 'jig-saw': in other words, how to program precisely the pattern of cell cohesion and cell death.

So, a feather is unique, there is nothing simple about its structure or development, and it cannot be understood as a modification of a reptilian scale.

Furthermore, to complicate matters, the body and the front and back limbs of a bird are *differently* covered. The body has downy feathers while the front limbs have large flight feathers, and the feet and back legs are largely covered by scales.

This is a most curious characteristic since in *all* other vertebrates

Above: Emerging barn owl feather in a nestling. A group of fluffy filaments emerge from the feather sheath and are continuous with the tips of the feather proper which emerges afterwards. By the time a bird is ready to fly most of the filaments will have been shed.
Left: Mature downy body feathers.

Python skin

Water monitor lizard skin.
Unlike birds, which moult their individual flight feathers over a period of time so as not to compromise flight, lizards moult their skin in large portions, while snakes moult their entire skin in one piece.

Slow worm moulting.

This budgerigar is growing a new flight feather. With just a couple of its feathers missing, the bird is unable to fly properly.

the body and both front and back legs have the same covering, showing that the genes responsible for the covering of the front and back limbs of non-birds are probably the same, or for practical purposes, inseparable. And a number of human diseases have a predilection for affecting both the hands and the feet, showing that many of the same genes are active in the skin at these sites of the body.

The distribution over the body of different feather types would have posed a serious problem for any dinosaur that wanted to change into a bird: we should find it surprising if mutations that affected the skin of the body of a dinosaur, such that it developed downy feathers, managed to arise *without* also producing downy feathers on the skin of both the front and back legs; or if mutations that affected the skin of the front legs of a dinosaur, such that it developed flight feathers, managed to arise without producing flight-type feathers on the skin of the body. Of course, any dinosaur with only downy feathers on the front legs and body would not be able to fly. And the *WYWYWG Principle* shows us that it is unlikely that a dinosaur that is completely covered in downy feathers would acquire the *very* mutations which would allow the skin of the body, front and back legs to develop differently, *and which would also* produce the highly specialised primary flight feathers, *only and exactly where they are needed*. The same problems and objections apply to the origin of the specialised feathers of the bird's tail, since unlike birds, in virtually all reptiles and mammals the skin of the tail is covered in the same manner as the skin of the body and legs.

While reptiles moult their skin either in a single piece or in large portions, the individual flight feathers of birds are moulted separately, and in an organised way. It is of prime importance that moulting birds do not lose *all* their feathers at the same time so that flight is not compromised. It is a common observation that a bird can be rendered almost incapable of flying at all by trimming just a few of its primary flight feathers. Similarly, any (supposedly) evolving proto-bird which lacked primary flight feathers would also be unable to undertake either gliding or flapping flight, and would most likely flutter and crash to the ground.

Mention must be made here of the curious *Sinosauropteryx prima*, a bipedal dinosaur found in Liaoning Province in China in 1996. The fossil contains impressions around the neck and along the back suggesting that this creature had a mane of downy filaments. It is not clear whether or not the rest of the body was also covered in the filaments, but this fossil raises the question:

> Could these strange fibers be the earliest examples of bird feathers? Jennifer Ackerman. 'Dinosaurs take wing'. National Geographic. July 1998. p76.

In fact, fossils have been found of a number of other dinosaurs which lived (supposedly) 120mya and which had down-like feathery coverings. For example, *Sinornithosaurus millenii* was a small bipedal dinosaur with long forelimbs and a long tail stiffened with curious thin bony rods. It had a shoulder girdle similar to that of archaeopteryx but:

> ...it could leap but not fly. Stephen Czerkas. 'Feathers for T. rex?' National Geographic. November 1999. p105.

And at 2m in length, *Beipiasaurus* is the largest dinosaur yet found with wispy fibres on its forelimbs.

There is, here, the obvious problem that if feathers evolved from hair-like precursors, why did not feathers evolve from the hairs of the shrew-like mammals (which presumably were warm-blooded and had an efficient four-chambered heart [see later]) that lived (supposedly) over 200mya? In other words, why didn't bats evolve feathers instead of membranes?

But going back to sinosauropteryx prima, sinornithosaurus millenii and beipiasaurus, these were clearly *fully* terrestrial dinosaurs and not birds. And since we know that dinosaurs were a group of animals quite distinct from the reptiles of today, we should not be surprised to find that some of them had what might seem to us, at first glance, unusual features not seen in other reptiles. Again, the *Chalicotherium Principle* reminds us that it is possible for sinosauropteryx to have downy feathers, *yet not to have any ancestor-descendant relationship to a bird*, and furthermore, for it to be a dinosaur, *yet not to have any ancestor-descendant relationship to any other type of dinosaur.*

Above: Hairy frog.
Below: Pangolin scales.
Bottom: Armadillo scales.

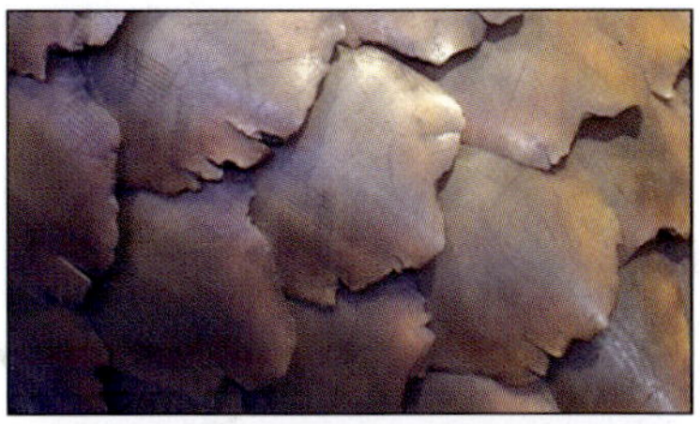

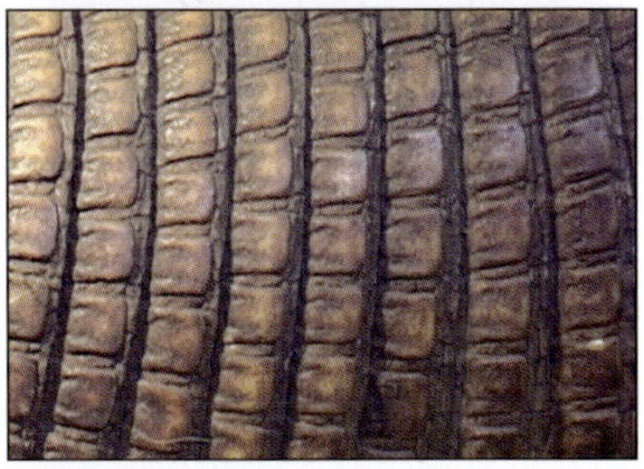

So dinosaurs such as sinosauropteryx more likely represent creatures with different sorts of skin to the lizards we are accustomed to seeing - simply creatures with filaments instead of scales. Indeed, there are living creatures which have different skin features from others of their kind, for example, the *hairy frog* (which has hair-like filaments on its legs and sides during the breeding season), the bizzare *tassled frog-fish* (which has hair-like fringes on its body that mimic a bed of red algae), and the *armadillo* and *pangolin*, both mammals with scales as well as hair. Actually, the bipedal dinosaurs such as sinosauropteryx were not the only ancient reptiles with hair-like fibres on their skin since they are also found in *pterodactyls*:

> Casts have been found showing that there was fur on the legs and wings. However, there is no evidence to suggest that they possessed feathers...
> J Z Young. 'The life of vertebrates'. 3rd edition. 1991. p320.

Of course, if we were to suppose that pterodactyls and sinosauropteryx *had* evolved from ancestors that had only scales, then we would still have the difficulty of explaining how their filaments could have originated. And if sinosauropteryx was some sort of bird ancestor, then its fossils ought to be found in sediments that are older than that of archaeopteryx. Yet sinosauropteryx is thought to have lived in the Cretaceous period 121mya: in other words, 30my *after* archaeopteryx, by which time modern birds such as confuciousornis had (supposedly) already evolved. In fact, the dinosaurs considered to have bird-like features are *younger* than archaeopteryx. Despite this undeniable evidence, authors still cling onto the belief that the *assumed*-more-primitive-looking-but-younger-than-archaeopteryx creatures represent forms ancestral to archaeopteryx. Hence the production of the misleading illustration in the *National Geographic* (July 1998) depicting a series of creatures - sinosauropteryx, velociraptor, unenlagia, caudipteryx and protarchaeopteryx (all of which are younger than archaeopteryx) – all (supposedly) representing traits ancestral to those of archaeopteryx:

> The family tree...is not a chronological progression but rather an illustration of how the traits of the modern wing evolved in different creatures in different locations at different times. Jennifer Ackerman. 'Dinosaurs take wing'. National Geographic. July 1998. p91.

This is no more than wishful thinking. The reason usually given for the anomaly that the majority of the dinosaurs considered to have bird-like features are *younger* than archaeopteryx is that:

> ...fossilisation is a rare event...you can take all the fossils ever collected relevant to this debate and fit them nicely in my office. Hou Lianhai. 'Dinosaurs take wing'. National Geographic. July 1998. p95.

However, such an explanation is misleading since for one (supposedly) 120myo-bird there are plenty of fossils. In fact there are:

> ...hundreds of specimens of *Confuciousornis*. Jennifer Ackerman. 'Dinosaurs take wing'. National Geographic. July 1998. p89.

And the *Intermediary Principle* reminds us that over immense periods of time, more fossils of intermediary forms should be found than of either any ancestor or any descendant alone. So, contrary to the evidence, if birds evolved from dinosaurs, many *non-bird-but-dinosaurs-with-bird-like* features *should* be older than archaeopteryx.

Before leaving feathers, one other issue must be addressed. It is an *assumed* rule that flightless birds (such as the ostriches and emus), with their *symmetrical* feather vanes, evolved from flying birds (which all have *asymmetrical* feather vanes). If the rule were true, in other words, if a mutation in a flying bird with asymmetrical feathers could result in non-flying progeny with symmetrical feathers, then the reverse mutation (a non-flying bird with symmetrical feathers changing into a flying bird with asymmetrical feathers) has an astronomical improbability in appearing since (except in specific cases such as in bacteria, as we have seen in chapter 5) spontaneous reversals of mutations are rarer than hen's teeth. This causes an insurmountable conflict for the evolutionist since it must then be accepted that archaeopteryx did not evolve from a non-flying feathered ancestor with symmetrical vanes.

Since we have no supporting evidence, we must reject the supposition that the ancestors of ostriches once flew or that the ancestors of archaeopetryx were once flightless creatures with symmetrical vanes. So, the finding of symmetrical vanes in either a living or an extinct feathered creature only means that the animal was most likely flightless. It does not mean that those symmetrical vanes were precursors to asymmetrical flight feathers.

Juvenile gannet beak. Like other divers, such as cormorants, gannets do not have nostrils.

### Beaks

All birds have beaks. A common suggestion is that having heavy jaws and teeth would make flight cumbersome or impossible, so birds evolved by replacing the bony jaws and teeth of their (supposed) dinosaur ancestors with lightweight beaks made of keratin:

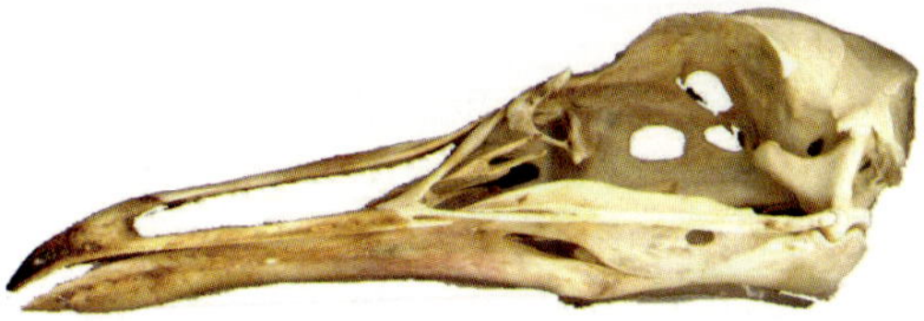

Archaeopteryx skull (right) showing its tiny teeth which, when compared to the herring gull skull with its large beak (above), would have been unlikely to have caused a weight problem during flight.
Note that the keratin part of the beak (above left) has been separated from the bone after death.
The beak of the merganser (left) has horny tooth-like projections to allow gripping of slippery fish.

> The overwhelming priority for every flyer is to keep weight to a minimum...a weighty jaw, laden with teeth, must have been a particular handicap for any creature trying to fly, for it would tend to unbalance the animal and make it nose-heavy... David Attenborough. 'Life on Earth'. Collins/BBC. 1979. p177.

But, it is clear that flight and teeth are *not* mutually exclusive: archaeopteryx was itself a flyer and had teeth; the extinct fully-flying bird icthyornis was about the size of a small gull and it too had a large skull lined with numerous small teeth; bats are extremely agile and have teeth which do not in any way hinder their manoeuvrability in the air; and modern gliding squirrels and gliding snakes are well able to become airborne (and also retain some degree of directional control) without their teeth and jaws being a hindrance.

Although beaks are very lightweight, nevertheless, in some birds this benefit is negated by absurd size. For example, pelicans and toucans have huge beaks, and the beak of the Indian hornbill can be 30cm long, yet all these birds are agile flyers. In fact, these birds do not need huge beaks at all. For example, the large beak of the pelican is not essential for catching fish since many diving birds of a similar size to

As with all birds, the beaks of the toucan (above) and the hornbill (below) are very lightly built and cause no problem to flight.

Above: Kingfisher. Most birds can fly while carrying a substantial proportion of their body weight.
Left: Pelican.

Sarah McAlpine

Ground hornbill (above) and a Tanzanian stork (below) with their large beaks.

Like these geese, most birds feed on the ground or in the water, and do not need to fly at all.

Green woodpecker feeding on the ground.

the pelican have much smaller beaks, yet manage to catch fish quite adequately. Other fish-catching birds often fly with relatively heavy fish held in their beaks or talons (or in their stomachs), and some eagles can fly while carrying prey weighing up to five times their body weight. Clearly then, the ownership of either jaws with teeth or an unnecessarily large keratin beak does not exclude powered flight and manoeuvrability. Furthermore, it may be surprising to learn that feathers are not so light - the weight of a flying bird's feathers can be greater than the weight of its skeleton. The penguin has horny spikes on the top of its tongue which may go some way to substitute for teeth. However, the penguin (which we might think would surely benefit from having teeth to catch fish with), despite having been around for (supposedly) 65my, in all that time has not managed to regain its (supposedly) lost teeth. Furthermore, it must also be noted that all living *non-flying birds* also have beaks, and none have managed to evolve their lost teeth by using (supposedly) dormant ancient teeth genes. Nor have any such non-flying birds evolved new genes and with them, new types of teeth.

So the need to reduce flying weight is not the reason why birds do not have teeth. Nor is diet. There are thousands of examples of animals with jaws and teeth which all survive to this day on *similar* diets (of seeds, nuts, insects, fish and meat) to those of birds, showing that there is no particular advantage of a beak over a toothed jaw.

As we have seen previously, the *variation* amongst Hawiian finches points simply either to the shuffling of, or to mutations within, *genes that were already present in the chromosomes of the parents*. Such observations of variation do not show us how genes coding for beaks could have arisen *when previously there were none*. So it is much more relevant to ask the question: *how* could a keratin beak and the loss of teeth evolve in a line of toothed reptiles? A beak is not just a modification of the scales around a reptile's mouth. A beak has definite form and function and requires, as with all things structural, a template to define it and to guide its construction. Since no beak could ever appear without there first arising the necessary genetic mutations to code for one, this presents a number of problems.

Firstly, how could mutations give rise to the growth of a beak *coincidentally* with either suppression or derangement of genes that cause teeth and jaws to grow? We know that regardless of diet or habit, beaks have no advantage over jaws with teeth, and reptiles that have teeth are not disadvantaged by their presence. Therefore, since in any toothed reptile *any* change towards a beak could confer no advantage, then any such reptiles would not survive and reproduce in greater numbers than their contemporaries which retained their full complement of teeth.

Furthermore, since in any toothed reptile *any* change leading towards losing teeth could confer no advantage, then in this case too, any such reptiles would not survive and reproduce in greater numbers than their contemporaries which retained their full complement of teeth. We have previously seen that a captive toothless Florida crocodile has normal jaws and remains fully a crocodile. We should expect such animals to quickly perish in the wild rather than for their progeny to gain a beak in order to help catch prey. Thus there will be a tendency for natural selection to always favour the status quo i.e. toothed reptiles; and any hypothetical trend towards a beak will probably never become established to the exclusion of teeth. Furthermore, if the whole variety of modern birds evolved from an ancestral population of primitive toothed birds, then we should see as many modern birds that have retained teeth as those that have evolved beaks. But it could be argued that since turtles have beaks and are reptiles, this shows that reptiles are capable of losing teeth and evolving beaks. However, this is not really evidence at all since there is no evidence to show that ancestors of beaked turtles once had teeth and no beaks.

Whilst it is possible for genetic manipulations to produce teeth in chick embryos, this only shows that the bird genome contains within it the capability of manufacturing teeth. Such evidence does not confirm that birds were once toothed dinosaurs but simply shows that creatures that are fully birds can have teeth.

Secondly, as we have previously seen when discussing both crocodilians and whales, teeth and jaws develop under the control of many different and *independent* genes. In other words, a whole series of genetic changes must be made to a toothed reptile to produce a beak. Therefore, since long before the (supposed) origin of any beak, *no* ancient reptile would have had *any* of the genes that could code for a beak, the *WYWYWG Principle* shows that it is most likely that no reptile could have accumulated *all* the necessary genes to produce a beak and to lose all the teeth.

Thirdly, if birds *did* evolve from toothed dinosaurs, the genes for beak production should just as likely have appeared in other unrelated creatures producing, for example, beaked snakes, frogs and mammals. But this is contrary to observation. Therefore, since beaks can confer no particular advantage over teeth, we should expect that in an animal that had teeth within a normal jaw, beaks could not have evolved at all.

There is no evidence to show that ancestors of beaked turtles once had teeth and no beaks.

## The bird heart and lungs

### The bird heart

In mammals, de-oxygenated blood returning from the body passes into the right atrium, from which it passes into the right ventricle and hence into the lungs via the pulmonary artery. Having lost much of its carbon dioxide and gained oxygen in the lungs, the blood returns to the left atrium from which it passes into the left ventricle. The left ventricle then ejects the oxygenated blood into the main outflow tract (the *aorta*) from which the blood is distributed around the whole body.

As in mammals, the heart of a bird also has four chambers - two atria and two ventricles (A2V2). On the other hand, the hearts of all living reptiles have three chambers - two atria and one ventricle (A2V1). All fishes, amphibians and reptiles have two aortic arches which join together to form a single main artery running backwards in front of the spine. Birds, on the other hand, have a single aortic arch on the right side of the body, while mammals have a single aortic arch on the left side of the body. Neither the mammalian nor the bird aortic arch has any advantage, one over the other. So, could a reptilian heart be modified to become like that of a bird?

The single reptilian ventricle receives both the de-oxygenated blood returning from the body (via the right atrium) and oxygenated blood returning from the lungs (via the left atrium), and previously it was thought that this arrangement caused the two streams of blood to mix together. However, in the lizard:

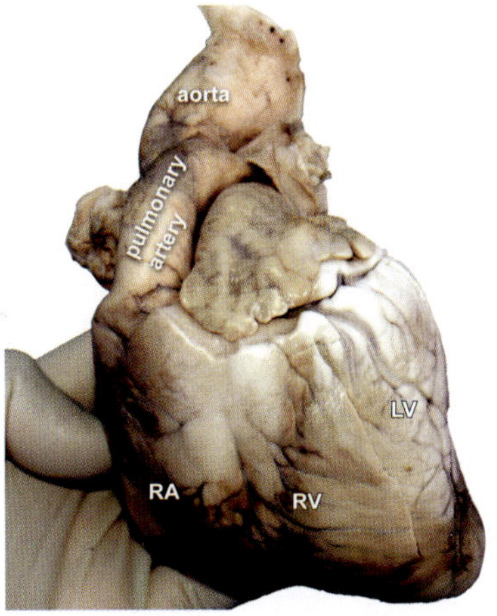


Above and below: Cow heart showing left-sided aorta and two completely separated ventricles.

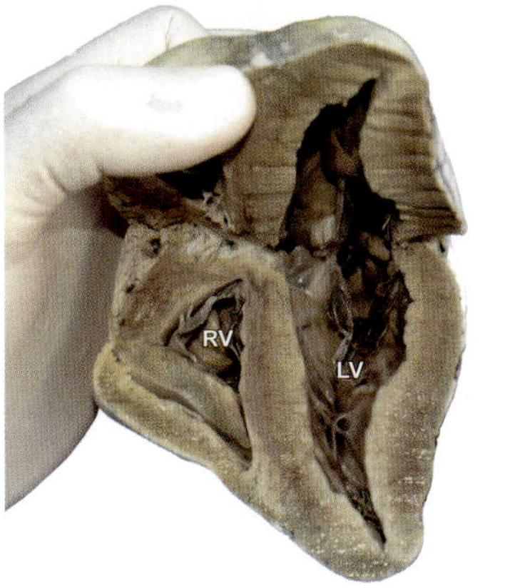

Right: The hearts of all reptiles have a single ventricle which is incompletely divided by a septum. In contrast, the hearts of birds and mammals have two completely separate ventricles. The circulation of reptiles comprises double aortic arches, whereas birds and mammals have a single aortic arch which lies on the right and left side respectively.
(Diagrammatic)

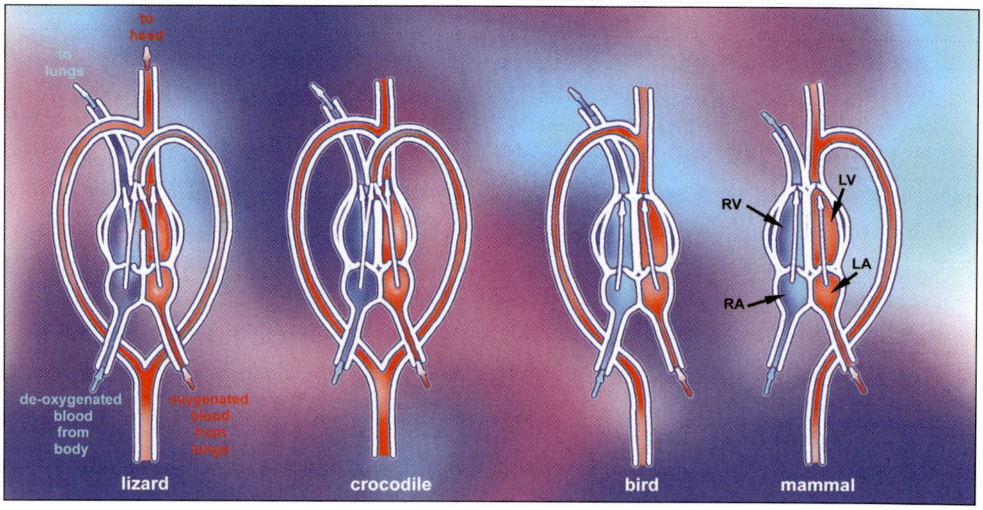


...radiographic studies have shown that when a reptile is breathing normally there is almost complete separation... J Z Young. 'The life of vertebrates'. 3rd edition. 1991. Clarendon Press. p281.

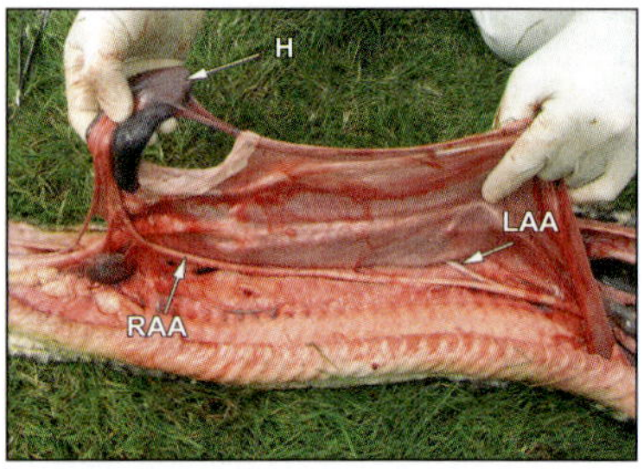


The heart (H), and right (RAA) and left (LAA) aortic arches of an Indian python.

In fact, the reptilian ventricle itself is *partly* divided into two chambers, the relationship of which to the pulmonary artery and to the two aortic trunks ensures that the flow of oxygenated and de-oxygenated blood is, to a great extent, kept separate. Since the reptilian heart and circulatory system is strikingly different from that of the bird or the mammal, the question then arises: could a series of mutations in ancestral reptiles have resulted in the evolution of the bird or mammalian system?* If such a change had been possible, it would have required (for simplicity's sake) at least the production of two separate ventricles associated with the loss of one aortic arch. The septum in the single ventricle of crocodiles almost completely divides it into two chambers - so it can be said that the crocodilian:

...blood circulatory system, with a heart featuring a divided ventricle, is physiologically more advanced than that of any other living reptiles... Rodney Steel. 'Crocodiles'. Christopher Helm. 1989. p32.

However, as with all reptiles, the crocodilian system has 3 main arteries - two aortic arches and the pulmonary (lung) artery - which straddle the outflow zone of both halves of the ventricle. This arrangement shows that the presence of a ventricular septum, even if almost complete, can co-exist with a double aortic arch. This in turn means that the development of the ventricular septum is *not genetically* linked *directly* to the formation of either one or two aortic arches. Therefore, there is no need for a bird to *lose* an aortic arch just because it happens to gain the A2V2 state. Furthermore, if the crocodile's heart and circulatory system is (seemingly) so close to that of the bird and the mammal, it is surprising that the crocodile has been unable to acquire the mutations necessary for a complete transformation into a two-ventricle-one-aortic-arch system, despite (supposedly) living on Earth for over 200my. In other words, in terms of their hearts and the genetic space that separates them, reptiles are unlikely to be a simple step apart from birds and mammals.
The notion of evolution also conflicts with the fact that in living lungfishes, from whose ancestors amphibians are supposed to have evolved:

* It might be argued that it is possible that the bipedal dinosaurs that (supposedly) evolved into birds may have been warm-blooded with four-chambered hearts, so the origin of the bird heart is not a problem. However, this only shifts the issue back in time since amphibians have only three-chambered hearts, and so the impossibility of the modification of a three- to a four-chambered heart remains.

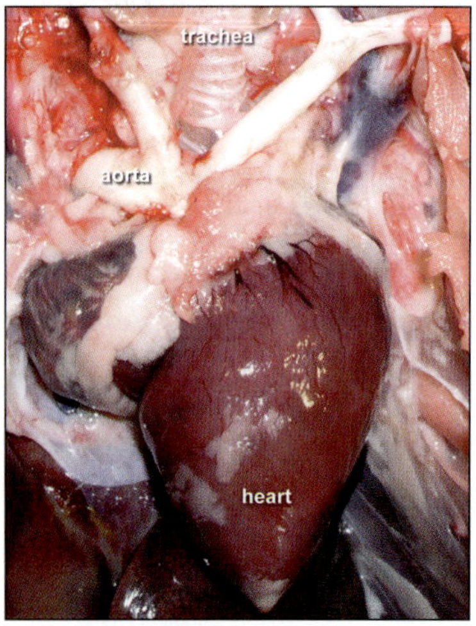


The heart of a turkey showing the right-sided aorta.

> ...the ventricle is partially divided...The lung-fish heart, in fact, shows an amount of structural division greater than that of any amphibian. Knut Schmidt-Nielson. 'Animal physiology'. 5th edition. Cambridge University Press. 1997. p129.

The fact that amphibians do not have a septum in their ventricles suggests that lungfishes and amphibians are unlikely to have had a common ancestor. Furthermore, after (supposedly) more than 300 million years of evolution subsequent to their separation from fishes, by now amphibians *should have* acquired a septum in their ventricles. The fact that they have not done so suggests that amphibians have not lived on Earth long enough (i.e. for millions of years) to have acquired such changes .

It has previously been noted that the large vessels of birds and mammals differ in that birds have a right aortic arch, whereas mammals have a left aortic arch. In about 1 in 100,000 humans, the aortic arch is found on the right-hand side, but this is also associated with a complete transposition of the heart (*dextrocardia*). Since the format of both the bird and the mammal heart work very effectively and there is no particular advantage, one over the other, then over time (contrary to observation) we should expect that some reptiles might have evolved into mammals retaining the right aortic arch, and some birds might have evolved retaining the left aortic arch. The embryological development of birds and mammals shows that neither the mammalian nor the bird heart develops first with a single ventricle which subsequently acquires a septum to convert it into two separate ventricles. Instead, the two ventricles develop *simultaneously* in a dynamic process. Furthermore, the ventricles are not simple muscle bags full of blood. For example, the muscle fibres in the right and left ventricles are continuous and are arranged in layers which form complex spirals, which end in the *papillary muscles* (the muscles attached to the heart valves which lie between the ventricles and the atria):

> The deep layers are three in number; they arise in the papillary muscles of one ventricle and, curving in an S-shaped manner, turn in at the interventricular groove and end in the papillary muscle of the other ventricle...The arrangement of these three layers ensures the synchronisation of ventricular systole* and the closure of the atrio-ventricular valves. 'Gray's Anatomy'. 34th edition. Longmans. 1969. p764. (*i.e. contraction).

If the crocodilian system *had* evolved from the basic reptilian plan, and if (supposedly) evolution has proceeded by small stepwise changes, then we should expect to find some other reptiles with a variety of septa producing lesser or greater degrees of ventricular separation than in the crocodile, along with variations in the aortic arches. Yet no reptile shows such features. Equally incongruous is that despite its (almost) four-chambered heart, except for short periods of rapid activity, the crocodile remains a relatively sluggish animal. Certainly, a medium-sized crocodile is no more active than the largest lizard, the *komodo dragon*. So, if it were feasible that genetic changes could result in the reptilian ventricle dividing into two parts, and in so doing confer major advantages in metabolism and speed, it is quite surprising that after (supposedly) millions of years, the four-chambered heart is not widespread amongst reptiles.

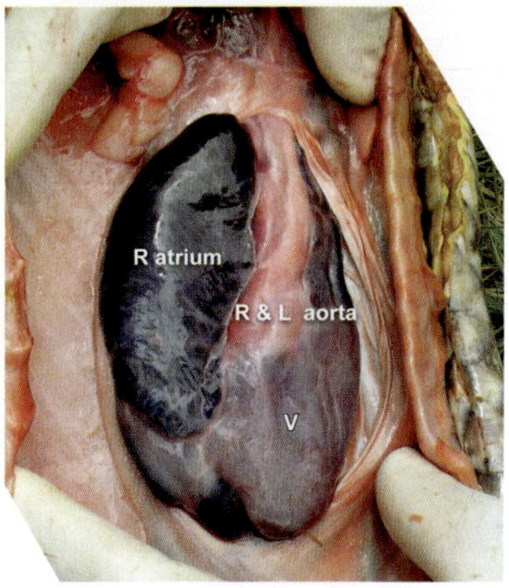


Above and below: Python heart showing twin aortas and a single ventricle (V).

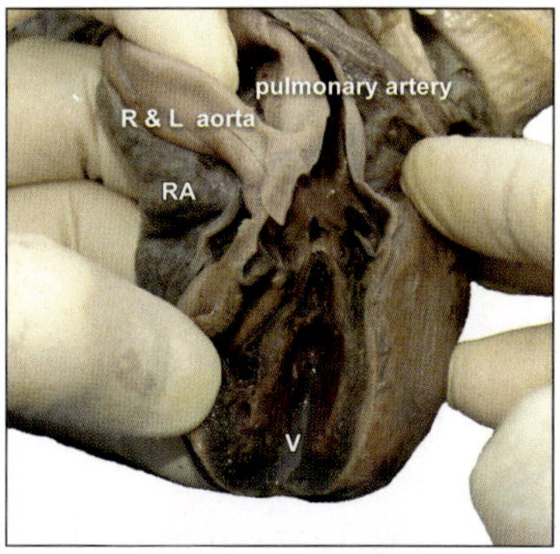


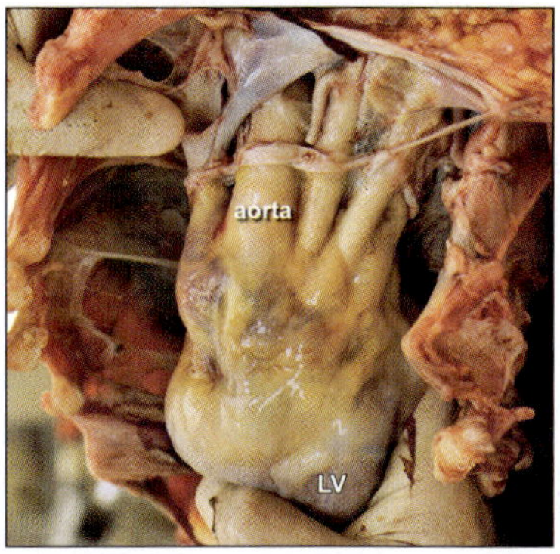


Above and below: Emu heart showing right-sided aorta and two completely separate ventricles. The inset below shows that the right ventricular wall (R) is thin compared to the left (L).

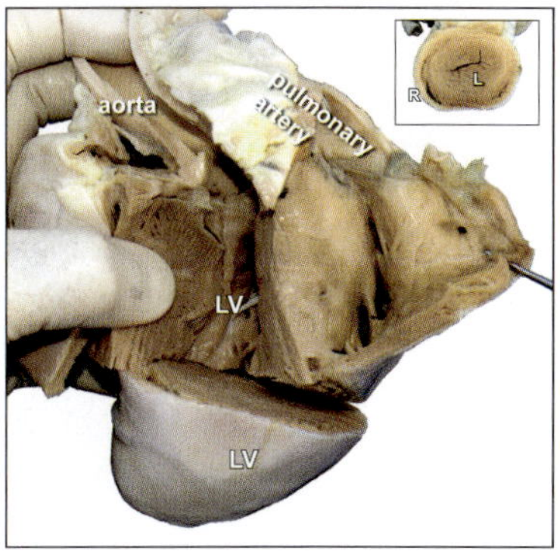

The water monitor (above) is a smaller version of the komodo dragon.

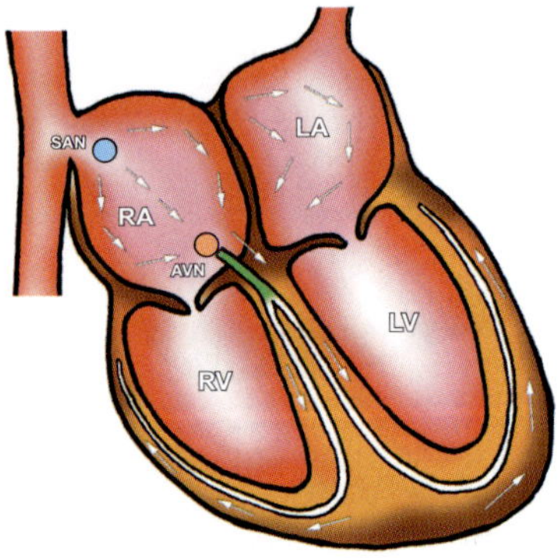


The heart conducting system.
Above: The electrical impulses pass from the sino-atrial node (SAN, blue) through the atrial muscle to the atrio-ventricular node (AVN, yellow), and from there to the right and left bundle branches (white) via the Bundle of His (green).
Below: The right side of a cow heart showing roughly where the SAN and AVN lie.

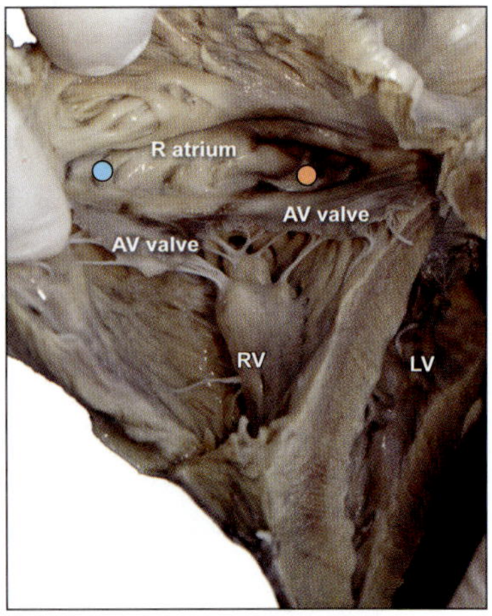


Since even some small alterations to a normal heart can have serious consequences, there is also the problem that a hypothetical series of simple changes – one that conforms to a *conceived* trend towards a four-chambered heart – might in fact face insurmountable difficulties. For example, in humans, when there is an abnormality and the septum develops incompletely (resulting in a *ventriculo-septal defect* or 'hole-in-the-heart'), this is *not* a reversion to the reptilian form with overlying of the three outflow tracts. Instead, the arterial arches are unaffected. However, a ventriculo-septal defect can cause unusual turbulence of the blood flow in its vicinity. This results in two serious problems.

Firstly, any bacteria that get into the blood (for example, via any tiny skin, mouth or intestinal abrasions) tend to settle around an area of turbulence and grow on the walls of the ventricles and the heart valves. This can cause a severe heart infection (*bacterial endocarditis*), which if left untreated can lead to death. And secondly, turbulence and irregular surfaces tend to attract blood clots on the inner lining of the heart. These blood clots can then break off in clumps of varying sizes, and get carried by the blood stream around the body as *emboli*. These emboli can block arteries in different parts of the body, and result in serious damage, for example, to the lungs, brain, kidneys and limbs. In the same way, any reptile with abnormal blood flow patterns, caused by mutations that alter the ventricular septum, is at risk of infection and emboli, and so is more likely to perish. Thus natural selection would tend to select for the status quo (the single ventricle) and vigorously oppose any change away from any design that gives rise to undue turbulence.

Nothing in the heart is simple, and the reader must not be misled into thinking that the bird and the mammalian ventricular septum is just a wall of tissue that some reptiles could have acquired by some of their genes gaining a few mutations. In birds and mammals, the contractions of the right and left ventricles are perfectly synchronised both with each other and also with the contractions of the two atria, so as to ensure a smooth and continuous blood flow from veins to atria to ventricles to arteries.

The triggering of atrial contraction is by impulses arising from specialised cells in the top of the right atrium, the *sino-atrial node*, which itself is subject to excitatory and inhibitory signals from the brain stem via *sympathetic* nerves and the *vagus* nerves respectively. A wave of electrical impulses passing from the atria to the ventricles causes the contraction of the atria, followed by the contraction of the ventricles. However, if the excitation wave from the sino-atrial node simply swept through atria and straight into the ventricles, there would not be enough time for the ventricles to fill properly before their contraction. So there has to be a delay before the ventricles are allowed to contract, and this is provided for by the *atrio-ventricular node*, which is situated just above the valve between the right atrium and right ventricle.

In fact, it is not normally possible for the atrial electrical impulse to pass directly to the muscle cells in the ventricles. Instead, the atrial impulses converge on the atrio-ventricular node, from which the impulses are directed along specialised conducting tissue (the *bundle of His*) which continue on into the ventricular septum as the right and left intraventricular bundles, which then spread out into the rest of the ventricular muscle mass. These specialised conducting fibres are arranged in specific patterns to ensure that the contraction of the muscle fibres in the ventricles proceeds in an orderly manner, starting from the tips of the ventricles.

Without this conducting system, the atria and ventricles would beat out of time (*complete heart block*) and cause severe problems to the blood flow. Such a complete heart block can occur in humans after a heart attack and may need treatment with a cardiac pacemaker (which is designed to stimulate the ventricles directly). The lizard heart has a single ventricle with a single ventricular conducting system. There is no reason to presume that a (supposedly) evolving ventricular septum would also acquire within it the nerve conducting system* necessary for the eventual proper synchronisation of the contractions of the fully-divided ventricle.

One final point regarding the heart: normal embryological development is not a series of separate events, but is instead a continuous process, with many of the genes which are operating at any one time having important effects on other often distant and seemingly unrelated aspects of the process of development. Thus, the genes involved with the development of the heart also have functions involving other aspects of embryological development, and so abnormal heart development is often associated with other diseases. For example, in *Trisomy 21* (Down's syndrome), in which there is severe mental impairment, perhaps 50% of affected individuals have some sort of congenital heart disease in addition to a high incidence of other major abnormalities.

So, it is implausable for a lizard's heart and circulation to become like that of a bird.

### *The bird lung*

Although there is some evidence to suggest that some dinosaurs may have had a bird-like lung system, the unique respiratory system of birds has no parallel in any other air-breathing animal, and it can be shown that the bird lung *could not* have evolved from the simple amphibian or reptilian form.

In amphibians, reptiles and mammals, the inspired air passes down the *trachea*, through the bronchi and into the lungs. The expired air then leaves the lung by flowing out in the exact opposite direction to the way it came in, as with a pair of bellows. In snakes and lizards the lungs are essentially simple balloons.

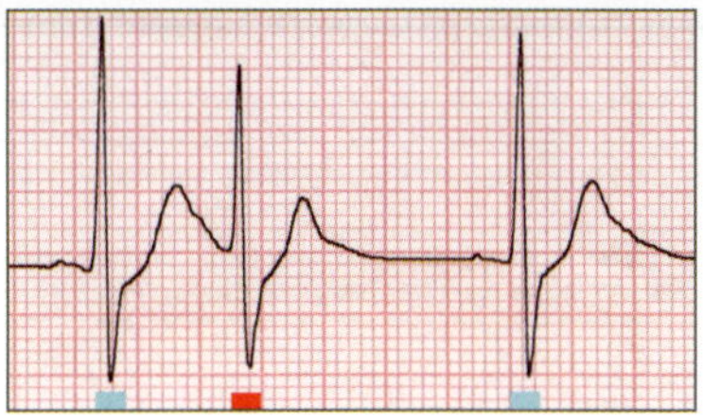

In the heart, ventricular contraction is normally triggered by an electrical discharge from the sino-atrial node which passes through the atria and into the ventricles. This finely balanced system can be disturbed by a number of factors.

In the above electrocardiogram (ECG), the normal electrical discharges (blue) are punctuated by abnormal and premature discharges (red) arising from the wall of the atrium.

***The reptile heart has no conducting fibres**

The hearts of cold-blooded vertebrates beat more slowly in proportion to their size than the hearts of birds and mammals, and they do not possess a special conducting system.

*Ref: Grays Anatomy. 34th edition. Longmans. 1969. p767.*

Below: The lungs of reptiles, exterior and interior views.

Iguana (A,B), slow worm (C,D), python (E,F,G, with inset showing close-up).

T: trachea, RL: right lung, LL: left lung, Li: liver.

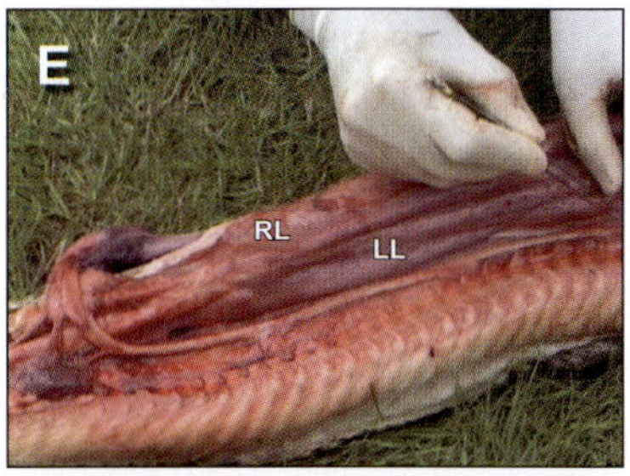


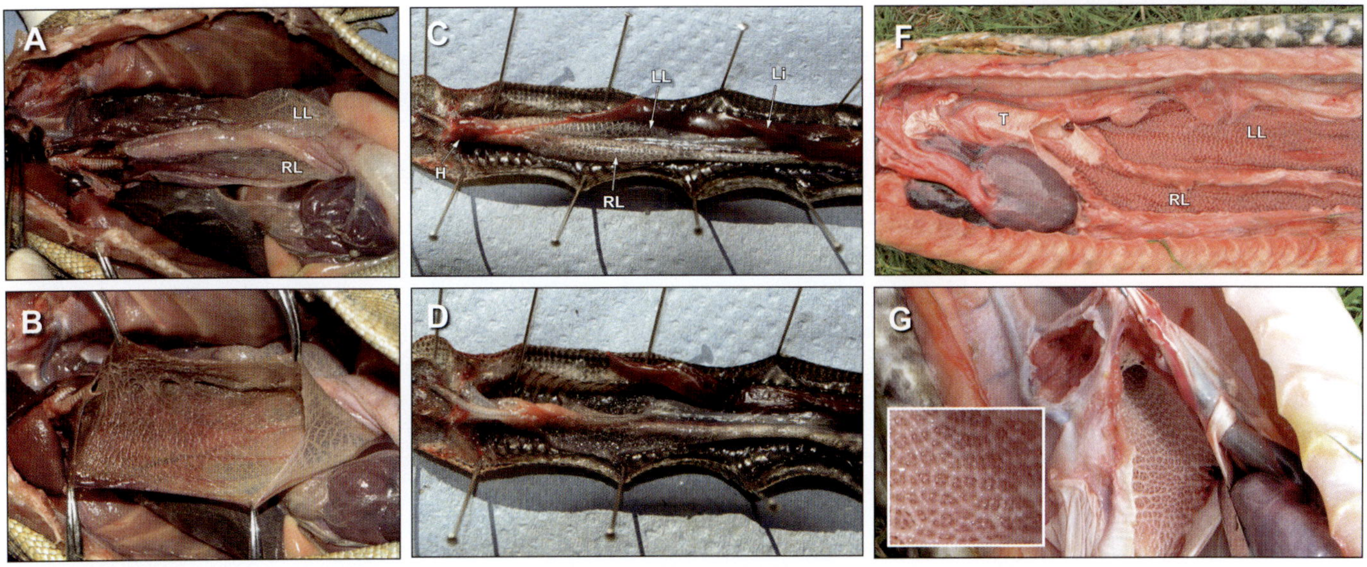

Right: The respiratory systems of mammals, birds and living reptiles.

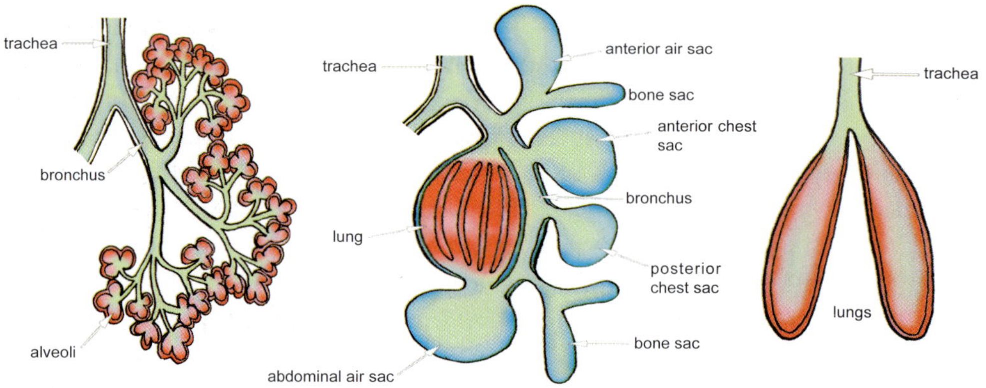


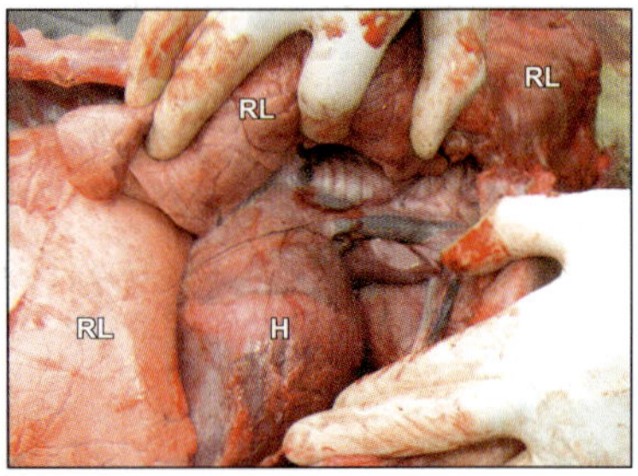


Cow chest cavity.
Above: Right lung (RL) and heart (H).
Below: Right and left lungs
T: trachea.

In mammals, gas exchange with the blood occurs in millions of tiny air sacs, the *alveoli*, which are lined with abundant blood vessels. Oxygen passes from the alveoli into the blood and carbon dioxide passes from the blood into the alveoli.

The air flow through the bird respiratory system, however, is not only fundamentally very different, but also far more complex. The lungs themselves are very compact compared to mammalian lungs, and are situated at the back of the chest, next to the spine. The bulk of the chest area is actually taken up by the heart and the chest air sacs. The bird lung has numerous connections with large thin-walled air sacs that lie in the neck, chest and abdomen, and which even extend into the limb bones and the spine. The air sacs themselves are not endowed with any significant blood vessels, so they do not take part in gas exchange, but probably act as bellows to help move the air in and out of the lungs.

From a practical point of view, the main bronchi do not convey incoming air directly into the lungs. Instead, the unique bronchi actually carry the incoming air straight *through* the lungs and directly into the posterior air sacs, which lie within the abdominal cavity. From these posterior air sacs the air then flows into the lungs. Unlike the lung of other land animals, the bird lung does not expand. Instead, it has fine, relatively rigid, parallel air tubes which are open at both ends so as to allow the through passage of air in a continuous tail-to-head unidirectional flow. From the lungs, the air then passes into the anterior air sacs, from which the air passes back into the main bronchus and is finally exhaled. This complex system is extremely efficient since air moves continuously in a tail-to-head direction through the lung tissue, allowing gas exchange during both inspiration and expiration.

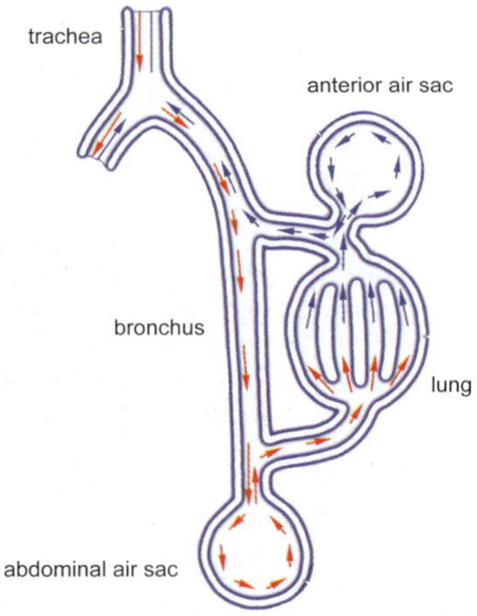


The air flow in the bird lung is a complex one-way circuit.

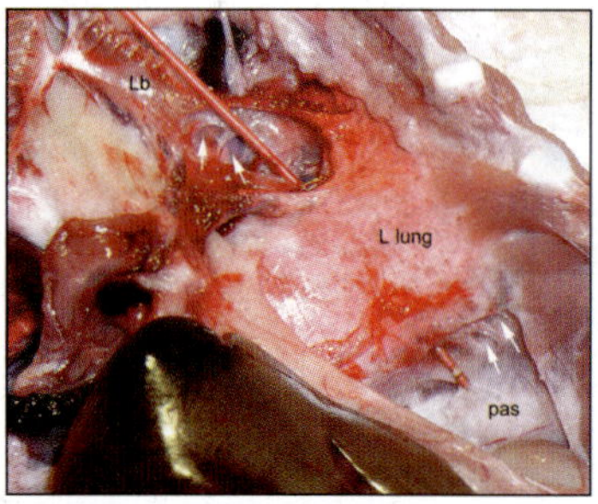


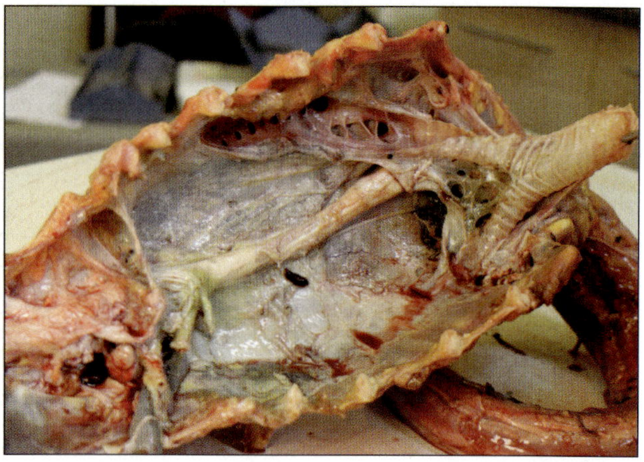

Above right: Turkey lung showing a probe in the left main bronchus which courses straight through the lung and opens into the posterior air sac (pas). The openings leading from the posterior air sac to the lung, and from the bronchus to the anterior air sacs, are arrowed.
Far right: Emu showing the small dense lungs situated in the back of the chest cavity, next to the spine, with multiple openings along the right brochus.

In addition to the unique structure of the air tubes, the blood vessels are also arranged in a unique way. In the bellows-type lung, the air spaces are enveloped by a random meshwork of blood capillaries. In the bird lung, however, the blood vessels lie in parallel to the air tubes and the blood flows in the opposite direction to the flow of air. This *counter-current* design ensures that the maximum amount of oxygen is extracted from the inspired air.

If we suppose for a moment that a series of stepwise changes could occur to the standard bellows-type lung of a reptile, the results would be of no benefit, and most likely would be detrimental. For example, suppose a sac (without any blood vessels) appeared as a ballooning out of one of the existing lung alveoli. Such a sac would tend to rupture easily when inflated, resulting in free air in the chest cavity itself *(pneumothorax)* and collapse of the lung. (In humans, a pneumothorax is often genetically related since it is more common in tall thin men.) In addition, a small sac would predispose to chronic infection (an example of this is the disease *bronchiectasis*, in which damage to the alveoli results in stagnation of air and mucus resulting in serious infections and breathing difficulties). Furthermore, if a sac appeared as a side branch to the main bronchus, this would reduce the proportion of air flowing to the lungs themselves and thereby *reduce* gas exchange by increasing the '*dead space*'*. Again, if the bronchus developed in such a way that it passed straight through the lung, this would result in death since the bronchus would have no communication with the lung tissue. And the presence of anterior and posterior air sacs without the alveoli changing into 'through tubes' would also just increase dead space.

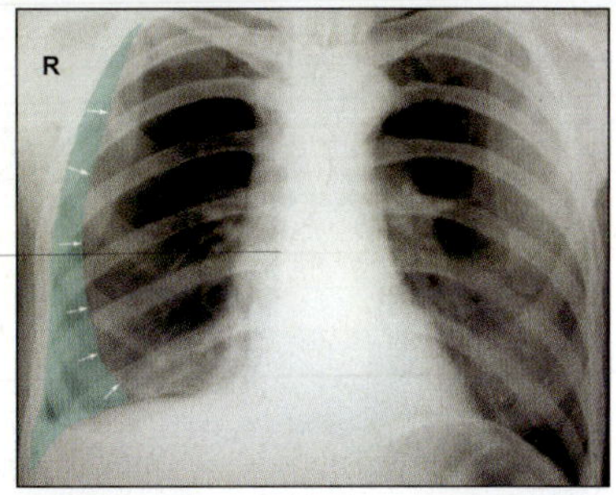


Pneumothorax. Air is present (pale blue) between the edge of the right lung (shown by arrows) and the chest wall.

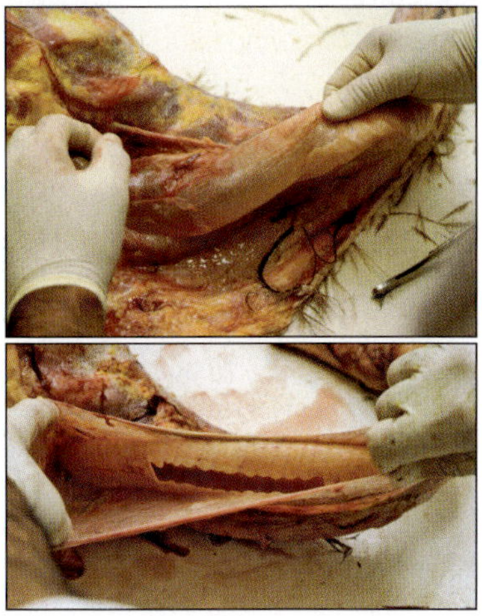

The emu throat bag has specific muscular controls, without which it would simply divert precious air away from the lungs.

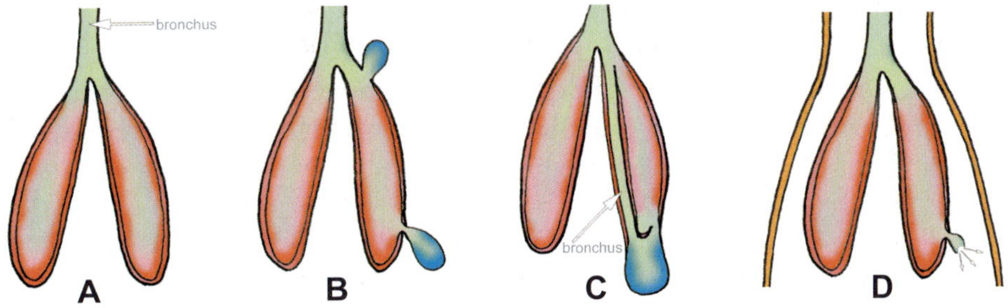


Mutations might result in a number of alterations to the reptile respiratory system, with air sacs arising in different sites (see main text).
A. The simple reptile lung.
B. Anterior and posterior air sacs with no other change.
C. A 'through' bronchus and posterior air sac.
D. Perforation of a small air sac would result in air in the abdominal cavity and deflation of the lung.

Furthermore, as we have seen many times previously, as a general rule, interference with the development of one aspect of an embryo will be associated with developmental problems elsewhere. Thus, whatever slight change to a reptile lung system we might imagine, it would either be detrimental, or have no beneficial effect. It is

* The term 'dead space' refers to those parts of the respiratory passages where no gas exchange can occur between the inspired air and the blood. In normal circumstances, dead space occurs only in the trachea and bronchi.
It is true that some snakes and lizards do have small sacs attached to their lungs. Their function (if any) is unknown, but what is certain is that sacs that have no blood capillaries can only increase dead space and reduce efficiency.

**Why a bird lung system could not have evolved**

1. A bronchus that opened into an abdominal air sac instead of into the lungs would result in death.
2. An expandable lizard-type lung that simply changed into a rigid, non-expanding bird-type lung would result in death.
3. If air sacs arose from the expandable lizard-type lung, this would result in a reduction in efficiency due to dead space, and any delicate air sac has the risk of bursting, causing air to escape from the lung and into the tissues.
4. Mutations would result in embryonic developmental derangements: the lungs could fail to develop *at all* (for example, see lungless frog, page 159), which would be fatal in any reptile or bird; alternatively, the lungs might be partially deformed, and this may be associated with multiple abnormalities affecting other sites of the body (for example, R-spondin 2, see page 508); furthermore, mutations affecting particular proteins may result in severe lung disease along with problems in other parts of the body (for example, cystic fibrosis, see page 107).

also notable that despite its undeniable efficiency, a bird does not need a bird lung in order to fly:

> This extensive and intricate respiratory system has been considered as an adaptation to flight. We can immediately say, however, that it is not necessary for flight, because bats (which have typical mammalian lungs) are good fliers and at times even migrate over long distances...during flight bats have an oxygen consumption very similar to that of birds of the same body mass. Knut Schmidt-Nielson. 'Animal physiology'. 5th edition. Cambridge University Press. 1997. p41-2.

Furthermore, most birds are about the same size (or smaller) than fruit bats, some of which can weigh as much as 1.5kg and can have a wingspan of 5 feet 5 inches (for example, the Bismarck flying fox from New Guinea). And whilst the bird lung system is demonstrably superior to the mammalian system at *high* altitude where the air is significantly thinner, the fact remains that most birds *do not* fly high in the sky. Indeed, in the routine day, very few birds do, or need to, fly much higher than tree tops.

So, we have seen that any reptile with a bellows-type lung, that was born with a mutation resulting in developmental structural changes which altered the air flow, would suffer the deleterious effects of increase in dead space, risk of infections, and other congenital abnormalities. It is therefore impossible that the functioning simple lung system of a reptile could ever have evolved into the bird system. This means that the bird lung can only have arisen by being built 'from the ground up', and not by the modification of an existing simpler lung.

It was acknowledged 150 years ago by its proponents that, if it could be shown that if any new organ or new creature could not arise by numerous stepwise changes to an ancestral form, the notion of evolution would simply fall apart. Therefore, having shown that it is impossible for the bird lung to have evolved, we can state the *Don't Be Fooled Again Principle*:

Since it is a fact that evolution is impossible, if we come across *any* data that *seems* to suggest evolution has occurred, we know that *that* interpretation of such data *must* be false.

### *Air sacs in birds and dinosaurs*

Living birds have air sacs in the neck, chest and abdomen, which extend into the neck, chest, and sacral vertebrae, and also into the humerus, pelvis and femur. It has also been shown that both the Berlin and the London archaeopteryx specimens have air sacs in:

> ...the cervical vertebrae of the Berlin specimen and in an anterior thoracal vertebra of the Eichstatt [London] specimen. This indicates the presence of a cervical air sac. We show that the London specimen also has pneumatized anterior thoracal vertebrae...the pelvis of the London specimen shows clear signs of the presence of an abdominal air sac... P Christiansen & and N Bonde. 'Axial and appendicular pneumaticity in Archaeopteryx'. Proc. R. Soc. Lond. B (2000) 267. p2501-5.

Fruit bat. Many bats are bird-sized and yet they manage with typical mammalian lung systems.

**Lung formation**

The proper development of branching structures is directed by the balance of a number of activating and suppressant triggers. Without this proper balance already in place, any changes to a lizard that was hoping to acquire a bird or mammal lung would be fatal:

> *Key gene families such as FGFs and BMPs are important mediators of branching morphogenesis. Canonical Wnt signaling negatively regulates branching morphogenesis, providing a balance to FGFs and BMPs which positively regulate this process. This multilayered control of growth and proliferation ensures that branched structures attain the morphology required to function efficiently.*
> C Dean et al. 'Canonical Wnt signaling negatively regulates branching morphogenesis of the lung and lacrimal gland'. Dev. Biol. 2005 Oct 1;286(1):270-86.

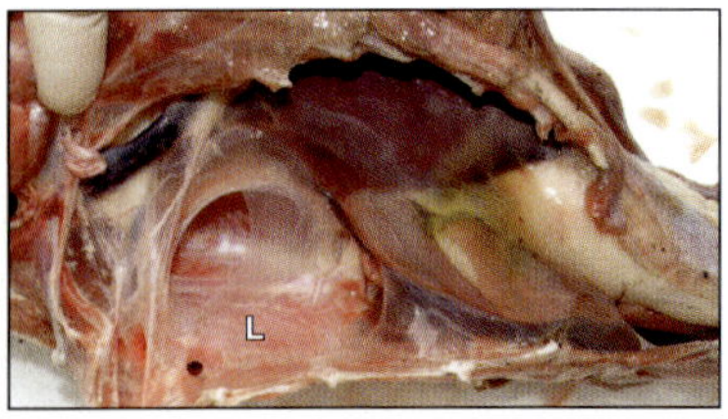


Above: Gannet right chest showing air sacs through which the lung (L) can be seen, which lies deep against the spine.

Below: Emu chest cavity showing the right chest air sacs (AS) opened up alongside the heart (H).

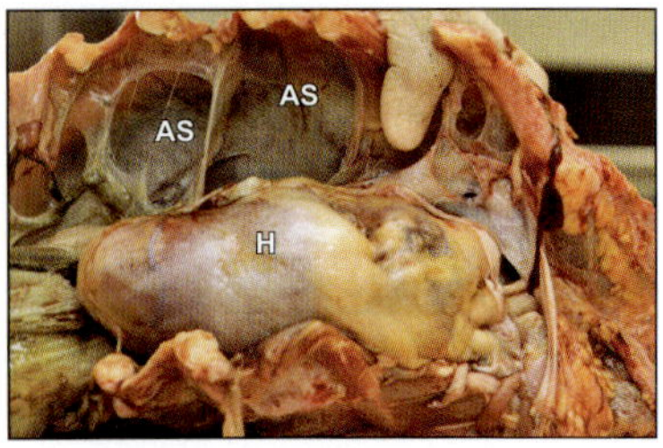

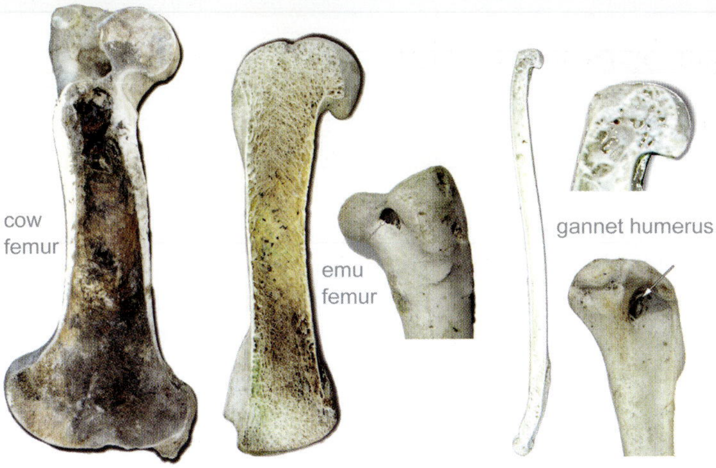


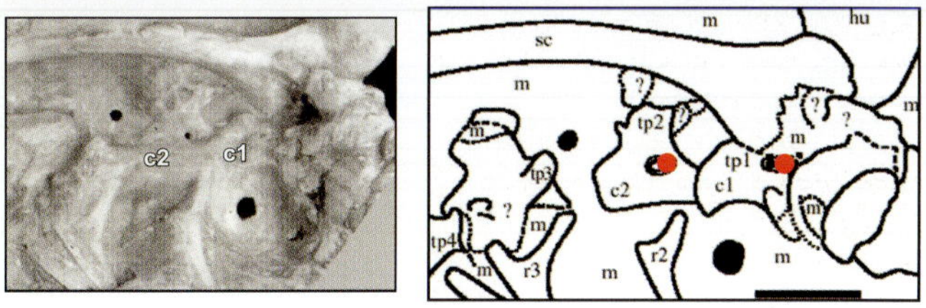


Air sacs in bones (not to scale).
Far left: Cow femur showing dense bone with no air cavities.
Middle left and left: Emu and gannet femur showing air hole (arrow), and honeycomb structure.
Above left and right: Archaeopteryx showing air holes (red dots) in first two neck vertebrae (*c1,c2'*).
*From Christiansen & Bonde. 2000.*

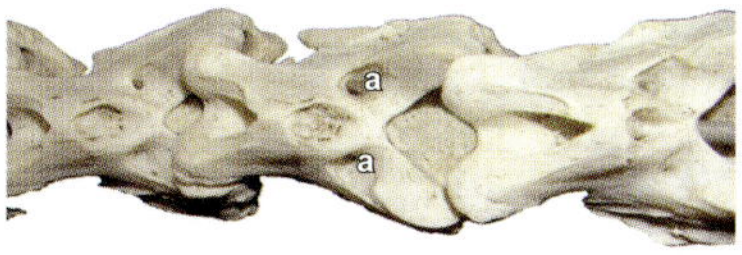


Above: Air holes (a) in emu lower neck vertebrae (from below).
Below: Air holes (a) in gannet lower neck vertebrae (from side).

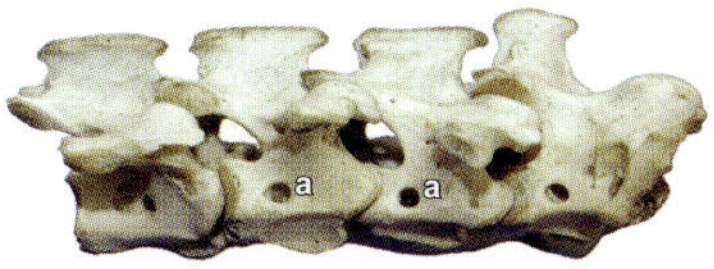


Some terrestrial bipedal dinosaurs* also show evidence of having had air sacs, and so it has been suggested that these dinosaurs had bird-like lungs:

> ...advanced theropod dinosaurs, such as dromaeosaurs, the group most frequently considered the sister group of birds...and tyrannosaurs, have vertebral pneumatic foramina on all presacral vertebrae. Oviraptorosaurs have pneumatization of the synsacral and proximal caudal vertebrae...indicating the possible presence of abdominal air sacs. P Christiansen & N Bonde. 'Axial and appendicular pneumaticity in Archaeopteryx'. Proc. R. Soc. Lond. B (2000) 267. p2501-5.

> Here we report...evidence for cervical and abdominal air-sac systems in non-avian theropods, along with thoracic skeletal prerequisites of an avian-style aspiration pump...these specializations imply that flow-through ventilation of the lung is not restricted to birds but is probably a general theropod characteristic. P O'Connor & L Claessens. Letter. Nature. Vol 436. 14th July 2005. p253-256.

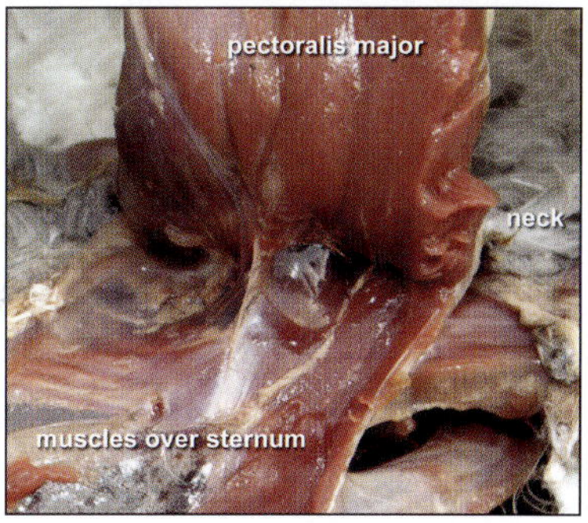


Barn owl: air sac underneath the pectoralis major muscle.

Most birds have special joints (*intracostal* joints) which divide each rib into two sections (one section attached to the spine, and the other attached to the breast bone), which enable the chest wall to move backwards and outwards:

> Birds utilize highly modified costal [rib] movement in association with a well developed sternum to ventilate their non-vascularized air sacs. The ribs of [living] birds possess unique synovial intracostal joints, the shape of which enables foreaft movement of the ribcage during lung ventilation. Consequently, during ventilation, avian ribs rotate so that the posterior end of the sternum is depressed upon inhalation, thus generating negative intra-abdominal pressure and filling of the large abdominal air sacs. J Ruben et al. 'Lung structure and ventilation in theropod dinosaurs and early birds'. Science. Vol 278. 14th November 1997. p1267-70.

Some argue that without this unique arrangement it is not possible for the abdominal air sacs to be inflated.

> Theropods lacked avian-like jointed or hinged ribs and an expansive sternum, structures without which proper ventilatory airflow cannot be maintained in the modern bird lung. J Ruben et al. 'Lung structure and ventilation in theropod dinosaurs and early birds'. Science. Vol 278. 14th November 1997. p1267-70.

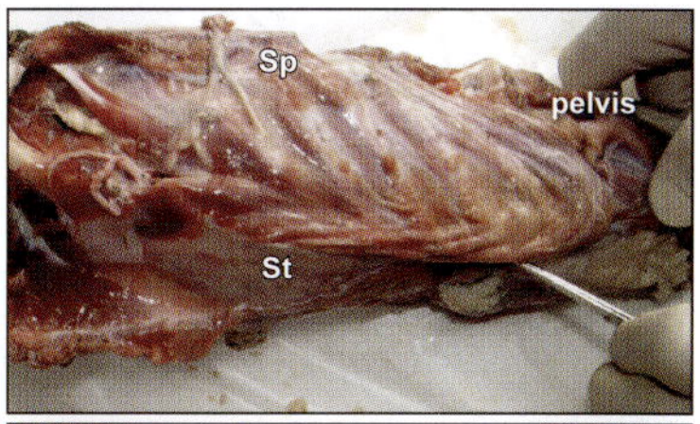


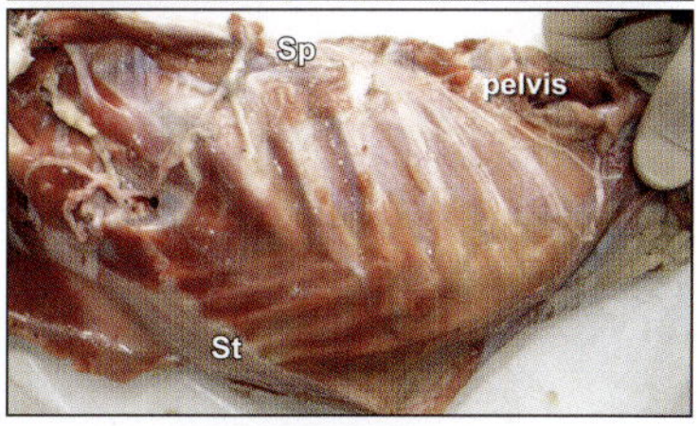


Gannet intracostal joints.
Above top: Chest compressed.
Above: Chest expanded.
Sp: spine, St: sternum.

*Pterosaurs are also known to have had air sacs in their bones.
(See New Scientist. 28th February 2009. p14; and R Butlet et al. 'Postcranial skeletal pneumaticity and air-sacs in the earliest pterosaurs'. Biol Lett. 1st May 2009)

*In kinematic studies of walking guinea fowl, the femur retracted little, only 5-15° compared with 60–80° retraction observed in the vertically oriented femur of walking alligators, while the avian knee moved through a 50-70° arc...Even during running the femur never passed vertical in birds...*
D Quick & J Ruben. 'Cardio-Pulmonary Anatomy in Theropod Dinosaurs: Implications From Extant Archosaurs'. Journal of Morphology. Published online: 20th May 2009.

None of the dinosaurs, including archaeopteryx, in which bone air sacs have been inferred, have such specialised rib joints.
It has also been suggested that without any rigid support any abdominal air sacs would simply be subject to paradoxical collapse as a result of the negative pressure caused by chest expansion. Such paradoxical collapse is avoided in birds since the almost horizontal position of the bird femur gives support to the side of the abdomen. The bird femur has relatively little movement, with the stride instead, being dependent on substantial contributions from knee and ankle movements*. In contrast, the femurs of bipedal dinosaurs were very mobile and could not have given the necessary support to the side of the abdomen. Therefore, it is argued that if dinosaurs had abdominal air sacs these could only have been small, and could not have functioned as bellows for the lungs. So it is unlikely that dinosaurs had a bird-like flow-through lung system.

Theropod dinosaurs probably lacked…enlarged abdominal air-sacs, and skeleto-muscular modifications consistent with their ventilation. In the absence of enlarged, functional abdominal air-sacs, theropods were unlikely to have possessed a specialized bird-like, air-sac lung. D Quick & J Ruben. 'Cardio-pulmonary anatomy in theropod dinosaurs: implications from extant Archosaurs'. Journal of Morphology. Published online: 20th May 2009.

Alligator.
Heart and lungs (HL)
Liver (Li)
Muscle between liver and pelvis (M)
Pubic bone (P)

In fact, the small bipedal dinosaur *Scipionyx* shows good evidence of having a crocodile-like *hepatic-piston*-type breathing mechanism. In crocodilians, the heart and lungs are separated from the abdominal contents by a large liver. Spanning the abdomen, and connecting the fibrous surface of the liver and the pubic bone, are a group of muscles which, when they contract, pull the liver tail-wards. The resultant negative pressure in the chest cavity then results in inflation of the lungs.
Scipionyx was discovered in (supposedly) 113myo deposits, and was the first dinosaur to be found in Italy. At 23.7cm in length, it was most likely just a hatchling, and clearly had a furcula. The specimen is virtually complete and, remarkably, soft tissues have been fossilised. Under ultraviolet light the colon and liver can be clearly seen.
Unlike in birds, where the trachea lies adjacent to the vertebrae of the neck (making a more direct entry into lungs which are situated just below the spine), the trachea of scipionyx lies well away from the spine. Also unlike birds, in which the abdominal air sacs displace the gut away from the spine, the colon in scipionyx is clearly next to the spine. This suggests that it did not have abdominal air sacs. Furthermore, the liver completely separates the heart from the colon as in crocodilians:

...the anterior border of the liver in Scipionyx is vertically oriented and completely subdivides the visceral cavity into anterior pleuropericardial and posterior abdominal regions.** J Ruben et al. 'Pulmonary function and metabolic physiology of therapod dinosaurs'. Science. Vol 283. 22nd January 1999. p514-6.

Scipionyx fossil showing preservation of the shape of a loop of colon (c). (Ruben 1999)

**Soft tissue features can also be seen in sinosauropteryx, but what was described by Ruben as being a diaphragm, is reported to actually be a crack in that fossil's matrix (see pages 454-5: Christiansen & Bonde).

In summary, evidence of bone air sacs is seen in a number of dinosaurs, but whether or not they had bird-like lungs is debatable, and examples such as scipionyx clearly show that some theropods most likely had a crocodile-like hepatic-piston ventilation system. Since it is impossible for a bird-like lung to have evolved stepwise from a simple reptilian lung design, then if dinosaurs had simple reptilian lungs, birds did not evolve from them. And if some dinosaurs had bird-like lungs, those dinosaurs in turn did not evolve from reptiles with simple reptilian lungs.

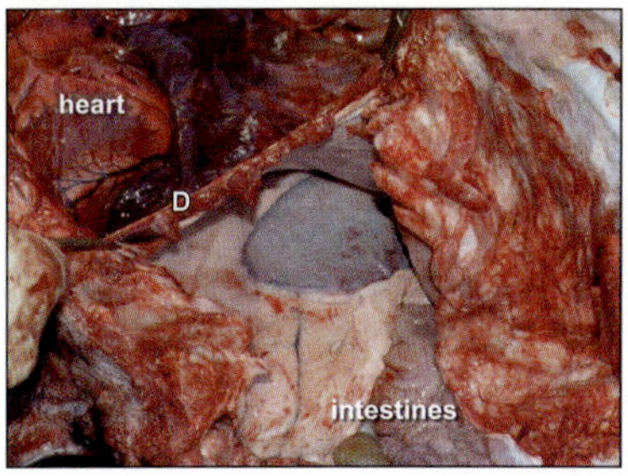


Cow (above) and squirrel (below) showing the diaphragm (D) completely separating the chest and abdominal cavities.

Below left: Iguana body cavity' (LL: left lung, Li: liver, H: heart), showing no separation between the heart, lungs and intestines.

### *Diaphragm*

Leaving the bird lung aside, the muscular diaphragm, which is only found in mammals, shows that it is also impossible for the *mammalian* lung to have evolved from some (supposed) ancestral reptilian design. Living amphibians, non-crocodilian reptiles and birds do not have a partition between the chest and abdominal contents, and the inflation of the lungs is achieved by a number of different mechanisms, including movements of the mouth floor (frogs), muscles which pull the liver towards the pelvic bones (some bipedal dinosaurs and crocodiles), and movement of the chest muscles (birds).

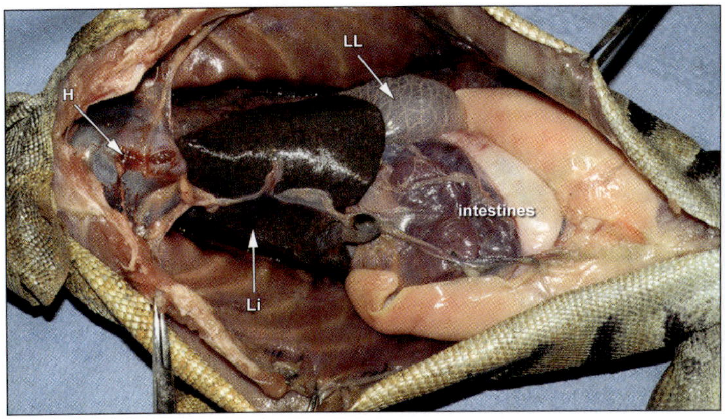


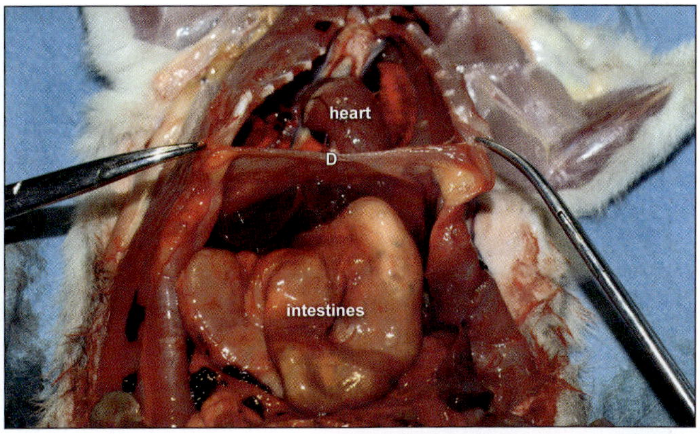


In contrast, both sharks and bony fishes have their hearts completely separated from the abdominal contents by thick fibrous partitions. It is impossible for the separate body cavities of a fish to have evolved into the single cavity of the amphibian or reptile, and then subsequently to have developed a diaphragm partition.

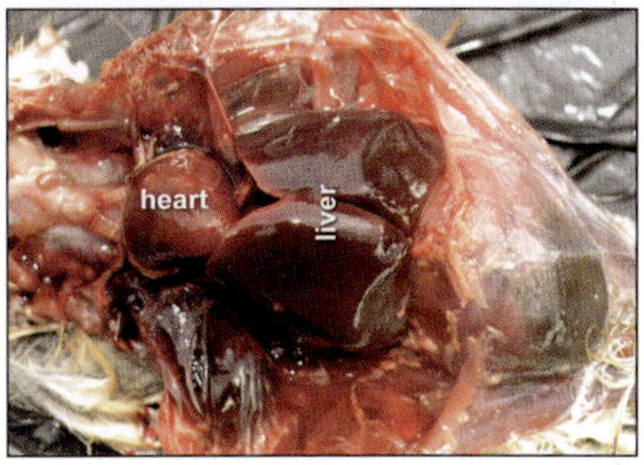


Barn owl: birds do not have a diaphragm between the heart and the liver to separate the chest and abdominal contents.

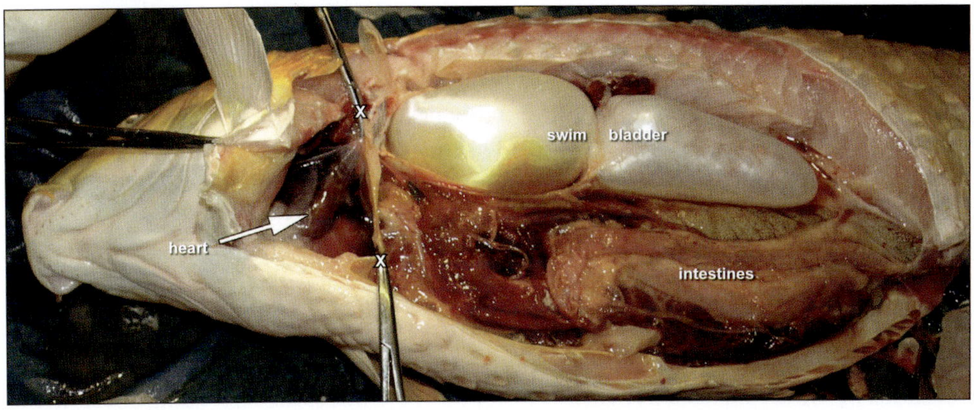


Above: Koi carp showing thick fibrous partitions separating the heart, intestines and swim bladder from each other. The partition between the heart and intestines is held between the forceps X and X.

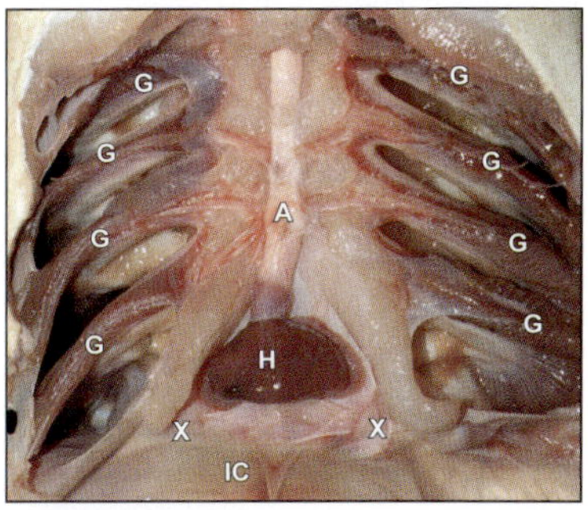


Right: Dogfish showing thick fibrous partition (between X and X) separating the heart (H) from the intestinal cavity (IC).
The dissection also shows the aorta (A) with its arterial branches going into the gills (G).

**Why a diaphragm could not have evolved**

1. A partial diaphragm would result in herniation of intestines into the chest, resulting in collapse of the lung, causing death.
2. A complete barrier separating the chest and abdominal contents could obliterate the main blood vessels and oesophagus, resulting in death.
3. Random migration of muscle tissue from the neck region of the embryo would result in useless diaphragm tissue anywhere in the body.
4. A diaphragm without specific involuntary nerves controlled by specific brain stem cells would be useless.
5. Muscle tissue that simply lies transversely across the body cannot function as a diaphragm since its contraction would pull the ribs inwards, resulting in deflation, rather than inflation, of the lungs. In fact, the diaphragm is specifically shaped as a dome with its perimeter attached to the tips of the lowermost ribs, and with special tendinous attachments to the anterior parts of the spine. As a result of this unique arrangement, when the diaphragmatic muscle fibres contract, the dome of the diaphragm is pulled downwards (or tail-wards), thus inflating the lungs.

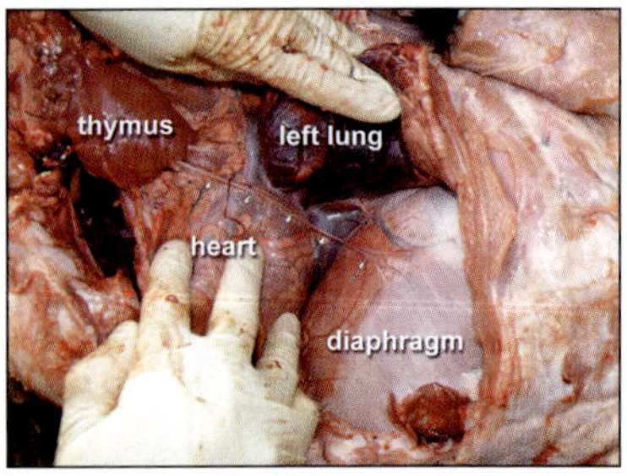


Cow: the left phrenic nerve is shown emerging from under the thymus and passing alongside the heart to the surface of the diaphragm.

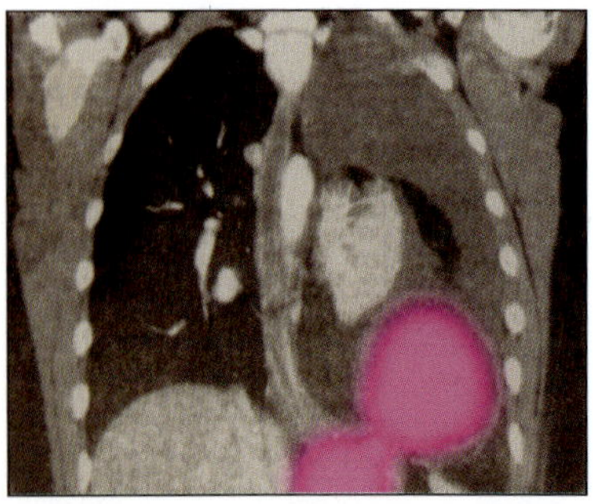

CT scan showing herniation of the stomach (colour applied) through the left half of the diaphragm, following a stab injury to the chest.
*(Modified from the British Medical Journal)*

It might be thought that the diaphragm could easily have evolved following mutations which caused muscles of the edges of the lower chest wall to migrate inwards and meet in the middle. However, apart from a very small part of its perimeter, the diaphragm *does not* develop out of local muscles. Strange as it may seem, the diaphragm develops from tissue (the *septum transversum*) which arises in the *neck* area in the developing embryo (adjacent to the tissue that is destined to become the fibrous sac around the heart, the *pericardium*), and which ultimately migrates down to where the bottom of the rib cage will be. Since a muscle always takes its nerve supply from the segment of the embryo in which it originated, this explains why the diaphragm is supplied by the *phrenic nerve*, the fibres of which originate mainly from the 4th spinal segment in the neck. Because of its long trajectory from the neck down through the chest, it is possible to damage the phrenic nerve at various places by a stab wound (or tumour), causing paralysis of the diaphragm.

Now, a stepwise origination of a diaphragm would be fatal to a reptile that was hoping to become a mammal, for the following reason. However small it may seem, a hole in the diaphragm can have devastating effects, because as the chest wall expands with breathing, the negative pressure in the chest would suck the stomach or intestines into the chest cavity. Such *herniation* would result in either compression of the lungs and their subsequent collapse, or strangulation of the herniated gut. This is a well-known complication in infants who are born with congenital defects in the diaphragm, and is also well illustrated by penetrating injuries to the chest which can cause a small laceration in the diaphragm (which, initially, can easily be over-looked).

Just such an example was the report of a teenager who was stabbed in the chest, and recovered very well, only to develop abdominal pain and vomiting, with a complete inability keep food down, one year later. When he was re-examined, a CT scan showed that he had just a 4cm perforation in the left side of his diaphragm, through which most of the stomach had herniated. Because the hole was so small, the blood supply to the stomach was strangulated, resulting in gangrene. Emergency surgery showed that:

> ...all but the pylorus [lower end] of the stomach had herniated through a 4cm central defect [in the diaphragm] and was incarcerated [strangulated] in the left chest with evidence of irreversible ischaemia [gangrene due to lack of blood supply]. J Ahmad et al. 'Penetrating trauma to the junctional zone needs aggressive management'. British Medical Journal. 3rd February 2007. p257-8.

Left untreated, this condition would have been fatal, but the patient made a good recovery after the stomach was removed surgically. As we saw with the descent of the testes, which can be arrested at any point, since the random migration of the septum transversum could cease at any point, it is impossible for it to have migrated to exactly where it is required. So, it is a fact that any incomplete diaphragm will have disasterous consequences. Therefore, a diaphragm could not evolve stepwise.

In addition, the diaphragm is of no use unless it is activated by an involuntary automatic nerve. It is implausible that by a random interference with the developmental cascades in some (supposed) primitive creature, the necessary nerve supply, and the necessary specialised cells in the brain stem (that subconsciously trigger the correct stimulus at the correct time), would appear exactly as and where needed.

### *Bird and dinosaur digits*

Digit formation is directed by combinations of highly complex triggers, about which we have but a very crude insight. Before digits are formed in the embryonic limb, a row of five condensations of tissue appear (cI, cII, cIII, cIV and cV), from which five digits (d1,d2,d3,d4 and d5) develop. The fourth condensation is the foundation to the digit production process, with the digits appearing in the order d4-d5-d3-d2-d1 from these condensations. However, the individual condensations do not actually define digit identity, which is directed by other controls. Each bone within a digit is called a *phalanx* (plural - *phalanges*) and digit 1 (which corresponds to our thumb) normally has two phalanges, the others normally having three phalanges.

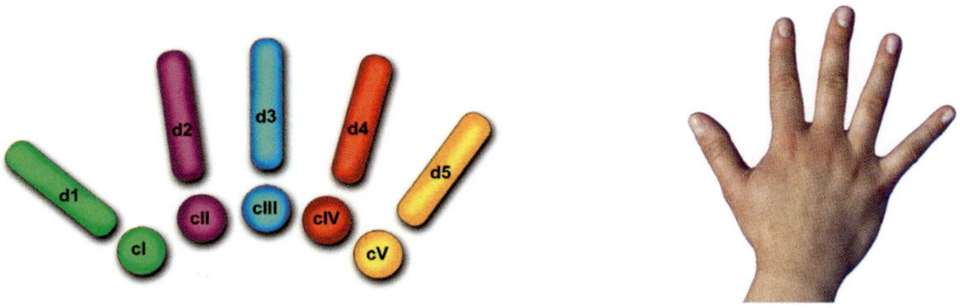

Condensations cI-cV normally give rise to digits d1-d5 respectively.

In different animals, some of the condensations dissolve away and do not produce digits at all. In the bird wing, it is well established that the digits develop from condensations cII, cIII and cIV, so birds have digits 2, 3 and 4.

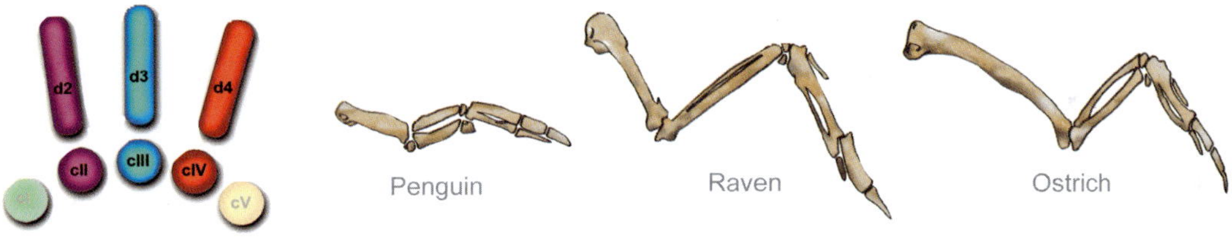

The wing bones of flying and flightless birds.
Birds have digits 2, 3 and 4.

However, looking at the bones of theropod dinosaurs and archaeopteryx, it is clear that their first digit is a 'thumb', so dinosaurs have digits 1, 2 and 3.

Dinosaurs have digits 1,2 and 3.

This is a problem for those who insist that birds evolved from non-flying dinosaurs since it appears that birds and dinosaurs have distinctly different digit identity. If non-flying dinosaurs evolved into birds, the latter would have had to regain d4 (which is lost in therapods) and lose d1 (which is present in therapods). This would seem a tall order. In order to force the evidence to fit the evolution paradigm, it has been suggested that theropod dinosaurs experienced a developmental *frame-shift* such that their digits 1-3 actually derived from cII-cIV, with the result that their bird descendants now also actually have digits 1-3 derived from cII-cIV.

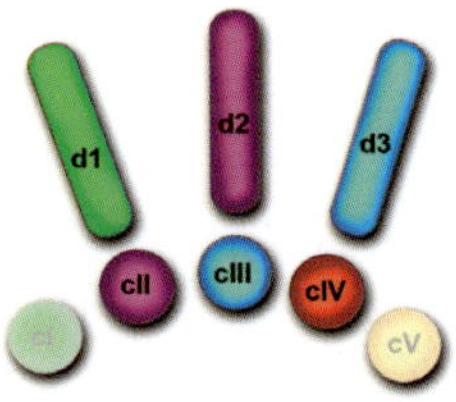

Frame-shift idea, which postulates that dinosaurs have digits 1,2 and 3, arising from condensations cII, cIII and cIV.

This might seem an elegant way to reconcile the apparently great differences between therapod and bird digits. However, there are at least two serious problems with the idea. Firstly, both therapods and birds have digits 1-4 in their feet and it is implausible that a frame-shift would affect only the forelimbs and not include the hind limbs:

> The homeotic identity shifts cannot be achieved by straightforward identity shifts in four consecutive digits alone. Galis et al. 'Hox genes, digit identities and the therapod/bird transition'. Journal of experimental zoology. Vol 304B. 2005. p198-205.

Secondly, even if a forelimb-only-frame-shift were possible, since any hypothetical frame shift would produce the same final form of the digits, this would have no evolutionary advantage, and so would not be selected for.
Another way of looking at digit development is to examine the activity of Hox genes. When chick wings and mice forelimbs were examined for Hox d12 and Hox d13, it was found that both genes are active in all the digits except the most anterior (front-most), with the latter developing in the absence of Hox d12. Since the mouse anterior digit is definitely digit 1 (equivalent to our thumb) and arises from condensation I, from this it has been claimed that the chick anterior digit is also digit 1 although it arises from condensation II, and therefore that birds really do have digits 1, 2 and 3 arising from condensations II, III and IV, that occured by a frame-shift in their (supposed) dinosaur ancestors.

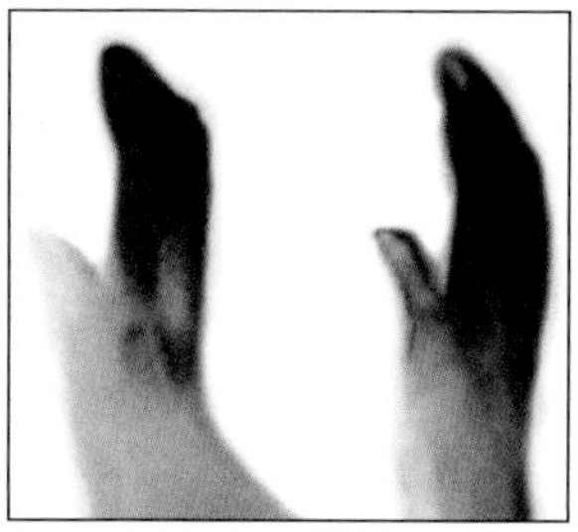

Chick developing wing showing that Hox d12 is not expressed in the first digit (left) while Hox d13 is expressed in all the digits (right).
*(Vargus & Fallon 2005)*

This seemed to support the notion that the bird digit arising from cII is the same as therapod digit 1. However, other research shows that mutant mice lacking Hox d12 can have normal digit identity in *all* their digits, and Hox d12 and 13 can be active in digits which can have one, two or three phalanges. So, the Hox d12/13 data alone cannot confirm that birds *have* digit 1 arising from cII. Therefore, the idea of a digit frame-shift:

> currently lack[s] convincing support. Galis et al. 'Hox genes, digit identities and the therapod/bird transition'. Journal of experimental zoology. Vol 304B. 2005. p198-205.

Finally, since we have seen unequivocal evidence to show that birds are not modified non-flying dinosaurs, even if birds and therapods *do have* digits 1, 2 and 3 arising from condensations II, III and IV, this *cannot* be evidence for common ancestry.

### *Bird DNA*

Despite the wealth of evidence that shows it cannot be so, it is still commonly asserted that birds evolved from dinosaurs:

> ...the chicken **is** a modern descendant of the dinosaurs. (My bold). International Chicken Genome Sequencing Consortium. 'Sequence and comparative analysis of the chicken Genome'. Nature. Vol 432. 9th December 2004. p695-713.

And so it is virtually impossible for scientific papers to remain objectively scientific. For example, one study involved comparing how antibodies react to T. rex soft tissue:

> In one experiment, antibodies that normally react in the presence of chicken collagen reacted strongly to the dinosaur protein, suggesting a similar molecular identity. National Geographic News, reporting on Science. 13th April 2007.

By suggesting that there is a 'similar molecular identity' (which is a valid conclusion), the report tries to give the message that this similarity confirms evolutionary relatedness. However, this is not a scientific conclusion at all. Antibodies that react strongly to proteins from the cell wall of the bacterium *Streptococcus pyogenes* also react strongly to the proteins on the surface of human heart valves. This is why antibodies produced as a response to some Streptococcus infections can subsequently attack the heart valves causing the serious disease *rheumatic fever*. We would not, from this knowledge, conclude that we are closely related to bacteria. Therefore, the finding of antibodies cross-reacting with T. rex and chicken allows no conclusion of chicken and non-flying-dinosaur relatedness.

Another analysis, which referred to a report that described the finding of the protein collagen (found in fibrous tissue) in a T. rex fossil, concluded that because the T. rex data conformed to amphibian rather than bird protein, the T. rex material was not authentic:

> We used authentication tests developed for ancient DNA to evaluate claims by Asara et al. (Reports, 13 April 2007, p. 280) of collagen peptide sequences recovered from mastodon and Tyrannosaurus rex fossils. Although the mastodon samples pass these tests, absence of amino acid composition data, lack of evidence for peptide deamidation, and **association of $\alpha$1(I) collagen sequences with amphibians** rather than birds suggest that T. rex does not. (My bold). M Buckley et al. Comment on 'Protein sequences from Mastodon and Tyrannosaurus rex revealed by mass spectrometry'. Science. Vol 319. No. 58594. January 2008. p33.

Of course, Buckley et al *assume* that T. rex *must be* related to birds and this prevents them from going where the evidence leads. Unless it is proved that there has been contamination by amphibian tissue, the similarity of the T. rex $\alpha$1(I) collagen sequences to that of amphibians means that this dinosaur was related to frogs, or that it was not related to birds, or that the similarity simply has no bearing, whatsoever, on (supposed) evolutionary relatedness.

Above: Emu claw.
Below: Extinct giant sloth.
The keratin proteins of birds are quite different from those of mammals.

Furthermore, the sequencing of the chicken genome has provided some contradictions to the evolutionary paradigm. For example, if birds and mammals do have a very distant common ancestor, we should predict that they should utilise the same keratin proteins, which would differ only as a result of minor mutational alterations over time. However, we find that:

> ...scales, claws and feathers of birds are formed using avian [bird]-specific family of keratins, whereas hair fibre formation in mammals involves a distinct keratin family. International Chicken Genome Sequencing Consortium. 'Sequence and comparative analysis of the chicken Genome'. Nature. Vol 432. 9th December 2004. p695-713,

which shows that birds and mammals *never* had a common ancestor. And yet the authors *start* the paper with the unscientific dogmatism that birds and humans once shared great...great grandparents:

> We present here...a draft sequence of the chicken genome, **which has** evolved separately from mammalian genomes for ~ 310 million years. (My bold). International Chicken Genome Sequencing Consortium. 'Sequence and comparative analysis of the chicken Genome'. Nature. Vol 432. 9th December 2004. p695-713.

The authors go on to describe evidence that shows that the typical evolutionary tree of life is pure fantasy, since the arrangement of genes on human chromosomes is more similar to that of chickens than mice:

> The synteny maps confirm an earlier observation that the human genome is **closer to the chicken** than to rodents in terms of chromosomal organisation of genes. (My bold). International Chicken Genome Sequencing Consortium. 'Sequence and comparative analysis of the chicken Genome'. Nature. Vol 432. 9th December 2004. p695-713,

In addition, the loss of a whole family of genes would be surprising:

> Loss of an entire domain family in the protein repertoire is unusual. Nonetheless, the SCAN domain family seems to be unrepresented in the chicken gene set and genome. International Chicken Genome Sequencing Consortium. 'Sequence and comparative analysis of the chicken Genome'. Nature. Vol 432. 9th December 2004. p695-713.

Rather than losing them, the probability is that birds never had SCAN genes in the first place. Finally, there is a clear absence of what evolution believers would call *'conserved'* genes. From an evolutionary perspective, conserved genes are those which are found very frequently in all types of living organisms, having been passed down from a common ancestor, often having vital functions. So birds should have their equivalent share of such genes. However:

> Overall, we find that among the predicted chicken genes, there is a notable under-representation of genes that are widely conserved and were presumably present in the mammal-bird common ancestor. International Chicken Genome Sequencing Consortium. 'Sequence and comparative analysis of the chicken Genome'. Nature. Vol 432. 9th December 2004. p695-713.

Losing so many genes that other creatures have (supposedly) retained so easily is downright negligent. Again, rather than losing them, since birds did not evolve from non-birds, the probability is that birds never had these genes in the first place. While teeth are found in a number of extinct birds but not in any living birds, the zebrafinch genome shows evidence of a fragment of an enamel-protein producing sequence:

> We have found AMEL [amelogenin] exon 2, with a deletion of 12 bases and a base substitution leading to a premature stop codon. (J Sire et al, unpublished data). J Sire, S Delgado & M Girondot. 'Hen's teeth with enamel cap: from dream to impossibility'. BMC Evol Biol. 2008; 8: 246.

Since birds lack enamel genes, they cannot produce teeth. However early tooth formation has been documented in the mutant chicken *talpid 2* (ta2):

> We show the initiation of tooth developmental programs as well as the formation of conical saber-like structures on the lower jaw of the ta2 chicken. The structures formed are similar to those seen in the first-generation teeth of the alligator in position, histological differentiation, and morphogenesis...
> The formation of teeth in the mutant is coupled with alterations in the specification of the oral/aboral boundary of the jaw... We propose that the oral/aboral boundary establishes a signalling center that, depending on its apposition to underlying competent mesenchyme, controls the initiation and suppression of teeth... M Harris et al. 'The development of archosaurian first-generation teeth in a chicken mutant'. Current biology. Vol 16. 21st February 2006. p371-77.

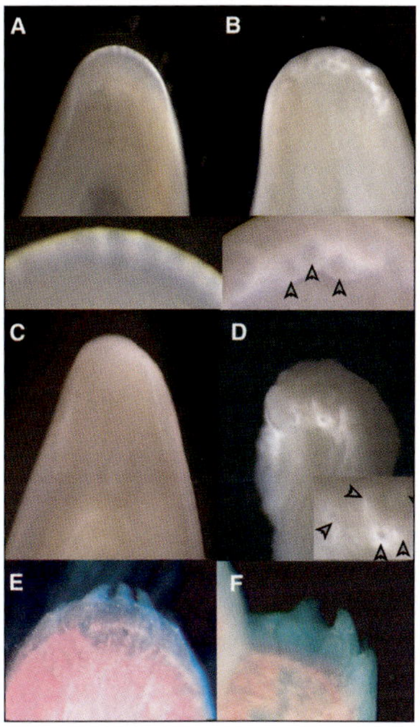


**Early tooth formation in talpid 2 chick**

B and D: Enamel-less teeth on the distal mandible and premaxillae of a ta2 chick.
A and C: Normal chick jaw for comparison.
D inset: Magnified view of the tooth formations.
E and F: Mandible of near-hatching ta2 jaw. The beak has been removed, uncovering underlying teeth at the tip.

*M Harris et al. 'The development of Archosaurian first-generation teeth in a chicken mutant'. Current Biology. Vol 16. 21st February 2006. p371-77.*

Since tooth formation requires *both* competent oral epithelium *and* competent underlying mesenchyme, the developmental cascades of which are both independently activated (in other words, one without the other has *zero* capability to form teeth), the *WYWYWG Principle* shows that no ancient *non-toothed-hoping-to-produce-a-tooth* creature *could* acquire the necessary cascades of signals. In addition, it is commonly claimed that birds lost their teeth a long time ago:

> Avian teeth were lost at least 70-80 million years ago. M Harris et al. 'The development of archosaurian first-generation teeth in a chicken mutant'. Current biology. Vol 16. 21st February 2006. p371-77.

Since (supposedly) 80mya a *variety* of fossil birds had teeth, all living birds cannot be descended from a single toothless pair. Therefore, since it is impossible for *all* of those diverse toothed birds to have lost their teeth, many living birds should have teeth. Therefore, there must have been some *global* population bottleneck that radically affected diverse bird types.
Furthermore, if birds lost tooth-forming potential 80mya, by now the genetic control of the formation of competent oral epithelium or competent underlying mesenchyme should be in *complete* disarray, so that neither epithelium nor mesenchyme should now retain any competence *at all*. But the *ta2* mutation shows that dental competence of the mouth mesenchyme is still functional. There are two conclusions that can be

made from this. One is that the systems for early tooth development have *additional vital functions* elsewhere in the chick embryo. This could only mean that anything less than a completely competent system would give rise to serious problems. And, this in turn, means that the system could not have arisen step-wise in a series of less-than-adequate components. The other conclusion is that the tooth development system *does not have vital functions* elsewhere in the chick embryo. Since mutations would certainly inactivate the system after millions of years, this means that chickens have not been around for millions of years. As previously discussed, the loss of teeth would not produce a beak. So, either way, the *ta2* mutation confirms that evolution is impossible.

So we are left with the problem as to how all the *mutally independent* and *unrelated* genes that are necessary to make a bird, could have accumulated in a (supposed) ancestral reptile and a line of its descendants. Suppose (although it is unlikely) that just a number of single mutations could result in completely different structures.

For example, suppose* a dinosaur had a mutation that caused it to gain a *beak*.

*Note that this example of a series of suppositions is being very, very generous to those who cling to evolution.
It must be understood that 'a mutation' would never produce a unidirectional lung or any of the other features mentioned. A whole suite of mutations would be required in each case, together with simultaneous changes in many regulative sequences in the non-coding DNA (which is now recognised not to be 'junk' at all).

Suppose a second dinosaur (geographically separated from the first) had a mutation that caused it to have a *unidirectional lung system*. Perhaps it might be able to function more vigorously than its fellows because of the benefits that the improved oxygenation might allow.

Suppose a third dinosaur (geographically separated from the first and second) had a mutation that caused it to develop *feathers*. It would be more likely to retain its body heat and thus be more active than its fellows.

Suppose a fourth dinosaur (geographically separated from the first, second and third) had a mutation that caused it to develop a *four-chambered heart*. This would allow for improved oxygenation of the tissues and most likely the animal would be more active than its fellows.

Suppose a fifth dinosaur (geographically separated from the first, second, third and fourth) had a mutation that caused it to develop hair. It too would be more likely to retain its body heat and be more active than its fellows.

In time, perhaps dinosaurs 2 and 5 could have gained a beak, and dinosaur 4 could have developed feathers. Now, if these five types of mutations all happened in the same lineage, it might be expected that eventually a bird would appear. But the *WYWYWG Principle* shows that the accumulation in a single lineage of such a series of mutations is implausible. It is far more probable that if the different mutations could appear, then they would appear and remain within *different* and geographically separate lineages. Thus, it is likely that no bird would or could emerge.

However, suppose, against all the evidence, that birds *did* evolve from dinosaurs. Then, if it *were* possible for some dinosaurs to acquire the necessary genetic changes, then it *must* have been possible for many of the changes to arise, stabilise in, and benefit a number of other vertebrates too. So, we should expect to find today, as well as true birds and true mammals, a number of hybrids, for example: some mammals with downy feathers instead of hair (both feathers and hair are efficient at minimising heat loss, and having downy feathers causes no problems to non-flying birds such as the kiwi or ostrich, so having downy feathers would cause no problem to a mouse or an antelope); some birds with hair on the body but feathers on the wings and tail, since the genes for flight feathers are separate from the genes for body feathers (such a combination would most certainly allow such animals to fly since bats manage to fly with hair-covered bodies); some bats with beaks; and some bats with a bird-type unidirectional lung system or, conversely, some birds with mammalian-type lung systems. But since we do not see *any* such hybrid forms, this goes against what the notion of evolution should predict.

So, going back to the subject of feathers, which is where most writers start and finish their discussion of the (supposed) evolution of birds from reptiles: even though the biological evidence says otherwise, it is curious that people are quite willing to accept that, by a random process, feathers evolved in animals that previously did not have them:

> The birds are characterised by this one unique feature, the possession of feathers, and it is very likely that these evolved only in a single lineage. John M. Smith. 'The theory of evolution'. Cambridge University Press. 1993. p304.

> All birds - and only birds - have feathers. Feathers are complex in terms of their structure and their development...The importance of all this in an evolutionary sense is that **feathers evolved (or originated) only once**. After all, **what are the chances of so complex a structure having evolved more than once**? Using parsimony, we must conclude that feathers evolved only one time... (My bold). David Fastovsky and David Weishample. 'The evolution and extinction of dinosaurs'. Cambridge University Press. 1996. p294.

Here Smith, Fastovsky and Weishample assume too much: that only birds can have feathers, and that birds came into being by an evolutionary process. Furthermore, since they are quite content to deny the possibility of a feather evolving more than once, then on that basis, they should also argue that an A2V2 system, or a unidirectional respiratory system, should each have evolved only just the once. That being the case, it should be argued that any two complex structures would most likely *not* have evolved fortuitously in the same lineage. So, from what we have seen, it should be logical to ask: how could so complex a *series* of alterations cause a reptile to ever evolve into a bird *at all*? Using parsimony, we should conclude that it never did.

The pattern that emerges is that all of the higher vertebrates have, or have had, flying members: reptiles had their pterosaurs; dinosaurs had their archaeopteryx; mammals have their bats; and most birds can fly. And we under no obligation to accept that any of these flying animals descended from non-flying ancestors.

The mental hurdle to overcome is the acceptance that feathers are not the sole preserve of birds. Just as the chalicotherium had a horse-like head but was quite unrelated to horses, so we should not be averse to the notion that although feathers are found in both dinosaurs and birds, this does not mean that feathered dinosaurs evolved into birds. Indeed, whether in dinosaurs or birds, we have seen good evidence to show that it is implausable for feathers to have evolved *at all*.

Thus, we should conclude that birds are unique creatures which, despite sharing a number of features in common with some dinosaurs, are nevertheless *so* different that it is implausible that birds are descendants of non-flying dinosaurs. So, in that context, we do not eat roast dinosaur at Christmas.

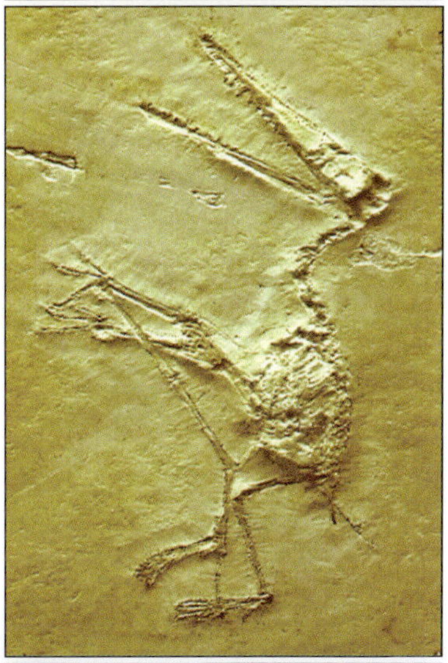

**Flapping vertebrates**
Top to bottom:
Flying dinosaur - archaeopteryx.
Fying reptile - Rhamphorhynchus.
Flying bird - pelican.
Flying mammal - fruit bat.

# CHAP

## The eye:

*image consciousness*

Elephant eye

PTER 9

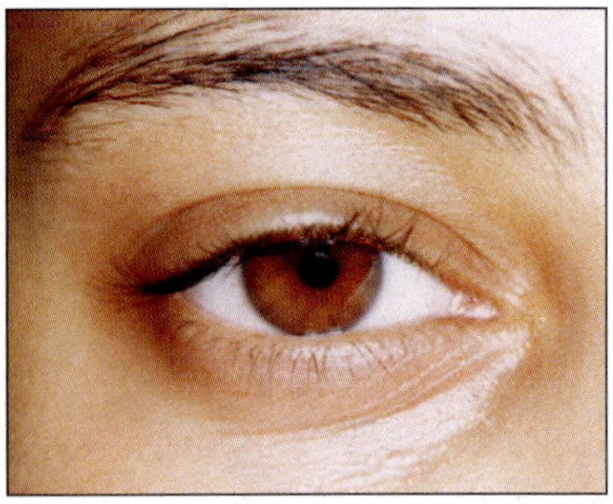

Human eye

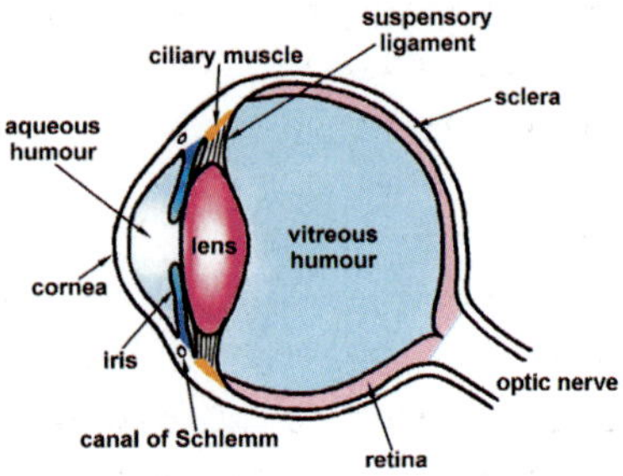


Mammalian eye

Cat with one blue and one brown eye.

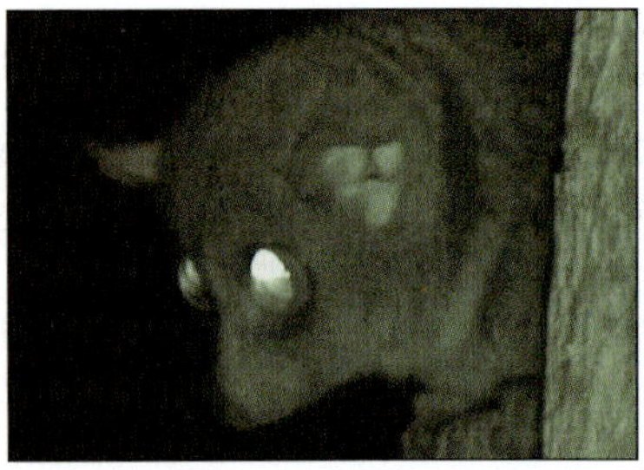

Eyes shining in torchlight show that many nocturnal creatures have a special layer beneath the retina which serves to reflect back any light that has passed through the retina, without stimulating the rods. This allows the rods a second opportunity to capture as much dim light as possible.

Whilst we must not let its sheer complexity cloud the issue of whether or not the human eye could, by chance processes, have evolved from nothing, the eye is not, as many a biologist might suggest, a structure based simply on modifications of the principle of the pin-hole camera model. Although we can name and have some considerable understanding of many of the large number of different facets that make up the eye, even so, this does no justice to the *supreme* complexity of such components and of the eye as a whole.

The human eyeball has a fibrous outer wall which is mainly opaque white (the *sclera*) but which has a transparent front part (the *cornea*). A complex arrangement of six individual muscles attached to the outside of the sclera allows for movement of the eyeball in any direction. However, absolute synchrony must exist in the activity of the muscles of the two eyes, since, if there is even a tiny difference in the direction in which the two eyes point, a squint will occur, resulting in double vision. In severe cases this can result in blindness in one eye since the brain will ignore the image coming from it, preferring to take note only of the image from the other.

On the inner side of the sclera lies a vascular layer (the *choroid*) which conveys blood to the whole of the eye. The eye has two main cavities, one in front of the *lens* which is filled with a watery liquid (the *aqueous humour*) and one behind the lens containing a jelly-like substance (the *vitreous humour*). After passing through the cornea, the light then passes through the *pupil* which is an opening in the middle of the *iris*, an elaborate system of muscles that can automatically increase or decrease the pupil size. The light then passes through the transparent *lens* which is held in position by the circumferentially-attached *ciliary muscle*. Contraction of the radial fibres of the ciliary muscle causes the lens to become thinner, and contraction of the circular fibres allows the lens to thicken by its own elastic recoil. After passing through the lens, the light passes through the vitreous humour and reaches the *retina* at the back of the eye. The retina lies on the inner side of the choroid and is composed of three main layers of cells: the innermost layer itself comprises a number of layers of special *nerve* cells which receive impulses from the light-sensitive *rod* and *cone* cells of the middle layer of the retina, and relay the impulses to the brain via the *optic nerve*. The outermost layer of the retina comprises *pigment cells*. In the condition *detached retina*, which is due to a defect in the cohesion between the pigment cell layer and the rod/cone layer, the retina tears away from the inside of the eyeball and this can result in serious visual impairment.

The rods and cones have light-sensitive pigments (*opsins*) located on tiny discs stacked within their outer segments. When photons hit these pigments, intricate chemical pathways transform the light energy into nerve impulses. The number of photosensitive cells in the retina can vary considerably: the *Nautilus* (a free-swimming mollusc that looks like a squid with a shell) has 4 million; humans have 130 million. In humans, the photosensitive cells are packed in at 200,000 per square millimetre, whereas the more sensitive eye of the buzzard has 1 million per square millimetre.

The eye is only one part of an elaborate system that results, ultimately, in vision. The impulses from the retina pass along the optic nerve to the *visual cortex* in the back of the brain. In mammals, the nerve fibres from the inner halves of the retina of each eye cross over on their way to the brain so that the information from the nasal side of the right retina is carried to the left side of the brain (and vice versa). Therefore, in order to produce a complete image, the two sides of the brain (the *cerebral hemispheres*) have to communicate with each other via a special system of nerves. Since, in both eyes, the light from the right visual field falls onto the left half of the retina (and vice versa), destruction of the visual cortex in the left cerebral hemisphere does not result in blindess, but instead results in the loss of the right half of the visual field of *both* eyes. There is no evolutionary trend in this unique set-up since, in all non-mammalian vertebrates, the visual fibres of each optic nerve cross over completely so that all the nerve fibres from the left and right eyes pass, respectively, to the right and left sides of the brain.

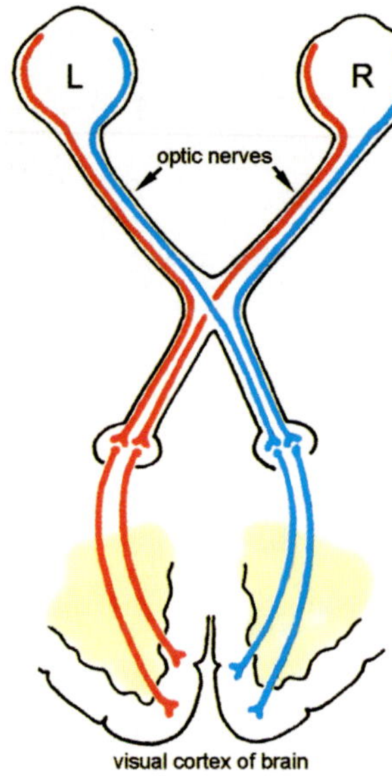


In mammals, nerves from the right half of each retina go to the right side of the brain (and vice versa). In other vertebrates, the nerves from each eye cross over completely so that all the information from the right eye is sent to the left side of the brain (and vice versa).

The nerve cells of the visual cortex communicate with other nerve cells in nearby parts of the brain such that ultimately, by a process itself completely unknown, the impulses from the eyes are *perceived* in the *mind* as a representation of the outside world.

In the eye, the image projected by the lens onto the retina is inverted and also transposed from left to right. In other words, the floor is projected to the upper part of the retina, the ceiling to the lower part, the right-hand side of the room to the left, and the left-hand side of the room to the right. However, in computing the retinal information, the brain actually corrects the orientation so that *in the mind* the virtual image is displayed in the correct orientation, so that the floor is seen where it actually is, below the feet.

To demonstrate the brain's ability to alter the orientation of the retinal image, there is a simple experiment that can be done using special prism spectacles. The prisms have the effect of reflecting the light so that the ground is projected to the lower part of the retina, and the ceiling to the upper part. Since, in the mind, the brain normally displays whatever is on the upper part of the retina as being towards the ground (and vice versa), to the wearer of the prisms it looks as if the world is upside-down. However, the brain does not accept without question the information coming to it from the eyes, but correlates the visual image with data from the other senses and computes what it thinks is the best view of the surrounding world. As a result, after continuously wearing the spectacles for a few days, the brain *corrects* the information coming from the eyes so that the apparently upside-down world is once again seen the right way up. After a few days the brain becomes accustomed to displaying whatever is on the upper part of the retina as being towards the ceiling (and vice versa) so that when the spectacles are removed, it looks to the wearer as if the world is again, upside-down. Fortunately, after yet another few days without the spectacles, the brain manages to re-compute its virtual image so that vision is once again back to the normal state before the experiment began.

This ability of the brain to interpret correctly the upside-down image received by the retina *is present from before birth*, and most certainly does not take a few days to become formulated. This is well demonstrated by the fact that many animals, within a few minutes of birth, manage to wobble onto their feet, and in only a short while longer, are able to run alongside the mother, without any problem in negotiating obstacles in their way. So their brains must be programmed for what is up and what

Nautilus

is down *before any light ever reaches the retina.*

It is immediately apparent, therefore, that in view of the tremendous complexity of the visual system, it is not possible to discuss the evolution of the eye without also considering, for example, the chemistry involved in converting light into electrical impulses, the embryological development of the eye, and even consciousness itself. Since it is a common belief that the human eye is the result, over time, of small gradual alterations arising in simpler, more primitive eyes, it is necessary to compare the eyes of a number of different creatures.

### *Different types of eyes*

Many different types of light-sensitive organs exist in the animal kingdom: at one extreme are structures that can just distinguish between light and dark; at the other extreme are the mammalian and bird eyes which can resolve minute differences in light intensity, colour, and form. However, from simple animals through to birds and mammals, whether we consider, for example, general eye shape, the pupil, the eyelids, the lens, or the rods and cones, there is no clear pattern to suggest any evolutionary trend from simple to complex eyes.

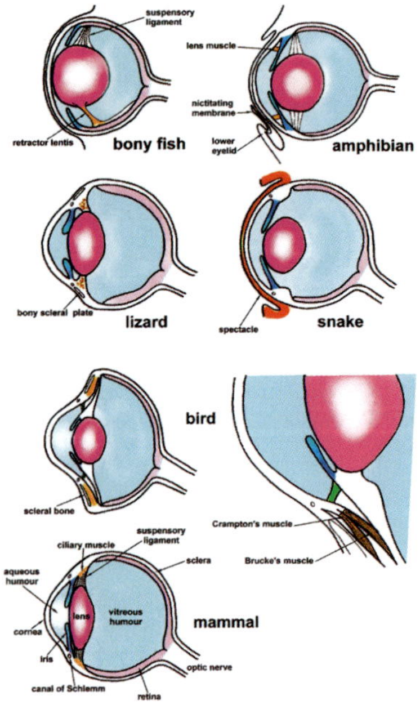


The structures of different vertebrate eyes show no evolutionary trend.

*Leeches* and *flatworms* have light-sensitive pigment cells around a simple bowl-like depression on the animal's surface, with which they can perhaps register a vague direction of incident light, or a shadow if something is close enough to cover the bowl. Clearly however, since the leech can move only extremely slowly it cannot use its eye to escape a predator, or to get a meal. It is likely therefore that in terms of vision, a normal leech has no significant advantage over a 'blind' leech.

The *Amphioxus* is a very simple fish-like marine animal about 5cm long. Its 'spinal cord' shows little in the way of thickening at the front end, and so it has little in the way of a brain. Instead of eyes, it has photosensitive cells in a pigmented 'eye' spot at the front end of the body. The amphioxus is found buried in the sand of coastal waters all over the world and lives on whatever particles come by in the water, so it clearly does not use its eye spot either to see food or to evade predators. Whilst leeches and flatworms are invertebrates, the (supposed) earliest vertebrates were creatures like amphioxus. Yet the eyes of leeches are more sophisticated than those of amphioxus, and moreover, amphioxus has managed to survive for (supposedly) millions of years without any further evolution of its 'eye'.

The lensless spherical eye of the nautilus is commonly described as being a simple pin-hole camera type of eye. It has a circular opening at the front which can range from 0.4mm to 3mm in diameter and which allows sea-water to pass directly through into the interior of the eye. Although the resolution improves as the size of the pin-hole gets smaller, the disadvantage is that less light gets through so the image becomes darker. So, if the pupil were to constrict down to its smallest size, this cannot be of any real use to it since:

Four-eyed fish. This unusual fish from Central America has prominent frog-like eyes which are divided into two parts, allowing it to see clearly both above and below the water simultaneously.

> ...most of the time its underwater home is not well-lit... John Downer. 'Supersense'. BBC Books. 1989. p37.

And since, without a lens to do the focusing, a blurred image is inevitable if the aperture is made larger than an actual pin-hole, any image formed on the nautilus

# Is the eye a poor design that must have evolved by trial and error?

In response to the idea that the eye is so complicated that it could not have appeared just by chance, here's a diatribe by one self-styled 'expert' on biological optics:

*"I'm taking my spectacles off and I can't see clearly. The eye is really very inefficient. If it were designed, He's a very bad designer. The eye evolved to be just as good as it needs to be. Why do we have microscopes? Why do we have telescopes?" Steve Jones. Professor of genetics. Jeremy Vine Show. BBC Radio 2. 2006.*

It is important to note here that Vine did not challenge such absurd logic, and unwary listeners may well have accepted Jones' assessment by virtue of his being a professor (albeit of a subject that, by itself, makes him no expert on origins). Of course, by saying that the eye is a bad design, or can't be the result of an Intelligent Architect and Manufacturer because sometimes there is faulty embryological development or tissue degeneration, is like saying that, overall, a space shuttle is a bad design or cannot have been the work of an intelligence because it has been known to disintegrate on re-entry as a result of some faulty insulating tiles. Of course, we all know that, despite an occasional fault, a space shuttle is *incredibly* well designed and manufactured.

It is a fact that although the human brain is not *conscious* of the arrival at the retina of less than nine or so photons (for reasons that would be, literally, blindingly obvious if we were conscious of every photon that hit the retina), the vertebrate retinal rod can detect the arrival of a *single* photon:

*This work examines the mechanism that enables retinal rod cells to register single photon absorptions with macroscopic signals of constant size and shape. F Rieke & D A Baylor. 'Origin of reproducibility in the responses of retinal rods to single photons'. Biophysical Journal. Vol 75. October 1998. p1836–185.*

Jones is not alone in promoting inane 'science' since he walks in the dark, hand-in-hand with the editors and writers of major scientific journals. For example, the unwary reader might not see through a recent New Scientist article with its *'the eye is badly-designed'* propaganda:

*The most famous flaw is found in vertebrate eyes. The light-sensing structure, the retina, is wired back-to-front, with the light-sensitive cells behind the nerves and blood vessels that support it...* ***Verdict: Back-to-front retinas are a mistake whichever way you look at them.*** *(My bold). C Ainsworth, M LePage. 'Evolution's greatest mistakes - blind spot'. New Scientist. 11th August 2007. p38.*

Not only was the New Scientist's science inane and flawed, it was published *over 3 months after* the paper in May 2007 which described the function of the retinal *Müller cells,* which lie between the retinal nerve cells and span the entire thickness of the retina. The authors of that article on retinal image-making rightly reasoned that although the 'inverted' retina might seem to be a problem, because incoming light should be refracted and scattered by passing through the nerve cell layer before hitting the rods and cones, nevertheless, since most of us can (literally) see that an eye with an 'inverted' retina *does* work *extremely* well, there *must* exist some previously unidentified mechanism to counter any tendency for light scattering. What they found was that Müller cells:

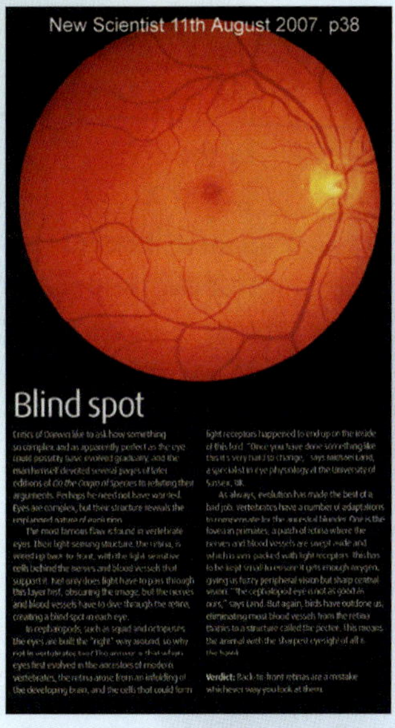


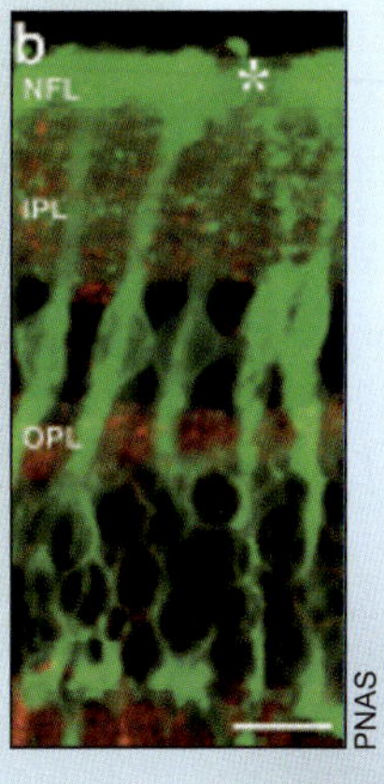


New Scientist's misleading declaration *'Verdict: Back-to-front retinas are a mistake whichever way you look at them'* was published *over 3 months after* the PNAS paper in May 2007, which described the function of the retinal Müller cells (right, in green), with which the editors should have been familiar, but failed to inform their readers.

*...have an extended funnel shape, a higher refractive index than their surrounding tissue, and are oriented along the direction of light propagation...[Müller cells] provide a low-scattering passage for light from the retinal surface to the photoreceptor cells... their parallel array in the retina is reminiscent of fiberoptic plates used for low-distortion image transfer... This finding elucidates a fundamental feature of the inverted retina as an optical system and ascribes a new function to glial cells. Franze et al. 'Müller cells are living optical fibers in the vertebrate retina'. Proc. National Academy of Sciences USA. 104(20). 15th May 2007. p8287–8292.*

Since without the 'optical fibres' a blurred image *would* result, and since an inverted retina without Müller cells *would* be a serious disadvantage, and furthermore, since it is implausible that the highly specific form of the Müller cells would arise by virtue of random mutations *exactly* where they are located, the new evidence conclusively confirms what we knew already: not only that the inverted retina is a design which works brilliantly and which cannot have arisen stepwise, but also that the eye *must* have been designed and manufactured 'from the ground up' in its complete and fully-functioning form.

See also pages 332-3.

retina cannot be very clear. Just how essential the eye of the nautilus is to its survival is not clear, but it could certainly improve its image-forming capabilities if it had a lens.

In the living world, a variety of lens-like structures can be found. For example, the worm *Nereis* has an eye filled with jelly, which acts as a poor lens. However, using a single large lens is not the only way that light can be focused, and the light-sensitive pigment cells of some jellyfish eyes each have a thickened transparent covering that acts like a feeble lens.

It is said that trilobites, a group of animals which (supposedly) evolved over 570 million years ago, were the first animals to develop high-definition eyes. For example, *Olenellus* had:

Close-up of head of a trilobite showing its complex eye.

> ...large crescent-shaped eyes...it is found in strata of the Lower Cambrian age... Chris Pallent. 'An illustrated guide to fossils'. Dragon's world. 1995. p82.

Even the earliest fossils of trilobites show evidence of eyes, the structure of which cannot be described as either simple or primitive. Many trilobite eyes were complex mosaic sensors comprising a cluster of perhaps 15,000 units, each with its own crystalline lens. Other trilobites had eyes with fewer units but with two-part lenses with a waved surface at the junction between the two lenses. One giant-eyed trilobite was able to see simultaneously in almost all directions, up, down, back and front. Incongruously, the much simpler eye of nautilus apparently evolved millions of years *later* since nautiloids:

> ...are first found as fossils in strata of late Cambrian age [just over 500mya]. Chris Pallent. 'An illustrated guide to fossils'. Dragon's world. 1995. p125.

> Nautiloids are primitive, marine cephalopods...They were most abundant in the Palaeozoic Era, 400 million years ago... Cyril Walker & David Ward. 'Fossils'. 1992. p134.

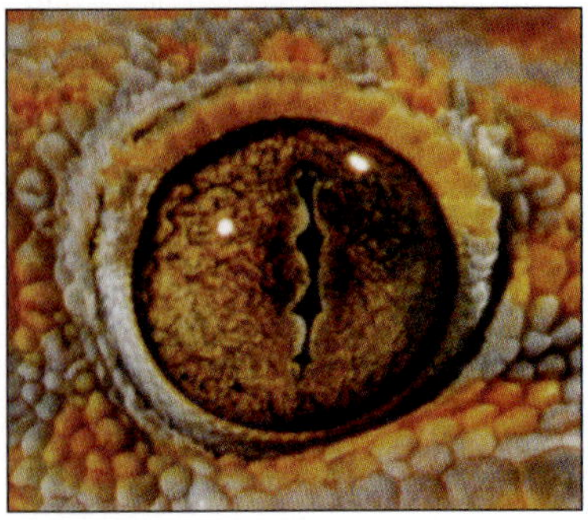

Close-up of gecko eye.

Although generally thought to be the more primitive, sharks, unlike most bony fishes, have pupils that can be contracted to protect against bright light. In the (supposedly more advanced) bony fishes, the pupil is essentially fixed in size. However, the pigment in the retina is able to migrate between the rods and cones, and the photosensitive cells have unique contractile elements within them, so that:

> In bright light the pigment expands, the cones contract forward, towards the light and the rods contract back, beneath the pigment. These photomechanical changes thus serve the same end as changes of pupil diameter in other vertebrates. J Z Young. 'The life of vertebrates'. 3rd edition. Clarendon Press. 1991. p177.

This makes no evolutionary sense because, since both sharks and bony fishes live in the same environments, we should expect that instead of evolving a completely different system to cope with varying light intensity, most bony fishes should have acquired the excellent adjustable pupils of their (supposed) ancestors, the sharks, and handed down the adjustable pupil to their (supposed) descendants, the amphibians and terrestrial vertebrates.

Pupils of other vertebrates show no particular pattern or trend. Most amphibians have circular pupils. Some lizards and snakes have vertical pupils while others have round pupils. Geckos have vertical pupils with a series of small circular openings. Crocodiles have vertical pupils. Chameleons have circular pupils and the eyes can turn independently in any direction.

Whilst these different pupils clearly satisfy their owners' visual needs, there is no particular evidence to suggest that these different pupils are essential for their survival or that they exist as a result of the natural selection of the circular pupil form. Take horses, for example, which have eyes with horizontal slits for pupils. It might be argued that a horizontal pupil would allow a horse to see predators over a wide horizontal area. However, if our human eyes with our circular pupils and 120° (per eye) horizontal field of vision were placed on the sides of our heads, our vision would be quite adequate in spotting predators in open grasslands. A field of vision of 120° (whether using both eyes, or only one eye) is certainly adequate for a UK driving licence. And, in fact, with largely frontal vision, baboons and humans have managed to survive very well in lion country. Furthermore, zebra and wildebeest rely to a large extent on smell, hence predators such as lions approach their prey from downwind. So to a horse there is no clear advantage in having horizontal pupils over the circular pupils of their supposed great...great grandparents, the sharks. Similarly, there is no clear advantage in the different pupils of other animals. For example, domestic cats have vertical pupils, yet other creatures with acute vision such as birds, big cats and many other mammals, including dogs and primates, have circular pupils.

Not only is there no clear evolutionary trend in the form of the pupil, but also the actual nervous control of the contraction and dilatation of the pupil shows no trend. The *sympathetic* and *parasympathetic* nervous systems act involuntarily to control many muscles and other bodily functions such as intestinal contractions, heart rate and pupil size. In flat fishes, the fibres of the sympathetic nervous system control the contraction of the pupil, and dilatation is effected by the action of fibres in the *oculomotor* nerve. This is the *opposite* arrangement to that found in mammals. In contrast, in sharks and in the eel, the contraction of the pupil is a result of the action of light directly upon the muscle of the iris itself, no nerve reflex being involved at all.

Unlike all bony fishes, many sharks have upper and lower eyelids. In addition, many animals also have a third eyelid, the *nictitating membrane*, which is a very thin structure that moves horizontally across the eye. It is found in blue sharks and hammerhead sharks but is not found in any bony fishes or amphibians. However, a nictitating membrane is seen in crocodilians and most lizards.

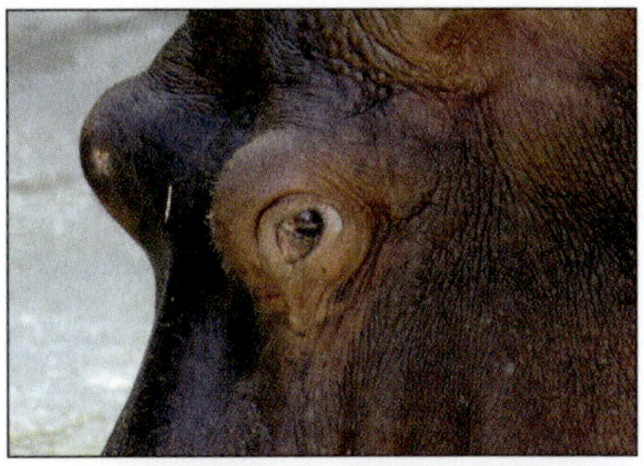

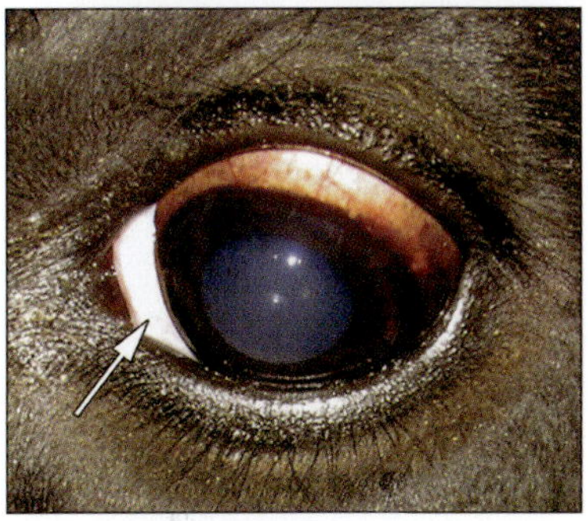

**Nictitating membranes**

Above top: Hippo.
Above: Cow.
Below: Cat.
Bottom: Dog.
Bottom left: Alligator. The nictitating membrane of the alligator underneath has completely covered the eye. The alligator on top is missing the tip of its lower jaw.

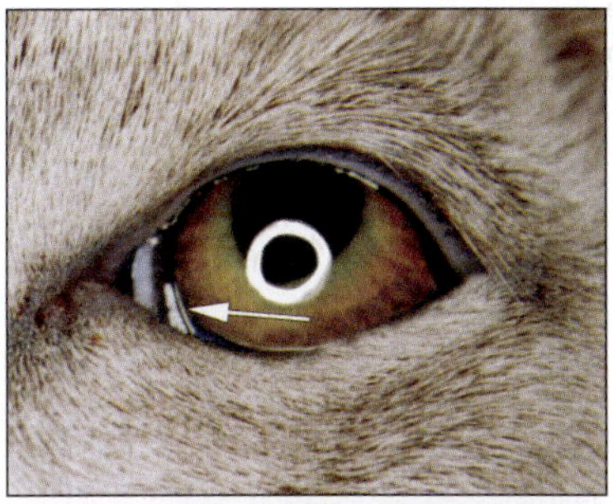

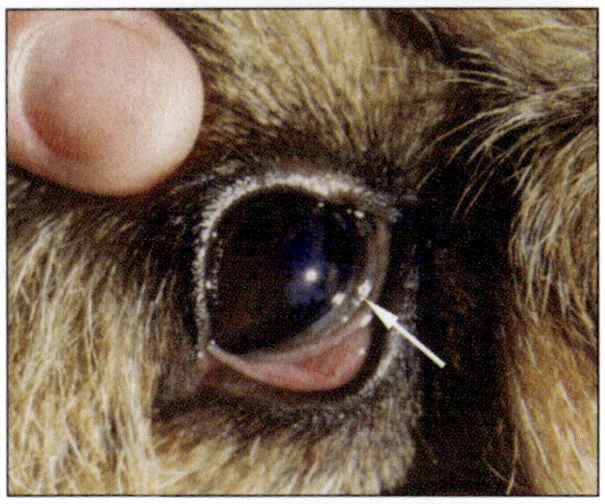

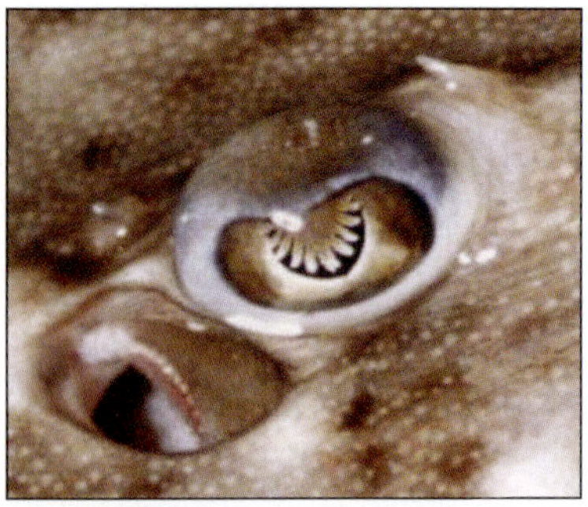

Above: Stingray eye.
Below: Dogfish eye.

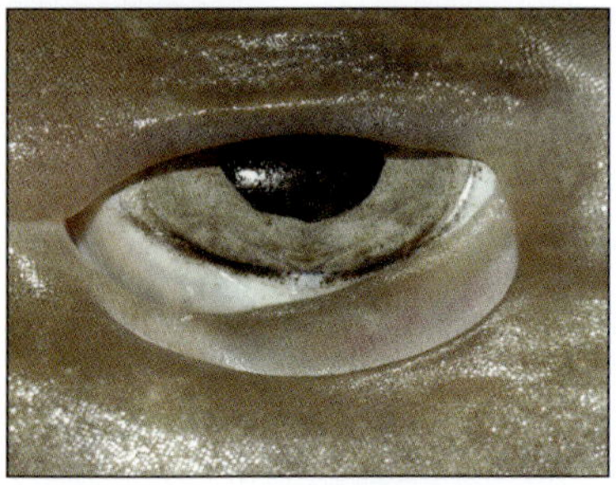

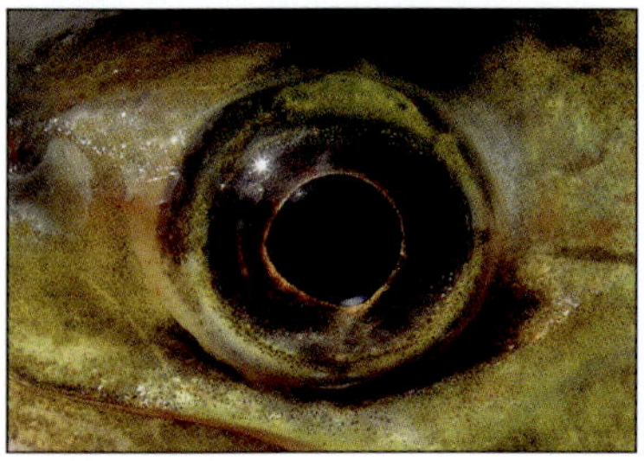

Above: Trout eye.
Below: Frog eye.
Bottom: Iguana eye.

Some mammals (such as dogs, cows and elephants) also have a nictitating membrane, yet others (such as humans) do not. Clearly, a nictitating membrane is not essential for vision either in water or in air.

The problem is that over time, the genes coding for any characteristic are liable to undergo deleterious mutations. If a characteristic is *not essential* to survival, then the lineage of creatures in which that characteristic exists will continue to live on, but eventually, that characteristic is likely to become degenerate or lost. So, as a result of the inevitable mutations that would have appeared after millions of years, sharks, dogs, cows and elephants should by now have lost their nictitating membranes. That they still retain them suggests that these animals have not existed for millions of years.

The term *accommodation* is applied to the ability of the lens to vary its focusing between near and distant objects. In this regard the lenses of fishes are distinctly different from those of all other higher vertebrates. Most fish eyes are spherical with a short focal length, and for accommodation the lens is moved as a whole, towards or away from the retina, by using special muscles. However, the muscular mechanisms are arranged differently in lampreys, sharks and bony fishes. In sharks, the active contraction of the *protractor lentis muscle* provides:

> ...accommodation for near vision by swinging the lens forward... J Z Young. 'The life of vertebrates'. 3rd edition. Clarendon Press. 1991. p144.

Elastic recoil allows for accommodation for distant vision. In contrast, in bony fishes, accommodation for distant vision is provided by the active contraction of the *retractor lentis muscle* and elastic recoil allows for accommodation for near vision:

> The eye is...accommodated for distant vision by pulling the lens nearer to the retina... J Z Young. 'The life of vertebrates'. 3rd edition. Clarendon Press. 1991. p178.

In contrast, the lens of the frog is unable to change shape:

> Accommodation is effected by protractor lentis muscles...These muscles move the lens forward for near vision whereas in teleostean fishes the muscles move the lens backwards... J Z Young. 'The life of vertebrates'. 3rd edition. Clarendon Press. 1991. p263.

Lizards are different again:

> Accommodation for near vision is usually produced by the ciliary muscles, so arranged that they cause the cilliary process to squeeze the lens, making its anterior surface more rounded... J Z Young. 'The life of vertebrates'. 3rd edition. Clarendon Press. 1991. p289.

In snakes, however, the ciliary body contains no muscle, and:

Above: Python eye. Snakes have a modified scale completely covering the front of the eye. Accommodation is attained by contraction of muscles attached to the spectacle, which results in backward movement of the lens.

Python dissection.
Top right: Spectacle in place over eyeball.
Right: Spectacle partially pulled away.

..accommodation is brought about in a manner unusual for reptiles, involving displacement of the lens... J Z Young. 'The life of vertebrates'. 3rd edition. Clarendon Press. 1991. p309.

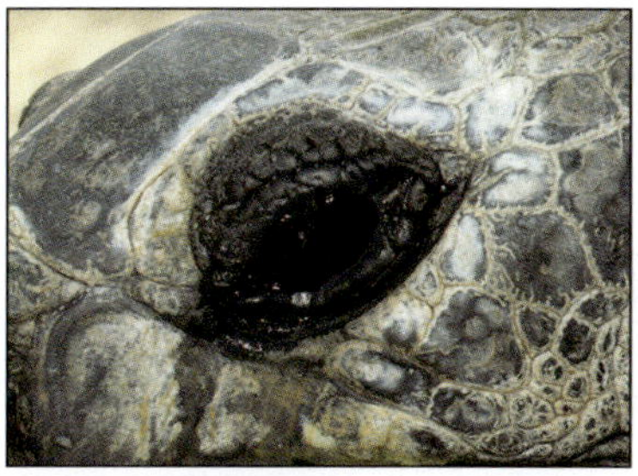

Giant tortoise eye.

The tuatara is mainly nocturnal and lives in burrows and has lived (supposedly) for more than 200my. Yet it has not lost either its eyelids or its nictitating membrane.

Snakes have a *spectacle*, a special thin tissue layer in front of the cornea. By the contraction of a circular muscle attached to its rim, the tightening of the spectacle pushes the lens backwards towards the retina.

Incongruously from an evolutionary viewpoint, except for a few lizards, a spectacle is not found in any other reptile, amphibian, bony fish or shark. The only other creature with a spectacle is the lamprey, (supposedly) a very primitive vertebrate, which clearly has no relationship to snakes. Whilst it is true that some lizards such as the *African striped skink* have a transparent window in the lower eyelid, that some skinks and geckos have a spectacle, and that a number of skinks vary considerably in the size of their limbs (for example, the three-toed skink has legs which are hardly noticeable on first glance), nevertheless, these reptiles bear no relation to snakes.

Furthermore, as we have seen, the truly limbless lizards such as the *slow worm* have normal lizard-like eyes with mobile eyelids and so are no closer to snakes than other lizards with normal legs. In addition, if snakes had evolved from lizards, we should expect to find that the eyes of the (supposedly) more primitive snakes (like the boas

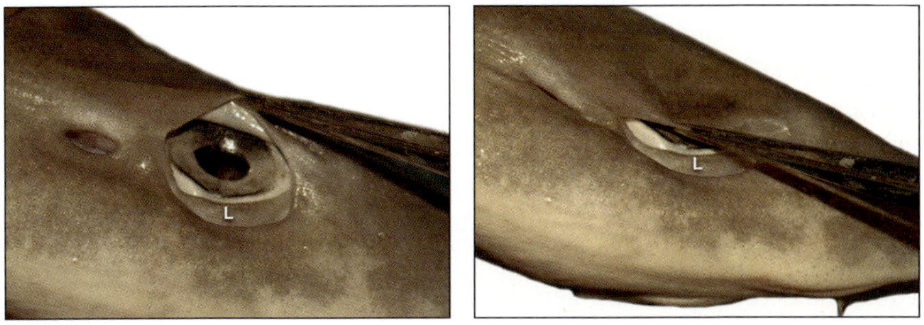


Dogfish.
The dogfish and other sharks have upper and lower (L) eyelids.

Gannet eyeball showing scleral ossicles.

and pythons) are more lizard-like than the eyes of the (supposedly) more advanced snakes (such as the rattlesnakes and vipers). But this is not so.

It is also notable that unlike lizards and birds, snakes do not have thin plates of bone (*scleral ossicles*) surrounding the cornea. Nevertheless, despite the unquestionable fact that the eyes of snakes:

> ...differ considerably in structure from any reptilian eye. David Attenborough. 'Life on Earth'. Collins/BBC. 1979. p167,

many people still believe that snakes were once burrowing lizards:

> If the snake's ancestors had been burrowers, then their eyes, like those of any other burrower, would have tended to degenerate. But if, before they were lost altogether, their owners had returned to life on the ground, then sight would once again be needed and the vestiges would redevelop. So the snake eye would have a structure peculiar to itself. This explanation is very persuasive but not yet universally accepted. David Attenborough. 'Life on Earth'. Collins/BBC. 1979. p167.

Basilisk

Actually, this explanation is not very persuasive at all. First of all, Attenborough offers no idea as to what environmental scenario, what time-scale, or what genetic mechanism would cause *all* snake ancestors, from *all over* the world, to start burrowing, and then to stop burrowing and live on the surface again. And secondly, as we have seen when discussing whales and leglessness, the characteristics that make a snake a snake, go far beyond just the eyes: snakes are by no means just legless lizards.

The fallacy in Attenborough's surmising is that if lizards evolved into snakes, they would not have lost their eyes *because* some lizards fancied burrowing. The *Galapagos Cormorant Principle* states that:

Disuse *per se* is not the cause of degeneration of a structure or function. Mutations occur as inevitable consequences of random chemical substitutions within DNA. However, if a mutation results in a degeneration of a structure or function which *is not essential for the survival of the affected individual*, then the individual might survive and pass on to its offspring that mutation. The progeny will then have degenerate structures and functions not because of disuse, but as a consequence of the prior mutation.

Therefore, the reality is that as with blind cave fishes, a mutant lizard born blind might only survive if it managed to hide or burrow, and a blind burrower that fancies crawling on the surface would not *as a result* develop eyes.

If a lizard gave rise to progeny which had no eyes (as with blind cave fishes), this would most likely be as a result of a mutation in some vital and key organising gene, resulting in the gene's failure to function. The rest of the eye-producing genes would not be *lost*, but would still be located within the chromosomes, and in their proper positions, but they would be incapable of being activated. It might be supposed that if the mutation in the key organising gene were to fortuitously become corrected in a subsequent generation so as to become fully functional again, this would simply trigger the rest of the dormant eye genes to enable the normal developmental processes to produce once again a normal *lizard*-type eye.

But there are two problems with this idea. Firstly, for a mutation in a DNA sequence to arise and then to become in some subsequent generation corrected back to the original sequence is unlikely: congenital diseases caused by genetic abnormalities do not rectify themselves in subsequent generations. For example, the defective gene for sickle cell anaemia *does not* in time and in subsequent generations become a normal haemoglobin gene; the countless generations of the axolotl *have not* resulted in 'corrected' mutations allowing progeny of these salamanders to undergo normal metamorphosis; and Galapagos cormorants do not give rise occasionally to normal-winged progeny.

Secondly, in the long time period of the supposed burrowing phase in the history of snakes, any dormant eye-producing genes would be liable to accumulate all manner of random mutations which would most likely result in failure of many of the necessary eye-producing genes themselves. So, even if the key organising gene were to re-acquire function, it would most likely *not* be able to trigger the formation of a normal eye but result instead in some form of *abnormal* eye. Most certainly, we should *not* expect any such progeny to develop a fully-functional and *quite different* eye such as that of the snake. So, it is improbable that the unique eye of the snake is the result of evolution amongst burrowing lizards.

Therefore, contrary to Attenborough's confident assertion that *'sight would once again be needed and the vestiges would redevelop'*, we should predict that the descendant of a genetically blind snake would most likely *not* regain its eyes.

But, for the sake of argument, even if *some* snake ancestors did burrow and by acquiring deleterious mutations did become blind, then not *all* snake ancestors would be expected to have evolved in the same manner. There would still have been *some* snake ancestors that always remained above ground, never became blind, and retained their normal lizard eyes. Such lizards ought to show at least some snake-like characteristics (such as very large numbers of ribs, elongated bodies, dislocatable jaws, etc.) so as to distinguish them from all other modern lizards. But this is not what we see today.

In fact there is no evidence to show that burrowing is associated with snake-like eyes. Indeed, the converse is true. There is at least one lizard, the *snake-eyed lizard,* which lacks eyelids and has a spectacle-like covering over the eye, but it is in all other respects completely a lizard, and shows no features that make it either a snake ancestor or descendant. Furthermore, despite its spectacle, it *does not* burrow and lives largely in arid open plains or bare ground. The *snake-eyed skink* also lacks eyelids but does not have a spectacle. It too does not burrow and specimens in Crete actually have relatively long limbs. Mention must be made of two reptiles that spend most of their time underground - the *European worm lizard* and *Wiegmann's burrowing lizard.* Whilst both these creatures lack legs, and have tiny eyes, the eyes are unlike those of snakes. In addition, the distinct grooves along the length of their bodies, and the arrangement of their scales in transverse bands or rings, also show that they are quite distinct from snakes. Neither creature represents an intermediary

Bird eye: the nictitating membrane in this emu is half-way across the eye.

between snakes and lizards. Thus, the evidence suggests that the snake eye is not a modified lizard eye, and that snakes are not modified lizards.
Nor is it plausible that the bird eye is a modified reptilian eye since both the shape and the movement of the bird lens are unique. The ciliary muscle is in two parts. One part pulls the cornea into a more globular shape, and the other part pulls the lens forwards so that its front end becomes not only more globular, but also bulges into the anterior chamber of the eye. In contrast, in mammals, the circular fibres of the ciliary muscle are attached to the circumference of the lens and their contraction allows the natural elasticity of the lens to resume a more globular shape.

### *Could the mammalian eye have evolved from a flat sheet of cells?*

So we have seen that the eyes of living creatures vary in complexity. However, the mere existence of different types of eyes with different degrees of complexity does not in itself constitute evidence that any one form of eye evolved from any other. Indeed, we can see no evolutionary trend from the (supposedly) primitive sharks through to the bony fishes, amphibians, reptiles, birds and mammals.
One thing that must be understood is that there is no such thing as a *simple* eye: even the 'simple' pin-hole camera type of eye is an extremely elaborate structure. Furthermore, it is totally inadequate to say that the human eye is a very complicated organ...it is a *phenomenally* complicated organ. Curiously, even though the human eyeball by itself is of no use without its retinal cells to capture the light and convert it into a nerve impulse, or without the brain to perceive the image into its consciousness, many people are quite content to confine any discussion about the supposed evolution of the eye just to the variation in the shape of the eyeball and the addition or not of a lens. So, to get a better perspective of the subject, let us look in turn at three essential components involved in the visual process: the eyeball, the retinal photosensitive cells, and the brain.
How could an elaborate structure such as the eye come into being? Perhaps the most commonly proposed idea is that starting with a simple flat sheet of photosensitive skin, the eye could have evolved over millions of years by a series of small sequential modifications. Surely, it might be thought, slightly changing something that is simple, time after time after time...will eventually result in something far more complicated. Superficially, the idea seems to be quite reasonable - a flat surface could become curved, then dish-shaped, and eventually become globular so as to form a pin-hole camera.
Some consider that even though the eye is so complex, there is now *no question* whatsoever that it evolved:

> ...the question is now one of process rate rather than one of principle.
> Dan-E Nilsson & Suzzane Pelger. 'A pessimistic estimate for the time required for an eye to evolve'. Proc.R.Soc.Lond. B. Vol 256. 1994. p53-58.

With the (unstated but implicit and unsubstantiated) assumption that each and every tiny *geometric* alteration to a flat sheet can be produced by tiny changes in the genes that determine embryological development, Nilsson and Pelger tried to work out how many 1% changes it would take to convert a flat sheet of light-sensitive cells into an eye with a lens. Of course, it is no surprise that if 1% is the minimum change that is allowed in any aspect, then not many *geometric* changes will be necessary for a flat sheet of light-sensitive cells to turn into a hollow sphere with a lens. In fact, they calculated that it might take less than 363,992 generations, and with a generation time of one year they assume that a camera-type eye could evolve in less than 364,000 years. This means that since the (supposed) origin of the first creatures with eyes 500mya, eyes could have evolved:

> ...more than 15000 times! Dan-E Nilsson & Suzzane Pelger. 'A pessimistic estimate for the time required for an eye to evolve'. Proc.R.Soc.Lond. B. Vol 256. 1994. p53-58.

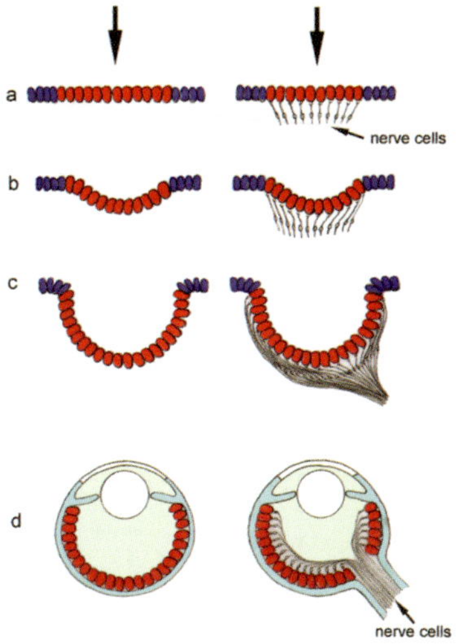


It has been suggested that the mammalian eye evolved from a flat sheet of photoreceptors in a sequence such as **a** to **d**. The vertical sequence on the left shows only simple geometric changes and this gives the superficial impression that such evolution is quite plausible. However, the sequence on the right includes the photoreceptors and nerve cells and this shows up a major problem.

The nerve cells associated with a shallow cup lie on the outside of the photoreceptor cells (shown in red), whereas in a mammalian eye the nerve cells lie on the inner surface of the retina. Since the two forms are the result of quite different embryological developments, there are no clear transitional states that the 'primitive' form would pass through to achieve the mammalian form.

Thus, in the supposed evolutionary scheme, the hypothetical step **c** to **d** is impossible.

But we *know* that starting with a flat sheet of light-sensitive cells, improvements in vision can be achieved by (a) gradually changing the flat sheet into a depression and then into a hollow ball, (b) narrowing the aperture at the front, and (c) introducing a lens. So the gradualistic model simply traces a sequence of changes which, when translated into a computer program, will inexorably lead to a globular eye with a lens, and it is no surprise at all that relatively few changes are needed to make the transition.

The point is that the step-by-step gradual change idea is a *false* model for a number of reasons. For example, the embryological development of an eye *does not* follow a series of sequential changes from a flat sheet of photo-receptive cells to a hollow ball; the mammalian lens does not arise simply as a blob of proteins which is then free to evolve its refractive index and shape, but is formed out of very specific modifications to whole cells; and we have no evidence to show that each and every tiny hypothetical *geometric* change to a flat sheet of cells is indeed an outward reflection of tiny changes in the genes that determine embryological development. Of course, the Nilsson-Pelger model does not discuss the hundreds of other eye components, each of which is in itself immensely complex and cannot readily be understood as being capable of origination by small stepwise changes. For example, consider the ciliary muscle which acts to alter the shape or position of the lens. The first (hypothetical) small changes to non-specific embryonic cells, such that they become in some tiny way like muscle cells, will not make a ciliary muscle in any shape or form, and so will not make any difference to the function of a lens, if one were present. And therefore there could be no natural selection for and evolution of such functionless cells.

It has also been suggested that since most of the structures within the eye are made up of proteins which have functions elsewhere in the body, evolution could simply recruit those existing proteins so as to produce *any* new structure that may be required by an evolving eye. But such a notion assumes too much, and is supported by too little evidence. For example, the fact that the iris is made up of virtually the same proteins as other parts of the eye does not explain how those proteins came to be *organised* in the way that they are. Again, the first (hypothetical) small changes to non-specific embryonic cells, such that they become in some tiny way like muscle cells, will not make an iris, and so will have no effect on the control of light reaching the retina. As before, there could be no natural selection for such functionless cells and hence no evolution.

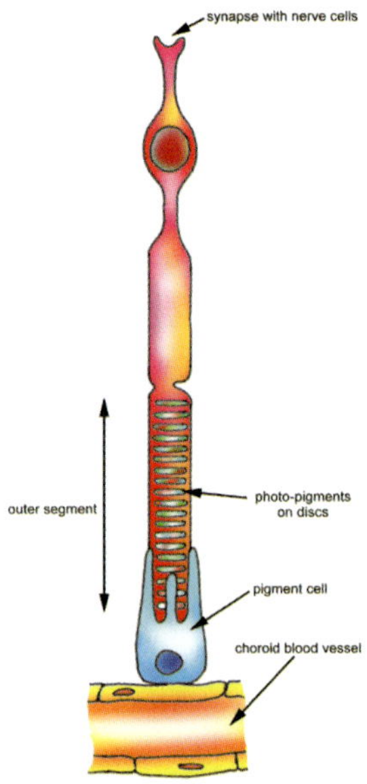


Diagrammatic relationship between a rod, a pigment cell and a choroid blood vessel.

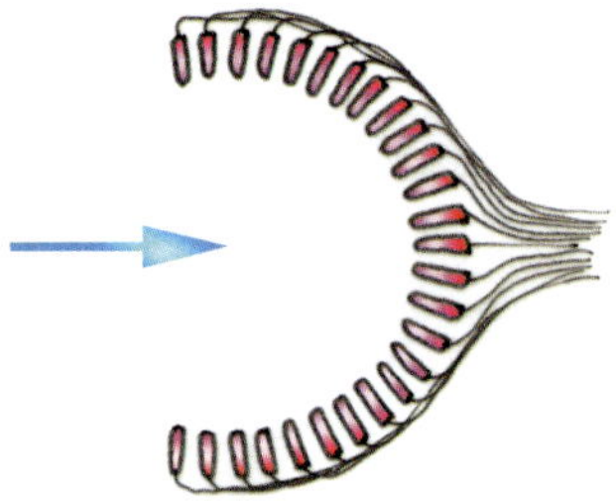

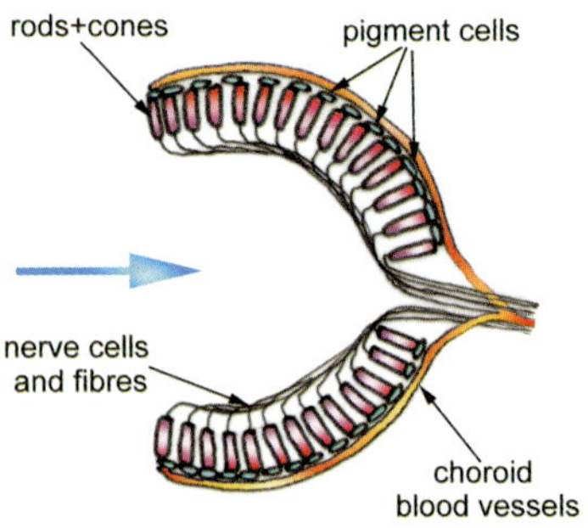


In a simple cup design (top) the light reaches the photoreceptor cells directly. Special Müller cells, acting like fibre-optic cables, conduct light between the retinal nerve cells directly onto the photoreceptor cells. This is a very efficient system (see also page 323). There is *no* evidence to show that this design is in some way imperfect and by implication that the eye *must have* evolved by a process of trial and error.

Furthermore, if any mutation could result in some 'proto-muscle' cells, it is implausible that such a mutation would affect the cells developing just where an iris ought to be, rather than (for example) interfering with the development of some of the retinal cells. So, the Nilsson-Pelger model says *nothing at all* about the plausibility of real biological events, and therefore, of real eye development.

During embryological development, the eyeball and the photosensitive cells arise from an outpouching (the '*optic cup*') of the developing brain, deep within the tissues. In contrast, the cornea and the lens arise from the skin cells overlying the developing eye. In a simple transplant experiment, if the developing optic cup is moved to another part of the embryo's body, not only will the skin at the new site start to produce a cornea and lens, but also, neither a cornea nor a lens will grow at the original (proper) site. This shows that the development of the lens and cornea is dependent wholly on the proper triggering function of cells within the developing optic cup, and is not an independent function of the overlying skin layer.

As the optic cup develops, two concentric and distinct layers arise from what was originally an homogenous group of non-specific embryological nerve cells. The outer layer of these cells is destined to become the dark pigment layer of the retina (which prevents internal reflections), and the inner layer of cells is destined to become yet two other layers, comprising the photosensitive cells (rods and cones) and the nerve relay systems that send signals to the brain.

The pin-hole camera-type and mammalian eye models have a number of important differences. In the pin-hole camera type of eye, the photosensitive cells lie innermost in the wall of the eyeball and incident light hits them directly. However, in the mammalian eye (and also in other vertebrate eyes), the incident light must first pass *through* the nerve relay cells before reaching the photosensitive cells. It might be suggested that this very design is illogical and inefficient, and points to a long and winding evolutionary process which has accepted numerous compromises along the way; and that if it *were* designed, the mammalian eye would have its 'wires' at the back so that the incident light could reach the rods and cones directly. However, the mammalian retina has special Müller cells, which act like fibre-optic cables, conducting light between the retinal nerve cells and directly onto the photoreceptor cells, without any distortion. This amazing system was only discovered in May 2007. What we *do* know is that the retina works *extremely* well, and in fact, there is evidence to show that the inverted design of the mammalian eye also has very important nutritional and detoxifying functions. Apart from initiating the chemical changes that produce nerve impulses, light also causes damage to the photo-pigment-containing discs within the outer segments of the rods and cones. In response to this damage old discs are broken down and replaced continually with new discs. Any damage also releases toxins such as *free radicals* which are continually neutralised by specific enzymes. Each melanin-containing cell of the pigment layer of the retina has fine processes (*microvilli*) that surround the outer segments of up to 20 rods and cones. These microvilli increase the surface area of contact, and toxins and nutrients pass respectively to and from the capillaries of the choroid through the pigment cells. The supreme importance of the blood supply for the proper functioning of the retina is shown by the fact that, weight for weight, the retina has the greatest blood

flow of any tissue in the body.
However, the choroid blood supply is actually much greater than the nutritional requirements of the retina so it must have other functions, one of which is protection from heat damage caused by light. It is found that if the blood flow in the choroid is experimentally reduced, the retina becomes more vulnerable to light-induced thermal injury. There is also evidence to show that the rate of choroid blood flow is controlled reflexly in response to changes in the intensity of the illumination. So, by being closer to the rods and cones than they would otherwise be in the simple cup design, the blood vessels of the choroid are better able to act as a protective heat sink.
Presumably, the simple cup design works adequately because of both its simplicity and its low energy requirements. However, the far greater functional complexity of the mammalian retina requires a more efficient support system. At every stage of evolution, for natural selection to operate, every form of intermediary eye must have been both functional, and more capable than its predecessor. Since any mutations that have (supposedly) caused an eye to evolve must have done so by some effect on embryological development, it makes sense to look at some of the sequences of events that occur in the developing eye. For example, we have seen that some specific activity within the cells of the developing optic cup is necessary for the developing lens to be induced to develop from the cells of the overlying skin tissue.
If the optic cup first existed without the ability to initiate the development of a lens, such initiating capability must

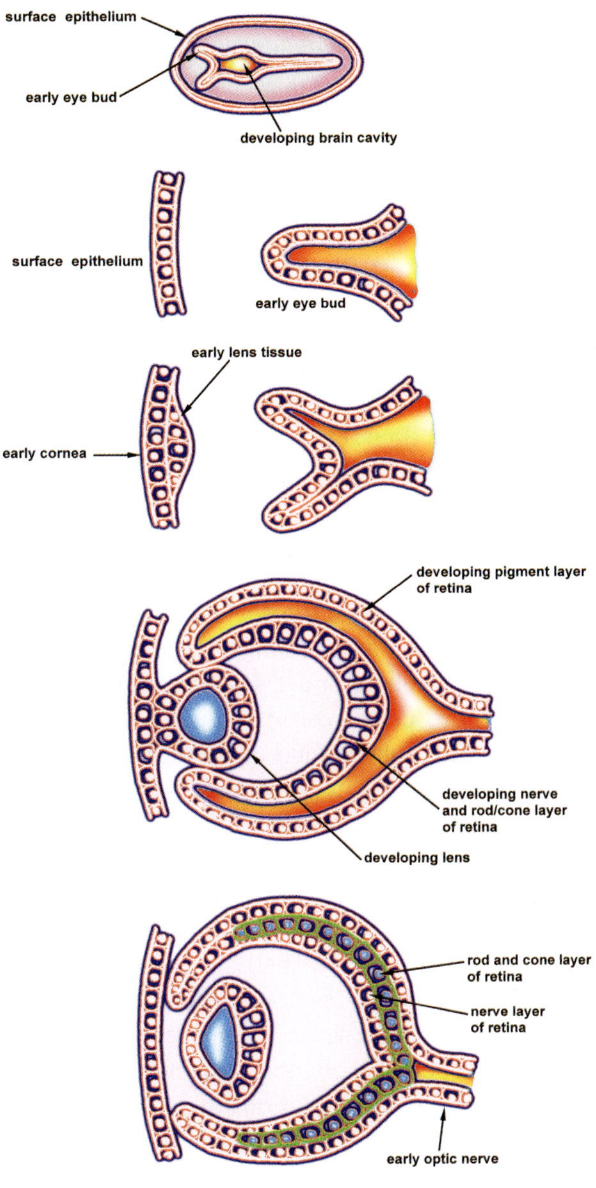


The vertebrate eye develops partly as an outgrowth of the embryonic brain and partly from the cells of the overlying skin. The vertebrate eye is not a modification of a flat sheet of photoreceptor cells.

Lens from a pig eye.

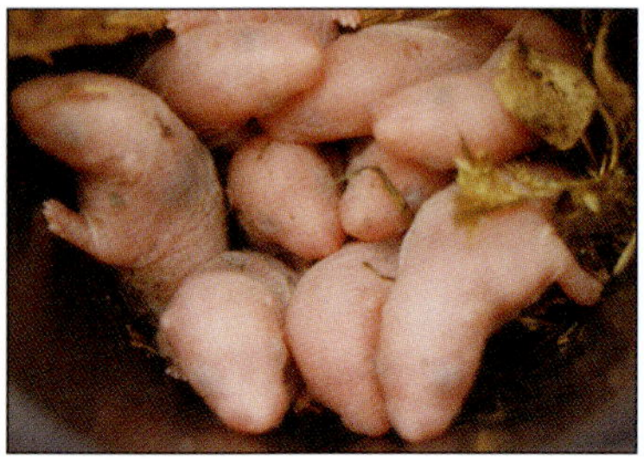
Like many other small mammals, rats are born blind with their eyes covered with skin.

A blind emu with a lens that has become dislocated so that it is now situated in front of the iris instead of behind it. Any supposed evolutionary development of a lens would most likely arise in the wrong place, resulting in blindness.

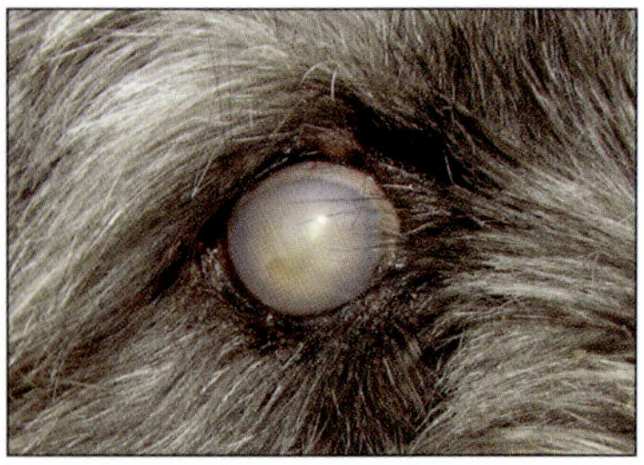
A blind dog which has an opaque cornea. Any supposed evolutionary development of a cornea would most likely incorporate opaque skin, resulting in blindness.

have evolved some time later as a result of one or more mutations. Furthermore, the *initiation* trigger for the formation of a lens is a separate process from the directives that control subsequently the lens *development* and *migration* into the developing optic cup. Even *within* the developing lens itself (the '*lens vesicle*'), the anterior cells (which ultimately form the anterior capsule of the lens) develop differently to the posterior cells (which become the lens proper). So, if in an embryo a mutation appeared so as to cause a lens vesicle simply to be *initiated*, the lens would develop haphazardly and would be likely to interfere with the development of the eyeball itself, leading to a deranged mass of tissue where in previous generations an eyeball had developed with a pin-hole opening at the front. This would result in the selection for the status quo and no evolution.

But suppose a crude lens did manage to arise without disrupting the development of the eyeball itself. Any blob of transparent jelly or crystalline substance which has a refractive index greater than that of water, and which has even a remote semblance of curved surfaces, can act as a feeble lens. So, no matter how crude, any cup which by chance developed any semblance of a lens within it would have been able to produce a sharper image than one with no lens at all. Those blobs producing the better images would have been selected for, and in time natural selection would have cumulatively refined the blob until it became increasingly like the familiar lens shape.

This may sound like a fine explanation for the origin of the mammalian lens but what is the evidence?

The lens is not a crystalline substance and, most certainly, congenital eye abnormalities do not provide evidence that any genes, altered here or there along their length, can be responsible for a variety of lens shapes that show even a ragged trajectory of changes from no lens to a biconvex lens shape made out of living whole cells, with each step being both functional and also an improvement over the previous state. When embryonic development goes wrong, it usually goes *badly* wrong. Problems with the development of the iris usually result in a section of the iris being absent (a coloboma). If the eyeball is defective, we do not see variations such as a 3/4 sphere or a 1/4 sphere etc. Instead, when the eyeball development is defective, this usually results in just a mass of poorly organised tissue.

In other words, when things go wrong with eye development, we do not see eyes with features which are a number of steps backwards along an hypothetical evolutionary pathway. Of course, for any particular organism that has a lens made of any blob, crystalline matter or living cells, it is questionable whether minor differences in optical performance can or do, in fact, reflect any (let alone a significant) difference in survival of the organism. Leaving aside the fact that survival is governed by multiple factors, as we shall see later, it is most likely that small differences in optical performance make no difference to survival.

Getting back to the lens, if, as a result of some mutations, any of the necessary controls responsible for the correct timing and location of the lens development appeared before the mutations that could cause lens initiation, then these controls would be of no use. This being so, the relevant genes would gradually accumulate further mutations, and end up with nonsense sequences long before any lens initiation signals could have evolved. But embryological development of the eye is even more complex than has been described so far:

> Under the influence of the lens the cells of the inner stratum proliferate and form a layer of considerable thickness from which the nervous elements...are developed. 'Gray's Anatomy'. 34th edition. Longmans. 1969. p181.

In other words, the development and organisation of the nerves that connect the rods and cones to the brain are triggered into being by the cells that comprise the developing lens. So, without the lens being first present the optic cup would develop into a functionless cavity. To put it another way, a (supposedly) primitive optic cup with distinct nerve and photosensory cells could never have existed and developed into a lens-less eye, since without the organising function of embryonic lens tissue (not yet evolved) there would have been no nerve cells to pass the signals from the photosensitive cells to the brain.

The mammalian eye is *such* a complex structure that we could cite component after component, unearthing the same problems (those of interdependent systems, and of the requirement for one system to be present and operational before another system can be introduced), time and again. For example, the liquid in the front part of the mammalian eye (the aqueous humour) is formed by the filtration of fluid from the blood. Removal of the fluid out of the eyeball and back into the veins is effected via a special canal system (the *canal of Schlemm*) that encircles the eye at the base of the iris. The correct pressure inside the eye is maintained by a balance between the rate of production and the rate of removal of the aqueous humour. If the outflow of fluid were obstructed in any way, the internal pressure would build up and damage the retinal cells. And this is what happens in the disease *glaucoma* which can result in blindness. Thus, a system for filling the mammalian eyeball with fluid cannot come into being without first there being a mechanism *already present* for the control of the fluid outflow.

So, the evidence suggests that for the eye to have evolved by a series of mutations resulting in gradual modification to a flat sheet of photoreceptor cells is implausible. Nevertheless, it is generally suggested that simple eyes evolved from no eyes at all, and that complex eyes evolved from simple eyes:

> Once primitive animals with muscles and nerves arrived, we can be confident, given a good molecular mechanism for rapid evolution, that a visual system will develop. The ability to see gives an animal a considerable selective advantage... Francis Crick. 'Life itself'. MacDonald. 1981. p109.

Crick certainly has faith that '*a visual system will develop*', but it is notable that he does not share with his readers any science with which to underwrite his confidence. We all know that the ability to see is useful. The question is: could a visual system arise from nothing? And is it true that *'The ability to see gives an animal a considerable selective advantage*'? Certainly, we all know that this is true when we compare the blind to the sighted, or to the partially-sighted, but the evolution of light receptors must have started when at first *no* organism could sense any light

**Mutations affecting the eye**

Mutations do not result in different degrees of eyeball curvature, but instead produce severe diverse abnormalities. For example, some result in defective lens development:

> *Laminins represent major components of basement membranes and play various roles in embryonic and adult tissues. [Laminin mutant] embryos exhibited an arrest in lens development and abnormalities in the body axis length and curvature.* Zinkevich et al. 'Laminin alpha 1 gene is essential for normal lens development in zebrafish'. BMC. Dev. Biol. 2006 Mar 7;6:13.

Other mutations result in defective retinal development:

> *[Mutations affecting] zebrafish nkx5-3 induced microphthalmia [small deformed eye] and disorganization of the developing retina.* Schorderet et al. 'Mutation in the human homeobox gene NKX5-3 causes an oculo-auricular syndrome'. Am. J. Hum. Genet. 2008 May;82(5):1178-84.

> *[Mutations affecting the] PITX3 gene...[resulted in] abnormalities in lens development...[and] retinas contained pyknotic nuclei accompanied by a reduction in the number of cells in different neuronal classes.* X Shi et al. 'Zebrafish pitx3 is necessary for normal lens and retinal development'. Mech. Dev. 2005 Apr;122(4):513-27. Epub 2004 Dec 10.

And as we have seen with blind cave fish, there can be various degrees of malformations.

at all. So any initial change could only have been from no light-sensitivity at all to a *very tiny* amount of light-sensitivity.

We know that irrespective of whether or not they have light-sensitive granules, single-celled animals manage to live and survive in the same pond. But does one have any *overall* selective advantage over the other? Further down the (supposed) evolutionary road, it is necessary to define whether the photoreceptors of the bowl-like organ of the leech can confer any advantage over a flat sheet of photoreceptors since, although they can both register light, neither organ can actually produce even a blurred image.

Euglena, a single-celled organism that has a light-sensitive granule near the base of its single flagellum.

The *euglena,* for example, is a single-celled organism which shows some features common to both plants and animals. It is important to the following discussion to note that the euglena reproduces only by simple *fission* (by one individual dividing into two daughter cells), and since there is no sexual reproduction, the genes inside one individual cell cannot be mixed with different genes from within a different individual cell. It is also important to note that the euglena is a complete and fully functional organism. By that is meant that it is fully capable of living in its environment and lacks no necesssary feature to ensure its survival. If it did evolve from something simpler it certainly has had no pressure to alter its form over (supposedly) millions of years.

The euglena swims with the aid of its flagellum and has a light-sensitive pigment granule (called a *photophore*) within its cytoplasm. Consider for a moment an hypothetical euglena ancestor that had neither a photophore nor a flagellum. This creature could therefore neither detect light nor move by its own volition. Suppose that following a mutation, this ancestor gave rise to a daughter cell that developed a photophore. But since it had no flagellum, the mutant still could not respond to any change in the intensity or direction of light any differently from its contemporaries. The *Galapagos Cormorant Principle* shows that, being of no practical value, in time the photophore would simply become degenerate and be lost. Similarly, suppose that following a mutation, the hypothetical euglena ancestor gave rise to a daughter cell that developed a flagellum. But since it had no photophore, the cell would not be able to 'detect' any change in light intensity and, with respect to incident light, its motion would be purely random. Again, the *Galapagos Cormorant Principle* shows that, being of no practical value, in time the flagellum would also become degenerate and be lost. Furthermore, we have already seen that even the humble flagellum is an incredibly complex structure, made up of a large number of proteins, each requiring specific genes in order to come into existence, and the 'simple' photophore itself is not so simple. The *WYWYWG Principle* shows that it is implausible for either a photophore or a flagellum to arise by a stepwise process.

There is another problem with the gradualistic origin of the eye, since in the (supposed) ancient scenario where all the organisms were very primitive, all the eyes would have been extremely simple, all the brains would have consisted of nothing more than a few neurones, and all the organisms would have been very slow-moving; so for any particular creature, any small change in either the 'eye', 'brain' or locomotion speed would most likely *not* have resulted in any significant survival advantage. To make a fair comparison, if any of its characteristics were to improve

only a tiny amount, most certainly a blind, deaf, toothless and clawless lion with a poor sense of smell would be unlikely to become a threat to a blind, deaf zebra which also had a poor sense of smell.
But the reader might protest: surely having an eye with only 1% of its normal visual acuity is better than being blind, and therefore can confer *some* advantage over having no eye? And some believe this to be true:

> Vision that is 5 per cent as good as yours or mine is very much worth having in comparison with no vision at all. So is 1 per cent vision better than total blindness. And 6 per cent is better than 5, 7 per cent better than 6, and so on up the gradual, continuous series. Richard Dawkins. 'The blind watchmaker'. Penguin. 1991. p81.

Well, 1% certainly is greater than nothing. And 2% is greater than 1% etc. etc. Yet, whilst *mathematically* this is true, nevertheless we all live in a real world and for practical purposes, any difference in visual acuity can only be translated into two groups: the first group having some *significant* benefit and the other group having *no significant* benefit. So, in the real world, and regardless of the mathematics - from a leech in a swamp to an ape swinging through the trees - we must ask: would an eye with 55% of normal visual acuity offer *significantly* more benefit than an eye with only 50% of normal visual acuity?
Imagine two men, both having poor vision, being in the same field as a lion. One man's visual acuity is 50% of normal, and the other's is 55% of normal. How will they fare with the lion? Most surely we should not expect the man with (in *mathematical* terms) the better vision to have any significant chance of survival over the other. So, in this case for practical purposes, 55% is *not* of *any* benefit over 50%.
But what of a primitive eye? You can perform a very simple experiment and easily get an idea of how limited a simple eye is by looking into a misty bathroom mirror. Even your human eye, with a human brain to interpret the image, will see only a misty *nothingness*. If the mirror is then wiped across just once with the teeth of a comb, even though a proportion of the mist on the mirror has been removed, you will still not be able to distinguish even your face. In fact, your own reflection will not become discernible as a face until a good number of passes have been made with the comb. Simple experiments such as this show conclusively that Dawkins' confidence in any small change in visual acuity being of some benefit in terms of survival is fallacious: a 1% or even a 5% improvement in a blurred image is *in practical terms* no improvement at all. And so, a 5% improvement confers no selective advantage.
It must also be noted that we have in this bathroom experiment discussed a *human* brain interpreting the image seen by a *human* eye. What Dawkins also fails to consider is that any primitive creature with a simple form of eye will most certainly have no more than a few neurones with which to interpret the image: in other words, a 'brain' that is very considerably less able to make sense out of a misty image than a human brain. Thus, starting with a totally misty image, if a human eye with its 200,000 photoreceptors per square millimetre connected to a human brain requires (say) a 15% improvement before it can discern its own face, then a (supposed)

Visual acuity chart. The full-sized chart is meant to be read at 6 metres, and at this distance most people with 'normal' vision can clearly read at least the 7th line (LNETHOA). Although a number of people will be able to read the 8th or 9th line, this confers no *significant* practical advantage.

primitive ancestor would have required an astronomical improvement in its eyes and 'brain' before it could have attained any selection advantage over any of its contemporary cousins.

So, we can see that only when there is a *substantial* improvement in visual acuity and perception that has practical relevance can the difference provide any significant survival advantage. No advantage means no natural selection. No natural selection means no evolution.

Yet Dawkins insists that:

> Gradual evolution by small steps, each step being lucky but not too lucky, is the solution to the riddle. But if it is not gradual, it is no solution to the riddle: it is just a restatement of the riddle...There will be times when it is hard to think of what the gradual intermediates may have been. These will be challenges to our ingenuity, but if our ingenuity fails, so much the worse for our ingenuity. It does not constitute evidence that there were no gradual intermediates. Richard Dawkins. 'River out of Eden'. Weidenfield & Nicholson. 1995. p84.

But it does not matter one iota whether or not we can *imagine* a theoretical series of small stepwise changes that might convert one structure to another. That will get us no closer to proving or disproving the *possibility* of the change, or more importantly, whether such a change actually occurred. And when he fails to offer any evidence to support his argument, Dawkins resorts to sharing his *feelings*:

> Is there a continuous series of Xs connecting the modern human eye to a state with no eye at all?...Has there been enough time for enough successive generations? We can't give a precise answer to the number of generations that would be necessary. What we do know is that geological time is awfully long...the number of generations that separate us from our earliest ancestors is certainly measured in the thousands of millions. Given, say, a hundred million Xs, we should be able to construct a plausible series of tiny gradations linking a human eye to just about everything!...**My feeling** is that **provided the difference between neighbouring intermediates in our series leading to the eye is sufficiently small, the necessary mutations are almost bound to be forthcoming.** (My bold). Richard Dawkins. 'The blind watchmaker'. Penguin 1991. p78-9.

But this assumes that *any* hypothetical small structural change *can* be the reflection of a small genetic change, and that any genetic change is, in time, *likely* to appear. The *WYWYWG Principle* states that the very mutation that is *required* or which *might be hoped for*, for the next evolutionary step, is unlikely to appear. So, if a pin-hole camera-type cup was present and if a lens would do very nicely as the next step to improve the efficiency of the eye, any such progression would be implausible.

It is quite clear that Dawkins has convinced himself that he is right. And even though he has merely expressed his *feelings* and has failed to produce any scientific evidence, he nevertheless has no time for those who might doubt his reasoning and pose the question: 'is it really plausible that thousands of lucky chance mutations happened coincidentally such that all the countless components of the eye evolved in synchrony?' Instead, he deals out the rebuttal that:

> 'this remarkable argument is frequently made, presumably because people WANT to believe its conclusion'. Richard Dawkins. 'The blind watchmaker'. Penguin 1991. p80.

In fact, Dawkins admits his confusion when he informs his readers that a creature such as nautilus, with its pin-hole type of eye, can exist for millions of years without acquiring any improvement in the structure:

> The eye is basically the same shape as ours, but there is no lens and the pupil is just a hole that lets in seawater into the hollow interior of the eye. Actually, Nautilus is a bit of a puzzle...Why, in all the hundreds of millions of years since its ancestors first evolved a pinhole eye, did it never discover the principle of the lens?...The system is crying out for a particular simple change...Is it that the necessary mutations cannot arise... **I don't want to** believe it... (My bold). Richard Dawkins. 'The blind watchmaker'. Penguin. 1991. p85-86.

One moment he rebukes anyone who might even *just question* the notion of evolution, but then he goes on to convince himself (and tries to slip it past his readers) that it is acceptable for *him* to ignore important evidence. The *denial* of evidence has no part to play in science, and it is clear that Dawkins' arguments are driven by his *beliefs* and not by facts. Since, as with all living things, it is likely that nautilus cannot be immune to random mutations of its genome, and if the eye of the nautilus has been around for (supposedly) millions of years without evolving into something *even slightly more* sophisticated, this must mean that nautilus is *unable* to improve on its pin-hole type of eye, or, of course, that it has not been in existence for millions of years. Or both.
Curiously, although it shares much of its anatomy in common with the nautilus, the octopus has a very sophisticated eye which is in many ways remarkably similar to a mammalian eye. Furthermore, if the human eyeball was the result of a large number of genetic changes which, over time, caused a flat sheet of photoreceptors to develop into a spherical eye, then we should expect to see some mutations in mammals giving rise to the failure of some of those genes, so as to result in abnormal eyes showing a variety of curvatures from a flat retina to a spherical ball, or other abnormalities which hint at ancestral intermediaries. However, no such congenital abnormalities occur. Thus, the nautilus illustrates the *Small Step Principle*:

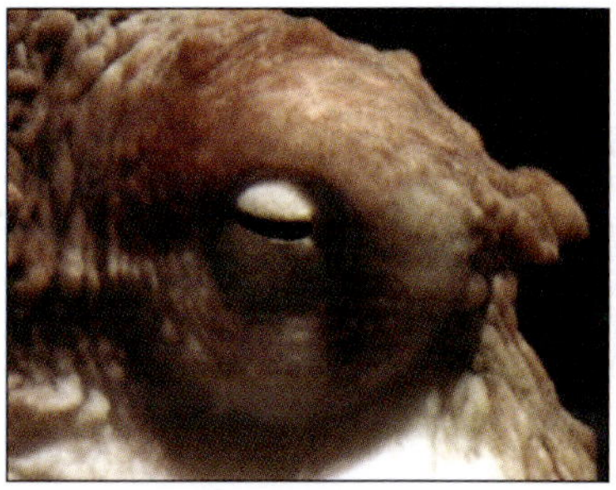

The octopus (above) and the cuttlefish (below) have sophisticated eyes which are very similar to those of mammals.

Any small step, even if infinitesimal, need not be *simple*, and even if a small transitional step may be *imagined*, this does not mean that that particular step *ever happened*, or is *possible*.

For it to have occurred, the evolution of living things requires that, for each and every biological structure or system which exists, or which in times past has existed, there must have been a *physical* precursor. We have already seen that different eyes do have widely differing components and systems, and yet we have no evidence on which to state that the mammalian eye is indeed the result of modifications to any physical precursors which link

it to a flat sheet of photoreceptors in some ancient organism. So, complex eyes are likely to have nothing more than *conceptual* links with less complex eyes.
It is curious indeed that some are content to accept that one creature can take so *long* to evolve relatively *minor* changes:

> After some **140 million years** of development, the nautiluses gave rise to a variant group with more flotation chambers to each shell, the ammonites. (My bold). David Attenborough. 'Life on Earth'. Collins/BBC. 1979. p44.

And yet at other times the same individuals are able to accept unreservedly that *sophisticated* changes *actually took place* in a very short time. It took (supposedly) one hundred and forty million years to develop a few more flotation chambers, and yet to produce the most sophisticated thing in the known universe - the human brain and the origin of humankind:

> The story starts (only) **five million years** ago on the plains of Africa. (My bold & parenthesis). David Attenborough. 'Life on Earth'. Collins/BBC. 1979. p293.

Furthermore, to be precise, we have no real evidence to show unequivocally that extra flotation chambers *could* or *did* evolve in 140my, or that the nautiluses *did* evolve into ammonites.
We have so far seen the implausibility of the notion that the mammalian eye evolved by small gradual changes from a primitive single photosensitive spot. Furthermore, we have seen that in discussing just one superficial aspect of the eye such as gradual changes in the eyeball shape, writers such as Dawkins deny their readers insight into the sheer complexity, for example, of just one retinal rod. There are, in fact, *so many* individual facets of the eye that we could discuss but due to the constraints of space and time, we necessarily have to be very selective. If a hollow ball is to become an eye it needs *light sensors* to detect the light in the first place. These sensors have to relay their signals to the *brain*, which in its turn has to display a virtual reality image in the *mind*. So it is natural to ask: how does the light which enters the eye create the nerve signals that go to the brain?

### *The conversion of light into a nerve impulse*

Central to the conversion of light into the electrical impulses is the molecule *rhodopsin*. Rhodopsin is localised to the 1000 or so special membrane discs found in the outer segments of the retinal rod cells and is made up of two component molecules: the protein *opsin* which is bound to the disc membranes, and *11-cis-retinal**, which is the light-absorbing pigment itself. A single human rod has about 4 x $10^7$ rhodopsin molecules, allowing the human eye to detect the arrival of as few as 5 photons. Incident light transforms the 3-dimensional conformation of *11-cis* retin*al* into *all-trans* retin*al*, and this forces a change in the 3-dimensional conformation of opsin to produce *activated opsin*. Since the opsin is bound to the cell membrane, its conformational change triggers a chain of chemical reactions which cause the closure of sodium channels in the cell membrane. We need not go into

* The terms '*cis*' and '*trans*' refer to different spatial orientations of atoms within a molecule. Whilst cis and trans forms of a molecule contain exactly the same numbers and types of atoms bonded together in exactly the same order, nevertheless as a result of subtle differences in some of the chemical bonds within the molecule, the cis and trans forms differ markedly in their three-dimensional configuration. This, in turn, can result in cis and trans forms of a molecule having markedly differing chemical activity. Any two molecules which contain the same numbers and types of atoms arranged in the same order but which have different three-dimensional forms are called '*isomers*'.

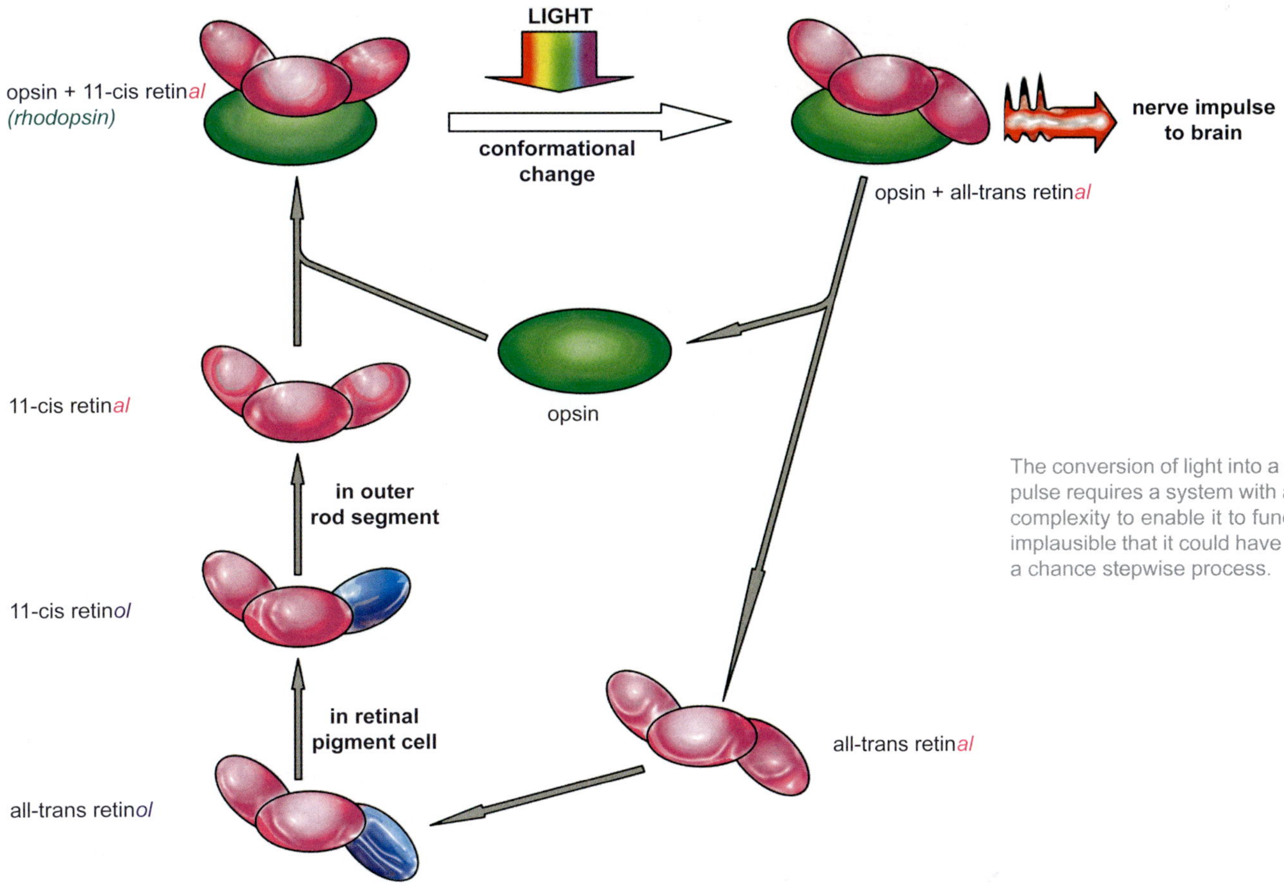


The conversion of light into a nerve impulse requires a system with a minimal complexity to enable it to function. It is implausible that it could have arisen by a chance stepwise process.

the complexity of the resultant movements of potassium and sodium ions across the cell membrane, and for our purposes it is sufficient to say that this last action results in the formation of the electrical impulses which ultimately are conveyed by the optic nerve to the brain.

The process of formation of rhodopsin begins with the naturally-occurring molecule *b-carotene* (found in high concentrations in carrots) which can be split into two molecules of *all-trans* retin*ol*.

The conversion of *all-trans* retin*ol* into *11-cis* retin*al* involves a number of chemical steps. First the *all-trans* retin*ol* is converted into *11-cis* retin*ol*. However, although *all-trans* retin*ol* and *11-cis* retin*ol* contain exactly the same numbers and types of atoms, the spatial orientation of the H-C=O ends makes the three-dimensional shapes of the two molecules significantly different. This seemingly simple difference is such that the change from one to the other *cannot* occur in one move by simply bending the molecule at one point. Instead, a number of intermediary chemical steps are necessary for the conversion to take place, and these require the presence of a specific enzyme to help the reactions proceed correctly. Since only *11-cis* retin*al* can be used to bind with the opsin, *11-cis* retin*ol* is of no use in the visual system. So the *11-cis* retin*ol* has to be converted into *11-cis* retin*al* with the help of another

specific enzyme.
The situation is made more complex by the fact the chemical reactions occur *within totally different cells* since the enzyme that catalyses the conversion of *all-trans* retin*ol* into *11-cis* retin*ol* is found:

> ...in the pigmented epithelial cell layer of the retina... Thomas Devlin. 'Textbook of biochemistry'. 3rd edition. Wiley-Liss. 1997. p945.

And the conversion of the *11-cis* retin*ol* into *11-cis* retin*al*:

> ...occurs in the rod outer segment. Thomas Devlin. 'Textbook of biochemistry'. 3rd edition. Wiley-Liss. 1997. p945.

Thus the conversion of *all-trans* retin*ol* into *11-cis* retin*al* and the latter's inclusion into the opsin molecule to form rhodopsin requires the presence of a number of specific enzymes, without which the molecular steps could not take place. Furthermore, the opsin protein itself has to have a precise structure for two main reasons: firstly, so that the *11-cis* retin*al* molecule can locate itself within the opsin molecule:

> The site on the chain of the protein [opsin] is one on which cis-retinal fits precisely. 'Solomon's organic chemistry'. 6th edition. Wiley. 1996. p611.

And secondly, so that when its attached *11-cis* retin*al* is converted to *all-trans* retin*al* by the action of light, the change in the opsin's 3-dimensional conformation is precisely that which is required to trigger the molecular steps necessary for the initiation of a nerve impulse.
Furthermore, whenever activated opsin is formed by the action of light, it spontaneously breaks up very quickly:

> Activated opsin is unstable and spontaneously dissociates, releasing opsin and all-trans retinal. Lodish et al. 'Molecular cell biology'. 3rd edition. Scientific American Books. 1995. p975.

Therefore, if activated opsin were to have arisen before a system was in place to initiate a nerve impulse, its spontaneous dissociation would serve no purpose. And for a system of reactions to appear which could ultimately produce a nerve impulse system but without the means of producing activated opsin, would also serve no purpose. Hence all the necessary molecules and systems must be in place for any function to manifest.
Once the activated opsin has dissociated into free opsin and *all-trans* retin*al*, the latter cannot be simply converted directly into *11-cis* retin*al* to be used again to produce more rhodopsin. Instead, *all-trans* retin*al* has to be converted into *all-trans* retin*ol* by another specific enzyme (*all-trans retinol dehydrogenase*). As described previously, a further series of reactions then converts the *all-trans* retin*ol* into *11-cis* retin*al*. Thus what appears as a simple hop from *all-trans* retin*al* to *11-cis* retin*al* is now seen in reality as a series of complicated jumps requiring a number of specific enzymes. So, we can see that the visual system requires a whole series of specific

enzymes to be present before the system can function.
So the biochemistry of vision demonstrates the *Irreducible Complexity Principle*:

If, for it to function, a system requires a minimum number of interdependent components, and if no component could have had any function outside that system, and if without all the components being present that system cannot function *at all*, then those components could not have evolved from simpler entities and that system could not have evolved from any fewer number of components.

It is therefore implausible that the enzymes necessary for the conversion of light into a nerve impulse could have arisen *de novo* by random mutations of DNA.

### *Colour vision*

The occurrence of colour vision shows no particular pattern amongst living creatures. For example, although even the (supposedly) more primitive shrimp can register colour, sharks see only in black and white. On the other hand, bony fishes can see in colour. Some bony fishes such as the piranha can also see infra-red and some fishes even see ultraviolet light. In fact, the common goldfish can see a range of ultraviolet greater than any other animal. In contrast, dogs do not have full colour vision: they have cones sensitive to only blue and green light, but not to red, while New World monkeys have cones sensitive to only blue and red light, but not to green. Most birds, Old World monkeys and humans have full colour vision with cones sensitive to blue, red and green light.
Two types of retinal cells are found in vertebrates and are named according to their shapes as seen under the microscope. The *rods* are found all over the retina (although they are more concentrated around the edges) and are sensitive to very dim light. Rods cannot register colour but can convey a grey-scale image, the shade being dependent (for our purposes) on the variation of the incident light *intensity*. The *cones* are found mainly around the centre of the retina, they function in bright light, and are colour sensitive. Commonly, three types of cones are found, each with a maximal sensitivity to a particular wavelength corresponding to the primary colours, red, green, and blue. Having only rods gives the greatest image definition, but obviously does not allow for the ability to distinguish colour. Having three types of cones, on the other hand, allows for colour vision, but at the cost of diminished definition. The relative numbers of rods and cones varies from animal to animal, and some have only rods while others have only cones:

> The retinas of pigeons contain only cones. Thus while pigeons have color vision, they see only in the bright light of the day. The retinas of owls, on the other hand, have only rods; owls see very well in dim light but are color blind. 'Solomon's organic chemistry'. 6th edition. Wiley. 1996. p610.

The rods (R), cones (C), nerve cells (N), and pigment cells (P) in the retina. (Diagrammatic)

Interestingly, many birds have arguably the most complex system of colour vision since they:

> ...not only have five pigments in their retinas but also each cone has an oil

> drop which filters the light, narrowing the band of wavelengths it responds to. Birds have five different filters available to combine with their five pigments, creating a powerful system for discerning hues... John Downer. 'Supersense'. BBC Books. 1989. p53.

Different birds have different arrangements of cones and filters according to their particular needs. For example:

> The pigeon has red and orange oil droplets concentrated on the view ahead and below, which is normally the ground. These filters probably enhance the colour variations in green vegetation. Seabirds which hunt on the surface of the water use red oil droplets to cut out blue scattered light... John Downer. 'Supersense'. BBC Books. 1989. p53.

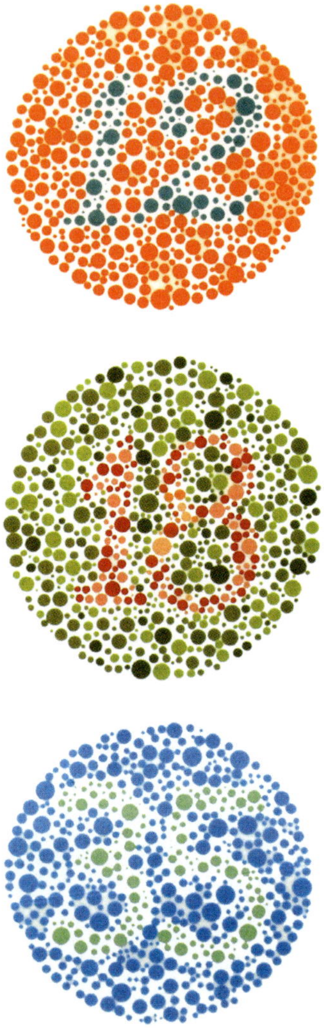

**Testing for colour vision**
A person with normal colour vision should be able to see different numbers within each of the three circles.

The sensitivity of a cone to different wavelengths of light is made possible because each type of cone contains a specific type of opsin which is maximally sensitive to blue (B), red (R) and green (G) parts of the spectrum. Each of the opsins B, R and G are coded for by different genes. In humans, the R and G genes lie adjacent to each other on the X chromosome and have a 98% similarity in their sequences. The B gene is located on chromosome 7. The retinal attached to the rod opsin and to each of the BRG cone opsins is the same. Thus, the difference in sensitivity to any particular wavelength of light is dependent entirely on the amino acid sequence of the opsin itself.

Since the genes coding for the R and G cones are very similar and are found adjacent to one another on the X chromosome, it is suggested that one gene arose from the other by duplication, followed subsequently by some mutations in the duplicated gene so as to alter its spectral sensitivity:

> The "blue" opsin is encoded on human chromosome 7, while the red and green opsin genes are located next to each other, head-to-tail, on the X chromosome. The red and green opsin genes **are** the product of an evolutionary recent gene duplication **because** they are 98 percent identical in sequence. (My bold). Harvey Lodish et al. 'Molecular cell biology'. 3rd edition. 1995. p977.

But such dogmatism must not be mistaken for science. Any similarity, however great, is *not* in itself evidence that one gene actually did develop into another. In fact, there are a number of problems with the notion that 3-cone colour vision evolved from 2-cone colour vision.

In terms of survival, 3-cone RBG colour vision is not necessarily superior to 2-cone BR colour vision. South American monkeys (such as *marmosets*, *tamarins*, *squirrel* and *capuchin* monkeys) have cones sensitive to blue and red light, but not to green. On the other hand, Old World monkeys (such as *colobus* monkeys, *macaques* and *baboons*) have full colour vision:

> ...it is widely agreed that primates emerged...becoming a distinct group from other mammals between 90 and 65mya...the New and Old World monkeys should not be equated, for they are monkeys in name only...their

common stem group began to divide into many descendant lines some 40 mya. Steve Jones, Robert Martin, David Pilbeam. 'Cambridge Encyclopaedia of human evolution'. Cambridge University Press. 1996. p199.

Top: Macaque, and Old World monkey.
Middle and bottom: Woolly monkey and lion tamarins from South America.

Since New World monkeys clearly survive and are well able to make their way through the trees and escape predators, it follows that 3-cone colour vision is not essential to life in the trees. And indeed in most circumstances, humans with BR vision (red-green colour blindness) are not significantly disadvantaged compared to those with BRG vision.

Suppose that the common distant ancestor to both monkeys and dogs lived 90mya and had only B and R genes, and that the Old and New World monkeys diverged 40mya. If it were significantly beneficial then it is implausible that, after 40my, no New World monkeys at all have acquired BRG 3-cone colour vision. In the same way it is implausible that after 90my some dogs have not acquired BRG 3-cone colour vision. And of course, if some monkey common ancestors (MCA) acquired a duplication of the R gene (resulting in BRR), the fidelity of DNA replication is such that it is likely that it would be copied *exactly*. This of course could not give rise to any change in the colour vision of the individual. But suppose that, some time later, the duplicated R gene underwent a mutation resulting in a change in one of the hundreds of amino acids in the opsin molecule. This would be most unlikely to result in the very amino acid sequence that is optimally sensitive to the green light. So after many millions of years, we should see a whole variety of amino acid sequence differences within the blue, red and green opsins of monkeys and apes, with these differences producing a continuous variety of subtle differences in the colour vision perceived. Yet the differences within the opsin amino acid sequences between different monkeys and apes are minimal.

Furthermore, if the R and G genes (which lie next to each other on the X chromosome) are the result of a duplication of an R gene, with subsequent mutations converting one of the genes into a G gene, then, contrary to observation, we should expect that, in a random manner, in some populations the G gene should have appeared as mutations within the new (duplicated) R gene, and in other populations the G gene

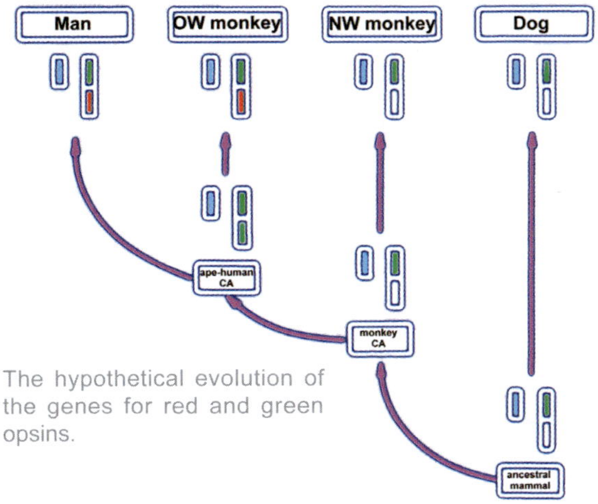


The hypothetical evolution of the genes for red and green opsins.

should have appeared as mutations within the old (original) R gene.
A major problem is also raised by the very fact that dogs do not have red cones and New World monkeys do not have green cones. We saw earlier that bony fishes (which are supposedly ancestral to all terrestrial vertebrates) have full 3-cone colour vision, and that many of their (supposed) descendants have only 2-cone colour vision. Since most bony fishes have full colour vision, then for any of their (supposed) descendants such as dogs and New World monkeys to now lack a red or green cone means that retaining that particular characteristic was unnecessary for their continued survival. Being non-essential, the genes coding for the red and green cones would, over time, have become degenerate as a result of random mutations. However, the genes themselves, showing marked alterations in their nucleotide sequences, would not have just disappeared, but would have remained in the chromosomes as genetic 'fossils'. In other words, if a dog or a New World monkey has indeed evolved from ancestors which were bony fishes, we should expect to find evidence of such degenerate genes. It is not clear from the literature if a degenerate red or green gene has been found in the chromosomes of dogs and New World monkeys respectively. If such genes do not exist, then it is implausible that either dogs or New World monkeys have evolved from fish ancestors.
Finally, since in reality there is no such thing as colour because it is within the *brain* itself that a colour image is perceived, simply developing 3 types of cones (B, R and G) where previously there were only 2 types (for example, B and R) cannot itself result in 3-cone colour vision. Light is essentially an energy form with a wide and continuous variation in wavelength. After light reaches the retinal cones and the resultant nerve impulses are transmitted down the optic nerves, the brain creates an image within itself (the original virtual reality) such that the different combinations of different wavelengths are *perceived* as different *colours*.
Thus, colour is merely an illusion. Indeed, each of the nerve impulses that emanate from each of the different types of cones are, in actual fact, indistinguishable from each other. In fact, the nature of the nerve impulses received by the brain from the cones is *exactly* the same as that of the nerve impulses sent to the brain from touch, vibration, temperature, smell, taste and hearing sensors. When *any* nerve is stimulated adequately the same process is triggered off: briefly, there is a shift of sodium ions into the nerve cell, very shortly followed by an eflux of potassium ions. These collective changes in the movement of sodium and potassium ions are then propagated along the length of the nerve as a *nerve impulse* (called an *action potential*) until the end of the nerve is reached. Here, there is a gap between the end of the nerve and the cell body of any other adjacent nerve. Obviously, for the signal to travel from one nerve to the other, somehow the gap has to be bridged. This is achieved by the action potential, on its arrival at the end of the nerve, triggering the release of a chemical that flows into the gap. Like a key in a lock, the chemical (called a *neurotransmitter*) then docks with specific sites (called *receptor sites*) on the second nerve's cell membrane. The functional entity which includes the junction between two nerves and the associated neurotransmitters and receptor sites is called a *synapse*. It is by interfering with this sort of process that the poison *curare* works on nerves outside the brain. Curare is found in plants of the genus *strychos*

A synapse showing a nerve signal passing from nerve **a** (left) to nerve **b** (right).
1. The arrival of the action potential at the end of nerve **a**, where neurotransmitters are stored in special pockets.
2. Once the action potential reaches the end of the nerve, the neurotransmitters are released into the synaptic gap.
3. The neurotransmitters lock into their specific receptor sites and an action potential is triggered in nerve **b**.
4. The neurotransmitters leak away from the receptor sites and nerve **a** has its neurotransmitters replenished while awaiting the arrival of the next action potential.

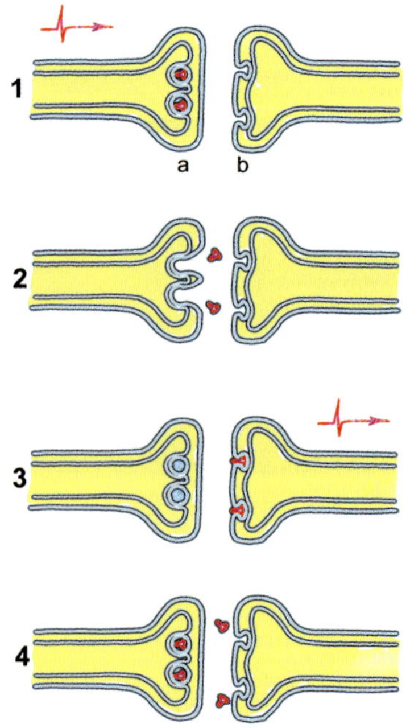

and was first used by native South Americans on the tips of their darts. When an animal is darted, the curare enters into the blood stream and permeates the whole body. The specific shape of the curare molecule allows it to fit into the receptor sites on the surface of muscles, and by so doing it prevents the normal neurotransmitter (*acetyl choline*) from stimulating the muscles to contract. The muscles are thereby paralysed and the animal dies of asphyxiation. A modified form of curare is routinely used in anaesthesia to 'paralyse' patients. The patient is not at risk since breathing is artificially maintained by a mechanical ventilator. Local anaesthetics such as *lignocaine* work on nerves in a quite different way by blocking the movement of sodium and potassium ions through the channels in the nerve cell membrane. This prevents an action potential being triggered by any stimulation of the area of tissue supplied by the anaesthetised nerve. Since the brain can receive no signal from the anaesthetised nerve, the affected area feels numb.

While on the subject of nerve impulses, we should certainly question whether or not the action potential/synapse system could have evolved by a series of stepwise changes. Clearly, if a (supposed) primitive cell generated an action potential but had no neurotransmitter present at the nerve ending, there could be no transfer at all of the signal to a second cell and there would be no function or advantage in an action potential being generated. In the same way, if the neurotransmitter appeared first, it would have no function without both the mechanisms for the production of an action potential to trigger it, and a receptor site to receive it, being in place. So, it is implausible that, starting with some lesser system, the action potential/synapse system could have evolved by a series of stepwise changes.

Since there is nothing 'out there' that has any colour, and given the identical nature of any action potential emanating from any type of cone (R,G or B), the mystery is not just how we see but also why we see any colour at all.

Without prior and appropriate programming within the brain to receive signals arising from cones with different spectral sensitivities, the brain could not distinguish between signals from R, G or B cones. In other words, unless the brain was first programmed to perceive full colour, simply having a number of different retinal cells with different spectral sensitivities would only result in grey-scale vision based on light intensity alone.

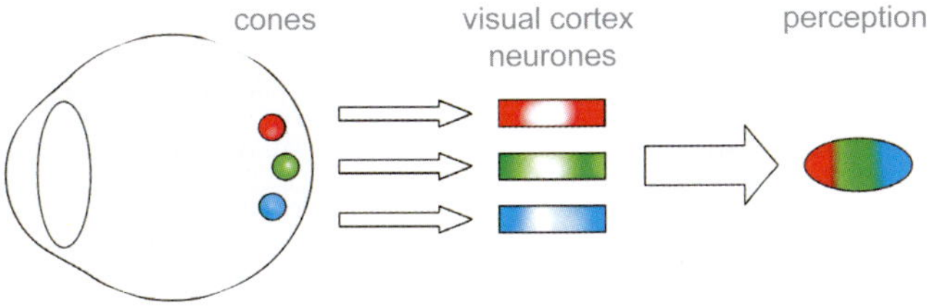


**Normal RBG colour vision**
The brain is able to differentiate between the different impulses from the R,B and G cones, so as to perceive full colour vision. Note, however, that whichever type of cone they originate from, *in their nature* the impulses from the eye to the brain are themselves identical. Therefore, colour perception must be dependent on the recognition by the brain of the spectral sensitivity of the type of cone that originates any particular impulse.

In individuals with *red-green colour blindness* the G gene is absent, and only R and B cones exist to send impulses to the brain. For practical purposes, any incident green light would weakly trigger either a R or B cone. In this case, although the brain might well have within it programming for the *perception* of full colour, nevertheless a non-full-colour image (let us call it a '2-colour' image) will be perceived by the individual.

**RG colour blindness**
Even though the brain might have G receptors, the absence of G cones results in '2-colour' vision.

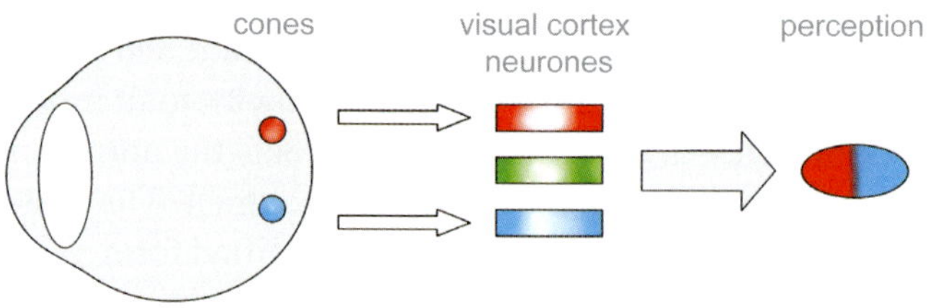


In the hypothetical ancient situation before there ever existed a G gene, a brain could only respond to cones containing R and B opsins resulting in a '2-colour' image.

In the (supposed) primitive situation with only R and B cones and no ability of the brain to process green light, only '2-colour' vision would exist.

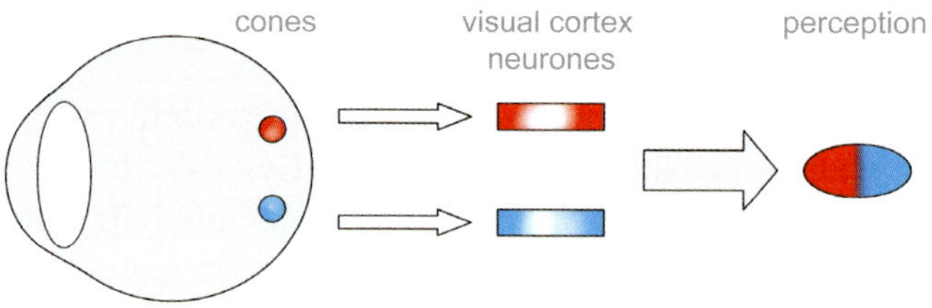


But suppose that subsequently some chance mutations resulted in G cones arising. The problem then would be that there would still have been no change in the neurones in the brain. Therefore, if it could be perceived at all, the signal from any G cone would only be recognised in the brain as if it originated from either a B or R cone.

Impulses from a newly-acquired G cone would either trigger no response from the brain or simply trigger R and B receptors.
So, without the brain being programmed to compute its impulses, the appearance of a G cone would still result in '2-colour' vision.

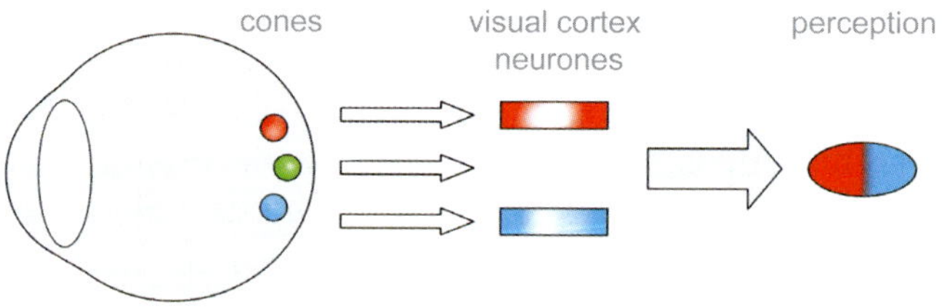


Thus, simply changing the genes for cone sensitivity could not result in colour vision unless other changes were to occur in genes that code for brain *recognition* of signals from R, G and B cones, and for the *perception* of colour. Although it may seem that a New World monkey with R and B cones and '2-colour' vision is a small step away from obtaining full colour vision by simply duplicating and slightly changing the R gene, in reality, the monkey may be *light years away* in terms of the possibility and probability space that separates it from 3-cone colour vision.

### *Perception and consciousness*

When light reaches the retina, the rods and cones fire off impulses to a number of nerve cells within the innermost layers of the retina itself. These nerve cells not only act as relay stations but are also involved in some preliminary processing of an image. The retinal nerve cells send impulses to the back part of the brain (the *visual cortex* of the occipital lobes) after which the signals are computed to make up a mental picture. However, the cells of the visual cortex are *not* arranged in neat grid patterns with co-ordinates corresponding to the layout of the rods and cones in the retina:

> ...functional mapping experiments do not reveal a topographically organised 'map' in relation to the retina. S. Zeki. 'The representation of colours in the cerebral cortex'. Nature. Vol 284. 3rd April 1980. p412-8.

Instead, the brain cells are organised in a very complex manner within the visual cortex, and their arrangement is in part dependent on their *function*. Thus, some groups of brain cells deal with colour, while others deal with horizontal or diagonal lines or with the movement of lines across the field of vision.

But what exactly is a mental picture and where is it projected into our consciousness as a virtual reality? Clearly, at the heart of the biology of vision is consciousness itself. Could consciousness *have arisen simply as a result of spontaneous chemical changes starting in some primordial pool of water?* Can consciousness be explained as simply the obligatory outworkings of some very complicated chemistry and nerve circuits? Most certainly, the brain is *not* like a computer with electric currents moving along fixed insulated wires or circuits. The total number of neurones in the human brain is about one million million. Any particular neurone can have up to around 100,000 other neurones making synaptic contact with it. Each such synapse can have a variable degree of stimulatory or inhibitory influence on the state of activity of that neurone which could itself be one of up to 100,000 neurones signalling to another neurone. This second neurone could be situated anywhere in the brain - either adjacent to the first neurone or even on the other side of the brain. The total number of connections between the million million neurones in the human brain is therefore astronomical. In addition to the numbers of interconnections, different synapses can have different neurotransmitters, each of which have different properties. The combination of the vast numbers of connections and the use of neurotransmitters allows for a flexibility in the manipulation of signals by any given neurone which would be unlikely to be matched by any silicon-based computer.

Any particular neurone like cell **A** can have up to around 100,000 other neurones making contact with it. The sum total of the multitude of signals from those neurones determines the signal that cell **A** will transmit to cell **B**, which itself may have up to 100,000 other neurones signalling to it. The total number of connections between the million million neurones in the human brain is therefore astronomical.

But does the brain achieve consciousness simply by virtue of its size and the number of its neuronal interconnections? Although some computers may simulate consciousness by responding to situations with a so-called artificial intelligence, the question is: could a far larger computer ultimately become *aware* that it is doing anything? Take an enormous computer and make it as large as can be imagined, and then enlarge it further. Imagine that it is initially programmed to respond to all manner of variations of possible input signals that it receives via pressure, light, sound, electromagnetic, gravitational and thermal sensors. Imagine that its programming allows it some freedom to vary its own response to the data coming into it via those sensors, and that it is programmed to say 'ouch' or to groan when its sensors are stimulated above a certain threshold level. However large you imagine the system to be, no matter how complex its actions or responses, the computer would still be dependent on its software and would operate simply on the basis of

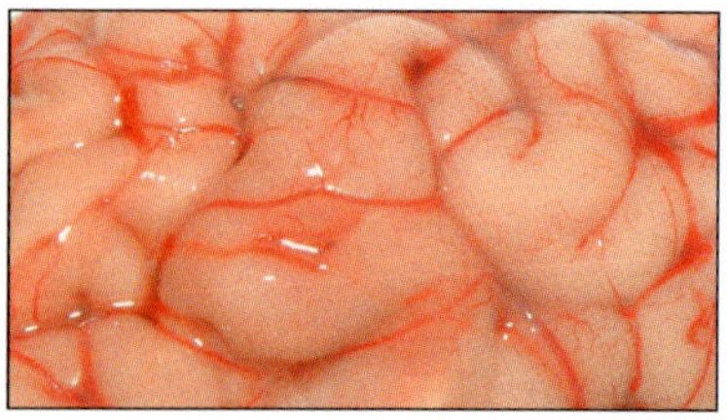

Cow cerebral cortex. The brain is not just a large computer functioning simply by billions of on/off triggers.

**Single cells make homes by themselves**

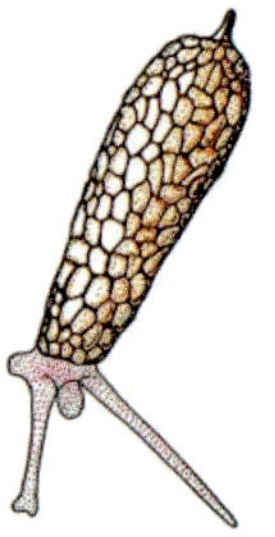

*The amoeba *Difflugia* can construct a home for itself out of sand grains. (*After Ford 2006*).

*'These phenomenal species collect granular material from the silt in which they live and glue the particles together to make a protective shell. The testate amoebae use specific components to make their shells. Some utilize minute grains of silica sand; others select centric diatoms and glue them carefully together. They do not confuse their raw materials, for each species has a preference of its own...It may be that the fundamental processes of thought are intra-, rather than exclusively interneuronal.'*

B Ford. 'Revealing the ingenuity of the living cell'. Biologist. Vol 53. No 4. August 2006. p221-4.

This has serious and profound implications for our understanding of how the brain works, and also of when the fertilised human egg becomes *human*.

yes/no commands without any *awareness* on its part. A computer does not understand anything and does not interpret anything - it just directs electrons according to how its programming dictates. In contrast, we are able to choose to do a number of things when confronted with the same situation in front of us. We can even choose an *illogical* response, or choose to do nothing at all.

There is a well-known mind experiment which involves a man in a room. A piece of paper is passed under the door into the room. On the paper is written some text, in a language A which is completely alien to the man in the room. However, his job is to translate the text into another language B, which is also completely alien to him. With him in the room, the man has a filing system which contains a large number of codes. All the man has to do is simply to look at the piece of paper given to him, and according to the shapes of the characters he sees, refer to an appropriate code for that character, from which he can define a character in the language B into which he is supposed to translate the text. Many such step-by-step references ultimately allow the man to write a complete and accurate translation without knowing or understanding either of the two alien languages. He then passes the translation back under the door. The person on the outside, not knowing anything about the man in the room or its contents, might therefore go away with the belief that someone in the room understood the first language and translated the meaning directly into the second language. But obviously, such a notion would be false, since the whole translation process was achieved by referring to a set of rules, which made no reference to the meaning of the words being translated. And so it is with computers and all things that work as they do. Their ability to solve problems can be impressive, but they function without understanding or awareness.

We know that for the brain to function, it is necessary on one hand for individual neurones to relay signals to other neurones via synapses, and on the other hand for the brain to be organised into specific collections of neurones scattered around in specific locations. The problem is that we have *no* idea how the microscopic activity of single neurones or simple nerve reflexes translates into the phenomena we are so intimately familiar with - thinking, memory and consciousness. Furthermore, while an intact brain is essential for consciousness to properly manifest, that *does not* mean that consciousness is simply and only the sum total of the neuronal intercommunications in the brain. It is far more mysterious than that and may well be an *epiphenomenon* inexplicable simply in terms of our current understanding of neuronal interconnections and brain chemistry*. At this point, our logic and our science leave us severely lacking and we can only say that we *have* consciousness. As to - in our minds - where and how we see and think, this is a *complete* unknown. So, how could consciousness have evolved? Since we have no understanding of consciousness (and therefore of whether it could or could not have evolved), there is no scientific basis to claim that it *is* the result of evolution. Furthermore, since we have already seen that the evolution of the eye is implausible and since our perception of vision requires consciousness, to claim that the eye is the result of evolution is a meaningless statement.

The subject of consciousness leads us to another associated question. Why do creatures need to experience pain? Since a molecule does not feel pain when it is broken down into its component atoms or when it undergoes a chemical reaction, and since evolution requires all animals to be simply and only massively complex chemical conglomerates, then it must be questionable whether natural selection *could* result in the evolution of the ability to experience pain and suffering. Many bodily actions occur merely as a series of reflexes for which the higher parts of the brain are uninvolved and unnecessary. For example, after having its head bitten off by a cat, a frog can continue to jump, swim, and dive! It can even climb out of the water and squat on a rock - properly balanced - all without its brain. In fact, a headless frog can continue to behave like this for several days, until its nerves and muscles finally decay and cease to function.

Bullfrog

Just as intriguing is the fact that the eye of the frog:

> ...has several types of nerve cells, each sensitive to different components of the picture. Some respond only to moving edges, while others are fly detectors triggered whenever anything of the right size moves into view. The frog relies on its fly detectors so much that it probably cannot see a motionless insect, and will starve to death rather than eat one...the frog only responds to stimuli that are of relevance to it, and **much of the visual information never reaches the brain**. (My bold). John Downer. 'Supersense'. BBC Books. 1989. p40,43.

A headless frog can still jump and swim for a few days until its muscles decay.

So, a frog will grab and eat its meal, swim and jump around without knowing that it is doing anything. In fact, humans, too, exhibit a considerable degree of automative responses to the world around. A good example is the degree to which we are unaware of our subconscious decision-making when driving a car, particularly when our attention is diverted by a conversation or other thoughts. All drivers will be able to witness to the fact that they can be driving along and realise that they have no conscious recall of any decision-making or activity on the road for perhaps the last few minutes. This shows how the eyes clearly deliver information about the world to parts of our brains which respond in pre-programmed and learned ways, and without that information reaching the parts of the brain involved in consciousness. Such subconscious responses to visual stimuli can also be demonstrated in some people with certain types of injury to the visual cortex, who, for example, can *know* that an object may be moving in a particular direction in front of their open (and normal) eyes but are nevertheless unable to consciously see anything at all. This so-called *blind sight* shows that there must exist specific nerve fibres that connect the eyes to deeper parts of the brain which are separate from the fibres connecting the eyes to the visual cortex.

So we should ask why *any* creature of *any* complexity could not exist and function perfectly well as a chemical conglomerate, showing varying degrees of sophisticated automatic responsiveness to various stimuli or circumstances...without any *pain* or *consciousness*. If life evolved from simple chemistry, then, if anything, we should expect that all creatures should behave simply and only like headless frogs.

So we see that, instead of being a simple open-and-shut case, the subject of vision opens up mystery upon mystery. If anything, this chapter is a gross simplification and only just scratches the surface of what exactly is vision and consciousness. For example, a whole book could be written purely on the mechanisms of the storage and release of neurotransmitters. Yet some people are willing to dive in blindly and claim that the origin of the eye is a simple extrapolation of natural selection. We shall certainly learn more in time, but whatever we learn is likely to show that vision, the mind and consciousness are more complex than we can imagine. Darwin himself considered it reasonable to ignore solid contradictory evidence that was well known from before he was even born:

> Reason tells me, that **if numerous graduations from a perfect and complex eye to one very imperfect and simple, each grade being useful to its possessor, can be shown to exist;** if further, the eye does vary ever so slightly, and the variations inherited, which is certainly the case; and if any variation or modification in the organ be ever useful to an animal under changing conditions of life, then the difficulty of believing that a perfect and complex eye could be formed by natural selection, though insuperable by our imagination, can hardly be considered real. (My bold).
> Charles Darwin. 'The origin of species'. Penguin Books. 1985. p217.

However, we have seen that evidence is as deficient as imagination. Some people try to make out that Darwin produced remarkably good science for a man of his time. However, regardless of his botanical observations, Darwin's studies of animals and anatomy were quite basic in most respects. Dr Henry Gray published the first edition of his masterly '*Gray's Anatomy*' in 1858, the year *before* the publication of '*The origin of species*'. The first *Gray's Anatomy,* which comprised 750 pages and contained 363 figures expertly drawn by Dr Henry Carter, shows the remarkable depth of anatomical knowledge 150 years ago, and in contrast, just how superficial and inadequate Darwin's anatomical studies actually were when compared to the studies made by his own contemporaries.

The fact is that had Darwin made the effort to do some simple anatomical studies of birds, reptiles and mammals, and conversed with the doctors of his time to discuss some basic anatomy and physiology, he would have seen how his idea of *small-stepwise-changes-leading-to-one-organ/animal-changing-into-a-completely-different-organ/animal* was not just implausible, but *impossible.*

High-powered microscopes and knowledge of molecular biology they did not have, but people in the 1850's had ample material at hand which completely disproved evolution. In 1848 Gray won a prize at the Royal College of Surgeons for an essay entitled '*The origin, connexions and distribution of the nerves to the human eye and its appendages, illustrated by comparative dissections of the eye in other vertebrate animals.*' There are six muscles (*extra-ocular muscles*) on the outside of the human eye: the *superior, medial, and inferior rectus* muscles and the *inferior oblique* muscle, which are all supplied by the 3rd cranial nerve; the *superior oblique* muscle which is supplied by the 4th cranial nerve; and the *lateral rectus* muscle which is supplied by the 6th cranial nerve. The 3rd, 4th, and 6th cranial nerves originate from quite distinctly different portions of the brain and take quite different routes to the eyeball, and show that the embryological development of these different muscles and their nerve supplies must be controlled by quite different triggers.

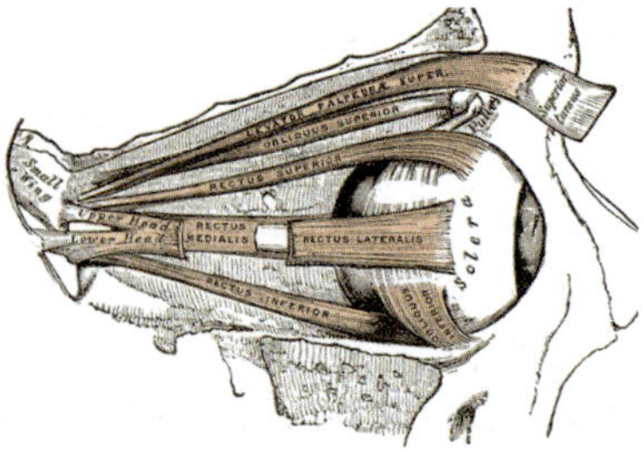
Detail from Gray's Anatomy, 1918.

Apart from the superior and inferior oblique muscles, all the other extra-ocular muscles are anchored on the inside of the back of the eye socket, and are attached

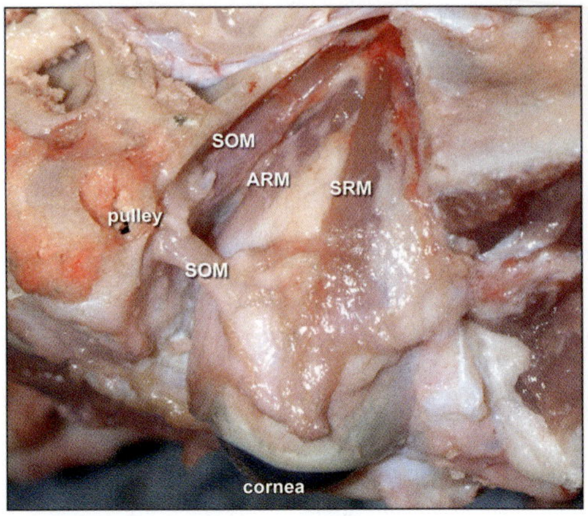

Pig left eye

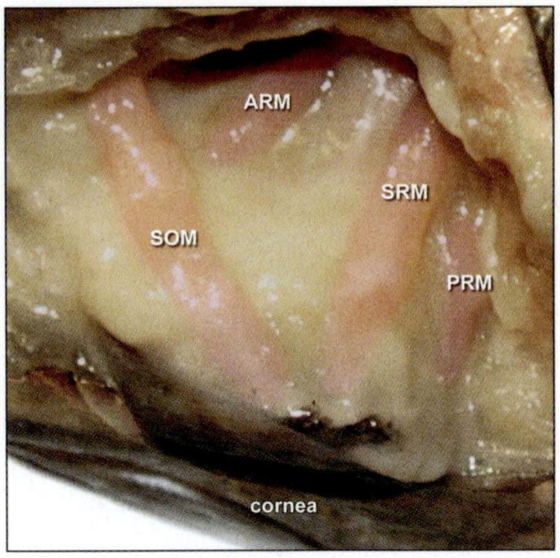

Salmon left eye

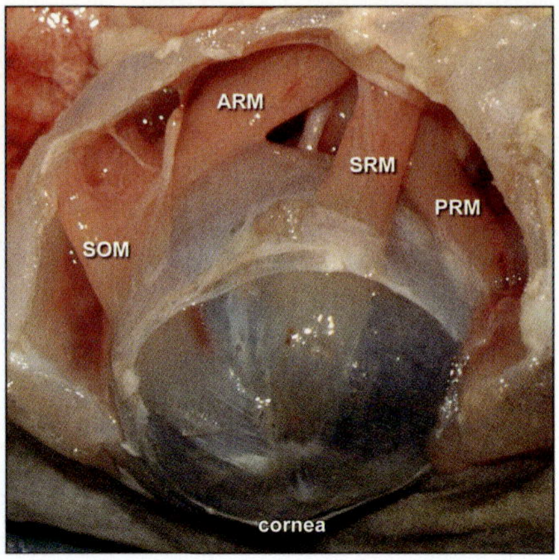

Dogfish left eye

in a straight line to the outside of the eyeball itself. In contrast, the superior oblique muscle, which is anchored to the bone at the back of the eye socket, runs not to the eyeball, but instead towards the front inner and upper corner of the wall of the eye socket. At this point its tendon passes through a fibrous *pulley* which alters its direction of pull by about 90 degrees in order to reach its insertion point on the top of the eyeball. As a result of this unique arrangement, the superior oblique muscle, the muscle belly of which essentially runs from the back of the eye socket to the front, is able to cause the eyeball to rotate in a manner so as to dip the gaze downwards and away from the nose. In addition to this remarkable arrangement, between the tendon and the pulley lies a *synovial pouch* containing lubricating fluid that facilitates the smooth running of the tendon, much in the same way that the tendons in the wrist and ankle slide within friction-reducing tendon sheaths. While mammals have a superior oblique muscle with a pulley system, in both cartilaginous and bony fishes the superior oblique muscle is a 'simple' structure arising from the front side of the orbit and attaches directly to the top of the eyeball. Since the is no functional intermediary between a superior oblique muscle that has no pulley system and one that does, the superior oblique muscle with a pulley system could not have evolved stepwise.

We have seen that while different types of eyes surely exist, this is not the same as saying that the '*numerous graduations from a perfect and complex eye to one very imperfect and simple'* exist. In fact, whether in living or extinct form, we have seen that intermediary forms both of eyeballs (such as 4/10 or 5/10 spheres) and of their myriad contents (such as rods, cones, the canal of Schlemm etc.) *do not exist*. Furthermore, the (supposed) evolution of a flat sheet of photoreceptors to a pin-hole camera-type eye, and ultimately to a mammalian eye, requires not only the sequential modifications to many existing genes, structures and biochemical systems, but also the origination of many new genes, structures and biochemical systems, many or perhaps most of which are independent of each other. So much so, that any such (supposed) changes could only be made by taking gigantic steps of a nature either chemical, embryological or statistical.

In other words, regarding the origin and evolution of the eye, what might from a *superficial* glance appear to be a simple, *seemingly* inevitable progression from simple to complex, is, in fact, impossible.

**External eye muscles**

Above and below:
SOM: Superior oblique muscle
SRM: Superior rectus muscle
ARM: Anterior rectus muscle
PRM: Posterior rectus muscle

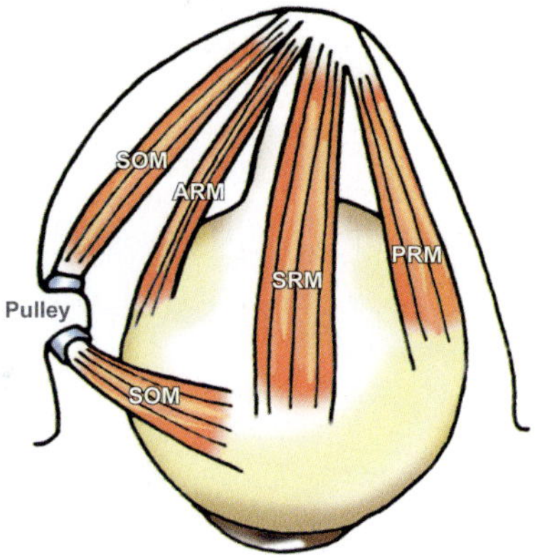

The superior oblique muscle of the mammalian eye has a pulley system.

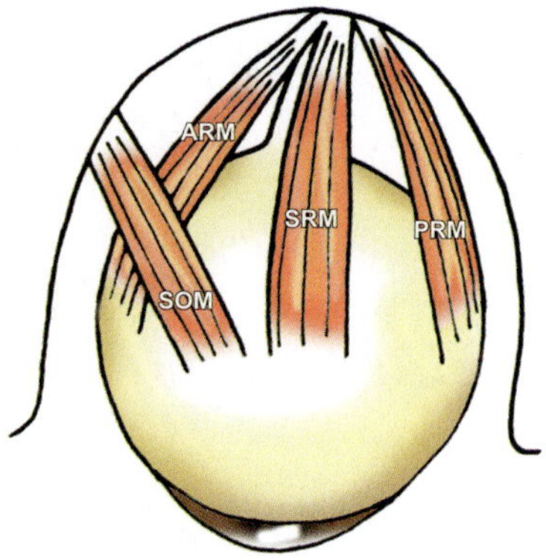

The superior oblique muscle of the fish eye does not have a pulley system.

# CHAP

## Human fossils:

*bone of ape bones?*

# SECTION FOUR
# The Man - one giant leap for apekind?

TER 10

Neanderthal skull

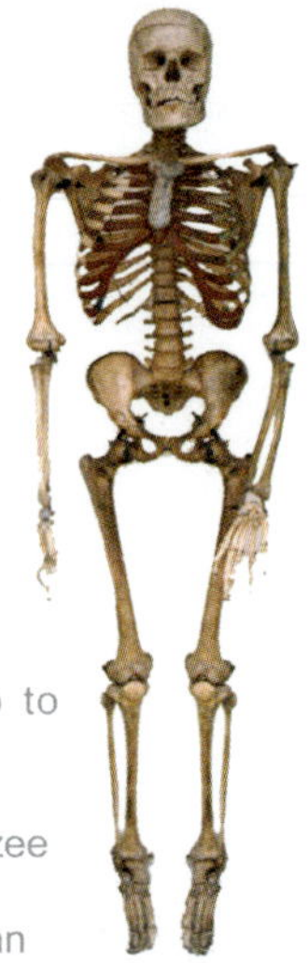

From top to bottom:
Human
Chimpanzee
Gorilla
Orang-utan

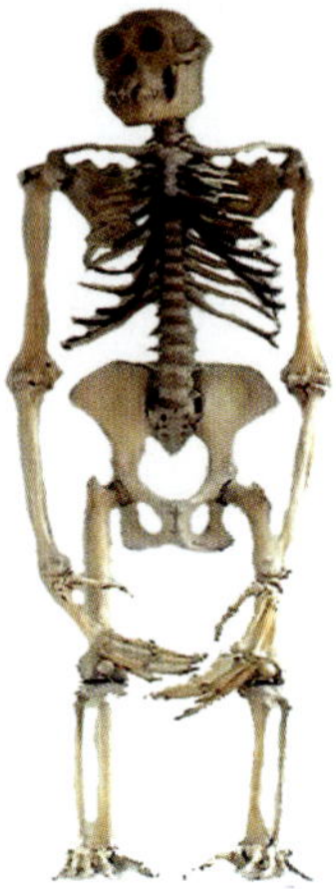

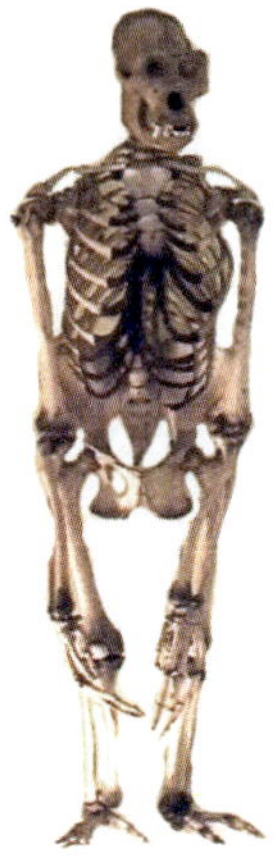

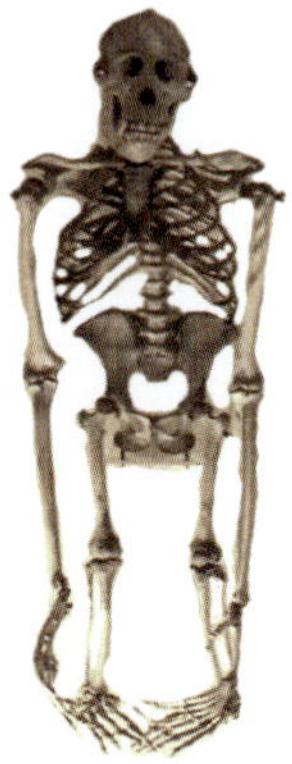

We have already seen that whether we look at the fossil record, cellular chemistry and structures, or living creatures, the evidence does not support the notion that there has been an evolution of organisms from simple chemicals to mammals. If this is indeed the case, then it follows that there cannot have been any evolution of Man from ape ancestors either, and therefore the following chapters are wholly unnecessary. However, the reader may protest that the concept of human evolution is well- supported by fossil evidence showing changes from ape-like to man-like skulls and skeletons. For this reason, our study cannot be complete without examining the evidence for the evolution of Man and seeing whether it correlates with our conclusions already derived from other evidence.

There are four main questions that we will address in the following chapters. Firstly, does the fossil evidence support the notion that humans evolved from apes? Secondly, is it plausible for an essentially quadruped ape to change into a bipedal human being? Thirdly, does the chromosomal evidence support the notion that humans evolved from apes? And finally, does the evidence support the notion that human intelligence evolved from something that was previously subhuman?

Although detailed models of human evolution are being constantly disputed and revised, nevertheless, most biologists have no serious questions about the general scheme of human origins. As Francis Crick, being somewhat economical with his science, explained:

> One branch, the primates, developed good colour vision and an enlarged cerebral cortex, eventually producing man. Francis Crick. 'Life itself'. MacDonald. 1981. p110.

Other books on fossils sing the same tune. For example:

> The fossil record of our ancestors, while not abundant, is complete enough to show **clearly** our common ancestry with apes and our evolution. (My bold). Arthur Busby III, Robert Coenraads, David Roots and Paul Willis. 'Rocks and fossils'. Harper Collins. 1996. p226.

Curiously, 'Rocks and Fossils' does not include *a single* human fossil. Indeed, information about human fossils is surprisingly absent from most books on fossils, and in actual fact, finding out just exactly how many human-like fossils are known and what they look like is not an easy task at all. Even the impressively-titled 300-page *'Encyclopaedia of Human Evolution'* by Jones, Martin and Pilbeam shows only a small sprinkling of human-like fossils, amounting to around just *twelve* in all. For a work purporting to be an encylopaedia on the subject, this constitutes lamentably little information. And surprisingly, the Natural History Museum of London's own exhibit on human evolution includes less than ten reproductions of individual human-like fossil skull specimens. Thus, members of the general public are given only a small amount of material, and a very sketchy and dogmatic story, which leaves little opportunity for formulating their own understanding.

So, what exactly is the evidence behind the notion that humans evolved from apes? Most of the supposed evidence for presumed links between ancient primates and human ancestors comes from differences in teeth. It is supposed that if primitive creatures became monkeys, some of which became apes, some of which became humans, then their teeth would show gradual transitions:

Elephant shrew

> In overall form, most Paleocene primates probably resembled modern tropical squirrels, and it may reasonably be asked why they should be classed as primates. The answer is in fact **equivocal**. Perhaps most important, their cheek teeth closely resembled those of later, undoubted primates. (My bold). Richard Klein. 'The human career'. Chicago University Press. 1981. p61-62.

Aegyptopithecus

But, as we have seen previously, the *Chalicotherium Principle* shows that the occurrence in two creatures of similar features does not in itself imply any ancestral or descendant relationship between the two creatures. Nevertheless, that aside, although there are a number of variations in the standard theory, basically it is thought that some 60mya a group of small ground-dwelling *shrew-like early mammals* took to the trees and became primitive monkeys. By 30mya some of these monkeys appeared in the fossil record as *Aegyptopithecus*, a creature considered to have looked more like a small tree-living dog than a monkey.

Then the fossil record becomes empty for an enormous *10 million years*. Between 17-20mya other ancient apes appeared, such as *Proconsul* - a tree-living ape whose limb bones suggest that rather than swinging with its arms, it may have instead walked quadrupedally on branches.

Also by 20mya more modern-looking apes had appeared in the form of the *Dryopithecus* family. Then, supposedly between 18 and 16mya, the Earth's climate began to change drastically and the forests began to decline. Many of the dryopithecines died out, but some that lived on the edges of the forests evolved into the ape-like *Ramapithecus* which flourished some 14-8mya in Asia, Africa and Europe. It is not possible to make any inferences about the body and limb form since all that is known about ramapithecus is a small collection of teeth and jaw bones.

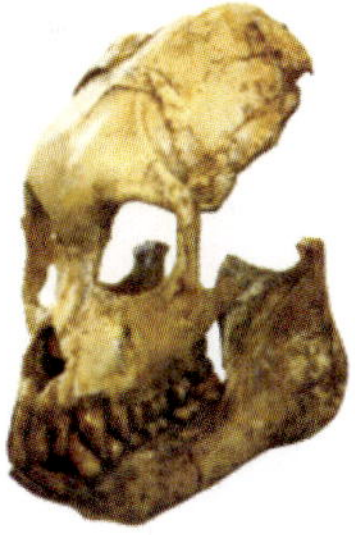

Above: Proconsul.
Below: Dryopithecus.

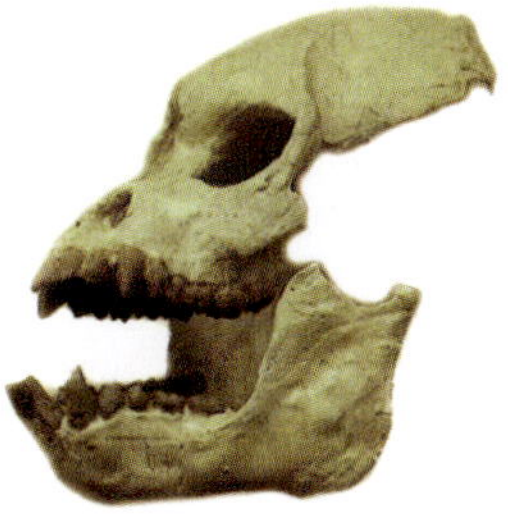

Primitive apes gave rise sequentially to *orang-utans, gorillas, chimpanzees* and *Australopithecines*. The oldest orang-utan fossils are only some 2myo. Orang-utans gave rise to gorillas, and later, to chimpanzees. Although orang-utan-like fossils have been found in Pakistan and the Far East, curiously, and importantly, the first

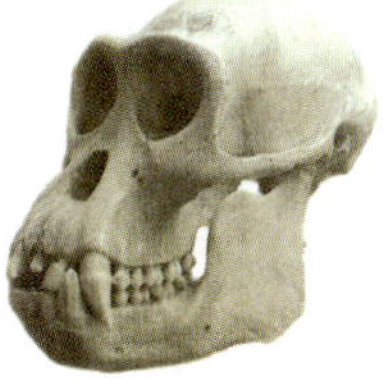
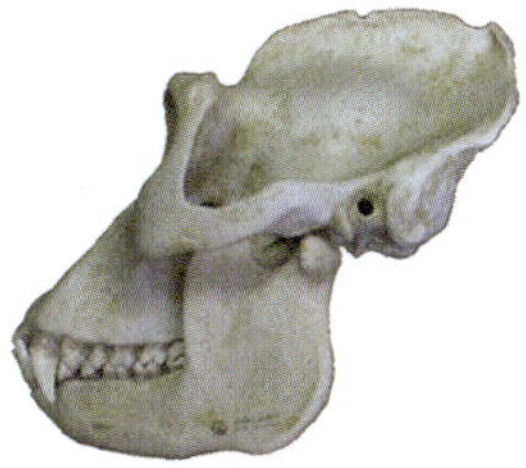
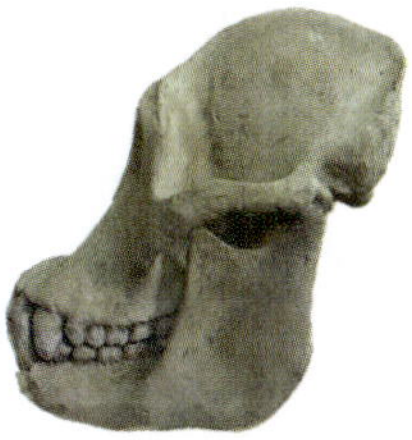

Above left: Chimpanzee.
Above: Gorilla.
Left: Orang-utan.

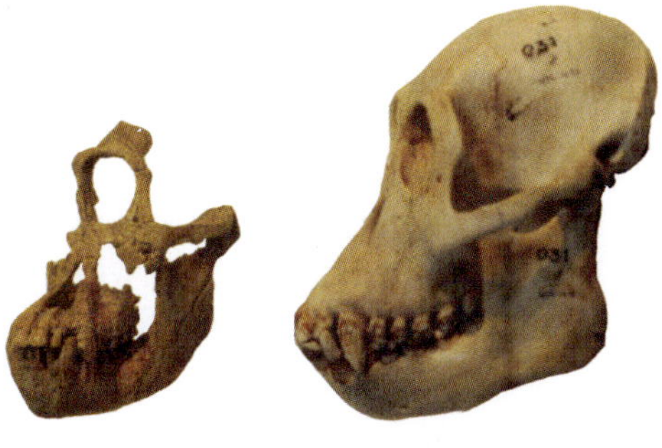

In many ways the skull of Sivapithecus (far left) is very similar to that of the modern orang-utan (left). Its fossils have been found in the Siwalik Hills of Pakistan.

Femur of Orrorin tugensis, (supposedly) 6 million years old.

fossil chimpanzee remains were recently identified from the Kapthurin Formation, west of lake Baringo, Kenya, dated around 545,000yo:

> The chimpanzee specimens comprise a minimum of three teeth. S McBrearty & N Jablonski. 'First fossil chimpanzee'. Nature. Vol 437. 1st September 2005. p105-8.

And the only known fossil of a gorilla is:

> ...a four-million-year old gorilla tooth. 'First fossil chimp'. New Scientist. 13th November 2004. p17.

This is curious since fossils of other forest-dwelling creatures such as proconsul, orang-utans, lemurs, sloths and monkeys *are* known and since (supposedly) chimpanzees and gorillas had been around for millions of years before both australopithecines and humans, there would have been significant opportunity for these great apes to become represented in the fossil record.

Recently a number of fossil bones were found in Kenya belonging to a new group of apes named *Orrorin tugensis*, and dated as being around 6myo. A partial femur suggests that Orrorin tugensis may have been bipedal. However, Orrorin also had curved finger bones which suggests that it spent a considerable part of its time in the trees. Nevertheless, the possibility that it might have been bipedal has been taken by some scientists to mean that Orrorin was a direct ancestor to humans.

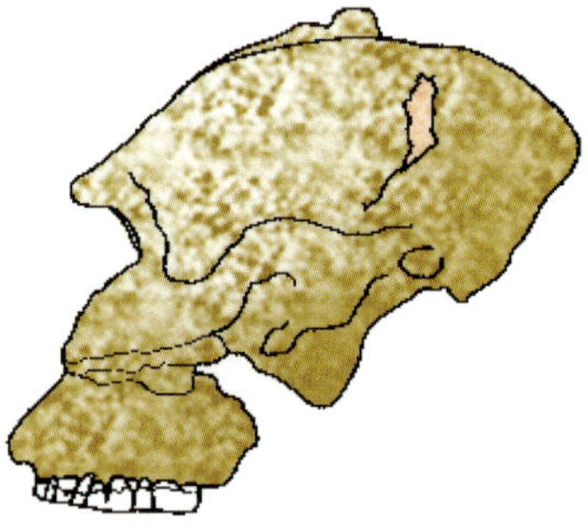

Robust australopithecine SK46.

Australopithecines are a distinct and extinct group of creatures that differ significantly from both living apes and humans, and it is widely suggested that australopithecines represent an ape-human intermediary. Despite the complete absence of any supportive evidence, it is supposed that australopithecines separated from their chimpanzee ancestors between 5 and 8mya. However, the story is constantly changing:

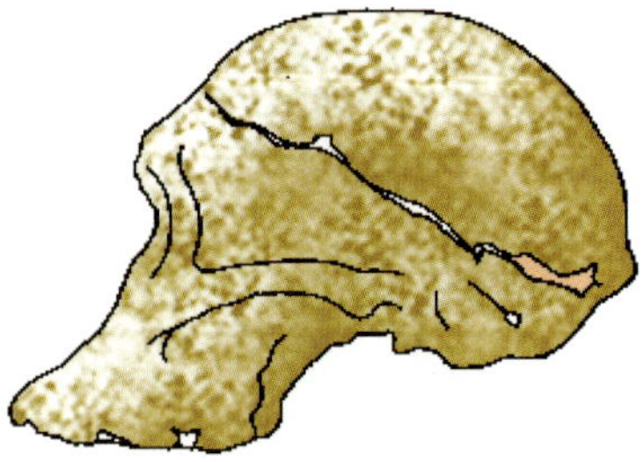

Gracile australopithecine Sts5.

> The recent discovery of a **surprisingly modern-looking ape from 22.6mya** in Africa, called *Morotopithecus bishopi*, has prompted some paleontologists to speculate a date of 6mya for the divergence between apes and the human lineage to be too recent. In fact, the presently accepted date is based largely on biochemical evidence. In the 1960s, when fossil evidence was the only clue, the split was thought to be at least 10mya, and possibly 15mya. (My bold). Ann and Patrick Fullick. 'The human story'. New Scientist. *Inside Science*. 7th June 1997. p2.

Australopithecines are conventionally divided into gracile (lightly built) and robust (thicker-boned) forms. The celebrated australopithecine fossil *Lucy* (AL 288-1) found at Hadar, Ethiopia, in 1974 is dated as being around 3.5 million years old, and australopithecines are thought to have lived until about 1mya. Although they could walk bipedally, the extent to which they could do so is subject to debate. Some consider that they walked like humans, while others think they walked on two legs in a limited manner, but with a greater ability than chimpanzees. Australopithecine fossils have only been found in the eastern half of Africa, from South Africa up to Ethiopia.

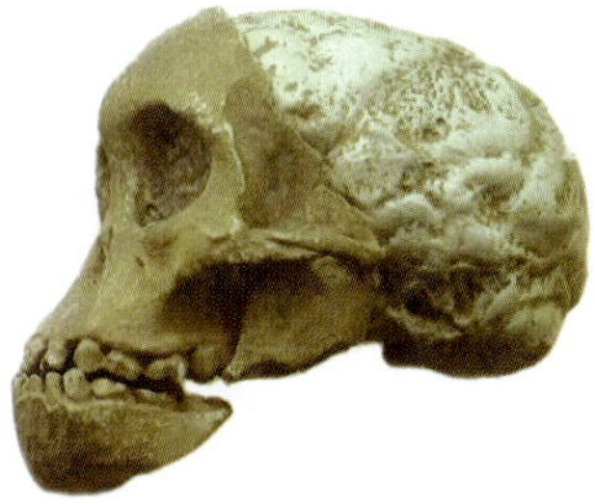

Juvenile australopithecine: the Taung specimen.

In time, australopithecines (supposedly) gave rise to *Homo habilis*, *Homo erectus, archaic Homo sapiens, Neanderthal man*, and modern Man. The first fossil of Homo habilis was found in the Olduvai Gorge in Tanzania, and fragments of partial skulls and partial skeletons (e.g. KNM-ER 1808 and 1500) have also been found in Koobi Fora in Kenya, all dated as being some 2 million years old.
Also dated as being from the same time period are the earliest known simple stone tools. Although the tools and the fossils were *not* found together, nevertheless:

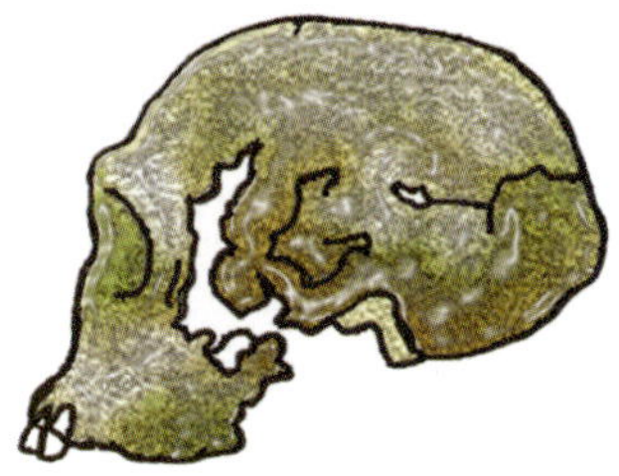

Small Homo habilis: KNM-ER 1813.

> The Olduvai hominid...was named Homo habilis meaning 'handy man' because of its large brain size and the possibility that it made tools. Michael Benton. 'The rise of the mammals'. Apple Press. 1991. p134.

As we shall see, fossils classified as Homo habilis consist of at least two distinct groups of creatures, whose brain sizes have been estimated to have been around 500ml to around 800ml.
Homo erectus first appeared in the fossil record over 1.8 mya. Typically, the erectus-type skull has coarse features with a low forehead, large eyebrow ridges, a prominent upper jaw and little in the way of a chin.
By 1mya Homo erectus was present in Asia and may have spread to southern Europe. Curiously, Homo erectus appears to have survived *without any change in his appearance for about one million years*, and it is commonly thought that he died out some 0.5mya. Homo erectus people were known to have made stone tools, had control of fire and buried their dead. Their fossils have been found in Europe, the Middle East, the Far East and Africa. It is generally assumed that Homo erectus was somehow replaced by a new form of human, archaic Homo sapiens:

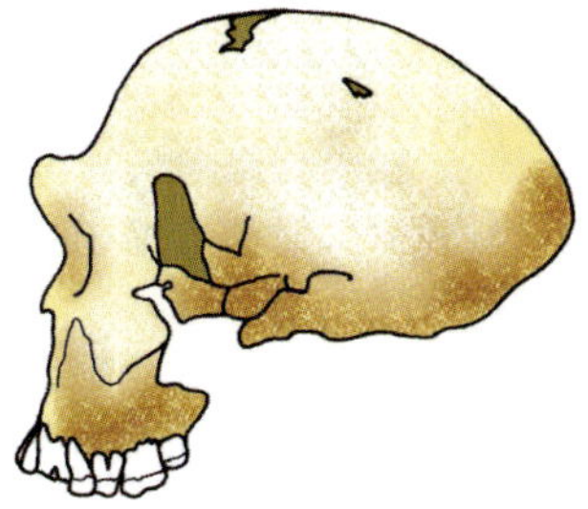

Homo erectus: KNM-ER 3733.

> By about 400,000 years ago there had been enough changes in certain human populations for a new species of early human to be recognised...The fossil remains concerned are usually known by the rather unsatisfactory term *'archaic Homo sapiens'*. Steve Jones, Robert Martin, David Pilbeam. 'Cambridge Encyclopaedia of human evolution'. Cambridge University Press. 1996. p245.

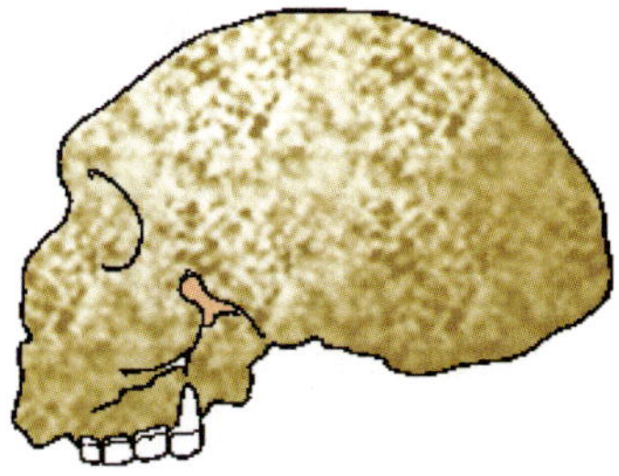

Archaic Homo sapiens: Petrolonica.

The term archaic Homo sapiens does not describe a species, but comprises a group of fossils which show features some way between the thick bones of Homo erectus and the slimmer bones of Neanderthals and modern Man. Archaic Homo sapiens fossils have been found in Europe, the Middle East, the Far East and Africa.
Neanderthal man first made his appearance in strata dated as being about 130,000 years old and apparently disappeared without trace some 35,000 years ago. The typical Neanderthal skull has thick features, but with less prominent brows than Homo erectus, and the back of the skull (the *occiput*) has a characteristic protuberance, the *'bun'*. There are features of the Neanderthal skeleton which are different from those of modern humans, but generally the skeleton was similar, but more stocky. Neanderthal people are known to have made flint tools, built huts out of animal skins, had control of fire, and also buried their dead. Fossils of Neanderthals are found restricted to the Middle East and Europe, although some specimens from Africa and the Far East exhibit some Neanderthal features.

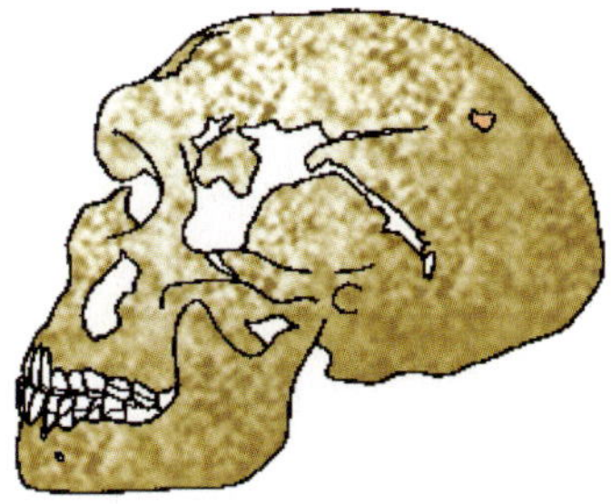

Neanderthal: la Ferrassie.

As the Neanderthals disappeared from the fossil record, the more modern-looking *Cro-Magnon Man* (modern Man) made his first fossil appearance in Europe. At the same time, seemingly without any prior graduality, sophisticated bone and antler tools appeared, along with complex artistry. Fossils of modern man dated as being (supposedly) up to 55,000yo have been found in the Skhul caves of Mount Carmel in Israel. The story gets complicated by the fact that fossils of modern-looking Man appear in the African strata of Ngaloba in Tanzania, and from Bodo and the Omo Valley in Ethiopia, that date back to 100,000ya. In addition, the fossil bones of modern humans have been found in Qafzeh near Nazareth, and are dated as being 100,000yo. As we shall see, the evidence points to ancient humans with modern features living much further back in time than that.

Not far away from Qafzeh, in the Kabara caves, have been found Neanderthal bones dating back 60,000 years. So Neanderthals and Cro-Magnons must have co-existed on the Earth for (supposedly) many thousands of years.

As to the distribution of human-like fossils world-wide, there are two main theories. The *multi-regional* theory proposes that Homo erectus people originally migrated out of Africa into southern Europe and Asia, and that over time the groups in different regions evolved separately into modern humans. The *out-of-Africa* theory proposes that Homo erectus people originally migrated out of Africa into southern Europe and Asia, and subsequently were eliminated by modern peoples as they too arose within and then migrated out of Africa.

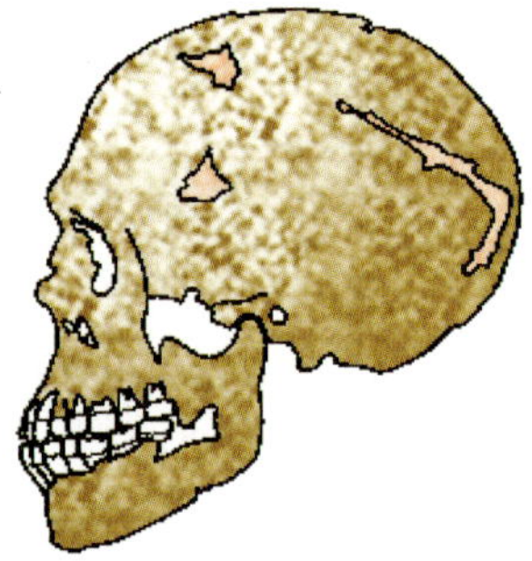

Modern Man.
Above: Jebel Qafezeh IX.
Below: Skhul V.

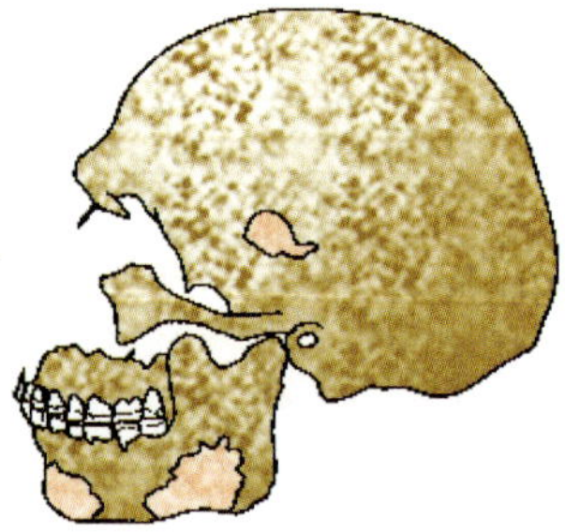

Recent petrified human skull from England.

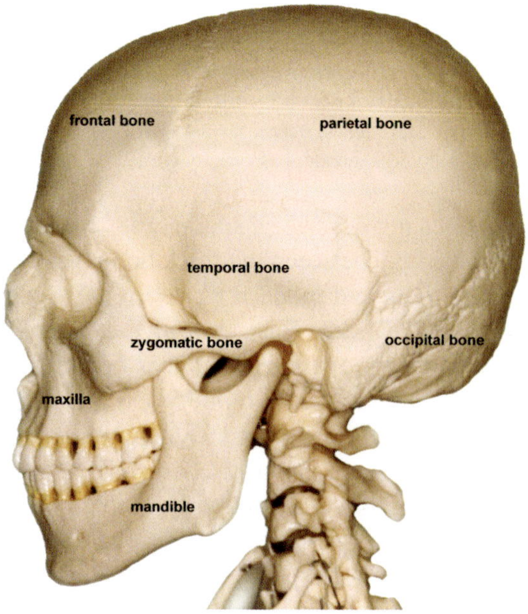


Modern human skull

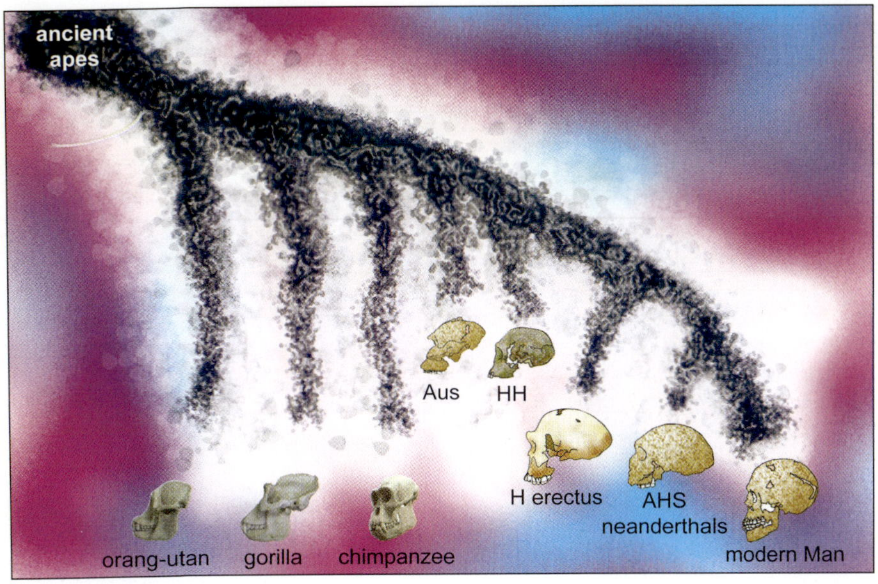


Standard scheme of human origins.
Aus: Australopithecines
HH: Homo habilis
AHS: Archaic Homo sapiens

And that is essentially the standard story of human evolution. Short of superficially attributing them to being in some way descendants of Homo erectus, the origin and disappearance of both Neanderthals and archaic Homo sapiens is usually left unexplained. Likewise, no real explanation is given for the initial appearance of either Homo erectus or modern humans.

Now that we have a broad outline of the candidates participating in the notion of human evolution, we can look at individual groups and specimens in more detail. Remember that when discussing ape-like fossils which have a few human-like characteristics, this in itself does not allow us to interpret those features as representing some evolutionary intermediate state between (supposedly) primitive apes and humans, just as the paleoparadoxia does not represent an intermediary between a terrestrial quadruped and the dugong. So we have to allow for the distinct possibility that fossils of extinct ape-like creatures represent, simply and *only*, apes that had different features to living apes. Similarly, when discussing supposedly ape-like features in essentially human-looking bones, this does not force us to interpret those features as representing some primitive intermediary between apes and humans. We have seen in previous chapters that within any given species, the external appearances can be very varied. This propensity for variation is also found in humans, so that without the benefit of a genetic analysis, even a *startling* difference in appearance cannot be interpreted unquestionably as representing a different species. Clearly, DNA analysis is not possible for most ancient bones, but nevertheless, that impossibility should not allow us to bend the rules of scientific enquiry and postulate or assume relationships when there is no clear evidence on which to do so.

So we have to be very careful when we come to interpret any individual fossil. To make things more complicated, uncertainties about the *age* of many fossils must also temper any dogmatism about any supposed evolutionary trends. For example,

Overleaf: Olduvai Gorge in Tanzania, where fossils of the supposed human ancestors, australopithecus and homo habilis, have been found. Hence the area has been dubbed 'the cradle of mankind'.

Olduvai Gorge, Tanzania

there are less than 30 known Tyrannosaurus rex fossils. Even though very large carnivores can never be numerous, nevertheless, in the course of their supposed existence over a span of 2 million years, 30 individuals could not possibly represent the variety of shapes and sizes that would have constituted Tyrannosaurus. How much greater, then, is the problem with primates. Out of the *millions* of extinct apes that have lived on this planet, only a minute percentage have been preserved as fossils. And of the millions of individual ancient humans that have lived, we have only the remains of precious few to study:

> It is accepted that the understanding of early Homo sapiens is hampered by...**a relatively sparse and poorly dated fossil record**. (My bold). Richard Klein. 'The human career'. Chicago University Press. 1981. p406.

There are also some glaring inconsistencies in the fossil record:

> There is **no fossil record of chimpanzees or gorillas** at all...There is **no fossil record of gibbons or orang-utans before the Pleistocene** [2 mya]. (My bold). Steve Jones, Robert Martin, David Pilbeam. 'Cambridge Encyclopaedia of human evolution'. Cambridge University Press. 1996. p22.

Clearly, if the fossil record of Neanderthals (i.e. no fossils dated earlier than say, 130 thousand years ago) is to be interpreted as meaning that Neanderthals did not exist before 130 thousand years ago, then by the same reasoning, we should also conclude that gibbons and orang-utans did not exist before 2 million years ago; and the complete absence of chimpanzee and gorilla fossils should force us to conclude that there were no chimpanzees or gorillas in existence before recent times. And although australopithecines were (supposedly) more recently evolved than orang-utans, the fossil dating also shows that australopithecines were (supposedly) alive at least 1.6 million years *before* the dating of the earliest orang-utan fossils. As we have previously seen, the *Coelacanth Principle* allows us to make no firm statement regarding the dates for the existence of either orang-utans, chimpanzees or gorillas... or of Neanderthals or modern Man...or of any creature living or extinct.

We would all like to think that scientific research is a search for truth and is purely objective and devoid of any bias. However, all theories necessarily have to start with some assumption as a foundation on which to base research. Thus, if it is assumed that humans *did* evolve from apes, that might lead us to look for, *and produce* from the fossil record, a line of skulls which might show an *apparent* gradual trend from ape-like to more human-like. But, as we know from the *Horse Principle*, even if such a supposed trend is construed, nevertheless, it might be quite erroneous. Therefore, if the truth is that no such evolution has in fact occurred, then any such supposed trend, even if seemingly convincing, would, of course, be purely imaginary.

So, let us look at the fossil evidence. On the stage of (supposed) human evolution, the relevant fossils fall comfortably into two main categories: those that are more ape-like than human, and those that are very human-like. Into the first group fall the australopithecines and the smaller Homo habilis specimens (the place of the larger Homo habilis specimens is open to question); and into the second group fall Homo erectus, archaic humans, Neanderthals and modern Man.

## *Ape-like fossils*

### *Australopithecines*

'Australopithecine' means, literally, 'southern ape', named after the first specimen found in South Africa, and the term covers a number of creatures with features that make them quite distinct from any living great ape. Since many workers cannot agree on how many australopithecine species are represented by the fossil record, it is reasonable here to describe four main types in order to simplify the discussion. Two types have been recognised in east Africa: the thicker-boned Australopithecus *boisei* and the thinner-boned Australopithecus *afarensis*. And in southern Africa another two types have been recognised: the thicker-boned Australopithecus *robustus*, and the thinner-boned Australopithecus *africanus*.

Australopithecines.
Right: SK46.
Far right: Sts5.
Below: OH5.
Below middle: KNM-ER 406.
Below right: StW5.

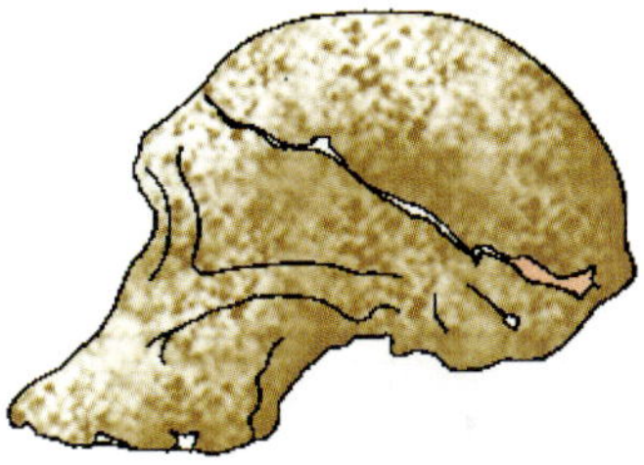

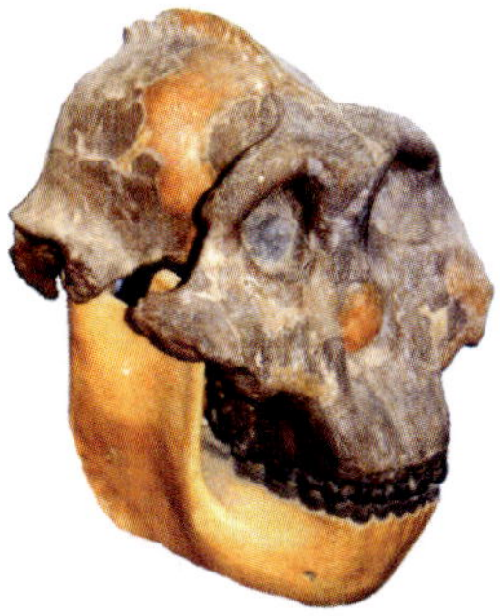

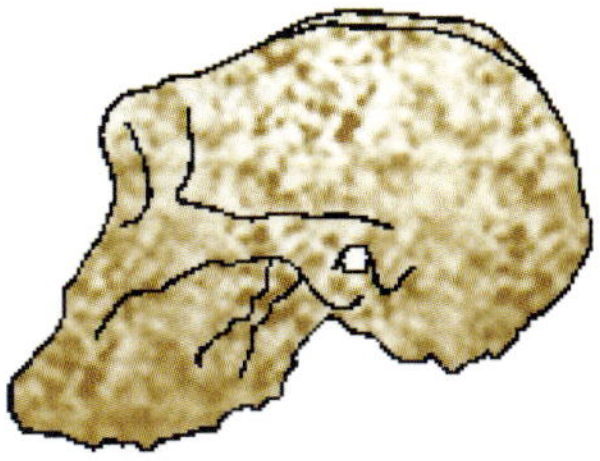

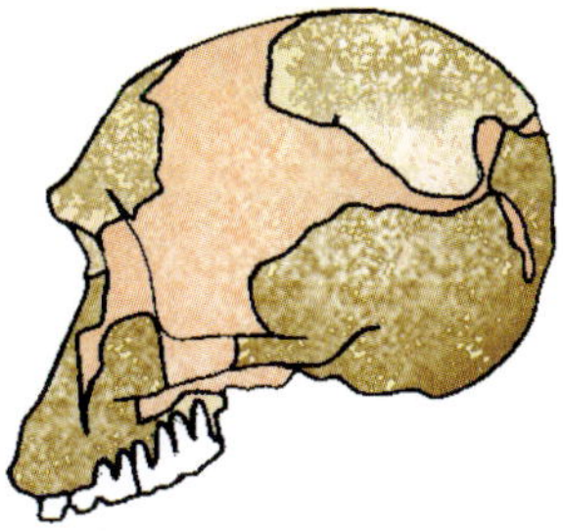

In fact, the number of intact australopithecine skulls found are very few in number and most of the fossils comprise only fragments of skull, jaw and teeth. But can we be certain that these few australopithecine fossils truly represent different species? We know that amongst many animals individual differences can be very marked - so much so that if they were named from fossils alone, a good number might be classified as different species. For example, the Siberian tiger can be up to 200kg heavier than the Bengal tiger. Forest African elephants may be only 2.5m tall compared with the savannah African elephant which may grow to about 3.5m in height.

Furthermore, within living species there can be considerable differences in both shape and size between the sexes *(sexual dimorphism)*. Some of the most marked instances of sexual dimorphism are found in the living great apes, mandrills and baboons, where the male is significantly larger than the female, with perhaps the greatest of all differences being shown in the gorilla. Apart from being much larger, the male gorilla skull also has a marked mid-line crest which, in contrast, is diminutive or absent in the female. Thus in the light of population and sexual variation, we must be cautious in ascribing creature-distinct status to the handful of known australopithecine skulls. In the absence of any genetic evidence, there

Above: Australopithecus afarensis.
Below: Australopithecus robustus.

always remains the possibility that they might simply represent different populations of the same animal.

A. afarensis, A. africanus, A. boisei and A. robustus fossils have been found in sediments assumed to be 3-4myo, 2.5-3myo, 1.4-2.2myo, and 1.5-1.8myo respectively. According to the *Coelacanth Principle*, we can infer nothing from this data regarding whether or not these creatures lived contemporaneously, or whether any one type was ancestral to another. What we do know is that australopithecines lived side-by-side with some animals which have remained unchanged up to the present day, and over a period of time in which the climate was also not very different from that of the present day. For example, the australopithecine skull SK54 (dated as being around 1.5-2.5myo), from Swartkrans in southern Africa, has two holes in its parietal bones which match with the lower canines of a fossil leopard jaw (which is just like a modern leopard's) from the same deposit; and many of the fossil animal footprints from Laetoli (see later) dated as being 3.5myo are identical to those of animals still living in the same area.

Since the australopithecines are generally considered to be *the* key intermediaries between apes and humans, it is necessary to examine certain aspects of their fossils in detail. In particular, it is thought that australopithecines walked fully upright, exactly in the manner of humans, and that almost this single feature *alone* confirms our ancestral relationship. So, did the australopithecines walk as we do, and if they did, does this mean that they were anything other than just apes that were slightly different from the ones living today?

The A. afarensis bones AL288-1 (found at Hadar in Ethiopia in 1974, and dated as being between 3 and 3.6myo) were, until recently, the most complete remains of any known australopithecine. Casts of the original bones can be seen in a number of museums, and photographs of the skeleton (commonly called by the nickname 'Lucy') can be found in many books on human origins. The bones comprising AL 288-1 suggest that it was an *adult* individual, and *if* it could stand upright it would have been no more than 1.07m (3 feet 6 inches) tall.

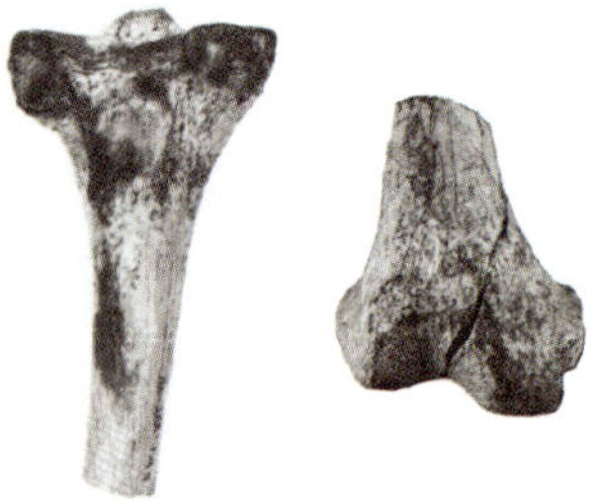

Australopithecine leg bones.
Left: AL 288-1. Upper right tibia.
Right: AL 333-4. Lower right femur.
(The AL 288-1 lower left femur was found badly fragmented.)

Curiously, the bones are always mounted in such a way as to give the impression of a creature looking much like a human being. However, this presentation is extremely deceptive since apart from the shape of the pelvis, knee and ankle joints, the bodily proportions are distinctly ape-like. The *humerus* (upper arm bone) is about 78% of the length of the *femur* (upper leg bone), whereas in humans the humerus is about 70% of the length of the femur. The *tibia* is about the same length as the femur, whereas in humans the tibia is about 85% of the length of the femur. Overall, the arms of AL288-1 are relatively long, and the legs are relatively short. In contrast to human elbows which reach down to the level of the pelvic bones, *in the position in which they have been mounted* AL288-1's elbows hang high above the pelvic bones. This is an unnatural posture and, in fact, if the spine were to be flexed, the long arms would allow a knuckle-walking gait - as in a chimpanzee or gorilla.

Some agree that australopithecines may:

> ... have fed and moved much like modern baboons, but with more emphasis on bipedalism*. Steve Jones, Robert Martin and David Pilbeam. 'Cambridge Encyclopaedia of human evolution'. Cambridge University Press. 1996.

*It is important to note that a stooping posture is such an inefficient form of locomotion that an evolutionary trajectory from knuckle-walking, to stooping, through to full bipedalism is impossible.

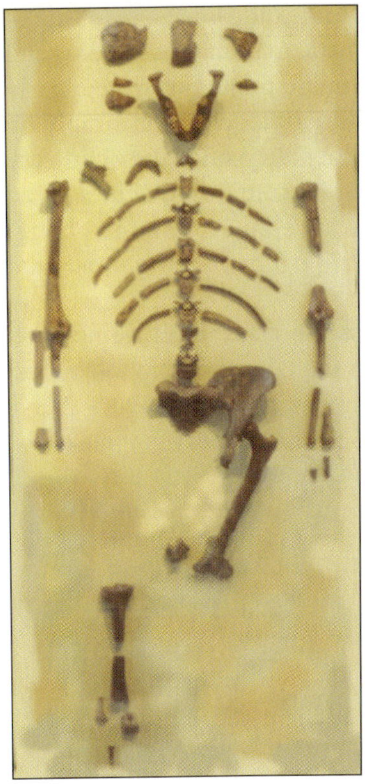

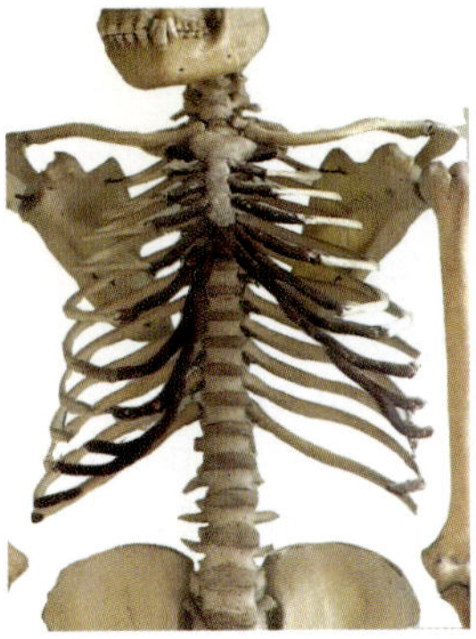

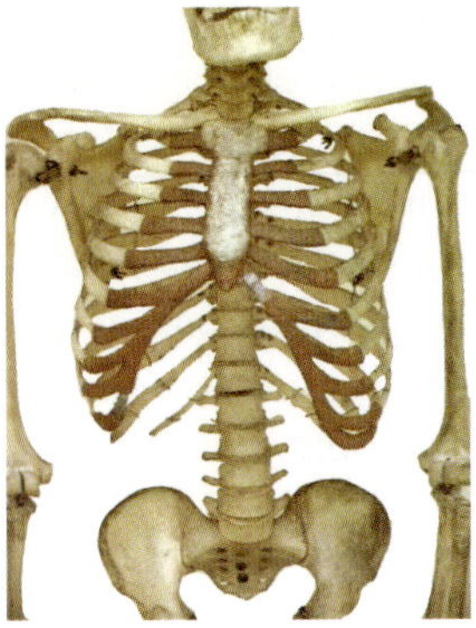

Left: The mounted fossil bones of AL288-1. Natural History Museum, London.

The initial appearance is one of a fully bipedal creature with human proportions. This is deceptive.

1. The legs are short and the arms are long, as in living apes.
2. The finger and toe bones are curved as in tree-dwelling apes.
3. When properly articulated, the rib cage would be conical (giving a 'pot-belly' appearance) as in the chimpanzee (top left), rather than being barrel-shaped as in humans (top right).
4. The elbows hang unnaturally high above the pelvic bones. If the spine is flexed, then a knuckle-walking form emerges - as in a chimpanzee or gorilla.

Above: Australopithecine model in the National History Museum, London, 2008. It falsely shows human-like feet when australopithecines definitely had thumbs on their feet. See page 369.

But others, looking only at the ankle joint, still insist that AL288-1 was *fully* bipedal:

> ...In every functionally significant feature examined the AL288-1 talocrural joint [ankle joint] is fully bipedal... Bruce Latimer et al. 'Talocrural joint in African hominoids: implications for Australopithecus Afarensis'. American Journal of physical anthropology. Vol 74. 1987. p155-175.

Certainly, detailed examination of the ankle joint shows features that are distinctly similar to the human joint, and both are different from those of living apes. In contrast to Jones et al., Latimer et al. conclude that living in trees played an *insignificant* part in australopithecine life, and that they were probably exclusively terrestrial creatures:

> Having made the adjustment necessary for habitual bipedalism, the hominid talocrural joint is no longer capable of functioning like the ankle of an arboreal pongid. Bruce Latimer et al. 'Talocrural joint in African hominoids: implications for Australopithecus Afarensis'. American Journal of physical anthropology. Vol 74. 1987. p155-175.

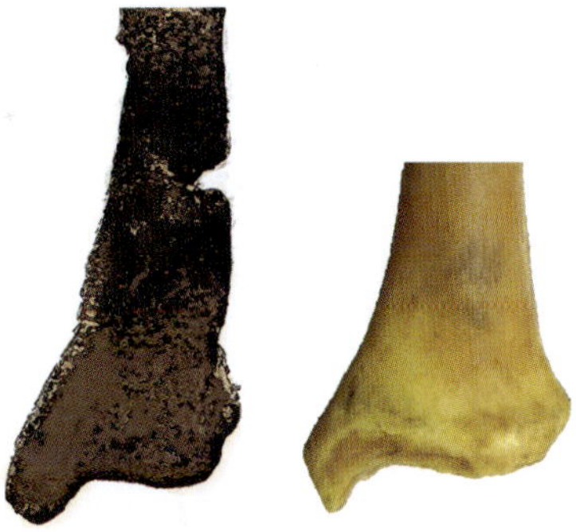

Left: Australopithecine left lower tibia AL 333-6.
Right: Human left lower tibia.

However, although the human ankle joint is less efficient than that of the ape ankle joint when it comes to climbing trees, it is nevertheless *quite possible* for humans with human ankle and knee joints to climb trees, *and also* to knuckle-walk quadru-

AL333w-4 AL288-1w

Australopithecine finger bones

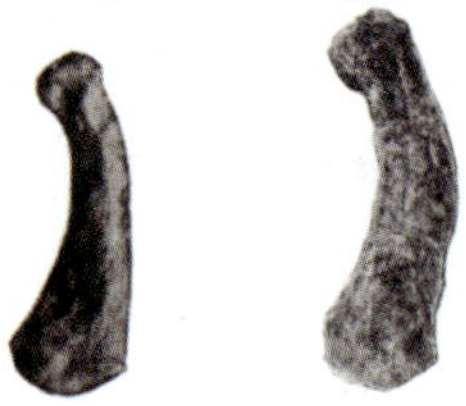

AL288-1y AL333-115

Australopithecine foot bones

Chimpanzee finger bones. Australopithecines had curved finger and toe bones as in chimpanzees, suggesting that they at least had a partly arboreal lifestyle.

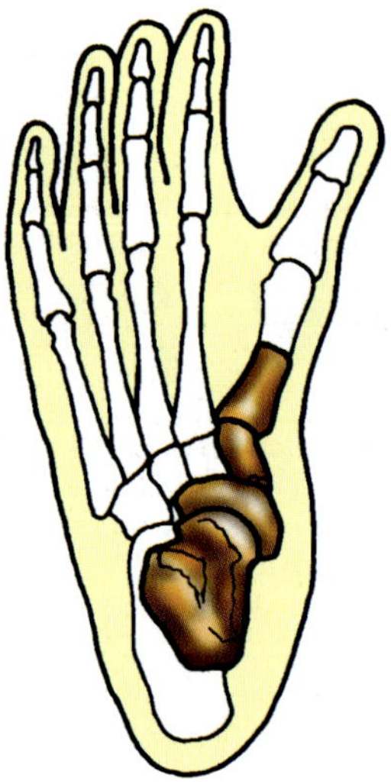

STW573 foot bones showing that while possibly being more bipedal than chimpanzees, nevertheless, australopithecines had ape-like feet and could not have made human-like footprints.

pedally - although our relatively short arms require us to flex the knees, resulting in a very clumsy gait. Furthermore, the lesson from the aquatic mammals artiocetus and rodhocetus (which had ungulate-like features in their ankle joints but were clearly unable to run) is that we are not at liberty to infer posture or locomotion simply on the basis of the anatomy of a few joints. Therefore, particularly since they had very long arms along with relatively short legs, the finding in australopithecines of an ankle joint or pelvis showing features that are somewhat similar to human bones *does not mean* that these creatures walked bipedally *like a human,* or that they did not knuckle-walk, or that they could not climb trees.

To clarify the issue further, it is necessary to discuss the australopithecine foot. However, AL 288-1 contains no foot bones other than the talus and two toe bones. Since all great apes, baboons and gibbons have a 'thumb' on the foot, and AL 288-1 was most likely only partially bipedal, it would be logical to expect that australopithicines, too, might have had ape-like feet. And in fact:

> The afarensis phalanges are arched, and proportionally a good deal longer than those in modern* feet. They might almost be mistaken for finger bones. Donald Johanson and Maitland Edey. 'Lucy, the beginnings of humankind'. Penguin. 1990. p349. (* Note: for 'modern' read 'human')

> The curvature of the hand and foot phalanges and the extreme robusticity...show that the species differed from the modern human condition. Ian Tattersall, Eric Delson, & John van Couvering. 'Encyclopaedia of human evolution and prehistory'. Garland Press. 1988. p77.

In other words, their fingers and toes were ideal for climbing trees. So, most certainly, australopithecine fingers and toes were ape-like. But what of their great toes - were they like human great toes, or were they like thumbs as in apes? Until only very recently, any reconstruction of an australopithecine foot was subject to pure speculation. However, some foot and ankle bones (STW573), which were found in a Sterkfontein cave in South Africa in 1980 and originally thought to be *antelope* bones, were identified in 1995 as being australopithecine bones, (supposedly) some 3.5myo.

In humans the geat toe metatarsal has a characteristic *facet* at the base where it articulates with the index metatarsal. This facet is not found in STW573, showing that, exactly as in living apes, the *great toe metatarsal* is *abducted* (i.e. projecting outwards, thumb-like). Clearly, and without any shadow of doubt, the australopithecine foot:

> ...possessed **an apelike great toe** that diverged from the other toes and was **highly mobile**...**departed to only a small degree from that of the chimpanzee**. It is becoming clear that Australopithecus was likely not an obligate terrestrial biped, but rather a facultative biped and climber... (My bold). Ronald Clarke & Phillip Tobias. 'Sterkfontein Member 2 foot bones of the oldest South African Hominid'. Science. Vol 269. 28th July 1995. p521-4.

Nevertheless, it is still suggested that the 'human-like' aspects of the ankle bones confirm the notion that the australopithecines represent intermediaries between apes and man:

> ...we have a real missing link in the evolutionary chain... Ronald Clarke quoted by Roger Highfield (Science editor). Daily Telegraph. June 1997.

But quite simply, no such conclusion can be made. The science of the matter can only say that this creature had feet that were different from both those of living apes and those of humans. Chimpanzees and gorillas and orang-utans all have feet which are excellent designs, well-suited for the purpose required of them. Similarly, our feet have their own unique features, which allow us to walk and run with ease. The australopithecine foot is yet another design – it is neither like the chimpanzee nor like the human foot. It is logical to presume, therefore, that its owners moved about in some unique way that mirrored neither the ape nor the human gait. In fact, we should not be surprised if australopithecines are just extinct apes which differed from living apes, since amongst all the very many other extinct and living creatures it is not uncommon to come across (seemingly) weird combinations of characteristics – the chalicotherium, paleoparadoxia and duck-billed platypus being excellent examples.

Thus, the fact that australopithecine feet shared some features found in both apes and humans allows us to conclude *nothing* about the ancestry or the descendants of these creatures. Most certainly, on the evidence, we are *not* at liberty to suggest that the australopithecine foot is some sort of intermediary structure representing an evolution between ape and human feet.

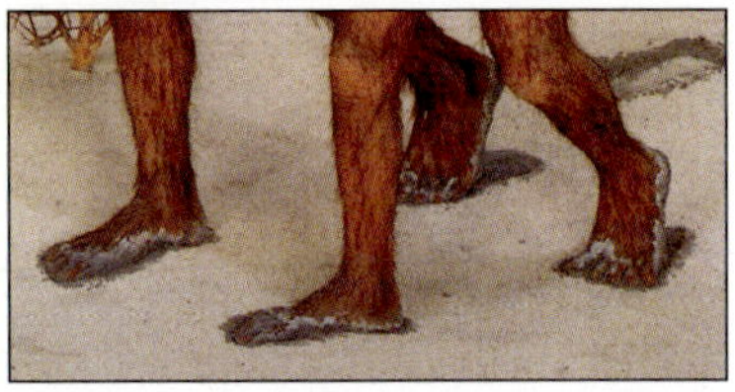

Although the STW573 foot bones were reported in 1995, nevertheless, in 2000, the BBC book 'Ape-man' (above) and the 'Walking with cavemen' TV programme (below) still depicted (incorrectly) that australopithecines had human-type feet.

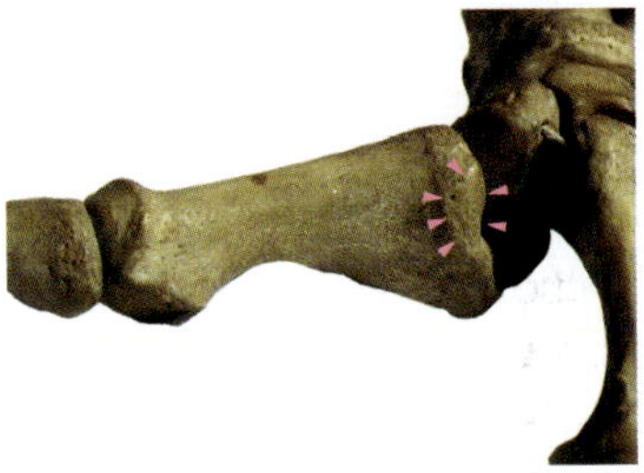

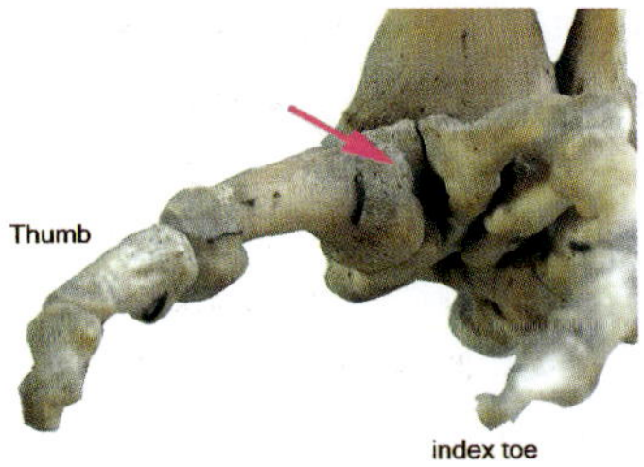


Far left, top and bottom: Human great toe showing facet (between arrow heads) at the base of the metatarsal which makes contact with the side of the index metatarsal. This facet is missing in both living apes and australopithecines since in these creatures the foot has a flexible thumb.

Left, top and bottom: Chimpanzee foot showing thumb. The side of the thumb metatarsal (arrowed) does not make contact with the base of the index metatarsal and so has no facet.

Below: Part of the foot showing the huge difference between the great toe of humans and the thumb of living great apes. The facet between the human great and index metatarsals is arrowed and shown in red.

(nav: navicular, mc: medial cuneiform, ic: intermediate cuneiform, lc: lateral cuneiform, mt: metatarsal).

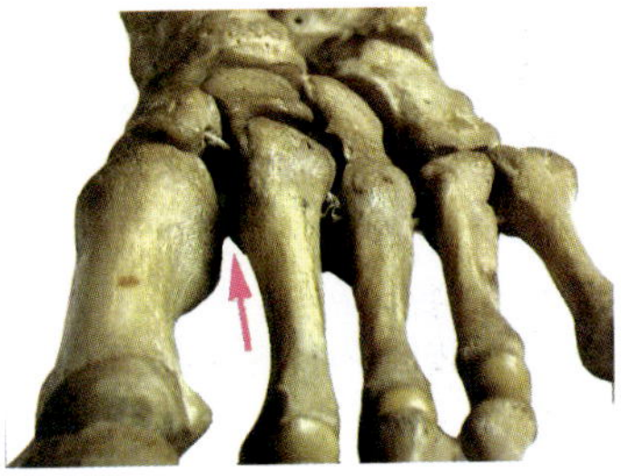

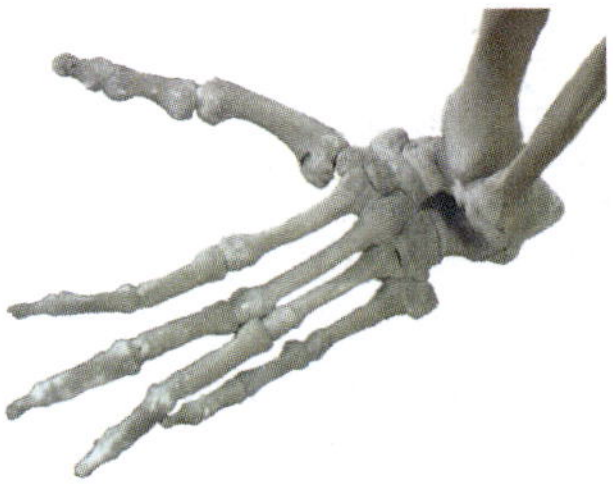

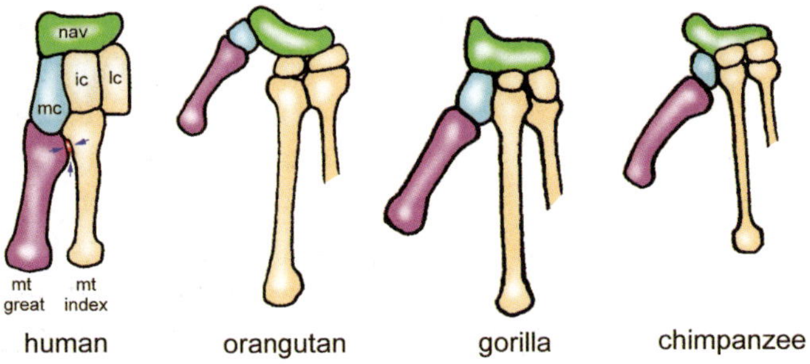

Ronald Clarke made a model based around the STW573 foot bones and suggests that it is shaped like the the smaller of the fossilised footprints in Laetoli, Tanzania, which are supposed to be 3.5myo (see main text later in this chapter):

*'I made a reconstruction of what the foot would have looked like, and it fits into the smaller of the Laetoli footprints…*

Ronald Clarke, 'Where do we come from?' Channel 5 Broadcast. April 2002.

However, it is quite evident that the little toe rests outside the edge (right arrow) of the footprint, showing that Clarke's model *does not* fit at all, and the left arrow points to the characteristic ridge that could only have been made by a foot in which the great and index toes touched each other. Rather than demonstrating that australopiothecines had human-like feet, Clarke's model confirms that australopithecines could not have made the Laetoli footprints.

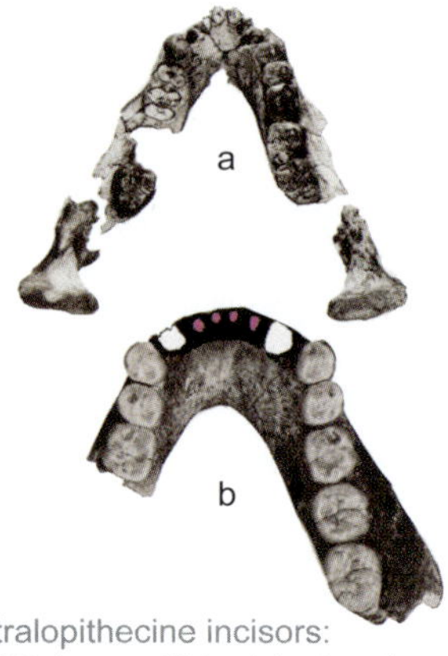


Australopithecine incisors: AL288-1 mandible (a) showing very small (broken) incisors and AL266-1 mandible (b) showing the tiny incisor sockets. (Not to scale)

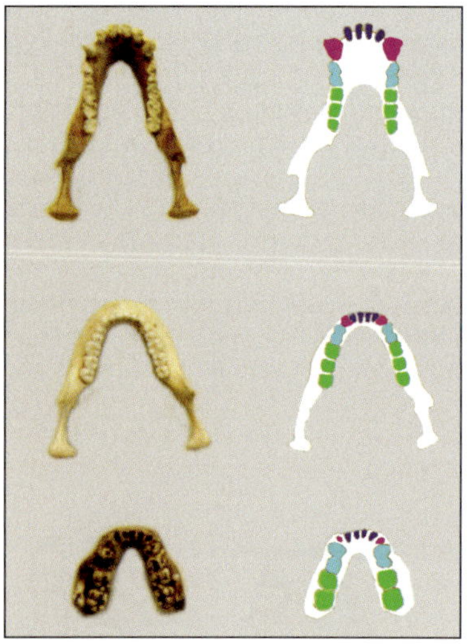

Different mandibles compared.
Top: Chimpanzee.
Middle: Human.
Bottom: Australopithecine.
Australopithecines had exceptionally small incisors and large molars. These features show that australopithecines are not intermediaries between humans and apes.

A number of other anatomical features must also be mentioned which demonstrate that there is no trend showing australopithecines to be intermediary between humans and great apes. In contrast to the living great apes and humans, australopithecines had very large molars and exceptionally small incisors in the lower jaw. The reduction in the size of incisors from apes to australopithecines and an increase again in size from australopithecines to humans does not make any evolutionary sense, since apes can eat anything that humans can. In the upper jaws of both living apes and australopithecines there is a gap (the *diastema*) between the lateral incisor and the canine tooth. This gap is necessary to accommodate the large lower canines and so a diastema is not found in humans. There are also important differences in the development of the permanent teeth: in both australopithecines and living apes, the first incisor erupts 2 years after the first molar, and the canines erupt after the second molar. In contrast, in humans the first incisor and first molar erupt together at about 6 years of age and the canines erupt before the second molar.

In addition, air sinuses are found in the zygoma (the side of the cheek) and the hard palate (the roof of the mouth) of both australopithecines and living apes, whereas there are no air sinuses in the zygoma or hard palate of humans. Australopithecines also had 6 lumbar vertebrae, in contrast to the 3-4 in living great apes and 5 in humans. The increase in the number of vertebrae from apes to australopithecines and the decrease again in number from australopithecines to humans does not make any evolutionary sense, since apes can climb trees with 4 lumbar vertebrae and humans can walk upright with either 4 (if one is congenitally absent) or 5 lumbar vertebrae. All these features show that australopithecines are not intermediaries between humans and apes.

Very recently, the whole of the rest of the skeleton (including the skull) belonging to the STW573 footbones was found in a limestone cave in Sterkfontein. This skeleton removes completely any doubt about the australopithecine form. In particular, the bones show a typical australopithecine skull with a brain case no larger than that of a chimpanzee or gorilla; foot bones with a typically ape-like grasping thumb; and hand bones which have long ape-like fingers but with a thumb that is larger than that found in living apes.

Recently, a new fossil, *Australopithecus garhi*, was found near the village of Bouri in Ethiopia, in sediments dated as being 2.6myo:

> *A. garhi* had relatively long, savannah-striding legs. His arms, however, were still apishly long... Peter Martin. 'History. The beginning'. The Sunday Times magazine. 12th September 1999. p18.

It is reported with unequivocal confidence that:

> *A. garhi* shapes up as being among the first accomplished makers of stone tools...He also practised large-scale systematic butchery. There an antelope's jaw with neat serrations where the tongue had been sawn out. Here, some leg bones chopped apart – some from a hipparion, a three-toed horse – with additional marks indicating a nice skill at filleting...In the use of stone tools, we see high orders of skill, of investment in time and labour, and of planning and of group effort. This is one very organised creature... Peter Martin. 'History. The beginning'. The Sunday Times magazine. 12th September 1999. p18.

Australopithecus garhri skull bones.

Limb bones found in the same sediment layer as the Australopithecus garhi skull bones, but separated from the latter by 150m. From left to right: humerus, radius, ulna and femur. Clearly, it cannot be stated unequivocally that the skull and limb bones belonged to the same creature.

The reader is given here the distinct impression that we have definite evidence of an upright-walking creature that manufactured tools, and there certainly seems to be no room for debate as to whether this creature was more human-like in form, or far more capable, than AL 288-1. But if we look at the report more closely, we see that the leg and arm bones were, in fact, found *over two years prior to* the discovery of the skull fragments and in a location *150m distant*. Clearly, such incidental associations are conclusive of nothing at all. In the same way that if the only bones found nearby were ostrich-like leg bones, one would not assume that *A. garhi* represented a bird-ape intermediary, so it is invalid to assume that the leg and arm bones belonged to the same individual or even to the same type of creature as the owner of the head bones.

Whatever type of creature the owner of the limb bones was, the absence of any foot bones makes it impossible to deduce whether or not the individual had an ape-like or human-like great toe. As to the notion that *A. garhi* used tools – there were no tools found with the fossil fragments at all! The reference to the use of tools was in fact inferred from the markings on animal bones found *elsewhere* in the:

> ...same-dated vicinity. Peter Martin. 'History. The beginning'. The Sunday Times magazine. 12th September 1999. p18.

So the head and limb fragments and butchered animal bones referred to were all found separated in time and space, with no direct evidence to link any of the ingredients in the recipe that made up a (supposed) bipedal, tool-making, sub-human creature.

Furthermore, from the reconstruction of the skull it is clear that the brain was no bigger than that of a chimpanzee:

> *A. garhi's* brain weight was still only a third the size of a modern human's. Peter Martin. 'History. The beginning'. The Sunday Times magazine. 12th September 1999. p18.

It is curious that there does not seem to be any concern regarding the questions: if australopithecus garhi had a brain no larger than that of a chimpanzee, and yet it (supposedly) could manufacture stone tools, then why did its brain have to enlarge any more (as we shall see later, the (unaided) manufacture of stone tools requires *fully* human intelligence); and why are gorillas (with their larger brains) not very much more intelligent than they actually are? There is no evidence, then, to suggest that A. garhi was anything other than another form of extinct ape.

### Kenyanthropus

Kenyanthropus. Although it seems that below the eyes this creature had a flatter face than that of australopithecines such as AL 288-1, nevertheless, its upper jaw protruded forwards in a manner not significantly different from that of australopithecines.

This fossil is of an ape-like creature that had a face significantly different from australopithecines. The specimen, consisting of numerous skull fragments (KNM-WT 40000), was found at Lomekwi on the western side of Lake Turkana, Kenya, in 1999. When the bones were reconstructed the creature had a flatter face (when measured horizontally below the eyes) than australopithecines, and clearly had a brain no larger than a chimpanzee. The other notable thing about this fossil is that the *foramen magnum* (the hole at the base of the skull through which the spinal cord exits) was underneath the brain case, strongly suggesting that Kenyanthropus walked upright. It was this feature that led to the claim that, rather than australopithecines, Kenyanthropus was a direct ancestor to human beings.

In the past it had been assumed that apes evolved into humans first by enlarging their brains and subsequently by becoming bipedal. It has now been suggested that Kenyanthropus shows that apes first became bipedal, and then their brains must have enlarged to the human form. However, it is important to note that regarding human origins, a number of invalid *assumptions* (these being central to the whole problem) are made here by evolutionists, namely: that humans *did evolve* from ape-like ancestors; that the finding of an ape with a large brain *confirms* it to be an ape-human intermediary; that the finding of an ape that walked bipedally *confirms* it to be an ape-human intermediary; that a small ape brain *can* become a human brain by acquiring a number of fortuitous mutations that would allow for both physical enlargement and the capability of speech and high intelligence; and that the foot, pelvis and spine structures of an ape *can* change into the human form by the acquisition of a number of fortuitous mutations. But, of course, the finding of an ape with a large brain, or one that walked bipedally, simply and only confirms that such creatures existed and *cannot itself* point to an ape-human intermediary. And, if humans *did not* evolve, any constructed trends from ape to human are simply imaginary. We shall deal in later chapters with bipedalism and the possibility of an ape acquiring changes to its brain size and form.

### Sahelanthropus

In 2001 new fossils comprising a skull (TM266-01-060-1), some teeth and a lower jaw fragment were found in Chad. The creature was named *Sahelanthropus tchadensis* and was heralded as the oldest known member of the human lineage. Its status as an intermediary between apes and humans was proposed mainly because this fossil has a shorter and less protuberant face compared to australopithecines and living apes; its molars are smaller (i.e. more human-like) than those of australopithecines;

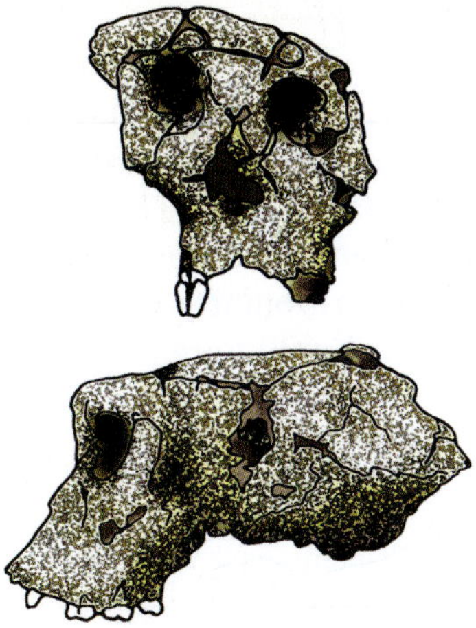

Sahelanthropus tchadensis (TM266-01-060-1), heralded as the oldest known member of the human lineage.
This fossil from Chad is seriously distorted with the front view showing the overlap of the bones on the outer rims of both orbits, and the marked displacement of the face to the left. Below the nasal opening this fossil has a shorter face than australopithecines and living apes. Although its molars are smaller than those of australopithecines, its palate is long and narrow, as in living apes. Unlike apes, the brow extends right across the head, and the foramen magnum lies more anteriorly. However, the side view clearly shows a distinctively ape-like face with an angulation of around 45 degrees, and a typical ape-like brain case which is no larger than that of a chimpanzee. In all probability this fossil represents simply an extinct ape that differed from australopithecines and living apes.
*Redrawn from: ' A new hominid from the Upper Miocene of Chad, Central Africa'. Michael Brunet et al. Nature. Vol 418. 11th July 2002. p145-151.*

and unlike apes, the brow extends right across the head and the foramen magnum lies more anteriorly. However, its palate is long and narrow, as in living apes; the nasal opening is wide as in apes; the side view clearly shows a distinctively ape-like face with an angulation of around 45 degrees; and it has a typical ape-like brain case which is no larger than that of a chimpanzee.

It is important to note that TM266-01-060-1 is seriously distorted with overlap of the bones on the outer rims of both orbits, marked displacement to the left of the right side of the face, and posterior displacement of the face and left brow. Correction for these deformations might result in a more protuberant face. Since there is no advantage in an ancient ape changing some of its dental and facial dimensions (australopithecines managed with large molars, chimpanzees manage with smaller molars; baboons have very dog-like faces yet tamarins and marmosets have relatively flat faces), there is is no scientific basis to presume that the 'mosaic' features were the result of evolution of a more chimpanzee-like creature.

Furthermore, the Sahelanthropus fossils were found in sediments that could not be dated by any radiometric method. Instead, the declared date of between 6 and 7myo was arrived at by comparing other fossil vertebrates (over 700 mammals including giraffe, hyena, gomphotherium and elephant) found in the same sediment layer with similar fossils in other areas. As we have seen earlier, dating fossils by comparisons with other sediments containing the same types of flora and fauna is a flawed and invalid technique. And the *Coelacanth Principle* shows that Sahelanthropus (and australopithecines) could have lived contemporaneously with ancient humans.

So, in summary, we have seen that australopithecines had relatively long arms and short legs, and although their ankle joints shared features with those of humans, their feet were more ape-like. They had ape-like skulls, and indeed, casts of the insides of the skulls show surface patterns and general features which suggest that they had ape-like brains (see later under KNM-ER 1470). Thus, since neither bipedalism, nor a flat face, nor small molars equates with human-ness, all we can say from the evidence is that australopithecines, Kenyanthropus, and Sahelanthropus represent, in all probability, simply extinct apes that differed from living apes and were not ape-human intermediaries.

### Homo habilis

Although many workers would state that Homo habilis can be clearly distinguished from both australopithecines and living apes on the basis of jaw and tooth structure, there has always been considerable dispute as to whether Homo habilis was just another form of australopithecine, or whether it was a human ancestor. The picture is complicated by the fact that fossils attributed to the group Homo habilis have cranial capacities that vary from around 500ml to around 800ml. For example, amongst the fossils classified as Homo habilis from Koobi Fora in northern Kenya, there are:

> ...**perplexing** finds...which complicate the picture of early human evolution. These include specimen KNM-ER 1813, a small-brained (510ml) skull...and KNM-ER 1805, which has teeth like 1813, a face like 1470, a brain capacity of only 580ml, and a braincase with a primitive crest along the midline... (My bold). Steve Jones, Robert Martin and David Pilbeam. 'Cambridge Encyclopaedia of human evolution'. Cambridge University Press. 1996. p243.

Homo habilis: KNM-ER 1813.

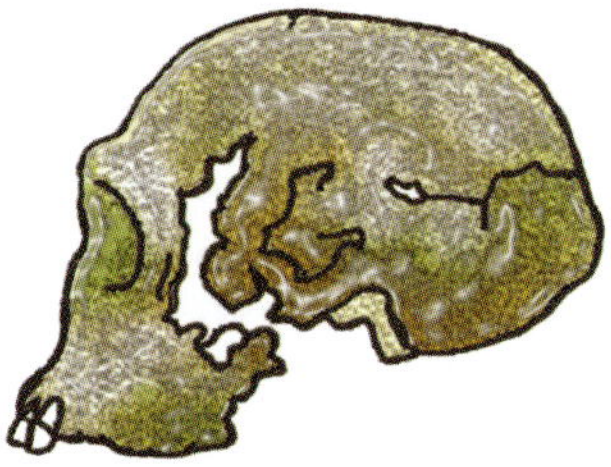

The upper jaw and limb bones of OH62. From left to right: Humerus, ulna, femur, and tibia.

Immediately, this raises the question as to whether or not a creature with a brain size of only 500ml (which is well within the range of the gorilla) could possibly be classed as *human*-oid. On the basis of the range of brain capacity, it is possible that the collection of fossils classified as Homo habilis could represent either just one, or a number of distinct creatures. For a long time virtually nothing was known of the body and leg bones of Homo habilis, but then in 1986, a new skeleton of Homo habilis (OH62) was found in Olduvai Gorge in Tanzania. Most importantly, included in the find were the arm and leg bones. This was the first time that any bones of the body of Homo habilis had been found and it was most surprising to find that:

> The total OH62 upper arm was almost certainly longer than that of 'Lucy'...this individual's femur was smaller and less robust than the AL288-1 femur...body size for this **fully adult individual** is estimated to be as small as or **smaller than that of any known fossil hominid**. (My bold). Donald Johanson et al. 'New partial skeleton of Homo habilis from Olduvai Gorge, Tanzania'. Nature. Vol 327. 21st May 1987. p205-9.

In actual fact, although fully grown, the Homo habilis individual OH62 was *less than 1.06m (3 feet 6 inches)* tall. So the evidence shows that both australopithecines and the small habilines had long ape-like arms, with relatively short ape-like legs. Some of the ankle bone features were similar to those in humans, but the feet were ape-like, with flexible great toes. Thus the evidence *only* tells us that australopithecines and Homo habilis were as different from orang-utans, chimpanzees and gorillas as baboons are, and as such australopithecines and Homo habilis cannot be classed together with the living great apes. More importantly, there is no evidence to suggest that the small habilines were in any way more human-like than living great apes. But what interpretation should be made of the larger Homo habilis fossils? The most famous Homo habilis fossil is KNM-ER 1470, which has a cranial capacity of about 750-800ml:

Since there is no evidence to show that the creatures called Homo habilis were ancestors to humans, or that they made tools, the misleading term 'Homo habilis' should be replaced by another classification that better reflects their ape nature.

> ...800 cubic centimeters is a closer figure. (My bold). Richard Leakey. 'The oigin of mankind'. Phoenix. 1994. p27.

This being so, the brain capacity of KNM-ER 1470 just falls within the lower range of modern humans, and yet it is not the largest ancient skull supposedly dating from between:

> 2.0 and 1.6 million years ago...one fragmentary skull probably had an **even bigger brain than 1470**... (My bold). Steve Jones, Robert Martin and David Pilbeam. 'Cambridge Encyclopaedia of human evolution'. Cambridge University Press. 1996. p242-3.

> KNM-ER 1590...A partial skull with **juvenile** dentition...the cranial volume **was large**... (My bold). Michael Day. 'Guide to Fossil Man'. Cassell. 1986. p207.

Homo erectus: OH9.

Whatever one's opinion is of the large habiline skulls, one thing is certain, and that is the *uncertainty* about what exactly they represent. Furthermore, there is no demonstrably valid morphological trend from the australopithecines, through Homo habilis to Homo erectus. We will discuss the skull KNM-ER 1470 in more detail later on.

### Human-like fossils

#### Homo erectus

Homo erectus: Peking Man.

Homo erectus first appeared in the fossil record (supposedly) around 1.8 mya and by 1mya Homo erectus was (supposedly) present in Asia and might also have spread to southern Europe. Curiously, Homo erectus appears to have survived without any change in his appearance for about *one million* years. While it is commonly thought that Homo erectus died out some 0.5mya, dating of the Homo erectus from Solo in Java has shown an age of only 35,000 years, and other specimens are even younger (see later, Kow Swamp and Cossack). Homo erectus people were known to have made stone tools, had control of fire and buried their dead.

There are three main views on the position of Homo erectus. Some consider that Homo erectus was a non-fully-human ancestor of modern humans. Others believe that Homo erectus was not ancestral to modern humans, but was a separate lineage that became extinct without leaving any descendants. Still others consider that Homo erectus was simply a fully human, earlier version of Homo sapiens.

Homo erectus: Solo Man.

Typically, the erectus-type skull has coarse features with a low forehead, large brow ridges, a prominent upper jaw and little in the way of a chin. However, there

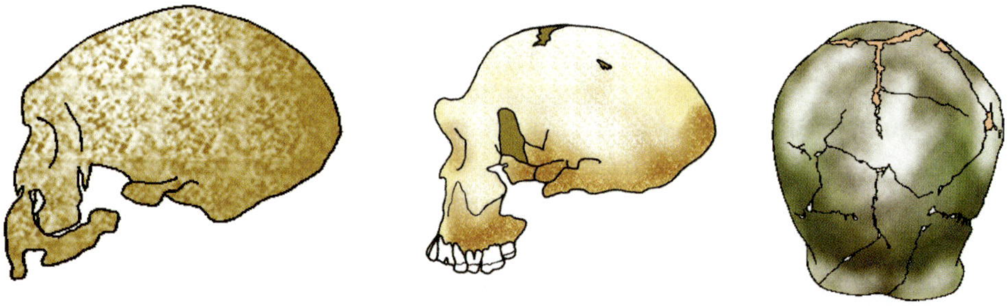

Homo erectus.
Right: Sangiran 17.
Middle: KNM-ER 3733.
Far right: Hexian.

Male and female mandrill skulls showing size differences.

is quite a marked variation in the shapes of the skulls commonly designated as being Homo erectus. This variation can be clearly seen in examples such as *Hexian* (0.24-0.4myo), KNM-ER 3733 (1.8myo), and *Solo* (35,000yo) from Java.
Even so, there are in fact more differences between the skulls of different breeds of dog, and between male and female *mandrills*, than there are between modern human and Homo erectus skulls. And all of the features seen in Homo erectus can be found within the normal variation of skull shapes of humans living today.
Although Homo erectus is commonly thought of as having a small brain, such as in the skulls from Trinil in Java (Java Man), or from Hexian in China:

> Java man...940 cc... Hexian man...1025 cc. Michael Day. 'Guide to Fossil Man'. Cassell. 1986. p341, 383;

in actual fact many Homo erectus skulls are almost as large as the average for modern human skulls. For example, the skulls from Ngandong in Java (Solo man) have all the features typical of Homo erectus yet the cranial capacity is:

> 1,035-1,255cc (based on the six best-preserved skulls). Michael Day. 'Guide to Fossil Man'. Cassell. 1986. p359.

Most certainly, there exists an enormous amount of variation in shape and size between many different living human groups, and the extreme differences in body build between, for example, African bushmen and Inuits are well known. Yet such differences in human form have appeared in *a relatively short period of time* and not over millions of years. For example:

> The ancient Japanese had long heads, broad faces, and wide flat nasal roots, and were prognathic. The later Japanese had rounded heads, narrower faces, and narrower and higher nasal roots, and they were less prognathic...significant differences have developed, over a short time span, between closely related and contiguous peoples, as in Alaska and Greenland... William Laughlin. 'Eskimos and Aleuts: their origins and evolution'. Science. 8th November 1963. p643-4. (The term *prognathic* refers to a prominent chin)

It is particularly relevant to note that Inuits, and Aleuts in particular, both have a number of features in common with Homo erectus and Neanderthal man. Some modern human skulls have particularly low vaults, and those of Aleuts:

...are among the lowest and most capacious in the world. William Laughlin. 'Eskimos and Aleuts: their origins and evolution'. Science. 8th November 1963. p633.

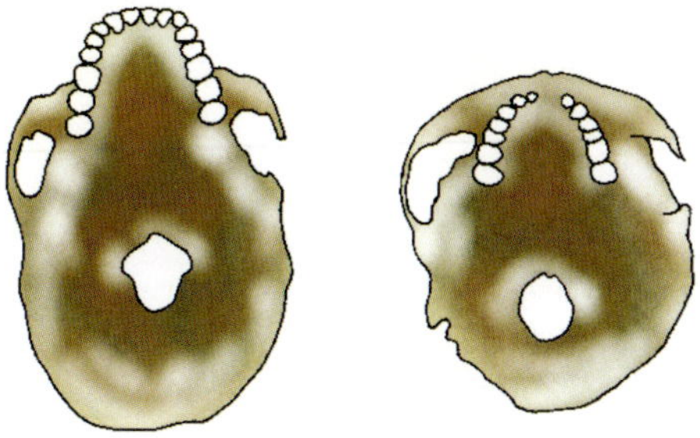

Long and short skulls of Aleuts (viewed from below).

The length and breadth of modern human skulls can vary dramatically:

> The relative narrowness of [the] cranium and the long occipital area are characteristic of the earlier Aleutian population...The great breadth of [the Neo-Aleut] cranium and the short occipital area are characteristic of individuals of this more recent population. William Laughlin. 'Eskimos and Aleuts: their origins and evolution'. Science. 8th November 1963. p637.

Along with these variations in general skull shape, there also exist variations in all the finer aspects. For example, the shape of the foramen magnum can be almost rhomboid, elliptical or almost circular. And along with such differences in the shape of the foramen magnum, the positions of the joints between the skull and the uppermost neck vertebra vary too.

Some *living* peoples have rather broad sides (*ascending rami*) to their lower jaws, which are even wider than those found in many fossil skulls:

> The enormously broad ascending ramus is characteristic of many Mongoloid groups. The breadth of this feature in Eskimos and Aleuts exceeds the breadth in Neanderthal man. William Laughlin. 'Eskimos and Aleuts: their origins and evolution'. Science. 8th November 1963. p639.

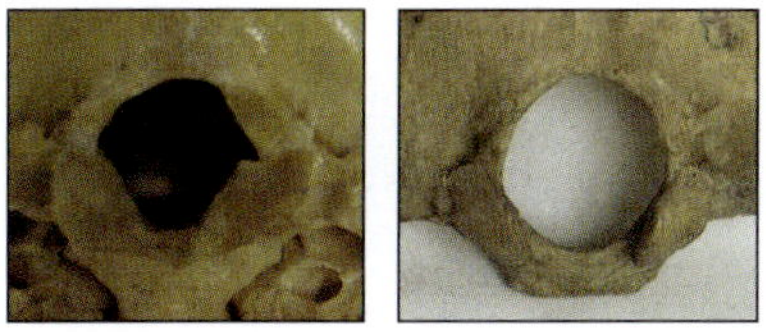

The foramen magnum of two modern human skulls showing marked variation in shape.

The lower jaws of Inuits and Aleuts have a bony protuberance (*mandibular torus*) on the inner aspects, but this is not unique to them:

> A form of this torus is found in Sinanthropus pekinensis [Homo erectus]. William Laughlin. 'Eskimos and Aleuts: their origins and evolution'. Science. 8th November 1963. p641.

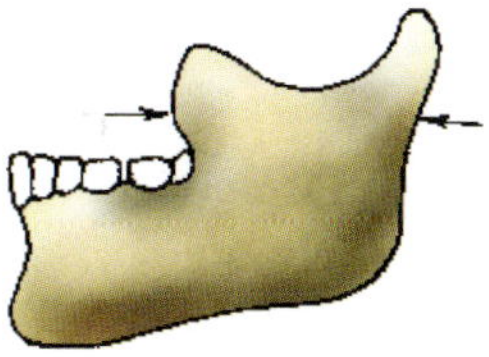

Aleuts: broad mandibular ramus.

The back part of the modern human skull is normally formed from one plate of bone (the *occipital* bone). Not infrequently, the bone plates may be divided into smaller sections. When the occipital bone is divided into two portions, the upper triangular portion is called the *inca* bone:

> This is found in Sinanthropus pekinensis [Homo erectus], Mongoloids, and American Indians in varying but of ten high frequencies. William Laughlin. 'Eskimos and Aleuts: their origins and evolution'. Science. 8th November 1963. p641.

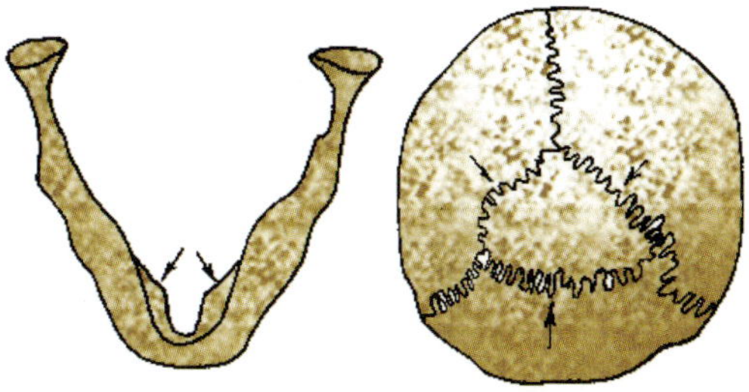

Aleuts: mandibular torus and inca bone.

Since 1967, the remains of more than 30 individuals showing Homo

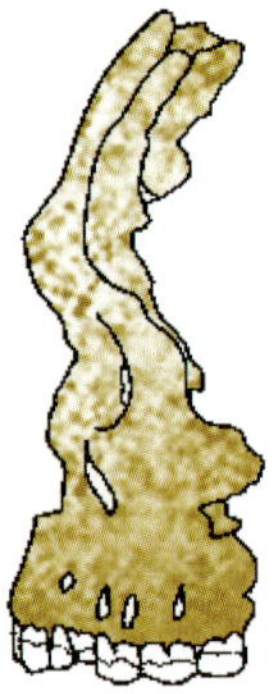

Kow swamp skull. Side view of face and frontal bone.

erectus features have been found in Kow swamp in south-east Australia:

> Analysis of the cranial morphology of more than thirty individuals reveals the survival of homo erectus features in Australia until as recently as 10,000 years ago...The frontal bones are particularly archaic... A.G. Thorne & P.G. Macumber. 'Discoveries of Late Pleistocene Man at Kow Swamp, Australia'. Nature. Vol 238. 11th August 1972. p316-9.

By 'archaic' is meant that the shape of the bones look primitive, but of course they are not old at all. And not being old means that that they cannot be primitive. Furthermore, regarding the Kow Swamp people, their forehead shapes fall within the modern range, since the:

> ...frontal curvature index values...overlap the ranges for near contemporary Aboriginal crania and homo erectus. A.G. Thorne & P.G. Macumber. 'Discoveries of Late Pleistocene Man at Kow Swamp, Australia'. Nature. Vol 238. 11th August 1972. p318.

The Cossack skull was found in 1972, near Cossack in north-west Australia. It has many Homo erectus features yet is dated as being *younger* than 6,500 years old:

> ...it does not seem as if burial was recent, for example within the last few hundred years, but it could be anywhere between that time and 6500 B.P. L. Freedman & M. Lofgren. 'Human skeletal remains from Cossack, Western Australia'. Journal of human evolution. 1979. p283-99.

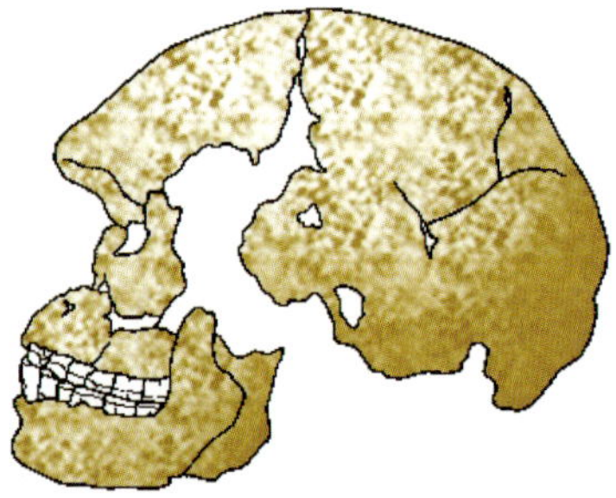

Cossack skull

This means that the owner of the (supposedly) primitive-looking Cossack skull *could* have lived after the death of Isaac Newton. Regarding the Cossack skull:

> The most striking feature is the markedly backward-sloping forehead. L. Freedman & M. Lofgren. 'The Cossack skull and a dihybrid origin of the Australian Aborigines'. Nature. Vol 282. 15th November 1979. p298.

In fact, the sloping forehead is found in living humans (see later), showing that the genes for many Homo erectus features are present in modern man. It is particularly important to note that the appearances of bare bones can look very different to the appearance of an individual *in life,* with all the soft tissues in place. For example, the X-ray of a skull may show a prominent brow, but the side view of the head may look unremarkable.

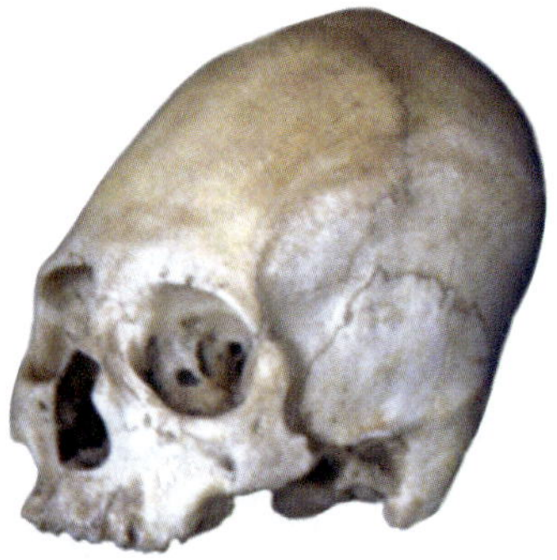

Ancient Peruvians often bound the heads of babies so as to mould the skulls into a more elongated form. This did not appear to affect their human-ness.

Furthermore, although there is no doubt that forensic reconstructions of facial appearances based upon facial and skull bones are helpful, the fact remains that such methods are still open to error. For example, the famous 2500-year-old 'Ice Maiden' was found frozen in the Alti Mountains of southern Siberia in 1993. Although the permafrost had preserved most of her body, the soft tissues of the face and skull had decayed leaving only bone. When one expert made a reconstruction of the face and declared that the features were fully:

> ...European... 'The ice maiden'. Horizon. BBC2 Broadcast. 7th December 1997,

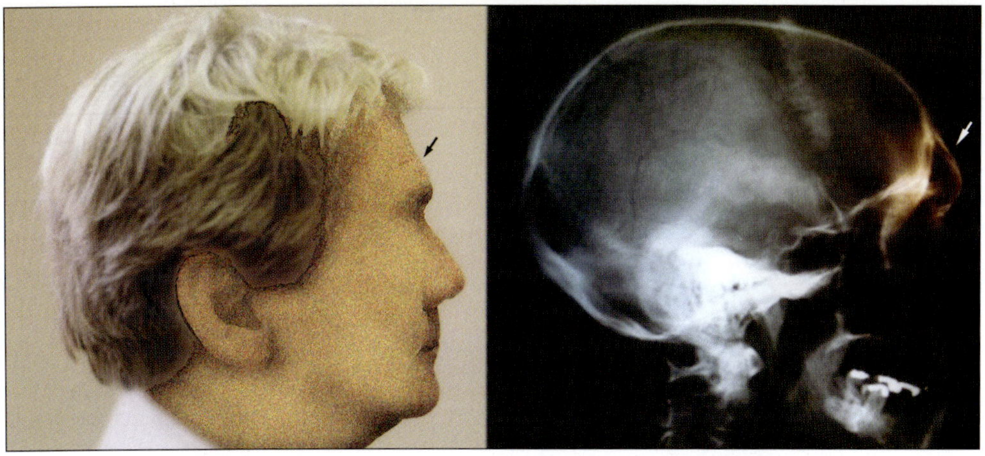

The skull X-ray (left) belongs to the man on the far left and shows that the brow prominence of a skull may not be mirrored in the living subject. So, whilst a number of Homo erectus skulls have very prominent brows, the living individuals would not necessarily have looked more ape-like than living humans.

this was immediately rejected by others. The skull was then examined by a forensic expert who declared the individual to have been clearly:

> ...Mongoloid... 'The ice maiden'. Horizon. BBC2 Broadcast. 7th December 1997.

To emphasise the point that it is not far-fetched to consider that human bones can be misinterpreted as being non-human, Owen Lovejoy's work on the bones in a 1000-year-old North American Indian (Amerindian) burial site demonstrates unequivocally that the degree of variation of the modern human form is enormous:

> The Amerindian collection undoubtedly represents a population belonging to the species homo sapiens, yet it includes many unusual bones that probably would have been assigned to **a different species**, or even **a different genus**, if they had been discovered as individual fossils... (My bold). John Reader. 'Missing links'. Collins. 1981. p232.

Whilst prominent brows are commonly seen in living humans of native Australian or African origin, there are living examples showing the typical characteristics of the Homo erectus skull form even in the UK. It is therefore quite clear that a skull with a large brow can house a fully human brain. This is not surprising since the prominence of the brow is largely related to the size of the frontal sinuses that reside within the bone, and has no bearing on the form of the underlying brain. It follows, therefore, that one cannot assume that the Homo erectus brain was in any significant way different from our own. What all this means is that the head and face, for example, of the individual to which the skull KNM-ER 3733 belonged *need not have looked dramatically different from us*. And the removal of the brow from the sloping forehead of Homo erectus skulls makes the front part of the profile comparable to the modern skull.

Living human (left) showing prominent brow and constriction at temples as in the Java Man skull cap (right).

Very few post-cranial bones have been found associated directly with Homo erectus skulls. However, a virtually complete skeleton (KNM-WT 15000) of a Homo erectus boy was found in the mid-

Section through skulls of chimpanzee (top) and gorilla (bottom) showing that the brow and frontal sinus have no bearing on the form of the brain case and brain.

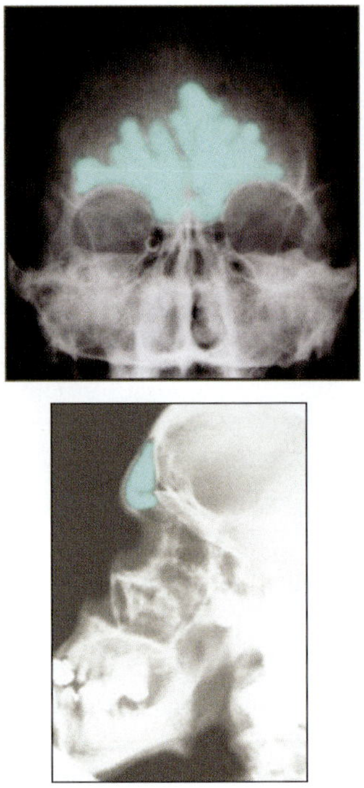

X-rays of living humans showing extent of the frontal sinus (shown in blue) within the brow.

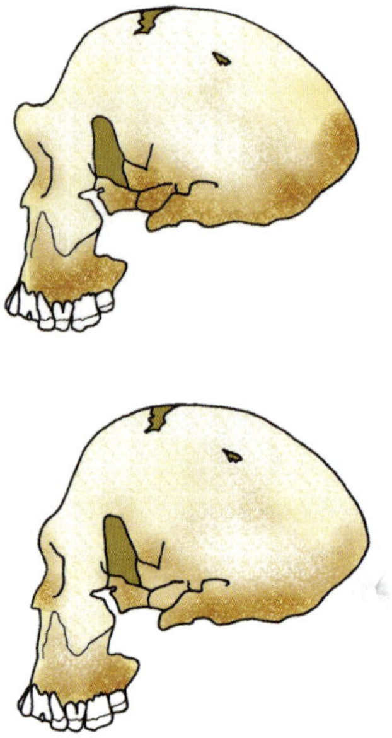

Removal of brow from KNM-ER 3733 results in a very modern appearance.

1980s on the western shore of Lake Turkana near Nariokotome in Kenya. Dated as being 1.6myo, he was clearly not fully grown since the growth plates of his bones had not fused. Except for the typical erectus features in the skull, the skeleton is indistinguishable from that of a modern human. It appears that the Turkana boy was between 9 and 12 years old and yet he was a heavily built individual, already 1.67m (5ft 6in) tall, with a cranial capacity of around 800-900 ml. When adult, he would have been over 1.83m (six feet) tall with the body proportions that we would expect of modern Kenyan bushmen.

The inner aspect of the front of the skull shows that the brain had a well-developed region called Broca's speech area, suggesting that he was well capable of human speech. However, the neck bones of KNM-WT 15000 show a spinal canal size which is smaller than that found in living adult humans of the same height. Because of this it has been suggested that his narrower spinal cord would not have had enough nerve fibres to allow him fine control of his breathing and also, since breath control is necessary for articulating speech, that the boy did not have the ability to speak. But this is pure speculation since the spinal cord contains a vast number of nerve fibres, only a small proportion of which are related to the control of the muscles of the rib cage and the diaphragm. We can be sure that a narrower spinal canal is of no relevance in defining whether or not the Turkana boy could or could not speak since 3-year-old children are well able to speak; and furthermore, despite having a brain smaller than a pea, no Broca's area, and a spinal cord only a few millimetres in diameter, even a budgerigar can mimic speech with fidelity. So, it is entirely

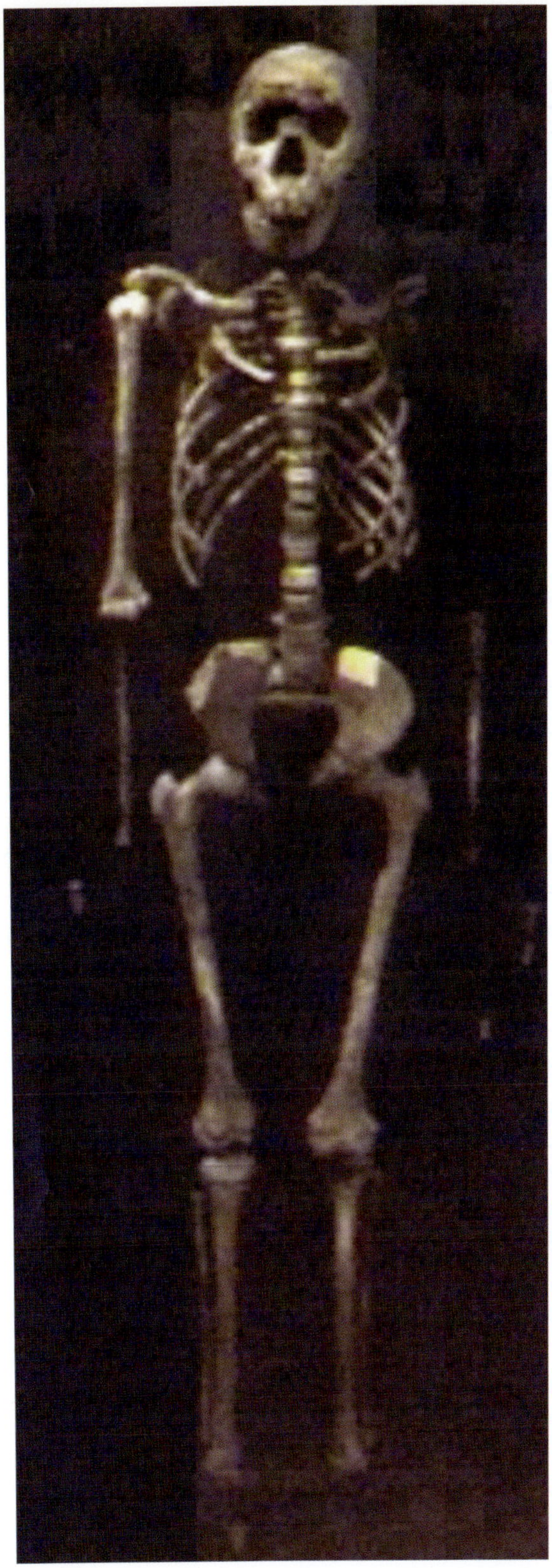

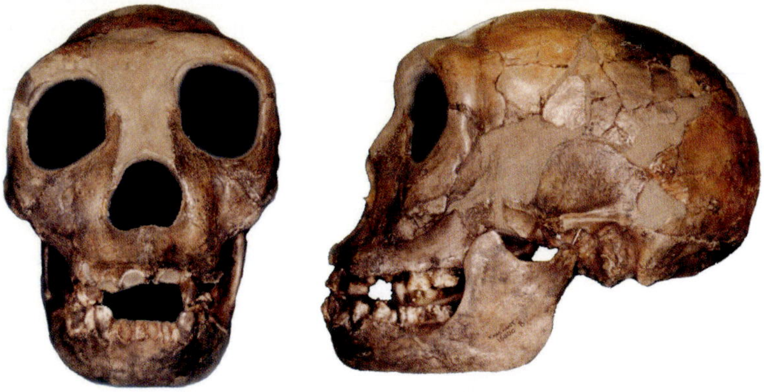

KNM-WT 15000 skull

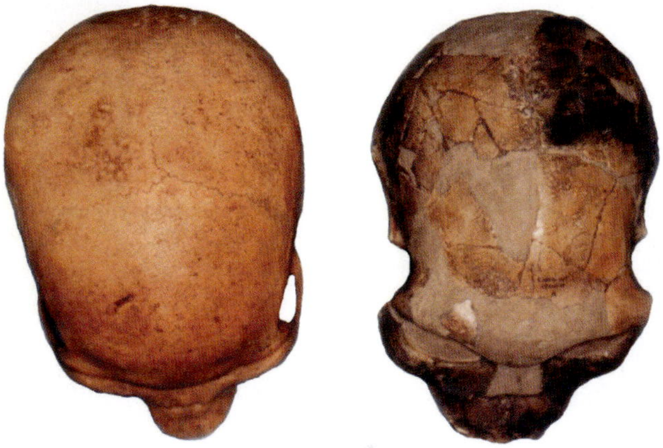

Left: Modern human skull.
Right: KNM-WT 15000 skull.

KNM-WT 15000 femur and tibia.

Although (on the basis of skull form) it has been proposed that from the neck upwards Homo erectus was not fully human, there is no convincing evidence to support such a notion (see main text).
The semicircular canals of the inner ears of humans differ from great apes in that humans have larger anterior and posterior canals and a smaller lateral canal than the great apes. When the inner ear bone structure in the skulls of living primates, australopithecines, Homo habilis and Homo erectus was examined by high resolution computerised tomography (CT scanning), it was found that:

> *The australopithecines show great-ape-like proportions and H. erectus shows modern-human-like proportions...The specimen StW 53, provisionally referred to as H. habilis, has semicircular canal proportions not seen in any of the other fossil or extant hominids or great apes.*
>
> Fred Spoor, Bernard Wood & Frans Zonneveld. 'Implications of early hominid labyrinthine morphology for evolution of bipedal locomotion.' Nature. Vol 369. 23rd June 1994. p645-8.

plausible that the Turkana boy could speak as well as we can.
However, it is immediately obvious that the boy had a backward sloping forehead and it has been suggested that this makes him non-human - that maybe he had the body of a man with the brain of an ape:

> ...he has hardly any frontal lobes so he has next to no forehead...above his eyes he goes straight back whereas in modern humans we **all** have this lump of frontal lobe...giving us this more vertical forehead... (My bold).
> Alan Walker. 'Ape-man'. BBC2 Broadcast. 7th March 2000.

But this is quite untrue since x-rays of living humans show conclusively that not only is the head form of living humans very variable, but also that a sloping forehead is not rare.

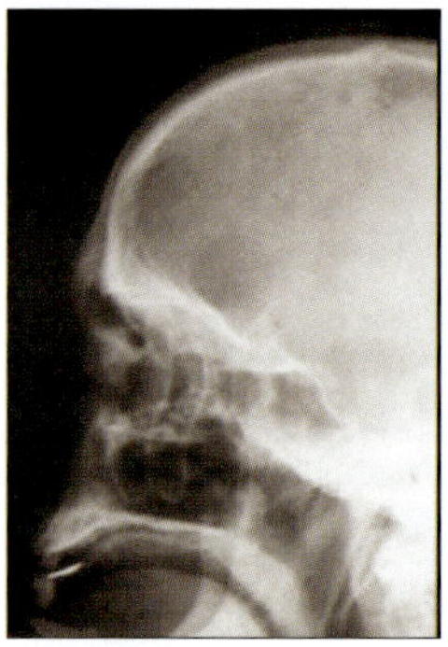

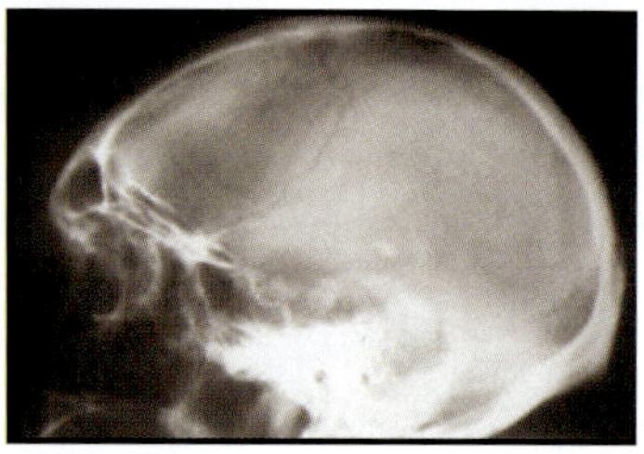

X-rays of living humans showing vertical and sloping foreheads.

Furthermore, the angle and appearance of the forehead in KNM-WT 15000 is itself deceptive, and can be altered significantly by changing the skull orientation, and by removing the large brow (see illustrations on left). Of course, if it is suggested that the Turkana boy had the brain of an ape and yet lived 1.6mya, this would have left an absurdly small amount of time for his ape brain to evolve into that of a human being. Furthermore, we have other good evidence to show that humans were alive even before the Turkana boy was born (see later), so this puts the supposed scheme of human origins into complete disarray.
So, in terms of brain size, skull shape and general skeletal form, Homo erectus falls comfortably within the modern human range. Furthermore, from the evidence of the bones alone, there is no valid reason to suppose that there was any difference in the genes or in the number of chromosomes between modern humans and Homo erectus. As we shall see, in terms of his tools, artefacts and navigational capabilities, Homo erectus must have had skills and intellect in line with fully modern humans. But since it is inconceivable that any capable man could possibly rest stagnant in his culture or technology for *one million years*, this point alone places a serious question mark over the dating of Homo erectus fossils. Since there is no scientific basis to consider those fossils commonly called '*Homo erectus*' as *anything other than* fully human, the term should be dropped. A more respectful and appropriate term would be 'sapien-*erectus*'.

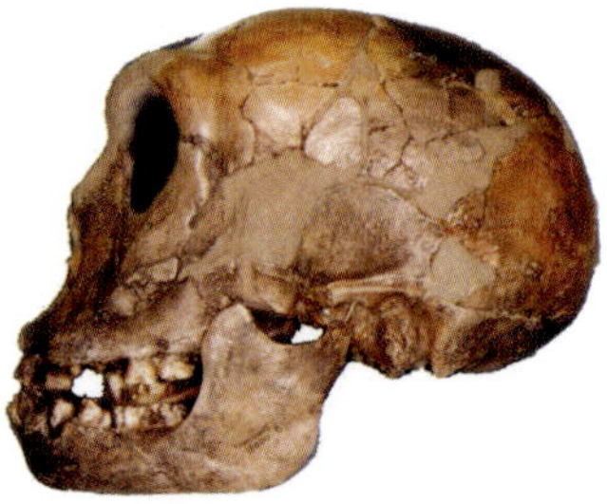

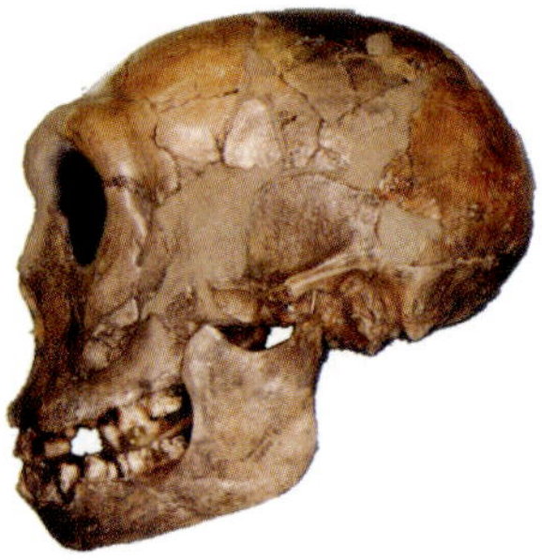

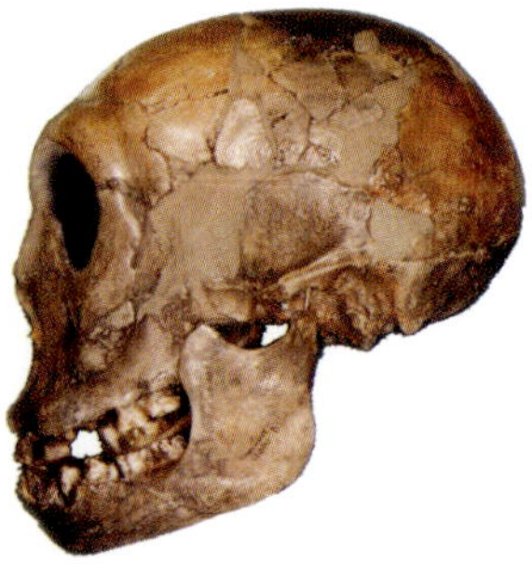

Top: The KNM-WT 15000 skull orientated with the forehead sloping backwards.
Middle: If the skull is rotated 20 degrees, the forehead starts to conform to a more modern shape.
Bottom: If the brow is reduced in size, the whole appearance conforms to the modern form.

### *Archaic Homo sapiens and Neanderthals*

Since the term *archaic Homo sapiens* describes a group of fossils with a wide-ranging form which overlaps considerably with the Neanderthals, these two forms of humans will be discussed together. But terminology can often do more to confuse than to clarify, since archaic Homo sapiens fossils:

> ...are most easily typified by characteristics belonging to other groups, such as neanderthals and homo erectus, rather than by their own unique characteristics. Ian Tattersall, Eric Delson, & John van Couvering. 'Encyclopaedia of human evolution and prehistory'. Garland Press. 1999. p54.

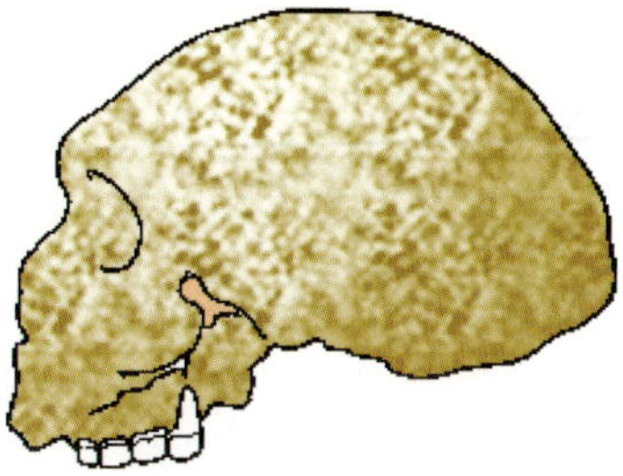

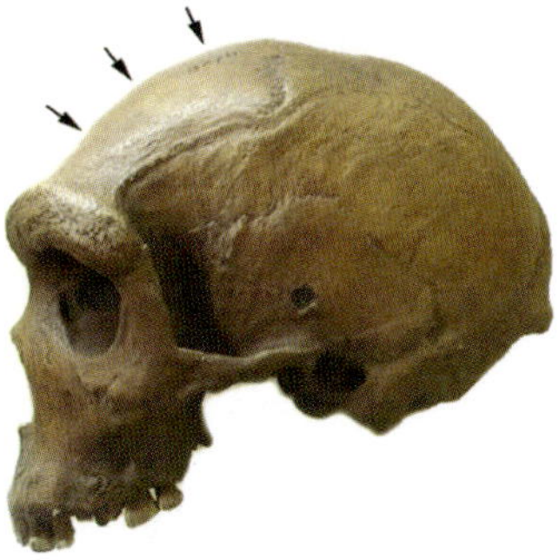

Archaic Homo sapiens.
Above top: Petrolonica.
Above: Broken Hill. The arrows point to the position of a very slight mid-line ridge which is not seen in this side view of the skull.
However, angular mid-lines are also found in skulls of living humans (below, arrowed).

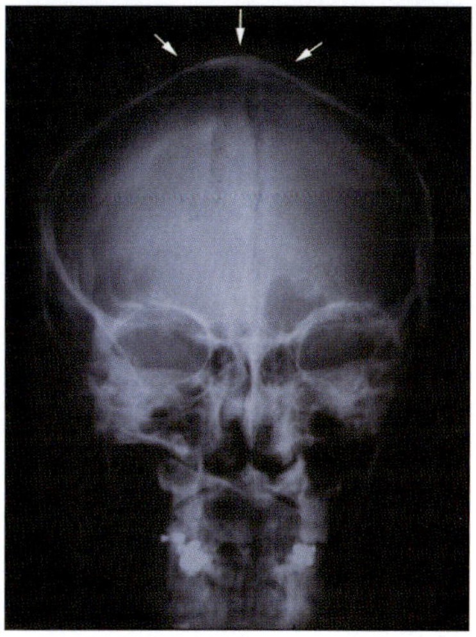

Archaic Homo sapiens fossils have been found in Europe, the Middle East, the Far East and Africa, examples being the Dali skull from central China (dated as being around 150,000yo). The *Petrolonica* skull (dated as being anything from around 200-700,000yo) is commonly ascribed to archaic Homo sapiens and shows features found both in Homo erectus and Neanderthals. The *Rhodesian Man* skull (also known as *Broken Hill Man*, after the mine/cave that it was found in) is dated as being anywhere between 40-125,000yo. However, it is accepted that in general shape, the Rhodesian Man skull is more modern-looking than some Neanderthal skulls, with a brain case volume of 1280cc which is well within the range of modern humans. An interesting feature is the slight mid-line ridge over the frontal bones which shows conclusively that a mid-line crest is in itself not representative of a primitive or ape-like feature. Also interesting is the fact that:

> The skull is in a remarkably fresh state of preservation, the bone having merely lost its animal matter and not having been in the least mineralised. Arthur Smith FRS. 'A new cave man from Rhodesia, South Africa.' Nature. Vol 108. 17th November 1921. p371-2.

Considering that other bones found in the same site show different levels of mineralisation, then it follows that the Rhodesian Man skull *could* be very much younger than commonly acknowledged.
Ever since the first fossil was found in the Neander Valley in Germany in 1856, disputes have continued as to the place of Neanderthal Man in human evolution:

> Exactly where Neanderthals belong in our family tree remains the subject of debate. In almost 140 years, Neanderthals have been cast in virtually every imaginable relationship to ourselves. They have been subsumed under our own modern species by some and thrust far out on the most remote branch of our family tree by others. Erik Trinkaus & Pat Shipman. 'The Neanderthals'. Jonathan Cape. 1993. p398.

In fact, there seems to be no evolutionary explanation at all as to Neanderthal Man's origins; and as for his demise, the highly speculative notions of displacement by or interbreeding with (supposedly) more advanced modern humans are commonly proposed. Neanderthal Man (supposedly) first made his appearance perhaps about 130,000-200,000 years ago, and apparently disappeared without trace some 30,000 years ago. Despite their stocky build and their stocky skull shape, there is no hard evidence to say that Neanderthals were, in any way, inferior intellectually to modern Man - they are known to have made sophisticated flint tools, built huts out of animal skins and had control of fire. They also buried their dead. Fossils of Neanderthals are found restricted to the Middle East and Europe, although some fossils from Africa and the Far East do exhibit some Neanderthal features:

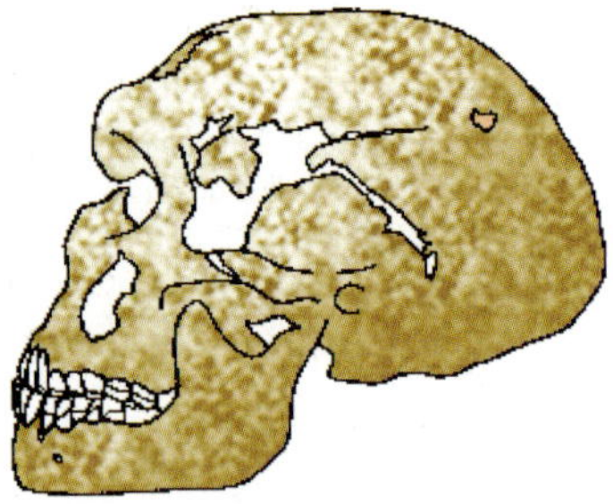
Neanderthal: la Ferrassie.

> The Asian neanderthals are less extreme than their European relatives when both are compared to modern humans. Ian Tattersall, Eric Delson, & John van Couvering. 'Encyclopaedia of human evolution and prehistory'. Garland Press. 1999. p371.

So, time and again we have seen that bodily form alone can be an inaccurate basis for making a biological classification. Although Neanderthals do have distinctive features, and are usually discussed separately, nevertheless, other than the observable differences in skull form, there is little evidence to consider archaic Homo sapiens and Neanderthal man to be different creatures:

> Estimates from the long bones of their skeletons suggest males averaged ca. 169cm (5' 6"), while females averaged ca. 160cm (5' 3") tall. Ian Tattersall, Eric Delson, & John van Couvering. 'Encyclopaedia of human evolution and prehistory'. Garland Press. 1999. p370.

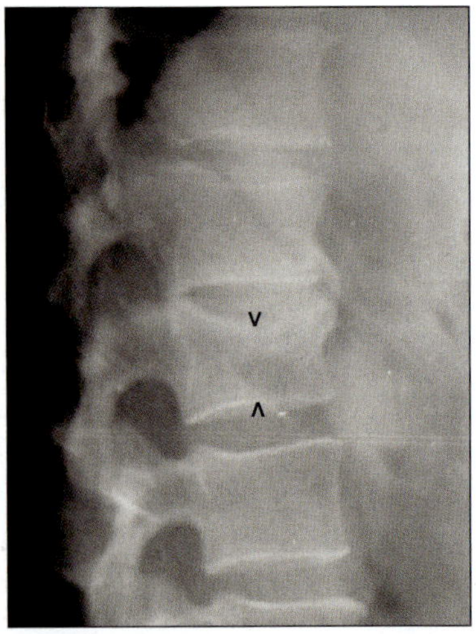
A compressed fracture of a lumbar vertebra.
The vertebra is deformed and wedge-shaped (arrowed). This commonly occurs in elderly people with *osteoporosis*, in which the bones are deficient in calcium.

However, if correction is made for the bowing of the leg bones, the Neanderthals would easily have been as tall as many modern Europeans. Yet even *as they are*, these heights are well within the range of modern humans since in the UK many living females are around 1.6m tall (see later, under Laetoli footprints), and even as recently as a few hundred years ago European males were significantly shorter than the men of today - as witnessed by the small size of their armour, and their skeletal remains. Incidentally, many Neanderthal long bones have evidence of fractures that have partly or completely healed showing that the bone injuries were sustained before death. It has been found that the patterns of the fractures (often involving the head, spine and upper arm) are in many ways similar to the injuries that are sustained by rodeo riders, which suggests that the injuries may have been sustained during hunting (or, of course, that they were into rodeos in a big way).
Both in the overall size, and in the size of the pelvic inlet, the Neanderthal pelvis was the same as a modern human pelvis. The adult pelvis from Kebara is dated as being 50-55,000yo, and is the most intact Neanderthal pelvis yet discovered:

> In general size, the Kebara pelvis does not differ significantly from that of modern Homo sapiens...the size of the inlet does not differ significantly from that of modern humans... Y Rak & B Arensburg. 'Kebara 2 neanderthal pelvis: first look at a complete inlet'. American Journal of physical anthropology. Vol 73. 1987. p227-231.

Although Rak and Arensburg suggest that the pubic bones of Neanderthals were relatively long and thin, their comparisons included only *four* modern human pelvices, which they admit is:

...a small sample (N=4). Y Rak & B Arensburg. 'Kebara 2 neanderthal pelvis: first look at a complete inlet'. American Journal of physical anthropology. Vol 73. 1987. p227-231.

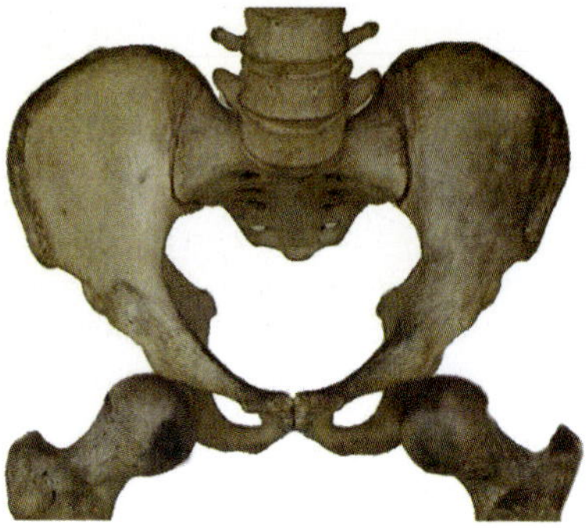

Modern human pelvis

Furthermore, comparisons with the human pelvis in standard texts on human anatomy show that any differences between the modern and the Kebara pelvis are minimal.

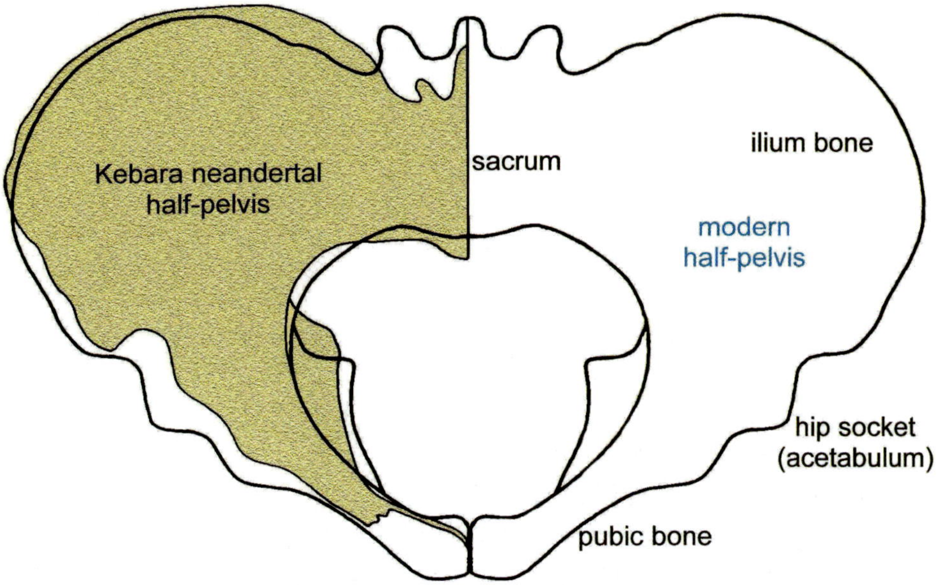


Shape of the Kebara Neanderthal pelvis with modern human pelvis superimposed.
*Human pelvis redrawn from Gray's Anatomy. 34th edition. Fig 509. p544. The Neanderthal pelvis is shaded.*

Any scheme describing the position of Neanderthals in relation to modern Man has to explain three main issues: firstly, the apparently sudden appearance of Neanderthals in a relatively short space of time (remember that Homo erectus supposedly remained unchanged for over one million years); secondly, their restrictive geographical distribution; and thirdly, the apparently abrupt disappearance of Neanderthals in a relatively short space of time.

The Neanderthal skull typically has a backward-projecting occipital bone (a *'bun'*) which is not seen in Homo erectus specimens or in the KNM-ER 1470 skull (see later). However, the shape of the occipital part of the skull of living humans is quite varied.

Since, as we shall see, modern-looking humans were living in Europe (supposedly) 800,000 years ago, long before any known Neanderthal, we therefore have to ask: how is it possible for Neanderthals to have evolved *from* such a modern-looking form? Clearly, if Neanderthals were inferior to modern humans, then most likely they would not have evolved in the first place. But in fact:

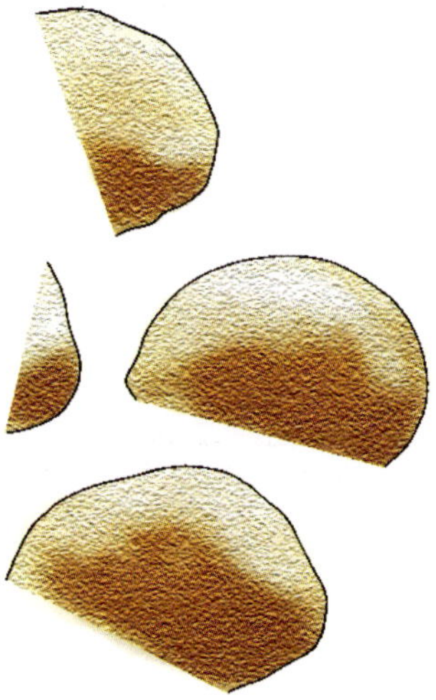

Variation in normal modern occipital shapes.
*Redrawn from Theodore Keats. 'Atlas of Roentgen variants that may simulate disease'. Mosby Year Book. 5th edition. 1992.*

Behaviourly, the neanderthals were strikingly modern... Richard Klein. 'The human career'. Chicago University Press. 1981. p407,

and since the evidence suggests that Neanderthals were not brutes, but were every bit as intelligent and capable as we are, then natural selection cannot explain their *complete* disappearance. If modern Man arose in Africa perhaps *one million years ago*, and subsequently displaced the Neanderthals, then we have the perplexing question as to why it took modern Man *so long* to displace the Neanderthals. At this point some authors just add pure supposition to pure supposition and clutch at anything:

> Perhaps the answer is that the bodily form [of Neanderthals] was only superficially modern and that some further neurological change was still necessary for full modernity... Richard Klein. 'The human career'. Chicago University Press. 1981. p410.

The fact is that the origin and disappearance of Neanderthals has not been explained by any supposed evolution, and contradictory theories live on:

> ...the neanderthals were direct ancestors of living Europeans...[or]...they represented a lineage of primitive hominids that had become extinct. Ian Tattersall, Eric Delson, & John van Couvering. 'Encyclopaedia of human evolution and prehistory'. Garland Press. 1999. p368.

Even the use here of the term *'hominid'* serves to complicate and muddy the waters since there is no convincing evidence to prove that Neanderthals were a different species to us. Could it possibly be that the Neanderthal form was the result of *disease*, perhaps caused by dietary and/or climatic conditions, rather than of some evolutionary quirk? In the middle of the 19th Century, Virchow, the famous pathologist, well-versed in the features of the diseases of his day, concluded that the Neanderthal bones represented individuals who had *rickets*, a condition in which, from an early age, the bones are abnormal and deficient in calcium. Nevertheless, the notion that rickets is the explanation for the Neanderthal form was soon rejected and buried. However, there remains sound evidence for rickets in Neanderthal bones:

> ...**every** Neandertal child skull studied so far shows signs compatible with **severe** rickets...I have examined the following Neandertal individual remains: la Chapelle-aux-Saints adult, Engis child, La Ferrassie I to VI adults and children, Gibralta adult and child, Neandertal, Peche-de-l'Ase, La Quina adult and child, Skhul IX adult, Tabun I adult. (My bold). Francis Ivanhoe. 'Was Virchow right about Neandertal?' Nature. Vol 227. 8th August 1970. p577-9.

Vitamin D is formed in the skin by the action of sunlight, and is also found in certain foods, such as fish, milk and eggs. Disease caused by vitamin D deficiency is now rare since treatment is so readily available, but deficiency can still be found in individuals living in temperate zones, particularly if they have dark skin; if their diets exclude fish, milk and eggs; or if they have limited exposure to sunlight, either as a result of wearing concealing clothing, or by being confined indoors:

> Vitamin D3 produced in the skin is by far the most important source, with smaller amounts supplied from the consumption of animal foods. Vitamin

> D status in the UK therefore depends greatly on exposure to sunlight during the months of April to October. There is no UV light of the appropriate wavelength in the UK during the winter months, when vitamin levels are dependent on the amount of vitamin D3 formed during the previous summer. Dietary vitamin D is important only when requirements are high, eg during growth, pregnancy and for those confined indoors. Eileen Manning. 'Management of vitamin and mineral deficiencies'. Prescriber. 5th November 1997. p76.

> We describe six cases of **florid** rickets in infants…all born…to parents who were postgraduate students or recent immigrants. All children had been breast fed for a prolonged period without vitamin D supplementation. Four mothers were…practising Muslims who wore concealing clothing. (My bold). Dr MZ Mughal et al. 'Florid rickets associated with prolonged breast feeding without vitamin D supplementation'. British Medical Journal. Vol 318. 2nd January 1999. p 39-40.

In the body, vitamin D is formed in a number of stages. First, and crucially, as it passes through the skin circulation, the precursor molecule *7-dehydrocholesterol* is converted into *cholecalciferol* (vitamin D3*)* by the action of ultraviolet light. This is why the formation of healthy bones is dependent upon adequate amounts of sunlight. Cholecalciferol is then converted into *25-hydroxycholecalciferol* in the liver. 25-hydroxycholecalciferol is then converted into the active molecule *1,25-dihydroxycholecalciferol* as it passes through the kidney circulation. The action of 1,25-dihydroxycholecalciferol is complex but in essence, a deficiency of it results in the removal of calcium from the bones, resulting in rickets.

Although ultraviolet light is a cause of skin cancers, indigenous people in hot countries have dark skin to protect against the harmful effects of UV light. However, since the intensity of sunlight is so great in such areas, enough UV light passes through dark skin so that the activation of 7-dehydrocholesterol is not a problem.

Bone is formed first by the construction of a fibrous and cartilaginous framework, followed by the laying down of calcium onto this matrix. In rickets, with the defective calcification of the fibrous matrix, the soft developing bones are subjected to various normal stresses, in response to which they become abnormally thick. However, the thickened bones are still frail and are susceptible to deformities and fractures. Rickets itself can vary in degree, and affected children can grow into adults who have characteristic bone deformities. Individuals who develop vitamin D deficiency in adulthood have *osteomalacia*, in which the bones look essentially normal in form, but contain less calcium.

Although most doctors practising today are unlikely to have seen a single case, the features of rickets are well described in older medical textbooks. Rickets can manifest itself very early in life:

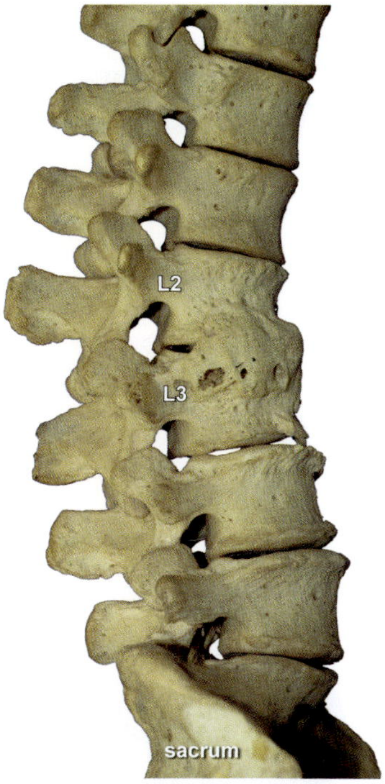

Arthritis of part of the spine with fusion of the 2nd and 3rd lumbar vertebrae. This is not due to rickets but gives an idea of how disease can damage the spine.

> An early sign, found sometimes as early as 3 months, is craniotabes: this is a curious softening of the occipital region...The skull shows striking changes. It is usually somewhat **enlarged**, **elongated**, and **flattened** on the vertex (box head), the anterior fontanelle remaining widely open long after the normal period of closure at the eighteenth month. There may also be pronounced thickening of the frontal and parietal eminences, leading

to the so-called 'hot-cross bun' or bossed head...The rickety head is apt to simulate hydrocephalus... (My bold). Price's Textbook of medicine. 9th edition. Oxford University Press. 1956. p465-6.

...the earliest sign of rickets is craniotabes...the bone becoming so thin that pressure with fingertips produces a dent which springs back when the pressure is released (as when pressing and releasing a tennis ball). John Forfar & Garvin Arneil. 'Textbook of paediatrics'. Vol 2. Churchill Livingstone. 1984. p1300.

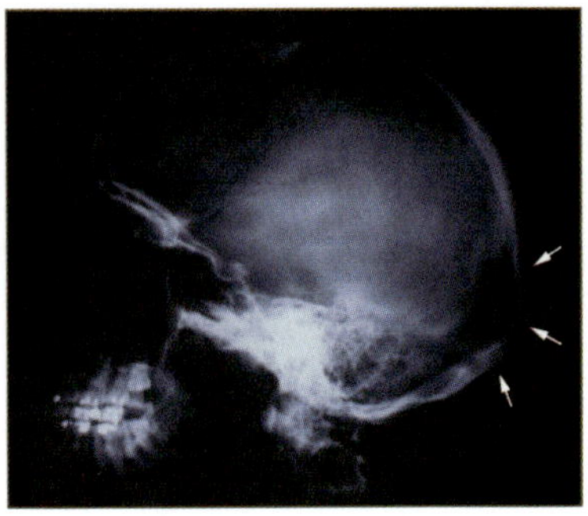

X-ray of normal living human showing bulge at the back of the skull (arrowed).

It is well established that any delay in the closure of the skull suture lines allows the skull to grow larger than normal before the skull bones fuse (this is also seen, for example, in *hydrocephalus*, which is caused by increased fluid and pressure inside the skull). So, rickets could in part explain the large head of the Neanderthal. Furthermore, localised expansion of the peculiar softened bone of the back of the ricketic skull could help explain the Neanderthal 'bun'. However, it would be a mistake to suggest that all the Neanderthal features were only abnormalities since, at the very least, the 'bun' lives on even today:

I think that my siblings and I may harbor Neaderthal genes. We all inherited an occipital bun, a bulge at the base of the skull typical of Neanderthals, from my father's side of the family, who came from Lithuania. Being of European extraction makes it possible for me to have Neanderthal ancestors. Bernard J Lane Sr. 'Neanderthals'. Forum. National Geographic. May 1996.

Getting back to rickets, however, the disease also causes abnormalities in tooth development and tooth decay:

Dentition is often delayed and not only may teeth erupt out of order but the enamel is hypoplastic and extensive caries is common. 'Price's Textbook of the practice of medicine'. 11th edition. Oxford University Press. 1973. p528.

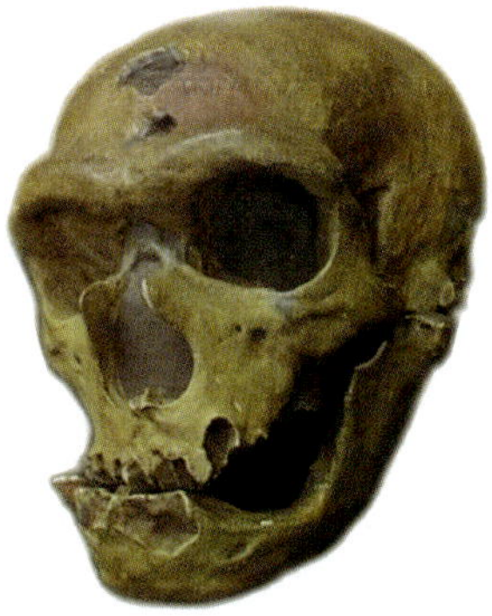

Neanderthal: la Chapelle.

And La Chapelle-aux-Saints Neanderthal demonstrates the loss of many teeth in a young individual:

...although this individual was only about thirty years old when he died, he had already lost most of his cheek teeth... Erik Trinkaus & Pat Shipman. 'The Neanderthals'. Jonathan Cape. 1993. p59.

Since we can be fairly certain that ancient peoples did not have large amounts of sugary intake, it is quite possible that La Chapelle-aux-Saints Neanderthal lost his teeth as a result of some disease process.

In rickets, an inward protrusion of the pelvis in the region of the hip joint (the *acetabulum*) is caused by the stresses of walking. This results in a:

...triangular pelvic brim as a result of weight bearing causing acetabular protusion... John Forfar & Garvin Arneil. 'Textbook of paediatrics'. 3rd Edition. Vol 2. Churchill Livingstone. 1984. p1300.

Although they did not consider a metabolic problem to be a potentially causative factor, nevertheless Rak and Arensburg also concluded that the Neanderthal pelvis shape was in some way caused by the stresses of walking, rather than something to do with giving birth:

> ...the uniqueness of the neanderthal pelvis may be attributable to locomotion and posture-related biomechanics rather than to obstetric requirements. Y. Rak & B Arensburg. 'Kebara 2 neanderthal pelvis: first look at a complete inlet'. American Journal of physical anthropology. Vol 73. 1987. p227-231.

Another deformity in rickets is that the sacrum is pushed forwards from its normal position:

> The pelvic inlet may be narrowed by forward displacement of the sacral promontory, or the outlet may be narrowed by forward movement of the lower parts of the sacrum and of the coccyx. James Hutchison. 'Practical paediatric problems'. 5th edition. 1980. p583.

Regarding the size of the Kebara Neanderthal's birth canal:

> ...the size of the pelvic inlet is comparable to that of modern Homo sapiens. Y Rak & B Arensburg. 'Kebara 2 neanderthal pelvis: first look at a complete inlet'. American Journal of physical anthropology. Vol 73. 1987. p227-231.

Since the Neanderthal pelvic inlet is the same size as that of modern humans, then the skulls of the newborn Neanderthals must have been roughly the same size as the skulls of modern human neonates. Therefore, the generally greater size of the Neanderthal skull must be the result of some enlargement *after* birth. This again is in keeping with what would be expected in rickets. Other features of rickets include spinal deformities and:

> ...bowing of the legs...thickening and distortion of the epiphiseal ends of the long bones...similarly at the costochondral junctions of the ribs a 'rickety rosary' may be visible... John Forfar & Garvin Arneil. 'Textbook of paediatrics'. 3rd edition. Vol 2. Churchill Livingstone. 1984. p1300.

Even a century ago, in Glasgow:

> ...rickets gave children 'legs so misshapen and soft...that they were helpless as babies as far as locomotion was concerned'. Rickets then was associated exclusively with urban malnutrition and atmospheric pollution... G Melvyn Howe. 'People, environment, disease and death'. University of Wales Press. 1997. p189.

Legs with rickets. Many children in Glasgow had rickets at the beginning of the 20th Century.

It is therefore quite possible that as those humans living in more temperate latitudes had to contend with the severe climatic changes associated with an ice age, *different groups would necessarily develop rickets to varying degrees.* Furthermore, those individuals that lived at higher latitudes would show the most severe characteristics of rickets. And this is exactly what the evidence points to:

NORMAL HUMANS

↓ Climate change

NEANDERTHALS — Combination of Rickets and Inbreeding

↓ Climate improvement

NORMAL HUMANS

> If the general environment at each dated site is considered, it is found that the extreme variability evident in the Neandertal populations can be related to two fairly straightforward conditions: latitude and paleoclimate...Thus if contemporaneous samples from Gibraltar, Tabun and Mapa [are compared]...it can be seen that a cline exists, the typical Neandertal traits decreasing towards the equator...It should be noted, however, that Sharidnar [Iraq] does not fit the gradient; the discrepancy may perhaps be accounted for on the basis of altitude. Francis Ivanhoe. 'Was Virchow right about Neadertal?' Nature. Vol 227. 8th August 1970. p577-9.

When the climate subsequently improved, the surviving Neanderthals would have given birth to children who did not develop rickets, and normal populations would have appeared again. Thus, the appearance and disappearance of Neanderthals can be explained by the development of rickets in vulnerable populations when the climate became particularly harsh, and the return to normal appearances and health in surviving populations when the climate became more favourable. Therefore, the peculiarities of the Neanderthal bones, their geographical distribution, and their sudden appearance and disappearance, can all be explained by *disease*, without the need for an evolutionary explanation. However, some have argued that the thickness of Neanderthal bones clearly goes against a rickets theory:

> ...the bones of neanderthals were extraordinarily stout, bespeaking of a muscular individual of supremely athletic habits, rather than the weakened and slender bones of an undernourished, calcium-poor rachitic sufferer. Erik Trinkaus & Pat Shipman. 'The Neanderthals'. Jonathan Cape. 1993. p59.

Icebergs breaking off the Breidamerkurjokull arm of the Vatnajokull glacier in Iceland.
The appearance and disappearance of Neanderthals may be explained by the development of rickets in vulnerable fully-human populations when the climate became particularly harsh, and glaciers extended southwards to include France and Spain. Modern human appearances and health could have been restored in surviving populations when the climate became more favourable.
Below: a chunk of 1000yr old ice from the Vatnajokull glacier.

But in rickets there is an increase in the fibrous matrix of bones:

> The bone shaft shows a coarse trabecular pattern due to the increase of non-calcified matrix. 'Price's Textbook of the practice of medicine'. 11th edition. Oxford University Press. 1973. p528.

And it is also notable that treated rickets can result in thicker and heavier bones than normal:

> ...bone is laid down in a more compact organized manner... 'Price's Textbook of the practice of medicine'. 11th edition. Oxford University Press. 1973. p528.

> Since the amount of osteoid laid down is not usually diminished in rickets...the bones may become heavier than normal when calcification does occur. 'Muir's Textbook of pathology'. 9th edition. Edward Arnold. 1971. p764. (The term 'osteoid' refers to the tissue which is the scaffold on which calcification occurs, so as to form bone.)

So, if the diet is improved, children with rickets will grow up to have thicker bones. And of course, despite treatment or a better diet, deformed bones will remain deformed:

> ...the child with healed rickets may have permanently bent long bones with bow legs, knock knees or other abnormalities. 'Muir's Textbook of pathology'. 9th edition. Edward Arnold. 1971. p764.

So, there is a good possibility that the Neanderthal skeletal form and its variation was to a large degree the product of varying degrees of rickets affecting populations of people who lived at varying latitudes and altitudes. If it were possible to compare, from a number of specific locations, enough Neanderthal specimens of different (and accurately determined) ages, then the disease theory should predict that such specimens should show varying degrees of deformities reflecting the onset, persistence and subsequent resolution of any adverse climatic change.
It must be noted that whilst Trinkaus and Shipman are content to dismiss Virchow's views as (supposedly) being biased due to his belief in the immutability of species and organisms, they would have it that their own prejudices are irrelevant:

> ...**our personal view is irrevocably stained with our prejudices** and convictions. The strongest of these is that human evolution indeed occurred...To us the fossils indicate that the earliest modern humans evolved out of Neanderthals... (My bold). Erik Trinkaus & Pat Shipman. 'The Neanderthals'. Jonathan Cape. 1993. p413-4.

Now, it could be argued that surely complete populations of Neanderthals could not have suffered from rickets for over 100,000 years. Most certainly, that would seem to be an unrealistically long period of time, since subsequent generations would most surely have developed paler skin and would have had opportunities to try different diets. Of course, the question then arises - could the dating of Neanderthals be in error? In fact, there is good evidence to suggest that they lived far more recently. The *Banolas mandible* is dated as being:

> 15,650 B.C...suspected to be from Neanderthal man found in 1887 near Gerona, Spain... Rainer Gerger & WF Libby. Radiocarbon. Vol 8. 1966. p480.

This report includes a comment that for a Neanderthal bone, the date is too recent to be correct. However, this is only on the basis of the *expectation* by the commentator that Neanderthals have not lived in more recent times. As we know from the *Coelacanth Principle*, no such dogmatism can be applied, and therefore in the absence of any *direct* and contradictory evidence, it is quite reasonable to accept that Neanderthals were living (supposedly) 15,650 B.C, and of course, it follows then by the same principle, that they could have lived even more recently than that.
Recently, the analysis of the *mitochondrial DNA* (mtDNA) from a Neanderthal bone has yielded some interesting and conflicting findings. We will deal with that issue in detail later when we come to discuss chromosomes.
So we have seen that in terms of skull form, brain size, skeletal form, and intellect, both Neanderthal man and archaic Homo sapiens fall within the modern human range. And as with Homo erectus, there is no valid evidence with which to infer any significant difference in either the genes or the number of chromosomes between archaic Homo sapiens, Neanderthal man and modern humans. Thus it is quite likely

that both archaic Homo sapiens and Neanderthal man represent, in association with a degree of inbreeding, *diseased* or *abnormal* fully modern human beings.
The whole notion of human evolution has become confused since there are now well-documented fossil specimens pointing to modern Man being present at a time (supposedly) *many hundreds of thousands of years* before it was previously thought.

### *Boxgrove Man*

The *Boxgrove Man*, from West Sussex in the UK, is supposed to have stood around 1.8m tall and weighed around 80kg. His remains are dated as being about 0.5myo and are associated with great quantities of animal bones and very sophisticated flint tools, suggesting that 0.5mya Boxgrove people were no less intelligent than living humans.

### *Trinil 1 femur*

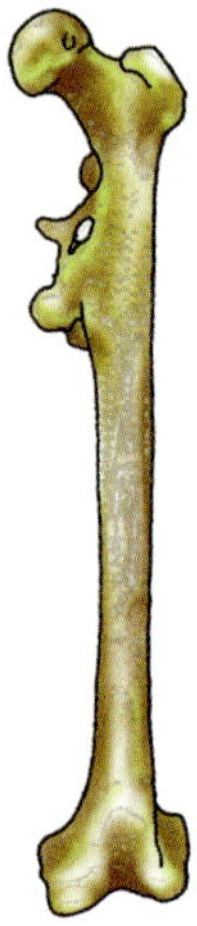
Trinil 1 femur

One complete femur and a number of femoral fragments have been found at Trinil in Java. There is some uncertainty about the exact dating of the Kabuh beds in which these fossils were found but:

> ...it seems that a date of 0.5-0.75m.y.b.p. is as good a working estimate as is available at present. Michael Day. 'Guide to Fossil Man'. Cassell 1986. p339.

The Trinil 1 femur has some disease-related overgrowth in the upper one-third, but otherwise the bone is indistinguishable from a modern human femur:

> The complete femur is remarkable in its general resemblance to that of modern man...the gross, radiological and microscopical anatomy of these bones does not distinguish them from modern human femora. Michael Day. 'Guide to Fossil Man'. Cassell. 1986. p340-1.

### *Homo antecessor*

The Daily Mail of 30th May 1997 reported the fossil bones of the newly-discovered *Homo antecessor* in Atapuerca in northen Spain as being aged 800,000 years:

> The discovery of skull and teeth fossils from six skeletons in a cave in Atapuerca, Spain, has astounded experts who previously assumed that humans with faces like ours evolved around 200,000 years ago. Daily Mail. 30th May 1997.

This report refers to an original paper which detailed 80 human fossil bones, amongst which was the bone ATD6-69, part of a face which:

> ...shows a completely modern midfacial topography...[and which] antedates other evidence of this feature by about 650,000 years. J.M. Bermudez de Castro et al. 'A hominid from the Lower Pleistocene of Atapuerca, Spain: possible ancestor to neanderthals and modern humans'. Science. Vol 276. 30th May 1997. p1392-5.

ATD6-69 clearly belonged to a juvenile since:

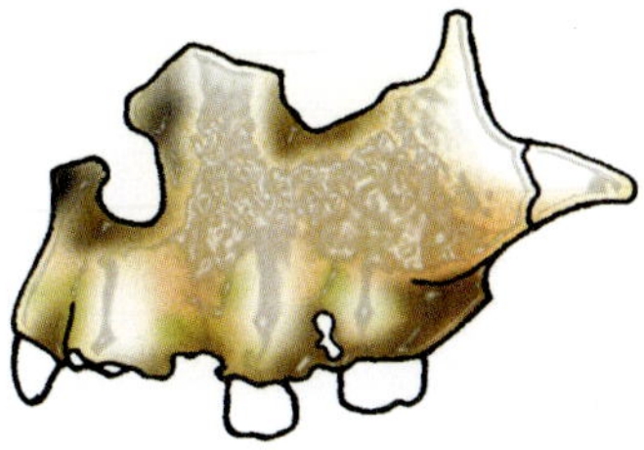

Atapuerca maxilla ATD6-69

> The eruption of the preserved teeth suggests that the age at death was 10 to 11.5 years for this individual. J.M. Bermudez de Castro et al. 'A hominid from the Lower Pleistocene of Atapuerca, Spain: possible ancestor to neanderthals and modern humans'. Science. Vol 276. 30th May 1997. p1392-5.

Although ATD6-69 has all the appearances of modern facial bones, nevertheless, there are some who are distinctly dubious:

> "...the Spanish team's theory rests heavily on an incomplete set of facial bones of an 11 year old child...It's always tricky to compare children with adults. The child's face might look completely modern...but only as a passing phase in growing up." Philip Rightmere, quoted in New Scientist. 7th June 1997. p16.

However, anatomical studies show that the appearance of a child's face when he or she is 11 years old is, in fact, a very accurate picture of what he or she will look like as an adult. Regarding the general shape of the skull:

> The growth of the vault proceeds rapidly during the first year and thereafter more slowly to the seventh year, by which time it has almost reached adult dimensions. 'Gray's Anatomy'. 34th edition. Longmans. 1969. p388.

And regarding growth of the face:

> The ethmoid bone, orbital cavities and upper part of the nasal cavities have almost completed their growth by the seventh year. 'Gray's Anatomy'. 34th edition. Longmans. 1969. p389.

By the age of twelve years, both the distance from the pupil of the eye to the lip line, and the distance from the lip line to the bottom of the chin, are only slightly less than the adult size. In other words, the shape of the face of an 11-year-old is indeed an accurate reflection of that of the adult-to-be. Most certainly, an 11-year-old human's facial bones can be easily distinguished from those of an ape. Since, therefore, it is unequivocal that ATD6-69 is entirely indistinguishable from modern human facial bones, there can be no doubt that modern-looking humans were alive in Europe (supposedly) 800,000 years ago.

### *KNM-ER 1470 skull*

The skull KNM-ER 1470 from Koobi Fora in Kenya is dated as being at least 1.9 million years old and there has been continued controversy over the interpretation of this fossil ever since its discovery in 1972. While some take KNM-ER 1470 to be the skull of Homo habilis, a (supposed) australopithecine-human intermediary, others are struck by the human-ness of its appearance:

> ...something that's so modern it makes your hair stand up. Donald Johanson & Maitland Edey. 'Lucy, the beginnings of humankind'. Penguin. 1990. p149.

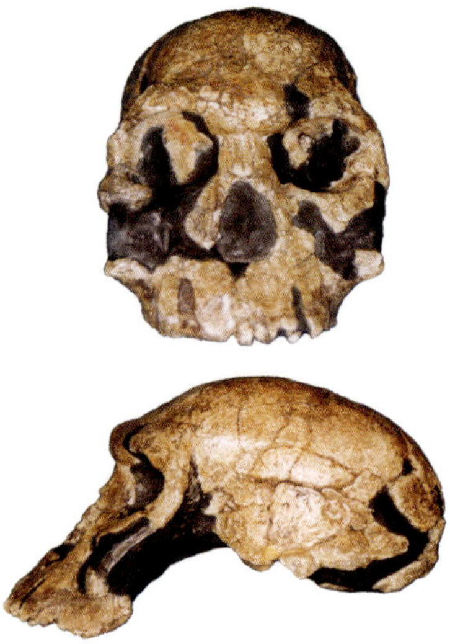

KNM-ER 1470

The controversy with the skull KNM-ER 1470 revolves around the fact that if it *were* human, then at 1.9myo, it is older than many australopithecine fossils. So, the generally accepted scheme of (supposed) human evolution would have to be discarded. Alternatively, if KNM-ER 1470 represents an ape and the fossil of the fully human Turkana boy (WT 15000) is 1.6myo, that would leave the *ridiculously* short time of only 300,000 years for apes to evolve into humans.

Since a significant part of the face is missing, there is no way of defining exactly at which angle the face and skull case should meet. So, in any reconstruction, the *maxilla* (the bone which makes up the middle of the face) can be made to join the skull at a variety of angles to the vertical. A more forward placement results in a more ape-like profile, whereas a more vertical angulation results in a more human profile.

Unless the reconstruction of the face has erroneously been made far too long, the base of the skull of KNM-ER 1470 lies very high up compared to the line of the teeth. This is very different from the human form.

As in apes, the face has very long cheeks and as in the australopithecine skull OH5, the bone between the nasal opening and the upper teeth is also very long. However, unlike in australopithecines, the zygoma (the bone at the side of the cheeks) does not seem to extend sideways beyond the line of the brain case.

To complicate the picture, from the way the skull has been reconstructed, the lower jaw would have been very large but there are no markings on the side of the skull to suggest the presence of large jaw muscles. Although unlikely, this might mean that the facial bones may have been put together incorrectly and that the face was not as long as in current reconstructions.

The appearance of KNM-ER 1470 from above resembles a smaller version of the Turkana boy's skull.

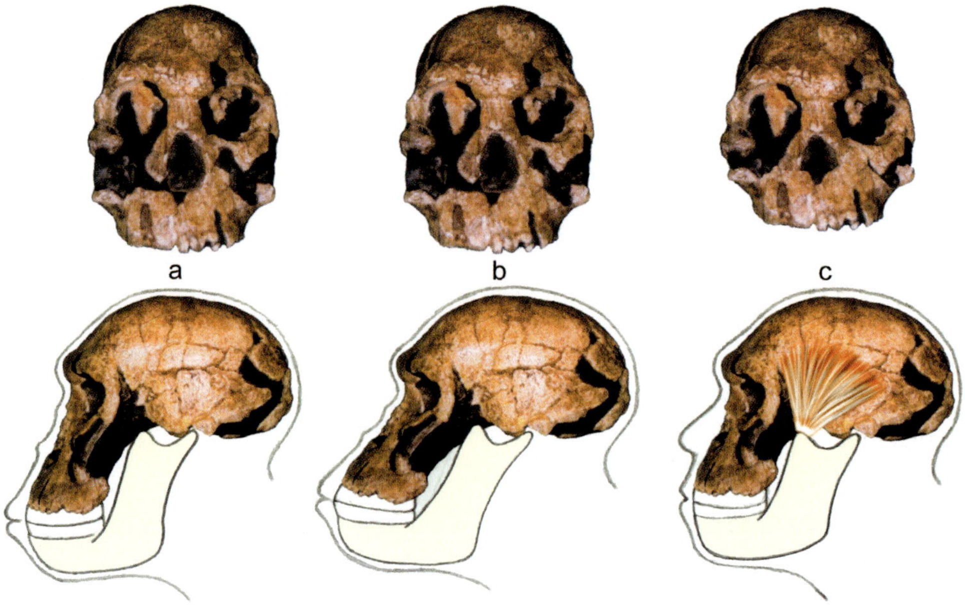


**Reconstructions of KNM-ER 1470**

Because the face is missing important parts, the appearance of KNM-ER 1470 can vary according to the way it is reconstructed.

The face can be relatively flat (a) or it can protrude markedly (b).

There is a possibility that the maxilla is too long in both (a) and (b) since it would require a very large lower jaw with very large muscles to move it. However, there is no mid-line crest (as in gorillas) and the side of the skull does not show evidence for the attachment of massive jaw muscles. If some of the small bones in the maxilla have been placed erroneously, and the reconstructed maxilla is shortened, a more human form appears (c).

However, in all three reconstructions the base of the skull is extremely high compared to the level of the teeth, making it unlikely that the fossil KNM-ER 1470 is human.

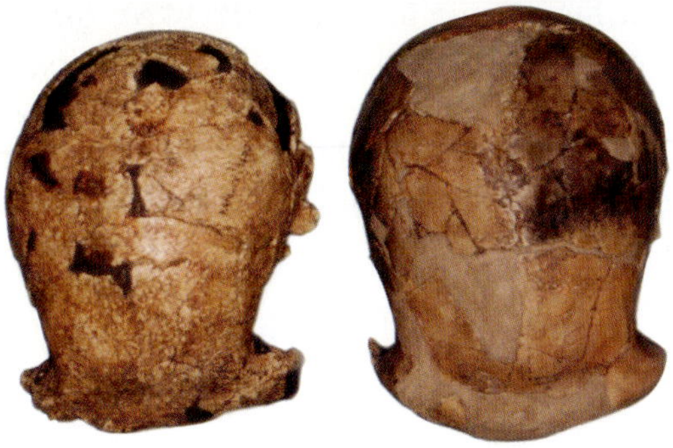

Skull KNM-ER 1470 (left) compared with Homo erectus KNM-WT 15000 (right). The difference in the skull cap is only one of size.

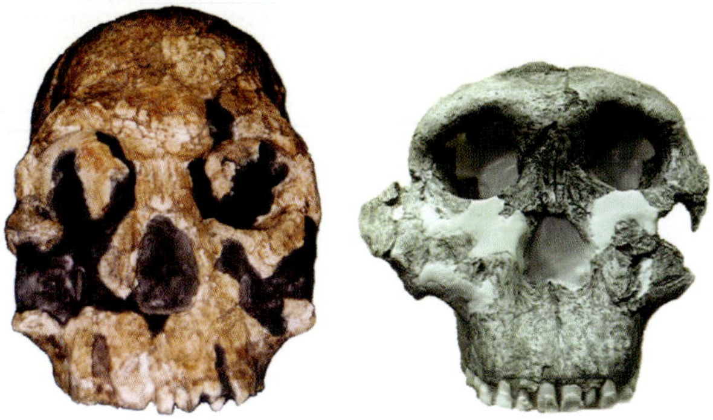

Unlike the australopithecine skull OH5 (right), the long ape-like cheek bones of KNM-ER 1470 (left) do not extend sideways beyond the skull sides.

However, the occipital flaring (the sideways protrusion at the back) resembles that of a typical australopithecine skull, although it is seen to a lesser degree in some Homo erectus skulls.

The teeth are missing but as in apes the front sockets are very large.

The cranial capacity of KNM-ER 1470 is generally accepted as being 750-800ml. In this respects KNM-ER 1470 is not unique, since the partial Homo habilis skull KNM-ER 1590 (which is also dated along with KNM-ER 1470 as being 1.9myo) also had a large brain. Although the individual to which it belonged was not fully grown, nevertheless, at the time of death, KNM-ER 1590 already had a brain case larger than KNM-ER 1470:

> KNM-ER 1590...A partial skull with **juvenile** dentition...the cranial volume **was large**... (My bold). Michael Day. 'Guide to Fossil Man'. Cassell. 1986. p207.

KNM ER 1470

Petro on ca

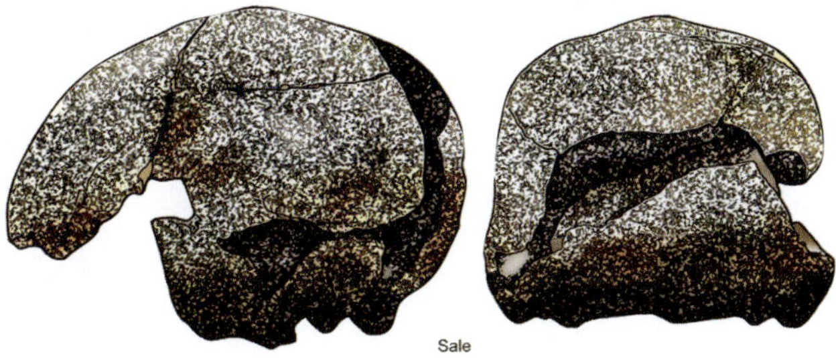

Sale

Skull KNM-ER 1470 (top) has flaring in the occipital region. This is seen to a lesser degree in the Homo erectus skulls from Petrolonica (middle) and Sale (bottom).

Depending on the estimate, the cranial capacity of KNM-ER 1470 is either the same or at most 50ml more than the largest gorilla brain, and between 80 and 130ml less than the smallest human brain. However, we know that intelligence is not proportional to brain size (see chapter 13). So from the brain size alone, KNM-ER 1470 could represent either a fully human or a fully ape intellect.

Endocasts of fossil skulls have been made and these allow the brain sizes and patterns to be compared with those of humans and of living apes. Apart from the human brain being much larger, the patterns of the *gyri* and *sulci* (the surface folds and gullies) are distinctly different. It is found that australopithecine brains resembled the brains of living apes since in the front lower aspect they had the same:

> ...fronto-orbital sulcus that characterises all extant ape brains... Dean Falk. 'Cerebral cortices of East African early hominids'. Science. Vol 221. 9th September 1983. p1072-4.

This particular feature is *not* present in humans or in KNM-ER 1470. §*Broca's area* is a specific part of the frontal lobe of the human brain and is *essential* for vocal communication. Damage to this area, for example, either by injury, stroke, or tumour, can result in a person losing the ability to speak whilst retaining the ability to understand language. It has been suggested that endocast of KNM-ER 1470 shows a Broca's area similar to modern humans:

> If Fig 1 illustrated an endocast of an extant human skull rather than that from KNM-ER 1470, one would conclude that the external gross morphology near and partially in Broca's area appeared normal and that the human in question had probably been capable of speech. Dean Falk. 'Cerebral cortices of East African early hominids'. Science. Vol 221. 9th September 1983. p1072-4.

However, living apes also have a Broca's area and, as with many animals, apes do communicate vocally to a limited extent. The *lunate sulcus* is a groove found at the back on the outer side of the human brain and is also seen in KNM-ER 1470, but it is not found in australopithecines or in living apes. So, the superficial brain markings in KNM-ER 1470 resemble the human form more than the ape form.

From the above mosaic of features, it is not possible to define whether or not KNM-ER 1470 represents an ape or a human being. The *Chalicotherium Principle* shows that we have no way at all of defining the body and limb forms that KNM-ER 1470 would have had. Furthermore, since the hypothetical evolution of australopithicines to humans requires a substantial sequential series of intermediaries, the *WYWYWG Principle* shows that such changes are implausible. Recent re-evaluation of the 1470 skull suggests that it is distinctly ape-like.*

In summary, if KNM-ER 1470 represents a large-brained ancient ape, that in itself does not mean that such a creature was more intelligent than living apes, or that it made stone tools, or that it could speak. If KNM-ER 1470 represents a small-brained human, that in itself does not mean that the individual was intellectually inferior to us, or that large-brained apes evolved into small-brained humans which in turn evolved into modern humans. Whatever it represents, KNM-ER 1470 should be set aside as offering no insight into human origins.

* *'Dr. Leakey produced an intrinsically biased reconstruction based on erroneous preconceived expectations of early human appearance that violated principles of craniofacial development,' said Dr Timothy Bromage, whose reconstruction, by contrast, shows a sharply protruding jaw and, together with colleague Francis Thackeray, Transvaal Museum, South Africa, a brain less than half the size of a modern human's…Because he did not employ biological principles, Dr. Leakey produced a reconstruction that could not have existed in real life,' Dr Bromage concluded.*
Man's earliest direct ancestors looked more apelike than previously believed. Science Daily. 27th March 2007.

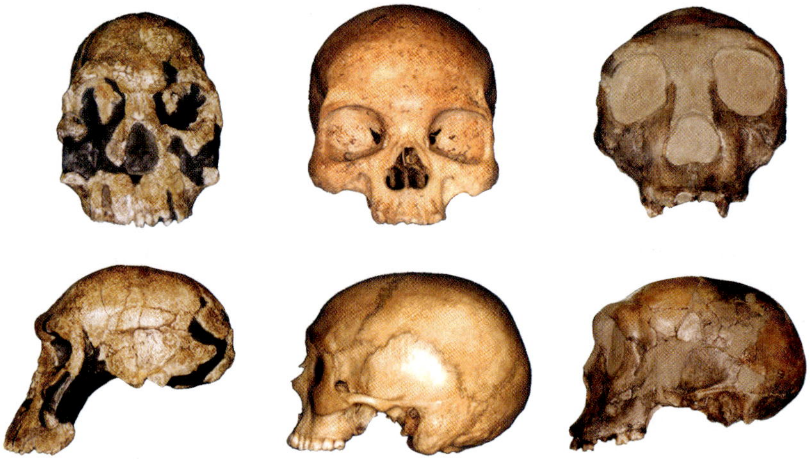

Skull KNM-ER 1470 (left) compared to the skulls of a modern human (middle) and the Turkana boy, KNM-WT 15000 (right). The difference in the levels of the teeth compared to the bases of the skulls is very marked.

### Ape-like and human-like features of KNM-ER 1470

| | Ape features | Human features | Features compatible with apes and humans |
|---|---|---|---|
| **Skull** | skull base high above teeth line | shape from above | occipital flaring |
| **Brain** | | no fronto-orbital sulcus lunate sulcus present | size of small human or large gorilla |
| **Face** | maxilla length nose-teeth distance | zygoma in line with skull side | angulation variable |
| **Teeth** | large size | | |

### *OH8 foot*

The OH8 foot from Koobi Fora in Kenya is dated as being 1.7myo, and there is no doubt that its form is indistinguishable from that of a human foot, as it:

> ...shows the characteristics typical of the bipedal human organ with an adducted hallux, a relatively large fifth metatarsal, arches, and several other features. Henry McHenry. 'Fossils and the mosaic nature of human evolution'. Science. 31st October 1975. p428.

> ...the (OH8) foot is non-prehensile and adapted for upright stance and bipedal gait... Michael Day. 'Guide to Fossil Man'. Cassell. 1986. p169.

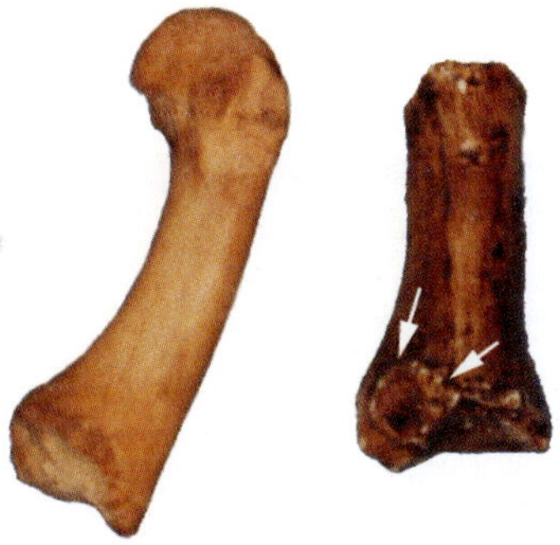

OH8 great toe metatarsal (above right) with the characteristic facet on the inner aspect of the base which is unique to the human foot, and which confirms that the bone was adherent to the index metatarsal and not mobile as in apes. In contrast, as in all apes, the chimpanzee foot's thumb metatarsal (above left) lacks this feature.

Despite its obvious human-ness and therefore its great importance, this ancient fossil is listed in the Natural History Museum of London's catalogue as simply:

> 0H8. Lt pes comprising 7 ossa tarsi and 5 metatarsalia. Kenneth Oakley, Bernard Campbell and Theya Molleson (editors). 'Catalogue of fossil hominids'. Natural History Museum of London. Part 1. 2nd edition. 1977. p59.

The specimen comprises a left foot which has lost all the toes along with all of the metatarsal heads. The posterior half of the heel bone (calcaneus) is also missing. The bones represent a foot that (including skin and soft tissues) would have been some 22.6 cm long. Although OH8 is generally taken to represent the foot of an ancient *non-human* adult, there are a number of points of detail that show it to have been the foot of a fully human *teenager*.

Examination of OH8 shows that the loss of the metatarsal heads was most certainly not as a result of the fracture of the bones. In humans

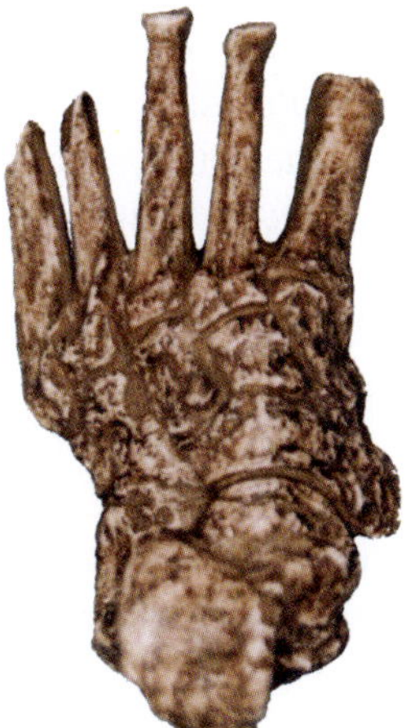

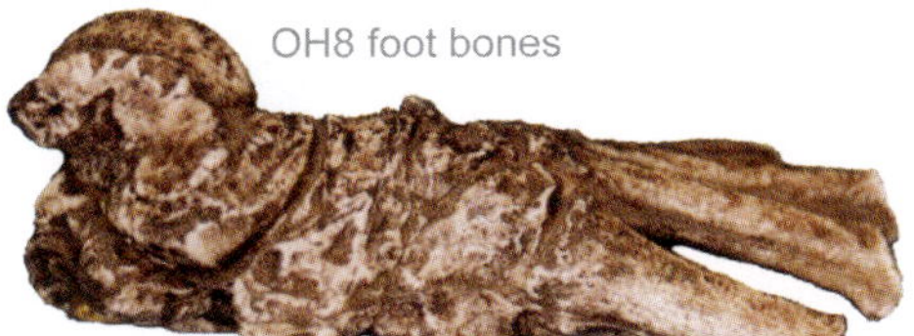

OH8 foot bones

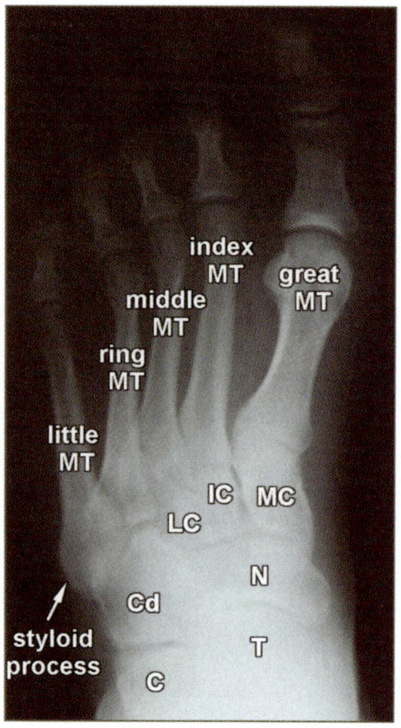


X-ray of a normal human foot.
MT: Metatarsal
MC: Medial cuneiform
IC: Intermediate cuneiform
LC: Lateral cuneiform
Cd: Cuboid
N: Navicular
C: Calcaneus
T: Talus

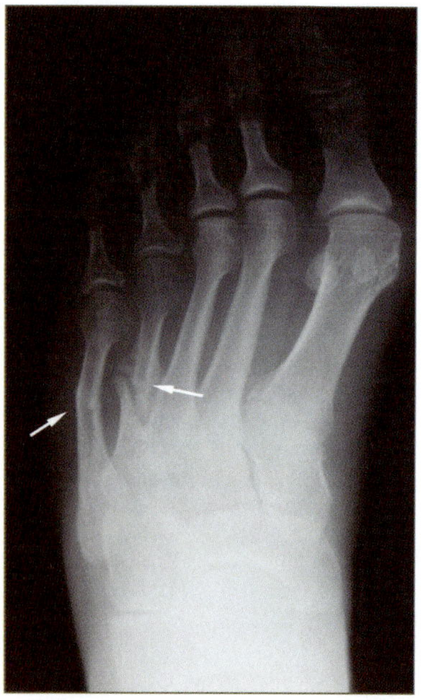

The commonest fracture in the foot occurs at the base of the little metatarsal. As in this case, other fractures involving the metatarsals almost always involve the shafts.

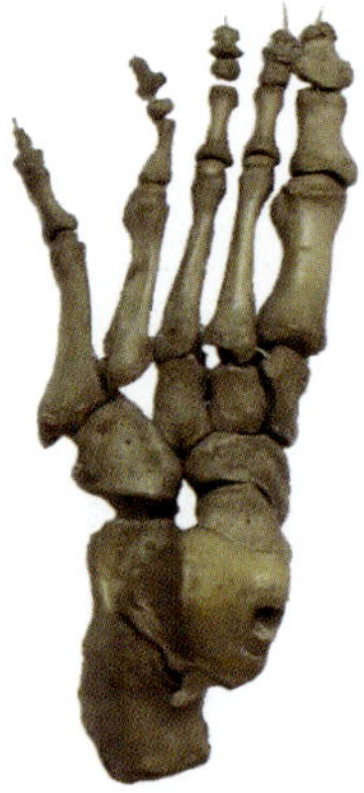

Modern human foot bones.

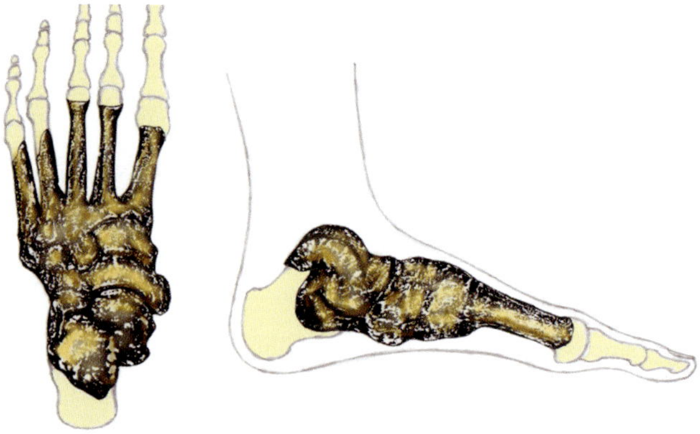

OH8 foot with missing bones and skin outline added.

it is very rare to find a fracture involving the head of a metatarsal bone, let alone a fracture of *every one* of the metatarsal heads. The commonest fracture by far is that of the *styloid process* of the base of the little metatarsal, which usually occurs after a twisting injury around the ankle. Such twisting injuries can also occasionally result in oblique fractures of the little metatarsal shaft. In relation to the stress of excessive activity, so-called *'march'* fractures may occur in the middle of the shaft of (usually) the index or middle metatarsal. Fractures caused by crushing also usually result in fractures involving the shafts of the metatarsals. Injuries in children commonly result in 'green stick' fractures in which the bones become bent or kinked rather than broken completely.

Before we continue, a few details about bone formation are necessary. The middle of a long bone is called a *diaphysis*, and each end of a long bone is called an *epiphysis*. In the growing foetus and before any hard material is laid down, the shapes of all the long bones are formed initially out of *cartilage*. This soft tissue is then converted into bone by the deposition of calcium phosphate, involving a process called *ossification*. Ossification commences whilst a baby is still within the uterus, and in large bones (such as the humerus, radius and ulna, and the femur, tibia and fibula), it proceeds from a *primary ossification centre* in the diaphysis. Over a period of time, *secondary ossification centres* appear, with (for the sake of simplicity) one at each epiphysis. In the small bones of the hand and foot, ossification usually proceeds from two centres, the primary one in the middle of the bone, and a single secondary ossification centre at one end of the bone. Bone elongation occurs as a result of continued bone deposition by the cells in a cartilaginous zone (called the *epiphyseal plate*) between the epiphysis and the diaphysis. The cartilage of the epiphyseal plate is not visible on x-ray but is represented by what appears to be a gap between the epiphysis and the diaphysis. Usually by the age of 17-18 years, ossification becomes complete as the cartilage of the epiphyseal plate also becomes bone and the primary and secondary ossification centres become fused together.

Cartilage is a tissue which is not normally preserved by fossilisation, and following the decay of a juvenile foot, the toes and the metatarsal heads could easily become separated from the rest of the foot bones, leaving behind the cupped ends to the metatarsals, exactly as seen in OH8.

The thumb metacarpal and great toe metatarsal usually have secondary ossification centres appearing at their *bases*, while all the other metacarpals and metatarsals have secondary ossification centres appearing at their *tips* (or heads).

Because the great toe metatarsal usually has its secondary ossification centre appearing at its base, it could be argued that the *shaft* of the great toe metatarsal of OH8 should have been lost rather than just the head. However, an ossification centre *can* sometimes be found in the great metatarsal head:

> ... in the first metatarsal there is sometimes an epiphysis for the head as well as for the base. 'Gray's Anatomy'. 34th edition. Longmans. 1969. p482.

In the adult, the base of the little metatarsal normally juts outwards as the *styloid process*. An ossification centre for the styloid process appears normally at around 10-12 years and fuses with the main shaft when growth is complete. The little metatarsal of the OH8 foot actually shows a smooth base which is in keeping with the expected appearance *prior* to skeletal maturity. Although it could be argued that the smoothness is due to erosion, this is unlikely since the tips of *all* of the other metatarsals have retained relatively sharp edges.

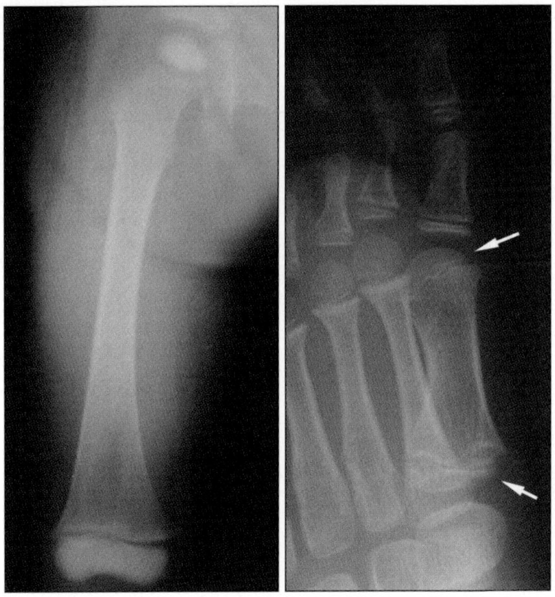

Left: The upper and lower epiphyses of the femur.
Right: This great toe metatarsal has two epiphyses.

The x-rays (below) of the left foot of a girl (LB) demonstrate well the development of the styloid epiphysis. When LB was aged 10 years and 5 months, the heads of all the lesser metatarsals could be seen to be separate from the shafts, and the styloid epiphysis has not yet appeared. When she was 11 years 8 months old, the heads of the lesser metatarsals could still be seen to be separate from the shafts, but the ossification centre of the styloid epiphysis had become clearly visible. At the age of 17 years 11 months, the x-ray showed adult features in which all of the ossification centres had fused. At the age of 10 years and 5 months, the length of LB's foot was 21cm, and by the time she had become an adult, her foot had grown to 24cm in length. It is obvious that had the OH8 individual grown into adulthood, the foot size could easily have been as long as that of LB. Since there is no anatomical evidence to suggest that OH8 belonged to any other creature, the OH8 foot most likely belonged to a human teenager, and could have grown considerably larger had the child not died. *Even as it is*, the size of the OH8 foot is within the normal range for small living *adult* humans.

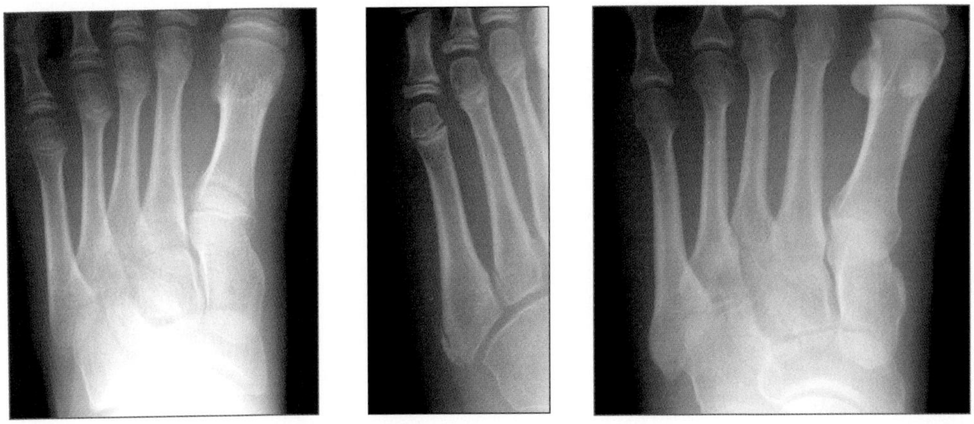

Far left: LB aged 10 years 5 months. The heads of all the lesser metatarsals can be seen to be separate from the shafts, and the styloid epiphysis of the little metatarsal has not yet appeared.
Middle: LB aged 11 years 8 months. The heads of the lesser metatarsals are still seen to be separate from the shafts, but in addition, the styloid epiphysis of the little metatarsal is now clearly visible.
Left: LB aged 17 years 11 months. Showing the features of an adult foot in which all of the epiphyses have fused.

### *KNM-ER 1481 and KNM-ER 1475 femurs*

There are a number of fossil femoral bones, found at East Rudolf in Kenya, which are accepted as being at least 1.9 million years old. The complete femur KNM-ER 1481 is indistinguishable from a modern bone, as are the upper femoral fragments KNM-ER 738 and KNM-ER 1503.

Femurs KNM-ER 1475 & 1481

In addition, the fossil bones KNM-ER 1475 comprise portions of a femur, tibia and fibula and:

> ...are **almost indistinguishable** from those of Homo sapiens. (My bold).
> Richard Leakey. National Geographic. June 1973. p820-1 & 828.

Both femurs KNM-ER 1481 and KNM-ER 1475 are approximately the same size, being about 400mm long, in other words, as long as a femur of a 1.7m-tall modern human. Given the normal human variation, these leg bones are, in fact, *completely* indistinguishable from those of living humans. Furthermore, they are completely different both in size and shape from the femurs of any ape or australopithecine. Not only are these femurs very much longer than in AL 288-1, but they are totally different in detailed anatomy around both the hip and knee joints. Since there is no evidence to the contrary, then the owners of these four femurs were in all probability fully human. Obviously, this throws into complete disarray any commonly held theory of the course of human evolution. So it is not surprising that KNM-ER 1481 is considered to belong:

> …to homo erectus...[or] attributed to...Homo habilis. Michael Day. 'Guide to Fossil Man'. Cassell. 1986. p206,

Left: KNM-WT 15000 femur.
Right: KNM-ER 1481 femur.

both of which are considered by Day to be sub-human creatures. But, we have seen that there is no convincing reason to consider Homo erectus as anything other than fully human. And as we have seen, the year after Day's book was published, the partial adult skeleton OH 62 was discovered, which clearly showed Homo habilis to be a very small long-armed ape-like creature. Since clearly, none of the bones KNM-ER 1481, 1475, 738 and 1503 belonged to either an australopithecine or any Homo habilis, there can be no doubt that the owners of these femurs were human beings living (supposedly) at least 1.9mya.

### *Laetoli footprints*

In 1975 some human-looking fossilised footprints in the form of a trail running for some 23 metres were found in Laetoli, Tanzania. The footprints had been made in mud that had been subsequently covered over by a few centimetres of volcanic dust. The main prints are made by two individuals, one larger than the other, who had been walking side by side. It appears that a third individual may be represented by some prints that seem to override the main prints. It seems that no paleontologist has any doubt about the age of these footprints which have been dated (supposedly) as being at least 3.5 million years old.

Within the same tuff, other fossilised prints can be seen which are *exactly* the same as those of *modern* animals and plants, which suggests that the climate at that time was much the same as today's. The animal tracks included those of hare, guinea fowl, baboon, giraffe and gazelle. Also present are the footprints of extinct animals such as a sabre-toothed cat, deinotherium and chalicotherium. There are also fossilised impressions of whistling-thorn leaves, which are identical to those found today. But having been dated as 3.5 million years old, these human-looking footprints posed a problem since humans were not supposed to have been alive at that time:

> "They looked so human, so modern, in tuffs so old", says footprint expert Dr Louise Robbins. Mary Leakey. 'Footprints in the ashes of time'. National Geographic. April 1979. p452.

*Unassociated* with the footprints, the partial remains of 22 individual australopithecines have also been found in the vicinity of Laetoli:

Above: Laetoli footprints. The main footprints belong to two individuals walking side-by-side. In addition, there are prints made by animals including hipparion, giraffe and guinea fowl.

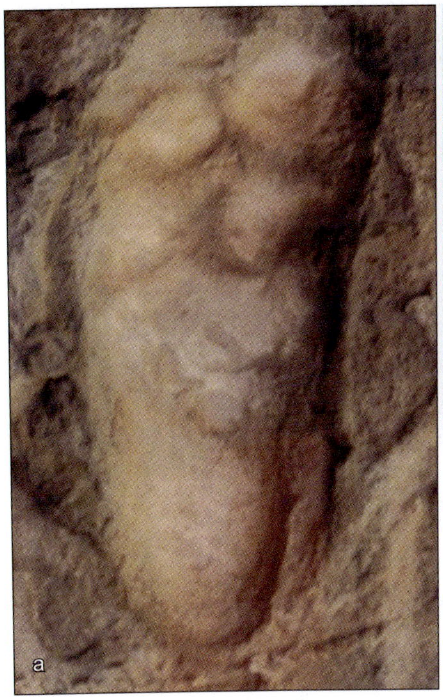

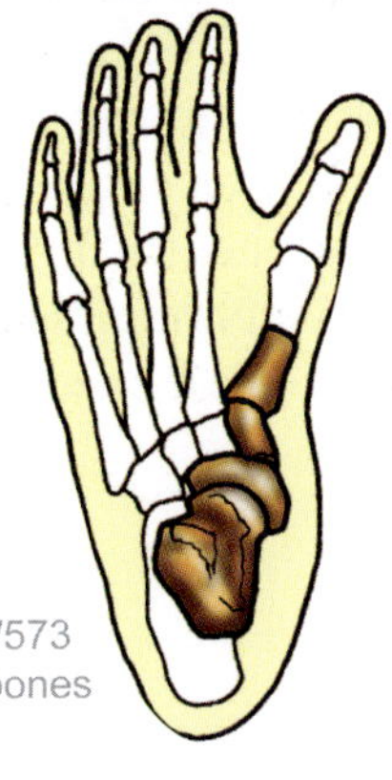

STW573 footbones

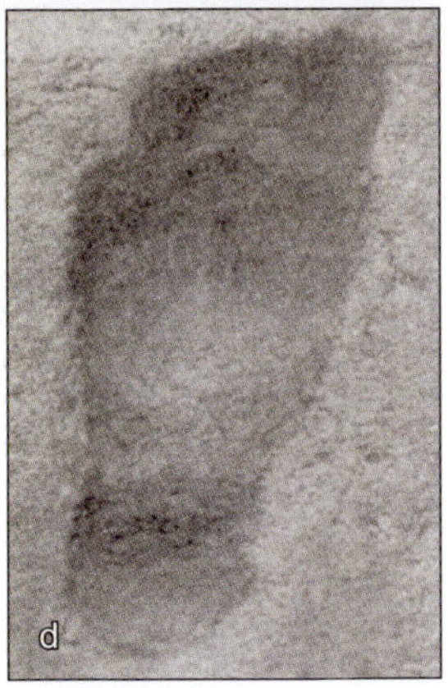


The Laetoli footprints are indistinguishable from those of humans. Furthermore, they could not have been made by australopithecines which had thumbs on their feet, as shown by the STW573 footbones. The impression (a) contains a small footprint which is very accurately aligned within a larger footprint. This feature points to human behaviour, as when a child deliberately steps into a footprint made by someone else.

The differences in the smaller footprints (b) and (c) were quite probably caused by a difference in the sloppiness of the soil combined with a degree of sliding, and cannot be taken as evidence that the small individual had thumbs on the feet.

Left: Specimen in ancient volcanic mud, Nicaragua. Peabody Museum. USA.
*E Bradford. 'On the compression of the front of the feet'. J Bone Joint Surg Am. 1896. s1-9. p110-23.*

Above: Guinea fowl.
Below: Whistling thorn tree.

> Teeth, jaws, and skull fragments from adults, as well as ribs and hand, arm, and leg bones from a 5-year-old... Mary Leakey. 'Footprints in the ashes of time'. National Geographic. April 1979. p452.

Despite such tenuous associations, Leakey speculated that the Laetoli footprints were made by australopithecines. Since then, the same interpretation has permeated the literature so that, over 20 years later, it has become standard *teaching*:

> Footprints 3.5 million years old from Laetoli indicate that Australopithecus walked fully upright, as do modern humans. Arthur Busbey III, Robert Coenraads, David Roots and Paul Willis. 'Rocks and fossils'. Harper Collins. 1996. p266.

But, of course, there is no reason to automatically associate the Laetoli footprints with any australopithecine bones that have been found in the vicinity. Even if a number of circumstances coincide in time or space, in the absence of any direct evidence to associate any two circumstances, it cannot be accepted that the circumstances are anything other than coincidental. Suppose a man is found standing over a dead body and has on his shirt blood stains that match the blood of the deceased, and his fingerprints are found on the murder weapon. This cannot be taken as unequivocal evidence that the man was the murderer. The *only* link between the Laetoli footprints and australopithecines is the *presumed* age of the tracks, coupled with the *assumption* that modern Man did not live contemporaneously with australopithecines. But this is pure supposition. It is curious that regarding the individuals responsible for the footprints, Mary Leakey writes:

> ...was the larger one a male, the smaller a female? Or was one mature and the other young? It is unlikely that we will ever know with certainty. For convenience, let us postulate a case of sexual dimorphism and consider the smaller one a female... Mary Leakey. 'Footprints in the ashes of time'. National Geographic. April 1979. p453.

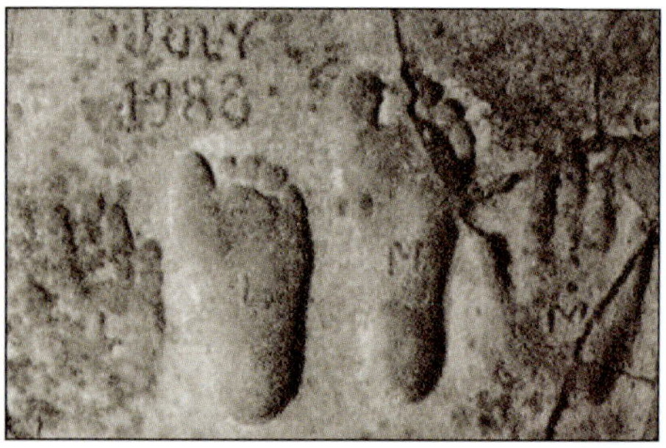

These footprints are of 3 and 7-year-old girls. The smaller print shows that the foot shape of toddlers is quite distinct from that of older children and adults.

But there is no reason *at all* to postulate either that the footprints were made by adults or that the smaller ones were made by an adult female. Although, in general, adults have bigger feet than children, and men have larger feet than women, from the age of 6 or 7 years all normal feet have the same basic form. However, the foot shapes of young children are distinctly less elongated than those of older feet, and the smaller Laetoli prints are quite in keeping with those of a human toddler. Furthermore, since the sex of an individual cannot be distinguished by footprint alone, there is no scientific evidence *whatsoever* to invoke any sexual difference in *these* footprints. And since neither set of footprints is large, if the individual who made the smaller footprints was deemed to be young, why not the other also?

Mary Leakey has a belief that Man has evolved from an ape-like ancestor, and she has allowed her belief to warp her science. More than that, in the absence of any evidence, she lets her imagination ramble as she *fabricates* ideas that the unwary reader might mistakenly take to be authentic science. So let's examine the evidence afresh.

The larger prints are about 21.5cm long and the smaller prints are about 18.5 cm long. Judging by the stride lengths, the larger individual has been estimated to have been around 1.4m tall, and the smaller individual around 1.2m tall. But, many living adult humans, even in the UK, are around 1.4m (4 feet 7 inches) tall. And regarding the pygmies in central Africa:

> Adults average about 4 feet 6 inches... Douglas Chadwick. 'The pygmy way'. National Geographic. July 1995. p35.

This means that many *living* humans are less than 4 feet 6 inches tall. Furthermore, as Leakey herself admits, height calculations based on the stride length are:

> ...far from exact... Mary Leakey. 'Footprints in the ashes of time'. National Geographic. April 1979. p435.

So the individual who made the large footprints could have been *more* than 1.4 metres tall. Furthermore, foot length can vary considerably even amongst individuals of the same height. Random measurements show that it is not uncommon for women to have small feet. In a random analysis, one woman whose height was 1.53m (just over 5 feet) had a foot size of just 22cm, only *5mm* longer than the larger Laetoli prints. Another woman had a foot size of only 20.2cm and a height of just over 1.42m (4 feet 8 inches). Yet another woman had a foot size of only 22.5cm but had a height of 1.65m (5 feet 5 inches). Furthermore, two women whose feet were identical in length (24.5 cm) differed by nearly 18cm (7 inches) in height.

A further complication is that the relative lengths of the trunk and legs can vary considerably. Therefore, not only could the Laetoli prints have been made by fully modern humans, but quite possibly, given the variation in the relationship between foot length and overall height, the individual who made the larger prints could have easily been 1.7m (5 feet 7 inches) tall. Furthermore, humans only a few hundred years ago were significantly shorter than we are today. In fact, there has been:

> ...a steady increase in the height of children which, since the start of the twentieth century, has amounted to half an inch per decade in five- to seven-year-olds and about an inch per decade in ten- to fourteen-year-olds. Today's five-year-olds are generally taller than five-year-olds in 1905. In eleven-year-olds and thirteen-year-olds, the difference is nearer **four inches**. (My bold). Melvyn Howe. 'People, environment, disease and death'. University of Wales Press. 1997. p189.

Since the Laetoli footprints are like modern human footprints, if the question of a presumed age of 3.5 million years was not part of the equation, then, most certainly, everyone would be happy to accept that these prints were most likely made by small human beings. Furthermore, not only did the Laetoli individuals have feet indistinguishable from those of modern humans, but they also walked just like us. Dr Louise Robbins makes the point that:

> 'The best-preserved print shows the raised arch, rounded heel, pronounced

These two ladies have the same foot length but differ in height by 18cm.

**Human feet vary considerably**

Foot dimensions...relating to shape characteristics of foot outlines taken for Japanese, Australian aborigines, Indonesians and French were analyzed...

There have been rapid increases in foot length for Japanese born after ca. 1950... Japanese born after ca. 1960 had a smaller foot girth and foot breadth for their foot length compared to Japanese born before ca. 1950... Elderly female Japanese born before 1930 had very wide feet and a weaker valgus [angulation towards the little toe] tendency of the first toe... Mongoloid populations including Japanese have a wider foot for foot length compared to Caucasoid and Australoid populations...East Asian populations including Japanese have a smaller foot length for height compared to Southeast Asians and Africans.

*M Kouchi. 'Foot dimensions and foot shape: differences due to growth, generation and ethnic origin'. Anthropol Sci. Vol 106. 1998. p161-88.*

**Other ancient footprints**

**Ileret, Kenya**

The footprints from Ileret, found in sediments (supposedly) 1.5myo, are essentially like those of modern humans. It is reported that in the Ileret prints, the angle between the big toe and the long axis of the foot ranged from 9° to 17°, compared to about 8° in living humans and 22° to 34° in the Laetoli footprints. However, this last measurement is patently false since the Laetoli print (a) on page 401 clearly shows that there is no gap at all between the great toe and index toe.

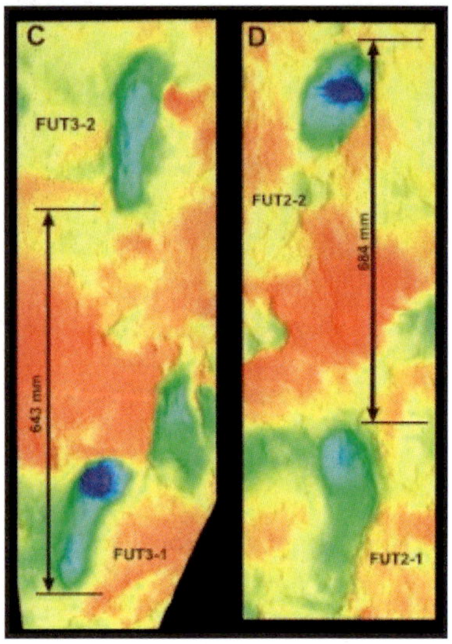


Color-rendered optical laser scan of footprints in the second and third trail at Ileret, Kenya.

*Bennett et al. 'Early hominin foot morphology based on 1.5-million-year-old footprints from Ileret, Kenya.' Science. Vol 323. 2009. p1197–1201.*

**Uskmouth, South Wales, UK**

One 10m-long set of footprints can be seen at low tide on the beach at Fornby Point near Liverpool, which is said to have been made about 5000ya:

> *"...one adult with swollen feet and a limping stride. One set of young male footprints show he had a deformed foot, with a missing second toe".*
>
> G Aaltonen talking to G Roberts. 'Sands of time'. National Trust Magazine. Autumn 2000. p36-9.

> ball, and forward-pointing big toe necessary for walking erect. Pressure exerted along the foot attest to a striding gait'. Dr Louise Robbins, quoted by Mary Leakey. 'Footprints in the ashes of time'. National Geographic. April 1979. p452.

However, as we have seen, an examination of the skeletons of australopithecines shows that these creatures were apes, and if they could move bipedally, just like gibbons or chimpanzees, perhaps they could do so for only short distances. And it is possible that they *did not* have a striding gait. Furthermore, since (as we have seen from the Sterkfontein and AL 288-1 evidence) their toes had long curved phalanges and abducted great toes, it is simply not possible for australopithecines to have made the Laetoli footprints.

So, the only things that we can objectively say about the Laetoli footprints are that the prints are like those of modern humans in size and stride length, and in detailed appearance; the larger footprints are consistent with those of a small modern adult; and the smaller footprints fall into the range for a child. *Nothing* else. As to what sort of individuals made the prints, there are only two candidates: a fully bipedal ape or a human being. But there is no fossil evidence of any fully bipedal ape with human-type feet. And the assumption that the prints could not have been made by humans, because humans are not thought to have lived at the (supposed) time that they were made, is just that: an *assumption*. The *Coelacanth Principle* shows that that possibilty cannot be excluded.

The Laetoli footprints are not the only ancient footprints that are like those of modern humans, and examples (presumed to be 10,000 years old) found in the caves at Niaux, in the south of France, contain some of the best preserved ancient human footprints in the world. Some of the prints were made by naked feet, others by individuals wearing shoes:

> En 1971, la dépression aux empreintes portrait 38 empreintes plus ou moins completes de pieds nus et 28 empreintes (au moins) de chassures modernes. [In 1971, there were found: *...38 more or less complete prints of bare feet and at least 28 prints of modern shoes.*] L.Pales. 'Les empreintes de pieds humains dans les cavernes'. Archives de l'Institut de Paleontologie Humaine. 1976. p157.

It is a coincidence that some of the Niaux footprints are the same size (21.5cm) as the larger of the Laetoli footprints, whilst the largest footprint is about 27.0 cm. With reference to a Niaux footprint that is 21.5 cm long, there seems to be no doubt that it was made by a human, and furthermore, that the individual was a *child*:

> Le sujet est **un enfant** francais de 9 ans; stature 130.1cm.
> [*The subject is a 9-year-old French* ***child****; height 130.1cm.*] (My bold).
> L.Pales. 'Les empreintes de pieds humains dans les cavernes'. Archives de l'Institut de Paleontologie Humaine. 1976. Caption to Fig 13. p54.

And when small human-like footprints are found elsewhere, common sense interprets them as most likely having been made by *human* children. Some *extremely* small human-looking footprints, thought to have been made 12,000ya, have been found in Monte Verde in southern Chile:

> Only about five inches long [12.7cm], it was probably made by a child. Rick Gore. 'The most ancient Americans'. National Geographic. October 1997. p98.

Since it was not thought that any people were in South America so long ago:

> The steps of that child insisted that someone like us had indeed reached the forests of Chile surprisingly long ago. Now the scientists have a lot of explaining to do. Rick Gore. 'The most ancient Americans'. National Geographic. October 1997. p98;

it is notable that Gore does not at least offer the possibility that this footprint could have been made by an australopithecine. I am not suggesting for a moment that the Monte Verde prints *were* made by australopithecines, but *if* it is to be suggested that human beings are not the only bipeds with human-like feet to have lived on Earth, then, however remote, logic must allow for such a possibility.

Clearly, both the Laetoli and the Niaux footprints are indistinguishable from those of modern humans, and some of the prints from both sites are exactly the same size. So, if Pales is allowed to (quite reasonably) submit that the small Niaux footprints were made by a human child of 9 years, and Gore is allowed to submit that the *12.7 cm* Monte Verde prints were also made by a human child, then we should be free to apply the same interpretation to the Laetoli footprints. So, if we were to be objective about the evidence, we should say only this: that the Niaux and Monte Verde footprints were made at a time when the only bipeds thought to be living were humans, hence it is *probable* that the Niaux and Monte Verde footprints were made by small humans. However, since the absence of evidence *does not exclude* the possibility of a non-human biped living 10-12,000 years ago, we cannot say categorically that the shoeless Niaux footprints or the Monte Verde prints *were not* made by non-humans.

Similarly, regarding the Laetoli footprints, the only scientific conclusion that can be made is that the Laetoli footprints were made at a time when the only bipeds thought to be living were non-humans, hence it is *possible* that the Laetoli footprints were made by non-humans. However, the (supposed) *absence of evidence* for humans living 3.5 million years ago *does not* exclude the possibility that humans were alive at that time, so we cannot say categorically that the Laetoli footprints *were not* made by humans. But since *there is* evidence (see below) that humans were living at the time that the Laetoli footprints were made; and since the Laetoli footprints are indistinguishable from those of modern humans; and since there is, in fact, *no evidence for any non-human creature (including australopithecines) ever having feet indistinguishable from a human foot*; then the most parsimonious conclusion must be that the Laetoli footprints *were* made by humans.

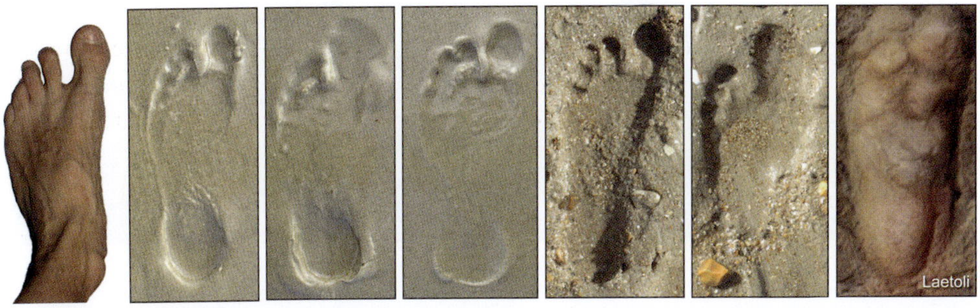


**Interpreting footprints**

Some claim that the great toes of the individuals who made the Laetoli and Ileret footprints are more divergent than those of living humans, and therefore that they were made by sub-human creatures (australopithecines and Homo erectus, respectively).

However, if we look at modern people who habitually walk barefoot, their feet are wider at the toes and the great toe diverges more than in shoe-wearing people. It has also been known for over 100 years that even in modern shoe-wearing individuals, walking barefoot with pressure placed on the forefoot creates a greater gap between it and the index toe than when the weight is placed on the heel, as shown in the photograph below.

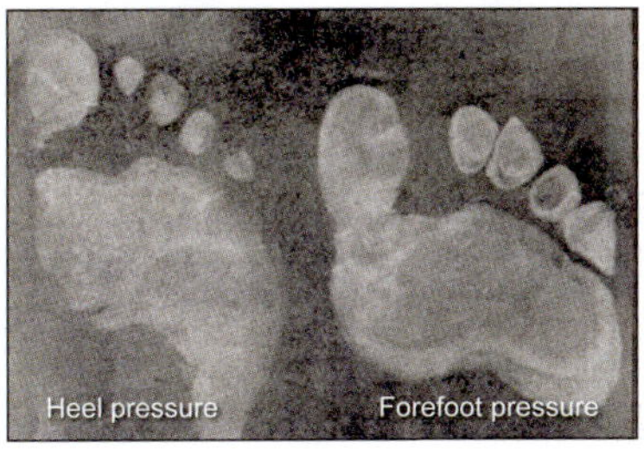


*E Bradford. 'On the compression of the front of the feet'. J Bone Joint Surg Am. 1896. s1-9. p110-123.*

Also, in ancient anatomically-correct sculptures and paintings, the great toes of barefooted or sandal-wearing people are not infrequently depicted as being divergent.

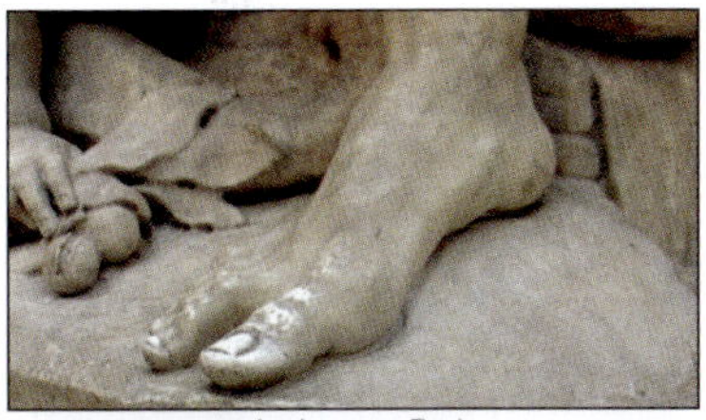

Le Louvre, Paris

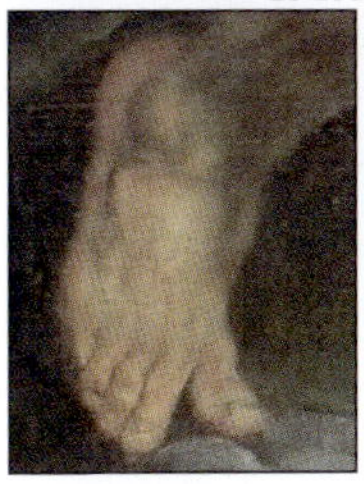

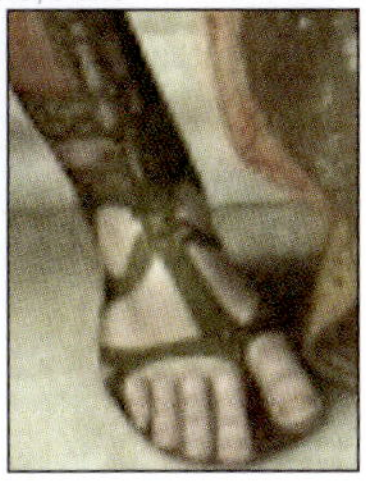

Furthermore, it is not unusual to see living shoe-wearing humans with quite divergent great toes.

Left: Individual (JD) with divergent great toe, along with a variety of his footprints made in sand of differing grain size, density, and dampness.

This shows that the Laetoli prints are entirely consistent with prints made by human feet.

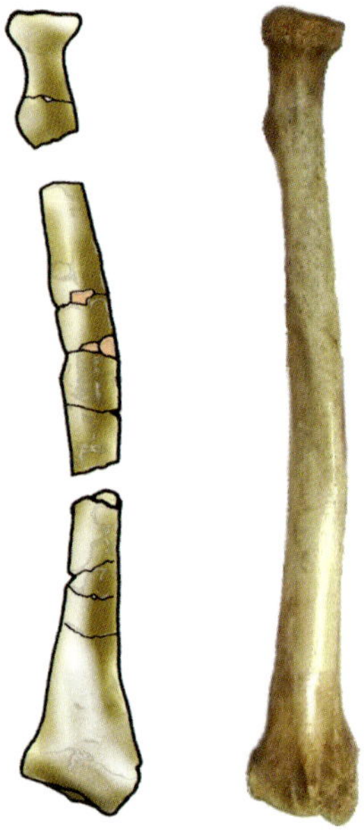

Left: Radius KNM-ER 20419.
Right: Modern human radius.

### *KNM-ER 20419 radius*

The KNM-ER 20419 radius is a forearm bone which was found in Allia Bay on the east side of Lake Turkana and is dated as being around 4 million years old. Very ancient human-like fossils are extremely rare, so it is surprising that it is given only a passing mention:

> In 1988 our field crew had discovered an unidentified hominid radius...
> Meave Leakey. National Geographic. September 1995. p51.

The fossil radius is almost complete, being essentially of three main parts, making an estimated total length of between 26.5 and 27.5 cm:

> In terms of overall morphological likeness, KNM-ER 20419 is most similar to the Australopithecus afarensis radii, AL288-1. The most notable difference is that of size; the estimated length of KNM-ER 20491 is probably **twice** that of AL288-1... (My bold). RE Heinrich et al. 'Hominid radius from the Middle Pliocene of Lake Turkana, Kenya'. American Journal of physical anthropology. Vol 92. 1993. p139-148.

In other words, most likely, the bone did not belong to an australopithecine. In fact, the size is typical of the radius of a modern human adult male, and furthermore, the thickness and general appearance show it to be quite different from the radius of any living ape. However, Heinrich argues that the upper and lower ends of the bone suggest some non-human features. He claims that the depression (the *'fovea'*) in the articular surface at the upper end of the radius is *centrally* placed in humans, but is *eccentric* in KNM-ER 20419 and in living apes. However, modern human bones show that the fovea can vary considerably in its position and *can* be eccentric.

Modern human radial heads.
Right: This pair of bones have eccentric foveas.
Far right: This pair of bones below have central foveas.

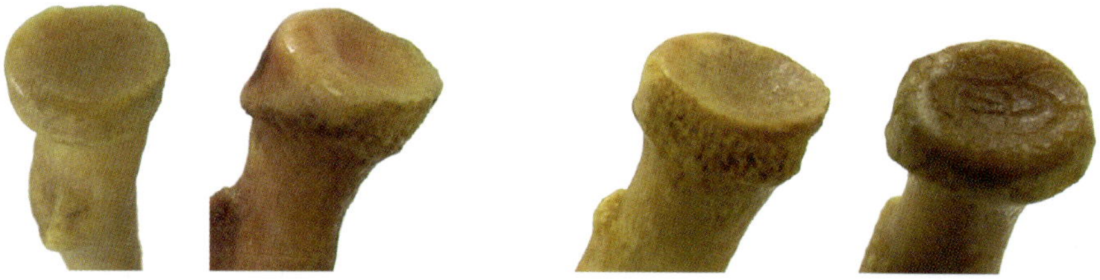

X-rays of the elbow joints taken from three different individuals, showing that the fovea of the head of the radius in the human can vary from being central (right: front view) to being eccentric (middle: front view; and far right: side view). The blue arrows point to the outside edges of the radial head and the pink arrows point to the edges of the fovea.

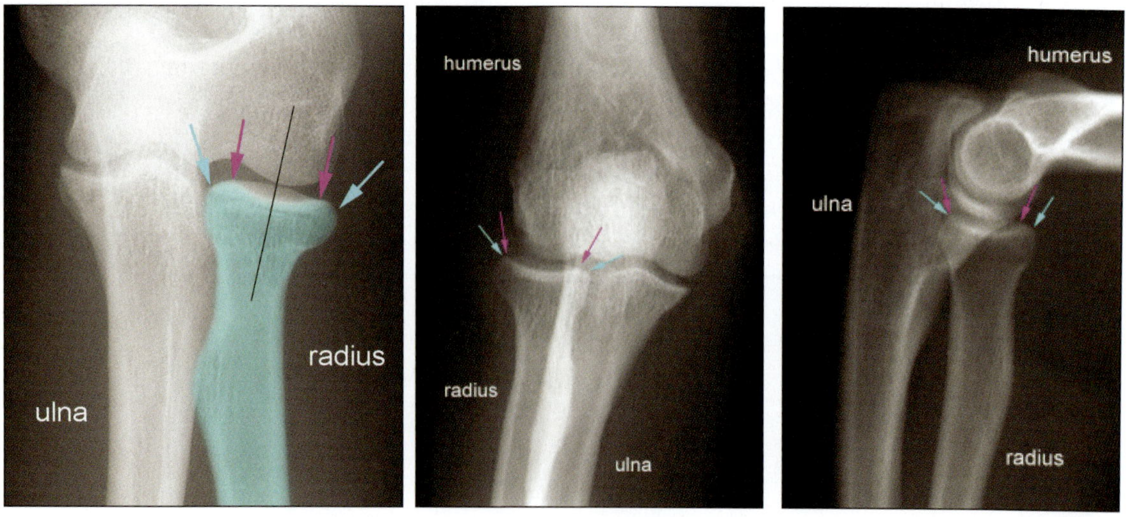

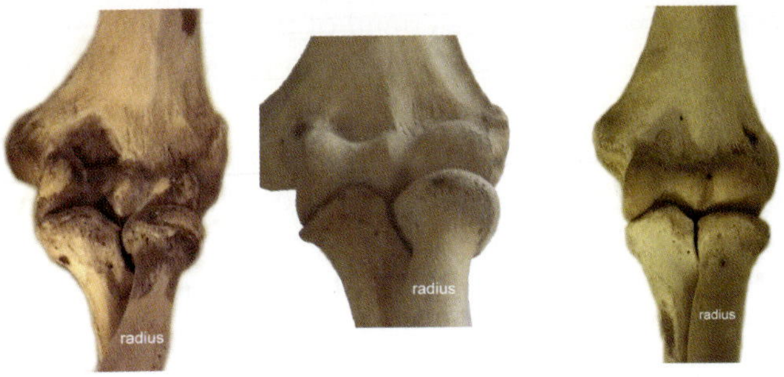


The radial heads of living apes are generally more stout than in humans.
Far left: Gorilla.
Middle: Chimpanzee.
Left: Orang-utan.

The lower end of the radius has two *articular facets* (joint surfaces) relating to contact with the *scaphoid* and *lunate* bones of the hand. Heinrich argues that in the human radius, the articular surface for the lunate is significantly smaller than the corresponding area for the scaphoid, whereas:

> The lunate articular surface of KNM-ER 20419 is larger than the corresponding area for the scaphoid... RE Heinrich et al. 'Hominid radius from the Middle Pliocene of Lake Turkana, Kenya'. American Journal of physical anthropology. Vol 92. 1993. p139-148.

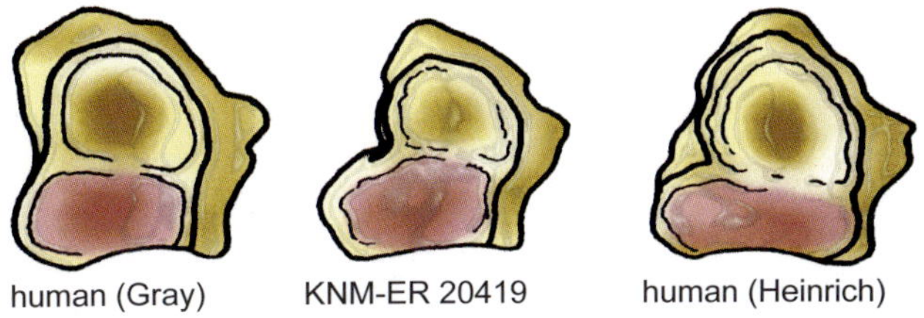


Distal radius.
Far left: Human (redrawn from Fig 39, Gray's Anatomy, 34th edition).
Middle: KNM-ER 20419.
Left: Human (Heinrich. Fig 6).
The scaphoid articular surface is above; the lunate articular surface (shown in magenta) is below.
The features of KNM-ER 20419 are fully consistent with the bone being human.

Although his analysis included only 10 human samples, Heinrich's Fig 6 shows a drawing supposedly representative of a human distal radius, which unlike that of KNM-ER 20419, certainly has a distinctly *small* lunate articular surface. But Heinrich's illustration is far from representative of the human form: Gray's Anatomy shows a significantly larger lunate articular surface. And even just a cursory look at modern human bones and wrist x-rays shows that there is, in fact, a great variation in the relative sizes of the lunate and scaphoid articular surfaces of the radius, and that the lunate and scaphoid articulations can be of roughly equal size.

These X-rays of adult human wrists show clearly that the lunate articulates with a sizeable portion of the radius.

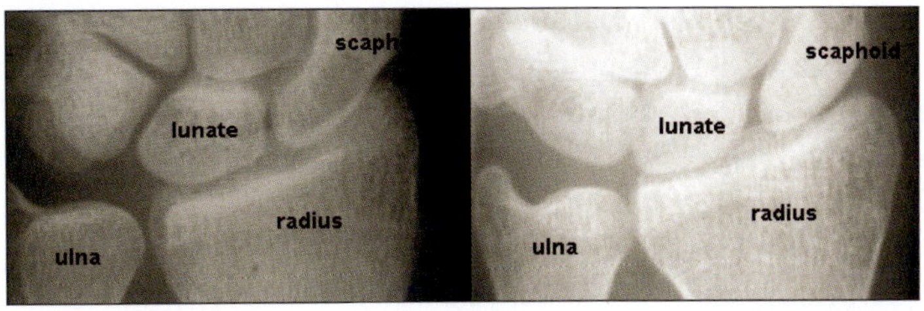


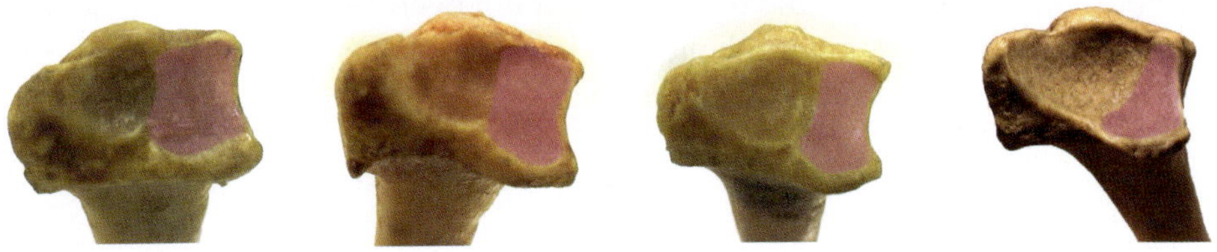

Lower radius. These modern human bones show that the articular surface for the lunate (shown in magenta) can vary from being large (far left) through to small (far right).

One final feature may be noted: where the lower end of the radius articulates with the lower end of the *ulna*, the articular surface is small in KNM-ER 20419, as it is in humans. This feature is distinctly large in living apes.
Therefore, in summary, KNM-ER 20419 is quite unlike the radius of any living ape, and is twice the size of the AL288-1 radius. Futhermore, it is indistinguishable from an adult human bone in size, in general appearance, and in its detailed features. Therefore, in all probability the radius KNM-ER 20419 belonged to a fully human individual living (supposedly) 4.1mya.

### *KNM-KP 29285 tibia*

The fossil KNM-KP 29285 comprises the upper and lower sections of a right tibia which is dated as being about 4myo, being found in Kanapoi on the south-west side of Lake Turkana. The middle section of the bone is missing. The bone is clearly not like that of any ape: as in humans, the upper articular surfaces of KNM-KP 29285 are roughly equal in size and perpendicular to the long axis of the shaft, whereas in chimpanzees they are distinctly asymmetrical and at an angle to the long axis of the shaft. Likewise, the lower end of KNM-KP 29285 is indistinguishable from a human bone.

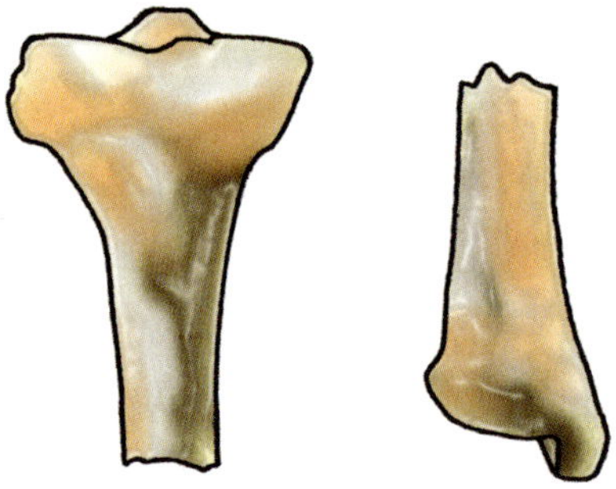

Kanapoi tibia KNM-KP 29285

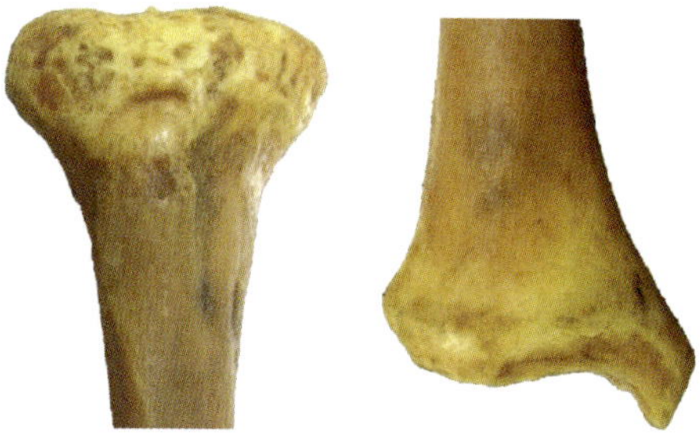

Modern human tibia

Thus the two sections of tibia:

> ...show clearly that its owner walked upright. Meave Leakey. National Geographic. September 1995. p51.

Curiously, the National Geographic article includes no dimensions of the fossil, and, without any scientific basis whatsoever, an illustration of the reconstructed tibia shows it associated with a foot which has a wide space between the great and index toes, suggesting that the individual was half-ape and half-human. This is pure fabrication, but an air of authenticity is suggested to the layman by the inclusion of a leg muscle (labelled with its Latin name *flexor hallucis longus*) and a few notes. It is surprising that Meave Leakey has so readily attributed the tibia to a bipedal ape, since even three years later the fossil had still not even been formally described in detail in any scientific journal, although a brief mention was made in *Nature* in 1995. However, no details are given in that article as to the exact geographical location of

the tibia in relation to the (obviously australopithecine) jaw fragments which were found in the same district. Certainly, there is no obligation to associate the australopithecine jaw fragments with the individual to which the tibia belonged.
In comparison to australopithecine bones, KNM-KP 29285 is a substantial size, being:

> ...larger than the largest from Hadar (AL333-42)...the estimated body weight of this individual is 55kg...somewhat higher than the mean body weight recently estimated for Australopithecus afarensis, and 1.7 times higher than that computed for presumed females. Meave Leakey, Craig Feibel, Ian McDougal & Alan Walker. 'New four-million-year-old hominid species from Kanapoi and Allia Bay, Kenya'. Nature. Vol 376. 17th August 1995. p566.

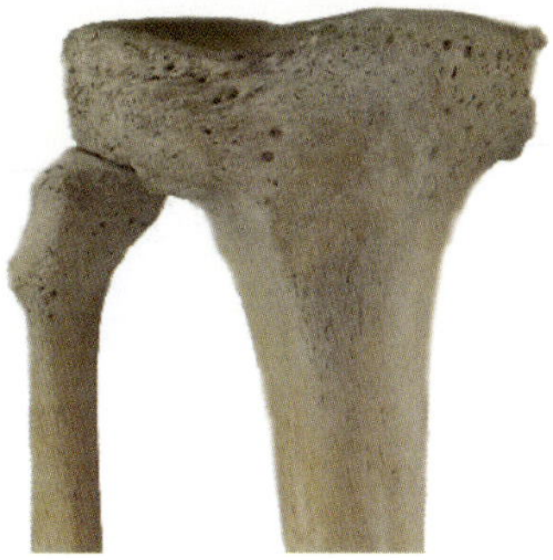

Above: Chimpanzee upper right tibia showing that a large portion articulates with the fibula.
Below: Upper right tibia of the australopithecine AL288-1 is significantly smaller than KNM-KP 29285.

So KNM-KP 29285 is certainly significantly larger than would have been expected had it belonged to an australopithecine. With reference as to where the upper part of the fibula forms a joint with the upper part of the tibia:

> Although the proximal fibular facet is broken away, the small size of the missing area shows that the articulation must also have been small, as in humans. Meave Leakey, Craig Feibel, Ian McDougal & Alan Walker. 'New four-million-year-old hominid species from Kanapoi and Allia Bay, Kenya'. Nature. Vol 376. 17th August 1995. p566.

Although no sizes are given in the article in *Nature*, judging by the 5cm scale included next to the illustration, the plateau of the upper tibia is about 6 cm in its transverse dimension and the whole bone may have been about 26.5 cm long (this is only a guestimate since the middle section of the bone is missing). These dimensions fall within the range of small modern adult humans. Thus, the features of Kanapoi tibia KNM-KP 29285 are consistent with it being a human bone.

### *KP 271 lower humerus*

KP 271 is a fossil of the lower part of a left humerus, which was found in 1965 in Kanapoi, near the southern edge of Lake Turkana. It is listed in the Natural History Museum of London's catalogue as simply:

> 11. KP 271. distal lt humerus. Kenneth Oakley, Bernard Campbell and Theya Molleson (Editors). 'Catalogue of fossil hominids'. Natural History Museum of London. Part 1. 2nd edition. 1977. p59.

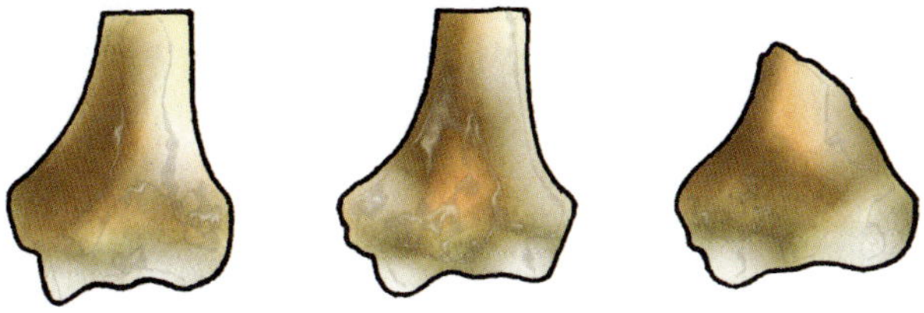

Chimpanzee Human Kanapoi 271

Lower humerus bones

Lower end of human humerus

There is no doubt that the Kanapoi humerus is considered to be extremely ancient, being (supposedly):

> ...4 to 4.5 million years old... Henry McHenry. 'Fossils and the mosaic nature of human evolution'. Science. 31st October 1975. p428.

> Northern Kenyan stratified sequence of early Pliocene age, between 5 and 4my according to K-Ar dating of associated basalts and faunal analysis...Included in the surface collection from the site is an adult hominid distal humerus... Ian Tattersall, Eric Delson, and Kohn Van Couvering. 'Encyclopaedia of human evolution and prehistory.' Garland Press. 1999. p295.

Although the fossil was found within surface erosions, the specimen seems to be genuinely very old:

> The fragment was found on the surface of exposures at the base of the west side of Narinangoro Hill. Color, hardness, and degree of mineralization agree with those of numerous specimens collected in situ in the sediments. The excellent state of preservation - the fragment shows no significant postmortem damage other than the break that separated it from the remainder of the bone - rules out the possibility of derivation from later deposits that may once have been present in the vicinity above the capping lava... Bryan Patterson & W.W. Howells. 'Hominid humeral fragment from Early Pleistocene of Northwestern Kenya'. Science. Vol 156. 7th April 1967. p64.

Furthermore, the specimen shows no feature that excludes it from being human. It is:

> ...indistinguishable from modern Homo sapiens. Henry McHenry. 'Fossils and the mosaic nature of human evolution'. Science. 31st October. 1975. p428.

The Kanapoi humerus is 60.2mm wide across its maximum width. The same measurement made of the humerus in 40 humans and 40 chimpanzees gave means of 58.02mm and 64.07mm respectively. When a series of seven dimensions were compared, it was found that:

> In these diagnostic measurements Kanapoi Hominoid 1 is strikingly close to the mean of the human sample. Bryan Patterson & W.W. Howells. 'Hominid humeral fragment from Early Pleistocene of Northwestern Kenya'. Science. Vol 156. 7th April 1967. p65.

*Also referred to as 'robust australopithecines'.

Furthermore, when compared to a humerus from *Paranthropus robutus** (whose maximal width was only 53.6mm) found in Kromdraai, South Africa, it was found that:

> ...the Paranthropus specimen is distinctly less like a hominine than is the specimen of Kanapoi Hominoid 1...To us the most interesting fact is the difference of form and size of the new fossil from the Kromdraai fragment identified as Paranthropus robutus...it is quite unlikely that Kanapoi Hominoid 1was a member of the same lineage; **although earlier in time it is more hominine**. (My bold). Bryan Patterson & W.W. Howells. 'Hominid humeral fragment from Early Pleistocene of Northwestern Kenya'. Science. Vol 156. 7th April 1967. p66.

In other words, although it is (supposedly) older than australopithecine fossils, the Kanapoi humerus looks human. However, the most surprising thing about Patterson and Howells' report is that although the Kanapoi humerus is indistinguishable from a modern human bone, for some extraordinary reason they conclude:

> Kanapoi Hominoid 1 may prove to be Australopithecus [africanus]. Bryan Patterson & W.W. Howells. 'Hominid humeral fragment from Early Pleistocene of North-western Kenya'. Science. Vol 156. 7th April 1967. p66.

But, if the Kanapoi humerus is physically indistinguishable from a modern human bone, then in all probability, *it must be more likely to be human than anything else*.

Thus, we have seen that there is, in fact, good evidence (according to standard dating) that modern humans were alive at least 4 to 4.5 million years ago. Furthermore, the fossil record shows no evidence for australopithecines having evolved from a common ancestor of chimpanzees, or for modern humans having evolved from australopithecines.

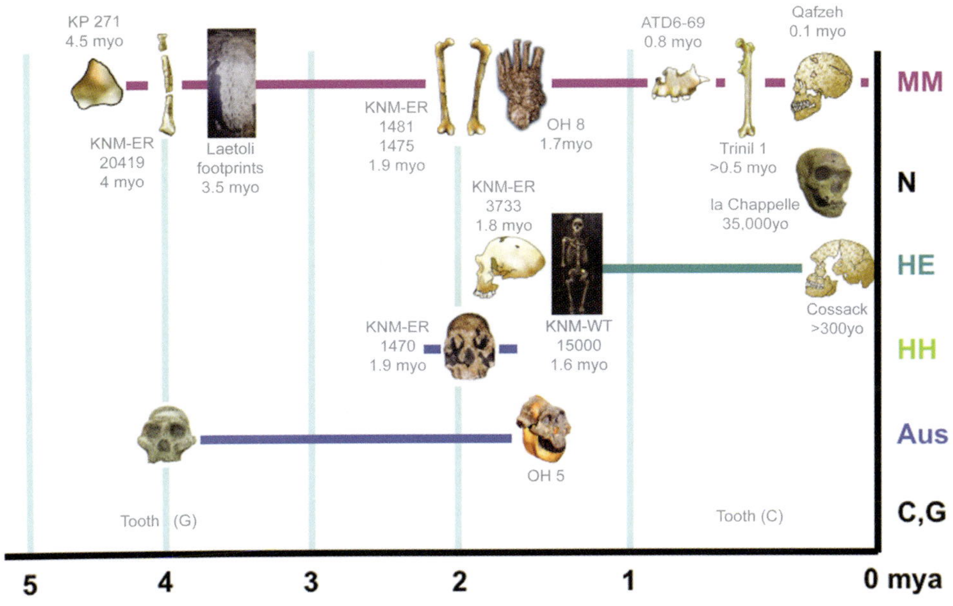


An overview of the fossil record of modern Man (MM), Neanderthals (N), Homo erectus (HE), Homo habilis (HH) and Australopithecines (Aus), chimpanzees and gorillas (C,G). The picture does not fit with the notion that humans evolved from apes via australopithecines. Human fossils are as old as the earliest australopithecine fossils. Note the *Coelacanth Principle* shows that all creatures could have lived contemporaneously in the past.

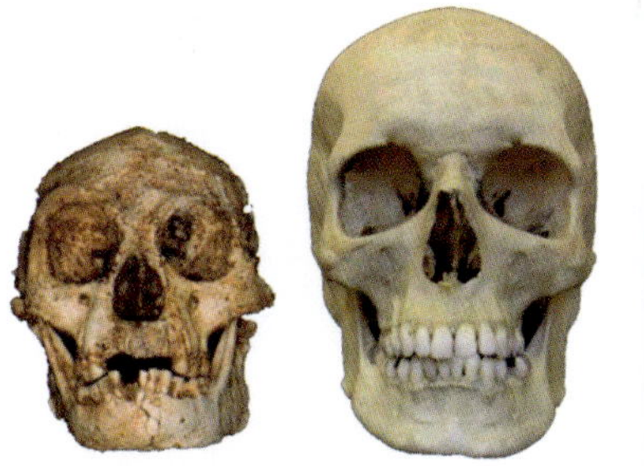

Left: LB1 skull from Flores.
Right: Normal human skull.

### *Flores LB1*

Despite the solid and comprehensive evidence against evolution, there are still those who promote the notion that humans are modified apes. And so in 2005, newspapers ran headlines about human-like fossils from Flores in Indonesia, dated at around 12-18,000yo, which had brain volumes in the region of just 380ml. One individual (LB1, a female) would have stood just 1 metre (3ft 3in feet) tall and weighed about 25 kg (55 pounds), and was around 30 years old when she died. LB1 was found in the same sediments that contain stone tools, evidence of the use of fire, and the bones of dwarf elephants, giant rodents, and komodo dragons*.

**M Morwood et al., 'Archaeology and age of a new hominin from Flores in eastern Indonesia'. Nature. Vol 31. 28th October 2004. p1087–1091. Also pages 1029, 1043 -1044 and 1055-1061.*

Note that since the Flores bones were not fossilised but had the '*consistency of wet blotting paper*', it has been suggested that the bones may be as little as 40 years old.

Pygmy forms are found in isolated populations all over the world, such as this pygmy elephant (above) and hippopotamus (below, with normal-sized skull in front) from Malta.

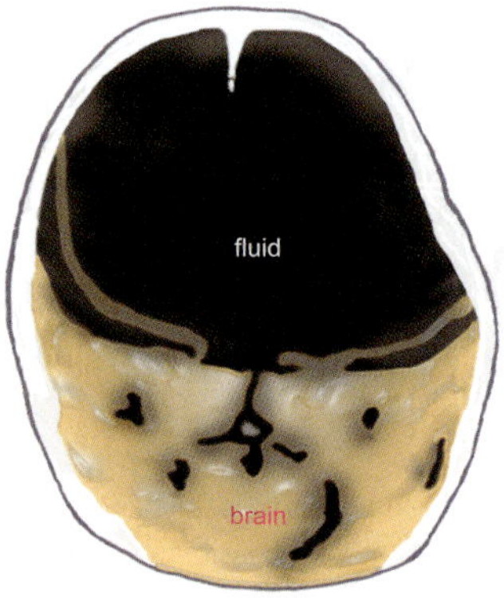


***A small brain is compatible with normal human intelligence**

Brain scan of a normal 55 year old man whose brain only occupies the back half of the skull. The large black area represents fluid which has compressed the brain and which may have been present from birth.

*Redrawn from British Medical Journal. Vol 327. 25th October 2003. p998.*

Because of the small skull size and the lack of a protruding chin (despite the fact that many people do not have a protruding chin), the Australian part of the research team originally suggested that LB1 represented some ancient pre-human ancestor. This was despite the fact that the casts of the skulls show a distinctly *non-ape* brain shape, and the face and cheek bones are distinctly human in proportion. Others thought the Flores fossils represented humans with small abnormal brains (*microcephaly).* However, the newspaper headlines were premature since the Indonesian members of the team claimed that it is not unusual to see very small Indonesians living in some isolated areas even today. Indeed, when a Daily Mail reporter went to the village of Rampapasa in Manggarai, Indonesia, he found that the women ranged in height between 4 feet 2 inches and 4 feet 6 inches:

> ...just a mile from the hobbit cave in Liang Bua where 80% of the 300 residents were 'mini humans'...Intriguingly, the children appeared of normal size. I was told that they stop growing in their teens. R Shears. Daily Mail. 7th May 2005.

Lee Berger et al have recently reported the finding of a number of human skeletons as small as LB1 in the Palauan archipelago in Micronesia, east of the Philippines. However this is old news:

> "The authors are apparently unaware of published data on living Filipino 'pygmies' that are just as small as these scrambled individuals from Palau", says William Jungers. Colin Barras. 'New bones suggest "hobbits" were modern pygmies'. Newscientist.com. 11th March 2008.

Despite all the 'is it human or is it ape-like' debate, there is simply *no scientific basis* to classify the Flores fossils as being subhuman just on the basis of their small size. If dogs can vary so much in size (the smallest chihuahua can be as small as 6 inches long and weigh just two pounds but is still the same animal as a wolf) then why not humans? In fact, a smaller-than-gorilla brain can be seen in some normal adult humans*. In addition, the tiny (2ft 9in, 9kg) but essentially-normally-proportioned 19-year-old man, Aditya Dev, from northern India, has a normal intellect yet has a brain that is *extremely* small indeed. Even smaller at 1ft 11in tall and weighing just 9 pounds is 15-year-old Jyoti Amge, who attends her local high school, in Nagpur, India, where she studies alongside classmates of her own age. However, to complicate matters, LB1's wrist bones seem to differ from *normal*-sized living humans:

> ...the hobbit's wrist bones looked just like those of a chimpanzee, or an early hominid such as Australopithecus, and had none of the specialisations for grasping that are seen in the wrist bones of modern humans. Bob Holmes. 'Hobbit wrist bones suggest a distinct species'. Newscientist.com. 20th September 2007.

However, it is pure conjecture to say LB1 was an ape simply from the appearance of some of the wrist bones without comparative analysis of the wrist bones of people with *microcephaly* or *cretinism* (see below). It has also been suggested that LB1's feet were:

> ...unusually large for its metre-high frame. H. floresiensis probably had to bend its knee further back than modern humans do, resulting in a sort of high-stepped gait. [The] big toe was quite short compared with the others...However, the shape of the toes, even the short big toe, is like modern human ones, says Bill Jungers, an anthropologist. E Callaway. 'Flores "hobbit" walked more like a clown than Frodo'. Newscientist.com. 16th April 2008.

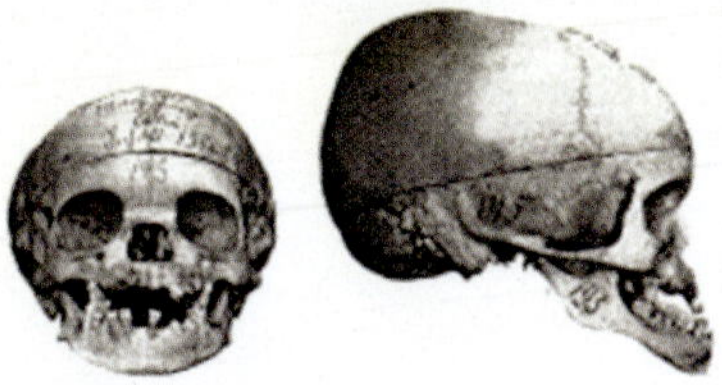

Above: Skull of a 25yo man of height 1.04m with cretinism (from Krantz, 1912).
Below: LB1 from Flores.

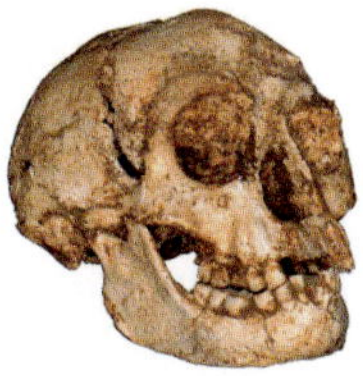

However, recall the fact that living shorter people can have the same size feet as significantly taller people (see page 403). And there is evidence from Indonesia of individuals and skeletal remains from the early part of the 20th Century which suggests that the Flores individuals were human beings with reduced function of the thyroid gland from birth (cretinism):

> We show that the fossils display many signs of congenital hypothyroidism, including enlarged pituitary fossa, and that distinctive primitive features of LB1 such as the double rooted lower premolar and the primitive wrist morphology are consistent with the hypothesis. P Obendorf et al. 'Are the small human-like fossils found on Flores human endemic cretins?' RSPB magazine. Vol 275. No.1640. 7th June 2008. p287-296.

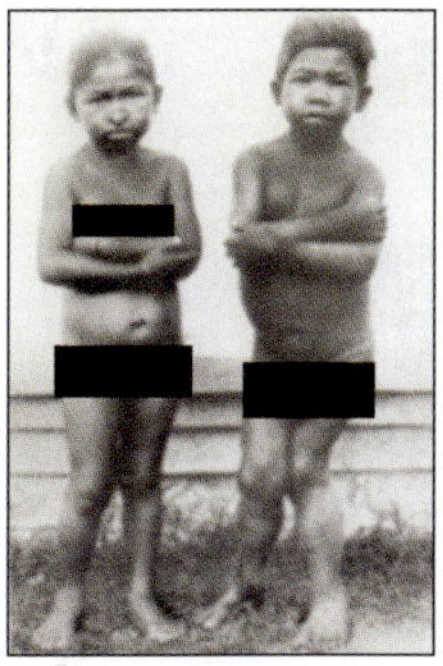

A young woman and man from Java with cretinism. Note the relatively large feet (compare the feet with the length of the tibia).
(Attributed to L Earland, 1932)

So, Indonesian individuals with cretinsim from the early 20th century showed clear evidence of very short stature, unusual wrist bones and relatively large feet. Incidentally, they had a normal intellect. Clearly, since humans *did not* evolve from apes, LB1 can only represent either a small human or a small ape.

In the absence of definitive clarification, the specimens should not be given the derogatory names 'Homo floresiesis' or 'hobbits' as this gives the misleading idea that these individuals *definitely were not* human beings. Since the possibility remains that the Flores individuals were fully human, it is more respectful to simply refer to the specimens by their catalogue numbers, e.g. LB1. Since living humans can be *smaller* (in both height and brain size) than the Flores individuals, and if, in the final analysis, the Flores individuals are deemed to be apes, the Flores findings are irrelevant to the matter of human origins.

In conclusion, we have seen that the fossil evidence actually paints a completely different picture from that which is commonly portrayed. Instead of man evolving from apes via crude-looking ancestors, the evidence points to populations of ancient human beings passing through some morphological changes (whether from inbreeding and/or disease) before gaining the modern human form.

To put it another way, the fossil record contains only human and non-human bones, with no evidence for any intermediaries. Apes have always been apes and Man has always been Man.

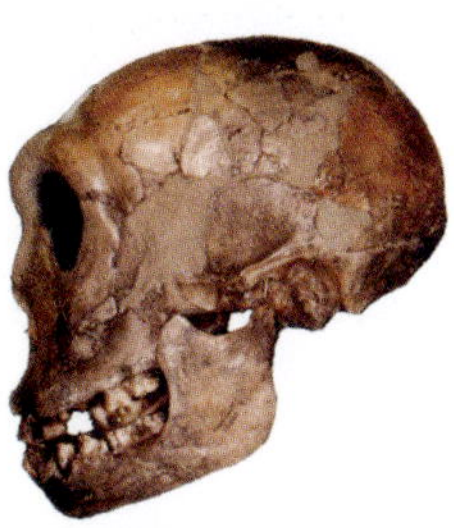

Homo erectus

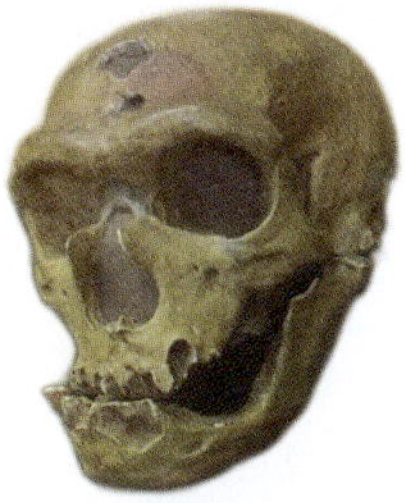

Neanderthal

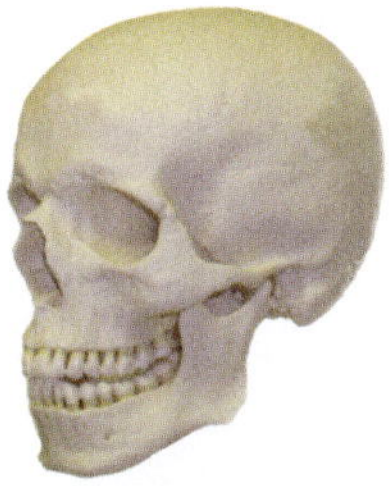

Modern Man

Fully human

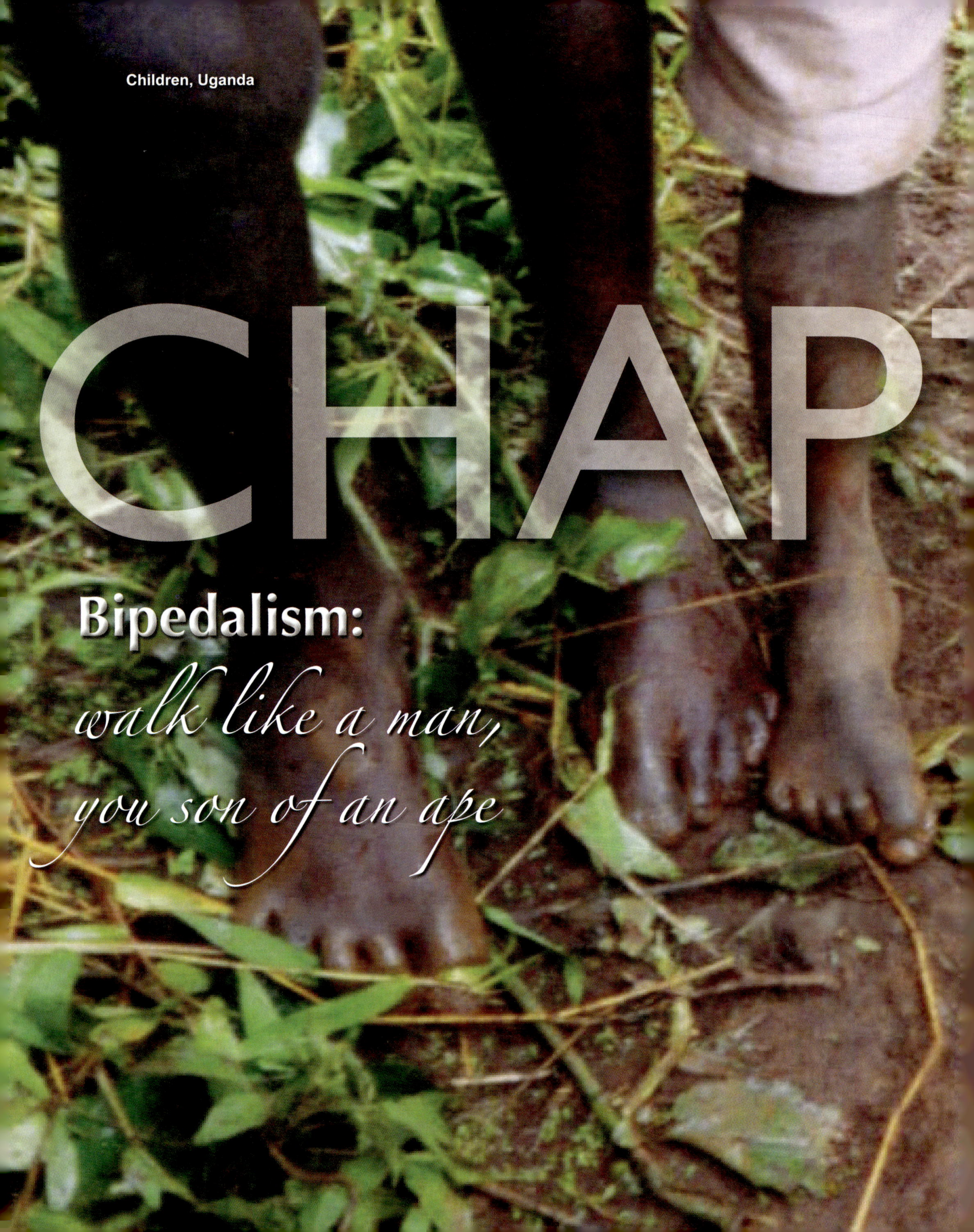

Children, Uganda

# CHAP

## Bipedalism: *walk like a man, you son of an ape*

have been held perhaps one metre high, since it would not have been fully upright in its gait. Such minor differences in height could have had no significant advantage in remaining cool.

Suppose an ape ancestor acquired a mutation that left it hairless. Before a lineage of hairless apes could subsequently appear, the hairless ape would have to mate increasingly, and eventually exclusively, with other hairless apes. But chimpanzee behaviour is such that virtually all members of a troop continually mate with each other, and there is no reason to expect that any supposed early human ancestor would have behaved any differently. This sexual activity would tend to *homogenise* the characteristics of the troop, rather than produce distinct offshoots. Since hairlessness offers no particular survival advantage, less hairy individual apes would simply mate with hairy apes, resulting in hairy offspring.

Furthermore, humans have distinctly different hair distribution patterns to apes. For example, unlike gorillas, which have a tendency to have *little* hair on their chests, men, in general, have hairy chests. Unlike men, male apes do not have much hair on the upper lip; in addition, any hair that apes do have on their heads, upper lips and face does not grow continuously as it does in humans. Furthermore, the eyebrows of humans are bushy but those of apes are relatively scant. Since none of the human patterns can offer any particular survival advantage over the ape pattern, they are unlikely to have arisen as a result of natural selection.

Proponents of the heat-loss theory assume that losing hair is a good way to deal with hot conditions. In fact, there is good evidence to show that this is not true. For example, hair is a very good insulator and is a major factor in reducing heat gain in camels which obviously manage very well in desert conditions. In one experiment in which a camel was shorn:

> …the water expenditure…was increased by about 50%. Knut Schmidt-Nielsen. 'Animal physiology'. 5th edition. Cambridge University Press. 1977. p274.

So, it would be more reasonable for a savannah ape to *retain* its fur as insulation against the sun's heat, rather than to lose it.

Could evolving bipedalism have given australopithecines an advantage in seeing and escaping from predators out in the open savannah? When lions stalk their prey in grasslands, they do not just walk or trot along, but they creep along on their bellies. In this manner they can be well hidden by grass that is less than one metre high. So, it does not follow that by standing upright an australopithecine would be any more able to notice a stalking lion than if it remained on all fours. Indeed, by walking quadrupedally an australopithecine might be *less conspicuous* to a predator.

Furthermore, being both much smaller than a man and not fully upright, any australopithecine most certainly could not have run as fast on two legs as humans can. In addition, whether running on all fours, or climbing trees, they would have been slower and less agile than baboons. Indeed, baboons can run on four limbs *faster* than any man can on two. Incidentally, whether in trees or on the ground, gibbons walk upright more than any other ape, yet they have not developed any great intelligence.

Therefore, since humans cannot run as fast as a baboon, are less agile in trees than

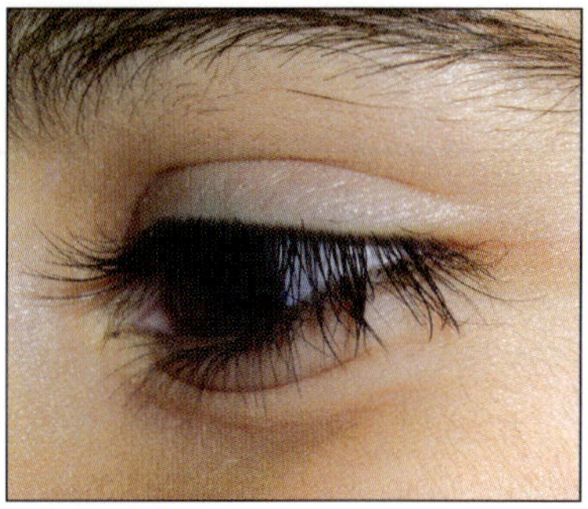

Bushy eyelashes in a normal child. Some humans carry mutations that result in their having a great excess of head and body hair (hypertrichosis). It has been suggested that this shows a reversion to our (supposed) ancestral ape form. However, affected children have very bushy eyebrows, very long eyelashes and moustaches - all characteristics which are not seen in living apes and which are unique to humans. Furthermore, hypertrichosis does demonstrate that hairiness is a feature which is independent of other features that characterise human-ness. There is no disadvantage in having hair and living in the savannah, and there is no reason to expect that a (supposedly) evolving upright ape should lose its hair as a side effect of it becoming (for example) bipedal, or developing a human-like larynx.

Camels would lose 50% more water if they did not have a coat of hair to protect them from the sun's rays.

Above: If they can, just like these wildebeest and zebra, savannah animals seek the shade to avoid the midday sun.

Right: Baboons manage very well in the savannah and their hair protects them from the sun.

Below: As with this baboon, there is no disadvantage in a *hoping-to-become-a-man* ape keeping its canines as a means of protection from predators.

Gibbons stand upright and walk bipedally in a more convincing manner than any other ape.

This gorilla demonstrated a premeditated display of annoyance against the author.
Above top: He picked up a piece of wood while he was still some distance away.
Row above: After approaching his target, the wood was then thrown with an outward flick of the wrist. Immediately on throwing the missile he ran away. In the knuckle-walking position, it is not possible for an ape to throw using an over-arm technique because of the limitation dictated by the shoulder joint. This means that although knuckle-walking apes can throw objects, they cannot do it with any significant force. A human throwing the same object would stand sideways, twist the hips, bend the spine sideways, and throw with an over-arm technique, utilising the additional leverage offered by hip and spine rotation.
Judging by the very limited throwing capability of the gorilla, it is implausible that in the open savannah a (supposedly-evolving) partially-upright-walking australopithecine would have been able to defend itself adequately against predators with sticks or stones. It would have been better off retaining large canines as in baboons.

An ape does not need to be bipedal to carry objects. This orang-utan is well able to walk with a tomato in its right foot, a lettuce in its left hand and another vegetable in its mouth.

any ape, and are weaker than a gorilla, if australopithecines did evolve bipedalism, this would have been a serious *disadvantage* to them. No chimpanzee or australopithecine could outrun a lion, leopard, hyena or wild dog. And when it comes to confrontation with a predator on the open savannah, a baboon, although a quadruped, has large canines and would be more able to hold its own against a hyena or leopard than any australopithecine. Thus, since there can be no disadvantage to having fierce canines, and we know that tooth size is not linked genetically to intelligence or strength, any (supposedly) evolving savannah-living ape would be likely to retain its fangs as additional protection. Furthermore, and very conveniently, fangs also allow carcasses to be torn apart without the need for stone tools.

What about bipedalism evolving so as to allow apes to carry tools? In fact, apes, baboons and monkeys can all manage to carry objects in one arm whilst running quickly on three limbs. And regarding a macaque which has been trained to climb up a tree and drop coconuts to the ground for its waiting owner:

> The fallen coconuts are now retrieved, and walking on two legs, it carries each one in its arms to a waiting basket... John Downer. 'Lifesense'. BBC Books. 1991. p192.

So, if walking on three or two legs while carrying objects is well within the capability of many primates without the necessity of true bipedalism, why could ancient apes not have learned to carry and use stone tools? The earliest tools would have been just stones and sticks, which can be easily carried by apes. However, although they can hammer things quite well, monkeys and apes cannot throw sticks or stones with any significant force or great accuracy. Apes use a relatively clumsy palm grip when throwing missiles, yet (although we generally use a more precise finger grip), if necessary, humans *can* manage to throw missiles quite accurately using a palm grip. Apes have relatively long fingers in relation to their thumbs but this does not make them less dextrous than humans. Indeed, macaques and apes can manage to manipulate objects as small as fleas while grooming each other. Humans with

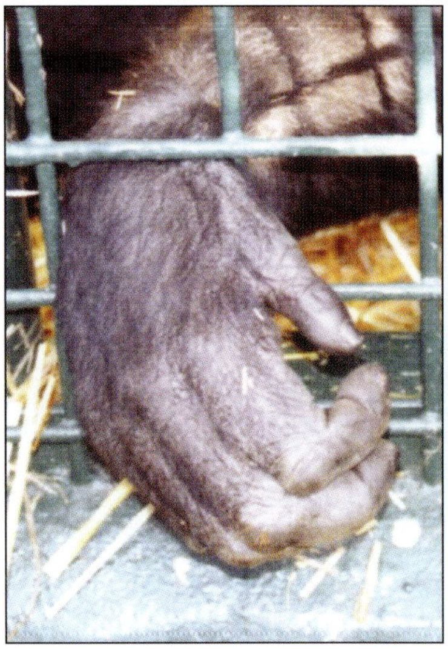

Right: Gorilla hand showing an effective opposition of the thumb and index finger.
Far right: Chimpanzee using a palm grip to hold an apple.
(The arrow points to its short thumb)

abnormally short thumbs are quite capable of using normal everyday tools. The illustration, below left, shows the hand of a man aged 72 years who had half his thumb amputated after it had been badly damaged in an explosives accident over 50 years previously during the Second World War. He has managed quite well over the years, despite having a thumb which is not only short but which is less flexible since it also lacks a joint along its length.

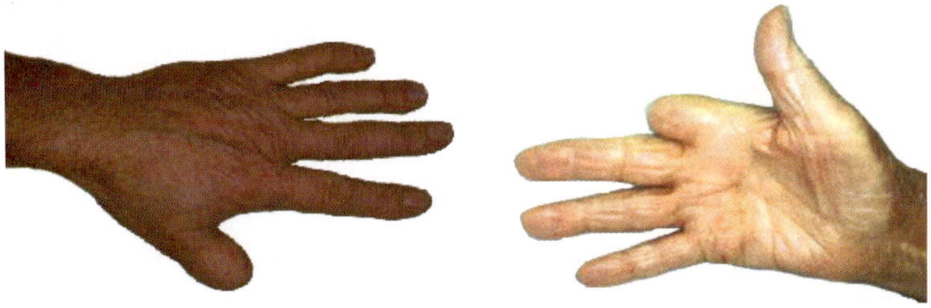

Left: A partially amputed thumb or index finger is not a problem to these men.

In the bush, a simple hand axe would have been manufactured whilst an individual was squatting or sitting, and the tool would have been used by the individual whilst squatting or kneeling over a carcass. Since apes are well able to accommodate such postures, they could both manufacture and use such a tool with their thumb and finger lengths unchanged, by modifying their brains and without having to become bipedal. So from this we can see that the key to the production and use of even a simple hand axe is not a human body, but a human *brain*. Therefore, the supposed evolution of bipedalism has nothing *whatsoever* to do with a (supposed) parallel evolution of the ape brain.

Could deforestation have forced the evolution of bipedalism in apes that were tempted into the newly-emerging savannahs? We saw when discussing the Laetoli footprints that the climate, plant growth, and animals in east Africa must have been largely the same (supposedly) 3.5mya as they are now. Furthermore, central Africa has a huge tropical rain forest, and in area, Zaire alone is the size of the whole of western Europe. So, more likely, any gradual deforestation in east Africa would have caused forest apes to move *further into* the abundant central African forests rather than to venture into open grasslands. And their survival in forests would have been more likely than in the savannahs, since this would have required no fortuitous genetic changes at all.

Having examined and found unsatisfactory a number of commonly proposed explanations for the origin of bipedalism*, we need to look at ape and human feet in more detail. Both chimpanzee and gorilla feet are enormously different from the human foot. Although in genetic terms the chimpanzee is commonly considered the closer of the two to humans, the chimpanzee foot is, if anything, less like the human foot than that of the gorilla. Yet the gorilla is effectively a quadruped and rarely walks bipedally, except for short distances. On the other hand, amongst all the apes, the gibbon's foot is the least like the human foot in appearance, yet the gibbon walks bipedally more often (and more convincingly) than any other ape. Despite the uniqueness of the human foot, some argue that the fossil evidence shows

Below: Demonstrating that fine manipulative movements are quite possible with a partially amputated thumb. Note that the presence of the index finger nail is more important than the length of the thumb when picking up very small objects such as this staple.

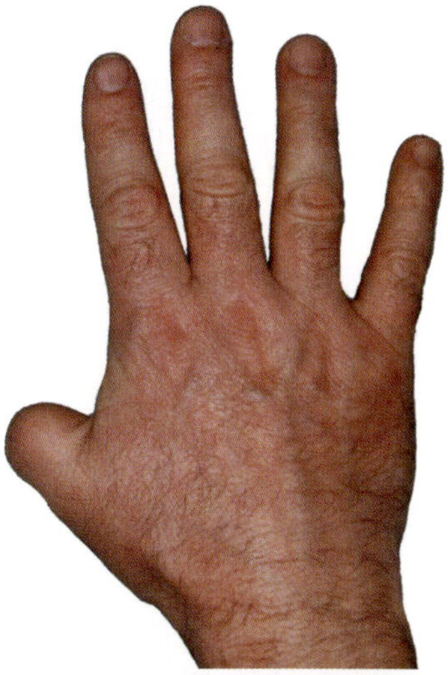

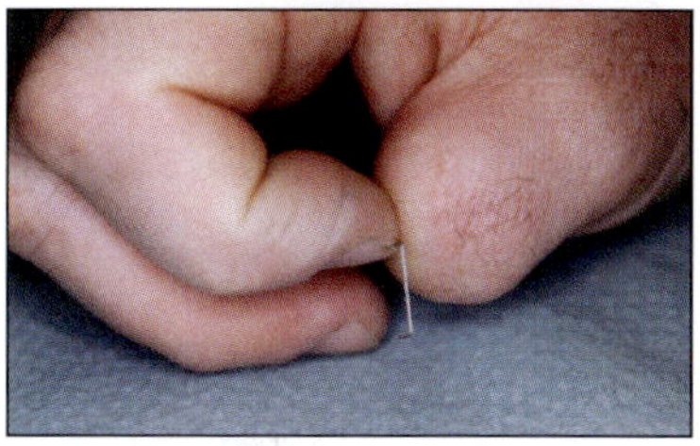

* Another theory is that apes became bipedal as a result of passing through an aquatic phase in their history. One serious problem with this theory, however, is that chimpanzees are generally afraid of stepping into water, and although they may wade, they are quite unable to swim. Furthermore, although macaques commonly swim and forage underwater, nevertheless, despite their aquatic capabilities, they have retained their fur and have not acquired any skeletal modifications. Indeed, macaques are no more bipedal than other monkeys.
More importantly, the aquatic theory still requires the evolution of major structural modifications, the plausibility of which is discussed in the main text.

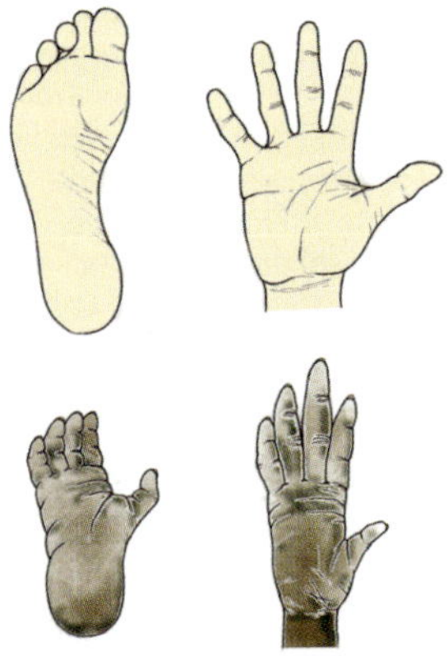

Human and chimpanzee foot and hand.

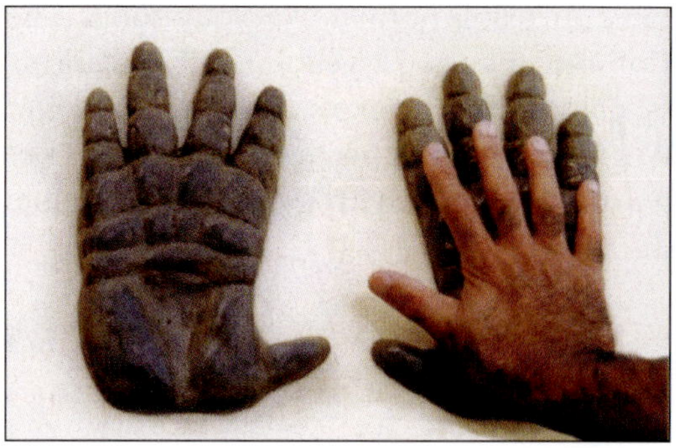

Casts of gorilla hands with author's hand superimposed.

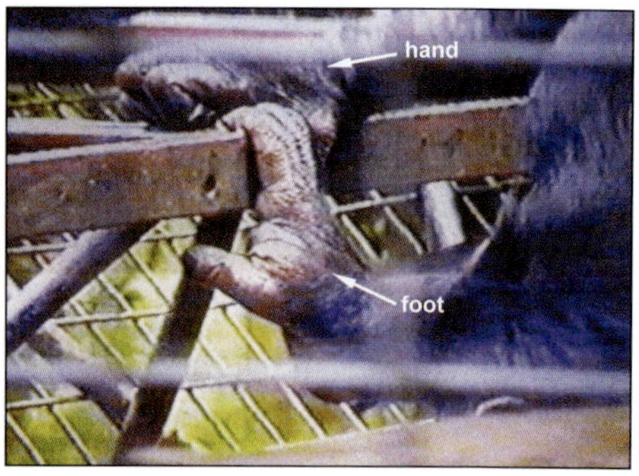


Chimpanzee hand and foot.

conclusively that the leg and foot bones of australopithecines were intermediate in form between modern apes and modern Man:

> Some of their foot bones were intermediate in form between ape and human and suggest some degree of aboreality, as well as terrestrial bipedality. Dr Ronald Clarke, quoted in the Daily Telegraph. June 1977.

Gibbon hand and foot.

The orang-utan foot is very much like its hand.

But although the australopithecine foot shows a mixture of human-like and ape-like features, this does not oblige us to accept that it arose as a *modification* of an ancestral ape foot, and that in turn it became modified to become a human foot. As we have already seen a number of times, any supposed modification to any biological structure or system should be *possible*, and should also be associated with a *reasonable degree of probability*. And furthermore, for natural selection to operate, any supposed modification should show a *significant* advantage over any (supposed) previous state.

It might be supposed that by just altering the position of the thumb of the ape foot, making it point forwards instead of sideways, and by making a few other alterations, it would be a simple matter to evolve the human foot from an ape foot. But the problem is that the human foot is a *very* complicated structure, and differs from the ape foot to a much greater degree than is commonly understood.

The joint at the base of the human great toe metatarsal (the *medial cuneiform-metatarsal joint*) allows very little movement, and the head of the great toe metatarsal is firmly fixed to the head of the index toe metatarsal by ligaments. However, the joint at the base of the thumb metatarsal of the ape foot is saddle-shaped, and the head of the thumb metatarsal is not held rigidly to its neighbour by ligaments - this allows for flexibility and grip in the same way as the thumb in the hand. Unlike in the ape foot, the bones of the human foot are shaped to form natural arches which are stabilised by special ligaments. These arches, which run both longitudinally and transversely, make for more efficient walking and running. All the features of the foot - such as the lengths of individual bones, presence or absence of ligaments, shape of joints, tendon attachments, tendon control etc. - are subject to a number of genetic controls, some of which are linked to each other, and some of which are completely independent of each other.

There are a number of hereditary conditions in humans which show that the devel-

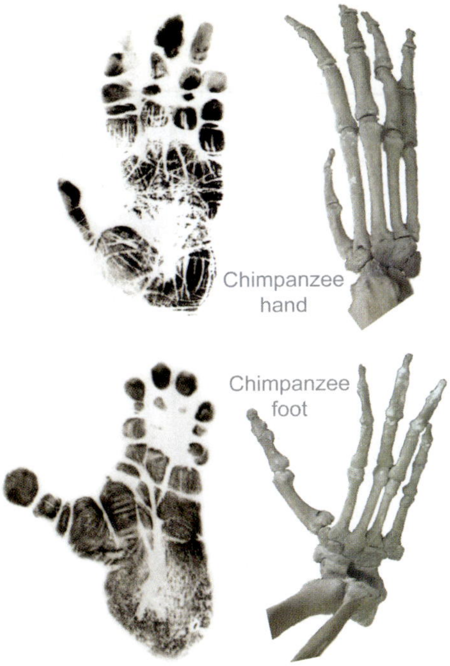


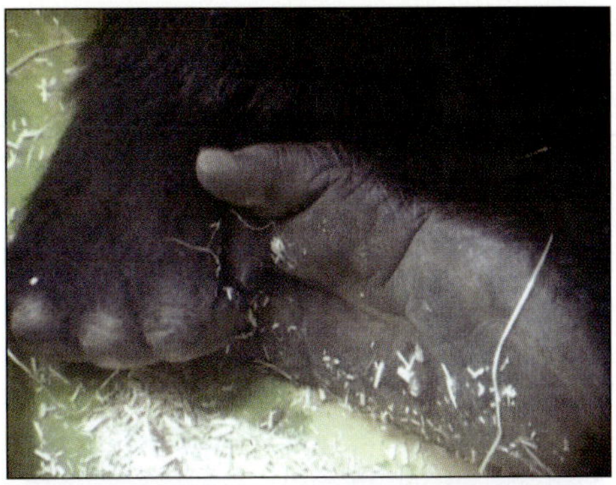

Gorilla foot

Gorilla feet. As in all great apes the great toes are highly flexible and extend sideways like thumbs. Unlike in human feet, the soles are flat and lack arches.

## Could an ape change into a human being?

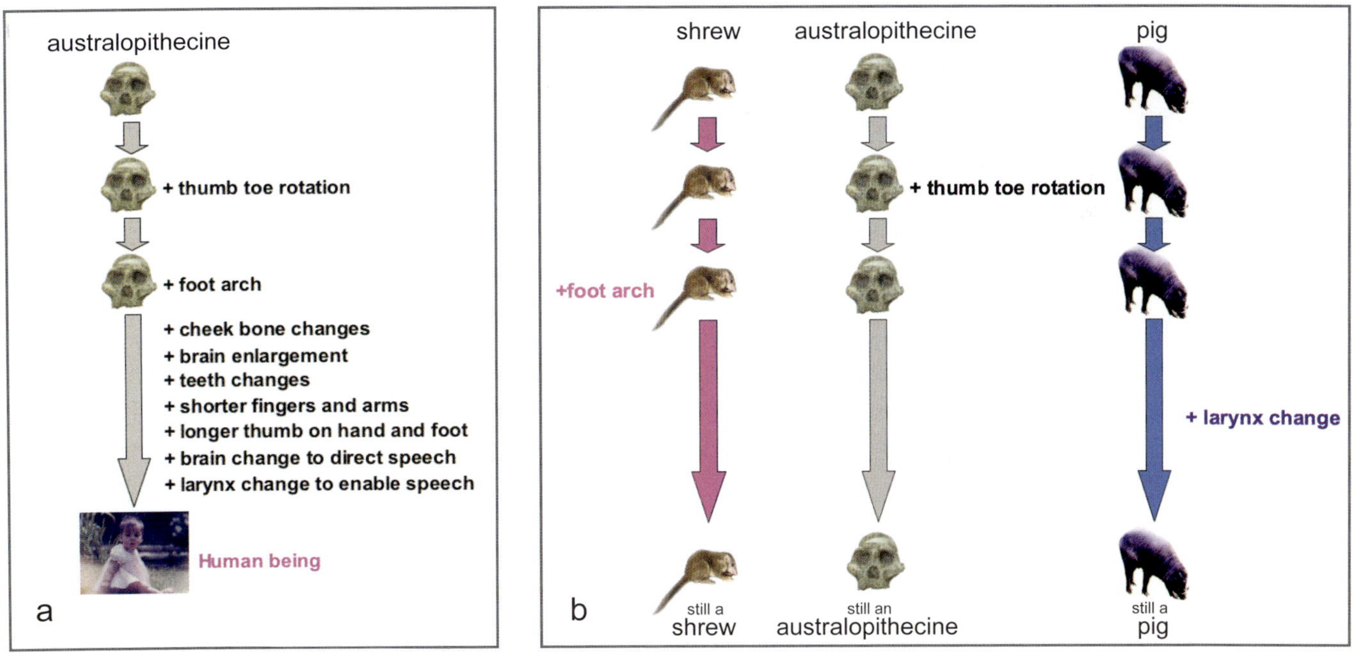


Any australopithecine that was hoping to become a human being would have had to acquire a whole series fortuitous mutations in its direct descendants (a). However, the *WYWYWG Principle* shows that this would be quite implausible. Since mice and humans have 99% of their genes in common and most mammals will have most genes in common, it would be more likely* that different features would appear in completely different creatures, at different times in history, and in different parts of the world (b), than for any one lineage of australopithecines to acquire a whole *series* of new features to become a human.

*But even this is implausible.

opment of different fingers and toes is under the control of different sets of genes. For example, in the presence of a ring D-group chromosome (in which the ends of a normal chromosome are joined together to form a circular structure), the thumb is lost altogether but the great toe is unaffected, showing that the development of the human thumb is controlled by different regulatory genes to those controlling the development of the great toe.

A number of different genes are responsible for the development of *short fingers*. Some conditions affect only certain fingers; some affect all fingers and toes; some only partially affect certain fingers; and some affect only the thumb and great toe. Some other gene defects cause abnormally short metacarpals and metatarsals.

Abnormalities of other genes result in the fusion of fingers and toes. However, not all digits need be affected. For example, the ring and little fingers can be fused without any abnormality of the toes; the middle and ring fingers can be fused in association with fusion of the index and middle toes; the middle and ring fingers can be fused in association with complete extra fingers; and the ring and little toes can be fused in association with complete extra toes.

Since we have a large proportion of our genes in common with apes, we should expect the same sort of patterns of abnormalities to affect apes too. Clearly then, for an ape foot to evolve safely into a human foot, any (supposed) mutations should not cause any abnormalities in the ape hand, or in the rest of the foot. But so many factors are linked in the hands and feet that for a *series* of *necessary* changes to be confined essentially to the foot *alone* would be implausible.

In addition, many genes have multiple actions at different sites and at different times during embryological development, so that following any particular mutation, even if innocent non-damaging effects are produced in one part of the body, damaging effects may occur elsewhere. In fact, most of the abnormalities described above are associated with a variety of generalised diseases, many of them serious. And there are many others. For example, in *von Waardingburg's syndrome* there is a harmless loss of colour in the hair on the front of the scalp (resulting in a *'white forelock'*) but this is associated with faulty gene expression elsewhere. A variation of the syndrome results in the absence of the metatarsal of the great toe, associated with an abnormally-shaped skull, asymmetrical eye sockets, squint, a long narrow nose, and short digits. *Tuberous sclerosis* is another congenital condition in which an innocent-looking abnormality of the skin of the cheeks is associated with mental deficiency, epileptic seizures and tumours at various sites in the body.

However, suppose, for the sake of argument, that the human foot *did* evolve from an ape foot; then, following some occasional mutation in humans, we might expect to sometimes see genetic 'throw-backs' resulting in some ape-like feature. But this does not happen - the mutations that do occur in humans result in *abnormal* hands and feet. For example, the *lobster hand (cleft hand) syndrome* is a dominant mutation resulting in a deep cleft dividing the hand into two parts. Another genetic abnormality, the *two-toed syndrome*, results in an abnormal bulbous foot which has a deep cleft dividing it into two parts, with severe malformation of the toes either side of the cleft. As a result of inbreeding, the *Wadomo* tribe in Zimbabwe and the *Kalanga* tribe of Botswana both have a high incidence of the condition. In these

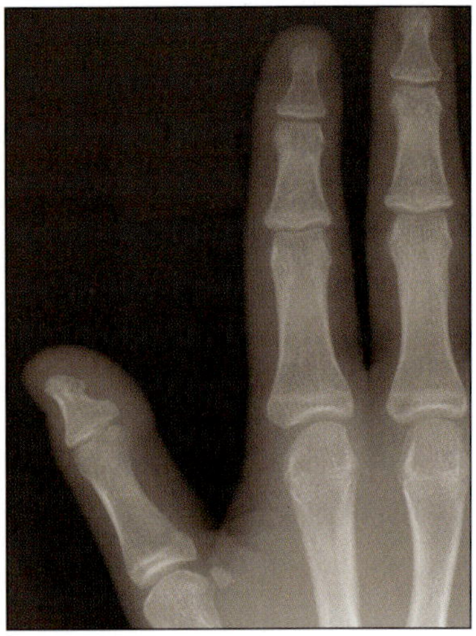

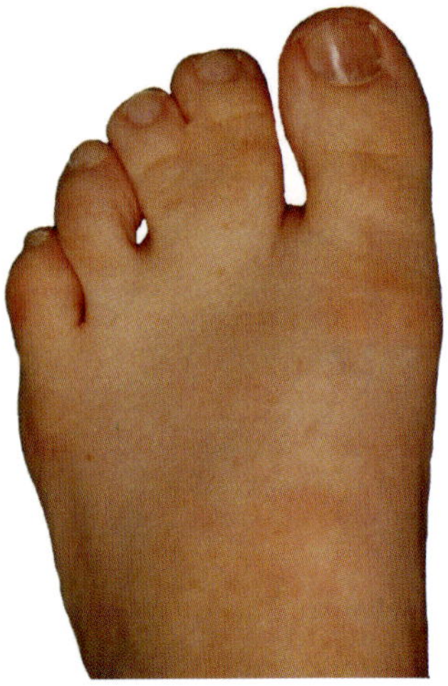

Above top: X-ray showing that the tip of this thumb has a short bone.
Above: It is not uncommon to see congenital fusion of the index and middle toes without the other toes or any fingers being affected.
Such examples demonstrate that the development of the different bones of the hand and foot are controlled by different genetic controls.

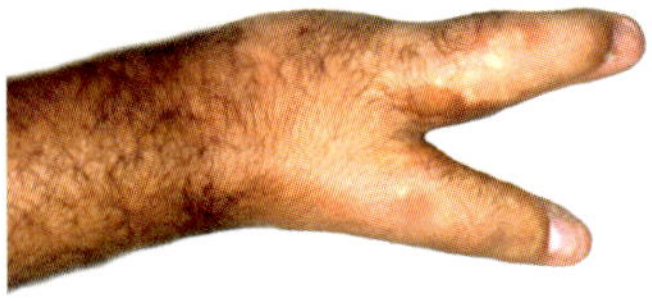

Congenital cleft hand

conditions, the hands may or may not be affected, showing that the genes involved have variable activity.

So, to put all this evidence into perspective, it becomes clear that the gorilla and chimpanzee feet are very different from those of humans, and are unlikely to be separated genetically from human feet by only a few simple gene alterations. If the ape form with thumbs on both the feet and the hands is supposed to be the more primitive state, then this means that *before the human foot existed*, genes existed that coded only for thumbs, and no genes existed for the coding of the human great toe. So, although a human great toe could *theoretically* be converted into a thumb by harnessing and activating the genes involved in the embrylogical development of the thumb of the hand, so as to apply their action to the development of the great toe, nevertheless, the same principle could not work in reverse for the modification of the thumb of an ancestral ape foot into a human-like great toe, since this would require the absurdly implausible acquisition (by a random process) of *a number of new* genes following a series of fortuitous and sequential mutations.

Furthermore, the *Galapagos Cormorant Principle* shows that since disuse could never be a mechanism for the appearance of a new, previously non-existing structure, the human foot could not have appeared as a consequence of the laziness of the ape foot thumb if ancient apes, being attracted into the savannah, no longer practised climbing. However, it is possible, for example, that as a consequence of some prior mutation, apes could give rise to progeny with weaker, shorter arms and with a reduction or degeneration in the thumbs on their feet. In time, this could result in apes being less able to climb and with the thumb being just a tiny (and also completely useless) appendage on the side of the foot. In fact, thumbs in the hands of primates tend to be *lost* rather than modified into different structures. For example, *colobus monkeys* in Africa have very reduced thumbs, and *spider monkeys* from South America have no thumbs at all. Furthermore, humans who have lost their great toes in accidents are still able to walk about quite well, although they cannot run as fast as they could before the accident. So, if anything, we should consider it far more likely that, by a deleterious mutation, an ape could lose the thumb on its foot, and that those of its progeny with the sturdiest index toes would give rise to four-toed

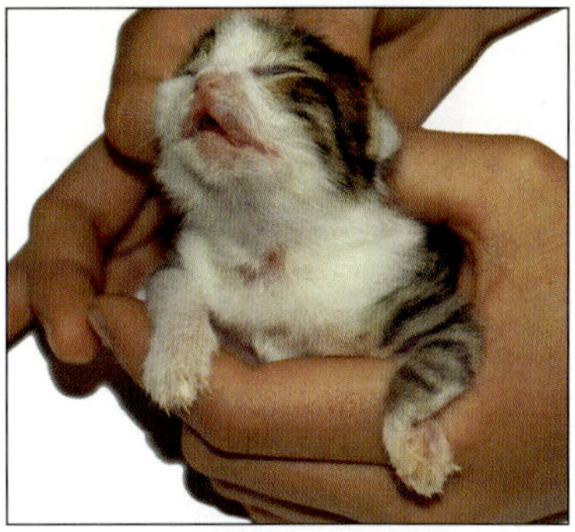

One-week-old kitten with deformed left foreleg.

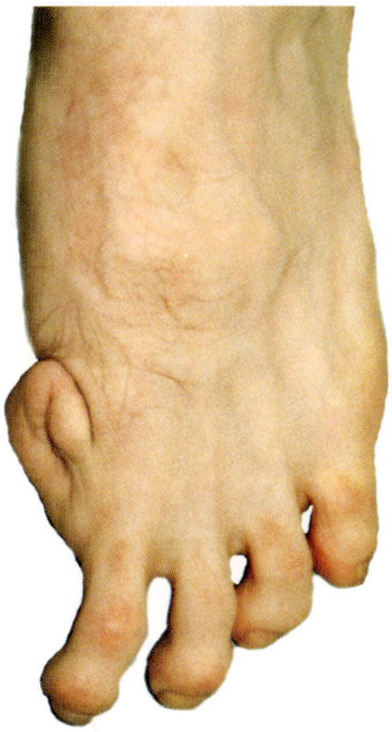

After an accident this man can still walk without his great toe. For an ape that wanted to walk bipedally, it would be simpler to lose the thumb on its foot than for the thumb to become a human-like great toe.

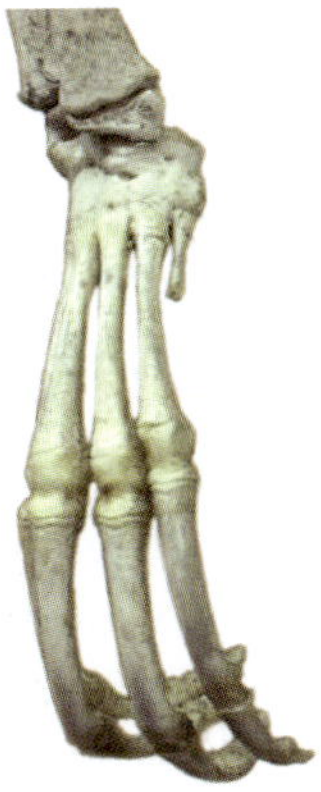

Far left: Spider monkey right hand bones showing that a tiny thumb metacarpal is present but there are no phalanges.
Left: Spider monkey showing absent thumb.
Spider monkeys have no thumbs on their hands. Colobus monkeys have only very short thumbs. Since gibbons have thumbs and are the most agile of primates, the absence of a thumb in some monkeys can be no advantage in the trees. Furthermore, the absence of a thumb is a distinct disadvantage when manipulating objects, and is likely to represent merely an inbred genetic abnormality.
Since the hands and feet of spider and colobus monkeys are otherwise normal, this shows that the genetic control of the development of the thumb is unrelated to the development of other parts of the hand. The same applies to the foot.

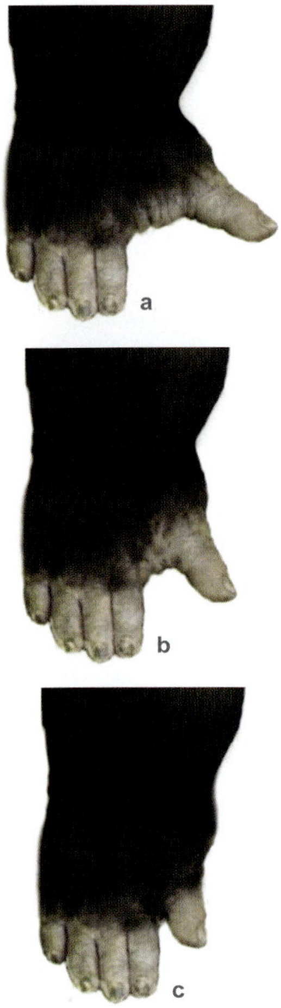


Gorilla foot: **a**. actual; **b**,**c**. modified.
The ape foot differs from the human foot in many fundamental and important aspects, such as the great toe length, direction and orientation, the little toe length and direction, and the shapes and sizes of the foot bones.
Unlike in apes, the human great toe metatarsal is rigidly joined to the index metatarsal joint and also to the medial cuneiform joint.
Unlike in humans, the ape foot has no longitudinal or transverse arches.
Simply altering the direction and rigidity of the thumb on the ape foot would not affect the lesser toes or the rest of the foot. Therefore, a mutant ape with a more forward-pointing and more rigid thumb would not be able to climb as well as its normal contemporaries and would be no better at running. There would be no selective advantage and so it is implausible that such a creature could have evolved towards the human form.

descendants. Of course, any such creatures would still be apes.

Most certainly, there is no reason why a particular ape that might possibly benefit from having a great toe *should* undergo a mutation to gain one. The *WYWYWG Principle* tells us that even if, at any time, a particular mutation would seem to be a logical thing to acquire in order to fulfil a *conceived* trend, it is nevertheless *improbable* that *that* particular mutation is just the one which will appear at that time.

So we have seen that on a genetic basis, it is improbable that a human foot could have arisen from an ape foot. Nevertheless, from a biomechanical point of view, could any modification to the ape foot be shown to have any significant advantage? If we were to suppose that following a mutation, the thumb on an ape's foot were to point a little more forwards instead of sideways, its metatarsal would be unlikely to become stabilised by ligamentous attachments to the index metatarsal and so it would still be mobile in all directions. Furthermore, simply altering its *orientation* would not affect the *length* of the thumb metatarsal. In humans, acceleration and agility are to the most part dependent on the stability of the foot and the flexion of the joint between the great toe metatarsal and the toe itself. Thus any ape with a flexible, short, more forward-pointing thumb on its foot is likely *not* to have any advantage in running, and is likely to be disadvantaged when climbing.

Furthermore, since we know that the human great toe is genetically independent of other structures, if there were to be any change in the thumb of the ape foot, the tendon alignments around the inside of the ankle would still be unlike the human form; there would still be no arches to the foot; and the pelvis, spine and leg bones would still be the same as before. Thus, such an animal would still have no increased ability to stand upright, or to run, and more importantly, it would be *less able* to climb trees than a gorilla with a normal ape foot. It would most likely, therefore, *not* be selected for in preference to the original ape form.

Alternatively, suppose, for example, that the *ape leg bones* were to change into more human-like bones. In this case there would be no change in the feet or the spine. And thus changes in simply the leg length, or hip, knee or ankle joints, could result in no significant advantage in attempting bipedal walking. However, such longer legs would make knuckle-walking more clumsy and thus be a disadvantage out in the savannah.

In the same way, if the *ape spine* were to change into a more human-like backbone with its unique double 'S' curve, there would be no change in the feet or the legs. So, the trunk could be held more upright, but the legs and feet would be the same, resulting in no overall advantage.

So we can show no significant advantage in a stepwise modification of the ape skeletal system. Instead, we actually see that solitary changes would produce disadvantages such that natural selection would tend to preserve the original ape form. By definition, the (supposed) evolution from fully-ape foot to fully-human foot necessitates that *all* intermediary forms should confer *significant* advantages on any ancestral form. And in accordance with the *Intermediary Principle*, since *all* (supposed) intermediary forms must have lived to produce at least as many offspring as any of their immediate ancestors, then if the australopithecine foot (with its human-like ankle joint, and ape-like thumb and long toes) represents an intermediary between

Fernando Revilla
Giant panda

the ape and the human form, there ought to be evidence of *other intermediaries* - for example, between the australopithecine and the human form; and in addition, fossils of these creatures with their half-australopithecine/half-human feet should be found in sediments younger than those of australopithecines. Of course, on the other side of history, we should also expect to see evidence of further intermediaries - between the australopithecine and the chimpanzee form; and in addition, fossils of these creatures with, for example, a half-australopithecine/half-chimpanzee pelvis should be found in sediments older than those of australopithecines. But the fossil record does not bear this out since, in general, it shows only three distinct forms - ape, australopithecine, and human, with no intermediaries at all.

It is curious indeed that while the evidence does not support the notion that apes stood upright and eventually became human beings, nevertheless, once the assumption of evolution becomes ingrained, logic and evidence are ignored. Regarding australopithecines:

> ...Although they were ape-like in many features, they were able to walk on two legs...and **because of this** we know that they **must have** played a very important part in the human story... (My bold). Prof. Leslie Aiello. 'Ape-man'. BBC2 Broadcast. 29th February 2000.

We have seen that there is no 'because of this' or 'must have' about it. Australopithecines were simply different apes, now extinct and of no relevance to human origins at all.

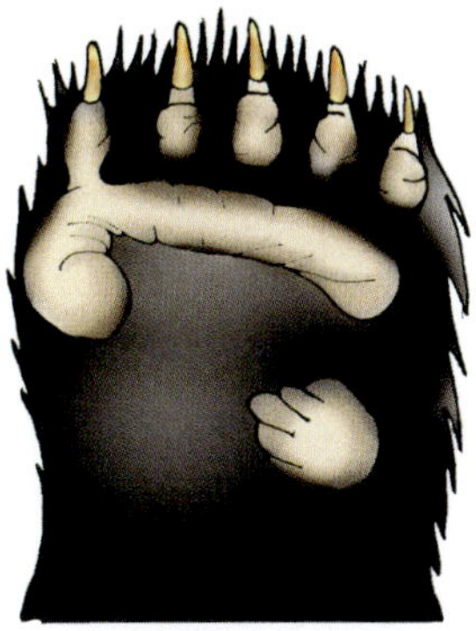
Giant panda left forefoot

### ***The panda's 'thumb'***

While on the subject of limb alterations, we should address the common mantra that the panda's 'thumb' is proof of the evolutionary process. The giant and red pandas are unique creatures in that, in addition to the normal five digits of the front paws, they both have a *sesamoid* 'thumb' (also called the '*radial sesamoid*') near the base of the first metacarpal, that acts functionally like a thumb. A sesamoid is a bone that lies *within* a tendon, and may act to modify pressure by the tendon, reduce friction, or alter the direction of pull of the tendon. This sesamoid-'thumb' feature is not something new since the fossil red panda relative *Simocyon batalleri* also had a sesamoid-'thumb', just like the living red panda. Simocyon was found near Madrid in sediments (supposedly) over 7myo, and was the size of a mountain lion.

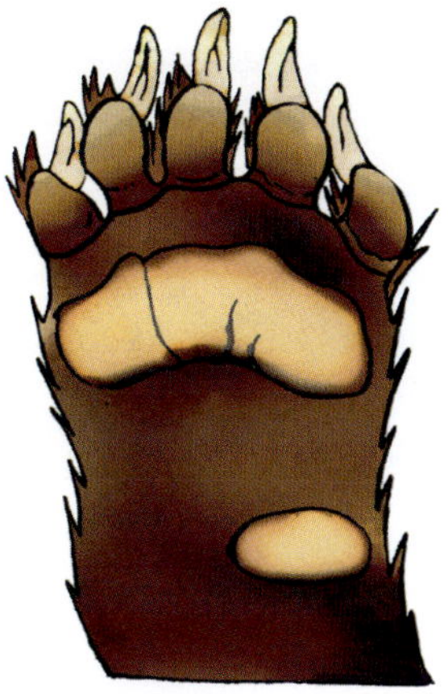
Brown bear left forefoot

A number of muscles have sesamoid bones embedded within the tendons. Examples in humans are the *pisiform* bone (near the wrist line on the opposite side to the thumb) which lies within the tendon of the muscle *flexor carpi ulnaris*, and the *patella* (or knee cap) which is the largest sesamoid in the body, and which lies within the tendon of the thigh muscle *quadratus femoris*.

The panda sesamoid-'thumb' lies within the tendon of the *abductor pollicis longus* muscle (APL, which, incidentally, in humans does not have a sesamoid bone) and is covered by a thick-skinned pad which joins the transverse pad of the palm at the base of the first true digit. In humans, the APL originates in the back of the forearm and is inserted into the base of the thumb metacarpal, and its action is to extend the thumb outwards. Instead of inserting only into the base of the first metacarpal, in the

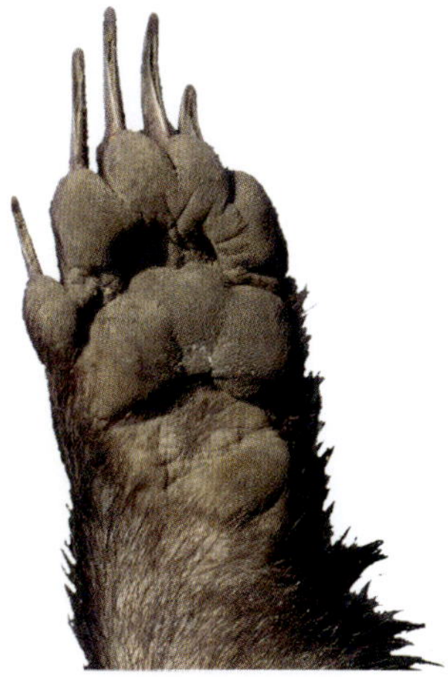
Badger left front paw

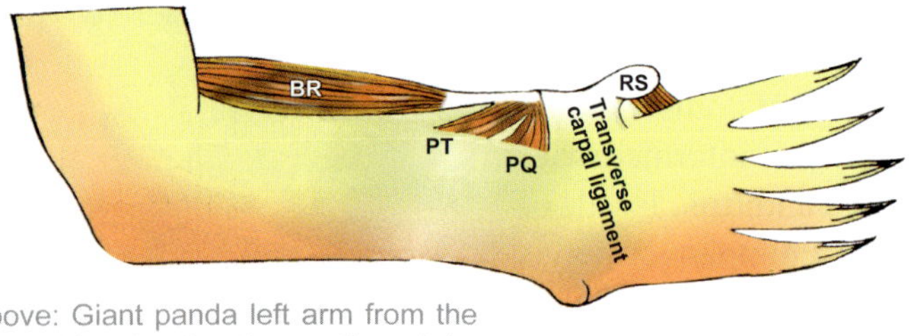


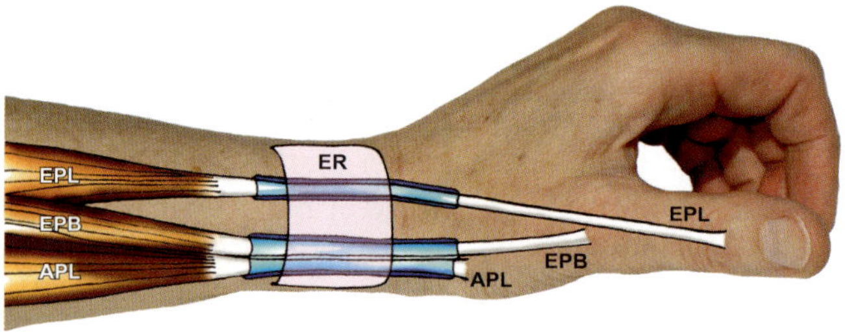


Above: Giant panda left arm from the underside, showing the position of the radial sesamoid (RS).
(*After Davis 1964*)
BR: brachioradialis
PT: pronator teres
PQ: pronator quadratus
Note that the abductor pollicis longus is hidden in this view since it lies on the back of the forearm.

Above right: The tendons on the thumb side of the human left wrist.
EPL: extensor pollicis longus
EPB: extensor pollicis brevis
APL: abductor pollicis longus
ER: extensor retinaculum

giant panda the APL tendon inserts partly into the sesamoid, to which two smaller muscles (*opponens pollicis*, OP, and *abductor pollicis brevis*, APB) also attach and act to pull the sesamoid-'thumb' towards the palm and the true digits.

Both the pandas (which are largely, but not exclusively, vegetarian) are able to manipulate bamboo and shear off the leaves by holding the bamboo in the gulley between the sesamoid-'thumb' and the first true digit. Bears also have a sesamoid in the APL and the general arrangements of the muscle attachments are the same as in the giant panda. However the bear's small sesamoid does not have an overlying pad and cannot be used for manipulation.

It is commonly said that the panda sesamoid-'thumb' is a crude organ, and since evolution can only operate blindly and randomly by modifications to existing designs, and so is likely to produce *sub*-optimal designs, the panda sesamoid-'thumb' is an excellent confirmatory example of the evolutionary process:

> The panda's true thumb is committed to another role, too specialized for a different function to become an opposable, manipulating digit. So the panda must use parts on hand and settle for an enlarged wrist bone and a somewhat clumsy, but quite workable, solution. S Gould. 'The panda's thumb'. Norton & Co. 1980.

Red panda

Incidentally, Gould perpetuates an old cliché – evolution can't change a 'specialised' structure so it must adapt generalised structures. That is not to say that evolution is possible, but Gould *assumes* (without offering any evidence whatsoever) that the bear APL/sesamoid/muscles unit *is* a less specialised structure, or one that is more easily modified during embryogenesis than is the bear first digit. However, the supreme complexity of the embryological development of the digits and limbs suggests that the digits and the sesamoid-'thumb' are both highly complex. Gould's statement also lacks any scientific credibility since humans have thumbs, which (supposedly) evolved ultimately from some ancestral shrew ancestor which would have had the first digit aligned like the others (i.e. no thumb), much the same as in bears. Therefore, if shrews (supposedly) could have ultimately developed a thumb, why not bears?

So, two issues need to be addressed here: is the panda sesamoid-'thumb' actually a sub-optimal design; and could it have appeared by modifications to the small sesamoid of a (supposed) bear/panda common ancestor?

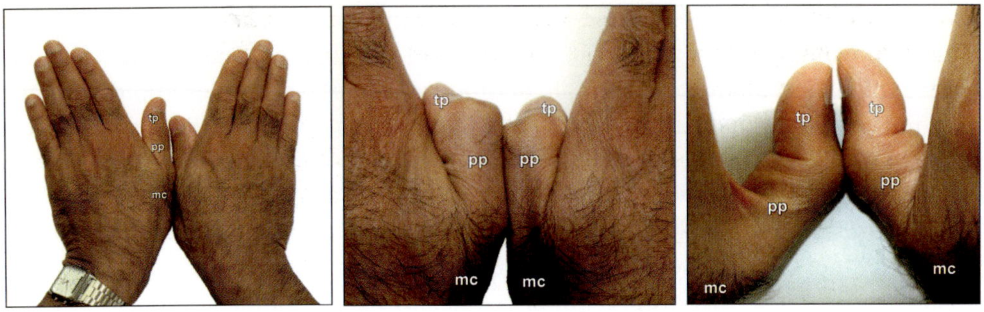


As a result of a childhood fracture, the right thumb is about 1cm shorter than the left, caused by damage to the metacarpal (mc) and the proximal phalanx (pp). As a result of the increased blood flow during the healing phase, the terminal phalanx (tp) of the right thumb grew larger than that of the left thumb.

Another example of enlargement is the *aye aye* which is a small lemur-like animal from Madagascar. It has longer than usual fingers with the middle finger being extremely long. With its keen hearing the animal can locate insects in crevices, and it uses the middle finger to pull them out. It is possible that an ancestor of the aye aye acquired a mutation that altered its digit growth pattern, and since the effect was not fatal or damaging, the aye aye has been able to make use of the result. In the same way, a guitarist born with a long, aye aye type of finger might be able to do some fancy fretwork that others would be incapable of. A more extreme example of enlargement is a tumour of the *pituitary* gland at the base of the brain, which can produce excess *growth hormone* resulting in gigantism (*acromegaly*), in which the whole body is affected, though normal proportions are generally retained. However, there can be serious ill-health in association with blindness.

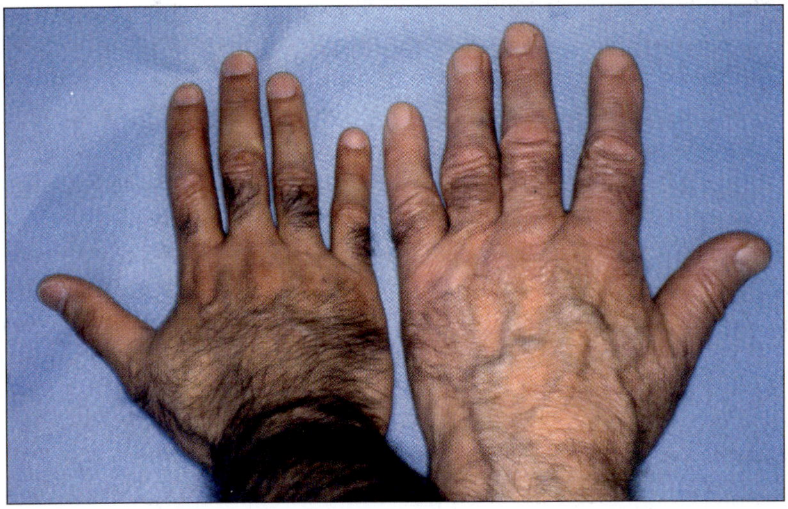

My hand (far left) is dwarfed by the hand of the patient (left) with acromegaly.

So, maybe the giant panda sesamoid-'thumb' *is* the result of enlargement caused by a change in developmental triggering in some ancient bear. However, there are a few problems with such a simple notion. Simple enlargement of the bear wrist would preserve its anatomical proportions rather than produce an elongated enlarged sesamoid precisely where and as required for use as a 'thumb'. For that to happen

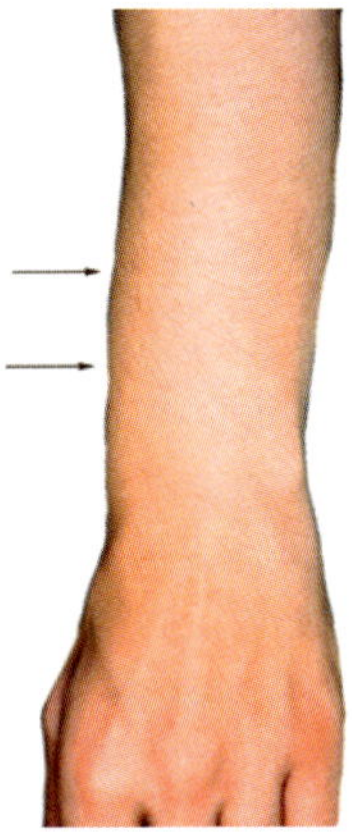

De Quervains tenosynovitis. The swelling is seen along the thumb-side of the wrist.

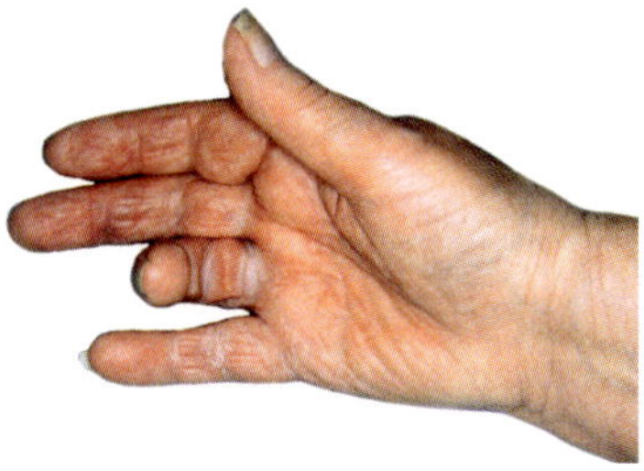

Trigger finger.
Above: The ring finger is locked and cannot extend.
Below: A swelling of the tendon gets jammed against the edge of the flexor sheath, preventing it from being straightened easily. When forced, the swelling suddenly flips under into the flexor sheath with a snap - the 'triggering'.

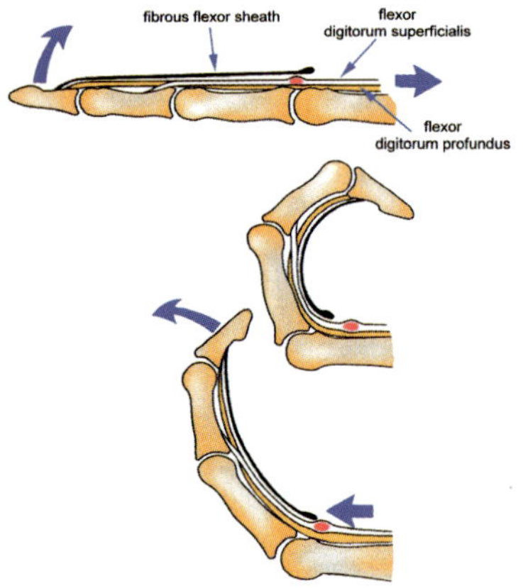


would be implausible. Furthermore, if the sesamoid-'thumb' did arise independently in the two pandas, twice upon a time, this should alert us to the possibility of some other explanation than an accident. In addition, the pad over the sesamoid-'thumb' does not develop as a result of friction and pressure (as does a guitarist's thickened finger-tip skin) but is present before the baby (not-so-giant) panda is born. Since no other sesamoid, not even the patella (the largest sesamoid), is covered by a skin pad, this means that the skin pad must be something that has to be coded for, *independently*, into the sesamoid-'thumb' itself.

To understand the matter properly, we need to leave the sesamoid-'thumb' itself for the moment and 'go back' to a *supposed* distant ancestor of shrews. This hypothetical creature would have had a 'basic' five-digit paw without a true thumb, with all the digits pointing forwards. The APL would have inserted into the base of the first metacarpal, contained no sesamoid, had no *synovial sheath* as it passed under the wrist *retinaculum* (a fibrous cuff that is wrapped around a group of tendons), and had no attachments from the opponens pollicis and flexor pollicis brevis muscles. At some time the ancient pre-shrew quadruped would have had to acquire synovial sheaths. And ultimately, long after leaving behind its shrew ancestors, the ancestral bear would have had to acquire all the other features mentioned above. Subsequently, the giant panda would have had to acquire a substantial elongated enlargement of the APL sesamoid, as well as the overlying skin pad.

Let's deal firstly with the synovial sheath. Whenever a tendon passes under a ligament it is enveloped with a synovial sheath, which is a thin tubular membrane containing the same sort of lubricating fluid as is found in the majority of joints. Without a protective synovial sheath wrapped around it, movement of a tendon would tend to cause friction, resulting in inflammation and pain. This is seen in the condition *De Quervain's tenosynovitis* in which the APL becomes inflamed as a result of excessive or unaccustomed use of the thumb.

So, an ancient tendon that has little friction has no *need* of a synovial sheath, and one that *is* subject to friction *cannot* do without one. Since a tendon that has no synovial sheath but is subject to friction will not, in time, conveniently evolve one, it will become inflamed, cause severe pain, or rupture. Similarly, if a tendon that slides under a fibrous tunnel develops any thickening, there is likely to be impingement or complete obstruction to its motion as the thickening jams against the edge of the tunnel. This sort of thing is seen commonly in the condition *trigger finger* in which a thickening of the tendon of the finger or thumb becomes jammed at the edge of the *flexor sheath*. Such a thickening can be caused by local friction or pressure.

So, a synovial sheath could not have evolved, and a tendon that slides under a fibrous tunnel cannot acquire a sesamoid without potentially restricting its movement. Therefore, the shrew or bear paw could not have evolved. The different parts *must* have been there, together, at the onset.

To get the full perspective we also need to highlight some important differences between bears and the giant panda. The external genitalia are significantly different in bears and pandas. Of particular importance is the *baculum* (the bone supporting the penis), which in the brown bear is long, straight and thin, in the giant panda is short and spoon-shaped, and in the red panda is small and thin with an upward-pointing, slightly curved tip. In contrast, the racoon, which is otherwise similar to the red panda, has a longer baculum which has a right-angled bend that points downwards.

In an attempt to reconcile the great differences between these creatures, it has been suggested that the giant panda genitalia are the result of some major alterations in the embryological development of the pelvic area. However, if the giant panda evolved from bears, its ancestors must have had bear-like genitalia. And since *significant* alterations in penile anatomy cannot support reproduction without corresponding *parallel* alterations to the female genitalia, it is implausible that any of these creatures have emerged from a common ancestor as a result of some random changes in regulatory gene activity during embryogenesis.

Although the limited genetic data involving mtDNA seems to place the giant panda much closer to bears than to the red panda, there is no obvious relationship when we look at the chromosomes: most bears have 74 chromosomes while the giant panda has 42 and the red panda has 36. And regarding its unique colouring, which as with the tiger has no special survival advantage but looks good, the giant panda is unlikely to have evolved from a plain-coloured animal.

If the giant panda's sesamoid-'thumb' did not evolve from enlargement of the bear APL sesamoid, there remain a few other possibilities. Firstly, it might still be argued that the red and giant pandas *must* have had a common ancestor since they both have a sesamoid-'thumb'. However, they are very different animals and the *Chalicotherium Principle* reminds us that many creatures have similar structures but no common ancestry. For example, both *micro*-bats (such as the vampire bats which use sonar) and *mega*-bats (such as fruit bats, which do not have sonar) have exactly the same wing form and structure, but are unrelated to each other. While the heads of fruit bats do resemble those of lemurs, fruit bats are unrelated to lemurs. And the heads of micro-bats resemble, well, nothing but the heads of other micro-bats.

Secondly, maybe the bear APL sesamoid is a mutational degeneration *of the giant panda's* sesamoid-'thumb'. However, this is unlikely because mutations generally result in deformities, yet bear paws are perfectly formed and have no signs of a sesamoid pad remnant.

To summarise, if the giant panda sesamoid-'thumb' *did* arise from the bear APL system by a sudden change in embryological growth triggers, we should expect the resulting animal to be a brown/black creature without eye patches, with no sesamoid-'thumb' pad, with the typical bear genitalia and chromosome number, and living *nowhere near* any bamboo forests in China. Instead, we find that in its looks, its paws, its genitalia, and its chromosomes, the giant panda is quite different from bears, and it hangs out in bamboo forests.

So, along with the fact that there are no fossils to show a non-panda gradually changing into a panda, the evidence points to the conclusion that (apart from natural variation) bears and the red and giant pandas have *always been* the way they are.

Finally, despite it being unlikely, even if the giant panda sesamoid-'thumb' *did* arise by enlargement of the bear APL sesamoid, this would *not be* evidence for evolution (the origination of a new structure, organ or creature), but would only demonstrate growth variation in a pre-existing structure.

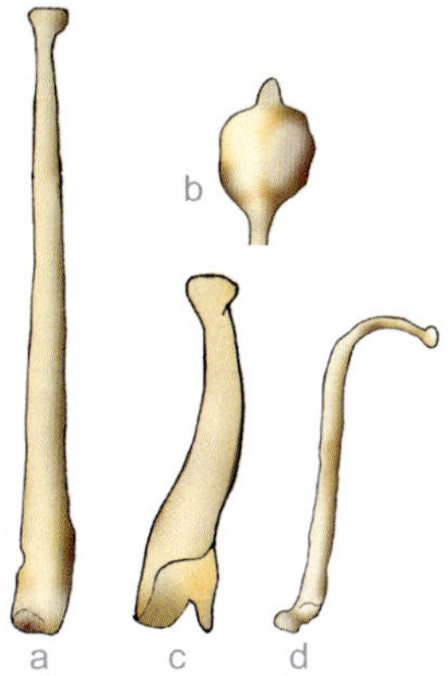

The baculum of: the brown bear (a), giant panda (b), red panda (c), and racoon (d).
*(After Davis, 1964)*

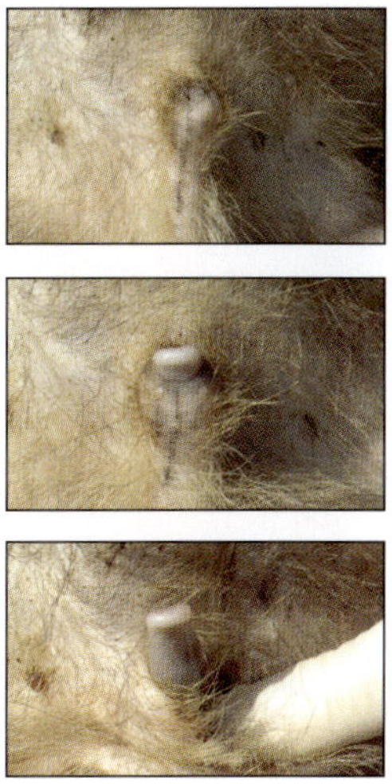
Above and below: The baculum of a badger.

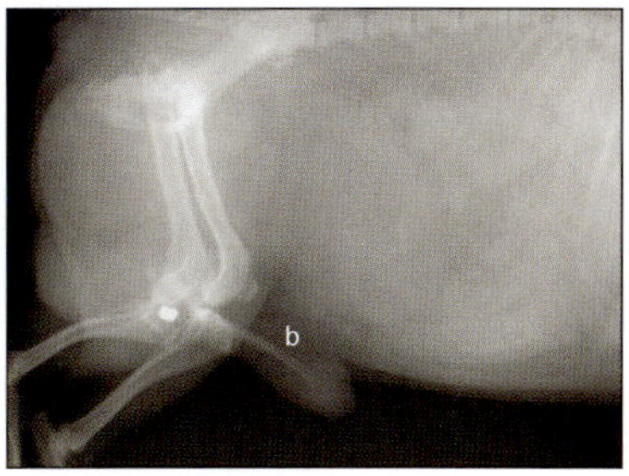

This X-ray shows the baculum (b) in the penis of a dog. The dense white spot is an air-gun pellet embedded in the back leg.

Vampire bat

### *Knee joint*

While on the subject of limbs, it is worth looking at the knee joint which, by virtue of its unique complexity, shows that it could not have evolved stepwise.

The human knee joint comprises the lower end of the femur (which has two bulges, the medial [inner] and lateral [outer] *condyles*) and the upper end of the tibia (which has a relatively flat surface). The two bones are connected on the inner and outer aspects by strong *collateral* ligaments, and in the middle by the *anterior and posterior cruciate* ligaments which cross each other, hence their name. The anterior cruciate is attached near the front of the tibia and passes upwards to the back end of the femur. The posterior cruciate is attached to the back end of the tibia and passes upwards to the front end of the femur. To understand the function of this arrangement, the anterior cruciate stops the femur sliding off the tibia when the knee is bent as in going upstairs, and the the posterior cruciate stops the femur sliding off the tibia when the knee is bent as in going downstairs.

In addition to the complex ligamentous arrangement, the knee joint also has specially-shaped semi-circles of cartilage (the *menisici*) that lie between the femoral condyles and the tibia. The menisci are important in relation to the rotation of the knee joint.

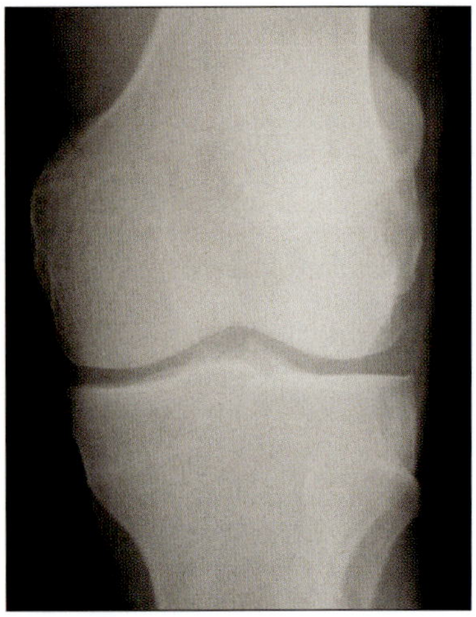

X-ray of a normal left human knee joint. The apparent space between the femur and tibia is taken up by the menisci.

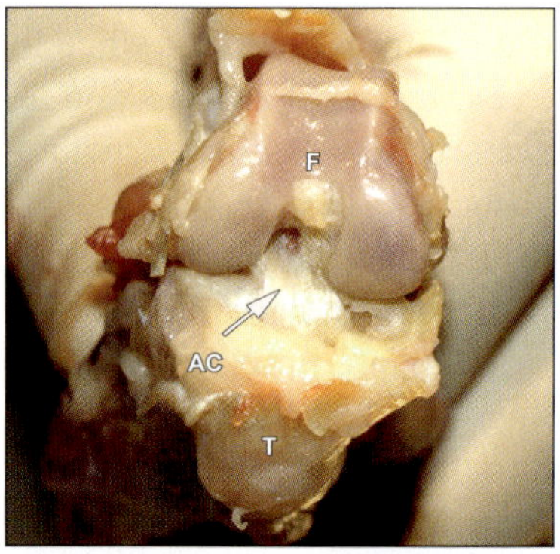


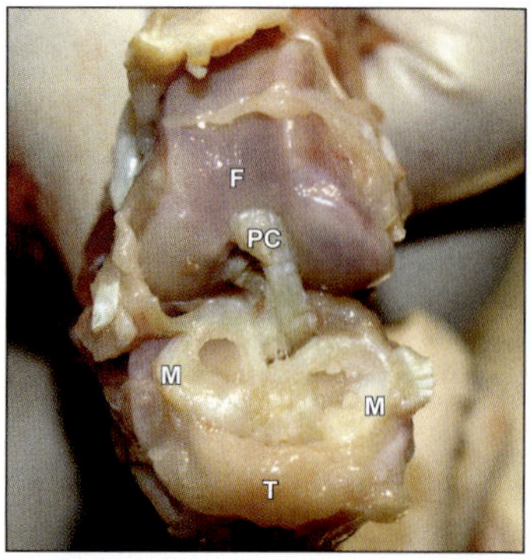


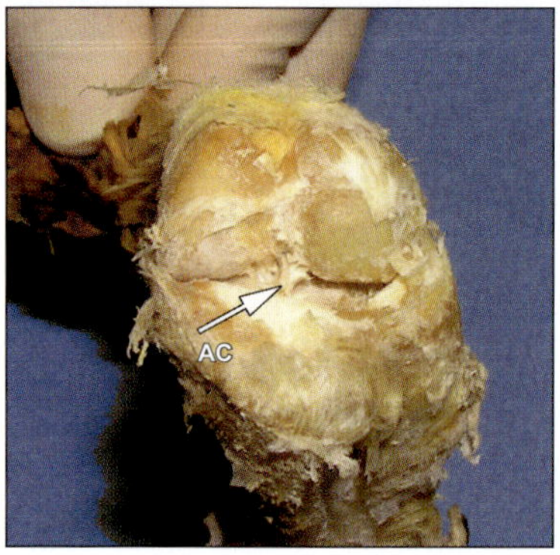


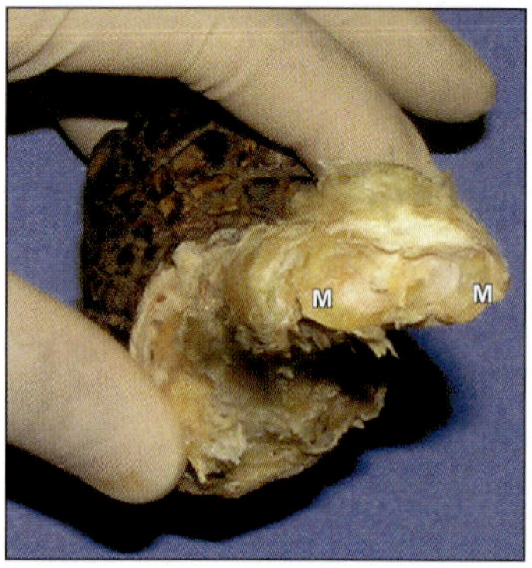


**Knee joints**
Above right: Cat anterior cruciate ligament.
Above far right: Cat menisci.
Right: Crocodile anterior cruciate.
Far right: Crocodile menisci.

AC: anterior cruciate ligament
PC: posterior cruciate ligament
F: femur
T: tibia
M: meniscus

(Crocodile is a preserved specimen)

When the knee is flexed, the lateral condyle of the femur slides backwards on the tibia and there is a risk of the back part of the lateral meniscus being pinched and torn in the process. This risk is minimised by the action of the *popliteus* muscle which lies behind the knee and which pulls the lateral meniscus backwards, out of harm's way. Interestingly, the fruit bat is the only mammal that is unable to rotate the knee and lacks both the menisci and the popliteus muscle.

The human knee joint is not just a simple hinge, but has a complex range of movements. When the knee is extended, the anterior cruciate ligament is the first ligament to become tight and this prevents further straightening. However, further contraction of the *quadriceps femoris* muscle causes the femur to be rotated inwards with the medial condyle of the femur sliding backwards on the top of the tibia, after which the collateral ligaments become fully tight. At this point the joint is completely rigid - *'locked'* in full extension, allowing us to stand for hours without any great muscular effort to maintain the posture. The knees of great apes do not have this facility and therefore apes can only stand, and walk inefficiently, with bent knees.

Not only are all the features of the human knee joint (collateral ligaments, cruciate ligaments and menisci) seen in mammals such as cats and sheep, but they are also found in crocodiles. So if humans share a common ancestry with crocodiles, the complex knee joint was in existence long before the earliest known crocodile (supposedly) 200mya.

If the cruciate ligaments and menisci evolved from a knee joint that millions of years ago did not have any, we would certainly not expect such complex mechanical systems to appear complete, and exactly *where they are needed.*

Although it is implausible, even if it is granted that just an anterior or just a posterior ligament did appear, and then, millions of years later, by an extraordinary amount of luck, the other appeared, we should certainly not expect to see the *hundreds* of lineages of (hypothetical) creatures, that had evolved from the one which had acquired one cruciate ligament or meniscus, *all* evolving the other also.

So, we should predict that there would be *plenty* of evidence of creatures with knee joints which have only one or other of the cruciate ligaments or menisci. However, we see none whatsoever. As the *WYWYWG Principle* shows, the complex knee joint could not have evolved *at all.*

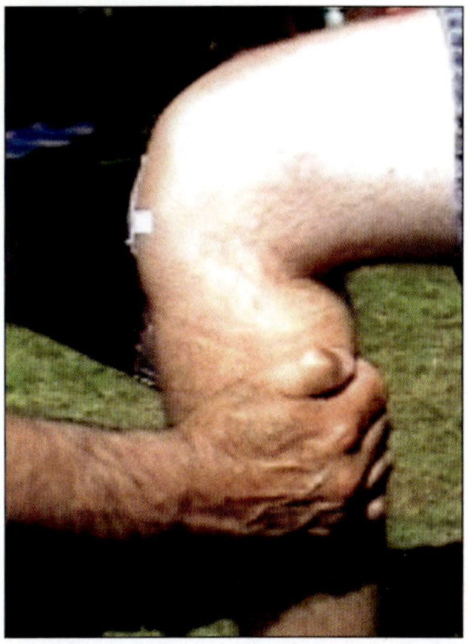

When the anterior cruciate ligament is torn, the lower leg can be pulled forwards a centimetre or so at the knee. Incidentally, it is not correct to say that absence of one of the cruciate ligaments will result in the complete failure of knee joint integrity. People with torn cruciates can still run, twist and jump, but function is just not as good as with a normal joint.

### *Bipedalism and intelligence*

We cannot leave this subject without addressing the supposition that bipedalism evolved in parallel with increasing intelligence. So, for the sake of argument, disregarding all the evidence against such a thing, suppose that somehow the feet, legs and spine of an ape *were* to all change such that a fully bipedal creature emerged. We would now have a bipedal ape. The shape of its body would certainly allow it to run bipedally instead of quadrupedally like a baboon, and its hands would be free to carry things. But any bipedal ape would still act and think like an ape. Other bipedal creatures have not evolved high intelligence. For example, the ostrich is fully bipedal, and its brain and forelimbs are free to evolve as with any other savannah-living creature. However, the ostrich is not intelligent, nor has it made use of its forelimbs for tool-making. Likewise, the bipedal dinosaurs were not intelligent, nor in all probability were they tool-users.

Suppose we put male and female apes into electric wheelchairs, with baskets attached so that they could collect and carry things, thus freeing their hands for manipulating whatever took their fancy in the savannah. Suppose we treated all of their offspring in the same way for millions of years, and those descendants were free to experiment with their hands and use more complex tools. Two questions arise. Firstly, should we expect this lineage of wheelchair-using apes to develop human intelligence? And secondly, should we expect this lineage of apes to develop human intelligence any more than any of their knuckle-walking quadruped cousins?

Since the genes controlling brain anatomy, function and intelligence are unrelated to the genes controlling the development of the feet, legs and spine, we should predict that freeing the arms alone should not promote or be associated with any fortuitous mutations that might affect the brain of *either* the wheelchair-tied lineage *or* the knuckle-walking quadruped apes. So we should conclude that human intelligence could not have been related to the acquisition of bipedalism by (supposed) ancestral apes.

Spider monkey demonstrating the grip of its prehensile tail.

Nevertheless, if we were to suggest that bipedalism was the key which freed the ape's hands for tool use, and which, as a result, allowed its brain to expand, then logically, we should not limit our discussion to changes in the feet and legs and spine, but go on to consider ways of freeing the hands *other* than bipedalism. And indeed, there are a few options which have at least as solid a biological foundation as bipedalism, and which would require *fewer* genetic changes than those required for altering the feet, legs and spine.

The simplest method of freeing a hand for carrying things is to run on *three* legs instead of four. In fact, apes, baboons and monkeys can all manage to carry objects in one arm whilst running on three limbs without any appreciable loss of speed or agility. A domestic cat or dog, or even a cheetah, that has had one of its legs amputated can still manage surprisingly well with only three weight-bearing limbs. If walking on three limbs while carrying objects in one hand is well within the capability of many primates, we should expect that ancient apes could have learned to carry and use stone tools without the necessity of bipedalism.

Carnivores often manage to carry quite large prey items in their teeth while retaining their quadruped gait. A leopard can even carry a whole antelope weighing as much as itself high into a tree. Macaques are also known to carry food in their teeth. So a little natural selection for individuals with the strongest teeth, jaw and neck muscles ought to have enabled ancient apes to carry food and tools in their teeth. Macaques and other monkeys also have large cheek pouches which they can fill with food that they can subsequently eat at leisure. So ancient apes could have simply developed large cheek pouches. Apart from cheek pouches, there are at least two other methods that could have been employed by ancient primates to carry food and tools. The first is a prehensile tail: spider monkeys from South America routinely use their prehensile tails to carry food items while walking on all fours.

Although a kangaroo is bipedal and has its forelimbs free, it is not particularly intelligent and it does not use tools.

The second is a pouch on the abdominal wall just like a kangaroo's. The reader might think this to be too absurd to take seriously, but in the kangaroo and other marsupials, evolution (supposedly) managed to produce a pouch where previously in some (supposedly) primitive mammal there was none present. So why could a pouch not appear in an ape? An abdominal basket could be very useful, either for carrying food back to a base camp to share with the rest of the troop, or for carrying stone tools. And all this whilst still retaining the quadruped gait. Not as absurd an idea as it first sounds. Such an ape could then (supposedly) get on with the business of accumulating mutations in its brain to make it more clever. However, it is a notable fact that although the fully bipedal kangaroos have had their forelimbs free *and* have had pouches in which to carry things for (supposedly) millions of years (and well before the origin of apes), they have not developed any higher intelligence or manual dexterity.

There is yet another option: fortuitous mutations could produce extra limbs. Individual animals with duplicated body parts are not unheard of. We know that some animals such as tortoises and snakes have a particular tendency to produce individuals with *two heads**. We know that in flies, only a simple mutation in a controlling gene system can result in a complete extra pair of wings or legs. We have also seen, when discussing whales, that some fishes can 'walk' using spines which are separate from the pectoral fins and which arise from the gill covers, showing that it is feasible that extra limbs could arise from the neck area. Extra fingers or toes are not rare in humans. So, could extra limbs occur in mammals? In fact, it has happened. In Dehli, in 1972, a free-roaming cow was seen which had *five* legs. The fifth leg was floppy, about one-third the normal size, and was attached at the left shoulder. The leg may have been a truly extra leg or it may have been a small representation of a 'Siamese twin'. For our purposes it does not matter since the point is that, genetically, the production of an extra limb *is* feasible.

So, it is reasonable to assume that, given enough time, by a mutation in a controlling gene, an ape could perhaps have developed a complete extra pair of arms. Again, this is not such a silly idea, since the underlying biological mechanisms already exist. A quadruped ape with an extra pair of arms would be able to run as fast as a normal ape, climb as well as an ape, and also be able to carry things. And as with the pouch idea, fewer genetic changes would be required to acquire an extra pair of arms than to change the feet, the legs, the spine, etc. With its extra hands free to manipulate things, and its gait unaffected, such an ape could get on with the business of accumulating mutations in its brain to make it more clever.

It would be difficult for anyone to say which of the options for the freeing of the hands of a knuckle-walking ape (i.e. bipedalism, tripedalism, cheek pouch, prehensile tail, abdominal pouch or extra arms) would be the most absurd, or to quantify which would be the most likely to occur. Certainly, from the evidence we have seen in this chapter, it seems that for a quadruped ape to become bipedal is implausible.

But the reader might say that surely the fact that the human genome is very similar to that of the chimpanzee leaves no doubt about a common ancestry - surely the genes confirm that evolution *must* have happened. In fact, recent work has shown that the *mouse* has 99% of its genes in common with humans, and that chimpanzees and humans share only 95% of their DNA. So chimpanzees and humans are not as close as was previously thought only a few years ago. Furthermore, as we shall see in the next chapter, the evidence of the chromosomes themselves does not support any evolutionary trend.

*Although the two-headed phenomenon is perhaps related to the egg being subjected to particular changes in temperature at critical times during embryological development, nevertheless, ultimately the resulting duplications must be reflected in some change in the chemical triggering of different processes during embryological development. Therefore, it is reasonable to assume that, given enough time, it would be quite possible that such an alteration in chemical triggering could also be caused by a mutation in a controlling gene. This would allow for the two-headed reptile to develop without any necessary temperature change, and for that two-headed animal to produce two-headed young.

This cow has an extra leg arising from its left shoulder.

# CHAPT

## Chromosomes: *Man, the deep end of the ape gene pool?*

Children, Uganda

**Different creatures vary in chromosome number**

| | Chromosomes TOTAL | Macro | Micro |
|---|---|---|---|
| Salmon | 54-56 | | |
| Trout | 84 | | |
| White perch | 48 | | |
| Lungfishes* | 36-38 | | |
| Coelacanth* | 48 | | 16 |
| Toad | 22 | 8 | |
| Moor frog | 24 | 12 | |
| Green frog | 26 | 10 | |
| Horned frog | 104 | | many |
| Spanish ribbed newt | 24 | 16 | |
| Chameleon | 36 | | 24 |
| Iguana | 38 | | 24 |
| S. American alligator | 42 | | 4 |
| Boa | 36 | | 20 |
| Anaconda | 36 | | 20 |
| Green cobra | 36 | | 20 |
| Rattlesnake | 36 | | 20 |
| Cobra | 28 | | 6 |
| Pigeon | 80 | | mostly |
| Ring dove | 68 | | mostly |
| Pheasant | 82 | | mostly |
| Razor-billed curasow | 82 | | mostly |
| Chicken | 78 | | mostly |
| Raven | 78 | | mostly |
| Humming bird | 74 | | mostly |
| Horned owl | 82 | | mostly |
| Rhea | 82 | | mostly |
| Cat | 38 | | |
| Dog | 78 | | |
| Hamster | 44 | | |
| Rabbit | 44 | | |
| Guinea pig | 64 | | |
| Domestic horse | 64 | | |
| Prezewalski horse | 66 | | |
| Grey zebra | 46 | | |
| Burchelli zebra | 44 | | |
| Ass | 62 | | |
| Onager (hemionus) | 54 | | |
| Tiger** | 38 | | |
| Lion*** | 38 | | |
| Great apes | 48 | | |
| Humans | 46 | | |

Note that the 'TOTAL' column includes 'average-sized', macro and microchromosomes.

*Data taken from 'Chromosome atlas'. Benirsche & Hsu' (Editors). Springer-Verlig. 1971.*
**Bogart et al. 'The chromosomes of the living coelacanth and their remarkable similarity to those of one of the most ancient frogs'. Journal of heredity. 1994: 85 (4). p322-5.*
***V. Thiagarajan et al. 'The karyotype of Panther (Pantheratigris tigris)'. Indian Vet. J. 69. July 1962. p651-652.*
****M. Geldenhuys. 'Die Kariotipering van die Leeu (Panthera leo)'. Tydskr. S Afr. Vet. Ver. 1989. 60 (1). p48-49.*

For the sake of simplicity, the various chromosome sizes between tiny (microchromosomes) and huge (marochromosomes) are referred to as 'average-sized'.

It is commonly stated that humans and chimpanzees have over 98.5% of their genes in common, and that humans and gorillas have around 97.7% of their genes in common. However, since in terms of anatomy and intelligence humans are *hugely* different from apes, these often repeated and currently fashionable statistics are meaningless without a broader perspective. So, before we can meaningfully discuss human and ape chromosomes with regard to any supposed evolutionary relationship, it is necessary to examine the numbers, shapes and sizes of the chromosomes (the *karyotype*), and the DNA content found in the cells of different creatures in general.

### *Chromosome number, size and DNA content*

If fishes and humans are *very* distant cousins, then amongst creatures along the supposed evolutionary pathway, the chromosomes should show arbitrary and random patterns. However, what is immediately apparent is that the major groups (fishes, amphibians, reptiles, birds and mammals) all show quite distinct patterns in chromosome number and size, and in DNA content.

Fishes have 36-84 chromosomes whereas their (supposed) descendants, the amphibians, generally have in the region of 22-26 chromosomes. Furthermore, the striking feature of amphibian chromosomes is that they are exceptionally large (*macrochromosomes*).

It has been thought that ancient ancestors of the coelacanth came out of the water and evolved into amphibians. However, although its karyotype resembles that of one particular type of frog (*Ascaphus truei*), the coelacanth has eight pairs of tiny chromosomes (*microchromosomes*), whereas, in contrast, most amphibians do not have *any* microchromosomes at all. Furthermore, the coelacanth and the frog are *completely* different creatures. Therefore, any similarity between coelacanth and ascaphus chromosomes does not necessarily mean that there is *any* relationship between the two animals, and, just as importantly, it does not explain what makes one a fish and the other a frog.

Lungfishes are the other candidates for the (supposed) ancestors of amphibians. Although lungfishes have the largest amount of DNA per cell of any living animal, and although they do not have microchromosomes and in that one sense they share characteristics with amphibians, nevertheless, apart from an exception like the horned frog, lungfishes have far more chromosomes than amphibians. Furthermore, in assessing relationships, the whole picture has to be considered. Although some biochemical studies have suggested that lungfishes and amphibians might be related: for example, comparisons of DNA sequences of the genes for 12s RNA and cytochrome b:

> ...supported a sister group relationship between lungfish and tetrapods.
> Bogart et al. 'The chromosomes of the living coelacanth and their remarkable similarity to those of one of the most ancient frogs'. Journal of heredity. Vol 85 (4). 1994. p322-5;

nevertheless, when both biochemical and morphological comparisons of the coelacanth, lungfishes and tetrapods were combined to produce a tree, it was discovered that:

> ...the position of the lungfishes on that tree was not significant. Bogart et al. 'The chromosomes of the living coelacanth and their remarkable similarity to those of one of the most ancient frogs'. Journal of heredity. Vol 85 (4). 1994. p322-5.

African lungfish

In contrast to their (supposed) ancestors the amphibians, reptiles on the whole have a larger number of chromosomes, comprising mainly average-sized chromosomes along with a number of microchromosomes, but they have no macrochromosomes. If reptiles were descended from amphibians, then we should expect that at least some reptiles - for example, crocodilians (supposedly, one of the most ancient of reptile groups) - should have retained some macrochromosomes from their (supposed) amphibian ancestors. But this is not the case.

In contrast to their (supposed) ancestors the reptiles, birds have a very large number of chromosomes (around 74-82), most of which are microchromosomes. If birds evolved from reptiles, and by so doing, accumulated their very large number of microchromosomes, then we should expect to find that the (supposedly) more modern reptiles should also have more microchromosomes than the (supposedly) more ancient ones. But no such trend exists. It is commonly assumed that the *constrictor snakes* constitute the most ancient lineage of snakes. But not only do the boa constrictor, the anaconda and the (supposedly) more advanced rattlesnake all have the same *number* of chromosomes (36), they also have roughly the same *forms* of chromosomes, since they all have 16 average-sized chromosomes and 20 microchromosomes.

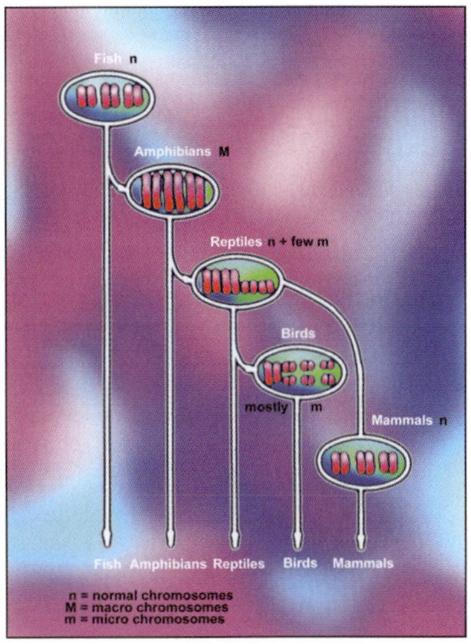


There is no evolutionary trend shown by the shapes of chromosomes of different animals.

If birds and reptiles had a common ancestor, it is implausible that there could have been any natural selection pressure for birds to acquire so many microchromosomes and, in contrast, for reptiles *not* to do so. And if there were some natural tendency for microchromosomes to accumulate by a break-up of larger chromosomes, and if we assume that both mammals and birds evolved from reptilian stock, then we should predict that mammals in general should also have a good number of microchromosomes to reflect their (supposed) reptilian ancestry. But this, too, is not the case. But, it might be argued that since microchromosomes are found in all monotremes, this confirms that placental mammals are distant descendants of egg-laying mammals which have simply lost their microchromosomes over time. However, if that were the case, then we should expect that birds should also have lost many of the microchromosomes of their (supposed) reptilian ancestors.

Although major animal groups do show quite distinct chromosome patterns, nevertheless, there can be marked differences in chromosome number *within* very similar animal types. For example, salmon have 54 chromosomes but trout have 84; the Reeves muntjac deer has 7 chromosomes but the Indian muntjac has 46; and the pigeon has 80 chromosomes while the ring dove has 68. Such marked differences between very *similar* creatures strongly suggest that it is unlikely that differences in chromosome number *per se* are responsible for the observable differences in appearance, anatomy and ability of animals that are markedly *dissimilar*.

To complicate matters, whilst chromosome numbers do vary markedly in some animals, in other animals they seem very stable. For example, considering the rate at which hamsters and rabbits breed, and the many millions of years they have (supposedly) been separately evolving, they all still have 44 chromosomes. This is in stark contrast to the notion that humans have evolved their 46 chromosomes from the 48 of the slow-breeding great apes - in only 8 million years.

Before leaving the subject of chromosome shape and form, we should note that amongst male and female mammals there exists a visible difference in one pair of the chromosomes, the *sex chromosomes*. Females have two X chromosomes (hence they are XX) whereas males have one X and one Y chromosome (hence they are XY), the Y being much smaller than the X chromosome. In birds, however, the situation is reversed in that it is the female that has one smaller chromosome. This is called a W chromosome, and the larger chromosome that pairs with the W chromosome is called the Z chromosome. So male birds are ZZ and female birds are ZW. If birds evolved from dinosaurs, and if crocodilians are close relatives of both birds and dinosaurs, then it is incongruous that there are no XY/XX or ZZ/ZW patterns in the male and female chromosomes of the alligator. Crocodilian gender is actually determined by the temperature at which the egg is incubated*. The female green cobra and the female rattlesnake have a smaller chromosome 4 than the male, yet in contrast to birds and mammals, reptiles such as the chameleon, anaconda and boa constrictor, along with frogs, toads, and fish, all have sex chromosomes which are visually identical. So the patterns of the sex chromosomes, too, are inconsistent with an evolutionary model.

*Crocodile eggs. Unlike in birds and mammals, the sex of crocodiles and alligators is determined by the temperature of the egg. Temperatures of around 32-33°C produce males, whilst a temperature slightly above or below that range produces females.
This shows that reptiles are unrelated to birds or mammals.

There are also marked and distinct differences in the *amount of DNA* contained in the chromosomes of different groups of creatures. For example, cartilaginous fishes are supposed to have evolved into bony fishes, yet sharks and rays have more DNA than bony fishes. On the other hand, amphibians have far more DNA than the bony fishes from which they were supposed to have evolved.

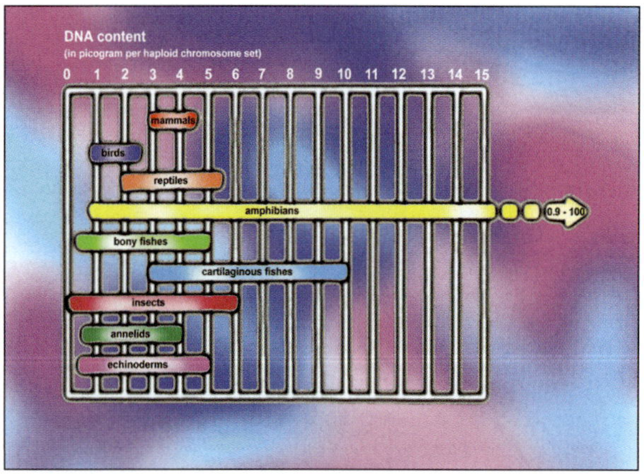


There is no evolutionary trend shown by the amount of DNA of different creatures.
Note: The *diploid* number is the total number of chromosomes in a cell. The *haploid* number refers to half that total number, i.e. the number of chromosomes in a sperm or egg.

Nor do the differences in DNA content reflect the complexity of the organism. For example, amongst their own kind, all insects and all amphibians would seem to be equally complex, yet the amount of DNA within each of these groups can vary by a huge amount. Similar degrees of differences are commonly found in plants. Many insects and worms have more DNA than many mammals. Amphibians and lungfishes have huge amounts of DNA, far greater than all other animals, yet they are not more complex than humans. Why these differences should exist is unknown, but we have seen already that, in times past, important organs such as the thymus and appendix were wrongly thought to be purely rudimentary, and that so-called 'junk' DNA was wrongly thought to have no biological function. Since the lack of understanding as to a purpose does not mean that no purpose exists, then their very existence and *persistence* suggests that the observed differences in the amount of DNA in different creatures may, in some way as yet unknown, be *necessary* to produce their different types.

Certainly, the differences in the DNA content seen in different animals do not suggest any particular trend of sequential step-by-step changes from simpler to more complex creatures. So, in terms of chromosome number, chromosome size, and DNA content, the observed patterns do not fit with an evolutionary model.

### *Chromosome alterations do not give rise to new creatures*

Considering the great length of the DNA molecule which constitutes even a small chromosome, it is surprising that breaks are, in fact, relatively rare. However, when breaks do occur, they can result in karyotype changes affecting chromosome number, shape and size. If a break does occur, any small broken fragment is usually lost (a *deletion*), since chromosome portions that lack a centromere cannot be captured and retained by the spindle mechanism during cell division. If the lost portion contains vital genetic information, then the result will be serious disease or death. Sometimes however, following a break, the fragments manage to reunite again, but not necessarily in the original orientation or on the same chromosome. A *translocation* occurs when one piece of a chromosome breaks off and relocates itself onto another chromosome. An *inversion* occurs when one piece of a chromosome breaks off and reunites again at the same site, but the other way round. A *reciprocal translocation* occurs when two chromosome pieces break off from two different sites and reunite by swapping places. In a *balanced translocation* (when one complete chromosome joins onto another chromosome) there is no practical loss or gain of any genetic material, so the individual retains all of its genes intact, but the total chromosome number is reduced by one.

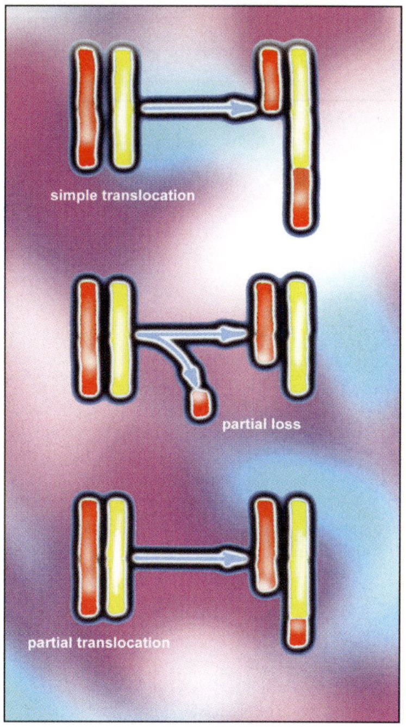


When chromosomes break up and rejoin, this can involve disruption or loss of genes.

Muntjacs are small Asian deer which are usually found in woodland. They have small pointed horns and the males have small tusks that extend down from the upper jaw. There are two commonly known types, the Reeves muntjac which is found in the southern part of China, and the Indian muntjac. Although they both have about the same total amount of DNA, the difference between the Indian and Reeves muntjac deer karyotypes is so great that it makes the difference between chimpanzees and human karyotypes look quite insignificant. The Reeves muntjac has only 3 pairs of large chromosomes and a Y chromosome, and the Indian muntjac has 22 pairs of chromosomes plus an X and a Y chromosome. However, despite the marked differences in the size and number of their chromosomes, the genes contained within these two animals are essentially the same, the only real difference being that the genes occupy different relative positions within the genome. Therefore, when we come to consider the notion of human evolution in chromosomal terms, we should likewise not expect any supposed process of chromosomal rearrangements to alter drastically either the genes themselves, or their functional outworkings in terms of appearance, behaviour and intelligence.

Reeves muntjac deer

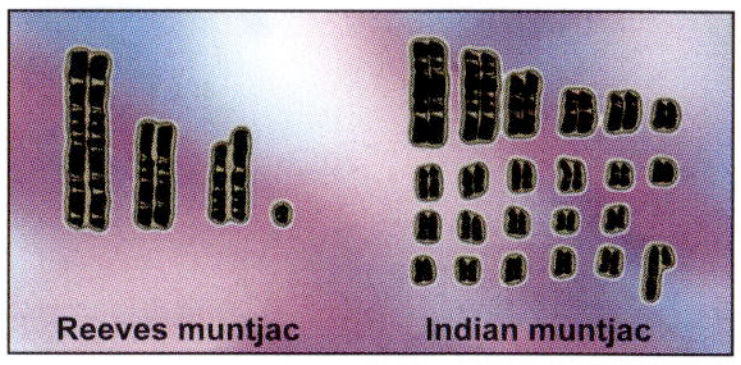


The Reeves and Indian muntjacs have vastly different chromosome numbers. However, the two animals are practically indistinguishable.

The muntjac deer is not an exceptional case and there are other examples of differences in karyotypes that do not result in different animals. Populations of Australian *wallabies* separated by as little as 70km have been shown to have significant genetic differences but they are all still wallabies. Populations of house mice usually divide into small family groups of 4-7 reproductively active individuals with one dominant male. Group territories are defended and there is little interchange between these social groups. Yet house mice populations in alpine valleys in Switzerland and Italy have been found to have a multiplicity of chromosomal rearrangements. Even in the confines of *a single barn*, karyotypically-distinct groups of house mice have been recognised. But for all their diverse karyotypes, all these animals are still house mice. In Israel, there are 4 known types of *mole rat*. The two northern types have 52 and 54 chromosomes; the central and southern types have 58 and 60 chromosomes respectively. However, they are all still mole rats, the genetic content, appearance and behaviour of which remain essentially identical. *Vervet monkeys,* which generally exist in small troops consisting of a single adult male and several adult females, are karyotypically diverse, but animals comprising these troops are still all recognisable as vervet monkeys.

In contrast, the segregation of other creatures into isolated breeding groups *does not necessarily* result in karyotype differences. For example, whilst inbreeding in dogs and in domestic horses results in the segregation of their genes, all dog and all domestic horse karyotypes remain the same, and furthermore, despite the diverse shapes and sizes of individuals, interbreeding is still possible. The African and Asiatic lion have no marked difference in karyotype, yet they have surely been separated into distinct breeding groups for as long as Reeves and Indian muntjac deer have existed. Similarly, despite its geographical separation, and despite the marked difference in its skin pattern, the extinct *quagga* has been shown to be genetically *identical* to the plains zebra. Or, to put it another way, the quagga was just a zebra with fewer stripes.

### *Break-up of genes lead to diseases and congenital abnormalities*

So far, we have seen that in any animal group, any reorganisation of chromosome sections that contain *whole* genes does not result in a different animal emerging. A good example in humans of a balanced translocation is when one chromosome 21 joins onto another chromosome. An individual with such an arrangement has 45 chromosomes instead of the normal 46, but is, in every respect, *entirely normal.* However, if rearrangements involve chromosome sections such that important genes are disrupted, then we should expect *non-viable* or *diseased* individuals to appear. And this is exactly what is seen in practice.

Not surprisingly, the loss of large amounts of chromosome material causes serious disease. But even *microdeletions* (the loss of tiny bits of chromosomes), when associated with the loss of parts of important genes, can result in very serious diseases and developmental abnormalities. For example, a microdeletion of chromosome 22 is found in 30% of children born with very serious abnormalities of the heart and of the blood vessels that arise immediately from the heart (such as the *tetralogy of Fallot* syndrome, where there is a hole in the wall between the two ventricles, and deoxygenated blood which should normally go to the lungs passes instead into the aorta and around the body). A microdeletion of chromosome 7 is found in most cases of *Williams syndrome* in which there are aortic valve abnormalities and learning difficulties. A microdeletion is found in the *Di George syndrome* (which includes heart defects, and parathyroid and thymus gland abnormalities). Similar microdeletions of chromosome 22 are found in the *velocardiofacial syndrome*, in which cleft palate, heart defects and learning disabilities occur. The *Cri du chat syndrome* results from the loss of a tiny fragment of chromosome 5 and this causes severe mental retardation. It is notable that in many diverse chromosomal abnormalities, severe defects in intelligence are commonly associated with other congenital abnormalities. Other examples of minor deletions include the eye cancer *retinoblastoma* which is caused by the loss of part of the long arm of chromosome 13. Obviously, the question here arises: how did any (supposed) ancient ancestral creature live before essential genes (such as those contained in the vital portions of chromosomes 5,7, and 22) ever evolved? Clearly any such hypothetical creature could not have survived to have ever left any progeny.

*Reciprocal translocations* can cause severe problems if important genes are disrupted in the process. For example, a subtle reciprocal translocation has been found in a patient with *recurrent miscarriages* (where a small portion of chromosome 5 was found to be attached to the long arm of chromosome 14, and a portion of chromosome 14 was attached to the short arm of chromosome 5).

The *Philadelphia chromosome* in humans is commonly found in the malignant cells of individuals with *chronic myeloid leukaemia*. It was originally thought that the Philadelphia chromosome represented a chromosome 22 which had lost a small portion of its long arm. However, it has since been shown that:

> ...the Philadelphia chromosome is actually involved in a reciprocal translocation with chromosome 9...the tip of the long arm of chromosome 9 has been joined to the body of chromosome 22, and the distal portion of the long arm of chromosome 22 has been joined to the body of chromosome 9. Peter Snustad, Michael Simmons and John Jenkins. 'Principles of genetics'. Wiley. 1997. p599.

This results not only in the disruption of the particular gene that lies at the breakage point on chromosome 22, but in addition:

> Chromosome 9 contains a gene that codes for a protein...that plays a role in cell proliferation. As a result of the translocation, one small end of the protein is replaced by about 600 extra amino acids encoded by a gene carried on the translocated piece of chromosome 22. This new, greatly enlarged protein apparently retains the catalytic activity of the original version but is no longer subject to the cell's normal regulatory mechanisms. Gerald Karp. 'Cell and molecular biology'. Wiley. 1996. p524.

This then results in the uncontrolled growth of a certain type of white blood cell to produce a malignancy - chronic myeloid leukaemia. *Burkitt's lymphoma*, a white blood cell cancer, is also caused by a reciprocal translocation. Other serious diseases caused by chromosomal rearrangements are being discovered all the time.

Therefore, on the evidence, we should conclude that it is highly likely that random loss of chromosome material, or breaks in chromosomes associated with random fusions of gene fragments, would result in damaging changes in the genes involved. In addition, it is implausible that any *new* enzymes, structures or organs could be so produced.

### Human and ape genomes

The notion that humans are evolved apes is so deep-rooted in many people's minds that, along with the *common chimpanzee* and the more slender *bonobo chimpanzee*, some people seem almost proud to class themselves as the *third chimpanzee*. So, what then should we make of the common statement that humans have over 98.5% of their genes in common with chimpanzees? In actual fact, this figure gives a warped perspective on (supposed) relationships since mice and humans have been shown to have *more than 99%* of their genes in common:

> The proportion of mouse genes without any homologue currently detectable in the human genome (and vice versa) seems to be less than 1%. R.H. Waterston et al. 'Initial sequencing and comparative analysis of the mouse genome'. Nature. Vol 420. 5th December 2002. p520-524.

Futhermore, recent work has shown that although single nucleotide substitutions account for a difference between chimpanzee and human DNA of around 1.4%, there are additional differences (of around 4%) as a result of whole sections of DNA being either present or absent from either creature:

*The term '*indel*' refers to a section of DNA that is present or absent in one creature but not the other.

> ...the DNA of both species was littered with indels* that add around another 4% to the genetic differences between us and chimps...The result is only based on about 1 million DNA bases out of the 3 billion which make up the human and chimp genomes...it's just a glance. Roy Britten. Proceedings of the National Academy of Sciences, DO1/10.1073/pnas.172510699. Reported in 'Frontiers'. New Scientist. 28th September 2002. p20.

**No similar examinations were made of gorilla or orang-utan genomes.

This means that chimpanzee and human genomes are likely to differ by more than 5.5%**. So, does that mean that *mice* are our closest cousins?

Actually, *all* mammals have similar amounts of DNA, distributed mainly over 40-60 chromosomes. This is not at all surprising, since a vast amount of all cellular chemistry is essential for life and therefore must *necessarily* be common to all mammals, whether mouse or elephant. Furthermore, all mammals have almost identical structures: two eyes, two ears, four limbs, similar head, chest and abdominal structures, similar hormonal and blood systems etc., etc. So, without knowing anything about biochemistry at all, we should predict that vast amounts of DNA *must* be common to all life forms, not just to mammals. And so it is.

If DNA replicates with extremely high fidelity (as it does); if all of an amoeba's biochemical activity is finely honed (as it is); and if all mammals evolved ultimately from an amoeba-like ancestor; this means that humans evolved as the result of a large series of *errors* in the replication of an ancient amoeba. Since errors reduce the amount of information in a system, and since compounding errors cannot continually introduce new and useful information, the evolution of single-celled organisms to Man by random mutations is impossible.

If the human genome contains 25,000 genes, then humans and mice could differ in something like 250 genes. If only 250 genes separate us from mice, and (although it is untrue) if by this we were to suggest that one gene reflects one facet, then there would be *only* 250 facets that need to be changed in a mouse to turn it into a human being. Then on the same basis (one gene equals one facet), that would mean that only 25,000 facets separate humans from a speck of dust. Is it really possible for those 25,000 genes *alone* to contain all the information necessary to produce a man from simple molecules? Clearly, this cannot be the case. It is more likely that the reason that the manufacture of either worms or elephants can be accomplished with so few differences in the genes is that, within the time and space available to a developing embryo, many gene products interact in a multitude of ways, with the number of *permutations* of possible gene interactions being incalculably huge. Since any new mutation could have effects in a bewildering number of permutations, then it is implausible that any ape lineage could have been so fortunate as to acquire *the very changes* necessary to cause it to become a human being. In other words, it is implausible that chimpanzees are simply a few genetic steps away from becoming human beings.

There still remains a common notion that the differences between the human and ape chromosomes can be interpreted as reflecting a series of changes from an orang-utan ancestor through gorilla and chimpanzee to Man. Chromosomes can be stained using various techniques which show up different patterns of bands that are unique to different creatures.

The *G-banding* technique shows up light and dark bands along the whole length of a chromosome:

> Chromosome banding analysis suggests the evolutionary history of these primates. A precursor of the ancestral hominoids and orangutan had the same chromosomes as the hominoid ancestor except for five chromosomes: 3, 7, 10, and Y are similar to those of orangutan, and chromosome 17 is like that of the rhesus monkey and the baboon. It appears that the orangutan diverged from this ancestor, with alterations in the structure of chromosomes 2q, 4, 8, 11, 17, and 20. The hominoid ancestor's chromosomes were similar to those of the human, chimpanzee, and gorilla. Before the divergence of humans and chimpanzees, their ancestor shared similar chromosomes 2p, 7, and 9. Human divergence from the chimpanzee is marked by the fusion of chromosomes 2p and 2q into chromosome 2, and by inversions in chromosomes 1 and 18. By contrast, the gorilla experienced nine chromosomal changes and the chimpanzee seven. Peter Snustad, Michael Simmons, and John Jenkins. 'Principles of genetics'. Wiley. 1997. p755.

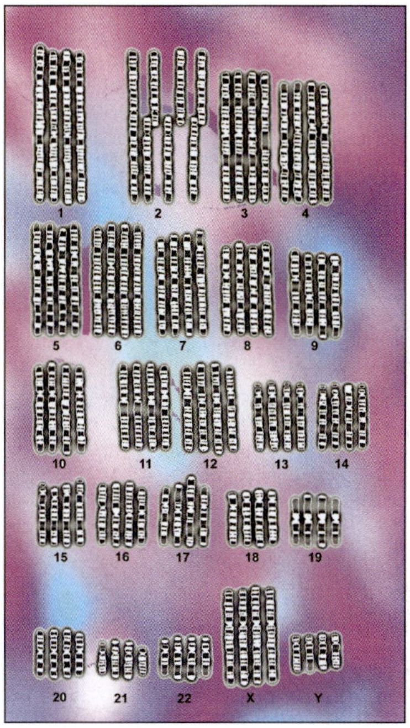

Human, chimpanzee, gorilla and orang-utan chromosomes compared (from left to right respectively).

Such an entry in a text-book of genetics certainly has an impressive air about it, but is it true? In actual fact, the evolutionary trend *presumed* to be suggested by the banding seen in the ape and human chromosomes is still based on the *prior assumption* that humans evolved from apes. So in essence, what is being described here is an *hypothetical* series of breaks followed by *hypothetical* re-orientations of portions of chromosomes. As when comparing fossil skulls, the *Horse Principle* shows us that even if the karyotypes can be shown to share similarities, and an *hypothetical* trend one to the other is conceived, this still does not constitute evidence that such a progression actually occurred.

In fact, if no evolutionary assumption is made, then based on the band patterns, a number of different evolutionary trees can be made up. For example, while the orang-utan X chromosome (OUX) has the same pattern as the human X chromsome (HX), the chimpanzee and gorilla X chromosomes (CX and GX) are both slightly different. From this we *could* suggest that HX gave rise to OUX, and that CX and GX arose from either HX or OUX.

Furthermore, it can be seen that of the 23 pairs of human chromosomes, eight (5,6,12,13,17,21,22,X) resemble the orang-utan chromosomes *more closely* than they do either the chimpanzee or gorilla chromosomes. And four human chromosomes (8,10,14,18) resemble the orang-utan chromosomes *as closely* as they do those of the chimpanzee, but *more closely* than they do the gorilla chromosomes. In a further seven (1,2,4,15,16,20,Y), the human chromosomes resemble the orang-utan chromosomes as closely as either the chimpanzee or the gorilla chromosomes. So, in fact, *it is only in three chromosomes (3,7,11) that humans are more similar to chimpanzees and gorillas than they are to orang-utans.*

With the *C-banding* technique (in which only small areas of chromosomes are stained), it is found that the gorilla pattern is closer to that of the human pattern, with the chimpanzee actually lacking completely the C-bands found on the human chromosomes 1, 9, 16 and Y. In addition, bands on the ends of the chromosomes (*terminal bands*) are:

> ...found in approximately half of all the chromosome arms from chimpanzee and in nearly all of those from gorilla, but they are conspicuously absent from man and orangutan. Jorge J. Yunis and Om Prakash. 'The origin of man: a chromosomal pictoral legacy'. Science. Vol 215. 19th March 1982. p755.

So, the notion that the human and ape chromosomes confirm a progressive evolution from orang-utan ancestors through to humans, and show that chimpanzees are very close relatives of humans, is invalid. In fact, we could interpret the banding patterns in all sorts of ways, and *if* we wanted to, we could make a case for the orang-utan or gorilla being man's closest relative, or for great apes having evolved *from* humans.

There are yet more problems raised by the chromosomes. Natural selection requires that new species arise by mutations in small isolated groups of individuals. If chimpanzee ancestors evolved into humans, then each difference in the chromosomes of humans and chimpanzees should represent distinct populations of intermediary creatures. This would mean (by one reckoning, as judged by the differences in the chromosomes) that at least 7 different creatures would have existed in the (supposed) lineage from chimpanzee to modern Man, of which just one might be the australopithecines.

If the differences that exist between chimpanzee, orang-utan, gorilla and human chromosomes do in fact reflect a series of translocations and deletions along with gene mutations over millions of years, then we should predict that australopithecine, Homo erectus and Neanderthal chromosomes *must* have shown some progressive trend towards the modern human karyotype. In addition, this changing karyotype trend *must* have reflected progressively increasing intelligence, bipedal gait, ability for speech, hairlessness, etc. However, at the present moment we do not have any way of knowing what australopithecine, Homo erectus, and Neanderthal chromosomes were like.

If living apes, australopithecines (Aust), Homo habilis (HH), Homo erectus (HE), Neanderthals (N), archaic Homo sapiens (AHS) and modern humans shared a common ancestry, then at some stage (suggested here after Homo habilis) the ape chromosomes would have reduced in number to attain the human format. But this is not as plausible as it might at first seem (see main text).
CA: common ancestor
OU: orang-utan
G: gorilla
C: chimpanzee

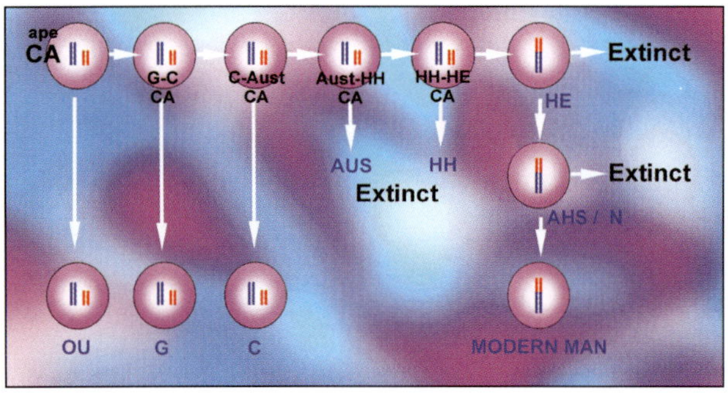


It is accepted that the isolation of small groups of animals and their subsequent inbreeding is an essential requirement for the evolution of specific lineages. This is a problem because inbreeding inevitably allows any recessive genes (which would otherwise be suppressed by dominant genes) to manifest, producing disease and abnormalities:

> ...Any time there is a departure from random mating in a population...rare recessive traits may appear with a frequency

> that is much higher than in the general population. Peter Snustad, Michael Simmons & John Jenkins. 'Principles of genetics'. Wiley. 1997. p683.

Incest in humans (involving father/daughter, brother/sister, or mother/son) has a high risk of genetic disorders appearing in the offspring:

> There is a 1 in 3 chance of mental handicap which can be severe; 1 in 15 for congenital malformations and 1 in 10 for autosomal recessive disorders... John Forfar & Gavin Arneil. 'Paediatrics'. Vol 1. Churchill Livingstone. 1984. p79.

It is well recognised that genetic variety within any breeding group is essential for the preservation of a healthy population. For example, regarding the white rhino:

> ...the rarest large animal on earth...with such small numbers there is a possibility of inbreeding that could weaken the stock... Alan Root. 'Survival Special: Garamba, the impossible elephants'. ITV Broadcast. 7th January 1998.

And regarding tigers:

> Now that areas where tigers can live are so much restricted and numbers are so much smaller, inbreeding will inevitably occur; over the next few decades this will weaken the genetic pool and have an adverse effect on the species as a whole. Valmik Thapar. 'The tiger's destiny'. Kyle Cathie. 1992. p86.

> Most tiger populations consist of fewer than 100 individuals, of which only about 40% constitute the breeding population. Inbreeding is inevitable and father-daughter and mother-son matings have been recorded...Genetic deterioration and reduced cub production will always follow from dramatically reduced population numbers. 'The tiger...will this be the last chapter?' World Wildlife Fund magazine. March 1998.

However, disregarding the potential for all sorts of congenital diseases in the progeny, how exactly could 48 chromosomes of a (supposedly) ancestral ape evolve into the 46 chromosomes of a human being? The answer commonly proposed is that two of the chimpanzee chromosomes merged to form the chromosome 2 of humans, thereby reducing the chromosome number.

### *Reduction of chromosome number*

Reduction of chromosome number would first of all require one chromosome to become translocated onto another chromosome, resulting in an individual with a single *balanced translocation* (BTC 1) and 47 chromosomes.

If two BTC 1 apes were to mate, then some of the offspring would be born with 2 balanced translocations (BTC 2) having gained one of the abnormal chromosomes from each parent. Such an individual would have 46 chromosomes (the other offspring of such a mating would either be normal or would not survive). As we have previously seen, such an individual would still have the normal *amount* and *quality* of genetic material, so it would remain a normal ape, and outwardly, it would be no different from any of its normal siblings.

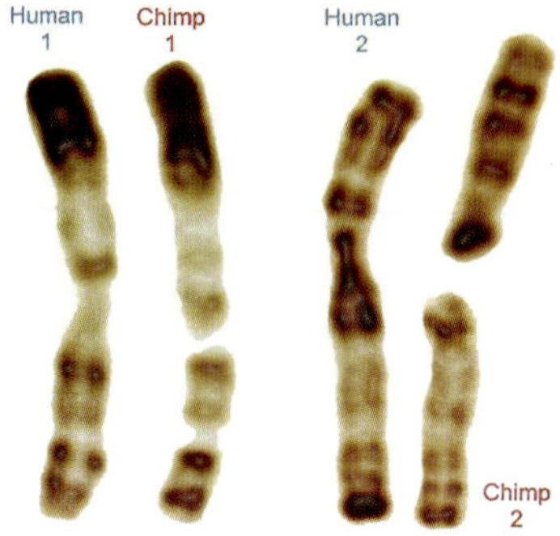

Human chromosomes 1 and 2 compared with three chimpanzee chromosomes. One common view is that two chimpanzee chromosomes fused together to produce the human chromosome 2.

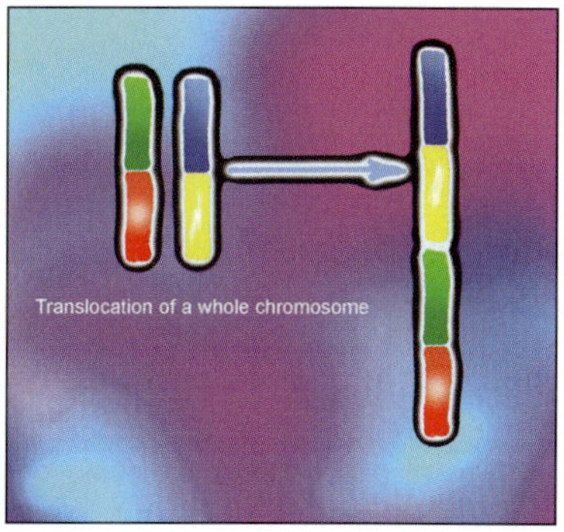

A balanced translocation occurs when one whole chromosome joins onto another chromosome. Since there is no loss or gain of genetic material, an affected individual will be entirely normal.

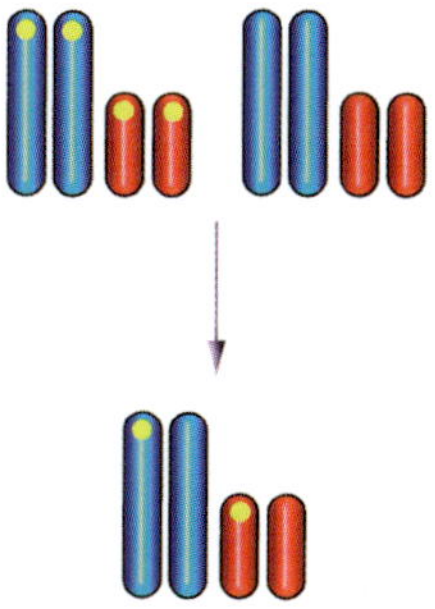
Diagram showing that the mating amongst normal individuals normally results in the equal distribution of chromosomes in the progeny, resulting in normal offspring.
For clarity, only two pairs of homologous chromosomes are shown. Maternal chromosomes shown with yellow dots.

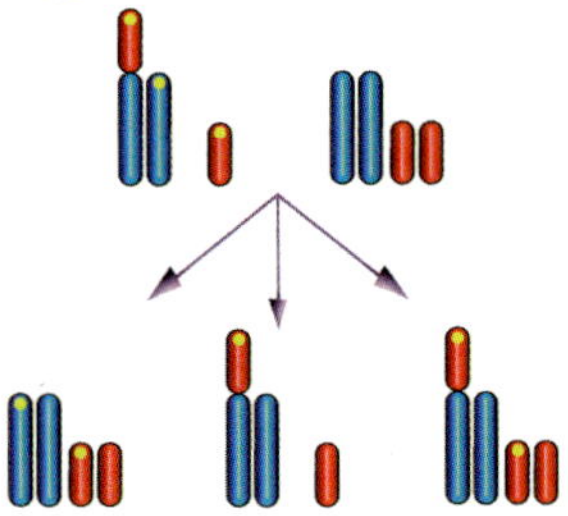

A mating between a BTC 1 and a normal individual results in 3 possible types of viable offspring.
The fourth possible type (monosomy, which has only one chromosome of an homologous pair) is incompatible with life and is not shown here.

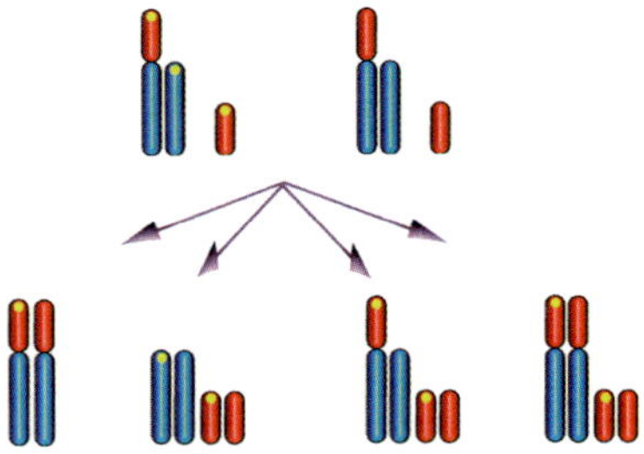

The mating of two BTC 1 individuals can result in some offspring being born as BTC 2, with a reduction in the number of their chromosomes. (The embryo with tetrasomy will be non-viable.)

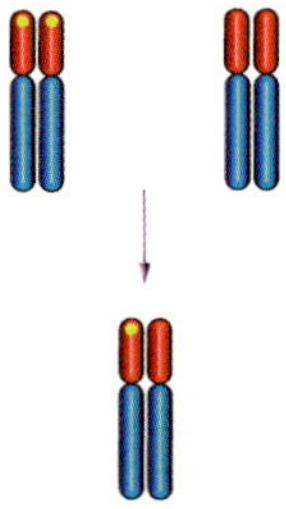

Mating between two BTC 2 individuals will result in all offspring being BTC 2.

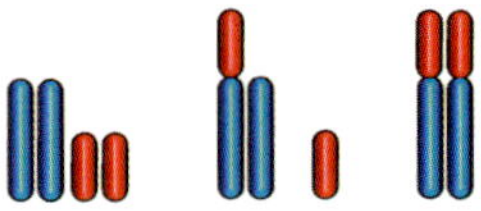

Since they all have essentially the same genes which are simply shuffled about in different arrangements on the chromosomes, all these individuals will look and behave the same. They will therefore represent the same creature.

Let's go through the whole process with the help of a few diagrams. A mating between a BTC 1 and a normal (N) individual will result in four possible types of offspring: one being monosomy (MS, which has only one of a pair of homologous chromosomes and will not survive), one trisomy (TS, which has one chromosome in addition to the two of an homologous pair: it will be severely disadvantaged and unlikely to survive), one N, and one BTC 1.

Once some BTC 1 offspring have grown up, interbreeding between two BTC 1 individuals will be possible; such a mating will result in the possibility of a number of types of offspring being born: one being MS, one TS, one BTC 1, and one double balanced translocation carrier (BTC 2). The latter would have a balanced translocation of both members of an homologous pair, resulting in a *reduction in chromosome number*.

Subsequent mating between two BTC 2 individuals will result in all offspring being BTC 2. In other words, offspring will have a reduction in chromosome number from 48 to 46.

But the problem here is that since the chromosomes of normal, BTC 1 and BTC 2 individuals would carry identical genes, arranged in virtually the same order, all three types of individuals would look and behave exactly the same as each other. Therefore, there would be no force that would cause them to become separated. They would all find each other equally attractive sexually, and then continue to interbreed, and as a result a number of different chromosome types would arise in the offspring. Any abnormal individuals would just be unlikely to survive, and all individuals, either Normal, BTC 1 or BTC 2, would be equally attracted sexually, one to the other.

In fact, chimpanzee behaviour itself would tend to prevent the formation of isolated breeding groups. Chimpanzees live in groups of up to 60 animals, in which both males and females will mate as often as possible and with as many partners as possible. In particular, Bonobo chimpanzees (which are supposedly closer relatives to humans than common chimpanzees) spend extraordinary amounts of time mating. Thus, since all individuals of a troop are likely to interbreed without any preference or prejudice, then, rather than any tendency for separation of Balanced Translocation individuals into isolated breeding groups, chromosomal mixing of the group is likely to be ensured. So, in chimpanzees, the isolation of individual groups resulting ultimately in a reduction in chromosome number is unlikely to occur. This is not to say that (as in humans) different subgroups of chimpanzees could not occur by geographical separation, since, in addition to the more distinct bonobo chimpanzee, at least four genetically-different groups of common chimpanzees have been identified.

The key issue is that different creatures do seem to have different propensities for undergoing changes in chromosomal number, and the reasons for such differences may be dictated largely by specific reproductive and social behaviours. Just because it is possible for muntjac deer and mice to undergo balanced translocations, and to separate into distinct breeding populations with varying chromosome number, it does not follow that such a thing is possible or likely in the great apes, *in the wild*.

If orang-utans have been separated from chimpanzees for (supposedly) more than

8my, we should expect that, if it were possible, orang-utans would by now have different numbers of chromosomes to chimpanzees. Furthermore, we should expect that those orang-utans that are separated into distinct breeding populations in different clumps of forest in different islands should by now have different numbers of chromosomes, one compared to the other. The fact is, however, that while the orang-utans of Borneo and Sumatra are as genetically distinct from one another as chimpanzees are from gorillas, those two populations of orang-utans have the same karyotype and represent very much the same animal. In addition, if the modern chimpanzee is separated from the modern gorilla by over 8 million years, then we should expect the two animals to be as different from each other (in chromosome number, external appearance and behaviour) as humans are from chimpanzees. But clearly they are not.

Also, since balanced translocations cannot alter the outward appearance or the internal physiology/biochemistry of an organism, then, if a BT 2 occurred in chimpanzees, some of which eventually gave rise to humans, we should predict that there should still remain some (or many) chimpanzees with 46 chromosomes (chimps[46]). We would not expect all chimps[46] to have become extinct.

### *Human chromosome 2*

This still leaves us with the problem of the human chromosome 2 (HC2), the sequence of which certainly closely matches that of the chimpanzee chromosomes 12 and 13. More than that, HC2 also has in its middle section a zone which contains repetitive sequences (TTAGGG/AATCCC) just like those that are found at both ends (*telomeres*) of chromosomes:

T T A G G G - C C C T A A
A A T C C C - G G G A T T

This is *exactly* what we would expect if HC2 is indeed the result of the fusion of two smaller chromosomes. But surely this is *absolute* confirmation of our ape ancestry? Actually, no, it can't be, because of the Big Picture: properly interpreted, all of the fossil, anatomical, physiological and embryological data unambiguously show that the spontaneous evolution of atoms to human consciousness could not, and did not, happen. So, we must recall the *Don't Be Fooled Again Principle* - there *has* to be another explanation. However, the search for the solution can be seriously hampered by the repeated brain-washing that we *are* evolved apes. Let's follow the trail logically. The evidence seems overwhelming that HC2 most likely *is* the result of the fusion of two smaller chromosomes. All the evidence shows that humans are not evolved apes, and have always been human. Therefore, HC2 could well be the result of the fusion of two smaller HUMAN chromosomes. So, 48 could be the number of chromosomes that people (Humans[48]) had in the very distant past.

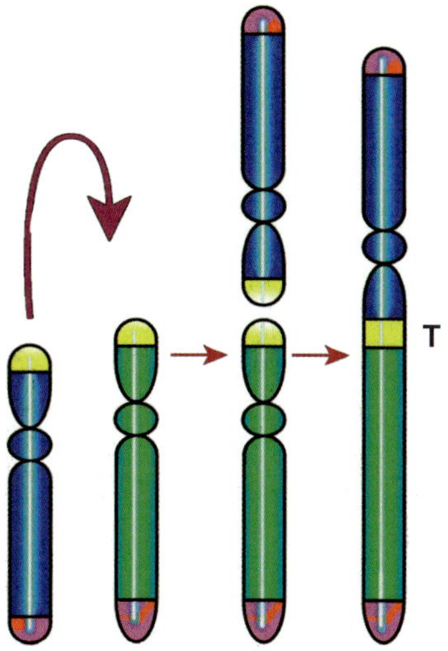


It seems certain that human chromosome 2 is the result of two smaller chromosomes having been joined together end to end, with a telomeric-like repetitive sequence (T) in its mid-zone.

Now, the same problem arises as when we discussed chimpanzees reducing their chromosome number. If humans originally had 48 chromosomes and if one group sustained a fusion resulting in people (Humans$^{46}$) with 46 chromosomes, then we should still see different populations of Humans$^{46}$ and Humans$^{48}$. But since that is clearly not the case, then there *must* have been a catastrophe that killed off *every* Human$^{48}$, with a massive bottleneck through which only Humans$^{46}$ passed and so survived. Furthermore, since *all* living humans are genetically *extremely* similar, all Humans$^{46}$ must have come from a *very* small gene pool *very* recently:

> Living humans are strangely homogeneous genetically, presumably because…their ancestors underwent a population bottleneck that wiped out many variations. P Kahn & A Gibbons. 'DNA from an extinct human'. Science. Vol 277. 1997. p176-177.

Since there would be no outward or inward difference between Humans$^{46}$ and Humans$^{48}$, this leaves the problem of just what sort of catastrophe would filter out *only* and *all* the Humans$^{48}$. It could not have been infective, or some sort of illness, because there would have been no difference in immunity and hardiness in the two groups. In addition, it could not have been geographically localised because people would have been widespread. So it had to be of *global* proportions.

At this point we need to make a little temporary detour, but be assured, we will not let go of the scientific method. However, we must be careful not to let prejudice or THE BOX* limit our reasoning.

* THE BOX: the flawed notion that evolution is a fact, and that we are modified apes.

Is there any evidence that gives us insight into the catastrophe? The first thing to clarify is that scientific evidence can take different forms. In law, the evidence of an (accepted to be honest) eyewitness is good evidence. If a person says he saw a green light in the sky, then unless there is good *evidence* to say that he did not, we basically have to accept that that is what he saw.

What about data in scientific literature – can we trust that? Most workers who depend on genome information have no direct involvement with gene sequencing and have to accept that the published human and chimpanzee data is true. Furthermore, in this case, even the gold standard of experimental repetition to confirm data accuracy is largely inapplicable since very few people have the time, equipment and money to verify the genome analyses that have been published. In other words, we have to take that information largely on trust. Now, it must be understood that for the scientific process to function properly, only the evidence can be allowed to speak. Professional qualifications or prejudice have no place alongside evidence. However, the scientific method is not what it might seem. For example, most authors on human fossil remains have probably not seen an *original* human fossil, but work from casts of the originals. Some paleontologists even seem to work from photographs:

> In one case (Ruben et al. 1999), the alleged identification of a crocodile-like, hepatic diaphragm in the basal coelurosaurian theropod Sinosauropteryx was not even based on a study of the fossil, but merely on examinations of photographs. In reality, the purported diaphragm is merely a longitudinal fracture in the slab, resulting in the breaking away of a piece

> of matrix (P. J. Currie, personal communication). The fossil shows no sign of a diaphragm. P Christiansen & N Bonde. 'Axial and appendicular pneumaticity in Archaeopteryx'. Proc. R. Soc. Lond. B (2000) 267, 2501-5.

Also, it is not uncommon for data or discussion that contradicts an author's conclusion to be entirely omitted. This is particularly important when we see that none of the genome papers offer or discuss any alternative scenarios to that of THE BOX. We saw an example of this when Jones gave the impression that the biochemical data connects hippos to whales, but he omitted to quote the second half a paragraph which contradicts that stance:

> ...three genes (the mitochondrial cytochrome b and two nuclear milk caseins) indicate an association between Cetacea and Hippopotamidae. Three others (the mitochondrial 12S rRNA and two nuclear hemoglobins) **do not** support such a relationship. (My bold). C Montgelard et al. 'Phylogenetic relationships of artiodactyls and cetaceans as deduced from the comparison of cytochrome b and 12S rNA mitochondrial sequences'. Molecular biology and evolution. Vol 14. 1997. p550-559.

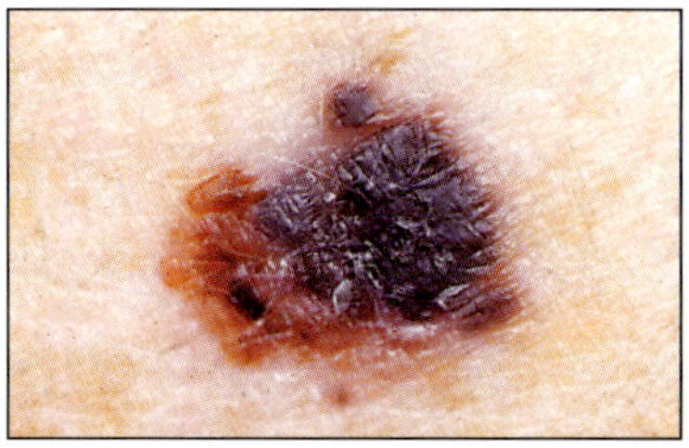

Malignant melanoma.
Experience does support the view that melanoma may be triggered by sudden exposure to intense sunlight: for example, when a person lives all year in a temperate climate and then flies off for a very sunny holiday for a couple of weeks.

No one is exempt from producing nonscience. For example, R Smith, a past editor of the *British Medical Journal* (BMJ), is recorded as saying in effect that only a *worringly small* percentage of articles submitted to the BMJ are statistically valid. In addition, the witness of experts in a particular field can be completely contradictory. For example, with the same world-wide data available for study, there is serious disagreement about a recent claim that melanoma is *not* related to sun exposure, and also about the concept that ultraviolet light may actually help *prevent* other cancers:

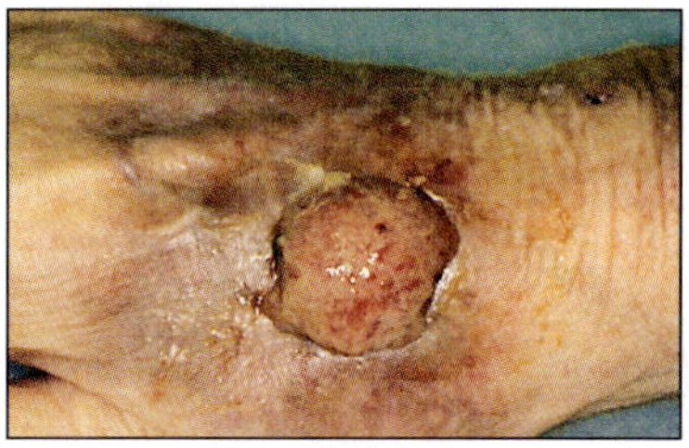

Squamous cell carcinoma.
This often arises within sun-damaged skin. Although this varies in its aggressiveness, it is fatal unless removed before it has spread.

> As a consultant dermatologist running two week wait cancer clinics and dealing with between four and eight melanomas a week...I find Shuster's arguments that melanoma is not related to sun exposure **irresponsible and beyond rational thought**. (My bold). A Cooper. BMJ letters. 9th August 2008;337:a1126.

> Cooper [previous letter]...professes to be unaware of evidence that ultraviolet light may protect against some forms of cancer...The evidence linking major cancers – including breast, prostate, pancreas and colon – to UVB light under-exposure is actually stronger than is the evidence linking melanoma...to UVB over-exposure. R Quinton et al. BMJ letters. 9th August 2008;337:a1130.

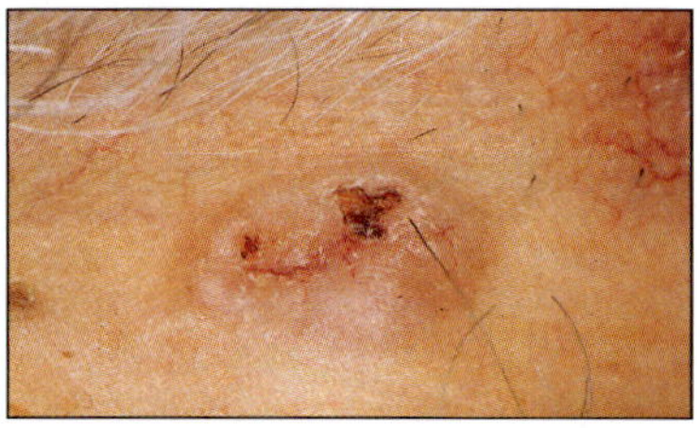

Basal cell carcinoma.
This often arises within sun-damaged skin but can also occur in normal skin. It is usually slow-growing and does not spread to other parts of the body. However, if not removed, it will grow progressively and cause severe local tissue destruction.

> If this [UVB] were as important as they [Quinton et al] suggest, Australians would be some of the healthiest around. An inspection of the readily available statistics shows that cancers of the breast, colon, and prostate kill with monotonous regularity in Australia and New Zealand as they do in western Europe. M Moncrieff & J Garioch. BMJ letters. 9th August 2008;337:a1131.

On a totally different subject, a recent report highlights the fact that while the UN's Inter-governmental Panel on Climate Change (IPCC) ruthlessly promotes the concept of a man-made global warming, the report "*Nature Not Human Activity Rules the Climate*" which is signed by scientists from 15 countries:

> ...presents a devastating analysis of the IPCC's case. Even on the IPCC's own evidence, the report concludes, 'anthropogenic greenhouse gases can contribute only in a minor way to the current warming, which is mainly of natural origin'. C Booker. 'Stop the $CO_2$ scare, before it's too late'. Sunday Telegraph. 20th April 2008.

A recent news report came out with the following statement, which might suggest (quite erroneously) to the unsuspecting public that there has been a significant* decline in children's school performances:

*Since from one year to the next, a reduction of 1% is of no statistical importance, and does not mean that there has been any meaningful change either in pupils' performance or in teachers' teaching, the statement '*fell by 1%*' is emotionally-charged.
The sound-bite that many listeners would come away with is that there has definitely been a decline in standards. However, on the data presented, such a conclusion is scientifically invalid.

> This year the number of students who passed their SATs **fell by one percentage point** compared to last year. (My bold). BBC Radio 4 News. 12th August 2008. (SAT = Standard Assessment Test, which relates a child's progress to same-age peers).

It is therefore clear that current scientific discussions and media reporting are by no means immune from being sloppy, of poor quality, completely flawed, or just plainly biased. All of the above shows that we have to be careful to be open and honest, not to mislead, and to follow where the evidence leads. The same has to apply to all reports in history. Clearly, *some* reports, such as those written in stone by egocentric kings, are likely to be either exaggerated or plain lies. However, lies and hyperbole are unlikely to arise without some motive. Since most people are generally honest, and if there is no significant evidence to suggest otherwise, it would generally be safe to assume that on matters that present no personal conflict of interest, their testimony would be reasonably sure. If that is not the case, then no one can trust anyone about anything. Clearly, that last statement is not true. People are generally trustworthy. So far, so good. Now, let's look at some ancient evidence.

In the cathedral in Carlisle, UK, there is a stone engraving of a *sauropod-like dinosaur*, which was made in 1496, hundreds of years before the outward form of any large dinosaurs became known from reconstructions of fossils in the 19th Century. So, contrary to commonly-held views, people over 500ya had enough knowledge of dinosaurs to accurately depict them. The question arises as to how they could have known what a sauropod looked like if no human being had ever reconstructed any dinosaur bones. One logical answer is that ancient people (even recent Humans[46]) had seen living dinosaurs and had passed on knowledge of their morphology.

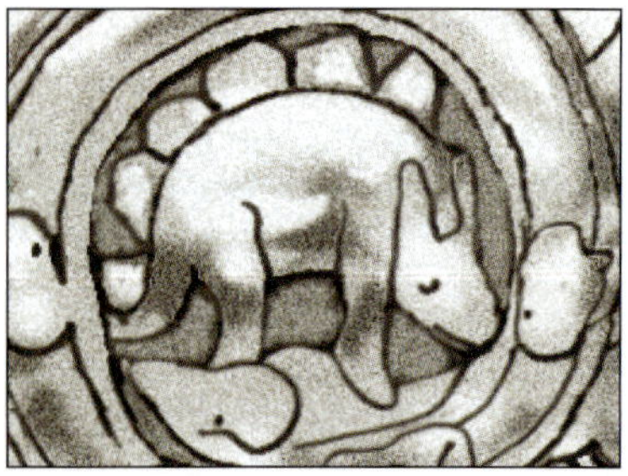

Drawing of a Stegosaurus stone carving from the temple at Ta Prohm, Cambodia, which was inaugurated in 1186 AD.

Although flying reptiles were not dinosaurs by anatomical definition, nevertheless, there is some provocative evidence suggesting that they lived until quite recently over the deserts of Arabia, as shown by the writings of the Greek historian Herodotus

In Carlisle Cathedral, UK, the brass engraving on the tomb of Bishop Bell (who died in 1496) depicts dinosaurs with spikes on their tails, along with fish, foliage and a hunting dog with a collar. Interestingly, the tomb is now covered with a mat.

(who lived around 484-425BC):

> [75] There is a place in Arabia not far from the town of Buto where I went to learn about the winged serpents. When I arrived there, **I saw** innumerable bones and backbones of serpents: many heaps of backbones, great and small and even smaller. This place, where the backbones lay scattered, is where a narrow mountain pass opens into a great plain, which adjoins the plain of Egypt. **Winged serpents are said to fly from Arabia at the beginning of spring**, making for Egypt; but the ibis birds encounter the invaders in this pass and kill them. [76]...The **serpents are like water-snakes...Their wings are not feathered but very like the wings of a bat**. (My bold). Herodotus. Chapters 75 and 76.

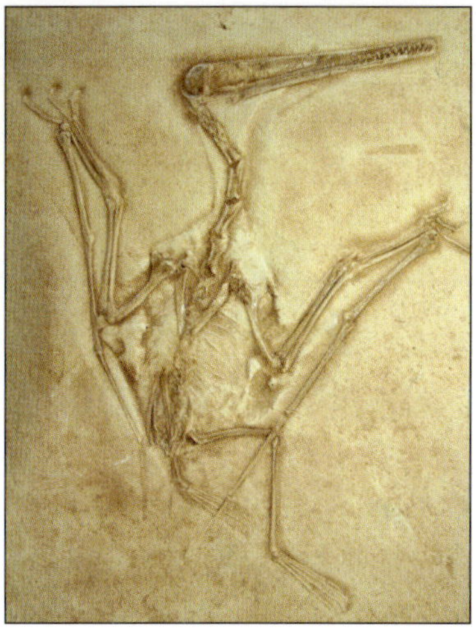

Above: Pterodactylus.
Below: Rhamphorhyncus.

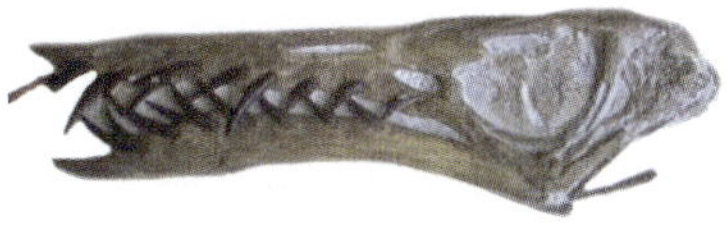

Flying reptiles are also mentioned in the Bible by Isaiah (around 700BC):

> **Isaiah 30**:[6] An oracle on the beasts of the Negeb. Through a land of trouble and anguish, from where come the lioness and the lion, the viper and the **flying serpent**...

Herodotus was quite wrong on some other points, for example, his description of a hippopotamus, but since he was often quite correct, we have to assess each of his observations individually.

The book of Job is probably one of the oldest texts in existence today, probably dating back to over 2000BC. It clearly describes large creatures such as the *behemoth* (which some commentators incorrectly say was a hippopotamus), which had a tail like a *tree*:

> **Job 40**:[15] Look at the behemoth, which I made along with you and which feeds on grass like an ox. [16] What strength he has in his loins, what power in the muscles of his belly! [17] His **tail sways like a cedar**; the sinews of his thighs are close-knit. [18] His bones are tubes of bronze, his limbs like rods of iron.

There is no doubt that the writer means to portray a picture of a massive animal with a *huge* tail, since the cedar tree is described elsewhere in the Bible as being used to make the *mast* of a ship:

> **Ezekiel 27**:[5]...they took a **cedar** from Lebanon to make a mast for you.

There is no known animal, living or extinct, which had a tail that in any form could be likened to a tree trunk - except for a sauropod.
Furthermore, since the Bible clearly relates in the *present* tense that Job was aware of such creatures living in the Jordan river, the inference is that they were living *alongside* mankind:

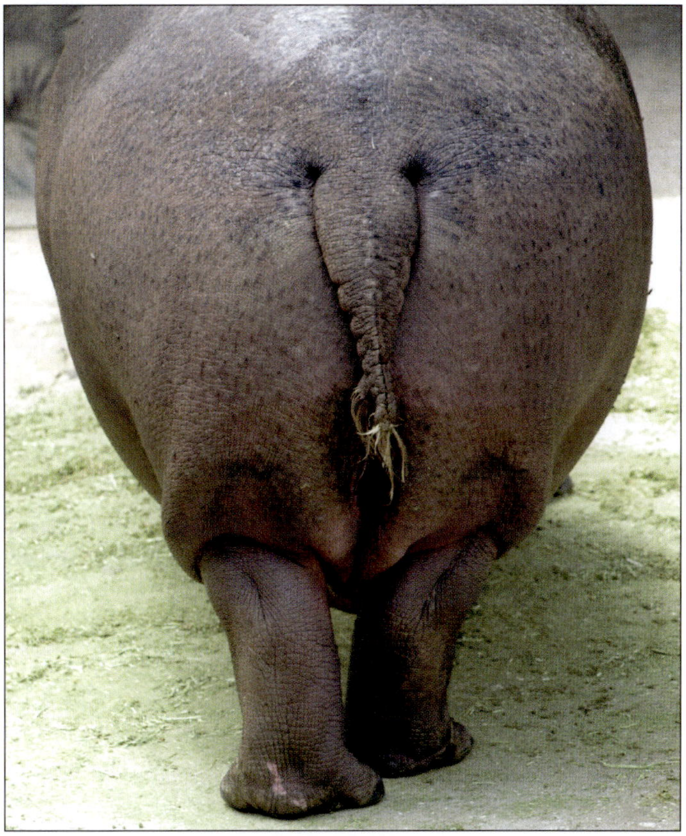

The hippo tail does not look like a cedar tree.

> **Job 40**:[23] When the river rages, he is not alarmed; he is secure, though **the Jordan** should surge against his mouth.

When we read such ancient accounts, we can only come away with the overwhelming conviction that these individuals were not stupid or naïve. So we have no choice but to give ancient writers the same attention as any living writer.

Now, what has all this to do with the extinction of all Humans[48]? We have seen diverse evidence that strongly suggests that humans and animals have not been around for millions of years. And we saw previously that the geological evidence points to a global flood that could have formed canyons and thousands of sedimentary layers in a matter of days, weeks or months.

It is a fact that there are a number of Flood stories from different civilisations from around the world, for example, the Hindu *Vedas* (unknown date, pre-1000BC), the *Gilgamesh flood epic* (pre-1500BC), and the *Astrahasis Epic* Babylonian flood (about 1635BC). Other flood stories from diverse peoples around the world (for example, from Inuits to native Australians) also contain elements that are similar to the Biblical account:

The 11th tablet of the Gilgamesh epic, narrating the flood story.

> These stories often have a common root that speaks of a real event, not just a myth. For example, every Amazonian society ever studied has a legend about a great flood. New Scientist. 31st January 2004. p44.

All these stories have such similarities that the most likely explanation for them is that they all came from a common original account. Whilst sharing a common root, some accounts have clearly taken on board fairy-tale content. In contrast, the Genesis version reads much like an historical account or a ship's log. It states categorically that the whole world was flooded, and that *all* humans and *all* land animals (apart from those on the ark) were killed:

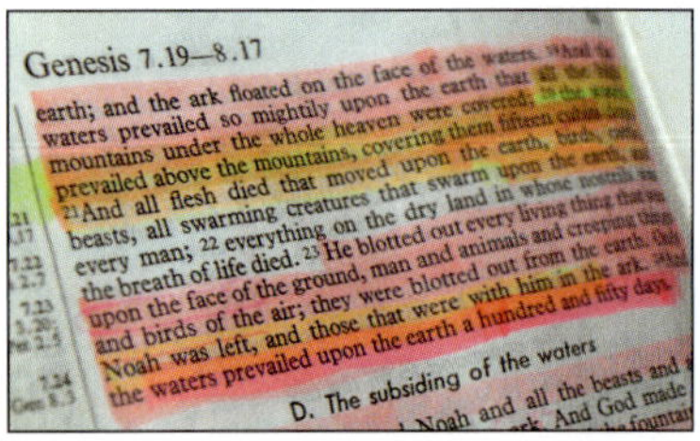


The Genesis flood account.

> **Genesis 7**:[21] And all flesh died that moved upon the earth, birds, cattle, beasts, all swarming creatures that swarm upon the earth, and every man…[23] only Noah was left, and those that were with him in the ark.

Since this left only the survivors that were in Noah's ship to subsequently populate the planet, we should predict: that all living animals can essentially be understood as being descended from the single breeding pairs of animal types that came off the ark; that all living people dispersed from one area on the Earth and are very closely related; and that all language, knowledge, religions, and technology would have arisen as those people reproduced and subsequently migrated away from the ark's landing site.

Now, we do have convergent data from multiple sources (geological, fossil, genetic and historical) that all support the notion of humanity passing through the most severe and recent population bottleneck. It is quite possible that marriages between close relatives had previously led to Noah and his tribe (amongst others) becoming Humans[46], with all Humans[48] being unnaturally filtered out by the flood bottleneck.

This would explain how only Humans[46] are alive today. Incidentally, this recent catastrophe model would predict that, genetically, all groups of land animals (e.g. cattle) and even marine creatures would be very homogenous.
Of course, ancient ignorance had led to all sorts of naïve concepts, but it is entirely proper that we give due consideration to the writings of ancient peoples. As we shall see in the next chapter, there is no evidence to show that humans were ever in some time past less intelligent than they are now. For example, it could be argued that the inspiration that sheep fur could be made into a poly-helix that could subsequently be woven into clothing, was as much an intellectual leap as was the elucidation of the DNA double-helix.

### *Chromosome differences do not define what makes a man a man*

So, from all the above we can see that the chromosome differences that exist amongst humans and the great apes cannot be responsible either for their physical and intellectual differences, or for the (supposed) evolution of humans from apes. Even amongst those who suggest that ape ancestors did evolve into human beings, and that in the process:

> Chromosomal inversions and translocations have moved genes into locations that have altered their activity patterns. Peter Snustad, Michael Simmons, and John Jenkins. 'Principles of genetics'. Wiley. 1997. p758,

there is the recognition that:

> ...the genes themselves have not changed much, nor have the amino acid sequences of the proteins they encode. Peter Snustad, Michael Simmons, and John Jenkins. 'Principles of genetics'. Wiley. 1997. p758.

So, we are all agreed that neither the arrangements of the chromosomes nor the genes that they carry can explain why apes are apes and human beings are human beings. But could the ape-human differences be explained by changes in the order of genes on the chromosomes causing alterations in the activities of *regulatory* genes? Without offering any *evidence* some people think that it can:

> ...researchers now **think** that divergence has largely been a consequence of changes in gene regulation. (My bold). Peter Snustad, Michael Simmons, and John Jenkins. 'Principles of genetics'. Wiley. 1997. p759.

But there are, for practical purposes, as many (supposed) translocations and inversions and deletions that separate the great apes from each other as there are those that (supposedly) separate apes from humans; and yet the opportunities for changes in regulatory genes amongst the great apes have not produced any significant differences in appearance, behaviour or intelligence amongst them. So there is no basis on which to propose that changes in the regulating genes of an ape could have produced a human being. And we have already seen, when discussing hox genes and embryological development, that the manufacture of living things requires degrees of organisation so astronomical that simple gene regulators and triggers are merely

Cave paintings of auroch. Lascaux, France.

Cattle.
All cattle most likely descended from the wild auroch, as depicted in the cave paintings from Lascaux. The cattle genome contains about 22,000 genes, and is more similar to that of humans than to the genomes of mice or rats:

> *Our high resolution examination of cattle shows that unlike the dog - which has restricted diversity and high levels of inbreeding - domesticated cattle had a large ancestral population size and that more aurochs must have been domesticated than wolves; reducing the severity of the domestication bottleneck.*
> The Bovine HapMap Consortium. 'Genome-Wide Survey of SNP Variation Uncovers the Genetic Structure of Cattle Breeds'. Science. 24th April 2009: Vol 324. No. 5926. p528-553.

It is unclear whether or not bison (below) and cattle had a common ancestor.

**Human-specific regulation of thumb development**

The gene HACNS1 contains 16 DNA nucleotides which are unique to humans. From chimps to chickens to frogs, this sequence is not found in any animal. When it is genetically engineered into mice:

> *The human DNA seems to activate genes in the budding wrist and thumb. Chimp and monkey versions, on the other hand, seem only capable of switching on genes in the developing shoulder…'There's a human specific function you don't see with chimp and rhesus'…*
> E Callaway. 'Junk DNA may have handed us a gripping future'. Newscientist.com. 4th September 2008.

It was therefore thought that HACNS1 represents one of the many regulatory gene changes that apes acquired on their (supposed) journey to manhood. However, since chimpanzees can manipulate things as well as we can, but lack the inspiration to do what we do with our thumbs, any *hoping-to-be-a-man* ape does not need to evolve any special embryological controls for its thumb. And the *Don't Mess With It Principle* shows that altering the developmental control of the ape shoulder would just result in an ape with a deformed shoulder.

single ice crystals on the tips of the icebergs that are embryological developmental processes.

Recently, the mouse genome has been sequenced and has been shown to have around 2.5 billion nucleotides compared to around 3 billion in humans. If mice and humans *did* have a common ancestry we could make a number of predictions. Firstly, if the human and mouse lineages split some 75mya, then as a result of random inversions, deletions, fusions etc., we should see in the current human and mouse karyotypes an arbitrary collection of chromosome shapes. But this is not found. Instead, we find that the chromosome patterns are quite distinct since *all* mouse chromosomes are *acrocentric* (i.e. the centromere lies at one end), whereas *almost all* human chromosomes are *not* acrocentric. This observed pattern actually suggests that mice and humans originated separately with different chromosome patterns, and have subsequently each acquired their own unique set of mutations, leading to their current forms.

Secondly, if mice have evolved to be so dramatically different from fishes (with which they (supposedly) shared a common ancestor that lived around 350mya) by progressively acquiring new genes and structures, then, after 75 million years since the (supposed) mouse-human split, we should now see significant differences in the genes of mice and humans. Instead, we find that mice and humans have 99% of their genes in common and that (supposedly) over time:

> …gene addition in the mouse lineage and gene deletion in the human lineage have not significantly altered the gene repertoire. R.H. Waterston et al. 'Initial sequencing and comparative analysis of the mouse genome'. Nature. Vol 420. 5th December 2002. p520-562.

Clearly, if only a 1% change has occurred in *75 million years*, such a slow rate of acquisition of new genes could not possibly have fuelled the (supposed) evolution of ancient fish into mice in 350 million years, nor could it account for the 5% difference between chimpanzees and humans, which is (supposedly) the result of the evolution of apes into humans in only 6 million years.

We have seen that two populations of modern mice living at two ends of a single barn, and different populations of muntjac deer, can have very different chromosome numbers and shapes. So, after 75 million years, we should expect there to have been an enormous number of changes in the karyotypes of both mouse and human lineages. This would make any assessment of evolutionary relationships, based on comparisons of gene locations, highly questionable. In fact, although the human chromosomes 17 and 20 contain genetic material equivalent to that found on portions of the mouse chromosomes 11 and 2 respectively, the patterns of the location of genes on the other chromosomes show no particular correlations:

> Other chromosomes, however, show evidence of much more extensive interchromosomal rearrangement… R.H. Waterston et al. 'Initial sequencing and John Jenkins. 'Principles of genetics'. Wiley. 1997. p759.

It is surprising then, that whilst it is admitted that by examining the genomes of only two species:

> It is not yet possible to recover the ancestral chromosomal order or reconstruct the precise pathway of rearrangements... R.H. Waterston et al. 'Initial sequencing and comparative analysis of the mouse genome'. Nature. Vol 420. 5th December 2002. p520-562,

nevertheless, for their calculations of evolutionary rates and other relational interpretations, the authors proceed illogically with three assumptions: that evolution *has* occurred, that mice and humans *did have* a common ancestor, and that this ancestor lived *75mya*:

> In the analyses below we use a divergence time for the mouse and human lineages of 75 Myr for the purpose of calculating evolutionary rates... R.H. Waterston et al. 'Initial sequencing and comparative analysis of the mouse genome'. Nature. Vol 420. 5th December 2002. p520-562.

It becomes obvious that, due to the omission of any discussion regarding predictions based on alternative models (such as no common ancestry and a different time-scale for their existence), the published interpretation of mouse-human relationships, being based solely on the comparisons of mouse and human genomes, has no scientific value or credibility whatsoever.

### *Evidence from human and ape genome sequencing*

Now that the complete human and chimpanzee genomes have been sequenced, it is *unequivocal* that the differences between apes and humans are *very* extensive and *very* profound.

Clearly, there are two fundamental models that may explain life on Earth. One model is that all life forms have arisen from a common ancestor, and have accumulated genetic variation and novel changes that are both responsible for, and account for, the observed differences in the genomes of different organisms.

The other model is that distinct groups of organisms *have always been distinct* (i.e. have never had a common ancestor) and have accumulated genetic variation and novel changes that are the observed differences in the genomes of different organisms, but which have *no bearing, whatsoever,* on relatedness between distinct groups.

However, despite this fundamental point, *all* the major genome studies commence with the assumption that evolution of all life from a common ancestor *has* happened and it is *only* on the basis of *that assumption* that the data are interpreted. What *none* of the studies offer is *any* analysis of the data in the light of what would be expected if humans *have always been distinct from chimpanzees.* Without this crucial assessment, *none* of the studies have *any* validity *whatsoever*. In other words, any such study is not science, which *must* address all potential options for any observation. Of course, since we have shown from multiple angles that evolution cannot and did not happen, then if a study is promoted as showing human-ape common ancestry, we know that there must be a *fundamental* flaw in that study. In fact, the data from the *chromosomal differences*, *gene differences*, *recombination hotspots* (RHs) and *TATA binding sites* (TBS), all confirm that humans and chimpanzees *did not* have a common ancestry.

### Chromosomal differences

#### Telomeres

*Telomeres* are the ends of chromosomes, and it is here that the differences in the chromosomes become more marked:

> The most striking regional pattern is a consistent increase in divergence towards the ends of most chromosomes. The terminal 10Mb [million bases/nucleotides] of chromosomes averages 15% higher divergence than the rest of the genome, with a sharp increase towards the telomeres... the cause of this effect is unclear. The Chimpanzee Sequencing and Analysis Consortium. 'Initial sequence of the chimpanzee genome and comparison with the human genome'. Nature. Vol 437. 1st Sept 2005. p69-87.

So the chimpanzee and human chromosomes are *very* different indeed. If apes and humans do not have a common ancestry then the cause of these findings would, obviously, be '*unclear*'.

#### Inversions and deletions

It has been known for *a long time* through banding techniques that chimpanzees and humans have *very great* and *important* differences in the way their chromosomes are organised:

> One of the significant unresolved differences between the karyotypes of humans and African apes is the presence of positively staining G-bands at the ends of many chromosome arms in the chimpanzee and gorilla but absence from human chromosomes. Furthermore, there is **no similarity** between other chimpanzee telomere-junction clones and human subterminal sequences, and therefore the organization of sequences adjacent to telomeres is very different between these closely related primates. (My bold). 'A subterminal satellite located adjacent to telomeres in chimpanzees is absent from the human genome'. Nature genetics. 1994. 6:52-6.

Despite this, the mantra of common ancestry ('*these closely related primates*') is repeated without question, even though the human and chimpanzee genomic sequences have confirmed that there are massive differences in the chromosomes:

> We searched the chimpanzee genome for information about large-scale alterations...Human and chimpanzee chromosomes differ by one chromosomal fusion, at least nine pericentric inversions, and in the content of constitutive heterochromatin...We also found suggestive evidence of many additional smaller inversions as well as older segmental duplications. The Chimpanzee Sequencing and Analysis Consortium. 'Initial sequence of the chimpanzee genome and comparison with the human genome'. Nature. Vol 437. 1st Sept 2005. p75.

The problem with inversions and deletions is that these have the possibility of seriously altering the way genes are regulated, causing disease. The number of such (assumed) events is staggering, and yet humans and chimpanzees are still generally healthy:

> The current study identifies 1,576 putative (supposed) inversion events. Thirty-three of these were larger than 100,000 base pairs... (My parenthesis). L Feuk et al. 'Comparative analysis of human and chimpanzee DNA sequence assemblies'. PLoS. Genetics. 28th October 2005.

### *Y chromosome*

The chimpanzee Y chromosome is much smaller than in humans and a number of the chimpanzee genes have been rendered inactive as a result of mutations:

> Here we find, by systematically comparing the DNA sequences of unique, Y-linked genes in chimpanzee and human, **which diverged** about six million years ago, evidence that in the human lineage, all such genes were conserved through purifying selection. In the chimpanzee lineage, by contrast, several genes have sustained inactivating mutations. (My bold). J Hughes et al. 'Conservation of Y-linked genes during human evolution revealed by comparative sequencing in chimpanzee'. Nature. Vol 437. 1st Sept 2005. p101-104.

Again this emphasises the fact that mutations cause *derangements* or inactivation rather than producing new genes with new functions. Note also that by using the terms *'diverged'* and *'conserved',* the study assumes chimpanzees and humans *did* have a common ancestor, rather than letting the evidence speak for itself.

### *Recombination hotspots*

Recombination hotspots (RHs) are discrete sites at which the maternal and paternal chromosomes swap genes during the '*crossing over*' at meiosis. More than 25,000 RHs have been identified, and:

> The mapping of recombination hotspots in the human and chimpanzee genomes reveals a surprising finding...**There is virtually no overlap between these two species** in the locations of their recombination hot spots. (My bold). L. Jorde. 'When we're hot we're hot, they're not'. Science. 1st April 2005.

These findings have been interpreted as showing that there have been rapid changes in the RHs of humans and apes. However, because the human genome is *very* stable, this conflicts with the interpretation of any rapid change. A more plausible explanation is that chimpanzees and humans do not share a common ancestry.

### *Genes*

It is found that there are around 35,000,000 nucleotide differences between human and chimpanzee DNA, and that around 53 human genes are partially or completely absent in the chimpanzee. The number of categories of substantial differences is large:

> 98 categories of rapidly evolving genes… Only 30 would be expected by chance, indicating that most (but not all) of these categories undergo significantly accelerated evolution relative to the genome-wide average… The Chimpanzee Sequencing and Analysis Consortium. 'Initial sequence of the chimpanzee genome and comparison with the human genome'. Nature. Vol 437. 1st September 2005. p78.

The term '*rapidly evolving*' is used because if chimpanzees and humans separated only 6mya, then these genes *must have* mutated rapidly. However, since such rapid mutations are simply not observed in humans, the evidence points to chimpanzees and humans not having a common ancestor.
It is found that the vast majority of the equivalent chimpanzee and human genes have different sequences:

> …20% of proteins are identical between the two species, and **80% of proteins are different.** (My bold). G Glazko et al. '80% of proteins are different between humans and chimpanzees'. Gene. Vol 346. 1st September 2005. p215-9.

Most differences do not result in any change in the functions of proteins, so they do not explain why apes are apes and humans are humans, but they are substantial enough to make chimpanzee and human tissues immunologically distinct (hence we don't transplant chimpanzee organs into humans). It is not surprising to find that those genes involved in fundamental cellular processes are virtually identical in apes and humans. For example:

> …cytoskeletal proteins show little divergence between human and chimpanzee. C Bustamente et al. 'Natural selection on protein-coding genes in the human genome'. Nature. Vol 437. 20th October 2005.

Note that as is always the case with genome papers, terms such as '*little divergence*' shows that ape-human common ancestry is *assumed* before any analysis is made. Since apes are very different from humans, it is not surprising to find that genes involved in protein regulatory processes are very different ('*rapidly-evolving*', in evolution-speak):

> …classes such as transcription factors, show an excess of rapidly evolving genes… C Bustamente et al. 'Natural selection on protein-coding genes in the human genome'. Nature. Vol 437. 20th October 2005.

### *TATA binding sites*

The sections of DNA between the TATA binding sites and the starting points of the protein-coding sequences are rich in elements that are intimately involved in the regulation of the transcription of DNA to mRNA. When these sequences are examined, it is found that chimpanzee and human sections are *entirely* different:

> Evolutionary conservation is particularly apparent upstream of coding sequences and in first introns, regions that are enriched for regulatory

Gene sequence showing TATA binding site.

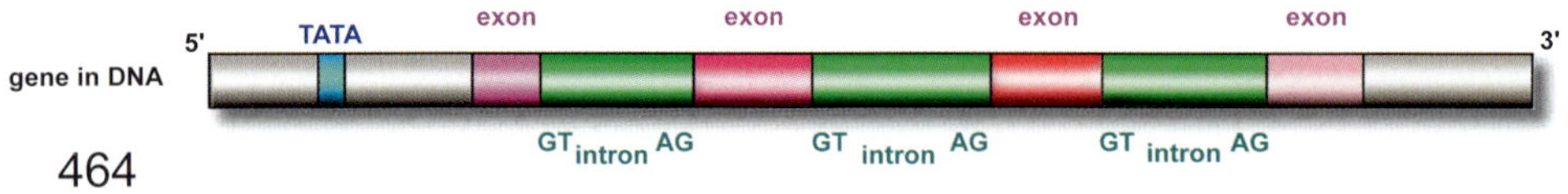

> elements. By comparing human and chimpanzee genomes, we show here that there is **almost no evidence for conservation** in these regions in hominds. (My bold). P Kneightley et al. 'Evidence for widespread degradation of gene control regions in hominid genomes'. PLoS. Biol. 2005. 3:282-288.

So, the fact is that the human and chimpanzee genes are subject to *entirely* different regulation. The issue is that gene regulation, which involves the TBS, TATA Binding Protein, DNA polymerase II, and a number of other proteins including activators and repressors which bind on the DNA upstream of the coding regions, is an incredibly complex phenomenon, and all these elements must be in place for transcription to proceed without error or hindrance. It is, therefore, implausible for even *one* (let alone virtually *all*) of the gene regulatory sites to undergo mutations that would safely and *completely* alter its sequence in a (supposedly) *hoping-to-become-a-human* ape ancestor.

On the evidence of the chromosomes, then, the most parsimonious conclusion is that the notion that the similarities between ape and human chromosomes reflect a common ancestry is implausible.

### *Neanderthal DNA*

We have shown - beyond all doubt - that a large body and variety of data converge on the conclusion that evolution from non-living chemicals to human beings cannot happen and did not happen. So, humans have no relationship with apes, and humans have always been human. Furthermore, all the Neanderthal evidence (from tools, use of fire, burying the dead, etc.) shows that these people thought and acted in ways no different from us today. Therefore, we have to address the issue of Neanderthal DNA in *that* light, and we should predict that the genetic evidence, properly evaluated, should support Neanderthal/Living Human unity. Furthermore, since, *categorically,* humans *do not* have a common ancestry with chimpanzees, the conclusions of any study that invokes human-chimpanzee common ancestry within its fundamental assumptions must be rejected as being *completely* valueless. For example:

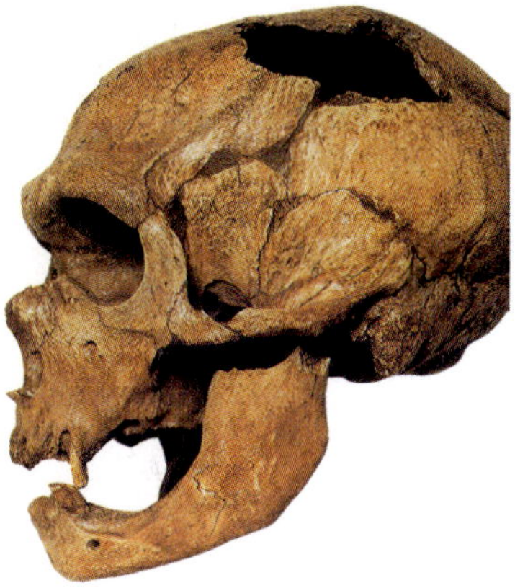
Neanderthal skull

> If the human–chimpanzee divergence time is set to 6,500,000 years, this implies an average human–Neanderthal DNA sequence divergence time of 516,000 years. R Green. 'Analysis of one million base pairs of Neanderthal DNA'. Nature. Vol 444. 16th November 2006. p330-336.

Since we are not modified apes, such statements are *meaningless*, and the effort and time spent on such blind-alley comparisons is pointless. Studying chimpanzee DNA can tell us nothing whatsoever about when Neanderthals lived and died out. Nevertheless, we can still make use of some observations. For example, the FOXP2 gene is thought to be involved in human speech and this gene is found to be identical in the chimpanzee, gorilla and rhesus monkey. The human gene codes for different amino acids in two places on exon 7 (*threonine* replaces *aspartic acid* at position 911, and *arginine* replaces *serine* at position 977), and the Neanderthal sequence is exactly the same:

**FOXP2**

The human differences in exon 7 are not found in any other genomes examined:

*...no amino acid substitutions are shared between the song-learning birds and humans. Furthermore, sequences from vocal-learning whales, dolphins, and bats do not share the human-unique substitutions.*
D Webb & J Zhang. 'FoxP2 in song-learning birds and vocal-learning mammals'. J. Hered. 2005 May-Jun;96(3):212-6.

But it is important to note that FOXP2 is extremely similar in all vertebrates from reptiles to humans, showing that it must have some important role in all non-speaking creatures:

*FoxP2 is extremely conserved... This suggests an important role for FoxP2 in vertebrate brains, regardless of whether they support imitative vocal learning or not.*
C Scharff & S Haesler. 'An evolutionary perspective on FoxP2: strictly for the birds?' Curr. Opin. Neurobiol. 2005 Dec;15(6):694-703. Epub 2005 Nov 2.

Since after millions of years we should see a random array of differences involving every non-essential nucleotide, reptiles and humans cannot have been around for millions of years.

> Here, we find that our closest extinct relatives, the Neandertals, share with modern humans two evolutionary changes in FOXP2... J Krause. 'The derived FOXP2 variant of modern humans was shared with Neandertals'. Current biology. Vol 17. 6th November 2007. p1908-1912.

So, at least with regard to FOXP2, living humans and Neanderthals are exactly the same. A small amount of DNA is also found in the mitochondria. *Mitochondrial DNA* (mtDNA) is a single circular molecule, which is itself separate from the DNA within the cell nucleus. Rarely, however, mtDNA can migrate into the nucleus where it can become incorporated into the nuclear DNA. At cell division, the mitochondria divide by simple budding and roughly equal numbers are found in each daughter cell. Since sperms have virtually no mtDNA, all of an individual person's mtDNA essentially comes from the mother. The entire human mitochondrial genome is found to contain 16,569 nucleotide pairs. Amongst living humans, there may be up to 8 mitochondrial nucleotide differences, and between chimpanzees and humans there are about 57 differences.

In contrast to the FOXP2 data, a number of mtDNA studies have concluded that living humans and Neanderthals are distinctly different. For example:

> The retrieval of mitochondrial DNA (mtDNA) sequences from four Neandertal fossils from Germany, Russia, and Croatia has demonstrated that these individuals carried closely related mtDNAs that are not found among current humans...**this excludes any large genetic contribution by Neandertals to early modern humans, but does not rule out the possibility of a smaller contribution**...samples considered as anatomically "transitional" between modern humans and Neandertals, such as Vindija analyzed here, fail to show any evidence of mtDNA admixture between the two groups. (My bold). D Serrel et al. 'No evidence of Neandertal mtDNA contribution to early modern humans'. PLoS. Biol 2(3). March 16 2004: e57 doi: 10.1371/journal.pbio.0020057.

However, it would be a mistake to think that producing family trees of human origins from mtDNA is foolproof. Most certainly, defining zoological relatedness and ancestry by making comparisons of different mtDNA is not an exact, reliable or even valid science. There are just too many questions regarding the meaning of mtDNA differences or similarities for any dogmatic interpretations to be made. For example:

> Analysis of DNA sequences has proven vexingly ambiguous in attempting to discern the two closest relatives among humans, gorillas and chimpanzees. Most analyses of mitochondrial DNA are so equivocal as to render a clear solution impossible... Steve Jones, Robert Martin, David Pilbeam. 'Cambridge Encyclopaedia of human evolution'. Cambridge University Press. 1996. p302.

> The phylogenetic position of the ancient DNA sequences recovered from Neandertal bones is sensitive to the phylogenetic methods employed. It depends on the **model** of nucleotide substituton, the branch support method, and the set of data used. (My bold). G Gutiérrez et al. 'A re-analysis of the ancient mitochondrial DNA sequences recovered from Neandertal bones'. Molecular biology and evolution. 2002. Vol 19. p1359-66.

In other words, depending on what you assume to be related to what, the branches can be attached in all manner of permutations. In fact, the Neanderthal mtDNA data can be interpreted so as to lead to the conclusion that living humans are closer to chimpanzees than are Neanderthals. So, things are not as cut and dried as is commonly made out.

Some have highlighted procedural deficiencies in a number of Neanderthal mtDNA studies, and when these are rectified, the results *do not* support the notion that Neanderthals and living humans are different creatures:

> Recent reports analyzing mitochondrial DNA sequences from Neandertal bones have claimed that Neandertals and modern humans are different species. The phylogenetic analyses carried out in these articles did not take into account the high substitution rate variation among sites observed in the human mitochondrial D-loop region and also lack an estimation of the parameters of the nucleotide substitution model. The separate phylogenetic position of Neandertals is not supported when these factors are considered... G Gutiérrez et al. 'A re-analysis of the ancient mitochondrial DNA sequences recovered from Neandertal bones'. Molecular biology and evolution. Vol 19. 2002. p1359-1366.

We also need to take note of an additional finding. Since an older Neanderthal (e.g. Feldhofer) would have been closer to our common ancestor than a younger Neanderthal (e.g. Mezmaiskaya), then the mtDNA of the older should be more similar to that of living humans. However, the reverse has been found:

> We also show that the most ancient Neandertal HVI region is the most divergent when compared with modern human sequences. However, the opposite would be expected if the sequence had not been modified since the death of the specimen. G Gutiérrez et al. 'A re-analysis of the ancient mitochondrial DNA sequences recovered from Neandertal bones'. Molecular biology and evolution. Vol 19. 2002. p1359-1366.

This suggests that there is something seriously wrong, since the supposedly older mtDNA *cannot* be more divergent. The only logical explanation is that the observed Neanderthal variation was present in individuals who lived *contemporaneously*, or who were separated by only a short period of time.

To illustrate the problems that are raised once the evolution paradigm is invoked, we can take a look at the mtDNA which was extracted from a (supposedly) 30,000-year-old Neanderthal bone. It was found that there were 27 differences compared to modern humans. However, it is thought that mtDNA:

> ...accumulates mutations at a reasonably **constant** and **rapid** rate. (My bold). Peter Snustad, Michael Simmons & John Jenkins. 'Principles of genetics'. Wiley. 1997. p758.

So this creates an impasse, not least because 57 changes in 8 million years is not *rapid*. For the sake of simplicity, if we assume that the (supposed) common ancestor of chimpanzees and living humans lived 8mya, then the observed 57 mtDNA differences between chimpanzees and living humans would obviously have accumulated

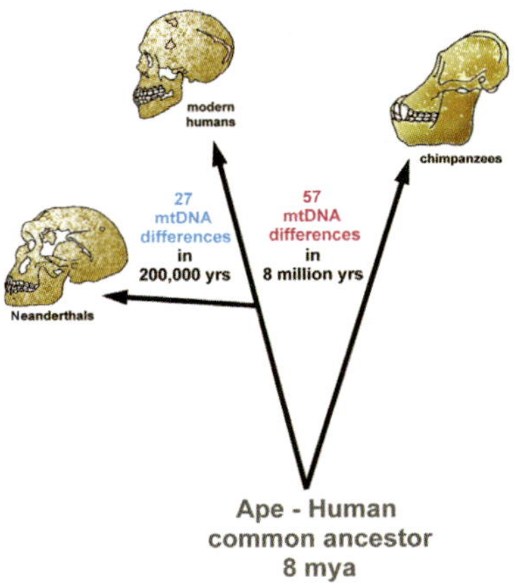


The human mitochondrial genome contains 16,569 nucleotide pairs. Amongst living humans there may be up to 8 nucleotide differences, and between chimpanzees and living humans there are about 57 differences. A (supposedly) 30,000-year-old Neanderthal bone was found to have 27 differences compared to living humans.

in 8my. This would mean a rate of 7 mtDNA differences per one million years, or 0.7 mtDNA differences per 100,000 years. At this rate of change, if the common ancestor of Neanderthals and living humans lived (say) 200,000ya, then there should be only 1.4 differences (not 27) between Neanderthals and living humans, and there should hardly be any differences *at all* amongst living humans. So something must be wrong here. Alternatively, if we assume that the 57 differences between chimpanzees and living humans *did* occur in 8my, then the 27 differences between Neanderthals and living humans would mean that the common ancestor of Neanderthals and living humans must have lived some 3.8mya, and not 200,000ya.

If, on the other hand, we assume that the 27 differences between Neanderthals and living humans *did* occur in 200,000 years, then this would mean a rate of 135 mtDNA differences per one million years, and therefore if chimpanzees and living humans *did have* a common ancestor 8mya, there should be around 1080 differences between chimpanzees and living humans. This is clearly not the case. Alternatively, if we assume that the 27 differences between Neanderthals and living humans *did* occur in 200,000 years, then the 57 differences between chimpanzees and living humans would mean that the common ancestor of chimpanzees and living humans would have lived only about 422,000 years ago.

So, from this we are left with two options, both of which are quite improbable: either Neanderthals lived a *very* long time ago or chimpanzees separated from humans relatively recently. However, we can escape from the impasse once we accept that the observed mtDNA differences between chimpanzees and humans are not related to common ancestry.

### Pseudogenes

Without question, the matter of *pseudogenes* might seem by many to be the most compelling evidence for common ancestry of apes and humans, and also for confirmation of evolution in general. An analogy is often made with plagiarism, in which one author copies another's work and, in so doing, unwittingly copies some errors also. Since the chances are that the second author would not write the exact same errors, a court would conclude that the second author's work was derived from that of the first. Similarly, since it is implausible that identical but random mutations would be found in exactly the same corresponding locations on the chromosomes of two different creatures (e.g. humans and chimpanzees), then identically mutated pseudogenes are strong evidence that those mutations originated in a common ancestor.

#### *What is a pseudogene?*

So, what is a pseudogene? Pseudogenes are DNA sequences that resemble known functional genes but seem to have no promotor with which to initiate transcription, and also seem to have disabling mutations including inappropriate stop signals embedded within the sequence. Because of this they have been thought to be non-functional. Pseudogenes can arise by a number of different mechanisms.

Firstly, a normal gene can simply acquire a mutation that makes it totally disabled

Transcription of mRNA from DNA.

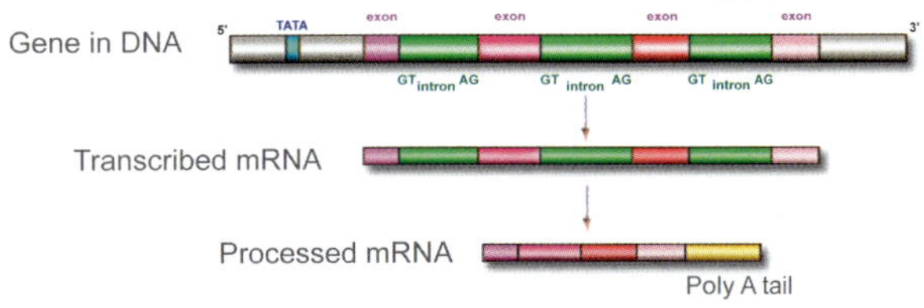

so it cannot result in a functional protein. Secondly, a normal gene may become duplicated, subsequent to which either the 'parent' gene or the duplicated gene may become a pseudogene by acquiring disabling mutations. A third mechanism is by a process of *reverse transcription,* in which a *processed RNA* can give rise to a characteristic DNA sequence which is randomly inserted at a site quite separate from the gene that gave rise to it. Reverse-transcripted DNA (a *retrotransposon*) can be recognised as such since it lacks any intron sequences (it has only the exons joined end to end) and it includes the tell-tale tail of *polyA*s of a processed RNA.

DNA sequences are generally accepted as being pseudogenes if they have greater than 70% similarity to known coding sequences. Some 20,000 pseudogenes have been identified in the human genome, hence the common notion that most of the DNA in the chromosomes is useless 'junk'.

For example, the human and chimpanzee *galactosyl transferase* pseudogenes have *identical* inactivating mutations. The *RT6* gene encodes a protein on the surface membrane of T lymphocytes (white blood cells) of rodents. Both the human and the chimpanzee RT6 pseudogenes contain *the same* three mutant stop codons. The *21-hydroxylase* gene encodes an enzyme involved in the metabolism of steroid hormones. In humans, the 21-hydroxylase gene is duplicated but the original gene has a section in which 8 nucleotides have been deleted. The chimpanzee has the *exact same* deletion*. The enzyme L-threonine aldolase (TA) has been shown to catalyse the conversion of L-threonine to yield glycine and acetaldehyde in bacteria, fungi and plants. The human TA gene is located on chromosome 17 and is non-functional. This is because it contains two single nucleotide deletions in exons 4 and 7 which cause frame-shifts, and also a premature stop codon:

> Expression of human TA mRNA was undetectable by RT-PCR... Mice, the western-clawed frog and the zebrafish have transcribed threonine aldolase/GLY1 genes, but the **human homolog is a non-transcribed pseudogene**. (My bold). A Edgar. 'Mice have a transcribed L-threonine aldolase/GLY1 gene, but the human GLY1 gene is a non-processed pseudogene'. BMC Genomics. 2005 Mar 9;6(1):32.

Perhaps the most well-known example relates to the enzyme *L-gulonolactone oxidase* (*GULO*) which is necessary for the final step in the metabolic pathway that converts glucose into *ascorbic acid* (vitamin C). Humans and a number of animals (e.g. apes, monkeys, guinea pigs, *all* bats** and some fish including trout and salmon) are unable to manufacture vitamin C since they lack GULO. Normally, rats can synthesise vitamin C but when their GULO is inactivated by a single nucleotide substitution, this results in scurvy. For example:

> [in the] mutant Wistar rat...the Cys to Tyr substitution is responsible for the L-gulono-gamma-lactone oxidase deficiency. T Kawai et al. 'A missense mutation of L-gulono-gamma-lactone oxidase causes the inability of scurvy-prone osteogenic disorder rats to synthesize L-ascorbic acid'. J. Biol. Chem. 1992 Oct 25;267(30):21973-6.

The functioning gene for GULO in the rat has twelve exons and is 70% similar to a sequence comprising five exons found on human chromosome 8. These five exons

**Non-viral retrotransposons**

Non-viral retrotransposons are classified into SINEs which are short interspersed elements, and LINEs which are long interspersed elements. LINEs comprise approximately 15% of the human genome.

Alu sequences are an abundant type of SINE which contain recognition sites for the DNA splicing enzyme Alu1.

LTRs (long terminal repeats) are found within viral retrotransposons and do not occur in LINEs.

LINEs and SINEs are particularly abundant in mammalian genomes.

*Two possiblitities for this are that the 8-nucleotide deletion could be due to a mutational hot-spot, or to a deliberate functional inactivation by the cell, in both humans and chimpanzees.

**The micro-bats, such as the long-eared bat (above), are unrelated to the mega-bats, such as the Egyptian fruit bat (below), yet both have an inactivated GULO gene.
*See: E Birney et al. 'Inability of bats to synthesise L-ascorbic acid'. Nature. 1976; 260: p626-8.*

are also found in apes. When a small section of the GULO sequence is compared in human, chimpanzee, orang-utan and macaque, all four sequences are seen to share a single nucleotide deletion that results in a frame-shift which completely inactivates the gene:

The actual human and primate GULO sequences (a) are the result of deletions of 'G's (b) followed by a frame-shift to the left.
*(After: 'Lesson B: Peseudogenes and common ancestry'. Indiana.edu/~ensiweb/lessons/psb.pdf.html)*

**a**

| | |
|---|---|
| Rat | **TACCCCGTAGAGGTGCGCTTCACCCGAGGCGATGACATTCTGCTGAGCCCC** |
| Human | **TACCTGGTGGGGGTACGCTTCACCTGGAGGATGACATCCTACTGAGCCCC** |
| Chimp | **TACCTGGTGGGGCTACGCTTCACCTGGAGGATGACATCCTACTGAGCCCC** |
| Orang-utan | **TACCCGGTGGGGGTGCGCTTCACCCAGAGGATGACGTCCTACTGAGCCCC** |
| Macaque | **TAACCGGTGGGGGTGCGCTTCACCCAAGGGATGACATCATACTGAGCCCC** |

**b**

| | |
|---|---|
| Rat | **TACCCCGTAGAGGTGCGCTTCACCCGAGGCGATGACATTCTGCTGAGCCCC** |
| Human | **TACCTGGTGGGGGTACGCTTCACCTGGAG GATGACATCCTACTGAGCCCC** |
| Chimp | **TACCTGGTGGGGCTACGCTTCACCTGGAG GATGACATCCTACTGAGCCCC** |
| Orang-utan | **TACCCGGTGGGGGTGCGCTTCACCCAGAG GATGACGTCCTACTGAGCCCC** |
| Macaque | **TAACCGGTGGGGGTGCGCTTCACCCAAGG GATGACATCATACTGAGCCCC** |

← frame-shift

From this data, it might be inferred that the human, chimpanzee, orang-utan and monkey genes must have gained their identical mutations by virtue of having a common ancestor. However, before we can accept that conclusion, we need to know three things: firstly, are pseudogenes largely non-functional; secondly, could seemingly random, seemingly useless sequences be deleted from or inserted into the *exact* same location in creatures that have no common ancestor; and thirdly, are there examples of pseudogenes that contradict the accepted evolutionary tree?

### *Are pseudogenes non-functional?*

How can we be certain that the majority of pseudogenes have no function? It is a fact of science that many previous views have been shown to be *entirely* wrong. For example, it was once thought that only one of the DNA double-helix chains is transcribed to RNA, but it is now known that both chains can be transcribed. It was once thought that the thymus was a useless relic, but that has been shown to be completely untrue - *thymic aplasia* (non-development) is fatal. It was once thought that the retina is a dumb design, but as we have seen, along with many other evidences, the Müller cells prove that notion to be completely wrong.

So, apart from those derangements caused by mutations, it is a fact that we continually discover that biological structures and molecules are there for a purpose. Therefore, we should be extremely *reluctant* to ascribe non-function to pseudogenes. Since inappropriate labels can disable proper understanding, and since some pseudogenes have been identified as having function, the logical thing would be to rename them. I will therefore substitute the term 'pseudogene' (meaning 'false gene', something which many examples clearly are not) with the term '*Query gene*' (*Qgene*) so that we remain alert to the possibility that they may have functions as yet undiscovered.

Since we have shown that the notion of evolution is unambiguously falsified, we have to address the issue of Qgenes in *that* light. Furthermore, as with mtDNA, since, *categorically,* humans *do not* have a common ancestry with chimpanzees, the conclusions of any study that includes human-chimpanzee common ancestry within its fundamental assumptions, or that uses the mouse or rat of (supposedly) 65mya as an assumed ancient and foundational human ancestor in its analysis of genetic sequences, must be rejected as being *completely* valueless. This is because, starting with the false premise of mouse-human common ancestry will lead, erroneously, to the invocation of diverse permutations of presumed insertions and deletions. Therefore, any interpretation of observations will be lacking if we do not, at all times, ask the question: what pattern would there be if humans, apes and mice did *not* have a common ancestry?

However, we must remember the *Don't Be Fooled Again Principle*. Since we have proved from diverse converging data that the evolution of simple chemistry to conscious thought did not happen and could not have happened, then, when we are confronted with evidence that might give some support to the notion that apes and humans have a common ancestry, we know that there must be another explanation, even if it is not immediately obvious.

Therefore, there must be an explanation for the *apparent* support that Qgenes give to the idea of evolution. Now, it is a fact that some Qgenes truly are non-functional, degenerative genes. We have seen examples of them already, but the example of *Mkrn1-p1* should be mentioned since it was originally reported (Hirotsune et al, 2003) to have regulatory activity on its 'parent' gene, *Mkrn 1*, but this was subsequently claimed to be erroneous:

> Here, we show that 5' Mkrn1-p1 is fully methylated on both alleles, a pattern indicative of silenced chromatin, and that Mkrn1-p1 is not transcribed and therefore cannot stabilize Mkrn1 transcripts...This study invalidates the data upon which the pseudogene trans-regulation model is based and therefore strongly supports the view that mammalian pseudogenes are evolutionary relics. T Gray at al. 'The putatively functional Mkrn1-p1 pseudogene is neither expressed nor imprinted, nor does it regulate its source gene in trans'. Proc. Natl. Acad. Sci. USA. 2006 Aug 8;103(32):12039-44.

However, despite the finding of sequences that are clearly non-functional, we have the problem that (perhaps most) Qgenes cannot be largely nonsense relics, since there is an undeniably *marked conservation* throughout their extensive sequences. After millions of years, *every* single nucleotide within *every* single *useless* 'junk' sequence within *every* individual should show a completely random pattern. That such randomness is clearly *not* observed, emphasises the fact that Qgenes *must* have diverse functions that are as yet unknown. Since a useless relic can continually acquire random mutations without affecting survival, and since a necessary gene will only tolerate a certain amount of alteration before it becomes inactivated, then one way of determining which sequences might be functional is to look for sequences which have apparently not changed much over time. This process has yielded some promising results:

> ...Two processed sequences are notable, their **conservation since the human-mouse split being as high as most protein-coding genes**; one

> is derived from the protein Ataxin 7-like 3 (ATX7NL3), and one from the Spinocerebellar ataxia type 1 protein (ATX1). (My bold). O Svensson. 'Genome-wide survey for biologically functional pseudogenes'. PLoS. Comput. Biol. 2006 May;2(5):e46. Epub 2006 May 5.

Furthermore, the different expression patterns point to different functions in mouse and man:

> ...processed pseudogenes, which are retrotransposed copies of mRNAs, have been regarded as junk or selfish DNA for a long time...the **expression patterns of transcribed pseudogenes in various tissues were quite different between human and mouse**... (My bold). H Sakai et al. 'Frequent emergence and functional resurrection of processed pseudogenes in the human and mouse genomes'. Gene. 2007 Mar 15;389(2):196-203.

Despite having only 80% similarity to ordinary functioning genes, deletions of sections or whole introns, ectopic stop codons, and frame-shifting, Qgenes can still have function. For example, regulation of RNA transcription can occur by pairing of mRNA from the Qgene to mRNA of the 'parent' gene. Such duplex formation limits the amount of protein coding mRNA that is free to be used for transcription in the ribosomes. This is exemplified by the *nitric oxide synthase* (*NOS*) gene and its Qgene the *anti-NOS*, which have sequences that are different by 20-24%. *Nitric oxide* (*NO*) acts as a neurotransmitter and is synthesised by the enzyme *neuronal NO synthase* (*nNOS*), and it is essential for the formation of long-term memory. In the snail it has been shown that *anti-NOS*:

> ...**functions as a negative regulator of nNOS** expression...during the critical period of NO requirement...We also provide the first example of the behavioral regulation of a pseudogene. (My bold). S Korneev et al. 'Timed and targeted differential regulation of nitric oxide synthase (NOS) and anti-NOS genes by reward conditioning leading to long-term memory formation'. J. Neurosci. 2005 Feb 2;25(5):1188-92.

In tomatoes, the Qgene *psiBCH* has a stop codon caused by a frame-shift, and 81-88% similarity to class I basic *chitinase* genes. Despite these apparently damaging mutations:

> ...psiBCH promoter-GUS analysis indicates that wound-response of the tobacco transgene was rapid and localized in the wounded area following mechanical wounding. Therefore, our results suggest that the **psiBCH promoter can provide targeted expression of genes**, such as protease inhibitors in response to pest attack. (My bold). U Baykal et al. 'A frameshift in the coding region of a novel tomato class I basic chitinase gene makes it a pseudogene with a functional wound-responsive promoter'. Gene. 2006 Jul 5;376(1):37-46. Epub 2006 Apr 6.

Placental and marsupial mammals do not lay eggs and it might be thought that they have no need of *vitellogenin* (egg yolk) genes, and that these genes have been lost during evolution:

> We here show that the three ancestral **vitellogenin-encoding genes were progressively lost during mammalian evolution**...in all but the egg-laying monotremes, which have retained a functional vitellogenin gene. (My bold). D Brawand et al. 'Loss of egg yolk genes in mammals and the origin of lactation and placentation'. PLoS. Biol. 2008 Mar 18;6(3):e63.

It has been found in the human and dog genomes that there are 'premature' stop codons and frame-shifting insertions/deletions in regions equivalent to the active vitellogenin genes found in the chicken. This would normally be interpreted as showing that mammals have acquired a damaging mutation to their vitellogenin gene (making it a pseudogene) but this is of no consequence because mammals have a placenta with which to nourish the embryo. However, the (*seemingly* defective) mammalian vitellogenin may simply reflect the *limited but necessary* requirement for yolk in all mammals in the very earliest stage of development (before the embryo is implanted in the wall of the uterus, and the placenta starts to provide nourishment):

> ...yolk is present in sufficient quantity to sustain the development of the embryo through the period of cleavage...During this period, the embryo draws upon the reserves of yolk in the dividing cells. (My bold). L Baggott. 'Human reproduction'. Cambridge University Press. 1997. p33.

Rather than representing a relic of mammalian evolutionary history, mammalian vitellogenin and other Qgenes may be genes that are active only for a short time after fertilisation or during embryological development, after which they are permanently inactiviated*.

So, in fact, the view is increasing that Qgenes are not inactive genetic fossils but are actually *vital* for the proper co-ordinated functioning of the whole DNA/RNA/protein synthesis system.

### *Can Qgenes be inserted or deleted within the exact same position in unrelated creatures?*

The second question is: can Qgenes be inserted or deleted within the exact same location in unrelated creatures? It is becoming increasingly apparent that yes, they can. The guinea pig GULO Qgene actually contains 47 out of 129 nucleotides that are *exactly* the same as in humans, but are different in the rat:

> A comparison [of] human exon sequences with the corresponding sequences of the guinea pig nonfunctional GULO gene revealed that the same substitutions from rats to both species occurred at a large number of nucleotide positions. Y Inai et al. 'The whole structure of the human nonfunctional L-gulono-gamma-lactone oxidase gene--the gene responsible for scurvy--and the evolution of repetitive sequences thereon'. J. Nutr. Sci. Vitaminol. (Tokyo). 2003 Oct;49(5):315-9.

Furthermore, the guinea pig GULO Qgene has four stop codons in the *exact* same place as in humans**. Since humans and guinea pigs are quite unrelated, the chance of this happening completely randomly is negligible, so there must exist hot spots for the insertion of stop codons.

***Qgenes function at specific times during embryogenesis**

*Lin et al described the activity of one embryonic stem (ES) cell gene that appears to be a pseudogene in the adult mouse[1]. They reported that ES cell-specific expression of Oct4 maintains the pluripotency and versatility of stem cells until they begin to differentiate into other tissues. They concluded that Oct4 may be a functional pseudogene with a unique and specific role in ES cells. It is important to note that all cells will carry this particular gene in their DNA, but the gene will be permanently switched off (and not expressed) in anything other than pluripotent ES cells. Maybe the once-functional gene in ES cells has become a pseudogene in differentiated tissue.*
*Another piece of evidence that suggests that pseudogenes might be functional at specific times in life is from the work of Choi et al. This research group extracted RNA from the ovarian tissues of 23 pre-menopausal patients and measured cyclin D2 and pseudogene cyclin D2 mRNA production in-vitro. They showed that the expression of mRNA from the gene and its pseudogene correlated with the age of the subject. In particular, they concluded that the expression of the pseudogene in the human ovary increases with age[2].*
'The changing face of pseudogenes'. www.truthinscience.org.uk. 6th September 2008.

*[1] H Lin et al. 'Stem cell regulatory function mediated by expression of a novel mouse Oct4 pseudogene'. Biochem. Biophys. Res. Commun. 2007 355: 111-6.*

*[2] D Choi et al. 'The expression of pseudogene cyclin D2 mRNA in the human ovary may be a novel marker for decreased ovarian function associated with the ageing process'. J. Assist. Reprod. Genet. 2001 18:110-3.*

**It would be interesting to know how this compares with the chimpanzee, or indeed with any animal that has an inactivated GULO. If this is a common and conserved inactivation, it speaks eloquently of a 'functional inactivation' and indicates that the gene might be functional at some stage in embryological development.

**Occipital condyles**

All the biological evidence, properly interpreted, confirms that no evolutionary trends exist. For example, all mammals have two occipital condyles (blue) at the back of the skull (where the skull articulates with the first neck vertebra), while all other land vertebrates have a single occipital condyle.

**Reptiles**

Diplodocus

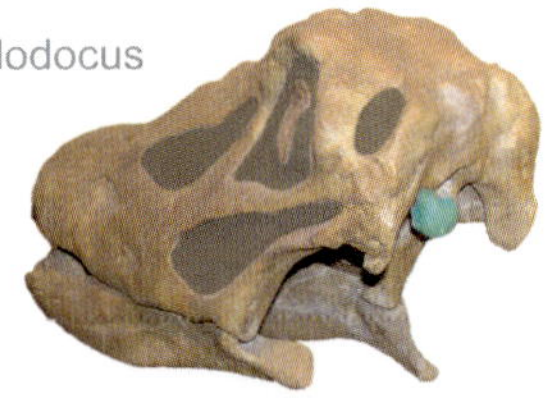

Crocodile

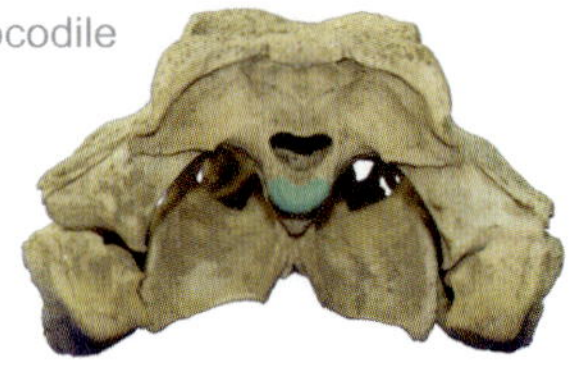

Python

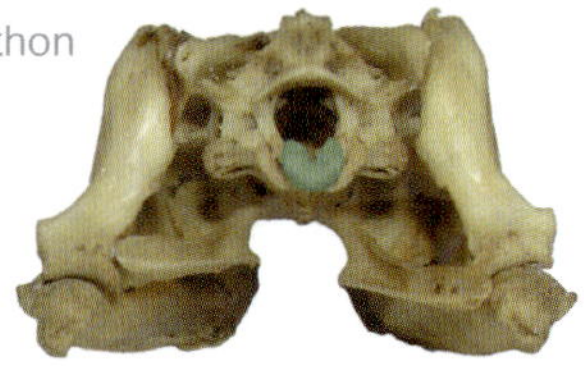

Tortoise

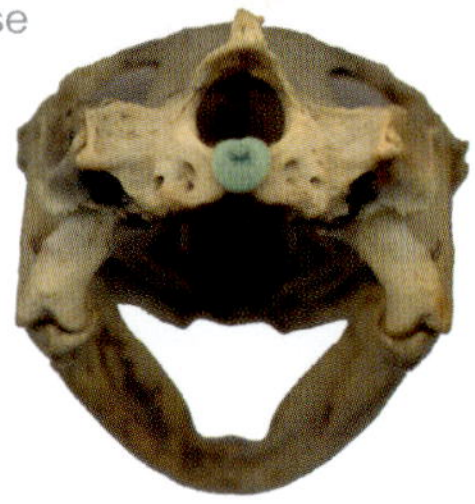

**Birds**

Emu

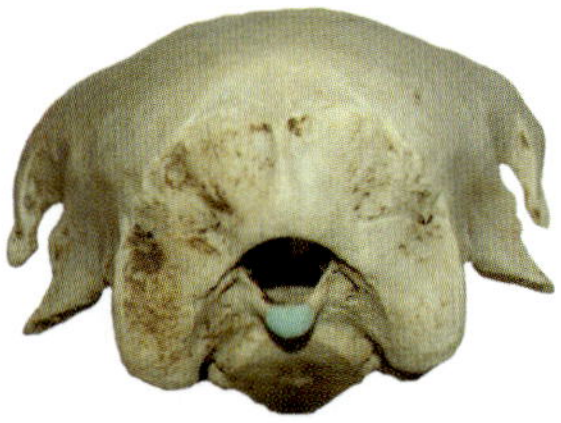

Gull

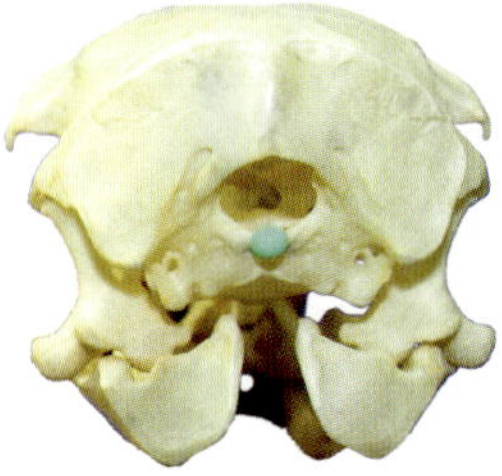

It is also clear that insertions of retrotransposons do not represent unidirectional genetic events. They can be *precisely* reversed:

> In this study, we show that retroelements can, in rare cases, be precisely deleted from primate genomes...**The deleted loci are indistinguishable from pre-integration sites, effectively reversing the insertion**. (My bold). L van de Lagemaat et al. 'Genomic deletions and precise removal of transposable elements mediated by short identical DNA segments in primates'. Genome Res. 2005 Sep;15(9):1243-9.

Another example is the intron between exons 4 and 5 of human blood group gene RhCE, which seems to have been precisely excised in apes:

> The CE intron was found in three of six gorillas, but in none of the seven chimpanzees. Sequence data suggested that the Alu element might have previously been present in the chimpanzee RH genes but was eliminated by excision or recombination. C Westhoff & D Wylie. 'Investigation of the RH locus in gorillas and chimpanzees'. J. Mol. Evol. 1996 Jun;42(6):658-68.

The last example is that of the *Mys* elements which are a type of LTR-containing retrotransposon found in *Sigmodontine* rodents from South America. It is found that unrelated animals have the exact same insertions:

> We have identified two hot spots for SINE insertion within mys-9 and at each hot spot have found that two independent SINE insertions have occurred at identical sites.
> These results have major repercussions for phylogenetic analyses based on SINE insertions, indicating the need for caution when one concludes that the existence of a SINE at a specific locus in multiple individuals is indicative of common ancestry. M Cantrell et al. 'An ancient retrovirus-like element contains hot spots for SINE insertion'. Genetics (In Review). 2001.

Therefore, when there are deletions of the same retrotransposon in different individuals, this cannot be confirmation of common ancestry.
So, there definitely seem to be hot-spots where identical insertions and deletions can occur, but identical insertions and deletions per se do not prove common ancestry.

### Some Qgenes contradict the expected evolutionary pattern

A number of Qgene studies show evidence that contradicts the expected evolutionary pattern. For example, when forty-three *human endogenous K family retroviruses* (HERV-K) were compared in great apes, two Old World monkeys, and two New World monkeys), it was found that the South American monkeys are closer to humans than are the African apes:

> ...five sequences from the *squirrel monkey* and three sequences from the *night monkey* are **more closely related** to human-specific HERV-K LTRs than they are to those of apes. (My bold). H Kim et al. 'Phylogenetic analysis of HERV-K LTR-like elements in primates: presence in some new world monkeys and

evidence of recent parallel evolution in these species and in homo sapiens'. Arch. Virol. 1999;144(10):2035-40.

The endogenous retrovirus *PTERV1* is found in chimpanzees, gorillas and monkeys, but:

> …is **absent from humans and Asian ape genomes**. (My bold). C Yohn et al. 'Lineage-specific expansions of retroviral insertions within the genomes of African great apes but not humans and orangutans'. PLoS. Biol. 2005 Apr;3(4):e110. Epub 2005 Mar 1.

In humans and in the gorilla, there are three *immunoglobulin C epsilon* genes: one active, one shortened, and one processed gene. However, the chimpanzee and Old World monkeys do not have the shortened gene:

> These results suggest that…**the gorilla is more closely related to man than the chimpanzee is**, unless the chimpanzee has lost the C epsilon 2 gene after the divergence of this species. (My bold). S Ueda et al. 'A truncated immunoglobulin epsilon pseudogene is found in gorilla and man but not in chimpanzee'. Proc. Natl. Acad. Sci. USA. 1985 June; 82(11): 3712–3715.

The *urate oxidase* (*Uox*) Qgenes contain a number of stop codons. Of these, the TGA codon at position 107 is absent in the human, chimpanzee and baboon, but:

> …occurs in the gorilla, the orangutan, and the gibbon, and therefore requires multiple origins of this nonsense mutation. Oda et al. 'Loss of urate oxidase activity in hominoids and its evolutionary implications'. Molecular biology and evolution. 19 (5) 2002: p640-53.

The evolution model requires that the Uox 107 stop codon appeared separately in the gibbon and orang-utan and subsequently reverted to a non-stop codon in the chimpanzee. In addition, Uox has a duplicated GGGATGCC segment in intron 4:

> Which is shared by the gorilla and orangutan... Oda et al. 'Loss of urate oxidase activity in hominoids and its evolutionary implications'. Molecular biology and evolution. 19 (5) 2002: p640-53.

Here, the evolution model requires that the non-duplicated GGGATGCC segment of the baboon became duplicated in the orang-utan and then subsequently reverted to a non-duplicated segment in the chimpanzee. This is better explained by human and ape genomes having independent origins and acquiring mutations independently. Sequences can also be found that are shared by quite unrelated creatures. For example, one identical intron is found both in cows and humans:

> [The **cow**] **P1 pseudogene** is very unusual. It appears to be derived from a partially processed transcript and contains an intervening DNA sequence of 861 bp that **corresponds in position with an intron in the human P1 gene**. (My bold). M Dyer et al. 'DNA sequences of a bovine gene and of two related pseudogenes for the proteolipid subunit of mitochondrial ATP synthase'. J. Biochem. 1989 May 15;260(1):249-58.

**Mammals**

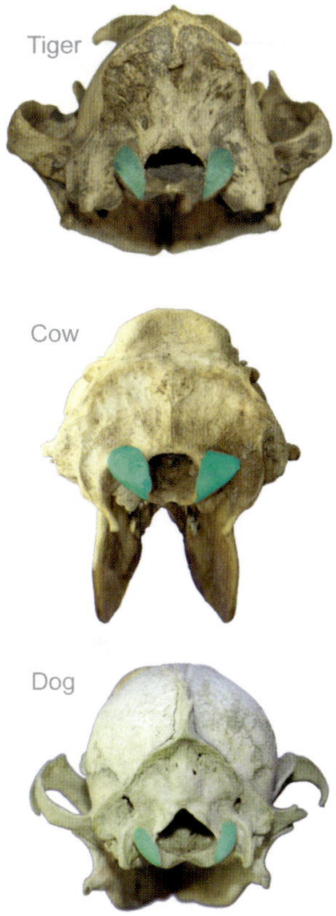


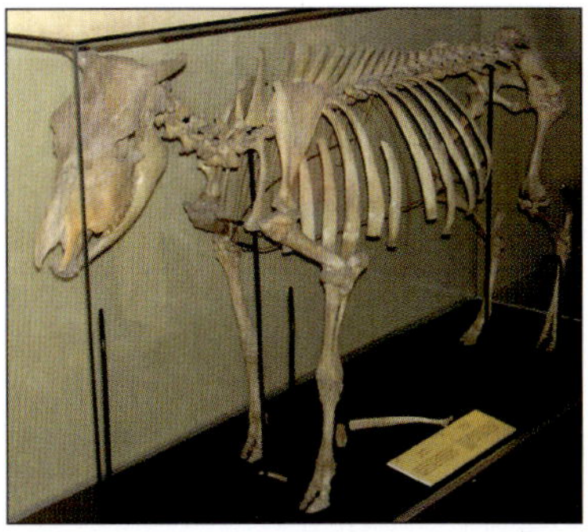

Dwarf cow (1m tall) from Stagelse, Denmark (1300 AD).
The cow and human P1 pseudogenes contain exactly the same 861bp sequence.

**Uox Qgenes in gibbon (Gib), baboon (B), orang-utan (O), gorilla (Gor), chimpanzee (C) and human (H)**

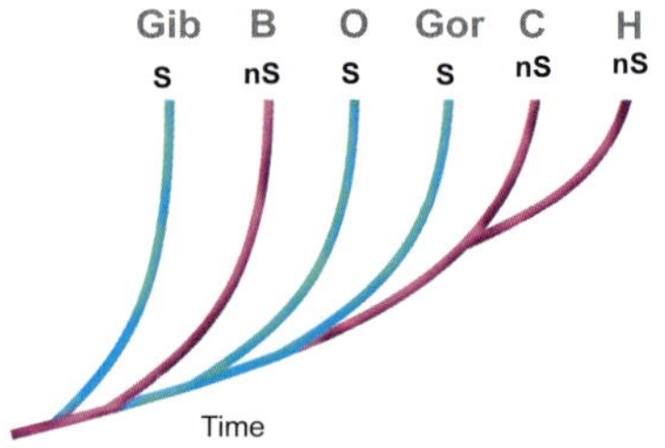


*Non-stop* (nS) and *stop* (S) distribution within the evolutionary paradigm. There is no evolutionary pattern.

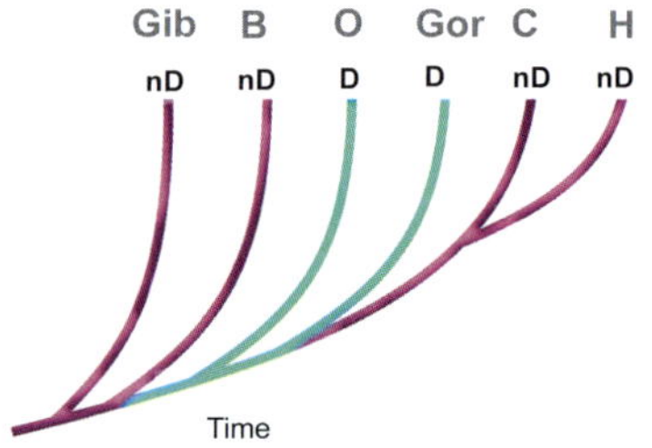


No *duplication* (nD) and *duplication* (D) distribution within the evolutionary paradigm. There is no evolutionary pattern.

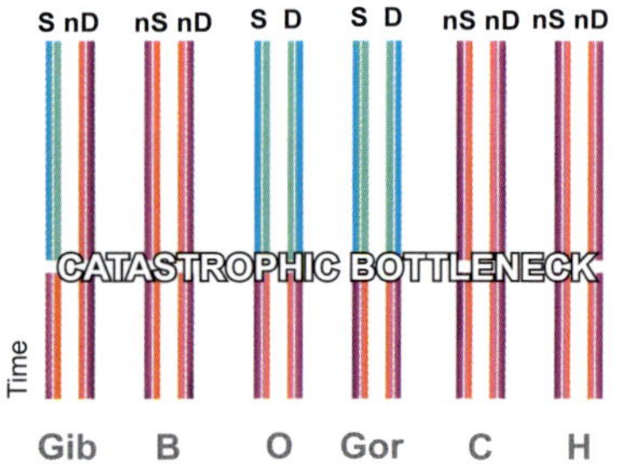


In contrast, if all non-great apes, great apes and humans were always separate creatures, which together passed through a severe population bottleneck, then the random pattern makes sense.

## The Uox Qgene: the problem and the solution

Baboons have a *non-stop* codon (TGA at position 107) and no *GGGATGCC duplication* in intron 4.
Orang-utans and gorillas have a *stop* and a *duplication.*
Chimpanzees and humans have a *non-stop* and no *duplication*.

From the evolution viewpoint, the original pre-great ape sequence would most likely have had a *non-stop* and no *duplication*, just like the baboon format. Therefore, the evolution model would require *all* of the orang-utan lineage to acquire a *stop* and a *duplication*, with no orang-utans surviving with a *non-stop* or a *duplication*, and no baboon individuals surviving with a *stop* and a *duplication*. After a few million years, the gorilla line would have arisen from the orang-utan/gorilla common ancestor, with *all* gorillas and *all* orang-utans retaining a *stop* and a *duplication*.
After another few million years, the chimpanzee line would have arisen from the gorilla/chimpanzee common ancestor, with *all* chimpanzees acquiring a *non-stop* and having no *duplication*, and with *no* gorilla individuals surviving with a *non-stop* and without a *duplication*. After yet another few million years, the human line would have arisen from the chimpanzee/human common ancestor, with *all* chimpanzees and *all* humans having a *non-stop* and having no *duplication*.

Clearly, this model does not make any evolutionary sense. If, for example, we look at the gorilla/chimpanzee branch point, we would predict that: if chimpanzees arose with a *stop* and a *duplication* acquired from the gorilla/chimpanzee common ancestor and subsequently acquired a *non-stop* and lost a *duplication*, then some chimpanzees should still have a *stop* and a *duplication*. Alternatively, if chimpanzees arose from within a group of *gorillas* that had acquired a *non-stop* and lost a *duplication*, then some gorillas should have a *non-stop* and no *duplication*. The observation just does not fit the prediction.

A solution to this seemingly bizarre genomic pattern is that all non-great apes, orang-utans, gorillas, chimpanzees and humans were always separate creatures, and all started off with a *non-stop* and no *duplication*. Then, independently, some lineages acquired a *stop* and a *duplication*. This would normally result in mixed populations: for example, some gorillas with a *stop* and a *duplication*, and some gorillas with a *non-stop* and no *duplication*. However, the observation that all individuals within the groups orang-utan, gorilla, chimpanzee and human have the same format, can be explained if all creatures passed through a severe and recent population bottleneck, *all* at the same time. As we have seen, there is convincing evidence of a global catastrophe that would have created such a bottleneck.

The vast majority of Qgenes occur in *unrelated* sites within apes and humans:

> We unambiguously map 287 retroviral integration sites and determine that approximately **95.8% of the insertions occur at non-orthologous regions between closely related species**. (My bold). C Yohn et al. 'Lineage-specific expansions of retroviral insertions within the genomes of African great apes but not humans and orangutans'. PLoS. Biol. 2005 Apr;3(4):e110. Epub 2005 Mar 1,

and show a random pattern:

> A human endogenous retrovirus (HERV-I; RTVL-I) has been located within the first intron of a haptoglobin-related gene. The evolutionary relationships within the HERV-I LTR family among hominoids and Old World monkeys showed **a random cluster**, indicating that HERV-I LTR elements have **evolved independently in primate evolution**. (My bold). J Lee & H Kim. 'Endogenous retrovirus HERV-I LTR family in primates: sequences, phylogeny, and evolution'. Arch. Virol. 2006 Aug;151(8):1651-8. Epub 2006 Mar 3.

And, although we have seen that the human GULO Qgene shares the same deletion that is found in chimpanzee, gorilla and orang-utan, nevertheless, up to 30% of the human sequence is different to that of apes, much of which shows *random* variation:

> The 164-nucleotide sequence of exon X of the gene was compared among human, chimpanzee, orangutan, and macaque, and it was found that nucleotide substitutions had occurred at random throughout the sequence... Y Ohta & M Nishikimi. 'Random nucleotide substitutions in primate nonfunctional gene for L-gulono-gamma-lactone oxidase, the missing enzyme in L-ascorbic acid biosynthesis'. Biochim. Biophys. Acta. 1999 Oct 18;1472(1-2):408-11,

which means that the identical deletions may well have functional significance. Other genetic evidence is also at variance with the chimpanzee-human cousinship mantra. For example, it is notable that malignant tumours are uncommon in monkeys and apes:

> Cancer is a major medical problem in modern societies. However, the incidence of this disease in non-human primates is very low. X Puente et al. 'Comparative analysis of cancer genes in the human and chimpanzee genomes'. BMC Genomics. 2006 Jan 26;7:15.

When the chimpanzee genome was examined for sequences corresponding to 333 human cancer genes, it was found that there was a great similarity in the genes:

> ...all examined human cancer genes are present in chimpanzee, contain intact open reading frames and show a high degree of conservation between both species. X Puente et al. 'Comparative analysis of cancer genes in the human and chimpanzee genomes'. BMC Genomics. 2006 Jan 26;7:15.

However, in between the identical nucleotides, the human cancer-generating *mutations* show patterns that are distinct to that of chimpanzees:

> ...**all** of the possible cancer-associated polymorphisms tested have human-specific alleles...suggesting a possible involvement of human-specific alleles in cancer susceptibility...(My bold). Y Maruta et al. 'Determination of ancestral allele for possible human cancer-associated polymorphisms'. Cancer Genet. Cytogenet. 2008 Jan 1;180(1):24-9.

This is quite incongruous since evolution would predict that many cancer-generating mutations would be identical in chimpanzees and people, and that human-specific malignant diseases would *not* be the norm.

### *Conclusion: Qgenes are not evidence for ape-human common ancestry*

While there is clearly much more to be learned than is known, the evidence shows that some Qgenes are simply mutated, inactivated regular genes and can be properly termed pseudogenes, and other Qgenes *often* exhibit features that point to functional roles involving gene regulation and generation of genetic diversity:

> Pseudogenes have been defined as nonfunctional sequences of genomic DNA (junk DNA) originally derived from functional genes...We describe some unexpected features of pseudogenes in diverse organisms that are inconsistent with this widely accepted point of view. Pseudogenes are ***often* evolutionary conserved** and **transcriptionally active**. (My bold). E Balakirev & F Ayala. 'Pseudogenes: structure conservation, expression, and functions'. Zh. Obshch. Biol. 2004 Jul-Aug;65(4):306-21.

Rather than Qgenes being evolutionary 'has-beens':

> A review of the evidence leads to the conclusion that **pseudogenes are important components of genomes**... (My bold). E Balakirev & F Ayala. 'Pseudogenes: structure conservation, expression, and functions'. Zh. Obshch. Biol. 2004 Jul-Aug;65(4):306-21.

In addition to the well-known process of gene transcription from the sense strand of DNA, it is now clear that transcription of Qgenes occurs from the anti-sense strand, and that insertions and deletions occur at hot-spots. It is therefore becoming apparent that the functional regulation of DNA transcription involves multiple layers of interdependent essential complexity that was previously *unimagined.*
Before finishing, it is necessary to clarify what can and what cannot be taken to be a proof. In the specific case of Qgenes, the logical argument would go something like this:

**Prediction**: evolution predicts that the same random useless genetic errors will be seen in creatures which have a common ancestor.
**Observation**: a number of creatures have the same useless errors.
**Conclusion**: those creatures must have a common ancestor.

This is not proof that the creatures share a common ancestor since the supposed errors may not be errors, may not be randomly generated, and may be the result of mechanisms unknown and unrelated to common ancestry. The fallacy of the conclusion of common ancestry may be seen with a different example:

**Prediction**: tomatoes that are red are ripe for eating.
**Observation**: red tomatoes are seen.
**Conclusion**: those tomatoes are ripe for eating.

The conclusion need not be true since the tomatoes may be unripe but may just have been stained red. So, a theory cannot be *proved* by predicting an observation since there may be other unknown causes for that same observation. In contrast, a sure way of *disproving* a theory is by observing something that the theory predicts will *not* be seen. For example:

**Premise**: chimpanzees and humans have a very recent common ancestry.
**Prediction**: chimpanzees and humans will not have widely differing inherited mutation-related cancers.
**Observation**: chimpanzees rarely have cancers and differ substantially from humans in their cancer-related mutations.
**Conclusion**: chimpanzees and humans do not have a common ancestry.

We have seen that rather than all Qgenes being non-functional, many have been shown to have function. Many have the features of functionality that is as yet undiscovered, and deletions and retrotransposons often appear at hot-spots. Therefore, the finding of identical mutations in diverse creatures is not evidence for common ancestry. Even those who are most aggressively evolution-minded admit that there is no genetic *proof* of ape-human common ancestry:

> Q. (from the audience): "Have there been, for example, insertions which one can track in humans in interesting locations, next to particular genes, which are going to suggest some affinity with chimpanzees because they might have similar insertions?"
> A. "That's a good question. I don't know the answer."
> Steve Jones. Royal Society Lecture, 2006.*

* Incidentally, the lecture did not prove evolution at all, since it dealt largely with the variation in HIV, and (erroneously) equated this with the spontaneous modification of dust into human brains.
It is worrying that, despite this, the MC congratulated Jones on a great presentation, and the audience applauded heartily.

Since the prominent paradigm has been that Qgenes are 'junk', few have been investigated for possible function. From what we already know, it is probable that, as more Qgenes are examined, less and less of them will be seen to be pseudogenes.

### *Qgenes are evidence against evolution*

Clearly, if the GULO Qgene in humans and apes was once functional, and has been inactivated in apes by identical mutations, this does not prove that we are modified animals. Neither is it evidence for evolution in principle, since such Qgenes can only represent a degeneration of a previously functioning gene, the genetic equivalent of the wings of the Galapagos cormorant - a degenerative feature that is not lethal.
Furthermore, Qgenes themselves falsify evolution: evolution requires that creatures would have lived happily for millions of years before GULO ever appeared. But, since we know that with a proper diet, creatures without functioning GULO can

not only survive but *flourish*, there is no reason for it to have evolved in the first place. Furthermore, if GULO had evolved at some point in time, evolution would predict that it would *not* have appeared in *most* creatures, because *every* creature at the time of its supposed appearance had managed to survive without it. Therefore, since GULO is widespread, yet is not essential, it did not arise by evolution. Also, the *Essential Sequence Principle* shows that easily-inactivated-by-a-single-nucleotide-change proteins such as GULO could not have arisen stepwise. And since most animals are vegetarian, and mutations are *inevitable*, then, after millions of years, *most* animals should have acquired inactivating (but not lethal) mutations of GULO. Since this is not observed, animals and humans cannot have been around for millions of years.

Baby alligators sitting on a turtle.
The genetic evidence suggests that animals have not been around for millions of years.

### *Conclusion*

We have already seen that since their cells have many similar functional requirements, it is inevitable that many genes will be found to be common to both bacteria and mammals. However, we should predict that if some bacteria ultimately evolved into mammals, the inevitability of mutations means that bacteria and mammals should have *no* genes that have remained *exactly* alike after billions of years. For example, human mtDNA contains 16,569 nucleotide pairs; if we assume that the 27 differences between Neanderthals and living humans did occur in 200,000 years, and if in general we assume (for the sake of argument only) that DNA within the cell nucleus accumulates mutations at the same rate as mtDNA, then this would mean that the human genome (which contains 3 billion nucleotide pairs) would accumulate some 2.35 million nucleotide changes per 100,000 years. If vertebrates have been around for (say) 450my, then since the time of the earliest fishes, the lineage leading to humans would have accumulated some 5 billion alterations compared to its ancient fish common ancestor. Obviously it is impossible to quantify just how many mutations have occurred in the history of life. Nevertheless, it would be safe to say that, by now, every single nucleotide that is responsible for representing *non-essential* amino acids in any given protein chain, should be different in *all* major classes of vertebrates. This is clearly not so. And if genes from very dissimilar organisms (for example, yeast and humans, which [supposedly] had a common ancestor over a billion years ago) are shown to be *identical*, for example:

> A gene that in humans causes an inherited disorder of the nervous system has an exact match in yeast... Steve Jones. 'Almost a whale'. Doubleday. 1999. p304,

then this confirms, firstly, that there has been no evolution of that gene, and secondly, that neither humans nor yeast have been around long enough for mutations to have become established. So, by implication, living things may not have been around for millions of years.

In summary, there is no demonstrable trend in chromosome number, size and shape from simple to complex animals; and rearrangements in chromosomes are not responsible for the observable differences either between different animals or between apes and human beings. Furthermore, the evolution of ape ancestors into humans requires there to have been a large number of lineages of isolated inbreeding populations, resulting in a number of distinct genetic intermediaries between humans, living apes, Homo habilis and australopithecines. There is no evidence to support such a notion. Finally, the evidence of Qgenes and mitochondrial DNA is ambiguous and unreliable, and lends no support to the notion that apes evolved into humans.

# Intelligence:

*When I was a child,*

*I thought like a child.*

*When my great, great...*

*great grandfather was a child,*

*did he think like an ape?*

CHAP

Chimpanzee

In many ways apes look a lot like us and indeed they have many mannerisms that remind us of ourselves. All the great apes show clearly that they are conscious beings, are self-aware in space and time, have emotions, and are capable of using tools. The question we have to ask in this chapter is: does the evidence from great ape intelligence, non-primate intelligence and human intelligence force us to the conclusion that the human brain is a modified ape brain? Because of the understandable temptation to anthropomorphise when dealing with apes, we have to be particularly careful to ensure that our objectivity is not compromised and not to presume beyond that which observation and logic allows.

Above: Gorilla.
Below: Orang-utan.

### *Comparisons of animal behaviour and intelligence*

There is the story of a male chimpanzee that would go regularly to a certain spot, stare at a waterfall for some time, and then get up and walk away. This story is interpreted by some as showing that chimpanzees have an interest in beauty, and furthermore, that our own appreciation of beauty is simply the result of mutational modifications within an ape brain:

> I always use that as an example of aesthetics in chimpanzees...a biological basis for appreciating beauty. Prof. Meredith Small. 'The Darwin debate'. BBC2 Broadcast. 28th March 1998.

But we have to be careful. Here, at the outset, evolution is *assumed* to be a fact, and then the observed biological world is interpreted in such a way as to fit in with that assumption. The fact that *we* might sit down and take in the vista of a waterfall does not mean that the chimpanzee was doing the same. It may simply have been looking blankly *through* the waterfall, with its brain in neutral gear. Actually, to be truly scientific, we ought to consider the possibility that the chimpanzee might have found the waterfall site a quiet place where it could get its brain around the meaning of life...or calculus...or quantum mechanics. Indeed, it is simply supposition and bad science to even suggest that this chimpanzee was thinking about anything in particular *at all.* The chimpanzee was never asked, and furthermore, it never made any attempt to convey its thoughts.

It is well recognised that the more we use our brains, the more agile our thinking processes become, and that this benefit extends into old age. We also know that the more stimulus is given, the more interconnections brain cells will develop. So, whilst the overall mental capability is determined by our genes, mental stimulation allows us to maximise our potential. So, in order to properly assess ape and other animal intelligence in relation to human intelligence, we ought to provide all the subjects with similarly intensive stimulation and encouragement.

If great apes are brought up from an early age to live within a human household, sharing the same food, playthings, furniture, bathroom, books, television, telephone etc. with human infants, it comes as no surprise that such apes manage to perform complex tasks, use sign language, and convey abstract thought by, for example, pointing

Chimpanzees can certainly convey a number of meanings through their seemingly small repertoire of grunts and hoots. For example, a particular sound may refer to a particular food item. However, are these chimpanzees (below) having a family discussion while they have their picnic? Are they communicating thought any more than a group of elephants would? The most important question is: does chimpanzee intelligence have any bearing on human origins?

at specific icons on the buttons of a purpose-made keyboard. However, there is a real danger in invalid anthropomorphism when watching apes manipulate and use cigarette lighters, play computer games, make and use stone flakes, drive electric cars etc.

Dolphins, too, recognise themselves in a mirror. Dolphins can easily learn sign language and possess sufficient understanding such that even if the position in a sentence of the words *left* and *right* are changed, they still respond correctly. For example, the commands 'get the basket on your *right* and put it under the water fountain on your *left*' and 'get the basket on your *left* and put it under the water fountain on your *right*' pose no problem.

Like us, apes and other animals learn initially by watching and imitating. At the Epcot Centre in Florida, dolphins have been trained to use a huge purpose-made underwater computer keyboard which has keys the size of the dolphin's head. Each key represents abstract thought (such as a place in the pool, a treat, or a toy) and words from an electronic voice are triggered when any particular key is touched. Using this system, the dolphins interact and communicate very well with the researchers but do not seem to be able to interact as well as apes. Does this mean that dolphins are less intelligent than apes? In fact, testing intelligence is no easy matter at all. In 6 years, the dolphins only had about 700-800 hours experience (far less than the contact that a human baby would have with its mother in only one year), and it is quite likely that the design of the dolphin keyboard itself seriously limits the dolphin's ability to communicate. Since any experiment can only provide data within the limits of the experiment, our ability to assess dolphin intelligence might be to a large degree hampered by our own ingenuity in devising the best equipment and experimental conditions. To put it another way, if we put a chimpanzee into an empty plain-sided tank and had a scientist interact with it for just 2 hours every day, the rest of the time leaving it to play by itself, it is likely that the chimpanzee, in being subjected to such an unstimulating environment, would perform rather badly in tests of its intelligence. However, we could not thereby conclude that the chimpanzee was unintelligent.

While it is true that dolphins and elephants are both very capable creatures, the important question is: are they less intelligent, and do they have lesser emotional responses, than apes? For a truly scientific study of comparative animal intelligence, both an orphan baby dolphin and an orphan baby elephant should be treated as part of a human family, with the same amount of cuddling, contact and nurturing as has been the privilege of a number of great apes. Obviously, it would be difficult to keep a baby elephant in the house, let it watch television, eat at the dining table, or sit on your knee while you read it a book, and it would be equally difficult to have such prolonged and intimate contact with a dolphin. However, that is not an excuse for imputing special and higher intelligence to the ape, without first giving other animals the opportunity to show what they can do and understand, under equally intensive contact and education. Such trials, under equivalent experimental conditions, are the cornerstone of scientific study. Indeed, it is because of their failure to comply with rigorous experimental and statistical protocols that many medical research studies (performed by highly intelligent people) are regularly rejected as

Top: Elephants playing football.
Middle: Dolphin.
Bottom: Orang-utan.
It is easy to bring up an ape in the home, but not so easy to do the same thing with a dolphin or an elephant.

being flawed. Clearly, the experimental conditions for comparing ape, elephant and dolphin intelligence are unlikely to be easily achieved. However, even if it were attempted, since the watery world of the dolphin is far different from a rain forest, the sort of contact/education that would maximise an ape's potential might be quite inappropriate for a dolphin.

What has all this got to do with evolution? Well, what we can establish is that other than humans, great apes may not be the only *very* intelligent creatures on this planet, and simply acknowledging the intellectual capabilities of apes does not in itself permit the giant leap of extrapolating that such intelligence is a reflection of our (supposed) ancestral state. Going back to the waterfall story, *even if* the chimpanzee was deliberately sitting there in order to absorb the aesthetics of the waterfall, that could just mean that some animals are able to appreciate more than that which has been previously ascribed to them. As it is with outward appearances, so it is with intelligence - it is simply not permissible for *any* similarity to be accepted as evidence for the evolution of one creature into another.

Orang-utan

However, since apes are generally thought to be very much more intelligent than other animals, it is not surprising that this has given rise to the notion that intelligence has increased over millions of years as monkeys evolved through apes and (supposed) primitive humans to eventually become modern Man. Inherent in this idea is the assumption that, as for every other biological characteristic, the natural selection of genetic mutations can, and did, produce human intelligence from within an ancestral creature that did not have human intelligence. Some would even say that evolution of intelligence was inevitable:

> It is possible that evolution, in the long run, will always produce a creature with a high degree of intelligence, because in the struggle for existence, intelligence always pays. Francis Crick. 'Life itself'. MacDonald. 1981. p111.

But, if a frog could think, it might not agree with Crick because frogs have managed very well for (supposedly) millions of years, apparently without any intelligence *at all.*

Whilst a number of people do assign to chimpanzees some special status above other creatures, nevertheless there are quite a few other animals which are considered to have feelings like ours.

Chimpanzee

We all know that dogs show emotions, and without it being pointed out to them, they certainly appear to understand when they have done something unacceptable (I hesitate to use the word 'wrong' as that might suggest a moral understanding which is not intended). We cannot be sure if dogs experience guilt or shame, but they do know what on their part is, and is not, acceptable behaviour. And when you watch elephants:

> ...for any length of time, you begin to realise that...elephants experience much the same sort of challenges and tragedies, and even maybe emotions as we do... David Attenborough. 'Echo of the elephants'. BBC2 Broadcast. 5th October 1997.

Young elephants are well looked after and protected by their mothers, older siblings and also other members of the herd. When confronted with the death of a herd member, elephants certainly show behaviour suggesting a personal loss. There is a film of a baby elephant which was trapped in a mud bank being helped out by other herd members. One elephant even succeeded in fashioning some 'steps' out of the bank up which the baby eventually managed to climb safely out of the mud. Anecdotal evidence points to wild elephants even showing concern for other creatures, such as a baby cow that was seen to be *carried* back to a water-hole when, having lost its mother, it attempted to follow the elephant herd.

It seems that anyone who spends a lot of time with any particular type of large mammal tends to believe that 'their' animals are endowed with special characteristics. So, after spending many years studying orang-utans in Borneo, some consider orang-utans to be more than just animals:

> You can't look into an orang-utan's eyes and say he's just some wild animal...there's something behind there...I'm always wondering what they're thinking... 'Animal zone'. BBC2 Broadcast. 26th October 1997.

Yet many people have the same feeling when they look into the eyes of their pet dog. The study of wild macaques in Sri Lanka shows that individual members of a troop of monkeys all have unique personalities and complex inter-relationships, and even seem to show signs of emotion. *Hagle* was a leader for many years and was then ousted from his position after a fierce fight with a rival:

> Hagle is badly injured...his old allies give comfort...and even clean his wounds... 'The natural world: the temple troop'. BBC2 Broadcast. 26th October 1997.

Hagle died and was not seen by the rest of the troop for three whole days. When, eventually, the troop did come across his dead body:

> Silently, the Temple Troop gather round their old leader...Hagle had been an ally to many of the males, a mate to most of the females, and a caring father to all his infants...The young males touch him with extraordinary tenderness...some of the females even stop the flies from landing on his face...what could they be thinking?...sights like this challenge the traditional view that animals do not have complex emotions... 'The natural world: the temple troop'. BBC2 Broadcast. 26th October 1997.

Orang-utans can appear to act like humans.

This is remarkable since these macaques have very small brains. Apes have much bigger brains so we might expect them to be much more capable animals. One seemingly impressive sight is that of wild chimpanzees 'fishing' for termites in a termite mound using a twig. It seems that these chimps have thought about a problem (how to get at the termites) and come up with a specific solution. What happens is that when the chimps come across a termite mound, they go to a nearby bush and break off a twig. While holding the twig with one hand, they then shear off the leaves by pulling the twig through the other hand. The twig is then passed into a convenient hole in the mound. If they are lucky, when the twig is withdrawn,

Chimpanzee using a stick to 'fish' for food in a log.

one or two termites might be walking on it. These are picked off with the lips and the twig reinserted into the hole to start 'fishing' again. It is not only wild chimps that go 'fishing'. Those in zoos regularly prod the holes in logs with sticks that they carry about with their lips. Until recently, orang-utans were not known to use tools in the wild, but it is now known that a population in Sumatra regularly uses sticks in order to 'fish' for termites and honey.

Now, superficially this looks like a very intelligent solution to a problem. But is it? In contrast, a gorilla has a different approach to the termite mound. It will simply rip the thing apart with its bare hands and eat termites by the mouthful. None of that dainty 'one termite at a time' method that the chimpanzee uses, after all, life's too short. The anteater also rips the termite nest apart and uses its tongue as a sticky spoon. Which animal, then, is showing the greater intelligence - the chimpanzee, orang-utan, gorilla or anteater? In terms of rate of termite catches, the chimpanzee's method is distinctly inferior to that of both the gorilla and the anteater.

The 'fishing' technique is clearly a learned behaviour, but despite having been observed for many years, individual chimpanzees do not show any signs of moving beyond their quite simple and rigid approach to catching termites. This shows that the chimpanzee is *not* particularly clever. If, for example, with the realisation that there must be a whole lot more termites in the mound beneath it, a chimpanzee stopped to think for a moment, dropped its stick and then ripped the termite nest open – now, *that* (possibly) would be a sign of intelligence that might separate a chimpanzee from a monkey. However, to complicate matters, even the tiny capuchin monkeys are known to use stones as hammers and anvils to break open nuts.

In fact, we should not be unduly impressed by watching chimpanzees fishing for termites since humans and apes are not the only creatures that use tools. *Egyptian vultures* break open large eggs by hitting them with stomes held in their beaks. *Shrikes* are birds which are well known to deliberately impale their prey (such as lizards, beetles, and stoats) onto the thorns of acacia trees in order to help tear off bits of flesh. While *woodpeckers* knock on hollow trees with their beaks so as to communicate with their fellows, the Australian *palm cockatoo* bangs its message using a stick held in its foot.

These crows on Vancouver Island regularly dunk their food before eating it.

Nor are humans and apes the only creatures that go fishing with sticks. The *Galapagos woodpecker finches* are known to use cactus spines held in their beaks to stab at insects in crevices beneath the bark of trees. Other finches use sticks as levers to get at grubs beneath bark. Finches are even known to lever at a piece of bark with a stick, hold the stick aside with a claw, try to get at the grub with their beak, and if unsuccessful, try again with the stick. *Crows* use sticks to 'irritate' grubs in deep crevices so that when a grub bites onto the end of the stick, the crow hauls out its 'catch'. Furthermore, crows are known, if necessary, to modify the shape of a twig by breaking bits off in order to make the twig more pointed, and even to carry about their 'favourite stick' with which they habitually go 'fishing'. Such behaviours are very similar to that of the 'fishing' chimpanzee and yet no-one anthropomorphises vulture, finch or crow behaviour.

During filming for a series of documentaries on Indian wildlife, a kingfisher chick was found helplessly entangled in some foliage a few feet above ground level. Very soon a *ratel*, a nocturnal carnivorous animal looking something like a cross between a skunk and a racoon, came across the chick, only to find that it was out of reach. However, without any hesitation, the ratel searched around and found a sizeable log, deliberately rolled it into position underneath where the kingfisher was, and climbed on top – but was still unable to reach the chick. Not easily beaten, it then found a larger log, climbed up, and this time reached up and grabbed its meal:

> ...extraordinary...this is truly remarkable...very few animals use tools in such a calculated way... Valmik Thapar. 'Land of the tiger'. BBC2 Broadcast. 8th December 1997.

Above: Weaver bird nests, Serengeti.
Below: Weaver bird.

Beavers demonstrate an ability to use tools that is often overlooked or underestimated. They go out of their way to cut down small trees and carry them away to a stream, they make a dam and build a nest within the newly-formed lake, with its entrance below the water level:

> Placing sticks on one another in such a way that they engage and are not readily dislodged is a skill that is easy to underestimate. David Attenborough. 'The trials of life'. Collins/BBC. 1995. p142.

Above: Beaver skull.
Below: Macaque.

Weaver birds are well known to make the most intricate nests by carefully entwining and interlocking different strands. However, other birds are also extremely deft. The Indian tailor bird is able to *sew* together a number of leaves using spider's silk:

> [It] creates a cup from living leaves...Holding a length of silk in its beak, it then pierces a hole in the leaf and pushes silk through it, tying a little knot in the end to prevent the thread slipping back through the hole and then doing the same thing on the other side so that the two leaf surfaces are secured to one another. David Attenborough. 'The trials of life'. Collins/BBC. 1995. p144.

Although it is not strictly tool use, other animals regularly appear to do things in anticipation of a desired result. For example, Californian *crows* drop walnuts from the tops of lampposts to crack them, and *gulls* often fly up and drop sea shells to break them. Eagles are known to drop tortoises from a height in order to break open their shells. Much of the diet of the *lammergeier* (*bearded vulture*) comprises bones with which it flies up and drops from a height, in order to break off fragments of a suitable size for swallowing. Macaque monkeys in Japan are known to separate seeds from gravel by throwing the mixture onto water - the gravel sinks and the floating seeds are gathered up and eaten. Of course, these sorts of activities can most likely be explained as *conditioned responses*. In other words, a gull that (for example) dropped a shellfish on its way back to its nest might find the shell open,

In some experiments in which the animal has to use a joystick to manipulate a cursor on a computer screen, pigs perform far better than dogs and, even more surprisingly, better than all but the most experienced chimpanzees.

and it therefore might be conditioned to *automatically* drop its next meal in the same way. It does not need to think about what it is doing any more than a dog does which salivates when trained to associate the sound of a bell with food.
More relevant to our assessment of animal intelligence is the fact that pigs can be trained to herd sheep, and in experiments involving specific movements of a cursor on a computer screen using a joystick, pigs have been shown to perform *much better* than dogs, and perhaps what is even more surprising:

> ...even faster than most chimpanzees... Prof. Stanley Curtis. 'Move over Babe'. QED. BBC2 Broadcast. 1997.

Clearly, assessing intelligent behaviour is not as simple a matter as it might first appear, and it is very important to interpret correctly any observations, since one can be very easily misled. Surprisingly, some people think that animals with brains even smaller than those of macaques are intellectually as capable as chimpanzees or dolphins. There are a number of important lessons to be learnt from taking a close look at Alex, the parrot who can talk, and (supposedly) count, hold a conversation and *read*:

> The 20-year-old bird is said to count up to six and to recognize and name some 100 different objects, along with their color, texture and shape; his ability to categorize **rivals that of chimpanzees**. (My bold). Madhusree Mukerjee. 'Scientific American News: interview with a parrot: 4/96.' p2. http://www.sciam.com/0496issue/0496scicit06.html. 14th January 1998.

But is this a sign of intelligence or just parrot-fashion conditioning?

> The parrots learn best when two trainers teach each other, one asking questions, the other 'learning' the right response. Alex listens and watches, and then breaks in, completing the model. This system is called two-part modelling, and it's the centrepiece of Pepperberg's method. Jake Page. 'Alex the talking parrot'. p4. http://www.discovery.com/DCo/doc/1012/world/nature/parrot/parrot l.html. 14th January 1998.

Macaw

Here is an example of one 'conversation'. When the parrot is presented with a metal key and a green plastic key, he is asked three questions:

> **Q.** **'What toy?'**
> A. 'Key.'
> **Q.** **'How many?'**
> A. 'Two.'
> **Q.** **'What's different?'**
> Q. 'Color.'
>
> (My bold). Madhusree Mukerjee. 'Scientific American News: interview with a parrot: 4/96.' p2. http://www.sciam.com/0496issue/0496scicit06.html. 14th January 1998.

On the basis of interactions such as this, Irene Pepperberg, the parrot's trainer, believes that the parrot *understands* questions, *thinks* about possible answers, and then gives the correct answers drawn from its own *knowledge*. But before we can

efit of any human intervention and teaching), stand out from other small-to-medium-sized mammals as a group of creatures *especially* endowed? Not necessarily.
So, regarding either intelligence, or the showing of emotion, or the expression of personality: if we were to say that a dog is on one level, and humans are on another far higher level, chimpanzees might simply be represented by a blip on the horizontal line representing the dog. *And they would not be alone there* – they might be virtually side-by-side with (amongst others) elephants, dolphins, and perhaps even pigs. So we see that when looking objectively at the behaviours and intellectual abilities of apes, there is no particular evidence that forces us to accept that chimpanzees are essentially human beings that have not yet evolved a larger brain.

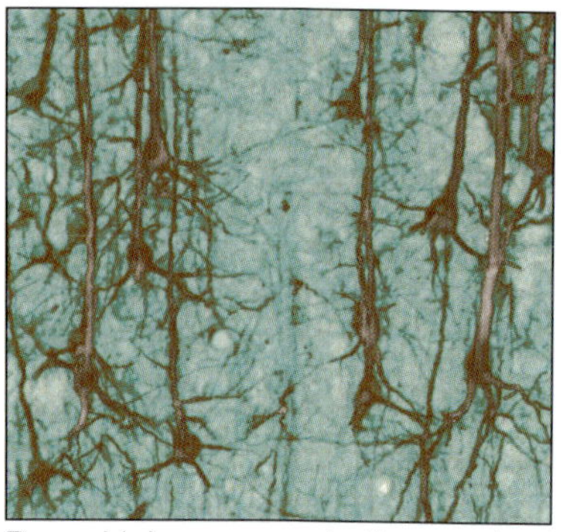
Pyramidal neurones in the cerebral cortex.

### *Comparisons of brain size*

Does the study of brain size itself give us any insight about our origins? Organisms vary considerably as to how many nerve cells (neurones) they have in their brains.

## Brain sizes

**Brain volumes**

| | |
|---|---|
| ***Sperm whale*** | 7820ml* |
| ***Elephant*** | 7500ml* |
| ***Bottle-nosed dolphin*** | 1600ml* |
| ***Human*** | 879-1850ml** (average: men 1430ml, women 1294ml***) |
| ***Gorilla*** | 340-752ml**** |
| ***Orang-utan*** | 276-540ml**** |
| ***Chimpanzee*** | 282-500ml**** |
| ***Cow*** | 500ml* |

* R Harrison & M Bryden. 'Whales, dolphins and porpoises'. Merehurst Press. 1990. p161.
** Gray's Anatomy. 31st edition. Longmans. p1060.
*** Gray's Anatomy. 34th edition. Longmans. p1112.
**** H McHenry. 'Fossils and the mosaic nature of human evolution'. Science. 31st October 1975. p429.

The human brain (seen from the left side).

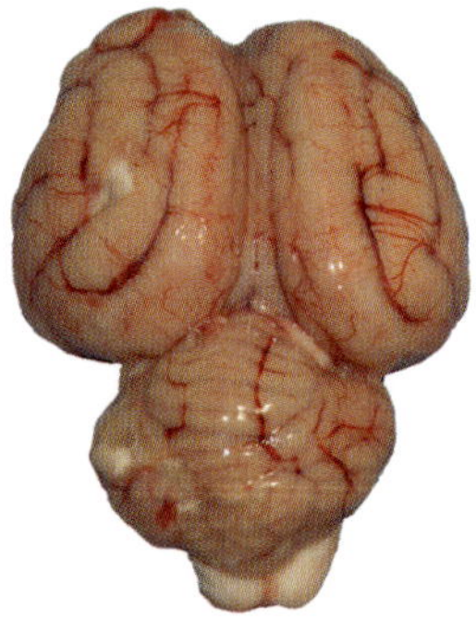
The brain of a domestic cat (seen from above, with the front of the brain at the top).

Some simple organisms have only a few neurones; a fly may have something like a million neurones; and a man has perhaps $10^{12}$ (a million million) neurones. Certainly, having more neurones makes for a more capable creature, but only up to a point. The huge diplodocus would have dwarfed the largest elephant, yet despite its brain being tiny, this was nevertheless sufficient for *all* of its activities, most of which were not significantly different from the activities of elephants. This shows that the bulk of an elephant's brain is *not* necessary for the control of basic bodily functions such as control of muscles, hormones, the heart and intestines, etc. By the same token, the 'wiring' in the brain stem and cerebellum of a mouse would be quite capable of handling an elephant's requirements for all the routine subconscious bodily functions and movements; and the relatively small brain of the spider monkey is as capable of handling all the complex movements in the three-dimensional world of the forest canopy as is the larger brain of a chimpanzee. The gorilla has the largest brain of any living ape, but it is not more intelligent than the chimpanzee or orang-utan. Furthermore, a 3-year-old human child is more capable intellectually

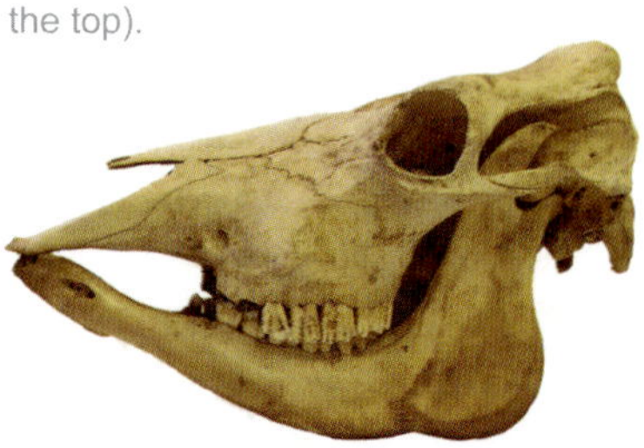
Cow skull.
It is noteworthy that the largest gorilla brain is only about 130cc less than the smallest normal adult human brain, and the cow brain is as large as the largest chimpanzee brain.

**Brain size and intelligence**

The largest gorilla brain is only about 130ml smaller than the smallest human brain. An elephant has a brain that is some 6600ml larger than the smallest human brain. The question is: would an increase in size of 130ml make the largest gorilla brain human? And why is an elephant not super-intelligent?

On average, the brain of a woman is some 136ml smaller than that of a man but this does not cause any compromise in intelligence. Nor is a man with a 1000ml brain intellectually inferior to one with a 1400ml brain.

There is therefore no advantage in a man having evolved a brain 10.5% larger than that of a woman, or even in humans having evolved brains larger than 1000ml.

Instead of wasting the extra energy required for maintaining a man's unnecessarily larger brain, evolution ought to have utilised the energy for other features, for example, to increase muscle bulk.

All the evidence points to the human brain having components that are unique and shows that the difference in intellectual capabilities between humans and apes is not attributable simply to brain mass or the number of neuronal connections. In other words, the notion that a human brain is simply a larger version of an ape brain has no scientific basis.

**The origin of the human brain**

Books on evolution do not give the reader a fair insight into just how complicated the human brain is. We have seen that an increase simply in brain mass does not explain the origin of human intelligence. Just as importantly, it must be understood that the human brain is not simply an homogenous mass of interconnecting neurones. If *any* part of the brain (for example, the pituitary gland, the brain stem or the cerebellum) is examined, both the naked-eye and microscopic views of the architecture of the brain show a *supreme* complexity. Whilst simplistic discussions about the human brain arising from an enlargement of an ape brain abound, there is little (if any) exposition on the (supposed) stepwise origin of the brain itself. The *WYWYWG Principle* shows that the stepwise origination of the myriad component parts of the human brain is implausible.

than a gorilla with an equivalent-sized brain. The female human brain has always been relatively smaller than that of the male but there is no evidence to show that it is intellectually inferior. Pigmy peoples have yet smaller brains but they have full human intelligence. These days, with better nutrition and medical care in childhood, people are generally growing larger. Hence people's brains are getting larger along with their bodies, but there is no evidence that this has affected their intelligence.
Since many animals seem to have extra brain tissue which is superfluous to requirements for their survival, and since such 'extra' brain tissue is not associated with a significant increase in intelligence, a large brain per se cannot be the cornerstone on which great intelligence rests.

Of course, it might be thought that it is *relative brain size* (the size of the brain in relation to the size of the body it is housed in) which is important. However, whilst in many cases relative brain size does correlate well with intelligence (for example, the brain is 0.2% and 2.3% of the body weight in elephants and humans respectively), this idea is not entirely satisfactory in explaining comparative intelligence, since the brain of a shrew is a substantial 3.3% of its body weight while that of a humming bird is a remarkable 4.2% of its body weight. In addition, there is the condition called *microcephaly* (which is caused by a number of genetic and developmental abnormalities) in which babies are born with very small brains. Usually, the brain is abnormal as well as being small, but in some cases a very small brain is known to be associated with normal intelligence. This all suggests that the capability of the human brain is not simply a factor of its size relative to an ape brain.

One distinctive feature of the human brain is its substantial continued growth after the neonatal period (all ape brains undergo very little increase in size after birth). Most of this enlargement is due to the increase in the number of *connections* between *existing* neurones that have been present from birth, rather than an increase in the number of neurones themselves.

One current theory proposes that as a result of mutations in ancestral apes, adult humans are essentially apes that have somehow managed to maintain the infantile ape head shape (round skull, flat face) into adulthood. This is so-called *neotany* - the brain continuing to enlarge beyond the time that normal ape-brain growth ceases:

> So we can see ourselves as sexually mature baby chimps...The timing of just a few control genes producing one of the great evolutionary leaps.
> 'Homeobox genes'. BBC2 Broadcast. 1998.

If size was all that mattered, then maybe if an ape brain was artificially enlarged and the animal brought up as a human baby and exposed to the same sort of stimuli and education as our children, then such an ape might, in terms of its intellect, become significantly closer to the human state. In fact, there is a simple experiment that could be done to enlarge an ape brain and so test the hypothesis. In the foetus, the individual plates in the skull are initially laid down as cartilage which is gradually mineralised to form bone. The growth of the skull continues after birth until the individual plates of bone have fused together. In the condition hydrocephalus, excess fluid is formed within cavities deep inside the brain and the resultant increase in pressure results in the skull expanding. A large skull results, but because the internal

pressure is damaging to the brain tissue itself, the brain becomes thin and reduced in mass. Intelligence can therefore be seriously impaired if the process is not treated by inserting a shunt to allow fluid to escape out of the skull. So, forcing the enlargement of the skull of an ape by injecting fluid into its brain cavities would not be a good idea. However, if we took a new-born chimpanzee and surgically interposed some inert material between the skull plates so as to prevent or delay their fusion, the skull and brain therein might continue to enlarge considerably. Would such enlargement, coupled with intensive affection, stimulation and teaching, be sufficient to make the animal more intelligent? In fact, there is good evidence to suggest that an ape with a large brain *would not* have an improved intellect.

If humans are apes which have somehow acquired delayed fusion of their skull bones, then perhaps we might expect that the premature fusion of human skull bones (*craniostenosis*) would cause some degree of reversal of our (supposed) evolution and result in a brain and skull that resembled more closely the ape form. However, this does not occur. In fact, when all the skull bones are involved (*acrocephaly*), premature closure gives rise to a more vertical forehead (rather than a sloping ape-like forehead). Furthermore, although the brain is smaller, its function is unaffected and there is no reduction in intellect (unless there is also an increase in pressure inside the skull – *raised intracranial pressure*). Finally, premature closure in humans can be associated with a number of other congenital abnormalities, such as *cleft palate* and fusing together of the fingers (syndactyly). So, the size and function of the human brain is unlikely to be simply the result of an ape brain being allowed to grow beyond the normal time of fusion of the skull bones.

Instead, could the size and function of the human brain be explained by the delayed fusion of skull bones in association with some active stimulus that would cause continued growth of all the body tissues, including the brain? In humans, this state of affairs can be produced by an abnormally high production of *growth hormone* (which is produced by the *pituitary gland* at the base of the brain). In infants, the condition gives rise to *gigantism* in which the head and body are greatly increased in size. However, (unless there is any associated raised intracranial pressure caused, for example, by a tumour) intellect is unaffected. Most certainly, intellect is not *improved* by the increase in brain size.

The reverse problem, an abnormally low production of growth hormone, results in reduced growth and small size (*infantilism*). Here the bodily proportions are retained, but the small brain size does not result in impaired intellect.

So from the above, the evidence suggests that forcing either a shorter or a longer period of brain growth in the human does not alter the brain's capabilities. In other words, in an ape, delayed skull bone fusion and a longer period of brain growth would most likely result only in an enlarged ape head and brain, with no change in intelligence.

Whilst we can demonstrate that the brain does not function like a desk-top computer, one analogy with a computer is valid: that an increase in processor size can lead potentially to an increase in computing speed. For example, if provided with the necessary software, spell-checking with a futuristic *Super duper 50*® processor would be quicker than when using a humble Pentium IV. However, without the

**Brain size and birth**

At birth, the human brain is significantly larger than that of a baby ape. A (hypothetical) mutation that might result in an ape foetus developing a larger brain *in utero*, would most likely result in impaction of the head during labour, and the death of both mother and baby.

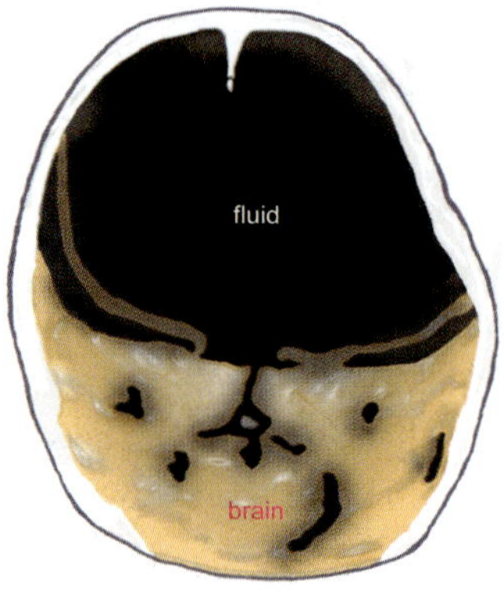


**A small brain is compatible with normal human intelligence**

Brain scan of a normal 55 year old man whose brain only occupies the back half of the skull. The large black area represents fluid which has compressed the brain and which may have been present from birth.
*Redrawn from British Medical Journal. Vol 327. 25th October 2003. p998.*

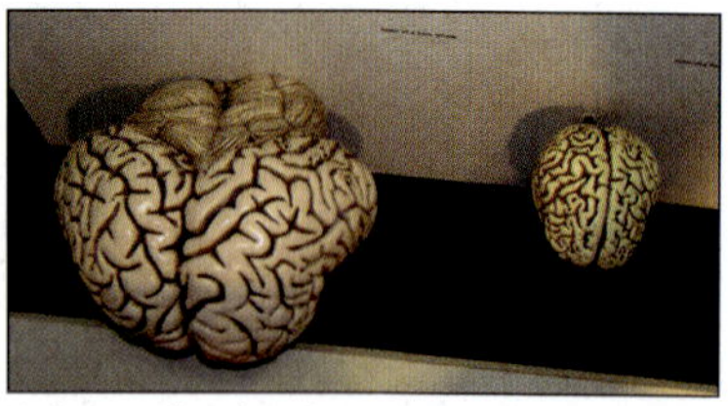

A large brain does not mean great intelligence.
Left: Blue whale brain cast.
Right: Human brain cast.

necessary software, simply being a fast processor per se would not allow the *Super duper 50®* to do any other task. So intelligence is dependent on how the brain is hard-wired and programmed as well as its size, and on how its connections are modified by experience and application to problem-solving. Thus the greater brain size (whether relative or absolute) of humans above that of the gorilla is not *the* cause of the phenomenon of human intelligence.

Most likely, what separates our human-ness from all other creatures is dependent on specific parts (such as Broca's area - see later) and on unique functional aspects of the human brain, not on its size. Nevertheless, despite this, increase in size is still presented as the major determining factor which facilitated the (supposed) journey from australopithecines to humans.

Some people believe that a large brain is an essential prerequisite for the manifestation of complex interrelationships and activities:

> The larger brain was **probably** an evolutionary advantage for social reasons as well as for technology. The ability to memorise social relationships and cooperate in foraging and hunting are important mental skills. (My bold). Ann and Patrick Fullick. 'Inside science: the human story'. New Scientist. 7th June 1997.

However, there is little persuasion in such notions since many small-brained savannah animals, such as a lone impala, can forage without planning the day's eating with relatives. Furthermore, chimpanzees, despite having a brain one-third the size of a human brain, are well able to organise a monkey-hunting party. However, lions, wolves and dolphins can also organise hunts, and many animals, such as wolves, big cats, monkeys, baboons and even meerkats, are involved in complex social interrelationships.

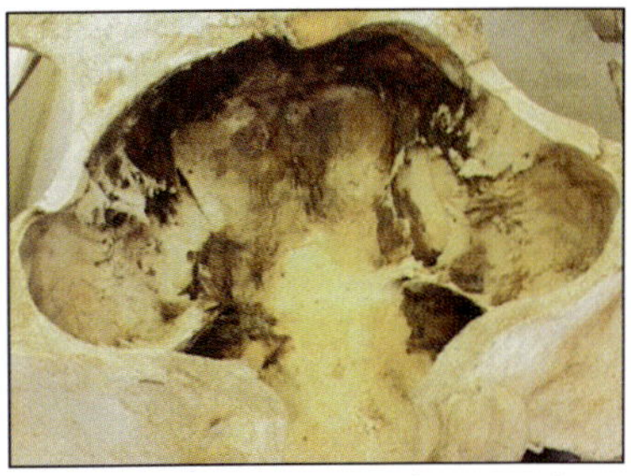

View into the brain case of an orca from above. The top of the skull has been removed and the back of the skull is at the bottom of the picture. The brain of an orca is significantly larger than that of a human brain.

Others suggest that enlargement of the ape brain was fuelled by a different diet once apes left the forests:

> Meat allowed us to free ourselves from the dietary shackles of our African homeland. It did more than that, however: **it made us brainy**. Easy to digest and rich in energy, meat provided the vital resources that our expanding brains demanded...says Leslie Aiello...**We started to eat meat, got smarter**, and thought of cleverer ways to obtain more meat. It was a loop...Breaking open the **marrow**-rich leg bone of an animal such as a wildebeest would have provided a **massive dose of calories** in a single swift meal. The calorific input would have provided the resources needed to fuel the swelling brains of humans, **an input that would - in turn - have improved their intellects** and therefore our ancestor's ability to find their own meat. (My bold). Robin McKie. 'Ape-man'. BBC Books. 2000. p109-113.

However, McKie and Aiello give the unwary reader a number of other false impressions: firstly, that somehow, eating meat or marrow can cause a brain to become larger and more capable intellectually than when eating a purely vegetarian diet. In fact, given an adequate and balanced diet, a meat-eating habit has no general advantage over a vegetarian habit. We have already seen that human vegetarians do

not have smaller or less capable brains, and that a bigger brain is not necessarily a better brain. Furthermore, the human digestive system *needs* a significant amount of vegetable roughage for it to function properly. Indeed, serious problems such as *chronic constipation, diverticular disease and colon cancer* can result from a high protein, low fibre diet.

A second false impression is that meat and marrow are significantly more nutritious than vegetarian food sources, particularly in terms of high and useful calorific value. In fact, meat contains much the same amount of protein as is found in many types of nuts, and in energy terms meat is no better than carbohydrate. So 100g of peanuts provide the same amount of protein and calories as 100g of meat.

## ENERGY CONTENT OF DIFFERENT FOODS

| FOOD | Energy content |
|---|---|
| Carbohydrate | 4kcal/g |
| Protein | 4kcal/g |
| Fat | 9kcal/g |

The daily energy requirement for physically active adults is 2500-3000kcal.
For sedentary adults the daily energy requirement is 1500- 2000kcal.

Although brain and marrow contain a substantial amount of fat, weight for weight these tissues contain *less* fat than there is in the fat layer found underneath normal skin. What McKie and Aiello may be trying to say is that it would be quicker to obtain the daily energy requirement by tearing off and eating a chunk of scavenged meat than by spending all day foraging for fruits, berries and tubers. However, such a view is questionable, since even when driving large distances in the Serenegeti it is uncommon to see a kill. So, an australopithecine might well have been better served by casually foraging rather than venturing out into the heat of the day looking for an uncertain dinner. Furthermore, scavenging left-overs from a lion or hyena is also a very risky business. If high-energy food was essential, the fat normally found underneath the skin of smaller mammals would have been more easily accessible to any (supposed) early ape-men (as shown by the monkey-hunting methods of modern chimpanzees) than that found in any carnivore left-overs.

Half-eaten buffalo, Serengeti. Not an appetising meal.

McKie and Aiello also make the assumption that savannah predators would leave significant amounts of a carcass, and would do so often enough so as to sustain a group of opportunistic australopithecines. However, having brought down a wildebeest or zebra, a group of lions or hyenas usually eats the chest and abdominal contents, and the limbs are usually severed and carried off by individuals to eat away from the main carcass. If they can at all avoid it, carnivores such as lions, leopards, hyenas and wild dogs *do not* leave significant portions of their kill. So, any meat, brain or marrow that might subsequently become available to a scavenging ape would be a *very* irregular occurrence, and so uncommon that any *hoping-to-evolve-a-big-brain* ape could not depend upon it even simply to survive. Indeed, baboons living in African savannahs today manage to survive very well on a largely vegetarian diet and do not depend to any degree on scavenging the occasional carcass.

Whilst it would seem implausible that any amount of enlargement could induce the ape brain to acquire human functions, it is still argued that there must exist a relatively small number of fundamental mutations that separate the human brain from its (supposed) ape ancestor, and that since there are thousands of genes operating in the brain, there must have been plenty of material for mutations and natural selection to work on over the years. Therefore, it is thought that it was *inevitable* that chimpanzee brains would evolve into human brains:

> There has been a **foolish tendency** for people...to deny the fact that evolution could affect the brain...well, **that's just silly**. There are 60,000 or so working genes in a human being and half of them are switched on at any one time in the brain...so clearly, there are many, many genetic changes in the brain which are open to natural selection. (My bold). Steve Jones. 'The Darwin debate'. BBC2 Broadcast. 28th March 1998.

It is now known (from the Human Genome Project) that we have only about 25,000 genes - not a great deal more than *roundworms*. Some have said that this shows that humans are:

> ...fundamentally simpler than we thought. Minerva. British Medical Journal. Vol 22. 24th February 2001. p500.

Actually, since the permutations of gene interactions available to the roundworm are astronomical, what this more likely shows is that roundworms are supremely complicated, and that the one-third more genes in a human being makes us *immeasurably more complicated* than some people are led to think.

Jones speaks very unkindly of intelligent people who might disagree with his dogmatism, but there are good reasons to consider his views invalid. It is all a question of the probability of random changes in a working system resulting in improvements in that existing system. Whilst we have ample good evidence from medicine to show that alterations in genes can result in *brain damage* and *malfunction*, we have *no evidence at all* to show (with the millions of possible mutations available to those 25,000 genes, and with the astronomical number of permutations in the way those genes could interact in a developing embryo) that there is even one mutation that could change the ape neurological connections or general brain form so as to become just one step closer to the human state.

Jones takes as his starting point the assumption that the evolution from dust to primate *has* actually occurred and so tries to force his views to fit in with that assumption. In contrast, what we have to do is to examine the evidence for what it is and to see what can and, just as importantly, what cannot, be deduced from it.

Since apes and humans have most genes in common, it has been suggested that perhaps the differences in brain capability might be explained by differences in the rate of *gene expression* (rGE) i.e. the rate at which mRNA and proteins are made from their genes. Although there are only very limited data available, nevertheless, if a greater rGE in the brain was responsible for human intelligence, we should be able to make a number of predictions.

Firstly, since there is no significant difference in intelligence between chimpanzees and orang-utans, there should be no significant difference between their rGE. However, this is not what is found. Orang-utans differ in their rGE as much from chimpanzees as the latter do from some humans. Secondly, since most humans of normal intelligence have roughly equivalent cognitive capabilities, the differences in rGE between them should be small. This is not found either, since the rGE in some humans is closer to that of chimpanzees than it is to the rGE of other humans:

> The variation in gene expression between individuals within the species is substantial, relative to the differences between humans and chimpanzees. For example, **one human brain sample differs more from other human samples than the latter differ from the chimpanzee samples**. (My bold). Wolfgang Enard et al. 'Intra- and interspecific variation in primate gene expression patterns'. Science. Vol 296. 12th April 2002. p340-3.

Thirdly, since most mammals share common anatomy and physiology, the rGE in organs in which there are no significant functional differences should be roughly the same in diverse mammals. However, this is not so. For example, the rGE in the livers of orang-utans and macaques, and in the white blood cells of macaques, differs markedly from that of chimpanzees and humans.
As with many researchers in biology, Enard et al. start their paper with the assumption that humans are modified apes:

> Striking differences in morphology and cognitive abilities exist between humans and **their closest evolutionary relatives**, the chimpanzees. (My bold). Wolfgang Enard et al. 'Intra- and interspecific variation in primate gene expression patterns'. Science. Vol 296. 12th April 2002. p340-3.

However, on the evidence so far, there is no justification for supposing that apes became humans by simple mutations that altered the rGE in their brains.
Could brain disease or injury give any insight into the human condition? Certainly, a human brain, damaged in just a few key places, will suffer the most severe loss in intellectual ability. But this does not mean that those few key areas are the result of evolutionary additions, bolted on to an ape brain. Nor does it mean that such key areas in the human brain are separated from the ape position by only a few genetic steps.
Look at it this way. We have previously seen that vision is a supremely complex phenomenon that includes both the conscious awareness of visual stimuli via the visual cortex, and also the unconscious registration of visual stimuli via other pathways (*blind sight*). Suppose we consider blind sight to be a primitive condition and that it existed at some time prior to the origination of conscious sight. We should certainly not consider that the supreme complexity of conscious vision is merely a few genetic steps away from a brain which has only blind sight, or that conscious vision could have evolved merely by an increase in *size* of a (supposedly) primitive brain that had no visual cortex or other accessory visual connections. In the same way, in the absence of any supportive evidence, it is unreasonable (perhaps even naive) to expect the human brain's capability for thinking and language to be related simply to mutations and enlargement affecting an ape brain.

Even if we suppose that the key areas in the human brain *are* separated from the ape position by any number of genetic steps, the question then arises: in order to increase intelligence, why did the ape brain not simply evolve changes in key areas, and retain its small size, and with it the ape face? Furthermore, if the chimpanzee has had (supposedly) 8 million years to develop since its ancestors (supposedly) separated from our own, we should expect the chimpanzee to have evolved far more intelligence than it has. Yet a chimpanzee is no more intelligent than the other great apes and is, in some ways, less intelligent than a pig.

It might be suggested that the living apes have evolved into a blind alley with regard to intelligence development. However, if both humans and the living apes shared a common ancestor, then they would have shared the same potential for mutations at those DNA locations which might allow for improvement in intelligence. Since there can be no deleterious consequences of having a better brain, then given enough time, apes would just as likely have obtained such beneficial mutations in the same way that humans are supposed to have done.

## *Speech*

### *Ape and human brains*

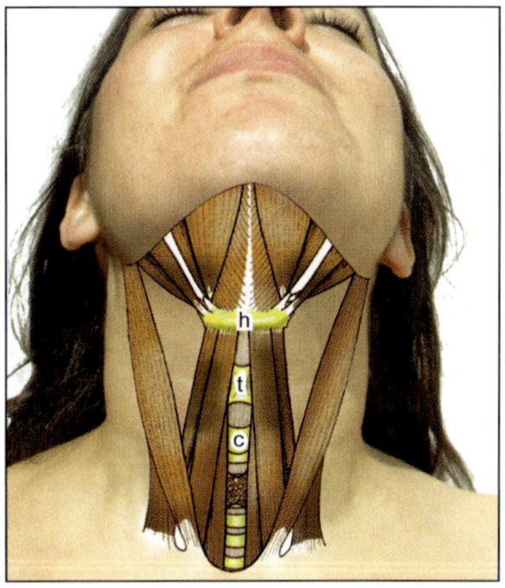


Front of the adult human neck.
h: hyoid bone
t: thyroid cartilage ('Adam's apple')
c: cricoid cartilage
The larynx is made up of the thyroid, arytenoid and cricoid cartilages. The vocal cords lie within the confines of the thyroid cartilage.

It might be argued that what has prevented chimpanzees, gorillas and orang-utans from entering into the spiral of increasing brain size and intellect, is that they do not have the right sort of throat to enable them to voice the consonants that allow the great flexibility of sounds that humans make. At birth in both humans and apes, the *hyoid bone* (which gives support to the muscles at the front of the throat) and the larynx lie high up in the neck. The high position of the neonatal larynx allows a watertight connection between the trachea and the back of the nasal cavity, and this permits babies to breathe and swallow liquids (which pass around either side of the elevated larnyx) at the same time. As a human baby grows, the hyoid bone and larnyx gradually descend so that by the age of about 3 months, they are situated in the more familiar position in the neck. Although the chimpanzee larynx also descends into the upper neck in the first year of life, the hyoid bone does not.

Certainly, the adult human larynx is better suited to speech, but why should the ape larnyx be *such* a handicap when it comes to making sounds? A great deal of communicating *can* be done with just *two* sounds since if the two sounds are arranged in a number of different repeats, then although the 'words' will be longer, nevertheless *thoughts* can be expressed. For example, many tribes convey thought across long distances by drumming. Only two drum sounds (producing a high and a low note) are necessary to convey complex messages. Each drum word is simply expanded into a stock phrase, each phrase being given its own unique tone and rhythm. Furthermore, humans who have had their larnyx removed due to cancer are able to belch intelligible sentences by controlled use of their chest and abdominal muscles. This shows that a human brain without a larynx is well able to communicate with speech. In other words, motor control of laryngeal muscles alone is not the key to human speech. Since apes have voluntary control over the timing, pitch and length of their grunts and squawks (in other words, they can choose when they want to make a sound and what sort of sound it will be), then we should expect that apes should be able at the very least to copy *some* human speech, however badly. Yet they seem incapable of even *trying* to do so. This is curious since apes can communicate with gestures and symbols without the need to speak at all (showing that they can think in abstract terms), and there is no reason why they should not be able to control their belching as well as any laryngectomy patient. Clearly, there must be more to speaking than just appropriating the ordinary voluntary control of the larnygeal, chest and abdominal muscles.

It might be argued that pivotal to the issue regarding the (supposed) evolution of speech is that both brain control of the larynx and the larynx position itself evolved together, with each feature facilitating the other's improvement. Could such a thing evolve suddenly, or slowly stepwise? Since significant alterations to the process of embryological development result in serious congenital *abnormalities*, it is implausible that a single mutation could result in a sudden and dramatic downward shift in the position and form of the adult ape larnyx without it being seriously damaged in the process. Could a series of slow, stepwise alterations make the change from ape to human form? Perhaps an ape with a larnyx that was positioned just a few millimetres lower in the throat might be able to grunt that little bit more intelligibly than its contemporaries. We know that the more any function of the brain is used, the more the corresponding neurones tend to increase their number of connections. So it might be argued that a slightly lower larnyx would allow a little more sophistication in grunting, and that any ape brain that made use of this greater vocal functionality would enlarge in those areas devoted to larnygeal muscle co-ordination. Such apes might be selected for because a slight increase in the ability to communicate vocally might give them a survival advantage. Their progeny might then develop an increasingly lower and more efficient larnyx, with more efficient neuronal control, eventually becoming able to speak in the same way as humans.

Such a progression might at first glance seem feasible but is there any scientific evidence that supports the idea? For there to be any evolution, any slight change must be significant enough for the individual ape concerned to have a survival advantage, such that its laryngeal and brain characteristics are passed on to its progeny.

However, the evidence is all heavily weighted against such evolution. We know that despite all the intensive human contact many apes have been subjected to, they do not speak *at all*. Being unable (but perhaps willing) to speak, such apes may well be frustrated inside the confines of their ape body. For our purposes here, this could in some way be compared to a human being undergoing an operation during which the anaesthetic renders him fully paralysed (and so unable to communicate) but awake enough to experience the whole procedure. No amount of thinking on his part will allow him to speak because of the chemical blockade of his muscles. However willing, it might be that an ape would not be able to speak even a little, since we know that a willing human brain which is damaged in key areas is *not* able to direct a normal larynx to produce spoken thought.

Furthermore, since the intensity of the mental stimulation that domesticated apes have experienced is astronomically greater than that which any (supposed) *hopping-to-become-a-man* australopithecine could ever have experienced, we can safely assume that any australopithecine that acquired a little more intellect would still be

Morphogenesis of the face and neck is driven by an intricate relay of signalling molecules and transcription factors organised into hierarchical pathways. The co-ordinated action of these pathways regulates the development of neural crest cells within the pharyngeal arches, resulting in proper spatiotemporal formation of bone, cartilage, and connective tissue:

> *Endothelin-1, a small, secreted protein…is crucial for patterning the neural crest cells that give rise to lower jaw and throat structures.*
> D Clouthier, T Schilling. 'Understanding endothelin-1 function during craniofacial development in the mouse and zebrafish'. Birth Defects Res. C. Embryo Today. 2004 Jun;72(2):190-9.

So, interfering with such finely-tuned regulatory pathways in a chimpanzee will result in a deranged larynx rather than a human one.

**Human and ape upper jaws have different structures**

> [A] *separate bone is found in most mammalian jaws, including those of primates. A separate premaxilla does not, however, occur in humans.*
> A Ten Cate. 'Oral histology'. Mosby. 1998. p45.

completely limited by its ape larynx and be unable to speak even a little. Since an australopithecine that could think more intelligently but could not speak would have no survival advantage over its contemporaries, its progeny would simply interbreed with its contemporaries and no evolution of the larynx could ensue.

But why should a willing intelligent ape not be able to speak, however crudely? One of the things that all ape brains lack is a part of the brain known as Broca's speech area, which comprises a unique bulge on the surface of the lower left side of the frontal lobe and which is essential for the vocalisation of human speech. There are a number of different areas of the brain which are known to be active during different forms of speech, and also during the process of thinking itself, so it is likely that Broca's area has widespread connections within the brain. People with a damaged Broca's area (for example, following a stroke) may have some associated paralysis but can be otherwise normal. Indeed, such people are fully capable of reading and understanding spoken language, and are also capable of communicating in sign language. However, they are incapable of speaking and are not able even to belch sounds as can a person who has had his larynx removed.

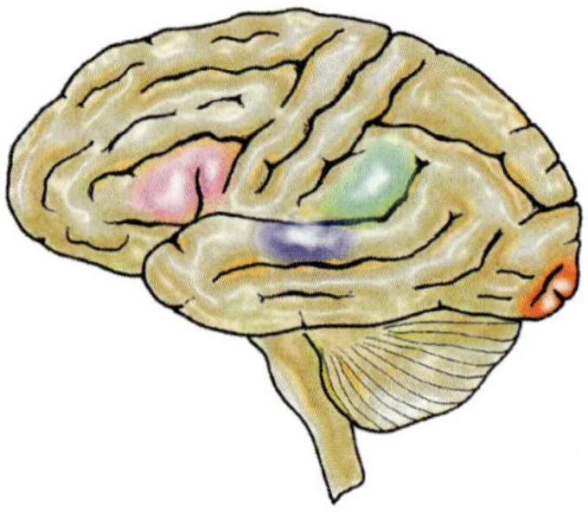

The human brain.
Pink: Broca's motor speech area
Green: Speech association area
Blue: Auditory area
Red: Visual cortex

So, the condition of the ape brain is likely to be in some ways similar to the stroke patient with a damaged Broca's area. For any form of speech (whether laryngeal, grunted or belched), the translation of thought into specific sounds requires thought to be converted into specific signals to the appropriate motor nerves and hence their target muscles. These speech-specific signals seem to be incapable of being mimicked by the mechanisms that otherwise exist to trigger the same muscle groups during voluntary breathing, breath-holding and belching etc. Perhaps this points to the human brain having unique 'wiring' and 'software' which is absent in the ape brain.

Nevertheless, against all the evidence, if, in some (supposed) subhuman ancestor, the genetic coding for the larynx and Broca's area managed to arise by some chance mutations that were beneficial, then we should predict that, given time, such modifications ought to have appeared in all other related creatures. Thus, after 8my of separation from the human lineage, for any of the great apes *not to have* a more human-like larynx or some semblance of a human-like Broca's area should be quite *unlikely*. The very fact that this has not happened suggests that the coding for the human larynx and Broca's area is likely not to have been a result of modifications to structures and systems existing in a (supposedly) ancestral ape.

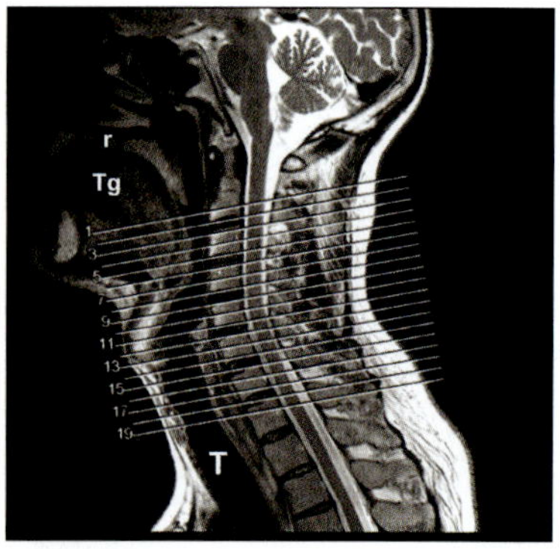


MRI scan of the human head and neck, showing the airway (black) from the back of the nose down to the trachea (T). The tongue (Tg) is touching the roof (r) of the mouth.

## *The larynx is not a dumb design*

### *Swallowing*

It is commonly claimed by believers in evolution that since it is a blind process, evolution will inevitably produce many designs that would be illogical and demonstrably inept. Since once in a while someone *does* choke on their food, this is taken by some to be proof that the larynx is one such dumb design:

> The list goes on…Windpipe next to gullet...mean[s] choking is not uncommon. Evolution's greatest mistakes. New Scientist. 11th August 2007.

Of course, there is nothing new here. Unscientific dogmatism has been around for a very long time:

> Our body is not the result of a straight-forward designing of the best possible machinery for our purposes but in many respects is such a patchwork that it seems sometimes a wonder that it functions at all. A Romer. 'Life of vertebrates 2'. (First published **1933**). Pelican 1963. p374.

So, what is the evidence that the human larynx is a bad design? In many non-human mammals (including the great apes), the larynx is situated high in the back of the throat so that the opening does not connect with the back of the throat but, like a periscope, fits snugly into the back of the nasal cavity. Soft tissue flanges to either side ensure a watertight fit such that food and water in the mouth can pass either side of the larynx and into the gullet without any risk of any liquid or solids entering the airway. In this way, many animals can breathe through the nostrils and drink *at the same time*.

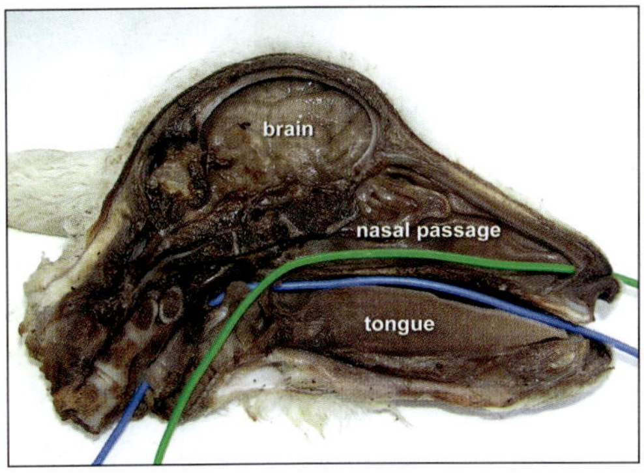


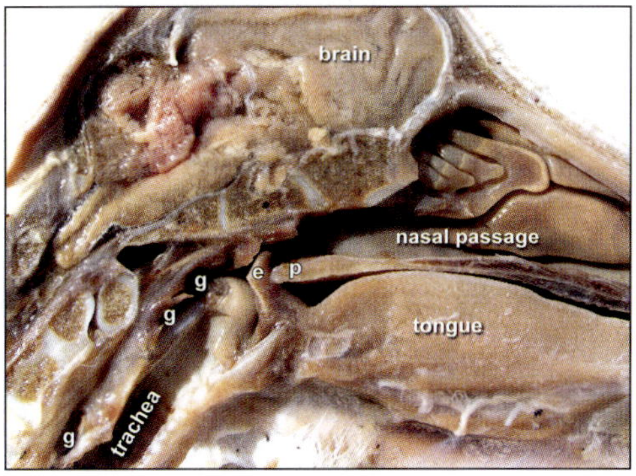


Far left: Lamb head showing the breathing (green) and digestive (blue) passages.
Left: Close-up of the larynx and the back of the nasal cavity. The tip of the epiglottis (e) makes a seal with the palate (p), allowing breathing through the nose to continue at the same time as food and milk passes either side of the epiglottis and into the gullet (g).

Newborn human babies have a similar arrangement, which allows them to breathe through the nose and suckle at the same time. However, by the time of weaning, the human larynx has started to move downwards, and eventually the infant becomes unable to breathe while swallowing.

With the larynx in the lower position, there is obviously a risk of food passing into the airway. However, choking is extremely rare because, at the very start of the swallowing process and before any material passes down beyond the back of the throat, there is a reflex closure of the airway. This reflex action is so strong that it can be used to good effect to inhibit a severe bout of coughing or hiccups, by simply taking a small sip of water whenever the urge to cough or hiccup occurs. So, in the absence of specific brain damage, it is simply not possible to cough or hiccup whilst swallowing.

Now, there are three specific mechanisms that operate to protect the airway automatically during swallowing. The first is the closure of the airway at the top of the larynx by

Babies can suckle and breathe at the same time.

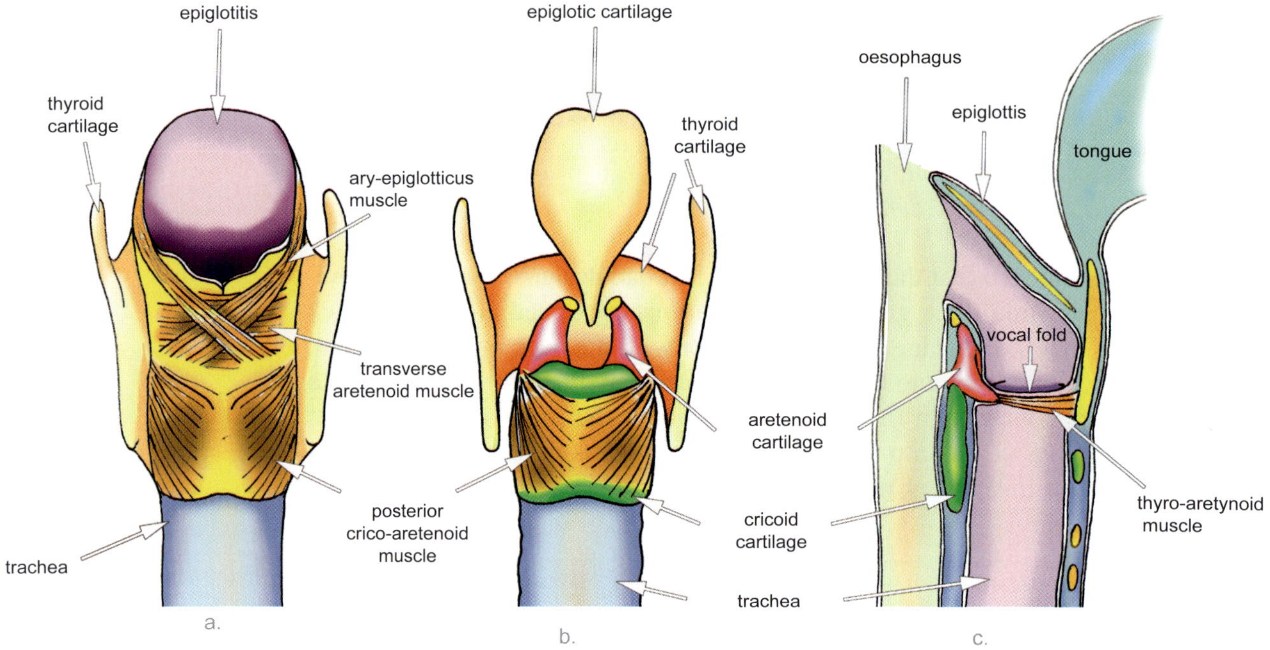


The human larynx is not a dumb design (see main text).
a. View from behind. b. View from behind with upper muscles and tissues removed. c. Side view.

Above: Dead python with tongue held in forceps, showing slit-like opening in the top end of the trachea (t).
Below: Crow.
Snakes and birds do not have a larynx, but the entry into the trachea is opened and closed by a muscular valve.

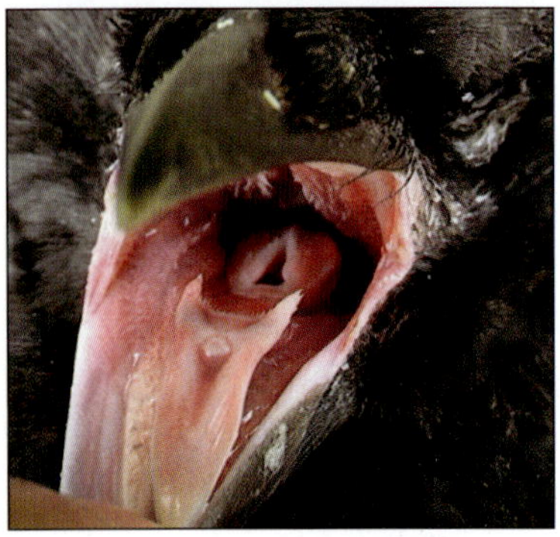

the contraction of the *ary-epiglottic muscle*. The second is the downward movement of the epiglottis to act as a lid on top of the larynx. The third is the upwards movement of the larynx so that it is covered by the backwards bulge of the base of the tongue.

The actual stages of swallowing demonstrate extreme precision. When the food passes over the back of the tongue, the larynx moves upwards and the epiglottis tilts backwards over the airway inlet. When the food passes down the back of the throat, the larynx moves further upwards as does the hyoid bone, and the epiglottis tilts further so that its tip points downwards.

So, while there *is* a risk of food being inhaled by having a common inlet to both the airway and the gullet, the fact is that choking during routine swallowing is as rare as hens' teeth (Talpid2 mutant excluded!). Furthermore, without having these reflex mechanisms *already in place* and the vital anatomical parameters precisely aligned, any (supposed) random alterations to the chimpanzee larynx and throat anatomy that could result in the larynx being sited lower down would be *fatal*.

However, because these mechanisms *are* in place in humans, the airway is well protected. Nevertheless, as Grandmother has always told you, there are two things you must never do: firstly, you must never mess about whilst eating, since laughing requires an open airway and so food can be *sucked* into the larynx, causing choking; and secondly, you must not jump about, swim or exercise for a good while after eating, as any regurgitation of food can also be sucked into the lungs, with potentially fatal consequences.

In reality, choking only occurs when Grandmother's advice is ignored, or when things like peanuts are given to babies and infants. Of course, choking can also occur when there is a degree of paralysis of the muscles in the neck and larynx, but

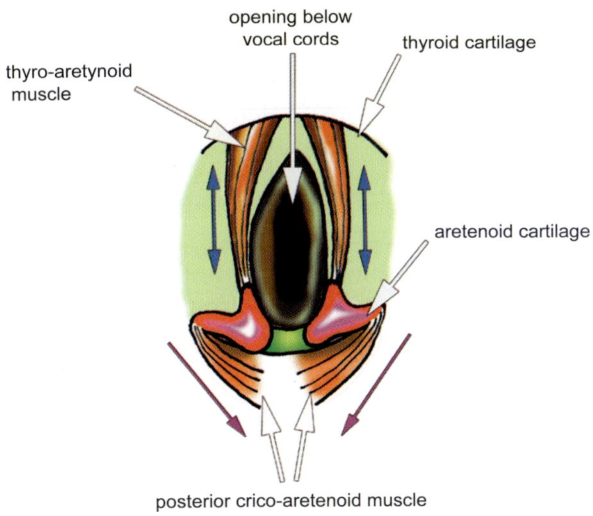


The vocal cords viewed from above.
Note: The vocal cords are made up of the thyro-aretenoid muscles and the overlying tissue (mucosa).
Contraction of the posterior crico-aretenoid muscles rotates the aretenoid cartilages, resulting in the opening of the vocal cords.
Contraction of the thyro-aretenoid muscles pulls the aretenoid cartilages forwards, resulting in relaxation of the vocal cords, and so reduces the pitch. Relaxation of the thyro-aretenoid muscles allows the aretenoid cartilages to move backwards by elastic recoil, resulting in tightening of the vocal cords, and so increases the pitch.

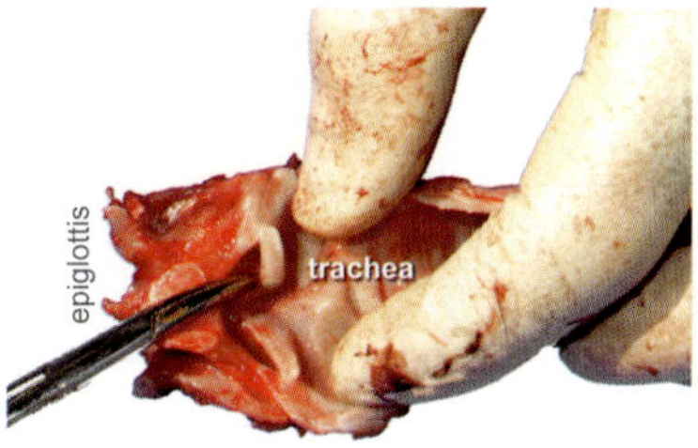


Badger vocal flap

such diseases actually demonstrate that a (supposedly) primitive larynx, lacking the necessary sophisticated nerves and muscles that control swallowing, could not have evolved stepwise. To emphasise this point, when the *recurrent laryngeal nerves* are damaged, this results in paralysis of the *posterior crico-arytenoid* muscles (see below) causing the vocal cords to remain only partially open. This would result in a severe reduction in the airflow, showing that without the crico-arytenoid muscles and their nerve supply *already in place* to open the vocal cords, the larynx could not have evolved.

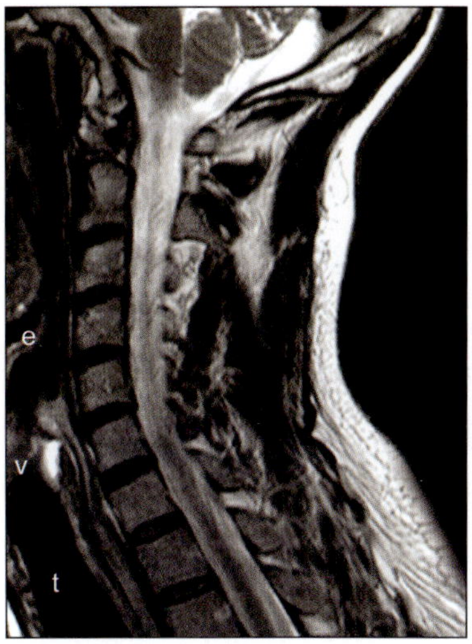


MRI scan of the human neck, showing the epiglottis (e), vocal cords (v), and trachea (t).

### *Phonation*

As to the mechanism of speech (*phonation*) itself, the human larynx demonstrates exquisite purposeful design, central to which are the arytenoid, cricoid and thyroid cartilages and their associated muscles and nerves.

The size of the opening between the *vocal cords* (or vocal folds) is determined by the opposing actions of the posterior crico-arytenoid muscles and the *lateral crico-arytenoid muscles.* The posterior crico-arytenoid muscles are attached to the back of the *cricoid cartilage* and also to processes on the outer sides of the *arytenoid cartilages* which sit on the sloping shoulders of the back edge of the cricoid cartilages. Contraction of the crico-arytenoid muscles causes outward sliding and rotation of the arytenoids cartilages resulting in the opening up of the gap between the vocal cords. Conversely, contraction of the lateral crico-arytenoid muscles causes inward sliding and rotation of the arytenoid cartilages resulting in the closing of the gap between the vocal cords.

The lengths of the vocal cords are determined by the opposing action of the *thyro-arytenoid muscles* and the *crico-thyroid muscles.* The thyro-arytenoids are thin muscles attached to the inside of the thyroid cartilage (behind the 'Adam's apple') and to the arytenoid cartilages. These muscles are covered by the internal lining (mucosa) of the larynx producing two folds running front to back. These are the vocal cords. The contraction of the thyro-arytenoids causes the arytenoid cartilages to be pulled forwards causing the vocal cords to shorten. In contrast, the contraction

of the crico-thyroid muscles causes the thyroid cartilage to be tilted forwards on the cricoid cartilage, causing the vocal cords to lengthen.
Some of the lower fibres of the thyro-arytenoids are attached to the internal lining of the larynx and their contraction determines the vertical depth of the opposing surfaces of the vocal cords.
The tissues of the vocal cords have a unique structure:

> The vocal folds consist of a felted membrane of fibro-elastic tissue whose strands run in all directions. Squares of this network are converted into diamonds by increased length of the vocal folds without a corresponding increase in tension. R Last. 'Anatomy, regional and applied'. Churchill Livingstone. 1973. p661.

As a result of this complex anatomy, the tension in the vocal cords is determined not by their length, but by the balance of the tensions in the opposing crico-thyroid and thyro-arytenoid muscles.
Regarding muscles involved in the actual control of phonation, they are all supplied by branches of the *vagus nerve*. But this makes no evolutionary sense since *all* the other functions of the vagus nerve (from intestinal contraction to heart rate control to stomach acid production etc.) are *automatic and subconscious*. Yet the most sophisticated function of the vagus nerve - speech - is a *conscious* one. However, just as when you think about moving your finger and the finger moves (but you have not consciously instructed any particular muscle to contract, and you have no *awareness* of *how* that happens), so, when you think about speaking a word (but you have not consciously instructed any particular muscles to apply a certain tension, and you have no awareness of how it happens), the sound that you wanted to express comes out of your mouth. Since it would require seriously complex alterations to the nerve fibres emanating from the cerebral cortex and subsequently issuing from the brain stem, it is implausible that conscious actions that previously did not exist could be added and transmitted alongside the subconscious signals carried by the vagus nerve. Therefore, as far as speech is concerned, without the necessary nervous controls, any anatomical changes to a non-human larynx would be completely useless.
So, we see that the human larynx is not a dumb design, but is, in fact, an incredibly sophisticated structure which, with its conscious control of the complex interdependent arrangement of opposing muscles, and specifically shaped and jointed cartilages, *cannot* be the result of blind random alterations to the 'simple' tube of a (supposed) ancient fish's gullet.

Development of the larynx is extremely complex.
For example, mutations that affect R-spondin 2 (Rspo2) result in fatal developmental changes in the larynx and lungs:

> *Rspo2(Tg/Tg) mice had severe malformations of laryngeal-tracheal cartilages, limbs and palate, and lung hypoplasia.*
> S Bell et al. 'R-spondin 2 is required for normal laryngeal-tracheal, lung and limb morphogenesis'. Development. 2008 Mar;135(6):1049-58.

### *Swim bladders and lungfish*

While on the subject of ancient fishes, there is the common notion that our lungs evolved by stepwise modifications to *swim bladders*. Most fishes have swim bladders which in many cases are connected to the gullet by a long tube. But the mistake must not be made that a swim bladder is a simple bag that could easily appear by random changes to the developmental trajectory of the gullet.

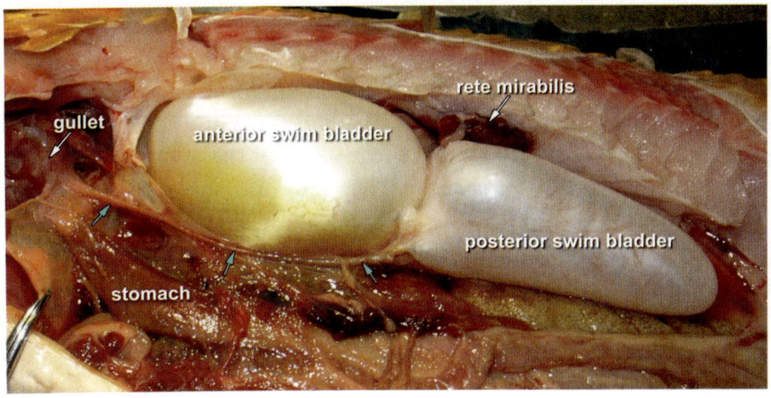


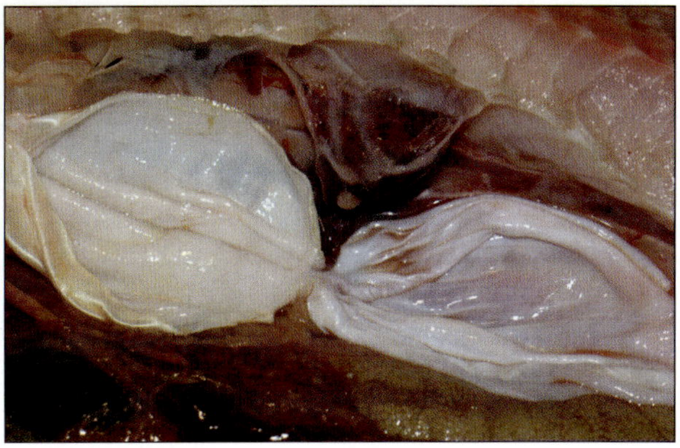

Above left: Carp showing the rete mirabilis, and the duct (blue arrows) between the gullet and the posterior swim bladder.
Above right: Close-up of the rete mirabilis and the swim bladders (opened up).

The amount of air in the swim bladder can be altered so as change buoyancy, to allow a fish to swim with neutral buoyancy at different depths. In order to ensure that contained air can be retained and concentrated, swim bladders have a special structure called the *rete mirabilis* which is a collection of long parallel capillaries which are arranged in such a way as to form a counter-current mechanism. It is a well-recognised fact that in cancers (which by their degenerative nature lack specific growth-guiding signals), the blood vessels are always found to be randomly-orientated. Therefore the parallel capillaries of the rete mirabilis are a *sure* sign of non-random arrangements. And the *WYWYWG principle* shows that it is implausible that the rete mirabilis would arise exactly where it is needed.

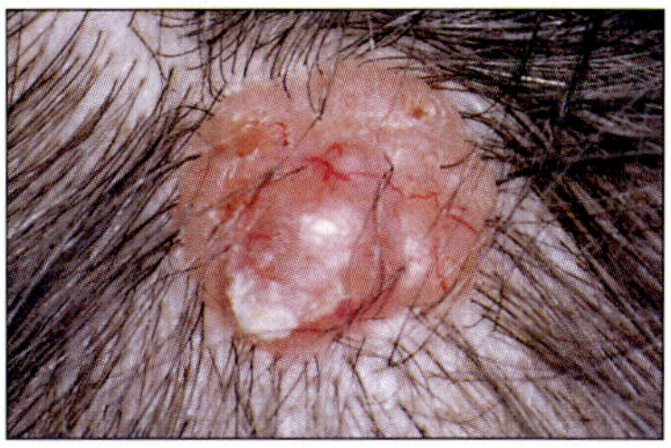

Basal cell carcinoma.
These slow-growing cancers typically show dilated capillaries over the surface. Cancers always demonstrate random orientations of their blood vessels.

Virtually all fishes breathe by using their gills to extract oxygen dissolved in the water. However, gills are of no use out of the water since they will quickly dry out and shrivel up. Many fishes (including the goldfish, garpike, and bowfin) do extract oxygen directly from the air when the oxygen level in the water is low, and lungfishes actually have simple lungs which are connected to the gullet. In the Australian lungfish, respiration is mainly via the gills. In the African lungfish, the lungs are the main organ for respiration and some use is made of the gills. In the South American lungfish, virtually all respiration is carried out by the lungs and only a tiny amount via the gills. However, the gills of the South American and African lungfishes are by no means redundant since:

> With carbon dioxide...both the African and South American lungfish exchange about half through the lung, the remainder through the gills. Hence, gills are important in carbon dioxide exchange, and the lung plays the dominant role in oxygen uptake. K Schmidt-Nielsen. 'Animal physiology'. Cambridge University Press. 1997. p40.

Australian lungfish

It is commonly promoted that swim bladders evolved into the lungs of lungfishes which in turn ultimately evolved into mammalian lungs. However, we have seen when discussing the bird lung and the diaphragm that neither the bird nor the mammalian lung system could have appeared from some simple reptilian structures. So we can already predict that we will find fatal flaws to the *swim bladder-to-lung* idea.
In the early embryo, the mammalian lung develops as simple expansion (the *lung bud*) from the underside of the tube that is to become the gullet (*oesophagus*). Then,

*starting from the tail end* of the lung bud, a wall begins to progressively separate the gullet from the lung bud. In time, the head end of the lung bud starts to form into a larynx. However, before any sign of the larnyx has developed, the simple lung bud becomes a pouch that splits into right and left halves, which subsequently and progressively divide to produce, ultimately, the alveoli of the two lungs. Now, since in the early embryological development of the lung bud there is no barrier or muscle valve between the oesophagus and the lung bud, a (supposed) early fish that developed a pouch arising from its oesophagus, as a *hoping-to-be-a-swim-bladder-or-a-lung*, would have no way of stopping food jamming in the pouch. In time this would stretch the pouch and ultimately cause pressure on the oesophagus so as to result in a complete obstruction to the normal passage of food, which would be fatal. This process can be seen in the disease *pharyngeal pouch*, in which a person develops a weakness in the wall at the top end of the oesophagus, which progressively enlarges to cause a visible swelling, usually on the left side of the neck. In time, as food progressively stretches and fills the pouch, a complete obstruction of the oesophagus can result which, if left untreated by surgery, will have fatal consequences.

A pharyngeal pouch (pp) starts off by stretching the gullet wall and, unless treated, will eventually block off the oesophagus (o) completely.

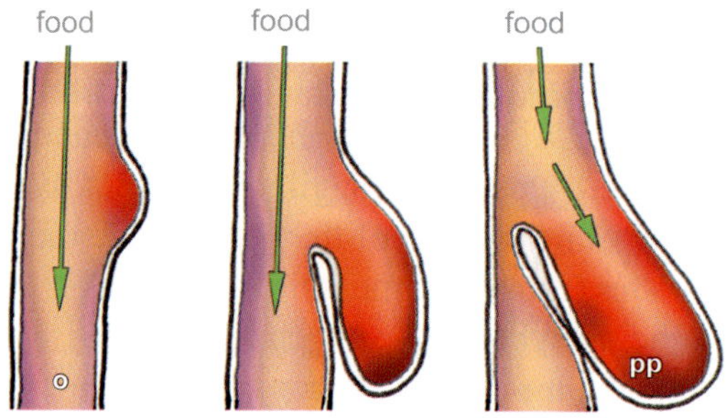


In effect then, any (supposed) very first lung bud would have been simply a blind pouch that had no advantage (it was not yet a swim blabber or lung), and would have been a serious disadvantage since, at its very inception, it would not have had a valve mechanism to prevent entry of food. Furthermore, it is implausible for fibrous and muscular tissue to have produced a valve *just where and when it was needed.* Therefore, since a functioning valve would have had to arise *simultaneously*, a swim bladder connected to the oesophagus could never have evolved stepwise from an out-pouching of the gullet.

So, the evidence from animal behaviour and intelligence, and from comparative brain size, offers no convincing support for the notion that there has been any evolution from apes to humans.

It may be obvious when you think about it, but the point must still be made that *even if* apes can use tools, can understand sign language and spoken words, are self-aware, show emotions, and can show appreciation of aesthetics – none of that in any way constitutes evidence that ape ancestors evolved into human beings. It simply and *only* shows what apes are capable of doing.

## *Human intelligence through history*

We must now look at the evidence for the evolution of human intelligence from the perspective of human history itself. We know that the further back we go in time, progressively less total *knowledge* would have been shared by humanity. It is, in fact, far easier for us to think and be creative, with the benefit of a library of knowledge being available to us, than it would have been for our ancestors who initially had, in terms of knowledge and experience, virtually nothing.

So, is there any evidence to suggest that human intelligence has changed *at all* throughout our history? In fact, over the millenia, everything points to a *constancy* which is quite remarkable. We all know of people who are, or were, from a generation or two before us, and it is clear that our parents, grandparents and great-grandparents thought their way through life *exactly* as we do today. Going back a little further, 500 years ago the Isaac Newtons of the world were no less capable than the best brains of today. For example, inside the Chateau Chambord in France is a twin stairway in the form of a double helix. The structure comprises two spiral stairways that start opposite each other on the same level, spiral upwards around each other, and finish opposite each other on the top floor. So, if one person walks down one stairway while another person climbs up the other stairway, they will reach the bottom and top respectively without meeting each other on the way. The twin stairway was designed by Leonardo da Vinci and fashioned beautifully out of large blocks of stone, the paper-thin joints between the stones having been cut impeccably. Anyone who has even tried to just *sketch* the stairway will understand that it is deceptively complex, and that both the penning of the working drawing and the construction of that stairway would be a major achievement for humans of any generation.

Going back even further, 3000 years ago the ancient Egyptians had mastered complex engineering and building methods. Five thousand years ago the Babylonians had understanding of complex mathematical models, had developed the wheel, had mapped out the visible star patterns and the phases of the moon, and had related them to the seasons and the time of year. Also from the same time period, African cave drawings show a degree of imagination, technical ability, and accuracy in life-like reproduction that is equal to that of any modern artist.

Double helix staircase at the Chateau Chambord, France.

Mesopotamian scroll and print-out 1500 BC.

Sumerian writing 2000 BC.

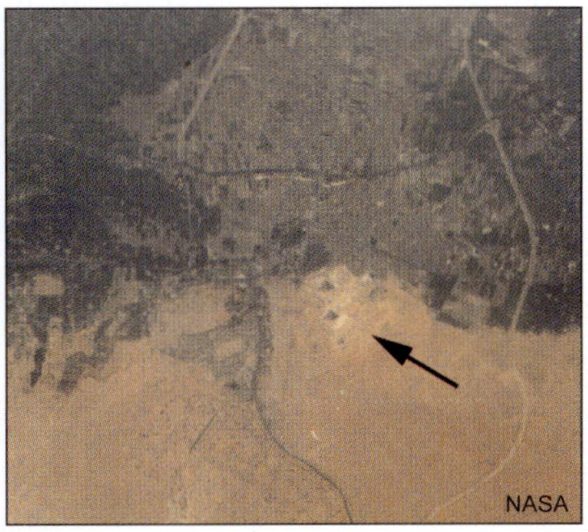


Sphinx and pyramid at Giza, Egypt.

Mesapotamian bronze knives 3000 BC.

The famous 'Iceman', dated as (supposedly) being 5300 years old, was found in 1991 up in the Alps between Austria and Italy, after his body had been exposed following a thaw. The Iceman's hair revealed excessive quantities of copper and arsenic, consistent with accumulation in the blood following years of copper smelting (malachite gives off arsenic vapour when heated). He had with him many items of tools and clothing. There was a leather quiver containing arrows. The arrow heads were made of flint and the flight feathers were held in place with tar and sinew. There were spare items of tools and threads in a 'survival' kit. Microfossil examination of the flint tools showed that the stones had come from a flint area some 120 miles south of the site of his death. There were also shoes which had grasses stuffed into them, presumably for insulation. The clothing contained corn grains of the type only obtainable by threshing. Clearly this man was as intelligent as humans are today.
There is excellent evidence from the Ukraine that (supposedly) about 8,000ya people undertook some form of surgery to the head. Although the reasons for the trephining are likely to have been erroneous and of no benefit to the patient, one skull has been found with a neat hole in it. The fact that the hole had healed over with bone shows that the individual survived for some considerable time after the procedure:

> ...cranium of skeleton No. 6285-9 from the cemetery of Vasilyeva II...the healed trephination is clearly visible as an area of remodelled bone on the left side... Malcolm Lillie. 'Cranial surgery dates back to the Mesolithic'. Nature. Vol 391. 26th February 1998. p854.

In 1994, over 300 superb paintings of animals were discovered on the walls of an underground cavern in the Ardeche, France. These are dated as being (supposedly) at least 30,000 years old and are:

> ...not only the oldest murals ever known but among the most beautiful...The paintings prove that man not only mastered drawing far earlier than was thought, but with great assurance, using shading and perspective to give depth and movement... Ben Macintyre. 'The cave of painted shadows'. The Times. 25th July 1998.

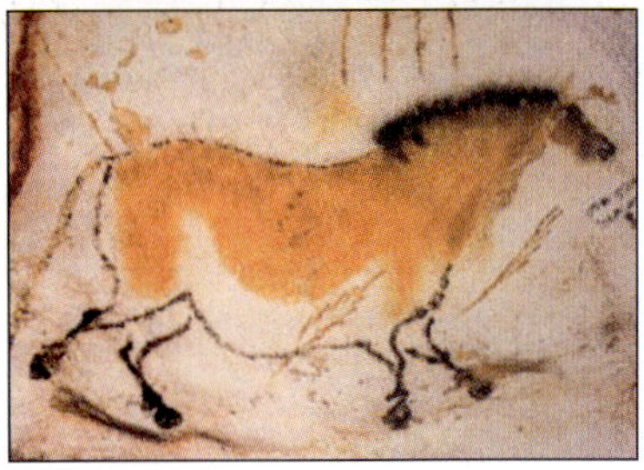

Cave paintings. Lascaux, France.

In July 1997, a section of a flute made out of the femur of a bear was found in a cave near Idrija in Slovenia, alongside other Neanderthal artefacts. Although the flute has been dated as being (supposedly) between 43-67,000 years old, the neatly-drilled holes show that the instrument is based on the same basic scale pattern used in modern Western music:

> In its original form it would have been about 15 inches long and capable of playing the entire scale...I can't imagine a group having conscious music without having language. Nigel Hawkes (Science editor). The Times. 5th April 1997.

Flute made out of bear femur

Of course, such complex use of musical sounds, and the making of an instrument to produce such notes, points to the use of language and breath control in every way as complex as ours is today. Evidence for the conquest and control of fire is seen in Neanderthal camps of (supposedly) 100,000 years ago. And even further back in time, there is evidence of fire having been used for hunting purposes:

> Fire...provided one of the most effective methods of hunting. Evidence of this was found at Ambrona, Spain, in sediments about 400,000 years old...At the end of 1981 evidence was found of a man-made fire 1,400,000 years old...near Lake Baringo, Kenya. LB Halstead. 'Hunting the past'. Hamish Hamilton. 1982. p168.

We all take so much for granted. How many of us these days, without the benefit of matches or cigarette lighters, could with confidence and competence go outside right now and be certain to light a fire to keep us warm through the night? Not many. Neanderthal people must have been just as able and as imaginative as we are. They just didn't have much knowledge or technology behind them. They were in the same situation as modern humans would be if, except for a few lone survivors, the whole of humanity were to be suddenly annihilated along with all our technology and libraries. Neanderthals used fire, they made skin-covered shelters, they hunted big game, they buried their dead, and the flowers in their graves showed that they cherished their loved ones.

So we are led to the conclusion that from us in the here and now, back to Neanderthals 100,000 years ago, there has been no change in the intellect or aspirations of Man. If there has been no change in 100,000 years, then it is probable that the 100,000 years before that also saw no change. Certainly, there is no evidence that forces us to *insist* that there *was* any change. One million years may seem a long time, but if we divide it by this block of 100,000 years-of-no-change, the result is only 10 such blocks of time, representing merely 10 opportunities (powered by any supposed mutations) for a slight (if any) change in intellect. In fact, we can go further back in time, and the evidence shows that the people of (supposedly) 800,000 years ago from Aterpuerca in Spain were much like we are today. Further back again, stone tools dated as (supposedly) being between 800,000 and 900,000 years old have been found in an ancient lake bed on the island of Flores in Indonesia. A number of these tools were clearly used for cutting up meat, chopping through bones and pounding vegetables and berries, since when they were examined under high magnification:

> ...four were...found to have edge damage, striations, polishing and residues indicating use in processing of plant material. MJ Morwood, PB O'Sullivan, F Aziz & A Raza. 'Fission-track ages of stone tools and fossils on the east Indonesian island of Flores'. Nature. Vol 392. 12th March 1998. p173-6.

The people who made and used these tools were clearly highly intelligent. Furthermore, since these individuals could only have got from the mainland to Flores across open waters, they must have made rafts or boats in which they navigated the seas. Therefore, the brains that successfully planned and undertook the colonisation of those islands must surely have been as capable and as intelligent as ours. So why do some people still use the term 'hominid'?

> These hominids are doing things that are the hallmark of modern humans. Alan Thorne, quoted by Tim Thwaites. 'Ancient mariners'. New Scientist. 14th March 1998. p6.

The term 'hominid' not only clouds the issue but also serves to deflect our understanding. Since there is no evidence that actually ties these activities and artefacts with australopithecines, and since (as we have seen) the evidence suggests that Homo erectus was fully human, the phrase should be more scientifically written as: '*these things are the hallmark of modern humans*', leaving out the red herring that suggests (without any evidence) that such activities were undertaken by non-humans. Nevetheless, it has been *assumed* that the individuals concerned could not have been modern-looking humans:

> The age of the artefacts...indicates that they were produced by Homo erectus rather than homo sapiens. MJ Morwood, PB O'Sullivan, F Aziz & A Raza. 'Fission-track ages of stone tools and fossils on the east Indonesian island of Flores'. Nature. Vol 392. 12th March 1998. p173-6.

However, no bone remains were found associated with these tools, and so without any direct evidence, it is pure speculation to say whether the owners of the tools were Homo erectus or Homo sapiens in form. In fact, any distinction is of no practical relevance since we are here discussing the *intellectual ability* of the individuals concerned. Furthermore, as we have seen already, the large brows and other minor bony differences of the Homo erectus people are evolutionary red herrings.

If we were then to assume that an intelligent person was living 1mya, we must assume that a generation before him, his parents must have been equally as intelligent. We must also assume that his great-great-great grandparents were also equally as intelligent. If 1 million years can pass without affecting human intelligence, we are then left with the puzzle: just when would an ancient man's ancestors *not* have been human - a thousand years before 1mya ... maybe 10,000 years before 1mya ... or perhaps 1 million years before 1mya? Actually, we have no factual basis on which to suppose that human intelligence has *ever* changed *at all*.

However, if we suppose that Homo erectus people of (supposedly) 1mya were just a little less intelligent than we are, meaning that human intelligence changes just a little bit over *very long* time periods, then a vast period of 8 million years from the (supposed) split between the chimpanzee and the human lineages would become only *eight* 1-million-year blocks of little-if-any-change time periods. But, since the Homo erectus boy KNM-WT 15000 died (supposedly) 1.6mya, and was most likely fully human, then 8 million years becomes only *five* 1.6-million-year blocks of little-if-any-change. Then again, if we accept that the Laetoli footprints (3.5myo) or the Kanapoi radius (4myo) are human, then 8 million years would leave space for little more than *two* blocks of little-if-any-change time periods.

## INTELLIGENCE: When I was a child I thought like a child. When my great, great...great grandfather was a child, did he think like an ape?

Leaving aside the large number of bodily differences between apes and humans, suppose that there were only a small number (to make it simple, say 10) of possible modifications to the DNA of ancient apes that could cause the change from ape intelligence to human intelligence. With the human genome containing 3 billion nucleotides, the chance of just one of those 10 necessary mutations occurring would be 1 in 3 billion. And the chance of all of the 10 necessary mutations occurring sequentially would be in the region of 1 in $10^{90}$. It is therefore implausible that the 10 *necessary* mutations could have occurred in the last 8my.

For many years, simple stone tools were rather naively and simplistically attributed to simple-brained ape-men, and more elaborate tools were thought to be the work of far more intelligent beings. However, stone tools cannot just be made out of any piece of rock (such as granite, igneous rock or slate) that happens to be lying around. The ideal stone is flint. Furthermore, a simple stone tool is not necessarily a reflection of a primitive or lesser mind, but might represent simply the best that the technology and accumulated knowledge of that time could allow. Nevertheless, it has been the norm to categorise ancient tools into groups according to their (supposed) age in antiquity, and according to the (supposed) degree of intelligence of the maker.

*Oldowan* tools are stones with relatively simple but deliberately-made chip marks, and first appear in the fossil record (supposedly) around 2.3mya. Some chips are clearly caused by simply knocking off a flake by using the stone as a simple hammer. Other stones bear additional chip marks to give one side of the flake a sharper edge. The *Acheulian hand axe* is the name given to a stone tool with two cutting edges and a rounded area for gripping. This tool requires a great amount of skill to produce, involving a carefully-planned series of blows using either a stout piece of wood or another stone as a hammer. It is usually taught that Homo erectus made these axes for about *one million years* without making any modifications *at all.*

Hand axe.
Whoever made it was incredibly skilled.

Anyone who has tried to make *any* stone tool will attest to the difficulty of finding the right sort of stone (one that will chip easily), of finding the right sort of hammer to hit the stone with, and of actually making the tool:

> To the uninitiated, Neandertal tools - mostly sharp-edged flakes of stone that can be held in the hand - look as if they have been casually knapped off a piece of convenient rock with another stone. Not so. Neandertals might have travelled many miles to procure just the right pieces of flint...'You need a lot of brains for flint knapping' says Jaques Pelegrin of the French Center for Archeological research...'You have to plan and organize how you are going to flake off each piece from the core rock ahead of time...Flint breaks under certain conditions. You have to learn those. **It takes months, if not years to learn to do it well...the techniques used by the Neandertals are no less difficult than those used later by modern humans'**. (My bold). Rick Gore. National Geographic. January 1996. p28.

Making a stone tool is much more difficult than it looks.

It might be argued that making stone tools is far easier than Pelegrin makes out since highly trained chimpanzees can make flakes from flint stones and use those flakes to cut dried skins. Perhaps, therefore, the making and use of stone tools does not require human intelligence, and perhaps it is quite plausible that ancient stone

Andrew Dickinson

Octopus changing from its spiky form (which it uses when it is camouflaged against coral) to its normal form.

Recently, a (supposedly) 95myo fossil of an octopus (below) was found in Lebanon. This specimen (*Keuppia levanteis*) is not significantly different from a living octopus, showing that it has not evolved at all.

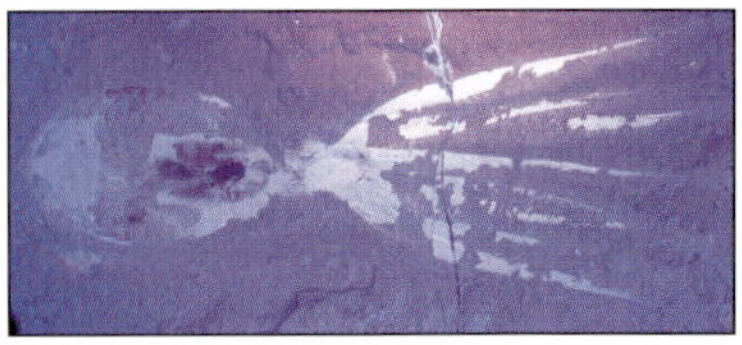

Dirk Fuchs
Freie University, Berlin

tools could easily have been made and used by australopithecines.

However, a trained ape is exactly that - it has been taught the complete method of knapping a stone flake, and how and where to use it, by humans, and has not had to learn and develop the technique stepwise by its own initiative. If an ape has the intelligence of a two- to three-year-old human, it is only to be expected that it *would* be able firstly to copy its human teacher in making a stone flake, and then to succeed in applying its own *limited* initiative so as to use the cutting ability of the stone flake for some purpose other than what was originally shown.

Furthermore, any such ability has to be put into context. For example, an octopus can get at some food that is inside a small screw-top bottle which is itself inside a larger screw-top bottle, by unscrewing both tops. A dolphin could not do that simply because it has no arms, but the experiment does not show that the octopus is highly intelligent, nor does it show that the dolphin is incapable of working out in its mind how to get at the food. So, *in context*, none of this ape activity has any bearing on whether or not apes evolved into human beings. It only shows what apes can do and nothing more. These apes have had artificial and accelerated learning imposed upon them, which they would be unlikely to have acquired without human teaching. In the same way, it is implausible that australopithecines could have acquired such learning on their own. Clearly then, it is most likely that the Oldowan stones and Acheulian hand axes were all the result of an intelligence at work no different from our own.

Now, another problem arises: one million years is a very long time period; indeed, it is so long that it is difficult even to fully appreciate just how long a time that is. For example, one million years is over *two hundred times* the time interval between the ancient Babylonians and the micro-chip. So, is it reasonable to assume that the Homo erectus individuals who conceived and made such complex tools continued to do so for *one million years without thinking about making a few modifications*? Only a few months, or at most years, after producing the first single-sided hand axe,

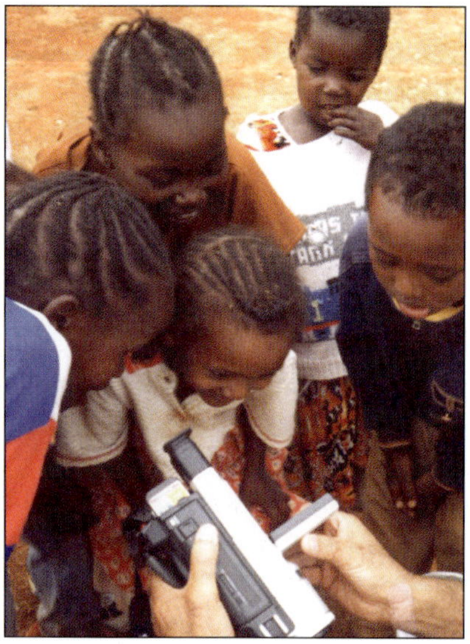

Human inquistiveness is never satiated

an individual must surely have been thinking about and trying out a few changes to the prototype. It is quite unreasonable to suggest that this very capable brain should take thousands of years just to progress from a single to a double-edged axe. Rather, we should expect that the process from thought to manufacture of a modified tool should take only a short time. But this raises a major problem: if we have moved from the Stone Age to the Space Age in just over 5,000 years, it is implausible that any people living one million years ago who could build bamboo rafts, navigate the seas and make sophisticated tools, would have remained in the Stone Age for the next *995,000 years*.

Therefore (all things being equal), an equally intelligent population of people living one million years ago would surely have moved into a Space Age 5,000 years later (i.e. 995,000 years ago).

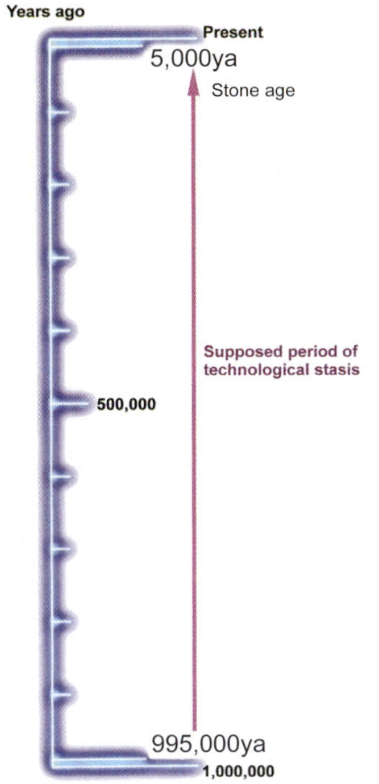


The total vertical distance represents a time period of one million years. The two horizontal lines near the top and the bottom depict periods of 5,000 years. If we have moved from the Stone Age to the Space Age in just the last 5,000 years, then (all things being equal) an equally intelligent population of people living one million years ago would have moved into a Space Age 5,000 years later (i.e. 995,000 years ago). It is implausible that people who could build bamboo rafts, navigate the seas, and make sophisticated tools could have remained in the Stone Age, technologically static, for the next 995,000 years.

Apollo 11

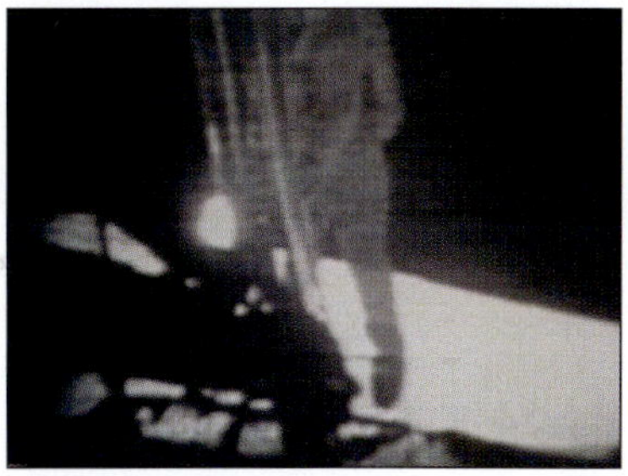

If 'Homo erectus' people could make stone tools one million years ago, humans would have landed on the moon around 20th July 997,000 BC.

We are left with two main conclusions from our study of intelligence. Firstly, since we have shown that there is no evidence to suggest that human intelligence is the result of modified ape intelligence, and that there is no evidence to show that human intelligence has ever been inferior to what it is now, it follows that there is no scientific basis for the claim that human intelligence has evolved from some intelligence which was previously subhuman. Secondly, since it is simply not possible for a thinking man *not* to relentlessly progress in his aspirations or his technology, it follows that the apparent 995,000 year period of relative stasis in Homo erectus stone technology may be *false*, and that the dating of the fossils may be *completely* wrong. This concurs with the substantial biological evidence we have seen which suggests that it is likely that life on Earth is not very old at all.

# CHAP

## Paradigm lost ~ *paradigm found*

# SECTION FIVE
# From one small speck to Man

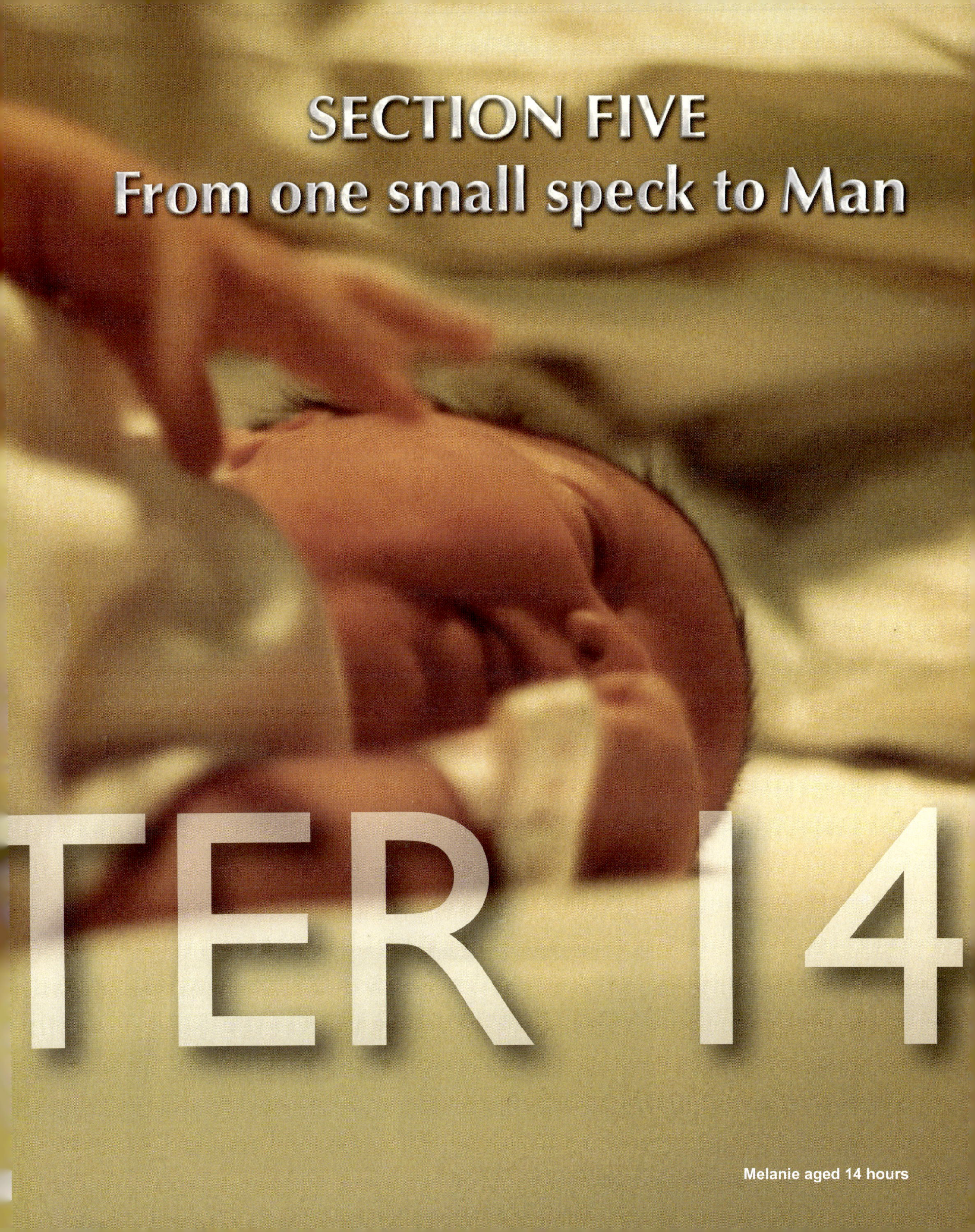


Melanie aged 14 hours

We saw previously (p225) that the Natural History Museum of London's 2008 book '*99% Ape: How evolution adds up*' claimed to:

> *"guide the general reader through the latest evidence".*

However, it shows pakicetus as a seal-like animal. There is a deliberate failure to inform readers that skeletal findings, *which have been known for more than 8 years*, confirm that pakicetus ran on its toes (and did not have paddle-like feet), had long legs, and a long tail.

**Evolution thought police**

In a letter to E Balls, the Secretary of State for Children, Schools and Families, a number of vocal evolution-believers have attempted to force the evolution myth on young children:

*Evolution*

> *We find it extraordinary that evolution and natural selection find no place in the section 'Science – life and living things' (page 6). The theory of evolution is one of the most important ideas underlying biological science. It is a key concept that children should be introduced to at an early stage so as to ensure a firmer scientific understanding when they study it in more detail later on. The wealth of new resources on evolution available for children of primary school age clearly demonstrates their ability to grasp its central concepts. We consider its inclusion vital.*
>
> T Blundell, P Braterman, R Dawkins, A Epstein, R Dunbar, R J Ellis, D Evans, J Gowans, A Hart-Davis, R A Hinde, T Hunt, H Huxley, S Jones, H Kornberg, H Kroto, J Lee, P Nurse, M Reiss, D Smith, K Stuart, F Stoddart, J Sulston, D Weatherall, J Williams, R Tallis, L Wolpert.
> 27th July 2009

## *Intellectual abuse and the Evolution Thought Police*

There are two things that I've found in my exploration of the living world. Firstly, living things are *mind-bogglingly* complex: the more deeply you look into the biological world, the more complexity you see. Secondly, whenever any biological stone is lifted up, underneath you find that any supposed evidence for evolution soon evaporates.

There are two things that most people (from all walks of life) say when it is explained to them that if they believe that their great…great grandparents were apes, then they *have to* accept the idea that the great...great ancestors of those apes were fish that lived around 300mya; that the great ancestors of those fish were worms that lived over 500mya; and that ultimately, over time, dust became human brains without any outside influence. Firstly, they had never thought about it like that. Secondly, they had never really believed that we came from apes.

You see, most people are not stupid. But a few people who control the education systems and the media insist (whether out of ignorance or otherwise) on trying to *force* the masses to accept a lie. Some people are so arrogant as to insist that since about half of the population of the USA is undecided about evolution, those people should be specifically targeted and encouraged to believe it:

> ..die hard creationists aren't a sensible target for raising awareness. What matters are those citizens who aren't sure about evolution - as much as 55% of the US population... Nature. Vol 451. 10th January 2008. p108.

However, we have shown that a huge body of diverse information all converges, beyond all doubt, on the conclusion that evolution from non-living chemicals to human beings *cannot* happen and *did not* happen. Therefore, since, *categorically,* humans *do not* have a common ancestry with chimpanzees, the conclusions of any study that invokes human-chimpanzee common ancestry within its fundamental assumptions must be rejected as being completely invalid. Furthermore, since false evolutionary concepts and conclusions often find their way into school textbooks (and even onto cereal and yoghurt cartons), this can result in the statutory obligation to learn falsehoods. If you repeat something often enough and loud enough, and if, in addition, you ridicule those who disagree with you, many people (including politicians) will tend to believe what you say. Such is brainwashing. So, teaching children that all life has a common ancestry and that we are modified apes constitutes *intellectual child abuse*.

Children should be allowed to think for themselves, not be told what to think.

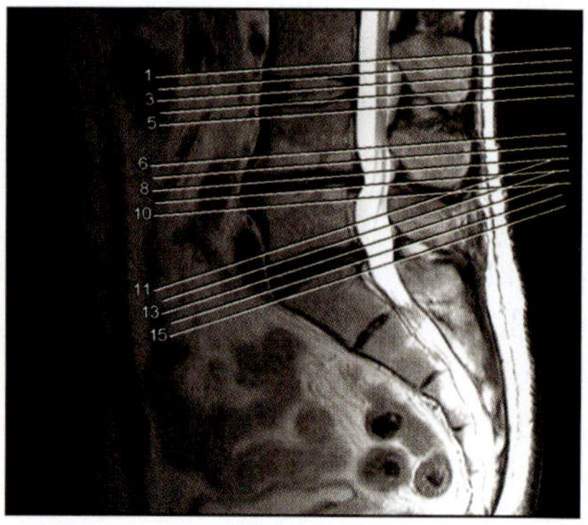


Right: Intellectual child abuse on a yoghurt carton. Our ancestors were never monkeys.
Far right: MRI scan of the sacrum and lower lumbar spine. The coccyx is the lowermost bone of the spine and has attachments of the muscles of the pelvic floor that support the anus and vagina.

As to the present situation – where courts accept evolution as fact (e.g. *Kitzmiller vs Dover** in the USA) when they are not competent to do so – this is a messy stain on common sense, the judicial system, and free thought. If you are told enough times by the media, by the justice system, by the Government, by your teacher, and even by your parents (who might – unwittingly or not – feed you evolutionary poison in the same spoon as your yoghurt) that you are a modified ape, then your synapses might just lose the freedom to even question the idea in the privacy of your own skull. In that case (biochemically, as well as legally), your freedom to think for yourself might become *obliterated*. Perhaps irreversibly. To those whose synapses are not (yet) blocked, beware: the Evolution Thought Police are already here. If you think that is over the top, let people speak for themselves:

> Spread the word: evolution is a scientific fact… Nature. Vol 451. 10th January 2008. p108.
>
> It's evolution, Stupid. New Scientist. 11th August 2007.
>
> There should be [a] term for the proponents of Intelligent Design. May I suggest IDiots? R Fortey. 'Why I hate this intelligent design story. It's simply IDiotic'. Daily Telegraph. 30th January 2007.
>
> …in America, ID is getting a hold only among those parts of the population who don't know anything. Richard Dawkins. Horizon TV Broadcast. BBC. 2007.

To put it bluntly: it is likely that those who are attempting to *prevent* children from even being *able* to question and challenge the validity of the evolution paradigm, have themselves been subject to intellectual child abuse in their formative years, and have now unwittingly become neuronal terrorists themselves. Either that, or else they find the alternative to the evolution myth wholly repugnant, and are just in wilful denial of the truth.

So we must preserve the facility to think outside THE BOX. In this case, THE BOX is the evolution myth, a coffin that inhibits the true scientific study of all things biological.

## *Evolution is not a theory*

Let's clear up a few things. Firstly, the argument about origins is not a matter of Science versus Religion, but of true science versus *nonscience* (non-science). Secondly, you don't have to have a belief in a Creator to question evolution – you just have to have an open and enquiring mind and to not be afraid of challenging dogma. And thirdly, although some anti-evolutionists would say that evolution is *just* a theory, this is quite untrue. Evolution is not just nonscience, it is not even a theory. Let me explain. In common language, the terms 'idea', 'hunch', 'supposition', or 'hypothesis' refer simply to starting points on the road of any enquiry. In non-scientific circles (and, regrettably, sometimes also in scientific circles), the word 'theory' is also commonly used to mean an 'idea'. However, in science, the technical term 'theory' is not *just* an idea, but it describes a concept which is supported by a large body of evidence and, importantly, which is not *fatally* invalidated by other solid evidence. So, we can say that the theory of the sun being orbited by the

*Note that although Padian, one of the witnesses, is said to have proved evolution to be true, he did no such thing. Padian used selective information and, in ignorance, the Court ruled in favour of evolution. Such is the process of law, which is often a game in which the truth can be clouded over.

**Beware 'scientific' reports**

*No fewer than 15 per cent of scientists at the National Institute of Health (the American Government's top health laboratory) recently admitted to bending data to fit their theories.*

Terence Keale. 'Peer review: the myth of the noble scientist'. Daily Telegraph. 19th February 2008.

**Unfounded scientific claims can give the impression of fact**

**WHAT THIS STUDY ADDS**

Distortions in the persuasive use of citation—bias, amplification, and invention—can be used to establish unfounded scientific claims as fact

One study looked at 242 papers that addressed the belief that β amyloid, a protein accumulated in the brain in Alzheimer's disease, is produced by and injures skeletal muscle. It concluded that:

*Through distortions in its social use that include bias, amplification and invention, citation can be used to generate information cascades resulting in unfounded authority of claims...*

*Primary data that weakened or refuted claims on which the belief was based were ignored (citation bias), and a small number of influential papers and citations exponentially amplified supportive claims over time without presenting new primary data (amplification). Certain related claims were invented as fact.*

S A Greenberg. 'How citation distortions create unfounded authority: analysis of a citation work'. British Medical Journal. Vol 339. 25th July 2009. p 210-213.

Similarly, the claim that evolution is confirmed, is repeatedly invented as fact.

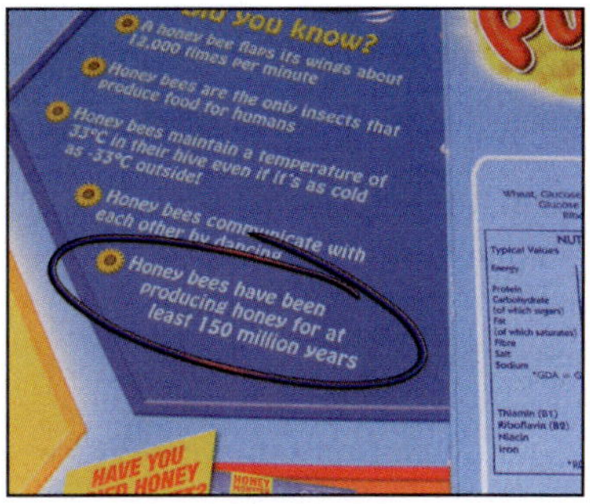


The back of a cereal packet. Children are given questionable assertions as dogma before they are even old enough to go to school.

planets is a true scientific theory, being well supported by evidence, and not fatally invalidated by any evidence. In contrast, we have seen that whatever biological stone you unturn, the solid evidence invalidates the idea that simple chemicals turned (spontaneously) into people. So the concept of evolution *is not a scientific theory*. It is simply a fatally flawed notion. Nothing else. So it should demand as much of our time as does the study of lunar green cheese.

It is understandable that everyone with deep-rooted beliefs will be seriously troubled by evidence that completely undermines their cherished ideas. However, those who object to unequivocal evidence against evolution out of dislike for the conclusion are not scientists. They not only delude themselves, but also lead astray from the truth countless innocents who do not have the ability to check out the evidence base for themselves.

False science should be exposed wherever it is smeared. For example, we have already seen that the Ape-Man BBC series in 2000 showed australopithecines with human-like great toes despite the fact that, 5 years previously, it had been shown conclusively by the fossil specimen STW573 that australopithecines had *thumbs* on their feet, not far removed from the chimpanzee form. Yet the Natural History Museum in London (14 years later) still has an exhibit showing an australopithecine with human-like feet. So, australopithecines could not have made the (supposedly) 3.5 million-year-old Laetoli footprints, but that myth is still told on TV programmes. We have seen that the finding of a lungless frog triggered excitement (and a call for its preservation) amongst 'researchers', yet all it represents is a congenitally-diseased animal, just like the Galapagos cormorant. More worryingly, as we have seen, the editors of the *New Scientist* tabloid declared the eye to be *'a mistake whatever way you look at it'* – a full *three months* after the fibre-optic properties of the retinal Müller cells were reported in *Nature* in May 2007. Clearly, those responsible for the *New Scientist* article were either ignorant of the *Nature* study (in which case they were incompetent), or else they *were* fully aware of the *Nature* study but deliberately chose to withhold that perspective (in which case they should be sacked).

### ***Pseudoscientist***

Talking of nonscience, for those who look up to Darwin as some sort of hero, his complete failure to address properly the myriad flaws in his ideas (which even in his time could have been easily identified) makes him, by definition, a very poor scientist. As we have seen, the fact is that had Darwin made the effort to do some simple anatomical studies of birds, reptiles and mammals (he would not have needed any sophisticated instruments - mainly just a scalpel and forceps), and to converse with physicians of his time to discuss some basic anatomy and physiology, he would have seen how his idea of *small-stepwise-changes-leading-to-one-organ/animal-changing-into-a-completely-different-organ/animal* was not just implausible, but *impossible.*

But human nature makes it difficult for a person to concede error, and so it is that deluded 'scientists' honour a man who has done more than most to stultify biological study, they mislead the innocent public, and they increasingly attempt to silence those who can discern and dodge the effusion of evolutionary landmines that are bluster and *non sequiturs**.

**Bluster*: loud arrogant or threatening speech or behaviour.
*Non sequitur*: a conclusion that does not follow its premisses.

## *Pseudoscience*

At this point it would be appropriate to address the bold claims of self-styled experts that evolution is true and that those who disagree are stupid.

### *New Scientist*

First of all, let's look at what the *New Scientist* (NS) has to say:

> Editorial: It's evolution, stupid. New Scientist. 12th January 2008.

You can almost hear the editor shouting out of his window at the dissenters down in the street. Hardly the language of intellectuals trying to establish the truth. The Deputy Head at my school ('Old Nick' - a bit scary at times, but he was OK - he just commanded respect) once told me: 'When you start to shout, you have lost the argument'.

The special NS article '*Evolution's greatest mistakes*' (11th August 2007) was intended to show just how dumb some aspects of the human body are, thereby confirming that these less-than-satisfactory designs can only owe their origins to the blind process of evolution. However, it will be obvious to anyone with at least one synapse, coupled with a nanogram of common sense, that the claims of bad design are unfounded and point directly to ignorance and/or arrogance on the part of the editors. Note that the '*straw man*' technique is used throughout, a straw man being a deliberately contrived (but inaccurate) state of affairs which is then pulled apart, giving the (false) impression that the argument has been won.

Here are verbatim quotations from '*The list goes on*' (NS, 11th August 2007. p39), with my corresponding comments below each one:

> **Female pelvis** *Human adaptation to walking upright has made giving birth more dangerous for women than for any other primate.* NS.

We have seen that there is no evidence, whatsoever, that the ancestors of humans were apes. Therefore, since neither the human pelvis nor the gait has changed, any problems with giving birth are not due to some mythical change from knuckle-walking to bipedalism.

In addition, as we discussed earlier with the giant panda, it is implausible that male and female genitalia could change in unison with one another: Hox gene mutations cause nothing but disturbances of normal tissue development and distortion of normal anatomy. So, interference with pelvic anatomy of an ancient ape would most likely give rise to infertile apes and their extinction. For example:

> Hypodactyly (Hoxa13(Hd)) mice have a 50-base-pair deletion in Hoxa13...they exhibit profound hypoplasia of the cervix and vaginal cavity...the penian bone is severely hypoplastic and misshapen. ... Our results indicate that infertility in Hypodactyly mutants is related to hypoplasia of the vaginal cavity and cervix in females and deficiency of the os penis in males. L Post & J Innis. 'Infertility in adult hypodactyly mice is associated with hypoplasia of distal reproductive structures'. Biol. Reprod. 1999 Dec;61(6):1402-8.

Incidentally, like most mammals, apes have a baculum – a special bone in the penis.

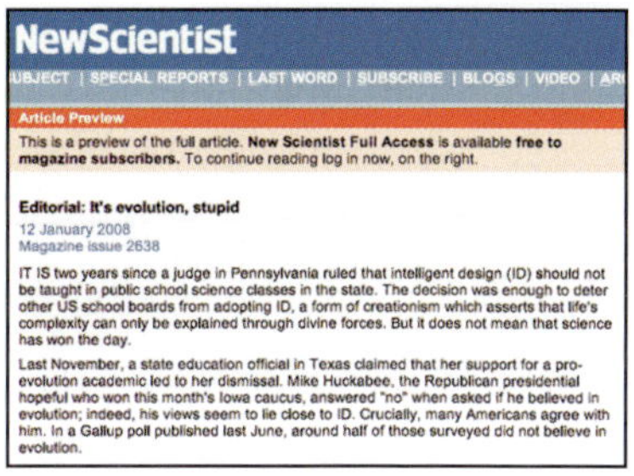

**NewScientist**

UBJECT | SPECIAL REPORTS | LAST WORD | SUBSCRIBE | BLOGS | VIDEO | AR

Article Preview

This is a preview of the full article. **New Scientist Full Access** is available **free to magazine subscribers.** To continue reading log in now, on the right.

**Editorial: It's evolution, stupid**

12 January 2008
Magazine issue 2638

IT IS two years since a judge in Pennsylvania ruled that intelligent design (ID) should not be taught in public school science classes in the state. The decision was enough to deter other US school boards from adopting ID, a form of creationism which asserts that life's complexity can only be explained through divine forces. But it does not mean that science has won the day.

Last November, a state education official in Texas claimed that her support for a pro-evolution academic led to her dismissal. Mike Huckabee, the Republican presidential hopeful who won this month's Iowa caucus, answered "no" when asked if he believed in evolution; indeed, his views seem to lie close to ID. Crucially, many Americans agree with him. In a Gallup poll published last June, around half of those surveyed did not believe in evolution.

A scientific journal should *only* follow the evidence, and should explore and consider *all* avenues of thought.
So, on that basis, *New Scientist* (above and below) and *Nature* (see p528), both of which indulge in intellectual abuse using statements which are intimidatory, derogatory, rude, and also plainly false, are periodicals which cannot be considered to be entirely scientific.

**The list goes on...**

New Scientist 11th August 2007. p39

THE FEMALE PELVIS Human adaptation to walking upright has made giving birth more dangerous for women than for any other primate

LINEAR CHROMOSOMES The ends of linear chromosomes erode as cells divide, something that cannot happen with circular chromosomes

EXTERNAL TESTICLES In harm's way

VAGINA AND URETHRA NEAR ANUS Leaves women prone to genital and urinary infections

WISDOM TEETH Many of us have jaws that are too small for these third molars

MUTANT GLO GENE Like most primates, humans cannot make vitamin C, rendering us vulnerable to scurvy unless we get plenty in our diet

THE APPENDIX No known function but if it gets infected it can kill you

WINDPIPE NEXT TO THE GULLET Means choking is not uncommon

ULNAR NERVE Runs behind the elbow, where it is unprotected (think funny bone), instead of in front of it

VULNERABLE BRAIN CELLS A few minutes of oxygen deprivation causes permanent brain damage in humans, yet an epaulette shark can survive for over an hour without oxygen

PARASITIC DNA Our genome is littered with "jumping genes" that can cause genetic diseases

ODONTOID PROCESS This extension of the last neck vertebra can easily fracture and damage the brainstem

FEET After coming down from the trees, we ended up walking on the "wrists" of our lower limbs, leading to all sorts of structural weaknesses

THE Y CHROMOSOME It is gathering mutations because it can't swap DNA with the X chromosome

VULNERABLE HEARTS A little heart damage triggers a disastrous cascade of events that causes further damage

HAIRY BOTTOMS Who needs them?

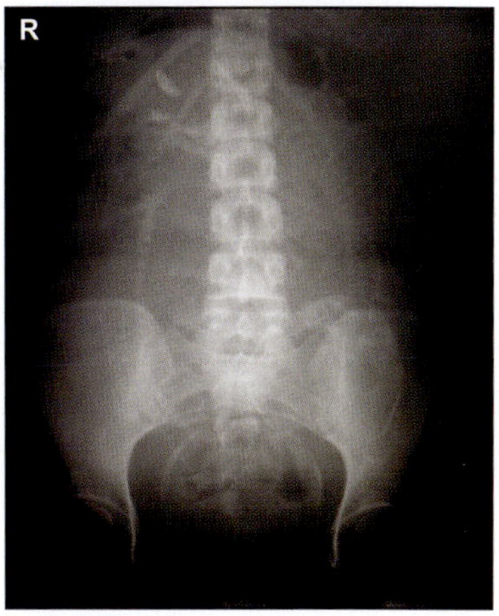


X-ray showing a twin pregnancy. Two spines can be seen, with one skull in the middle of the pelvis, and the other on the left overlying the pelvic brim.
Humans did not evolve from apes, and no other mammals have any particular problem giving birth. Therefore, the only logical explanation for the (otherwise natural) process of labour in women being both painful and protracted, is that there must be a *unique*, non-evolutionary reason behind this particular state of affairs.

Although humans do not have a baculum, a single case of congenital ossification of the penis was described by Champion and Wegrzyn in 1964. The child also had a cleft scrotum, so the ossification was not a normal feature. In any case, this has no bearing on human origins since the differences between humans and apes is extensive, and there is no evolutionary pattern in the anatomy within primates:

> Chimpanzees and Bonobos lack a true glans penis...chimpanzees and bonobos do not [have] a bulbous glans penis while rhesus monkeys and baboons do...In contrast to chimpanzees [which do not have paired corpora cavernosa], **humans**, rhesus monkeys, and baboons [do] have paired corpora cavernosa. (My bold). G Denniston, F Hodges, M Milos (editors). 'Male and female circumcision'. Kluwer Academic/Plenum Publishers, New York. 1999.

Furthermore, as with other animals, the ape baculum is situated in the *glans* area. When ossification occurs in the human penis, the calcification is always abnormal, being the result of injury or inflammation, and is found in the *shaft* or *base* of the penis, not in the glans. So, human penile calcification bears no relationship to the baculum normally present in many other mammals.

**Linear chromosomes** *The ends of linear chromosomes erode as cells divide, something that cannot happen to circular chromosomes.* NS.

We have seen that cell division mechanisms could not have evolved, and that the systems that exist are phenomenally complex and efficient - so there's nothing bad about any aspect of the entire replication system. However, things can go wrong with any system, and to prevent the telomeres from being eroded, linear chromosomes have specific mechanisms to protect and replicate DNA ends:

> Mammalian telomeres contain a specific protein complex, shelterin, that functions to protect chromosome ends from all aspects of the DNA damage response and regulates telomere maintenance by telomerase. W Palm & T de Lange. 'How shelterin protects mammalian telomeres'. Annu. Rev. Genet. 4th August 2008.

It must be remembered that (supposed) evolution can only proceed by an advantage. So, whether you understand them or not, linear chromosomes *must* have some *significant* advantage over circular chromosomes. However, if linear chromosomes had evolved, they would have appeared at a time when shelterin and telomerase did not exist. Therefore, the *WYWYWG principle* shows that those first linear chromosomes would have broken up, and so linear chromosomes could not have safely evolved at all. Therefore, the linear chromosome/sheterin-telomerase system must have appeared as a fully functional unit from the beginning.

**External testicles** *In harm's way.* NS.

The fact is that testicular trauma is very rare. Furthermore, in humans, if the testicles do not descend, and remain high in the groin or inside the abdomen, there is a real risk of malignant change occurring. So there are good reasons for testicles being in a scrotal sac (see page 110).

**Vagina near anus** *Leaves women prone to genital and urinary infections.* NS.

The fact is that despite the kidneys (via the bladder and ureters) and the abdominal cavity (via the vagina, uterus and fallopian tubes) being potentially in *direct* contact with bacteria from the outside world, the majority of women do not develop kidney infections or peritonitis from germs passing upwards in those two channels. Therefore, there exist *extremely* efficient mechano/muscular and immune systems to prevent potentially fatal infections. Occasional urinary infections are not evidence for a bad design since the evidence for an excellent design is overwhelmingly obvious. Genital infections rarely occur outside situations of promiscuity.

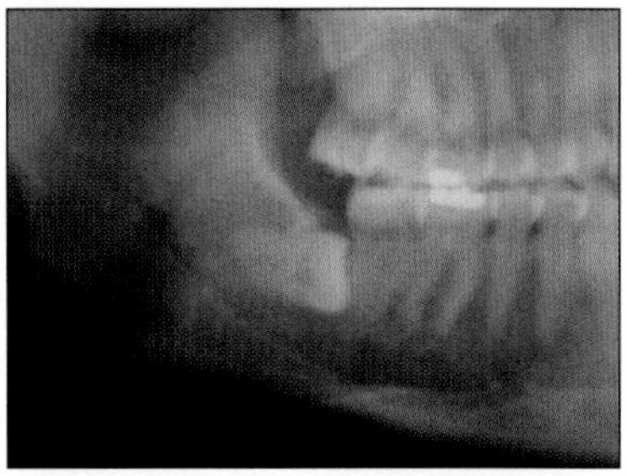

This lower right wisdom tooth is impacted in a horizontal orientation. Interestingly, the wisdom tooth is so wise that if the 2nd (normal) molar is removed, the inbuilt developmental programming will cause the horizontal 3rd molar to rotate through 90 degrees and erupt normally.

**Wisdom teeth** *Many of us have jaws that are too small for these third molars.* NS.

Paleontological evidence shows that modern human jaws are not deformed relative to ancient jaws, but in many cases they are smaller, and this is mirrored in the increase in wisdom tooth impaction. However, the developmental growth of the jaws is subject to many variables, the most important being diet and genetics. If foodstuffs that require a lot of chewing are not eaten, then – to put it simply – as a result of reduced muscular forces on the bones, the jaws do not develop as large as they should, and this results in increased risk of impaction of the third molar. This is not evidence of bad design, but of degeneration of the developmental process as a result of an unsatisfactory diet. It is also found that people with squarer faces (e.g. in some areas of eastern Europe and Africa) have larger jaws and so have less incidence of impaction. Of course, in addition to having genes that result in larger jaws, it may be that many of these same people may have a coarser diet than people in the West.

**Mutant GLO gene** *Like most primates, humans cannot make vitamin C, rendering us vulnerable to scurvy unless we get plenty in our diet.* NS.

We have shown that inactivated mutant genes simply demonstrate *degeneration* in previously perfect genes, and the *Essential Sequence Principle* shows that complex proteins could not have originated stepwise from inactive precursors.

The fact is that, given a reasonable diet, scurvy is a rare condition. Getting scurvy during long sea voyages without fresh food is like getting dehydrated after walking through the Sahara for a day without water. It does not need to happen and it can be easily avoided. As discussed previously, Qgenes (pseudogenes) *do not* confirm common ape-human ancestry.

**Appendix** *No known function but if it gets infected it can kill you.* NS.

This is plainly untrue. As we have discussed previously, the appendix contains numerous lymphoid cells, which are involved in protecting against invasion by bacteria from within the gut. Incidentally, it is only permissible to say it has *no* function if it is *known* that it has no function. Yet on the basis of what they don't know (no *known* function), the authors are willing to make the unscientific leap of declaring that the appendix is of no use and is a liability. It is important to note that appendicitis is a *very* rare condition in countries where people eat unrefined foods and plenty of roughage. So, problems with the appendix are related largely to eating an unnatural (highly refined, low fibre) diet.

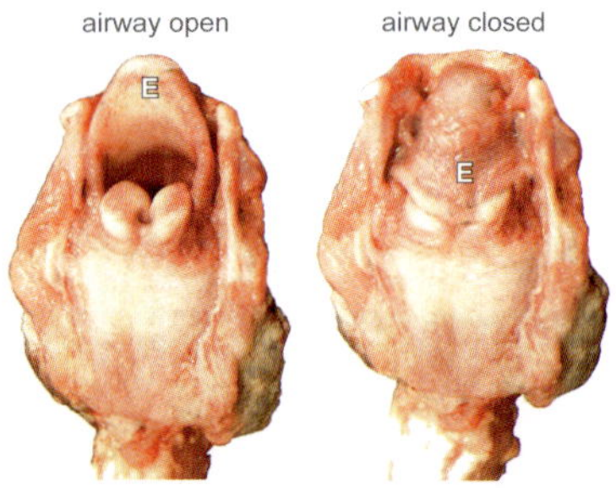


Calf larynx from behind: the epiglottis (E) is very effective in protecting the airway from inhalation of a foreign body.

Similarly, constipation is the single most common cause of abdominal pain in the West, but this is not the result of some design flaw in the intestines, but is associated with a reduced intake of unrefined, fibrous food. To be blunt, the digestive system is designed to take in plenty of roughage and to eliminate semi-soft material, which should simply plop out upon squatting. Anyone who goes to Africa or India and eats a rural diet will quickly realise that anyone who has to strain or wait *at all* when going to the toilet, is constipated.

**Windpipe next to gullet** *Means choking is not uncommon.* NS.

This also is plainly untrue. Choking is *extremely* rare, and as we discussed earlier, the larynx is an amazing, highly complex structure that could not have arisen stepwise from a simple tube.

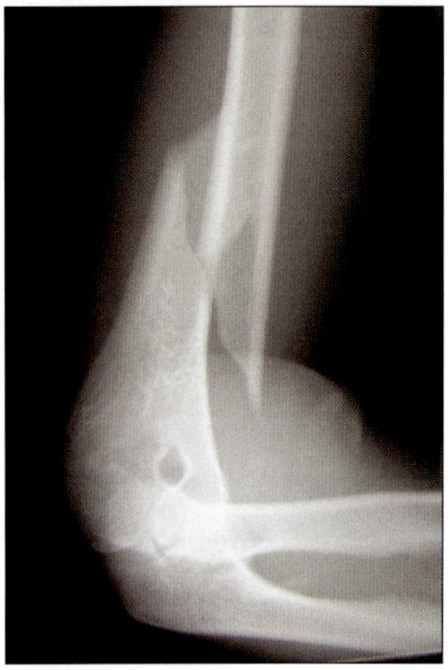

A fracture of the lower third of the humerus. This can be associated with damage to the main artery in the arm. The ulnar nerve is not in danger.

**Ulnar nerve** *Runs behind the elbow, where it is unprotected (think funny bone), instead of in front of it.* NS.

Actually, this statement (think funny joke) shows incredible *ignorance* and *arrogance*. It is one thing to ruminate over an issue, but it's another thing to *teach* in an international (supposedly) scientific journal on something you have no clue about. It is true that the ulnar nerve runs behind the elbow, but so does the radial nerve, which runs a little higher up and on the outer side of the humerus. It is *extremely rare* for either nerve to be damaged around the elbow, but children falling off climbing frames do have a habit of breaking the humerus just above the elbow. In such cases, the upper fragment usually protrudes forwards and can damage the median nerve and brachial artery, which run above and *in front* of the elbow joint. Of course, you can get tingling in your fingers when you knock the back of your elbow (hitting the ulnar nerve) but this very rarely causes anything other than transient discomfort. So the position of the ulnar nerve behind the elbow is *not* a bad design.

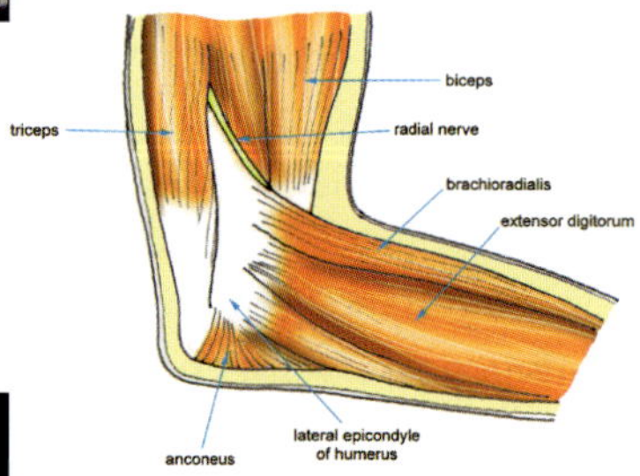


Outer aspect of right elbow

Right elbow.
The ulnar nerve runs behind the inside edge of the elbow. The radial nerve runs just above the outside of the elbow. It is rare for either nerve to be damaged.

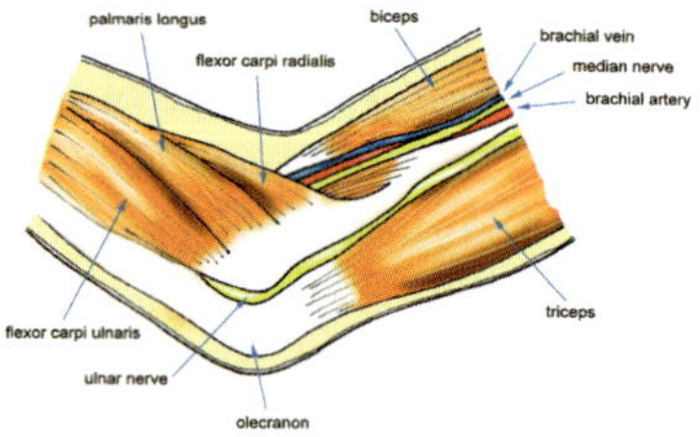


Inner aspect of right elbow

**Vulnerable brain cells** *A few minutes of oxygen deprivation causes permanent brain damage in humans, yet an epaulette shark can survive for over an hour without oxygen.* NS.

However vulnerable it might be in certain circumstances, we must not lose sight of the fact that the human brain is the single most complex structure in the universe. Furthermore, most people do not die from simple oxygen deprivation of the brain, but from many other causes that ultimately result in cessation of the blood circulation. The very fact that humans can read and write, and sharks can't, shows that even the editors of nonscience articles have far more capable neuronal circuits than do sharks.

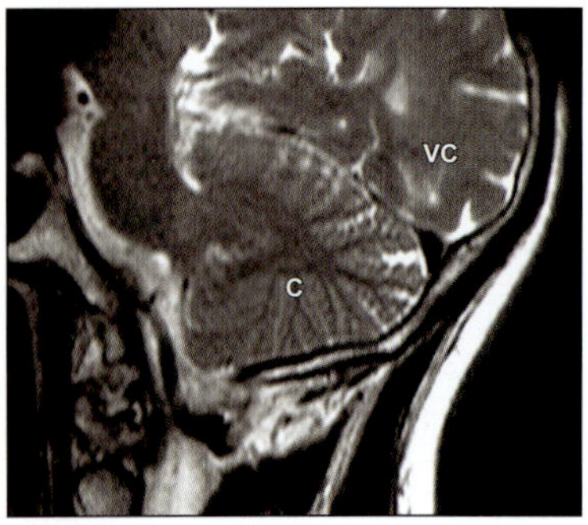


MRI scan of the back of the human brain showing the cerebellum (C) and part of the visual cortex (VC).
The human brain is the most sophisticated structure in the universe.

**Parasitic DNA** *Our genome is littered with 'jumping genes' that can cause genetic diseases.* NS.

All the genetic evidence points to mutations - copying errors - causing disease. The same evidence also shows that before any errors, the genes were functional and good. So, everywhere, genomes do not present a case for bad design, but instead demonstrate *degeneration* in a previously excellent design. Pathogenic viruses, bacteria and plasmids etc. probably all reflect damaged goods that now cause trouble.

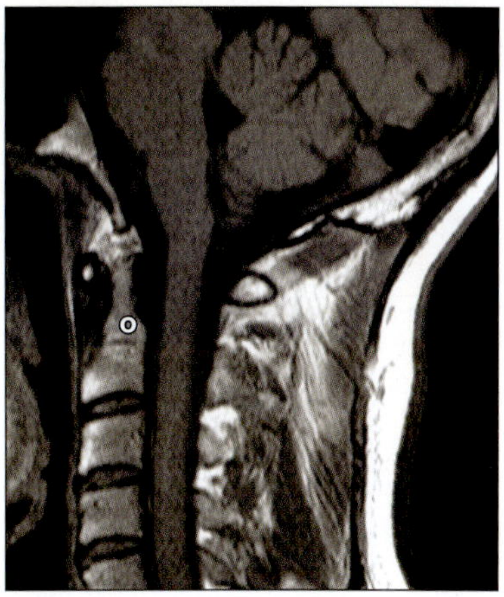

MRI scan of the neck and base of the skull showing the odontoid peg (o).

**Odontoid process** *This extension of the last neck vertebra can easily fracture and damage the brainstem.* NS.

Actually, the odontoid process does not easily fracture even if you fall on your head from a cliff (which usually causes a fracture of the middle or lower neck vertebrae). But the odontoid process *is* liable to fracture if you hang by the neck in a noose. The reality is that fractures of the spine (whether from injury or bone disease) almost always involve vertebrae *other than* the odontoid process of the axis. Incidentally, the odontoid process is part of the *second* neck vertebra (also called the *axis*), not the last. Also, fracture of the odontoid process damages the *spinal cord* just below the brain stem (which lies within the skull).

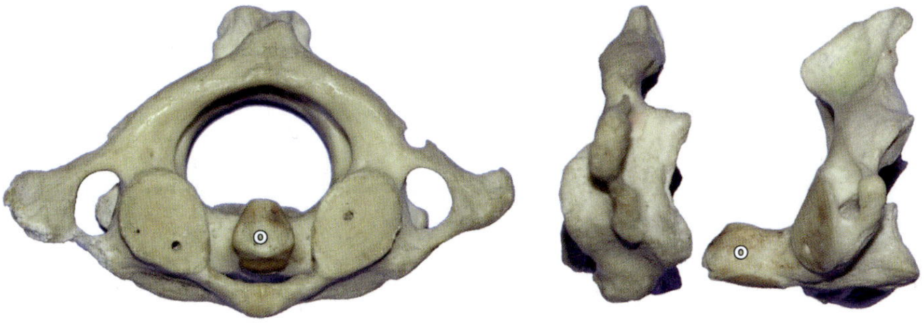

The first (atlas) and second (axis) human neck vertebrae from the left side and from above.
The odontoid process (o) is part of the axis but it actually represents the body of the atlas. Turning of the head occurs at the joint between the odontoid process and the atlas. Nodding of the head occurs between the atlas and the base of the skull.

**Feet** *After coming down from the trees, we ended up walking on the 'wrists' of our lower limbs, leading to all sorts of structural weaknesses.* NS.

This is plain nonscience, since in addition to the fact that our ancestors never lived in trees, most people live their lives with no problem, whatsoever, with the bones of their feet. Obviously, arthritis can cause foot problems, but far more people have arthritis of the hands, hips and knees than anywhere else. Anyone who has ever used their feet for walking about, running, or for sport will know how good the design is.

**The Y chromosome** *It is gathering mutations because it can't swap DNA with the X chromosome.* NS.

If evolution happened, then the Y chromosome could only have come into existence because it had an advantage. So there must be benefits that the authors are ignorant of, which override any deficiencies. But evolution did not occur. And since we all know that mammalian cell reproduction is a fantastically complex and efficient system, it cannot be a dumb design.

**Vulnerable hearts** *A little heart damage triggers a disastrous cascade of events that causes further damage.* NS.

A heart can normally work very well, beating at 60-80 times a minute for 70-80 years and longer, so there is no question that it is extremely efficient. If your car runs out of oil (even if was built in Japan or near where archaeopteryx used to hang out), the engine will enter into a cascade that will seriously weld the pistons. The heart is both far more complex and far more reliable than any car.

**Hairy bottoms** *Who needs them?* NS.

Clearly, the writers of the article must have patted each other on the back, being well pleased to have come up with what they thought was a humorous little snipe to finish with. Obviously, they did not realise that their whole article would be well suited for publication on April 1st*.

* April Fool's Day, UK.

### *Nature Journal*

Secondly, let's look at the claim of the journal *Nature* that evolution is true and that this 'truth' must be instilled into everybody's mind:

> Spread the word: evolution is a scientific fact... Nature. Vol 451. 10th January 2008. p108.

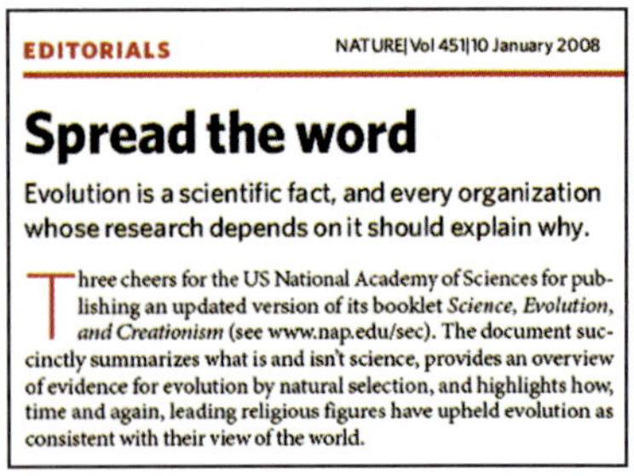

EDITORIALS NATURE|Vol 451|10 January 2008

**Spread the word**

Evolution is a scientific fact, and every organization whose research depends on it should explain why.

Three cheers for the US National Academy of Sciences for publishing an updated version of its booklet *Science, Evolution, and Creationism* (see www.nap.edu/sec). The document succinctly summarizes what is and isn't science, provides an overview of evidence for evolution by natural selection, and highlights how, time and again, leading religious figures have upheld evolution as consistent with their view of the world.

The *Nature* editorial recommends that the '*Science, evolution and creationism*' publication of the *National Academy of Science and Institute of Medicine* be used to explain to everyone why evolution is true, so it is important to take a look at what this document says. As before, I have given verbatim quotations, with my corresponding comments directly below each one:

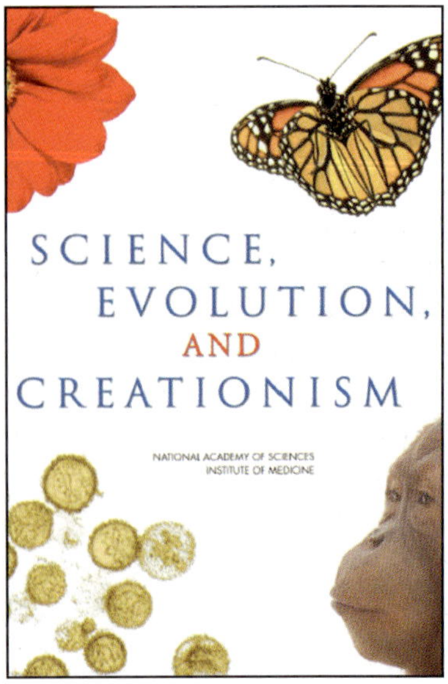


**The Fossil Record** *The concept of evolution is supported by fossil findings in rock layers from different ages of Earth's history... Scientists have found an overwhelming number of fossils in rock layers of different ages that repeatedly confirm the changes in life forms that are predicted by the theory of evolution.* Nat. Ac. Science & Inst. Med.

This is untrue. The fossil record shows that highly complex creatures are found even in the lowest sedimentary layers. The fossil record also shows that all the major types of creatures appear 'suddenly' without any hint of gradual evolution from a previous form. We have also seen from the *Coelacanth Principle* that all creatures could have lived contemporaneously in the past, and that it is completely invalid to make an evolutionary tree invoking common ancestry from the fossil record. And it is also a fact that fossilisation of wood, bones and shells can occur in a very short space of time - in only a relatively *few* years rather than *millions* of years. This should at least make people rethink about the age of the earth.

**The Theory of Evolution Has Been Repeatedly Tested and Confirmed** *We all know from our experience that biological traits pass from parents to offspring. This is the basis of evolution.* Nat. Ac. Science & Inst. Med.

This is untrue and misleading. It is obvious that variation occurs between individuals in a family and within populations, but parent-offspring-sibling variation simply and *only* involves juggling with whatever genes already exist. Simple heredity does not produce anything new.

However, for *Life* to have come into being, and subsequently to have (supposedly) evolved, would have required not derangements of existing proteins and structures, but the *de novo* origin of the very first genes, and subsequent improvements in their function along with information and templates by which new structures and organs and functions could become manifest. In some great distant past, for a (supposedly) very short, primitive and ancestral protein (having initially a nonsensical sequence of amino acids and, by definition, having at one time *no* function) to have *become* a functional protein, and hence to have facilitated the manifestation of a new structure or a new organ (i.e. one that previously never existed), would have required a truly magical alteration to the specific sequence of nucleotides that coded for that protein. In other words, evolution is a system that would have required countless mistakes in existing functioning systems to have produced all life today. *All* the evidence shows that any such change of *this* nature is not possible nor has it ever happened.

Furthermore, when the genes of an offspring are altered in any key areas, the *Don't Mess With It Principle* shows that the result is usually a mess. So, in fact, there is a *limit* to the differences that an embryo can acquire compared to its parents, since any significant alteration to genes affecting developmental processes results in disease or death. This is exemplified by the Galapagos cormorant, the lungless frog, the sloth bear, blind cave fishes, Tanzanian bats, cystic fibrosis etc. Instead of genes and systems 'getting better' or 'improving' over time (evolution), observation shows the reverse to be true: that genes and systems which in times past were good, have progressively degenerated.

It has been claimed that polio virus has been manufactured from scratch*. However, a virus is not a living thing and is not comparable to a bacterium or a single-celled protozoa. Furthermore, the virus was not manufactured *de novo*, but was manufactured using existing complex molecules such as groups of nucleotides.

**J. Cello et al. 'Chemical synthesis of poliovirus cDNA: Generation of infectious virus in the absence of natural template'. Science. 9th August 2002. Vol. 297. p1016-1018.*

*Nobody knows how a mixture of lifeless chemicals spontaneously organized themselves into the first living cell.*
New Scientist. 12th July 2003. p24-39.

**DNA Research** *Molecular biology and genetics have revealed how evolution works at the level of molecules.* Nat. Ac. Science & Inst. Med.

This is untrue. To make an analogy with house-building, DNA is just a self-replicating library of information about how to make the materials that a house is made of, such as the plastic and copper of which the gutters and water pipes are made. The genes do not contain any information about how to use those materials to arrange and construct the guttering or the plumbing. To put it in the context of real biology, DNA is *not* the template for Life: for practical purposes, it only codes for proteins. Looking at DNA with its 'mile after mile' of nucleotide repeats, and the myriad particles entering (seemingly) randomly into chemical reactions within a cell, there is no evidence for any information (molecular or otherwise) that says 'build an eye'...or even 'build a toenail'. And yet it is assumed that different genes, interacting in different permutations, are alone sufficient to provide all the information necessary for the making of living bodies. Other than for their own 3-dimensional configuration, proteins do not carry any inherent *information* about any biological structures or systems. For example, the protein collagen is the main component of tendons, but the gene for collagen does not carry any information for making a nerve-muscle-tendon-bone-joint complex. So whilst it is true that a substitution of one amino acid for another can change the shape of a protein (for example, to produce the abnormal Haemoglobin HbS), or change its activity so as to alter some aspect of an embryo's

**Not a human ancestor**

A (supposedly) 47myo fossil lemur (nicknamed 'Ida') was unveiled at the American Museum of Natural History in New York on 19th May 2009, and hailed as the 'eighth wonder of the world'. Generally, there is nothing special about the fossil since in its overall proportions and anatomy, it just looks like a lemur. However, living lemurs have curious fused and forward-pointing elongated lower incisors (called a 'tooth comb'), and also a claw on the index finger (the other digits have nails) with which they groom companions. Since the fossil lacks both of these structures, it is thought by some that the fossil represents a side branch from ancient lemurs that led to apes and humans. However:

> *In order to establish that connection, Ida would have to have anthropoid-like features that evolved after anthropoids split away from lemurs and other early primates. Here, alas, Ida fails miserably.*
> C Beard. 'Why Ida fossil is not the missing link'. New Scientist online. 21st May 2009.

*(Continued on opposite margin)*

Lemur fossil, nicknamed 'Ida'. Since it lacks a baculum, it is thought to be a female.

Alex Dunkel

Lemur grooming claw.

Lemur skull showing long protuberant lower incisors. Incidentally, lemurs also have a small cartilagenous 'secondary tongue' (lying underneath the true tongue) which is used to remove hairs from between the lower incisors.

development (for example, to produce a leg instead of an antenna on a fly's body), we have seen that it is likely that genes and their corresponding proteins only act so as to reflect in structural or functional form the existence of some other unseen, unknown and separate organisational system/s. In addition, the *Don't Mess With It Principle* shows the evolution of mollecular biology to be impossible.

> **DNA Research** *DNA studies support findings from other branches of science. For example, species that appear to be more distantly related from their positions in the fossil record are found to have correspondingly greater differences in their DNA than species that appear more closely related in the fossil record.* Nat. Ac. Science & Inst. Med.

An example is given of the gene that codes for *leptin*:

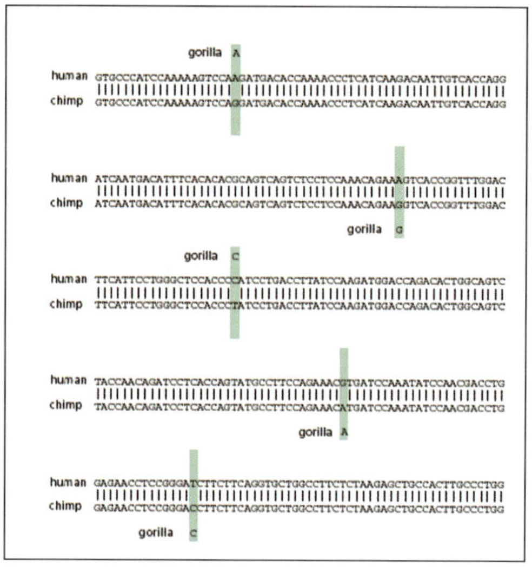


Human and ape sequences for leptin, an enzyme involved in fat metabolism.

> *Comparison of the human and chimp DNA sequences for the gene that encodes the hormone leptin (which is involved in the metabolism of fats) reveals only five differences in 250 nucleotides. Where the human and chimpanzee sequences differ, the corresponding nucleotide in the gorilla (shaded bars) can be used to derive the nucleotide that likely existed in the common ancestor of humans, chimpanzees, and gorillas. In two cases, the gorilla and human nucleotides match, while in the other three cases, the gorilla and chimpanzee sequences are the same. The common ancestor of the gorilla, chimpanzee, and human is most likely to have had the nucleotide that is the same in two of the three modern-day organisms because this would require just one DNA change rather than two.* Nat. Ac. Science & Inst. Med.

At first glance, this all sounds extremely sensible and valid. Yet all it's really saying is that humans have *this* in common with chimpanzees and *that* in common with gorillas, so our common parents most likely had this or that. The text gives the impression that a common ancestor has been proved, but actually, it's just *assumed*. However, we have seen that there is no fossil evidence of any ape-human intermediary, and diverse evidence shows that it is impossible for humans to have evolved from apes. So, apes and humans are *not* related. Therefore, *all* analyses of this type, based on an *assumption* of ape-human common ancestry (including, for example, analyses of mtDNA of Neanderthal and living humans) are scientifically invalid. This is seen in all the major genome studies, for example, of mouse, chicken, and chimpanzee. The mouse genome project starts with the assumption that mice and humans *did* have a common ancestor, and that this ancestor lived 75mya:

> In the analyses below **we use a divergence time for the mouse and human lineages of 75 Myr** for the purpose of calculating evolutionary rates... (My bold). R Waterston et al. 'Initial sequencing and comparative analysis of the mouse genome'. Nature. Vol 420. 5th December 2002. p520-562.

The chicken genome project starts with the assumption that the chicken *did* evolve from bipedal dinosaurs and that the chicken *is* a distant cousin to humans:

> ...the chicken ***is*** a modern descendant of the dinosaurs...**the chicken genome...has...evolved separately from mammalian genomes for ~ 310 million years.** (My bold). International Chicken Genome Consortium. 'Sequence and comparative analysis of the chicken genome provide unique perspectives on vertebrate evolution'. Nature. Vol. 342. 9th December 2004. p695.

Chick

The chimpanzee genome project starts with the assumption that human features *are* acquisitions from ape ancestors:

> Human-chimpanzee comparative genome is **essential** for narrowing down genetic changes involved in the **acquisition of unique human features**... (My bold). International Chimpanzee Chromosome 22 consortium. 'DNA sequence and comparative analysis of chimpanzee chromosome 22'. Nature. Vol 429. 27th May 2004. p382.

*(Continued from opposite margin)*

The 'Ida' fossil simply and only represents a lemur which has different lower incisors and index nails to those of most living lemurs. In fact, the living aye-aye is an example of a bizzare lemur that does not have a 'tooth comb' and has claws on all its digits.

Therefore, since all these genome studies invoke evolutionary relatedness as *the basis* for making any analyses, they are nonscience, being simply attempts to make the facts fit in with the assumption of common ancestry.

Ring-tailed lemur

> **Common ancestry** *There are common structures and behaviors among many species... The fossil record, DNA research, the evidence that species have common ancestors, and other findings add up to overwhelming evidence that evolution by natural selection is how life on Earth arose and became diverse.* Nat. Ac. Science & Inst. Med.

It is true that fossils, comparative anatomy, biochemistry and genetics show that there is a great amount of similarity between all living and extinct organisms. It is easy, therefore, to see how a theory of evolution was initially conceived and subsequently developed. And with the notion that life on Earth started a few billion years ago, it is easy to see how it might be assumed that such huge amounts of time could easily accommodate a slow step-by-step evolution. However, we have seen that, contrary to common belief, the evidence shows that neither human beings nor any other creatures, whether living or extinct, have lived on earth for millions of years.

It is often claimed that much of modern medical advances owe their success to the evolutionary paradigm, without which progress would have been unfruitful. However, while it is true that many medical advances are made by utilising an understanding of *how* human, bacterial and rodent cells function, it is quite wrong and deceptive to claim that evolutionary relationships *per se* have *any* bearing on such research.Yes, animal experiments do yield useful medical information, but this is simply because many animals share common anatomy and physiology with humans. As we have seen, not only is commonality of structure and function not proof of common ancestry, but also, the evidence (for example, from the fish fin/tetrapod limb, dinosaur front limb/bird wing, the ape foot/human foot,) shows, *unequivocally*, that different but *seemingly*-easily-intertransformable structures are separated by chasms that are *impossible* to bridge. So, a (supposed) stepwise evolution has no scientific basis. The evidence from all angles points directly to the conclusion

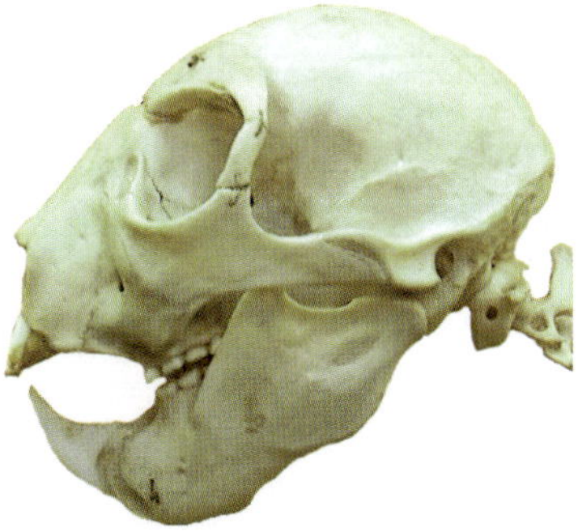

Aye-aye skull.
The aye-aye is a type of lemur which does not have a tooth-comb, and has claws on all its digits. Like rodents, its incisors grow continuously during life. Interestingly, it has a curious elongated middle finger which it uses to get at grubs under bark.

that all those similar-but-different structures could *only* have been put together as complete and separate entities from the beginning.

> **Some creationists claim that certain features of living beings are too complex to have evolved through natural processes.**
> *They claim that structures such as a bacterium's flagellum...the human eye, or the immune system are "irreducibly complex" and must have been created intact by an "intelligent designer." But biologists have discovered that components of the flagellum have their own individual functions and also have found intermediate forms of flagella. Both of those findings support the idea of the flagellum evolving from existing structures over time.* Nat. Ac. Science & Inst. Med.

The bacterial flagellum keeps popping up in any discussion of complexity, but it is only one of countless examples showing that a stepwise evolution from *random atomic motion* to *thought* is a '*Just So*' story. We have seen that when structures (such as the eye, flagellum, bird lung, bird wing, feather, knee joint and larynx) are looked at critically and in diverse detail, all the evidence points to those structures having been put together - 'from the ground up' - as complete functioning entities, without any evidence of a stepwise progression from simpler structures. Regarding the bacterial flagellum, we saw that the TTSS is not an ancestor to the bacterial flagellum. So, the biological world contains countless examples of *proof* that evolution does not deserve the status of a scientific theory.

> **Some creationists argue based on scripture** *that the Earth cannot be old enough for the diversity of life to have emerged through evolution. Yet measurements from geology, astronomy, and other fields have repeatedly confirmed the ancient age of the Earth (approximately 4.5 billion years).* Nat. Ac. Science & Inst. Med.

Evolution is a myth, not because the Earth might be far younger than $10^{17}$ seconds old, but because, as we have seen, *no* amount of time will cause the impossible to become possible.

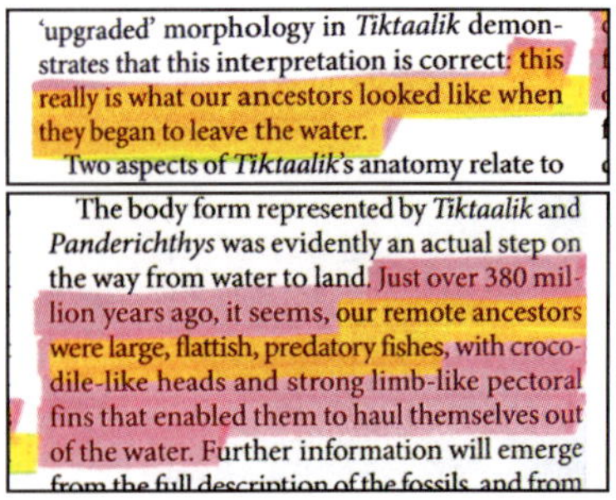
'upgraded' morphology in *Tiktaalik* demonstrates that this interpretation is correct: this really is what our ancestors looked like when they began to leave the water.
Two aspects of *Tiktaalik*'s anatomy relate to

The body form represented by *Tiktaalik* and *Panderichthys* was evidently an actual step on the way from water to land. Just over 380 million years ago, it seems, our remote ancestors were large, flattish, predatory fishes, with crocodile-like heads and strong limb-like pectoral fins that enabled them to haul themselves out of the water. Further information will emerge

Tiktaalik: our great ancestor...according to *Nature*.

> **By understanding evolution, scientists were able to predict what type of creature existed and in what geologic layer it would be found.**
> *The discovery of Tiktaalik fills another gap in the fossil record.* Nat. Ac. Science & Inst. Med.

As we have seen from anatomical and embryological evidence, tiktaalik cannot be a transitional form between fishes and tetrapods. Those who continue to claim a transitional position for tiktaalik simply think in too narrow a framework, and give little consideration to the the diverse anatomical and embryological aspects which give rise to the *Don't Mess With It Principle*, and which make the fish-to-tetrapod transformation impossible. Furthermore, we have seen that the *Coelacanth Principle* shows that tiktaalik could have lived both before and after all other extinct creatures, nullifying any supposed evolutionary trees.

Clearly, when it comes to biological origins, the scientific journals are full of plain nonscience.

## ***Purposeful design***

We have seen that evolution is impossible, and that everywhere, the biological world points unambiguously to non-random, non-trial-and-error, non-natural design. However, for those who are still sceptical, we will now take a look at the changes that occur in the blood circulation of the human baby at birth. Not only does this example deal with familiar structures (heart and lungs) and so makes the subject easy to appreciate, but it also demonstrates exquisite precision, with a beauty befitting a final example of purposeful design.

We will first look at the circulation of the baby in the womb and the blood circulation in the adult, and then look at the changes that happen at birth.

*The ductus arteriosus shunt also prevents the right ventricle from pumping against the high resistance within the (collapsed) lungs. If the ductus arteriosus closes before birth, this results in increased pressure in the right ventricle and can lead to right-sided heart failure, which can be fatal for the baby in the womb.

### *Blood circulation in the womb*

While the baby is in the womb, the lungs are non-functional because oxygen supply and carbon dioxide removal is undertaken by the mother's lungs, via the placenta and umbilical cord. Since only a minimal blood flow is required simply to sustain the foetal lungs and to enable their growth, the lungs are by-passed, with most of the blood in the pulmonary arteries being directed into the aorta through a special tube, the *ductus arteriosus**.

Oxygenated blood and nutrients are carried from the placenta to the baby via the umbilical vein inside the umbilical cord. Inside the baby's abdomen, the umbilical vein passes into the liver and joins the left *hepatic vein*, which then carries the oxygenated blood to the *inferior vena cava*, the large vein that takes the blood to the right atrium. The arrangement of the plumbing is such that blood from the inferior vena cava is directed straight into the left atrium through a special opening in the inter-atrial septum, the *foramen ovale* (oval hole). From the left atrium, the oxygenated blood passes into the left ventricle and then into the aorta. Most of the oxygenated blood from the left ventricle is directed through to the large arteries which branch off the first part of the aorta and take the blood to the head and arms. This blood then returns in its de-oxygenated state to the right atrium via the *superior vena cava*, and because of the physical arrangement, this de-oxygenated blood is channelled into the right ventricle without much mixing with the incoming oxygenated blood of the inferior vena cava. Having passed into the pulmonary trunk, the de-oxygenated blood is conveyed through the ductus arteriosus into that part of the aorta which supplies the lower part of the body. At the level of the pelvis, the *umbilical arteries* branch off from the *internal iliac arteries* and take the relatively de-oxygenated blood and waste back to the placenta.

The cycle repeats itself, with the baby's heart beating at around 150 times a minute.

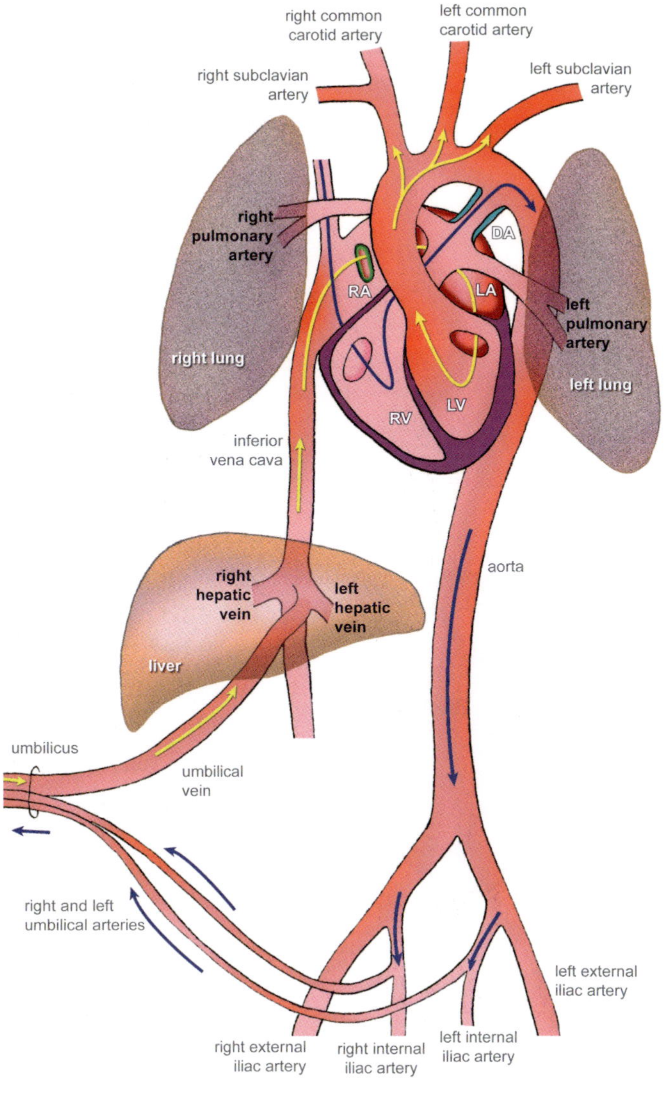


The blood circulation of the baby in the womb, with the foramen ovale in green, and the ductus arteriosus (DA) in blue. Yellow arrows represent oxygenated blood coming from the placenta; blue arrows represent de-oxygenated blood from the baby's tissues, going to the placenta.

### *Blood circulation in the adult*

In the adult, the veins deliver blood to the right atrium of the heart, from where it passes into the right ventricle. It is then pumped out into the pulmonary arteries and through to the lung capillaries. Here the blood picks up oxygen from the lung alveoli, and carbon dioxide in the blood is discharged into the alveoli. After passing through the lung capillaries, the bright red oxygenated blood is collected in the pulmonary veins and returned to the heart, but this time to the left atrium. From here the blood passes into the left ventricle, which then pumps the blood through the aorta and subsequently into arteries throughout the body. The arteries divide into smaller and smaller tubes and eventually become the capillaries, which are just a single cell thick. Oxygen and nutrients pass from the blood across the capillary walls into the various cells of the body, and in the reverse direction, carbon dioxide and waste materials pass from the cells into the capillaries. The dark red *de*-oxygenated blood from the capillaries is then collected in the veins and directed back to the right atrium, where the cycle repeats itself, with the adult heart beating some 60-80 times a minute.

### *Circulatory changes at birth*

At birth, the baby starts to breathe for itself, and the placenta is normally separated from the mother within 5-10 minutes. This necessitates three fundamental changes to the blood circulation to take place, *in a very short space of time*.

One: the umbilical vessels start to constrict. This prevents bleeding from the umbilicus (so that obstetrical clamping of the cord is not essential). The mechanism for this involves *Wharton's jelly,* a special tissue (comprising unique cells and a number of *mucopolysaccharide* molecules) which occupies around 60% of the cross-sectional area of the umbilical cord at birth. Wharton's jelly has the property of becoming less viscous with movement (*thixotropic*), and temperature change at birth causes physical changes to the molecules, resulting in automatic and permanent clamping of the umbilical veins and arteries within a few minutes of birth.

Two: the reduction in blood volume in the umbilical vein and in the veins leading to the heart results in a reduction of the pressure in the right atrium. At the same time, as an increased volume of blood flows into the lungs and returns to the left atrium, this results in an increase of the blood pressure in the left atrium. This causes a flap (the *septum primum*) to close over the foramen ovale, with the result that all the right atrial blood is forced to pass into the right ventricle and into the pulmonary arteries.

Three: the special muscle fibres (*ductal tissue*) in the ductus arteriosus constrict, and so blood in the pulmonary arteries is directed into the lungs.

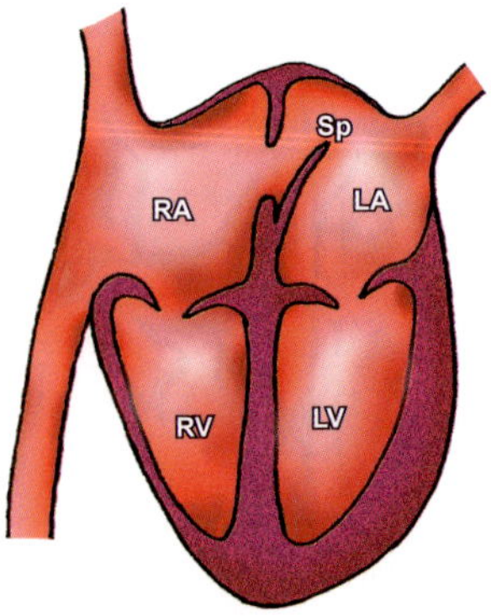


The septum primum (Sp) is a flap of tissue which closes off the foramen ovale when the left atrial pressure rises at birth.

### *Circulatory problems*

Serious or fatal consequences are inevitable if the necessary circulatory changes do not occur precisely at the *right time*, and in the *right place*.

#### *Umbilical vein*

Wharton's jelly provides *vital* support to the blood vessels in the cord. If it is absent in just a small segment, this can predispose to a fatal constriction in association with

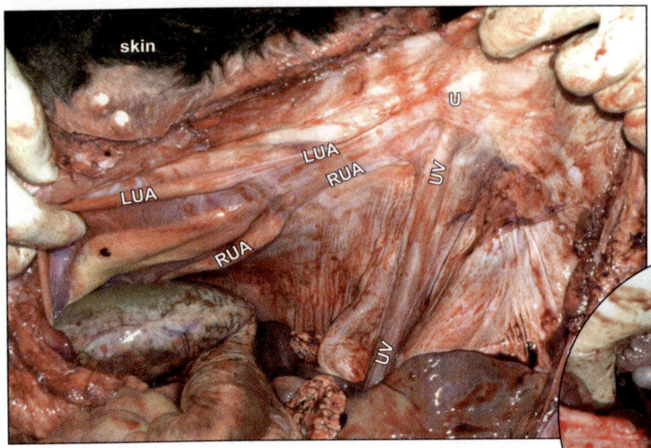


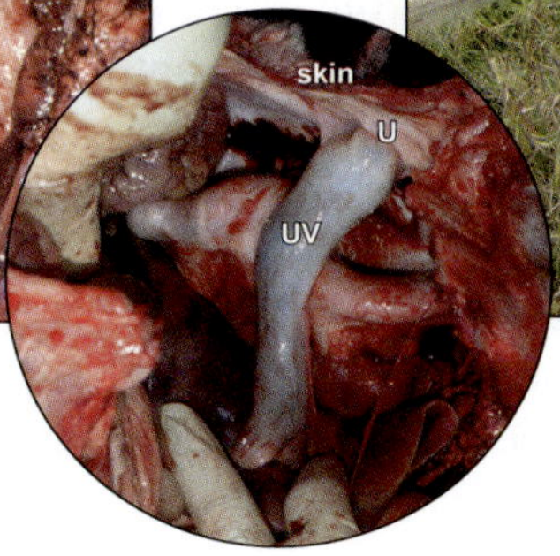


Above: One-day-old calf showing the umbilical vein (UV) and the right and left umbilical arteries (RUA, LUA) converging on the umbilicus (U), as seen from inside the abdomen. Right: The engorged umbilical vein before the blood was drained out.
NB. The head is to the right.

Sarah Juhasz

A cow giving birth.

*torsion* (knotting of the cord), while the baby is still in the womb:

> The stricture of the cord is usually accompanied by torsion and characterized by a fibrosis of the Wharton's jelly and a thickening of the vascular walls which obstructs the fetoplacental circulation, leading to anoxia and fetal death. J Tavares Fortuna & M Lourdes Pratas. 'Coarctation of the umbilical cord: a cause of intrauterine fetal death'. Int. J. Gynaecol. Obstet. 1978;15(5). p469-73.

New-born calf with the umbilical cord transected and still hanging out of its mother. If the cord did not constrict spontaneously, the calf would die from haemorrhage.

Furthermore, if Wharton's jelly were completely absent and a baby survived to term, the cord would not constrict *at all* following birth, and as soon as the cord was severed, the baby would die from haemorrhage. This means that *every* hypothetical pre-mammal that evolved an umbilical cord, but did not have the special Wharton's jelly, would bleed to death at birth .

Of course, it is implausible that any animal that did not have Wharton's jelly in the umbilical cord would acquire it exactly where it is needed, and, just as importantly, nowhere else. More likely, it would occur in the wrong place and cause serious or fatal blockages to blood flow in random locations such as in the eyes or the brain, or in the main vein of the liver (the *hepatic* vein, which is a direct continuation of the umbilical vein). Such inappropriate locations would result in automatic constriction at birth, resulting in blindness, brain damage and liver failure, respectively.

Furthermore, if the whole system was not specifically designed to prevent it, clotting of the blood inside the hepatic vein would result in extension of the clot into the inferior vena cava, with fatal embolisation to the heart and lungs (pulmonary

**Umbilical cords**

Below far left: Still-born spider monkey showing no signs of constriction in the cord.

Below middle: 7-day-old kitten. The cord has detached, leaving just a shrivelled stub.

Below: Baby boa that died some time after birth, by which time the cord had shrivelled. Note that although snakes such as boas, vipers and garter snakes do not lay eggs, the young are not connected to the mother by a placenta. Instead, the umbilical cord is connected to the yolk sac.

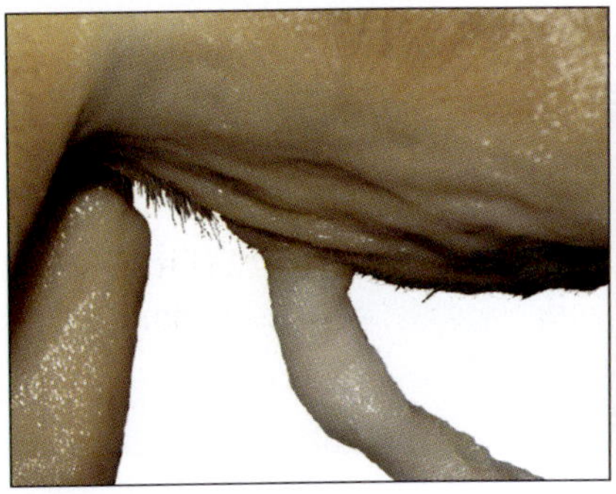

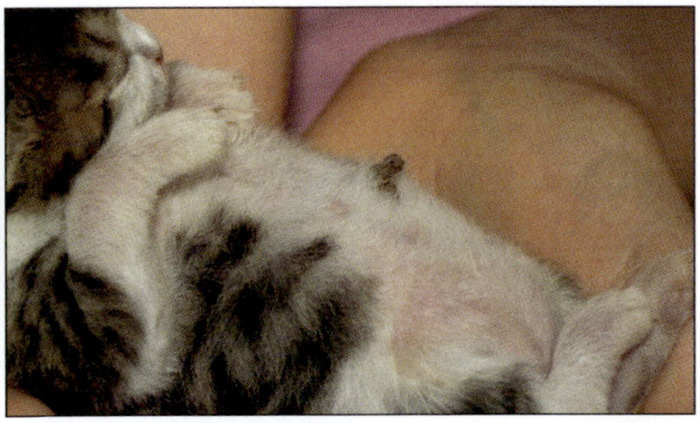

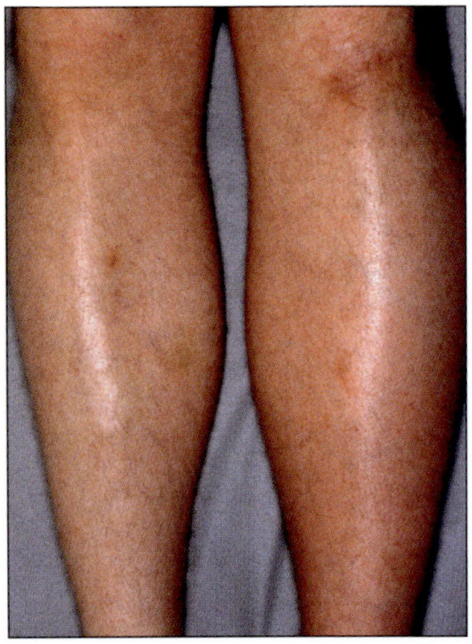

Deep vein thrombosis of the left lower leg. The blocking of the blood flow back to the heart results in marked swelling.

embolism), as can happen, for example, with deep vein thrombosis (DVT) in the legs following major surgery.

The *WYWYWG Principle* shows that the evolutionary origination of the Wharton's jelly constriction mechanism is impossible.

### *Patent foramen ovale*

Although the flap of the septum primum closes off the foramen ovale at birth, the hole is not permanently sealed until about the end of the first year. When the flap fails to block off the foramen ovale, the condition is called *patent foramen ovale* (*PFO*). It is true that over 20% of healthy people have a PFO which causes no problems, whatsoever, throughout life. However, in most of these cases, the PFO is usually small. On the other hand, if the PFO is large, the blood passing from the left atrium into the right atrium will result in fatigue, and eventually in an increased pressure in the right side of the heart which can result ultimately in heart failure*.

*Heart failure is a medical term to describe the situation where the heart becomes incapable of pumping effectively. This results in increased pressure in the venous system, with swelling due to retention of fluid. Without treatment, the condition is ultimately fatal.

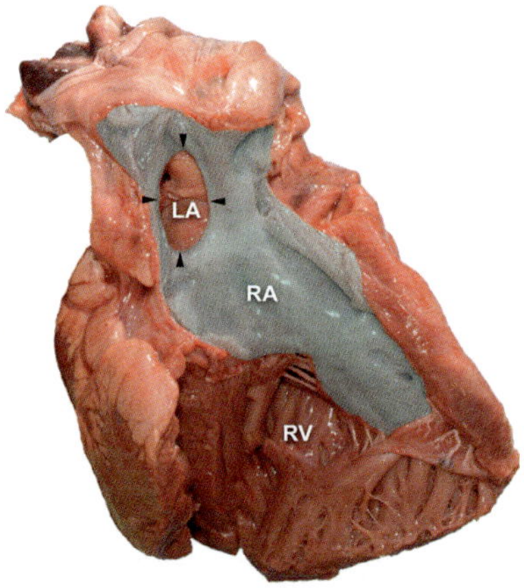


One-day-old calf heart seen from the right side: The right atrium (RA, blue tint) and the right ventricle (RV) have been opened up, and the cavity of the left atrium (LA) can be seen through the large (2.5cm high) patent foramen ovale (arrowed).

### *Patent ductus arteriosus*

The ductus arteriosus usually closes within the first 10 days of life. If the ductus arteriosus fails to close (*patent ductus arteriosus*, *PDA*) but has only a small diameter, it may cause no problems at all. However, a large PDA will result in blood being shunted from the aorta into the lungs resulting in fatigue, poor growth and, eventually, high blood pressure in the lungs with the risk of heart failure.

This means that *every* hypothetical creature that did not have a closed-off ductus arteriosus and was *hoping-to-become-a-mammal* would have had problems because of a PDA, since none would have had the special ductal tissue in the wall of the ductus arteriosus.

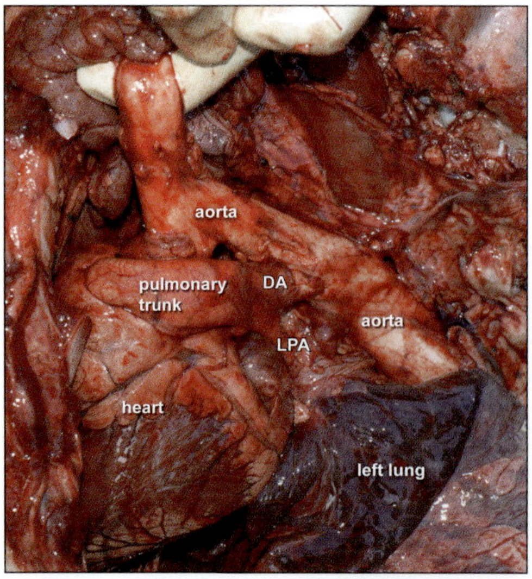


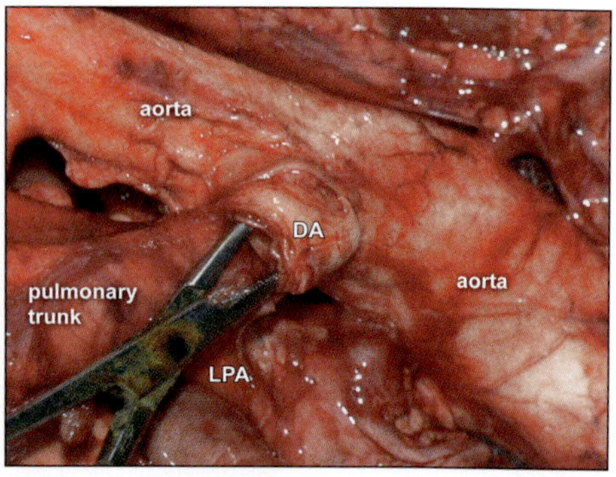


One-day-old calf.
Right: The ductus arteriosus (DA) is seen connecting the aorta to the pulmonary trunk.
Far right: Close-up of the ductus, which is over 1 cm in diameter.
LPA: left pulmonary artery.

### *Coarctation of the aorta*

Of course, as with Wharton's jelly, it is implausible that any animal that did not have the special ductal tissue, would acquire it *exactly* where required, and nowhere else. More likely, it would occur in the wrong place (*ectopic*) and cause serious or fatal blockages to blood flow in random locations, such as in the arteries of the ear, brain, kidneys, or lungs. Such ectopic locations would result in automatic constriction at birth, resulting in deafness, brain damage, kidney failure or stillbirth, respectively.

An example of this sort of thing is *coarctation of the aorta*, a condition where ectopic ductal tissue ends up in part of the aorta alongside or opposite the ductus arteriosus. Contraction of these fibres occurs sometime after birth and results in narrowing or complete blockage of the affected segment of the aorta.

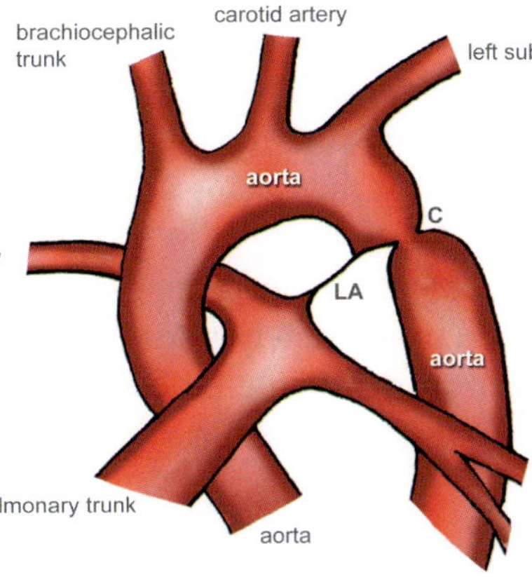


Coarctation of the aorta (C) adjacent to the ligamentum arteriosus (LA), which is the remnant of the fibrosed ductus arteriosus.

Coarctation of the aorta to the right of the ductus arteriosus is often severe and can be fatal at birth, whereas the prognosis for a coarctation to the left of the ductus arteriosus varies in severity depending on the degree of blockage.

Survival with coarctation will often result in severe high blood pressure, with the risk of heart failure.

### *Ductal tissue*

Not only does the ductal tissue have to be in the right place and not in the wrong place, but the muscle fibres must contract at the *right time* – immediately *after* birth, and *not before*.

The patency of the ductus arteriosus is subject to complex interactions. In the womb, the ductus arteriosus is normally kept open by the dilating effect of circulating chemicals called *prostaglandins*, and as the pregnancy progresses, the sensitivity of the ductus tissue to prostaglandins reduces. Normally, the ductal tissue constricts in response to the rise in the blood oxygen level at birth. If the ductus arteriosus were to close *before* birth, the result would either be fatal, or would seriously impair growth and development. Intra-uterine closure of the ductus arteriosus can occur without any identifiable cause. In addition, *non-steroidal analgesics* (such as aspirin, metamizole or ibuprofen) can inhibit the patency-maintaining effect of the prostaglandins, and if taken during pregnancy, this can result in intra-uterine closure of the ductus arteriosus:

**The ductus arteriosus muscle fibres have complex properties**

*We report that O2 reversibly inhibits a 58-pS voltage- and 4-aminopyridine-sensitive potassium channel, causing membrane depolarization, an increase in intracellular calcium through L-type voltage-gated calcium channels, and constriction of the ductus arteriosus. We conclude that the effector mechanism for O2 sensing in the ductus arteriosus involves the coordinated action of delayed rectifier potassium channels and voltage-gated calcium channels.*
M Tristani-Firouz et al. 'Oxygen-induced constriction of rabbit ductus arteriosus occurs via inhibition of a 4-aminopyridine-, voltage-sensitive potassium channel'. J. Clin. Invest. 1996. 98(9). p1959-1965.

> We report on four cases with prenatal DC [ductal closure], presenting at 34, 35, 36 and 37 weeks of gestation. They were referred to fetal echocardiography because of abnormal routine echo scans with unexplained signs of right heart decompensation [early heart failure]...Any application of non-steroidal anti-inflammatory drugs in pregnancy requires close fetal follow-up due to their potentially life-threatening effect. B Schiessl et al. 'Prenatal constriction of the fetal ductus arteriosus - related to maternal pain medication?' Z. Geburtshilfe Neonatol. April 2005. Vol 209. p65-8.

Although, throughout life, the muscles in the walls of blood vessels normally contract and relax in response to changing demands, the ductus arteriosus must contract and *stay contracted.* As a result, over time, the ductus arteriosus eventually becomes just a thin solid fibrous cord.

Finally, any special muscle fibres in the ductal tissue must be orientated correctly (ie. circumferentially), such that when they contract, the blood vessel is constricted. If the muscle fibres were orientated in a longitudinal or a stretched-out helical arrangement, the vessel would not constrict properly, with adverse consequences.

With all these variables and inherent complexities, the *WYWYWG Principle* shows that the evolutionary origination of the ductus arteriosus constriction mechanism is impossible.

So, we see that the absolutely essential transformation of the circulation of the baby in the womb to the circulation of the newly-born could not have arisen stepwise. Furthermore, the instructions for the whole process must have been present (and complete) *from the beginning*, making this an undeniable example of purposeful design.

### *Imperfection arising from previous perfection*

As we have seen from the erroneus articles in diverse scientific journals, it is commonly taught that faulty structures and diseases confirm the blind process of evolution. However, the reverse is true.

Congential heart disease and other diseases are not evidences for evolution, the blind process by which imperfect systems (supposedly) appeared from non-existent systems, and (supposedly) subsequently became rather good, but retained some bad points. Instead, these diseases point unambiguously to previously perfect systems acquiring errors, resulting in imperfect systems and diseases.

## ***Information and intelligence***

The evidence categorically shows that we are not modified apes. There is no evidence to show that the stuff of which the universe is made (matter or energy) can, *of itself*, create any living thing. For its manifestation, the organisation that we observe in the biological world requires *information.* But information is a reality that is separate from time and space, energy and matter. Naturally, the question arises: can information exist in the absence of *intelligence*?

For example, *1+1=2* is an *absolute* truth. It cannot be violated. It cannot *not* exist. 1+1=2 is true now and always has been so. Without mathematical principles, the physics upon which the (supposed) BIG BANG was dependent could not have given rise to real cosmological events. It follows then that the principle of 1+1=2 must have existed '*before*' the origination of time and space (for want of a better label, let us call it Time Zero, $T^0$), in a dimension that we cannot comprehend: a dimension in which there 'once' existed no time, space, energy or matter (NoTSEM), yet from which time, space, energy and matter could and did appear. In other words, within this NoTSEM dimension, the information and principles that *allowed* the *origination* and *maintenance* of time, space, energy and matter existed 'before' $T^0$. And this information and these principles still continue today to exercise their controlling effects on the universe.

To put it another way, there would have existed within that NoTSEM dimension (and there still exist) the mathematical and physical principles by which, for example, a feather could be made and by which it could function for flight. Of course, if, prior to the origin of the universe, the information by which the universe could come into being, or the principles of feathered flight, already existed, then this suggests the existence of an intelligence *prior to* the origin of the universe.

We have seen that, fundamental to the issue of evolution, there is *not* just the problem of a chimpanzee ancestor being modified so as to become a human being. The problem is that of the speck of dust in front of you *becoming* you. We must accept that some things are possible, and some things are *not* possible....*no matter how much time you allow*. Therefore if, after studying all the evidence, you do not consider it a likely possibility that, spontaneously, a speck of dust could become human, then neither can you consider it plausible that a chimpanzee 'ancestor' (which is only one link in the [supposed] evolutionary pathway linking dust to humans) could have given rise to *you*. Actually, we can take the argument even further back than the dust. Before the origin of the universe, there was NoTSEM. So the question is: is it likely, from within the NoTSEM dimension, that the *nothingness*, left to itself, could ultimately become conscious and conceptualise about the NoTSEM dimension from which it (supposedly) came?

**Many scientists do not accept the Big Bang theory**

Dr E Lerner, along with 33 other astronomers, declared:

*The big bang theory can boast of no quantitative predictions that have subsequently been validated by observation.*

New Scientist. 22nd May 2004. p20.

*Big bang predictions are consistently wrong and are being fixed after the event...Take the most distant galaxies ever spotted, for example. According to the accepted view, when we observe ultra-distant galaxies we should see them in their youth, full of stars not long spawned from gas clouds. This is because light from these faraway galaxies has taken billions of years to reach us, and so the galaxies must appear as they were shortly after the big bang. But there is a problem. "We don't see young galaxies," says Lerner. "We see old ones...They are pretty much the same range of stars as present-day galaxies".*

New Scientist. 2nd July 2005. p30.

### *The conclusion of the matter*

Many will still cling on to flawed nonscience, and claim that the truth is that we are just evolved and particularly clever apes. But, whatever your own conclusion, let it be based on reason and knowledge and possibility and probability. The molecular biology of DNA and genetics only explains the variations and mutations in existing genes. And natural selection simply and only explains the death of the not-so-fit. *Nothing* more. Natural selection does not even dent the armour of questions such as: exactly how did your actin and myosin molecules become the motor power to move your index finger as it lifts that speck of dust?

By embracing evolutionary thinking, it is commonly believed that any *order* in the natural world exists by chance and functions *for no purpose*. However, since all of the biological world shows features of deliberate design and living things did not come into being spontaneously, we must accept that the natural world and everything contained within it *did not come into being without purpose*. If a purpose seems to anyone incomprehensible, unimaginable or even distasteful, this does not deny that a purpose exists.

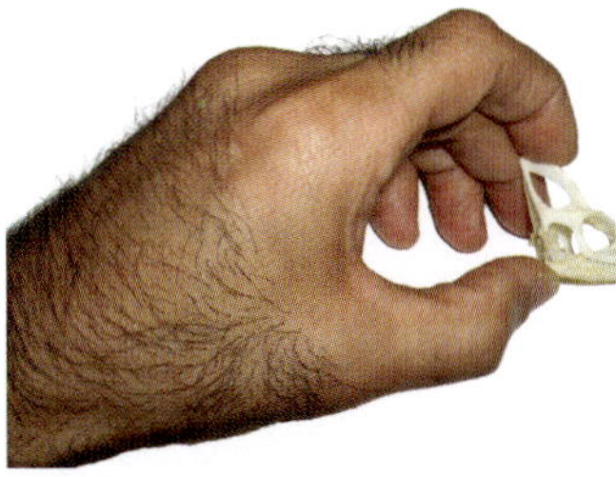

Tendons move the fingers and thumbs holding the chameleon skull. But the gene for collagen does not carry any information for making a nerve-muscle-tendon-bone-joint complex.

**Paradigm found**
Since evolution is a myth, *all* research based on the assumption that diverse living organisms such as worms and humans share a common ancestry is looking into biological blind alleys, and is a total waste of time and money. Biological processes can only be properly investigated and understood once it is realised that bacteria did not give rise to eukaryotes, and apes did not give rise to humans.

The evidence that we already have is more than enough to state unequivocally that the spontaneous, undirected origin of life, and its evolution by natural selection, is an imagined unreality. A myth. It is axiomatic that *even if millions of people accept as truth something that is imaginary, then that thing will still be imaginary.*
So, paradigm lost, and paradigm found: terrestrial quadrupeds did not and could not have become whales; bipedal dinosaurs did not and could not have become birds; and apes did not and could not have become human beings. Therefore, whales and birds and frogs and humans, along with all other types of living creatures, have always been the way they are...from the beginning.
So, the fact is that our great, great…great grandfather was not an ape.

### *A final word*

We have seen that there is no evidence for evolution, and that we are far more than just molecular complexity. Everywhere, the biological evidence points to deliberate, purposeful design. As sure as 1+1 can only equal 2, the biology unequivocally shows that our being can only be the work of an Intelligent Architect and Manufacturer. Having said that, no doubt die-hard evolution-believers will continue to repeatedly present 'another missing link' or another 'proof' of evolution. However, do not be misled...always remember the *Don't Be Fooled Again Principle*:

Since it is a fact that evolution is *impossible*, then if we come across any data that *seems to suggest* that evolution has occurred, then we know that *that* interpretation of such data *must* be false.

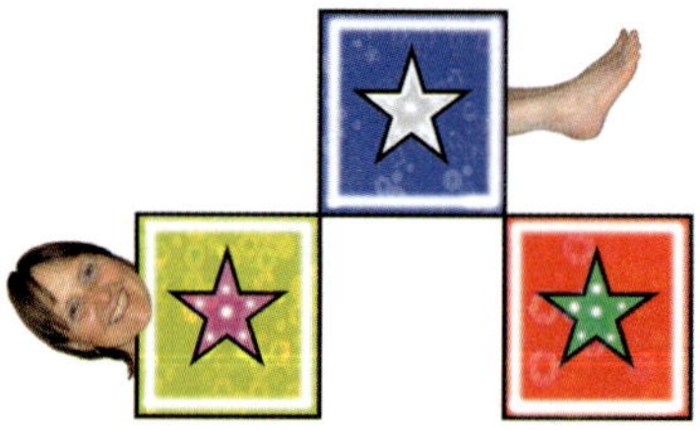

**Don't Be Fooled Again**
Even if you don't know how the trick was done, you know that the face and the feet cannot belong to the same person.
Similarly, even if you may not always see the flaw, you know that any suggestion that, spontaneously, hydrogen gas turned into rock and water, which ultimately turned into monkeys and human beings, is false.

Indian python

## Anyone who has no doubt
that chimpanzees are our distant cousins,
or that fishes and worms share a common ancestry with humans,
either has not studied the wealth of unequivocal evidence to the contrary...
or is living in a state of denial.

Gorilla
Not a distant cousin

# Appendix

## List of Principles

### *CHALICOTHERIUM PRINCIPLE:*
Purely on the basis of its teeth or skull, for any fossil: (i) The nature of the body skeleton cannot be predicted or assumed, and (ii) No animal can be assigned to any ancestor-descendant or distant cousin relationship.

### *COELACANTH PRINCIPLE:*
In the absence of any direct evidence to the contrary, any creature could have lived at any time before or after the time of its earliest or latest known fossil.

### *DON'T BE FOOLED AGAIN PRINCIPLE:*
Since it is a fact that evolution is impossible, if we come across any data that seems to suggest evolution has occurred, we know that that interpretation of such data must be false.

### *DON'T MESS WITH IT PRINCIPLE:*
If you mess with a working complex multipart system controlled by multi-hierarchical activatory and suppressant triggers – you get a mess.

### *ESSENTIAL SEQUENCE PRINCIPLE:*
No enzyme (nor any of its hypothetical, shorter ancestral molecules), missing any of the amino acids that are essential for the integrity and minimal function of the enzyme's active site, could possibly function at all.

### *GALAPAGOS CORMORANT PRINCIPLE:*
Disuse per se is not the cause of degeneration of a structure or function. Mutations occur as inevitable consequences of random chemical substitutions within DNA. However, if a mutation results in a degeneration of a structure or function which is not essential for the survival of the affected individual, then the individual might survive and pass on to its offspring that mutation. The progeny will then have degenerate structures and functions not because of disuse, but as a consequence of the prior mutation.

### *HORSE PRINCIPLE:*
a. Even if different fossil specimens can be arranged in an order suggesting a presumed evolutionary trend, nevertheless no such evolutionary trend need be true.
b. In the absence of any direct evidence, if any number of creatures share any number of characteristics, it cannot be assumed that for any of those creatures the shared characteristics reflect any ancestor-descendant relationship.

### *IRREDUCIBLE COMPLEXITY PRINCIPLE:*
If, for it to function, a system requires a minimum number of interdependent components, and if no component could have had any function outside that system, and if without all the components being present that system cannot function at all, then those components could not have evolved from simpler entities and that system could not have evolved from any fewer number of components.

### *SMALL STEP PRINCIPLE:*
Any small step, even if infinitesimal, need not be simple, and even if a small transitional step may be imagined, this does not mean that that particular step ever happened, or is possible.

### *WYWYWG PRINCIPLE:*
If a number of genes are together essential for the manifestation of a particular structure or characteristic; and if in the absence of any one of those genes the structure or characteristic could not manifest; and since, if the necessary individual genes were to appear, it is likely that they would appear haphazardly, and within unrelated lineages that would be separated in both time and space: then that structure or characteristic is likely never to manifest - in any lineage.
In other words, it is implausible that a creature will acquire the very mutation that it might hope to acquire so as to continue its (supposed) evolutionary journey:
What You Want You Won't Get.